AIRCRAFT GAS TURBINE ENGINE TECHNOLOGY

SECOND EDITION

IRWIN E. TREAGER

Professor, Department of Aviation Technology
Purdue University

Gregg Division

McGraw-Hill Book Company

New York	Düsseldorf	Paris
St. Louis	Johannesburg	São Paulo
Dallas	London	Singapore
San Francisco	Madrid	Sydney
Auckland	Mexico	Tokyo
Bogotá	Montreal	Toronto
	New Delhi	
	Panama	

LIBRARY OF CONGRESS CATALOGING IN PUBLICATION DATA

Treager, Irwin E
    Aircraft gas turbine engine technology.

    Includes index.
    1. Aircraft gas-turbines.   I. Title.
TL709.5.T87T73   1979   629.134'353   78-15881
ISBN 0-07-065158-2

**Aircraft Gas Turbine Engine Technology,
Second Edition**

5 6 7 8 9 10 11 12   WCWC   8 9 8 7 6 5

The editors for this book were D. E. Gilmore and George
McCloskey, the designer was Jackie Merri Meyer, the art
supervisor was George T. Resch, and the production
supervisor was May E. Konopka. Cover photo by Hue
and Eye/Landskroner. It was set in Bembo by Waldman
Graphics, Inc.
Printed by and bound by Webcrafters, Inc.

CONTENTS

PREFACE

Although the gas turbine, or jet engine, is a relative youngster to aviation, its growth and refinement have not only given new life to this industry, but also have been so rapid that keeping abreast of developments in this area has become very difficult. A great deal has been written about the gas turbine engine from an engineering viewpoint, but there is relatively little consolidated information treating this type of power plant at the technical level. The second edition of *Aircraft Gas Turbine Engine Technology* continues in the attempt to correct this deficiency.

Extensive changes have been made to several sections of the book, especially Chaps. 2, 10, 12, and 18, in order to reflect radical advancements in:

1. Design and usage of the turbofan and other gas turbine engines
2. New materials and methods of construction
3. Improvements in fuel controls
4. Changing maintenance and overhaul procedures and philosophies

Although the original text was designed primarily to provide a source of information about the gas turbine engine for aircraft technicians, it has been found to be very useful to other students at all levels, including those in engineering, who wish to study this form of prime mover.

The author has tried to follow a logical presentation and to use the type of approach that does not assume a great deal of technical information on the part of the reader. Beginning with the background and development of the gas turbine engine, the book ends with a discussion of several modern engines of this type. The section of the book dealing with mathematical relationships, which are an integral part of any study dealing with this type of engine, has been simplified without, it is hoped, sacrificing clarity and completeness to any great degree. As the heading for Chap. 3 implies, all who can add and subtract should have little difficulty reading and understanding this part of the book. The part of Chap. 12 dealing with the fuel control is slightly more detailed than other sections because fuel metering is a critical factor in correct engine operation and because the fuel control is probably the most complicated and difficult to understand unit on the entire engine. The appendixes are devoted to a compilation of appropriate mathematical formulas, a glossary of terms related to the gas turbine engine, and several pages of applicable conversion factors and tables.

ACKNOWLEDGMENTS

The information contained in this book has been collected from a great number of sources. It represents the work of many people and organizations, without whose contributions this text could not have been completed. I wish to express my gratitude to the following organizations for their efforts in providing technical information, pictures, charts, tables, and other materials.

Airbus Industrie, Blagnac, France

AiResearch Mfg. Co., Div. of Garrett Corp., Phoenix, Arizona

Air Training Command, U.S.A.F., Randolph A.F.B., Texas

American Petroleum Institute, New York, N.Y.

American Society of Mechanical Engineers, New York, N.Y.

American Society of Tool and Manufacturing Engineers, Dearborn, Michigan

Avco Corp., Lycoming Division, Stratford, Connecticut

Aviation Power Supply Inc., Burbank, California

Aviation Week and Space Technology, New York, N.Y.

Beech Aircraft Corp., Wichita, Kansas

Bell Helicopter Textron, Division of Textron Inc., Fort Worth, Texas

Bendix Electrical Components Division, Bendix Corp., Sidney, New York

Bendix Energy Controls Division, Bendix Corp., South Bend, Indiana

Boeing Aerospace Co., Seattle, Washington

Boeing Commercial Airplane Co., Renton, Washington

British Aircraft Corp., Weybridge, Surrey, England

Chandler Evans Corp., West Hartford, Connecticut

Columbus Aircraft Division, Rockwell International, Columbus, Ohio

Convair Division, General Dynamics Corp., San Diego, California

Curtiss-Wright Corp., Woodridge, New Jersey

DeHavilland Aircraft of Canada Ltd., Downsview, Ontario, Canada

Detroit Diesel Allison Division, General Motors Corp., Indianapolis, Indiana

Department of the Air Force, Washington, D.C.

Douglas Aircraft Co., Long Beach and Santa Monica, California

Ex-Cell-O Corp., Detroit, Michigan

Extension Course Institute, U.S.A.F., Gunter A.F.B., Alabama

Falcon Jet Corp., Teterboro, New Jersey

Fairchild, Fairchild Republic Co., Farmingdale, Long Island, N.Y.

Fairchild Industries, Germantown, Maryland

Federal Aviation Agency, Washington, D.C.

Gas Turbine Publications Inc., Stamford, Connecticut

Gates Lear Jet, Wichita, Kansas

General Dynamics Corp., Fort Worth Division, Fort Worth, Texas

General Electric Co., Aircraft Engine Group, Cincinnati, Ohio, and Lynn, Massachusetts

General Laboratory Associates, Inc., Norwich, N.Y.

Grumman Aircraft Engineering Corp., Bethpage, N.Y.

Hamilton Standard Division, United Technologies Corp., Windsor Locks, Connecticut

Hawker Siddeley Aviation Limited, Kingston Upon Thames, Surrey, England

Holley Carburetor Co., Warren, Michigan

Hughes Helicopters, Division of Summa Corp., Culver City, California

Industrial Acoustics Co., Inc., New York, N.Y.

International Nickel Co., New York, N.Y.

Investment Casting Institute, Chicago, Illinois

Israel Aircraft Industries Ltd., Lod Airport, Israel

Kaman Aircraft Corp., Bloomfield, Connecticut

Kelsey Hayes Co., Utica, N.Y.

Koopers Co., Inc., Sound Control Dept., Baltimore, Maryland

Lockheed California Co., Division of Lockheed Aircraft Corp., Burbank, California

Lockheed-Georgia Co., Division of Lockheed Aircraft Corp., Marietta, Georgia

Magnaflux Corp., Chicago, Illinois

Materials Systems Division, Union Carbide Corp., Indianapolis, Indiana

McDonnell Douglas Corp., St. Louis, Missouri

Misco Precision Casting Division, Howe Sound Co., Whitehall, Michigan

Mooney Aircraft Inc., Kerrville, Texas

Naval Air Training Command, Washington, D.C.

Northrop Corp. Aircraft Division, Hawthorne, California

North American Aviation Inc., Los Angeles and El Segundo, California

Pan American World Airways, New York, N.Y.

Pesco Products Division, Borg-Warner Corp., Bedford, Ohio

Phillips Petroleum Co., Bartlesville, Oklahoma

Philosophical Library, Inc., New York, N.Y.

Pilatus Aircraft Ltd., Switzerland

Piper Aircraft Corp., Lock Haven, Pennsylvania

Pratt and Whitney Aircraft of Canada Ltd.

Pratt and Whitney Aircraft Group, United Technologies Corp., East Hartford, Connecticut

Rolls-Royce Ltd., Derby and Bristol, England

Ryan Aeronautical Co., San Diego, California

Shell Oil Co., New York, N.Y.

Sikorsky Aircraft, Division of United Technologies Corp., Stratford, Connecticut

Society of Automotive Engineers, New York, N.Y.

Socony Mobil Oil Co., Inc., New York, N.Y.

Solar Division, International Harvester Co., San Diego, California

Stalker Development Co., Bay City, Michigan

Sundstrand Aviation Division, Sundstrand Corp., Rockford Illinois

Swearingen Aviation Corp., Subsidiary of Fairchild Industries, San Antonio, Texas

Teledyne CAE Turbine Engines, Toledo, Ohio

Texaco, Inc., New York, N.Y.

Thermal Dynamics Corp., Lebanon, New Hampshire

Utica Division, Bendix Corp., Utica, New York

Vertol Division, Boeing Airplane Co., Morton, Pennsylvania

Vought Systems Division, LTV Aerospace Corp., Dallas, Texas

Welding Journal, American Welding Society, New York, N.Y.

Westinghouse Research Laboratory, Pittsburgh, Pennsylvania

Williams Research Corp., Walled Lake, Michigan

Woodward Governor Co., Rockford, Illinois

Wyman-Gordon Co., Worcester, Massachusetts

A special note of thanks to my wife for all her help and encouragement.

IRWIN E. TREAGER
Professor of Aviation Technology
Purdue University

PART ONE
HISTORY AND THEORY

CHAPTER ONE
BACKGROUND
AND
DEVELOPMENT

Long before humans appeared on earth, nature had given some creatures of the sea, such as the squid and the cuttlefish, the ability to jet propel themselves through the water (Fig. 1-1).

There have been many examples of the utilization of the reaction principle during the early periods of recorded history, but because a suitable level of technical achievement in the areas of engineering, manufacture, and metallurgy had not been reached, there was a gap of over 2000 years before a practical application of this principle became possible.

THE AEOLIPILE

Hero, an Egyptian scientist who lived in Alexandria around 100 B.C., is generally given credit for conceiving and building the first "jet engine." His device, called an *aeolipile* (Fig. 1-2), consisted of a boiler or bowl which held a supply of water. Two hollow tubes extended up from this boiler and supported a hollow sphere which was free to turn on these supports. Attached to the sphere were two small pipes or jets whose openings were at right angles to the axis of rotation of the sphere. When the water in the bowl was made to boil, the steam shooting from the two small jets caused the sphere to spin, very much like the lawn sprinkler is made to spin from the reaction of the water leaving its nozzles. (This phenomenon will be explained in Chap. 3.) Incidentally, the aeolipile was only one of a number of inventions credited to Hero, which include a water clock, a compressed-air catapult, and a hydraulic organ. He also wrote many works on mathematics, physics, and mechanics.

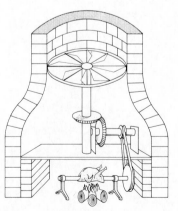

FIG. 1-2 Hero's aeolipile.

LEONARDO DA VINCI

Around A.D. 1500 Leonardo Da Vinci described the *chimney jack* (Fig. 1-3), a device which was later widely used for turning roasting spits. As the hot air from the fire rose, it was made to pass through a series of fanlike

FIG. 1-3 DaVinci's chimney jack.

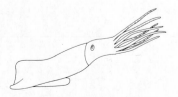

FIG. 1-1 The squid, a jet-propelled fish.

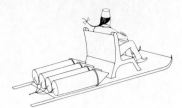

blades which, through a series of gears, turned a roasting spit, thus illustrating another application of the reaction principle.

ROCKETS AS A FORM OF JET PROPULSION

The invention of gunpowder allowed the continued development of the reaction principle. Rockets, for example, were constructed apparently as early as 1232 by the Mongols for use in war and for fireworks displays. One daring Chinese scholar named Wan Hu intended to use his rockets as a means of propulsion (Fig. 1-4). His plan was simple. A series of rockets were lashed to a chair under which sledlike runners had been placed. Unfortunately, when the rockets were ignited, the blast that followed completely obliterated Wan Hu and the chair, making him the first martyr in humanity's struggle to achieve flight. In later times rockets were used during several wars, including the Napoleonic Wars. The phrase "the rockets' red glare" in our national anthem refers to the use of rockets by the British in besieging Fort McHenry in Baltimore, during the war of 1812. And, of course, the German use of the V-2 rocket during World War II, and the subsequent development of space vehicles is contemporary history.

BRANCA'S STAMPING MILL

A further application of the jet-propulsion principle, utilizing what was probably the first actual impulse turbine, was the invention of a stamping mill built in 1629 by Giovanni Branca, an Italian engineer (Fig. 1-5). The turbine was driven by steam generated in a boiler. The jet of steam from a nozzle in this boiler impinged on the blades of a horizontally mounted turbine wheel which, through an arrangement of gearing, caused the mill to operate.

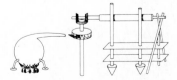

SIR ISAAC NEWTON

It was at this point in history (1687) that Sir Isaac Newton formulated the laws of motion (discussed in detail in Chap. 3) on which all devices utilizing the jet-propulsion theory are based. The vehicle illustrated in Fig. 1-6, called Newton's wagon, applied the principle of jet propulsion. It is thought that Jacob Gravesand, a Dutchman, actually designed this "horseless carriage," and that Isaac Newton may have only supplied the idea. The wagon consisted essentially of a large boiler mounted on four wheels. Steam generated by a fire built below the boiler was allowed to escape through a nozzle facing rearward. The speed of the vehicle was controlled by a steam cock located in the nozzle.

THE FIRST GAS TURBINE

In 1791 John Barber, an Englishman, was the first to patent a design utilizing the thermodynamic cycle of the modern gas turbine and suggested its use for jet propulsion (Fig. 1-7). The turbine was equipped with a chain-driven reciprocating type of compressor, but was otherwise basically the same as the modern gas turbine in that it had a compressor, a combustion chamber, and a turbine.

SIR FRANK WHITTLE

During the period between 1791 and 1930, many people supplied ideas which laid the foundation for the

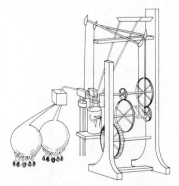

modern gas turbine engine as we know it today. When in 1930 Frank Whittle submitted his patent application for a jet aircraft engine, he had the contributions of many people to draw from. For example:

Sir George Caley—Invented the reciprocating hot air engine. This engine (1807) operated on the same cycle principle as the modern closed-cycle gas turbine.
Dr. F. Stoltz—Designed an engine (1872) approaching the concept of the modern gas turbine engine. The engine never ran under its own power because component efficiencies were too low.
Sir Charles Parsons—Took out many comprehensive gas turbine patents (1884).
Dr. Sanford A. Moss (Fig. 1-8)—Did much work on the gas turbine engine, but his chief contribution lies in the development of the turbosupercharger. Credit for the basic idea for the turbosupercharger is given to Rateau of France; it is in reality very similar to a jet engine, lacking only the combustion chamber (Fig. 1-9).
Dr. A. A. Griffith—Member of the British Royal Aircraft establishment who developed a theory of turbine design based on gas flows past airfoils, rather than through passages.

The work of many others, in addition to those mentioned, preceded Whittle's efforts. In addition, there were several jet-engine developments occurring concurrently in other countries. These developments are discussed on the following pages. Whittle is considered by many to be the father of the jet engine, but his contribution lies mainly in the application to aircraft of this type of engine which, as indicated previously, was already somewhat refined.

In 1928, at the time that Dr. Griffith was involved in his work with compressors and other parts of the gas turbine, Whittle (Fig. 1-10), then a young air cadet at the Royal Air Force (R.A.F.) College in Cramwell, England, submitted a thesis in which he proposed the use of the gas turbine engine for jet propulsion. It was not until 18 months later that this idea crystallized, and he began to think seriously about using the gas turbine

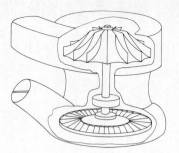

FIG. 1-9 A turbosupercharger.

engine for jet propulsion. By January of 1930, Whittle's thinking on the subject had advanced to the point that he submitted a patent application on the use of the gas turbine for jet propulsion (Fig. 1-11). In this patent were included ideas for the athodyd, or ramjet, but this was removed from the specifications when it was determined that the ramjet idea had already been proposed.

The period between 1930 and 1935 was one of frustration for Whittle and his co-workers. During this time his idea had been turned down by the British Air Ministry, and by several manufacturing concerns because the gas turbine was thought to be too impractical for flight. In 1935, while he was at Cambridge studying engineering, he was approached by two former R.A.F. officers, Williams and Tinling, with the suggestion that Whittle should acquire several patents (he had allowed the original patent to lapse), and they would attempt to raise money in order to build an experimental model of Whittle's engine. Eventually, with the help of an investment banking firm, Power Jet Ltd. was formed in March of 1936.

Before the new company was formed, the banking firm had placed an order with the British Thomson-Houston Company at Rugby for the actual construction of the engine, minus the combustion chamber and instrumentation. Originally Whittle had planned to build and test each component of the engine separately, but this proved to be too expensive. As planned, the new

FIG. 1-8 Dr. Sanford A. Moss.

FIG. 1-10 Sir Frank Whittle.

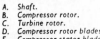

A. Shaft.
B. Compressor rotor.
C. Turbine rotor.
D. Compressor rotor blades.
E. Compressor stator blades.
F. Radial blades.
G. Diffuser vanes.
H. Air collecting ring.
J. Combustion chamber.
K. Fuel jet.
L. Gas collector ring.
M. Turbine stator blades.
N. Turbine rotor blades.
P. Discharge nozzle.

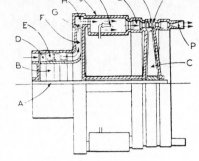

FIG. 1-11 Whittle's patent drawing. (G. G. Smith, *Gas Turbines and Jet Propulsion*, Philosophical Library, New York, 1955.)

FIG. 1-12 Whittle's first experimental engine—1937. (G. G. Smith, *Gas Turbines and Jet Propulsion*, Philosophical Library, New York, 1955.)

engine was to incorporate specifications beyond any existing gas turbine. As Whittle explains in his book, "Jet—The Story of a Pioneer":

Our compressor was of the single stage centrifugal type generally similar to, but much larger than, an aero-engine supercharger (or fan unit of a vacuum cleaner). The turbine was also a single stage unit. Thus the main moving part of the engine—the rotor—was made up of the compressor impeller, the turbine wheel and the shaft connecting the two. It was designed to rotate at 17,750 revolutions per minute, which meant a top speed of nearly 1500 feet per second for the 19 inch diameter impeller and 1250 feet per second for the 16½ inch diameter turbine. [Author's note: feet per second = $(\pi D/12)$ (rpm/60) see Chap. 5.]

Our targets for performance for the compressor, combustion chamber assembly and turbine were very ambitious and far beyond anything previously attained with similar components.

The best that had been achieved with a single stage centrifugal compressor was a pressure ratio of 2.5 with an efficiency of 65 percent (an aero-engine supercharger). Our target was a pressure ratio of 4.0 with an efficiency of 80 percent.

Our designed airflow of 1500 pounds per minute (25 pounds per second) was far greater in proportion to size than anything previously attempted (that was one of the reasons why I expected to get high efficiency). For this pressure ratio and airflow the compressor required over 3000 horsepower to drive it. Power of this order from such a small single stage turbine was well beyond all previous experience. Finally, in the combustion chamber, we aimed to burn nearly 200 gallons of fuel per hour in a space of about six cubic feet. This required a combustion intensity many times greater than a boiler furnace.

The design and manufacture of the combustion chamber, which was let to an oil burner firm, Laidlaw, Drew and Company, proved to be one of the most difficult design problems in the engine. But by April 1937, testing on the first engine began, and although the engine's performance did not come up to specifications and there was much heartbreaking failure, the machine showed enough promise to prompt the official entry of the Air Ministry into the picture (Fig. 1-12). With new funds, the original engine was rebuilt and the combustion chamber design was improved somewhat.

Testing was continued at a new site because of the danger involved.

The original engine was reconstructed several times (Fig. 1-13). Most of the rebuilding was necessitated by turbine blade failures due to faulty combustion. But enough data had been collected to consider the engine a success, and by the summer of 1939, the Air Ministry awarded to Power Jets Ltd. a contract to design a flight engine. The engine was to be flight tested in an experimental airplane called the Gloster E28. On May 15, 1941, the W1 Whittle engine installed in the Gloster E28 made its first flight, with Flight Lieutenant P. E. G. Sayer as pilot. In subsequent flights during the next few weeks, the airplane achieved a speed of 370 miles per hour [595 kilometers per hour (km/h)] (mph)] in level flight, with 1000 pounds (lb) [4448 newtons (N)] of thrust. The Gloster/Whittle E28/L39 is shown in Fig. 1-14. Sayer was later killed flying a conventional aircraft.

FIG. 1-13 Early Whittle designs.

FIG. 1-14 The Gloster E28/39, which flew in 1941.

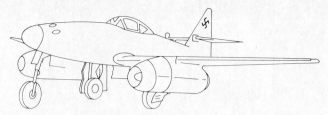

FIG. 1-16 The ME262 German operational jet fighter.

GERMAN DEVELOPMENT

Work on the gas turbine engine was going on concurrently in Germany with Whittle's work in Britain. Serious efforts toward jet propulsion of aircraft were started in the middle 1930s. Two students at Gottingen, Germany, Hans von Ohain and Max Hahn, apparently unaware of Whittle's work, patented, in 1936, an engine for jet propulsion based on the same principles as the Whittle engine. These ideas were adapted by the Ernst Heinkel Aircraft Company and the second engine of this development made a flight with Erich Wahrsitz as pilot on August 27, 1939, which is now generally considered to be the *earliest date* of modern jet propulsion. The HE178 was equipped with a centrifugal-flow jet engine, the Heinkel HeS-3b, which developed 1100 lb [4893N] of thrust and had a top speed of over 400 mph [644 km/h] (Fig. 1-15).

FIG. 1-15 The HE178 was the first true jet-propelled aircraft to fly (1939). (J. V. Casamassa and R. D. Bent, *Jet Aircraft Power Systems*, 3d ed., McGraw-Hill, New York, 1965.)

Subsequent German development of turbojet-powered aircraft produced the ME262, a 500-mph [805 km/h] fighter, powered by two axial-flow engines. (The terms centrifugal flow and axial flow will be examined in Chap. 2.) More than 1600 ME262 fighters were built in the closing stages of World War II, but they reached operational status too late to seriously challenge the overwhelming air superiority gained by the Allies (Fig. 1-16). These engines were far ahead of contemporary British developments and they foreshadowed many of the features of the more modern engine, such as blade cooling, ice prevention, and the variable-area exhaust nozzle. An interesting sidelight to the German contribution was that on September 30, 1929, a modified glider using Opel rockets was the world's first airplane to achieve flight using a reaction engine.

ITALIAN CONTRIBUTION

Although strictly speaking not a gas turbine engine in the present sense of the term, an engine designed by Secundo Campini of the Caproni Company in Italy utilized the reaction principle (Fig. 1-17). A successful flight was made in August 1940, and was reported, at the time, as the first successful flight of a jet-propelled aircraft (Fig. 1-18). The power plant of this aircraft was not a "jet" because it relied upon a conventional 900-horsepower (hp) [671-kilowatt (kw)] reciprocating engine instead of a turbine to operate the three-stage compressor. Top speed for this aircraft was a disappointing 205 mph [330 km/h], and the project was abandoned in late 1948.

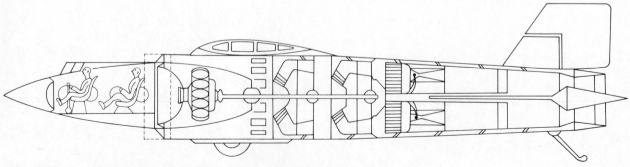

FIG. 1-17 The CC-1, a proposed Italian design never flown. This illustration shows the compressors being driven by a reciprocating engine.

FIG. 1-18 The Caproni-Campini CC-2 flew using the engine configuration shown in Fig. 1-17. (G. G. Smith, *Gas Turbines and Jet Propulsion*, Philosophical Library, New York, 1955.)

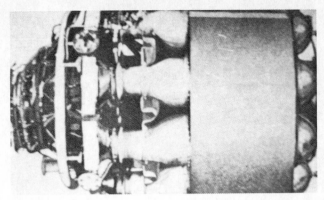

FIG. 1-20 General Electric I-A, the first jet engine built in the United States.

DEVELOPMENT IN AMERICA

America was a late comer to the jet-propulsion field, although it must be said that at the time it was felt that the war would have to be won with airplanes using conventional reciprocating engines.

In September of 1941, under the auspices of the National Advisory Committee for Aeronautics, the W.1X engine, which was the forerunner of the W.1, and a complete set of plans and drawings for the more advanced W.2B gas turbine, were flown to the United States under special arrangements between the British and American governments. A group of Power Jets engineers was also sent. The General Electric Corporation was awarded the contract to build an American version of this engine because of their previous experience with turbosuperchargers and Moss' pioneering work in this area.

The first jet airplane flight in this country was made in October 1942, in a Bell XP-59A (Fig. 1-19) with Bell's chief test pilot Robert M. Stanley at the controls. The two General Electric I-A engines (Fig. 1-20) used in this experimental airplane were adaptations of the Whittle design. But while the Whittle engine was rated at 850 lb [3781N] of thrust, the I-A was rated at about 1300 lb [5782N] of thrust and with a lower specific fuel consumption. (Specific fuel consumption will be defined later in the book.) To make the story even more dramatic, both engine and airframe were designed and built in one year. A project of similar proportions would take several years at the present time.

General Electric's early entry into the jet-engine field gave the company a lead in the manufacturing of gas turbines, but they were handicapped somewhat by having to work with preconceived ideas, after having seen Whittle's engine and drawings. Now, the N.A.C.A. Jet Propulsion Committee began to look about for a manufacturer to produce an all-American engine. Their choice was the Westinghouse Corporation, because of this company's previous experience with steam turbines. The contract was let late in 1941 by the Navy, but it was decided not to inform the Westinghouse people of the existence of the Whittle engine. As it turned out, this decision was a correct one, for the Westinghouse engineers designed an engine with an axial compressor and an annular combustion chamber. Both of these innovations, or variations thereof, have stood the test of time and are being used in contemporary engines.

Shortly thereafter, several other companies began to design and produce gas turbine engines. Notable among these were Detroit Diesel Allison, Garrett AiResearch, Boeing, Teledyne CAE, Avco Lycoming, Pratt & Whitney Aircraft, Solar, and Wright. Of these, Boeing, Westinghouse, Solar, and Wright are no longer manufacturing jet engines. Additionally, many of these companies have undergone name changes. The several companies currently in production offer a variety of gas turbines to the user, most of which are discussed in Chap. 2, along with the airplanes in which these engines are installed.

FIG. 1-19 America's first jet airplane, the Bell XP-59A, powered by two General Electric I-A engines.

REVIEW AND STUDY QUESTIONS

1. How old is the idea of jet propulsion?
2. Describe the first practical device utilizing the reaction principle.
3. What was Leonardo Da Vinci's contribution to the development of a jet engine?
4. Who were the first people to use rockets? Give an example of the use of rockets in war.
5. Between the years 1600 and 1800, who were the contributors to the development of the gas turbine engine? What were those contributions?
6. What was Sir Frank Whittle's chief contribution to the further development of the gas turbine engine?
7. Give a brief outline of the efforts of Whittle and his company to design a jet engine.
8. Describe the German contributions to the jet engine.

9. Which country was the first to fly a jet-powered aircraft? What was the designation of this airplane and with what type of engine was it equipped?

10. When considering who was first with the development of a jet engine, why should the Italian engine be discounted?

11. What American company was chosen to build the first jet engine? Why?

12. Describe the series of events leading up to the first American jet airplane. Who built this plane?

13. List several American companies who manufacture gas turbine engines.

CHAPTER TWO
TYPES,
VARIATIONS,
AND
APPLICATIONS

INTRODUCTION

The author has included in Chap. 2 almost every American engine produced currently, or within the last several years, and most of the American and foreign aircraft in which the engine is installed. Also included are many foreign engines that are used in American aircraft. In addition, several special-purpose gas turbine engines, such as auxiliary power units, missile power plants, etc., are also listed where these engines incorporate unusual or interesting design features. The engines are arranged alphabetically by manufacturer and within the major classification, by compressor type. It should be kept in mind that the specifications which accompany each of the engines only approximately reflect actual engine parameters, such as thrust, airflow, specific fuel consumption, etc. This is due to the fact that several configurations (dash numbers) are possible for each model engine. All values are given for sea level, static conditions, at maximum power. It is hoped that readers will find this section a useful and valuable reference throughout their studies of this form of prime mover.

THE GAS TURBINE ENGINE

Gas turbine engines can be classified according to the type of compressor used, the path the air takes through the engine, and how the power produced is extracted or used (Fig. 2-1).

Compressor types fall into three categories:

1. Centrifugal flow
2. Axial flow
3. Centrifugal-axial flow

while power usage produces the following divisions:

1. Turbojet engines
2. Turbofan engines
3. Turboprop engines
4. Turboshaft engines

Compression is achieved in a centrifugal-flow engine by accelerating air outward perpendicular to the longitudinal axis of the machine, while in the axial-flow type, air is compressed by a series of rotating and stationary airfoils moving the air parallel to the longitudinal axis. The centrifugal-axial design uses both kinds of compressors to achieve the desired compression.

In relation to power usage, the turbojet engine directly uses the reaction resulting from a stream of high-energy gas emerging from the rear of the engine at a higher velocity than it had at the forward end. The turbofan engine also uses the reaction principle, but the gases exiting from the rear of this engine type have a lower energy level, since some power has to be extracted to drive the fan. (See pages 79–81 for a more detailed explanation of the operating principles of the fan engine.) Turboprop and turboshaft engines both convert the majority of the kinetic (energy of motion), static (energy of pressure), and temperature energies of the gas into torque to drive the propeller in one case, and a shaft in the other. Very little thrust from reaction is produced by the exiting gas stream.

From these basic types of gas turbine engines have come the literally dozens of variations that are either in actual service, or are in various stages of development. Many combinations are possible, since the centrifugal- and axial-flow compressor engines can be used for turbojet, turbofan, turboprop, or turboshaft applications. Furthermore, within the major classifications are a host of variations, some of which are discussed on the following pages.

CENTRIFUGAL COMPRESSOR ENGINES

Variations of this type of compressor include the single-stage, two-stage, and single-stage double-entry compressor (Fig. 2-2). The centrifugal design works well for small engines where a high compression ratio (pressure rise across the entire compressor) is not too essential, or where other design or operational considerations may take precedence.

The principal advantages of the compressor are:

1. Light weight
2. Ruggedness, and therefore resistance to foreign object damage
3. Simplicity
4. Low cost
5. High compressor ratio *per stage* (but number of stages is limited)

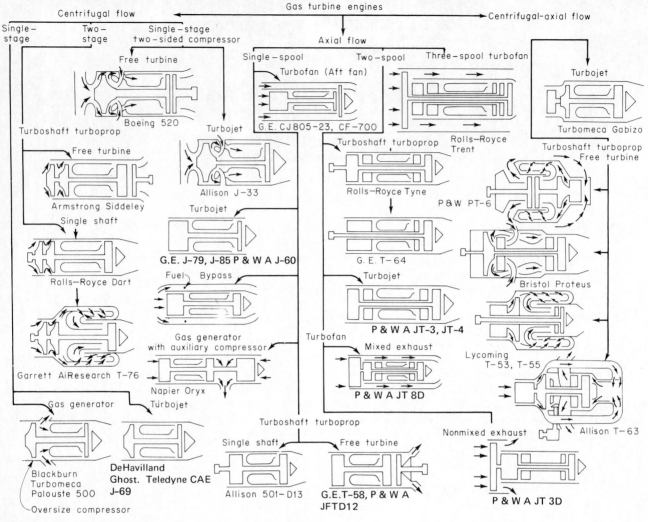

FIG. 2-1 The family tree. Note: The fronts of all the engines are to the reader's left.

Probably the most famous example (historically speaking) of this type of powerplant is the Detroit Diesel Allison J33 (Fig. 2-4), used in the first U.S.A.F. jet, the Lockheed P-80. Newer versions were used in the T-33, which was a training version of the P-80. Centrifugal compressors have found wide acceptance on smaller gas turbine engines. Examples of this application are the Teledyne CAE (Continental Aviation and Engineering Co.) J69 (Fig. 2-55), and the auxiliary power units based on this engine's series (Fig. 2-56), (figure 2-57 represents further developments of the basic J69 design), the Williams Research Corporation WR24-6 and WR27-1 (Fig. 2-58), the Detroit Diesel Allison model 250 series III (Fig. 2-5), and Avco Lycoming LT series engines (Fig. 2-11). Two other examples of engines equipped with a form of the centrifugal compressor are the Rolls Royce Dart (Fig. 2-47), and the Garrett AiResearch TPE331 (Fig. 2-18). These turboprop engines incorporate a two-stage compressor and integral propeller reduction gearbox. (See Fig. 2-8 for an example of an engine equipped with a separate propeller reduction gearbox.) Interesting features to note on some of these engines is the radial-inflow gas-producer tur-

bine shown in Fig. 2-14(a) and (b), and the "free-power" turbine shown in Figs. 2-14 and 2-15. The radial-inflow turbine is essentially the opposite of the centrifugal or radial compressor. It receives the hot gases from the combustion chamber at its periphery where they then proceed to flow inward toward the center, causing the turbine wheel to turn.

The free-power turbine used on many different forms of gas turbines has no mechanical connection to the primary or gas-generator turbine which, in this situation, is used only to turn the compressor in order to supply high-energy gases to drive the free-power turbine. The design lends itself to variable-speed operation better than the single shaft, and it produces high torque at low free-power turbine speeds. In addition, this type of powerplant has the advantage of requiring no clutch when starting or when a load is applied.

Some centrifugal and axial engines incorporate a heat exchanger called a regenerator or recuperator. The purpose of the regenerator or recuperator is to return some of the heat energy that would normally be lost with the exhaust, to the front of the combustion chamber. By doing this, less fuel needs to be added to reach the tur-

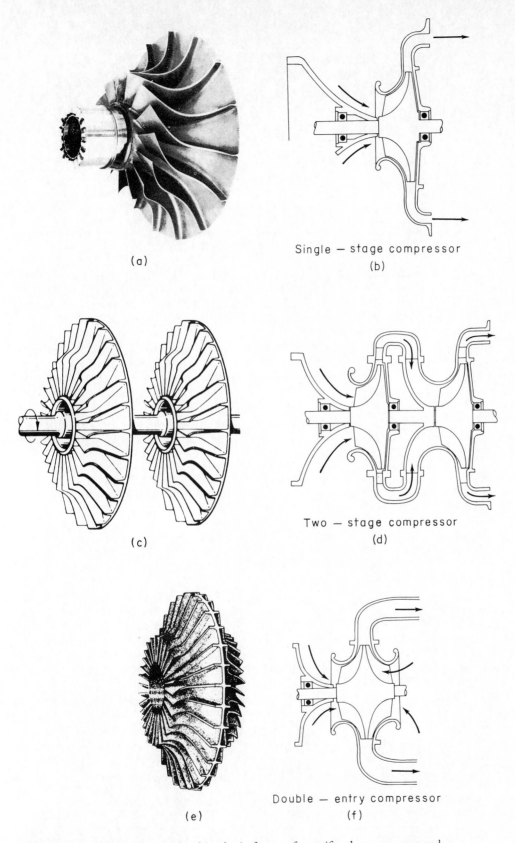

(a)

Single — stage compressor
(b)

(c)

Two — stage compressor
(d)

(e)

Double — entry compressor
(f)

FIG. 2-2 Drawings showing the three basic forms of centrifugal compressors and schematics showing the airflow through each.
(a) and (b) The single-stage centrifugal compressor.
(c) and (d) The two-stage centrifugal compressor (compressors in series).
(e) and (f) The two-stage or double-entry centrifugal compressor (compressors in parallel).

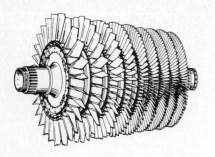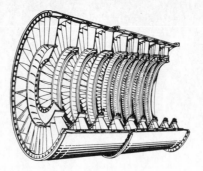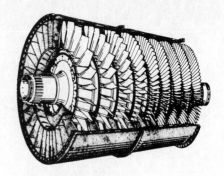

FIG. 2-3 The axial-flow compressor rotor and stator assembly.

bine limiting temperatures and this will result in high thermal efficiency, low specific fuel consumption, and low exhaust gas temperature. Although regeneration has been used on a number of ground-power engines, at the time of this writing, no aircraft engines use this method of power recovery because of excessive weight and/or regenerator air-sealing difficulties. Two basic regenerator or recuperator types are the rotary drum shown in Fig. 2-6, and the stationary or nonrotating type designed for the Allison T78 turboprop shown in Fig. 2-10. The latter was never put into production.

AXIAL-FLOW COMPRESSOR ENGINES (*FIG. 2-3*)

Engines using this type of compressor may incorporate one, two, or three spools (a spool is defined as a group of compressor stages, a shaft, and one or more turbine stages, mechanically linked, and rotating at the same speed; see Figs. 2-48, 2-40, and 2-52 for examples of a single-spool, two-spool, and three-spool engine respectively), forward or rear fans, afterburners, and free-power turbines, and may be used in a variety of applications such as turbojet, turbofan, turboprop, and turboshaft engines.

Most large gas turbine engines use this type of compressor because of its ability to handle large volumes of airflow at pressure ratios in excess of 20:1. Unfortunately, it is more susceptible to foreign-object damage, it is expensive to manufacture, it is very heavy in comparison with the centrifugal compressor with the same compression ratio, and it is more sensitive to "off design" operation. (See Chap. 5 for aerodynamic and thermodynamic considerations relating to the axial flow compressor.)

Examples of axial-flow machines are even more numerous than those of the centrifugal-flow types and include all of the uses to which gas turbines may be put. Two of the most widely used axial-flow engines are the Pratt & Whitney JT3 (J57) and JT4 (J75) series powerplants (Fig. 2-40). These engines were used in early-model Boeing 707s and 720s and Douglas DC-8s, and except for dimensional and thrust values, are essentially the same in construction. Newer designs are installed in several military aircraft. A forward-fan version of this engine, the JT3D (Fig. 2-42), is currently powering later versions of the 707, 720, and DC-8, and some

military aircraft. Other Pratt & Whitney Aircraft forward-fan engines are the JT9D (Fig. 2-44) (a high-by-pass-ratio design; see page 81 for a discussion of bypass ratio) used on the Boeing 747, the JT8D (Fig. 2-43) used on the Boeing 727 and 737 and Douglas DC-9, and the TF30 (Fig. 2-45) installed in the General Dynamics/Grumman F-111, the Navy Grumman F-14, and Vought Corsair II. Typical of the two-spool design used by Pratt & Whitney Aircraft is the J52 (Fig. 2-38), a straight turbojet engine used in the Douglas Skyhawk and Grumman Intruder. One of the few supersonic-cruise engines, the J58, is also manufactured by this company (Fig. 2-39).

Pratt & Whitney Aircraft makes the JT12 (J60), a small axial-flow engine in the 3000-lb [13,344-N] thrust class (Fig. 2-36). One of these engines is installed in the North American Buckeye, two JT12's are installed in the North American Sabreliner, while four are used to power the earlier-model Lockheed Jetstar. Note the placement of the engine(s) on these aircraft. This company also manufactures an axial-flow turboprop, the T34 (Fig. 2-41), which is used in the Douglas C-133, and a free-power turboshaft engine, the JFTD12 (Fig. 2-37), two of which are used in the Sikorsky Skycrane helicopter.

Pratt & Whitney's latest engines include the F100, an augmented (afterburner) two-spool turbofan (Fig. 2-46), for the McDonnell Douglas F15 and the General Dynamics F16, and the JT15D used on the Cessna Citation (Fig. 2-35). The JT15D is an interesting design incorporating a reverse-flow combustion chamber constructed to keep the engine as short as possible. (See Figs. 2-12, 2-13, 2-19, and 2-34 for other engine designs using the reverse-flow burner concept.)

Another major manufacturer of both large and small axial-flow gas turbines in this country is the General Electric Company. One of their most highly produced machines is the J79 series (Fig. 2-22) used in many supersonic U.S. Air Force and Navy airplanes. A commercial version of this engine is called the CJ805-3 (Fig. 2-23), and an aft fan counterpart, the CJ805-23 (Fig. 2-24) is used in the Convair 880 and the Convair 990, respectively. Two points worth noting about these engines are the variable-angle inlet guide vanes, the variable-angle first six stator stages in the compressor, and the location and the method of driving the fan in the CJ805-23 engine. The fan, located in the rear, is "gas coupled" to the primary engine as opposed to the me-

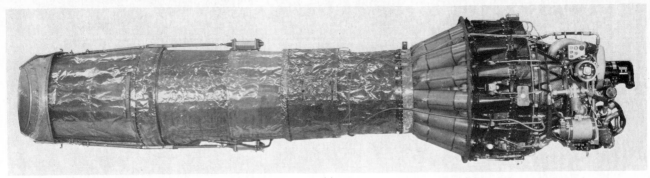

(a)

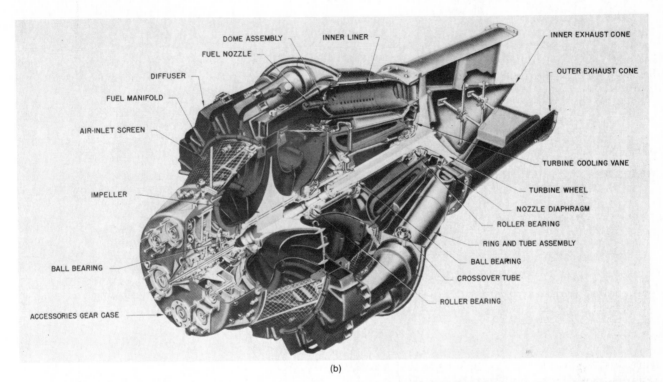

DOME ASSEMBLY — INNER LINER — INNER EXHAUST CONE
FUEL NOZZLE — OUTER EXHAUST CONE
DIFFUSER
FUEL MANIFOLD
AIR-INLET SCREEN
TURBINE COOLING VANE
IMPELLER
TURBINE WHEEL
NOZZLE DIAPHRAGM
ROLLER BEARING
RING AND TUBE ASSEMBLY
BALL BEARING
BALL BEARING
CROSSOVER TUBE
ACCESSORIES GEAR CASE
ROLLER BEARING

(b)

DETROIT DIESEL ALLISON DIVISION J33 (*FIG. 2-4*)

Although this engine is no longer in production, it is included here as an example of a centrifugal flow turbojet with a double-sided compressor. Since several models were produced, engine operating parameters are given as ranges and/or approximations. The compression ratio is approximately 4.5:1, the compressor flows 90 to 110 pounds per second (lb/s) [41 to 50 kilograms per second (kg/s)] at 11,800 rpm, and is driven by a single-stage turbine. Fourteen can-type combustion chambers are interconnected by cross-ignition tubes. Specific fuel consumption is about 1 pound/pound of thrust/hour (lb/lbt/h) [102 grams/Newtons/hour (g/N/h)]. The engine weighs about 1900 pounds (lb) [862 kilograms (kg)] and produces thrusts up to 6000 lb [26,688 N].

(c)

FIG. 2-4 The Detroit Diesel Allison Division J33, now out of production, is included here as an example of a large double-entry centrifugal-compressor engine.
(a) External view of the Allison J33 equipped with an afterburner.
(b) Cutaway view of the Allison J33 turbojet.
(c) The Lockheed T-33, training version of the F-80 Shooting Star, is powered by one Allison J33 engine. The T-33/F-80 is now removed from the U.S.A.F. inventory.

TYPES, VARIATIONS, AND APPLICATIONS   13

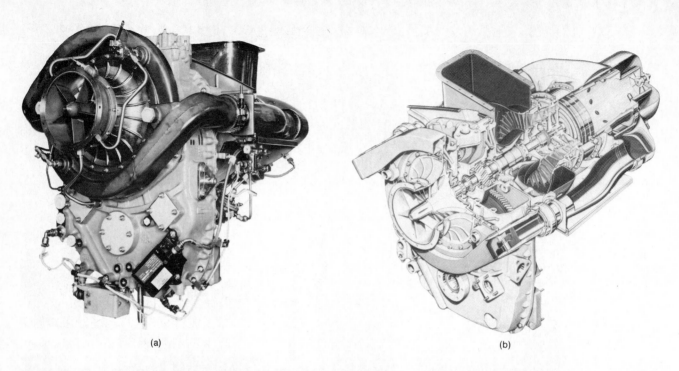

(a)

(b)

(c)

FIG. 2-5 The Detroit Diesel Allison Division of General Motors Corp. model 250–C28 series III engine capable of producing 500 to 650 hp.
(a) External view of the Allison model 250-C28 engine.
(b) Sectioned view of one version of the model 250 series III engine.
(c) The Sikorsky S-76 with two Allison model 2-C28 or -30 engines.

14 HISTORY AND THEORY

chanical coupling used in many of the Pratt & Whitney Aircraft designs and others. Placing the fan in the rear and having it gas-coupled is claimed to compromise basic engine performance to a lesser degree. In addition, the engine can be accelerated faster and the aft fan blades are automatically anti-iced by thermal conduction. Forward, mechanically coupled, fan designers claim fewer problems resulting from foreign-object damage, since most of the foreign material will be thrown radially outward and not passed through the rest of the engine, and furthermore, claim that the forward fan is in the cold section of the engine for highest durability, reliability, and minimum sealing problems.

In addition to their aft-fan designs, General Electric also produces a *high-bypass*-ratio forward-fan engine called the TF39 (Fig. 2-26) which powers the Lockheed C5A Galaxy, the largest airplane in the world.

From the TF39, General Electric has developed a series of engines utilizing the same basic gas generator (core) portion of the engine, but has changed the fan and the number of turbines needed to drive the fan. Such an engine series is the CF6 (Fig. 2-27), installed in the McDonnell Douglas DC-10, A300 air bus and other wide-body transports. The Rockwell International B-1 Bomber uses one of the newest General Electric designs, the F101, *medium bypass* turbofan (Fig. 2-30). A *low bypass* General Electric turbofan engine is the F404 used in the McDonnell Douglas Northrop F-18 (Fig. 2-31).

Like Pratt & Whitney Aircraft, General Electric manufactures a series of smaller gas turbine engines. The CJ610, or J85 (Fig. 2-20), is used in the Gates Lear Jet, Northrop Talon (F5), and the early Jet Commander. (The Jet Commander, now made in Israel, is called the Westwind 1124, and is powered by the Garrett Ai-

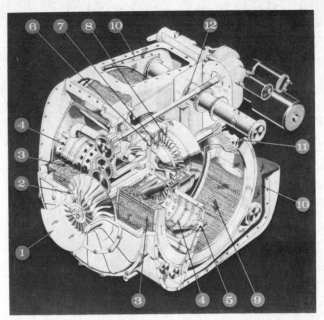

FIG. 2-6 The Detroit Diesel Allison Division GMT-305 regenerative gas turbine. (See specification inset for description.)

Research TFE731 engine.) As might be expected, General Electric has developed an aft fan version of this engine called the CF700 (Fig. 2-21), two of which are installed on many models of the Falcon fanjet. (Text continues on page 34.)

DETROIT DIESEL ALLISON DIVISION 250-C28 SERIES III (*FIG. 2-5*)

The Allison Series III represents a nearly total redesign from previous model 250 (T63) engines, (See Fig. 2-9.) principally through the elimination of the multi-stage axial compressor and development of a high compression ratio (7:1) centrifugal compressor with an airflow of 4.33 lb/s [1.96 kg/s], rotating at 51,005 rpm. Output shaft speed is 6000 rpm in the shaft version, while the turboprop has an additional gear reduction. Air from the compressor flows through external pipes to the single can-type combustion chamber, through a two-stage gas producer turbine and then through a two-stage free-power turbine which powers the load through the drive gears. Specific fuel consumption is .64 pounds/shaft horsepower/hour (lb/shp/h) [290 grams/shaft horsepower/hour (g/shp/h)]. The engine weighs 200 lb [90.7 kg] and produces 500 shp.

DETROIT DIESEL ALLISON DIVISION GMT-305 WHIRLFIRE (*FIG. 2-6*)

The engine is designed to be used as a prime source of power for land vehicles, but it is included here to show the rotating regenerator. The GMT-305 Whirlfire is shown in the cutaway with some of the frame parts omitted for clarity. The arrows show the airflow through the engine. Air enters the inlet (1), is compressed to over 3 atm by a centrifugal compressor (2), and absorbs exhaust heat as it passes through two rotating regenerators (3). The heated compressed air then enters the combustors (4) where the fuel nozzles (5) inject fuel for combustion. The combustion gases pass through the turbine vanes (6) and drive the gasifier turbine (7). The gases then drive the power turbine (8) which is not mechanically connected to the gasifier shaft. Hot exhaust gas is cooled to 300-500° F [150-260° C] as it passes through the self-cleaning rotating regenerators (9) and is directed out the exhaust ports (10). The power turbine drives the power output shaft (11) through a single-stage helical reduction gear. The gasifier turbine drives the accessory shaft (12) through a set of reduction gears. (See Fig. 2-10 for a description of a fixed regenerator engine.)

(a)

DETROIT DIESEL ALLISON DIVISION J71 (FIG. 2-7)

The 16-stage axial-flow compressor of this turbojet engine flows 160 lb/s [73 kg/s] at a compression ratio of 8:1 at 6100 rpm. The can-annular combustor has ten interconnected flame tubes. A three-stage turbine drives the compressor. Specific fuel consumption is 1.8 lb/lbt/h [183.5 g/N/h] with the afterburner in operation. The engine weighs 4900 lb [2223 kg] and produces 10,000 to 14,000 lbt [44,480 to 62,272 N] under normal and reheat (afterburner) operation.

(c)

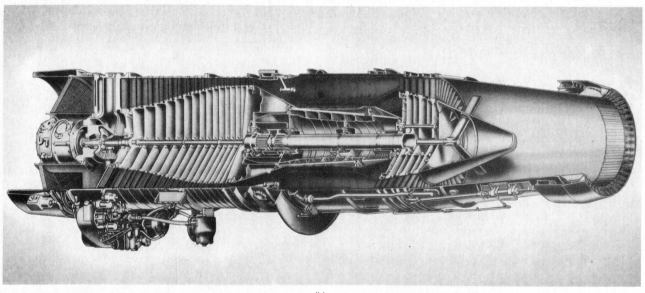

(b)

FIG. 2-7 The Detroit Diesel Allison Division J71 turbojet engine.
(a) External view of the Allison J71 engine.
(b) Cutaway view of the Allison J71 engine.
(c) The McDonnell B-66 has two Allison J71 turbojets.

16 HISTORY AND THEORY

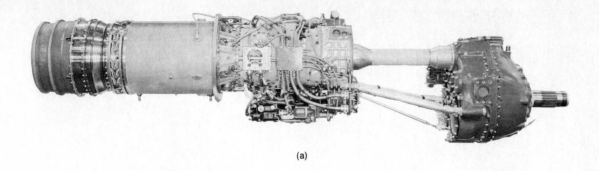

(a)

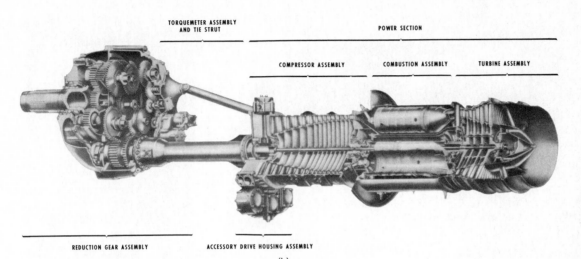

TORQUEMETER ASSEMBLY
AND TIE STRUT

POWER SECTION

COMPRESSOR ASSEMBLY   COMBUSTION ASSEMBLY   TURBINE ASSEMBLY

REDUCTION GEAR ASSEMBLY   ACCESSORY DRIVE HOUSING ASSEMBLY

(b)

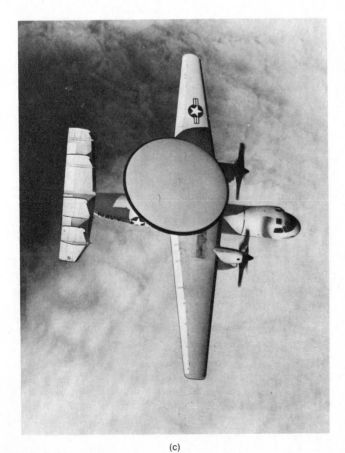

(c)

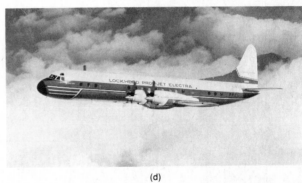

(d)

FIG. 2-8 The Detroit Diesel Allison Division 501-D (T56) series engine.

(a) External view of the Allison 501-D13 turboprop engine. Note that the gearbox can be offset up or down.

(b) Sectioned view of the Allison 501-D13 with the reduction gearbox offset up.

(c) The Grumman E-2A Hawkeye with two Allison T56 engines.

(d) The Lockhood Electra with four Allison 501-D13 engines. (Continued on following page.)

(e)

FIG. 2-8 (continued)

(e) The Allison 580 Convair Conversion uses two Allison 501-D13 engines.

(f) The Lockheed—Georgia Division C-130 powered by four Allison 501-D15 turboprop engines. Several versions of this aircraft are available.

(g) The Lockheed—California Division P-3C Orion. The anti-submarine warfare (ASW) weapons system is equipped with four Allison T56-A-14 engines and is based upon the Electra transport aircraft.

(h) The C-130SS Hercules. This stretched C-130 is powered by four Allison T56-A-15 engines having 4591 ESHP each. (See Chap. 3.)

(f)

(g)

DETROIT DIESEL ALLISON DIVISION 501-D SERIES (T56) (*FIG. 2-8*)

This is a single-shaft turboprop engine equipped with a separate propeller reduction gearbox which reduces the propeller rpm to 1020. The 14-stage axial-flow compressor is driven by a 4-stage turbine. Compression ratio is 9.5:1, and airflow is 33 lb/s [15 kg/s] at 13,820 rpm. The combustion chamber is the can-annular type with six flame tubes. Specific fuel consumption is .53 lb/eshp/h [241 g/eshp/h], and the weight of the engine is approximately 1800 lb [817 kg]. The power produced ranges from 3750 eshp (3460 shp plus about 725 lb [3225 N]) of thrust to almost 5000 eshp, depending upon the model.

(h)

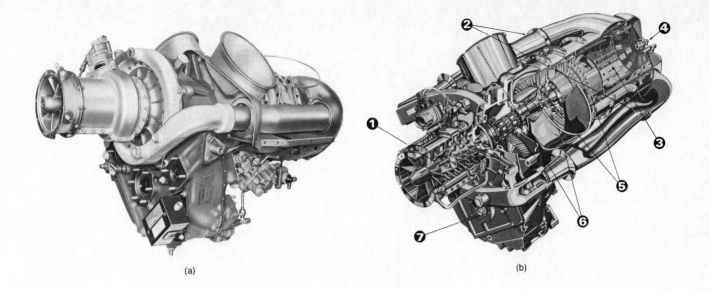

(a)

(b)

(c)

FIG. 2-9 The Detroit Diesel Allison Division Model 250 Series engine (T63).
(a) External view of the Allison Model 250-C-20B turboshaft engine.
(b) Cutaway view of the Allison Model 250 engine. (1) Compressor: air enters the inlet and is compressed to over 6 atm by the six axial stages and one centrifugal stage of the compressor. (2) Air-transfer tubes: the high-pressure discharge air from the compressor is transferred rearward to the combustion section through the two air-transfer tubes. (3) Combustor: the single combustor regulates and evenly distributes the engine airflow. (4) Fuel nozzle: fuel is injected through a single duplex-type fuel nozzle. (The fuel is ignited by a single ignitor plug adjacent to the fuel nozzle and used only during the starting cycle.) (5) Turbines: the hot combustion gases pass forward through the first two-stage axial turbine which drives the compressor and thence through the second two-stage axial turbine which drives the power output shaft. (6) Exhaust: after passing forward through the turbine section, the gases are exhausted upward through twin exhaust ducts. (7) Power output shaft—6000 rpm: the energy of the turbine section, after passing through appropriate gearing in the accessories gear case, is available from an internally splined shaft at either the front or rear output pad.
(c) The Detroit Diesel Allison Division Model 250-C18 powers the Model 206B Jet Ranger helicopter. Other models of this engine are installed in the Bell Long Ranger 206L, a stretched version of the Bell 206B, the Hiller UH12, and the Agusta A-109A. (Continued on following page.)

DETROIT DIESEL ALLISON DIVISION MODEL 250-C18 (T63) (*FIG. 2-9*)

This is an axial-centrifugal flow, free-power turbine turboshaft/turboprop engine. The five-stage axial and single-stage centrifugal compressors provide a compression ratio of 6.2:1, and a airflow of 3 to 3.6 lb/s [1.4 to 1.6 kg/s] at 51,600 rpm. There is a single reverse-flow combustor at the rear. The compressor is driven by a two-stage turbine and the load is driven by a two-stage turbine which turns at 35,000 rpm. Specific fuel consumption is .7 lb/shp/h [318 g/shp/h]. The engine weighs approximately 170 lb [77 kg], and produces from 317 shp to over 400 shp, depending upon the model. Shaft rpm is 6000, and propeller rpm is 2200.

(d)

(e)

FIG. 2-9 continued

(d) The Hughes Helicopter Division OH-6A Light Observation Helicopter (LOH) is powered by one Allison T63 (civil model 250-C18). A commercial version of this helicopter is called the Hughes 500 C/D, and is powered by an Allison 250-C20 400SHP engine.

(e) The BO-105C, manufactured by MBB (Messerschmidt-Boelkow-Blohm), and marketed and supported in North America by Boeing Vertol, is driven by the Allison model 250-C20B engine.

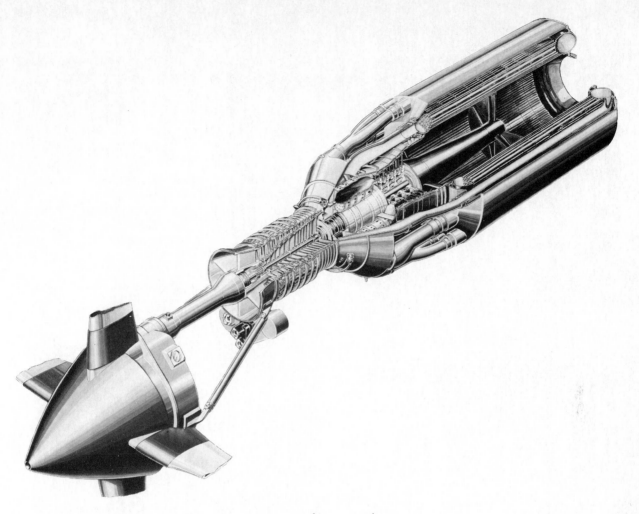

FIG. 2-10 The Detroit Diesel Allison T78 regenerator turboprop engine.

DETROIT DIESEL ALLISON DIVISION T78 (*FIG. 2-10*)

Based slightly on the Allison 501 (T56) designs, the Allison T78 regenerator turboprop incorporated a variable stator compressor and a fixed recuperator or regenerator which promised a 36 percent reduction in specific fuel consumption over a comparable non-regenerative turboprop engine. The engine had a relatively unusual side-entry burner. It was never placed in production, but is included here as an example of an engine incorporating a fixed regenerator. The function of the regenerator is to cycle some of the heat energy normally lost through the exhaust back into the engine at the front of the combustion chamber. In this way less heat energy, in the form of fuel, needs to be metered before reaching turbine temperature limits. (See Fig. 2-6 for a description of a rotating regenerator engine.)

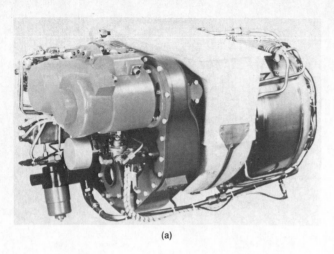

(a)

GEARBOX MODULE   GAS GENERATOR MODULE   COMBUSTOR/POWER TURBINE MODULE

(c)

(b)

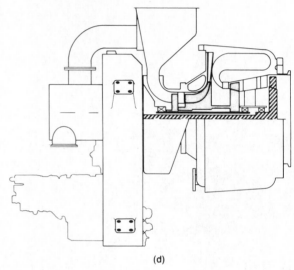

(d)

(e)

FIG. 2-11 The Avco Lycoming Corporation LTS/LTP series turboshaft/turboprop gas turbine engines.

(a) The Lycoming LTS101. In addition to being installed in the Bell Model 222 (e), this engine is also scheduled for use in the:

   1. United States Sikorsky S-55T-2 twin conversion helicopter.

   2. Japanese Kawasaki KH-7 light twin helicopter.

   3. French Aerospatiale AS-350 Sunbird helicopter.

(b) The Lycoming LTP101, turboprop version of the LTS/LTP series engines for use in the Italian Piaggio P-166-DL3 and the British Britten-Norman Turbo Islander.

(c) The Lycoming LTS101 engine showing the modular design which allows initial lower cost and easier maintainability.

(d) Schematic cross section of the Lycoming LTS101 turboshaft engine.

(e) The Bell model 222 commercial light twin-turbine helicopter powered by two Lycoming LTS101 engines.

AVCO LYCOMING LTS/LTP SERIES (*FIG. 2-11*)

The Lycoming LTS/LTP series turboshaft/turboprop are the smallest of the company's aircraft engines, but they can produce more than two horsepower for each pound of engine weight. The single-stage axial and single-stage centrifugal compressor is driven by one turbine wheel and has a mass airflow of 5 lb/s [2.27 kg/s] with a pressure ratio of 8.5:1. The single-stage power turbine drives the load through a gearbox. The LTP version has an additional reduction gear stage so that the propeller will turn in a range of 2000 rpm. Specific fuel consumption is .551 lb/shp/h [250 g/shp/h]. The engine weighs 290 lb [132 kg] and produces 610 eshp or 587 shp plus 57.5 lbt [256 N].

(a)

1 ANNULAR INLET
2 VARIABLE INLET GUIDE VANE
3 FIVE-STAGE AXIAL COMPRESSOR
4 SINGLE CENTRIFUGAL COMPRESSOR STAGE
5 RADIAL DIFFUSER
6 AREA SURROUNDING COMBUSTION CHAMBER
7 COMBUSTION CHAMBER
8 VAPORIZER TUBES (IN EARLIER VERSIONS) OR ATOMIZERS
9 COMBUSTION TURNING ZONE
10 AIR-COOLED FIRST TURBINE NOZZLE
11–12 TWO-STAGE COMPRESSOR TURBINE
13 FREE-POWER TURBINE NOZZLE
14–15 TWO-STAGE FREE-POWER TURBINE
16 THROUGH-SHAFT
17 PLANETARY REDUCTION GEAR
18 INLET HOUSING

(b)

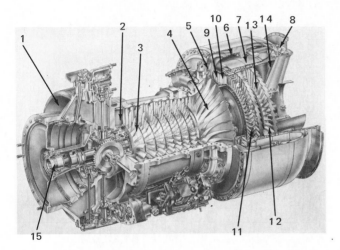

1 ANNULAR INLET
2 INLET GUIDE VANES
3 SEVEN AXIAL STAGES
4 SINGLE-STAGE CENTRIFUGAL COMPRESSOR ASSEMBLY
5 RADIAL DIFFUSER
6 AREA SURROUNDING THE COMBUSTION CHAMBER
7 ANNULAR, REVERSE FLOW, COMBUSTION CHAMBER
8 ATOMIZER OR VAPORIZER TUBES
9 180° TURNING AREA
10 AIR-COOLED FIRST TURBINE NOZZLE
11 COMPRESSOR-TURBINE ROTOR ASSEMBLY
12 FREE-POWER TURBINE NOZZLE
13 TWO-STAGE FREE-POWER TURBINE ASSEMBLY
14 STRUT
15 CONCENTRIC OUTPUT SHAFT

(c)

(d)

FIG. 2-12 The Avco Lycoming Corporation T53/T55 turboprop-turboshaft series engine.
(a) External view of the Lycoming T53 turboshaft engine.
(b) Cutaway view of the Lycoming T53-L-13 with a two-stage compressor turbine.
(c) Cutaway view of the Lycoming T55-L-7 with a single-stage turbine to drive the compressor.
(d) The Bell 204B, civil version of the military UH-1 Iroquois helicopter, uses one Lycoming T53 turboshaft engine. (Continued on the following page.)

(e)

(f)

(h)

(e) The Kamen HH-43B Huskie is powered by one Lycoming T53 engine.
(f) These three Bell helicopters use the Lycoming T53 engine (top to bottom: UH-1D, UH-1B, and the Huey Cobra).
(g) The Lycoming T53 turboprop version powers the Grumman OV-1A Mohawk.
(h) The Boeing Helicopter Vertol Division CH-47 is equipped with two Lycoming T55 turboshaft engines.

AVCO LYCOMING T53 (T55)
(FIG. 2-12)

This engine incorporates a five-stage axial and single-stage centrifugal compressor. The compression ratio is 6:1 to 7:1 and mass airflow is 11 to 12 lb/s [5 to 5.5 kg/s] at 25,400 rpm. The combustion chamber is a folded or reverse-flow annular type. Some models are equipped with a single gas generator turbine and a single free-power turbine, while others have two of each kind. The load (turboprop or turboshaft) is driven at the front of the engine by means of a concentric shaft. Specific fuel consumption, weight, and power ratings all vary with the model engine but run from approximately .6 to .7 lb/shp/h [272 to 318 g/shp/h], 500 to 600 lb [227 to 272 kg] of weight, and 1100 to 1400 shp plus 125 lbt (556 N) respectively. The T55 is a growth version of the T53 and is capable of producing over 3700 shp.

(g)

AVCO LYCOMING ALF502
(FIG. 2-13)

This engine is derived from the Avco Lycoming T55 turboshaft engine. The high bypass ratio fan and single-stage low-pressure compressor is driven by the last two stages of a four-stage turbine through reduction gears. Fan bypass ratio is 6:1. Total airflow is 240 lb/s (109 kg/s). The combustion chamber is of the reverse flow, or folded annular type for short engine length and turbine blade containment in case of failure. See Figs. 2-11, 2-12, 2-18, 2-19, 2-34, and 2-35 for additional examples of this type of combustion chamber. Specific fuel consumption is .42 lb/lbt/h [42.81 g/N/h]. The engine weighs 1245 lb [565 kg] and produces 5500 to 6500 lbt (24,464 to 28,912 N) depending upon the model.

(a)

(b)

SINGLE-STAGE FAN
WITH A SINGLE
LOW-PRESSURE
COMPRESSOR STAGE

FOLDED ANNULAR
COMBUSTOR

TWO-STAGE
LOW-PRESSURE
TURBINE

TWO-STAGE-COOLED
HIGH-PRESSURE
TURBINE

HIGH-PRESSURE COMPRESSOR
7 AXIAL + 1 CENTRIFUGAL STAGE

REDUCTION GEAR

(c)

FIG. 2-13 The Avco Lycoming ALF502 turbofan.
(a) External view of the ALF502 turbofan engine.
(b) Two Lycoming ALF502 turbofans installed in the Canadair CL-600 Challenger.
(c) Cutaway view of the Avco Lycoming ALF502, high-bypass-ratio geared fan engine. Notice that the core is basically the Lycoming T55 engine.

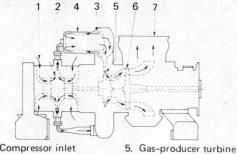

1. Compressor inlet   5. Gas-producer turbine
2. Compressor   6. Output turbine
3. Compressor collector   7. Exhaust
4. Burner

(a)

(b)

FIG. 2-14 The Boeing 520 (T60) turboshaft engine.
(a) Schematic showing the arrangement of the internal parts and airflow through the Boeing T60 engine.
(b) Notice that the gas-producer turbine is of the radial-inflow design. (Most piston engine turbosuperchargers use this type of turbine.)

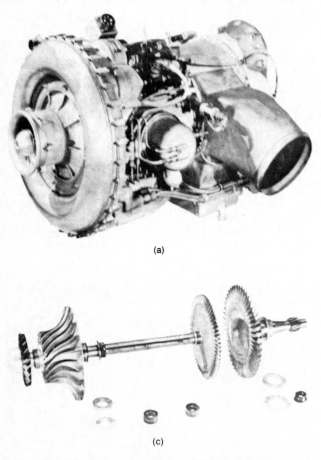

(a)

(c)

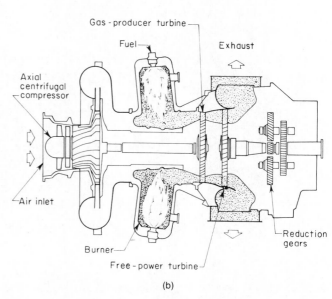

(b)

FIG. 2-15 The Boeing Model 550 (T50) turboshaft engine incorporating an axial-centrifugal compressor and a free-power turbine.
(a) External view of the Boeing T50-BO-10 drone helicopter engine.
(b) Sectioned view of the Boeing T50 engine showing the arrangement of parts and airflow.
(c) The Boeing T50 (Model 550) rotor system.

BOEING MODEL 520 (T60)
BOEING MODEL 550 (T50)
(FIGS. 2-14 AND 2-15)

Although these engines are no longer in production, they are included here as examples of engines equipped with double-sided centrifugal compressors and radial-inflow and axial-flow power turbines. Since the engines are obsolete, no specifications are included, except to note that they were rated at approximately 300 to 450 shp.

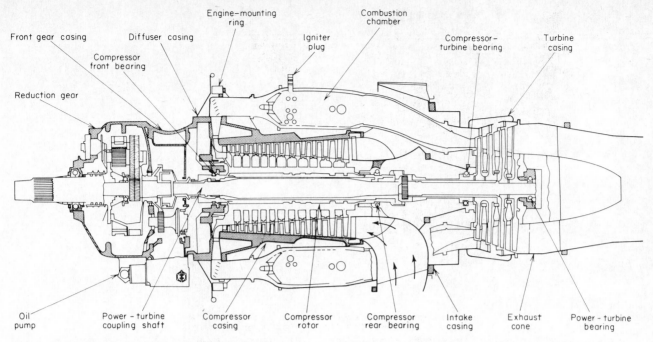

Front gear casing
Compressor front bearing
Reduction gear
Diffuser casing
Engine-mounting ring
Igniter plug
Combustion chamber
Compressor-turbine bearing
Turbine casing

Oil pump
Power-turbine coupling shaft
Compressor casing
Compressor rotor
Compressor rear bearing
Intake casing
Exhaust cone
Power-turbine bearing

FIG. 2-16 Sectioned view of the Bristol Proteus. Notice the reverse (forward) flow of air through the compressor.

BRISTOL PROTEUS (*FIG. 2-16*)

The unusual design and arrangement of the parts in this engine warrant its inclusion in this section. Air enters toward the middle of the engine and flows forward through a 12-stage axial compressor and a single-stage centrifugal compressor. The compressor is driven by a two-stage turbine and has a compression ratio of 7.2:1, a mass airflow of 44 lb/s [20 kg/s], and rotates at 11,755 rpm. Eight can-type combustion chambers are located on the outside of the compressor. A two-stage free-power turbine drives the propeller by means of a concentric shaft through the engine. An integral gearbox reduces the propeller rpm by a ratio of 11.593:1 in relation to the free-power turbine rpm. Specific fuel consumption is .48 lb/eshp/h [218 g/eshp/h] in cruise. The engine weighs 2900 lb [1315 kg] without the propeller, and produces 4445 eshp, or 3960 shp plus approximately 1213 lb [5395 N] of thrust. It may be equipped with water injection.

(a)

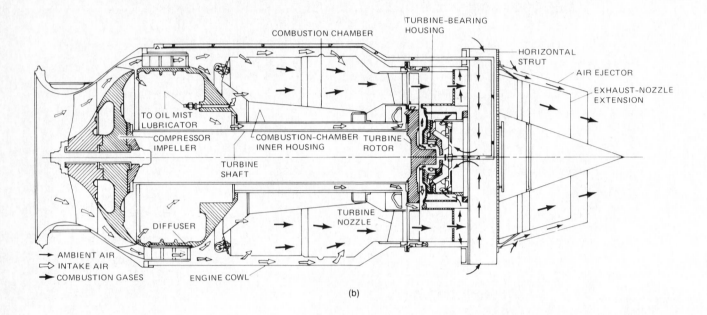

COMBUSTION CHAMBER

TURBINE-BEARING
HOUSING

HORIZONTAL
STRUT

AIR EJECTOR

EXHAUST-NOZZLE
EXTENSION

TO OIL MIST
LUBRICATOR

COMPRESSOR
IMPELLER

COMBUSTION-CHAMBER
INNER HOUSING

TURBINE
ROTOR

TURBINE
SHAFT

TURBINE
NOZZLE

DIFFUSER

→ AMBIENT AIR
⇨ INTAKE AIR
➡ COMBUSTION GASES

ENGINE COWL

(b)

(c)

FAIRCHILD J44 (*FIG. 2-17*)

Although this engine is obsolete, it is included here as an example of a mixed-flow compressor. In this type of compressor the air transitions from a centrifugal to axial flow over a single stage. The compression ratio is 2.7:1, and airflow is 25 lb/s [11.4 kg/s] at 15,780 rpm. The engine has an annular combustion chamber and a single-stage turbine. Specific fuel consumption is 1.4 to 1.6 lb/lbt/h (142.7 to 163.1 g/N/h). Engine weight is 370 lb [168 kg] and it produces 1000 lb [4448 N] of thrust.

FIG. 2-17 The Fairchild J44 turbojet engine.
(a) External view of the Fairchild J44 engine.
(b) Sectioned view of the Fairchild J44 turbojet showing the arrangement of parts and the airflow.
(c) Drawing showing the mixed-flow compressor design.

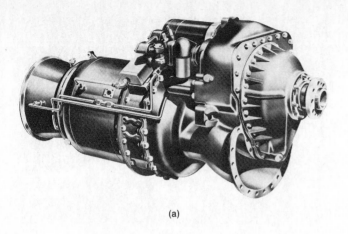

(a)

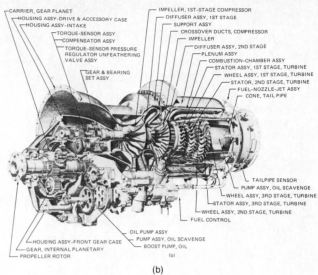

CARRIER, GEAR PLANET
HOUSING ASSY-DRIVE & ACCESSORY CASE
HOUSING ASSY-INTAKE
TORQUE-SENSOR ASSY
COMPENSATOR ASSY
TORQUE-SENSOR PRESSURE
REGULATOR UNFEATHERING
VALVE ASSY
GEAR & BEARING
SET ASSY

IMPELLER, 1ST-STAGE COMPRESSOR
DIFFUSER ASSY, 1ST STAGE
SUPPORT ASSY
CROSSOVER DUCTS, COMPRESSOR
IMPELLER
DIFFUSER ASSY, 2ND STAGE
PLENUM ASSY
COMBUSTION-CHAMBER ASSY
STATOR ASSY, 1ST STAGE, TURBINE
WHEEL ASSY, 1ST STAGE, TURBINE
STATOR, 2ND STAGE, TURBINE
FUEL-NOZZLE-JET ASSY
CONE, TAIL PIPE

TAILPIPE SENSOR
PUMP ASSY, OIL SCAVENGE
WHEEL ASSY, 3RD STAGE, TURBINE
STATOR ASSY, 3RD STAGE, TURBINE
WHEEL ASSY, 2ND STAGE, TURBINE
FUEL CONTROL

OIL PUMP ASSY
PUMP ASSY, OIL SCAVENGE
BOOST PUMP, OIL

HOUSING ASSY-FRONT GEAR CASE
GEAR, INTERNAL PLANETARY
PROPELLER ROTOR

(b)

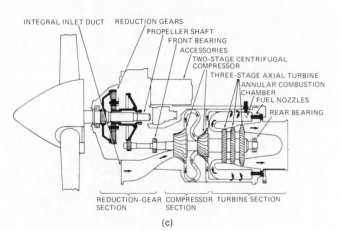

INTEGRAL INLET DUCT   REDUCTION GEARS
                        PROPELLER SHAFT
                        FRONT BEARING
                        ACCESSORIES
                        TWO-STAGE CENTRIFUGAL
                        COMPRESSOR
                        THREE-STAGE AXIAL TURBINE
                        ANNULAR COMBUSTION
                        CHAMBER
                        FUEL NOZZLES
                        REAR BEARING

REDUCTION-GEAR   COMPRESSOR   TURBINE SECTION
SECTION   SECTION

(c)

(d)

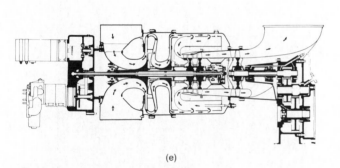

(e)

(f)

FIG. 2-18 Garrett AiResearch 331 series engine.

(a) External view of the Garrett AiResearch TPE331 turbo-prop engine.

(b) Cutaway view of the Garrett AiResearch TPE331 (T76) single-spool turboprop engine.

(c) Sectioned view of the TPE331 (T76) engine showing airflow.

(d) The Garrett AiResearch TSE331-7 with compressors, turbines, and load (through reduction gears) on the same shaft (single-spool engine).

(e) The TSE331-50 model incorporates a free-power turbine.

(f) The Turbo II Aerocommander is equipped with two Garrett AiResearch TPE331 engines.

(g) Two Garrett AiResearch TPE331 engines are installed in the Mitsubishi MU-2 turboprop aircraft (Continued on following page.)

(g)

TYPES, VARIATIONS, AND APPLICATIONS 29

GARRETT AIRESEARCH TPE331 (T76) (*FIG. 2-18*)

The Garrett AiResearch TPE331 is a single-shaft (spool) turboprop engine. The compressor has two centrifugal stages in series that are driven by three turbine wheels. Compression ratio is 8:1 and mass airflow is 5.8 lb/s [2.63 kg/s] at 41,730 rpm. Propeller rpm is reduced to 2000 by means of a 20.86:1 integral gearbox equipped to sense torque. A reverse-flow combustion chamber is used. Specific fuel consumption is .66 lb/eshp/h [300 g/eshp/h], and the engine weighs 330 lb [150 kg]. Power is 600 to 700 eshp approximately, depending upon the dash number. Other forms of the 331 Series are shown in Fig. 2-18 (d) and (e).

(h)

(i)

(j)

(k)

(l)

FIG. 2-18 (continued)
(h) The Volpar Super Turbo 18 Conversion using two Garrett AiResearch TPE331 engines.
(i) The Cessna model 441 Conquest Propjet is powered by two Garrett AiResearch TPE331-8-401 engines.
(j) The North American OV-10A counterinsurgency (COIN) aircraft with two T76 engines.
(k) Two Garrett AiResearch TPE331 engines are installed in the Beech King Air B100 turboprop aircraft.
(l) The Swearingen Aircraft Metro II with two Garrett AiResearch TPE331 engines.

(a)

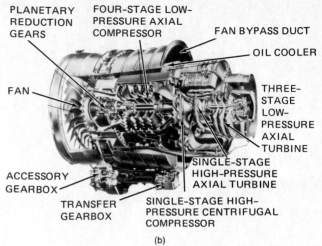

PLANETARY
REDUCTION
GEARS

FOUR-STAGE LOW-
PRESSURE AXIAL
COMPRESSOR

FAN BYPASS DUCT

OIL COOLER

FAN

THREE-
STAGE
LOW-
PRESSURE
AXIAL
TURBINE

ACCESSORY
GEARBOX

SINGLE-STAGE
HIGH-PRESSURE
AXIAL TURBINE

TRANSFER
GEARBOX

SINGLE-STAGE HIGH-
PRESSURE CENTRIFUGAL
COMPRESSOR

(b)

(c)

(d)

(e)

FIG. 2-19 The Garrett AiResearch TFE731 is an exceptionally quiet, high-bypass-ratio, two-spool, geared-fan turbofan
engine equipped with a reverse-flow combustor.

(a) External view of the Garrett AiResearch TFE731 turbofan engine.

(b) Cutaway view of the Garrett AiResearch TFE731 turbofan engine.

(c) Two TFE731-3 engines are installed in the Israel Aircraft
Industries Westwind 1124.

(d) The Gates Lear Jet 25 fitted with two Garrett AiResearch
TFE731 engines instead of the previous General Electric
CJ610s [see Fig. 2-20(d)] almost doubles the range of this
aircraft.

(e) The TFE731 installed in the latest Gates Lear Jet 35.
(Continued on following page.)

GARRETT AIRESEARCH TFE731 (FIG. 2-19)

The Garrett AiResearch TFE731 is a two-spool, geared front fan engine. Use of the geared fan increases operational flexibility and gives better performance both at low and high (50,000 ft [15 km]) altitude. A three-stage turbine drives the four-stage axial compressor at 19,728 rpm and the geared down (.555:1) fan, while the first-stage turbine drives the centrifugal compressor at 28,942 rpm. The total compression ratio of the engine is 19:1 with an airflow of 113 lb/s [51.3 kg/s] and a bypass ratio of 2.66:1. Specific fuel consumption is .49 lb/lbt/h [50 g/N/h]. The engine weighs 710 lb [322 kg] and produces 3500 lbt [15,568 N].

(f)

(g)

(h)

FIG. 2-19 (continued)

(f) The Falcon Jet Corporation 10 with two TFE731 engines.

(g) Two Garrett AiResearch TFE731 engines power the latest Hawker Siddeley HS-125-700 business jet.

(h) The Lockheed Jet Star II retrofitted with four Garrett AiResearch TFE731 engines. (See Fig. 2-36.)

(a)

(b)

FIG. 2-20 The General Electric J85 (CJ610) series turbojet engine.

(a) External view of the General Electric J85-21. Notice the variable compressor stator and inlet guide vanes.

(b) Cutaway view of the afterburner-equipped General Electric J85 engine. Note the variable-area exhaust nozzle (VEN). (Continued on following page.)

GENERAL ELECTRIC J85 (CJ610) (FIG. 2-20)

The General Electric J85 is a turbojet engine which has an eight-stage axial-flow compressor. The compressor handles 44 lb/s [20 kg/s] of air and has a compression ratio of 6.5:1 to 7:1 at 16,500 rpm. Variable inlet guide vanes and compressor bleed valves are linked together to allow for rapid acceleration. The engine has an annular combustion chamber and a two-stage turbine drives the compressor. Specific fuel consumption is .9 to 2.2 lb/lbt/h [92 to 224.2 g/N/h] and thrust is 2400 to over 4000 lbt [10,675 to 17,792 N], depending upon the model and whether the engine is equipped with an afterburner. Weights range from 400 to 600 lb [182 to 272 kg].

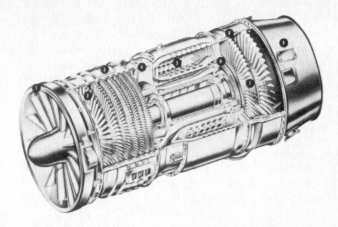

1 FORWARD FRAME
2 EIGHT-STAGE AXIAL-FLOW COMPRESSOR
3 COMPRESSOR CASING
4 MAIN FRAME
5 COMBUSTOR
6 TURBINE NOZZLE
7 TWO-STAGE AXIAL-FLOW TURBINE
8 TURBINE CASING
9 EXHAUST SECTION AND JET NOZZLE

(c)

(d)

(e)

(f)

(g)

FIG. 2-20 (continued)

(c) Cutaway view of the General Electric CJ610 turbojet engine.

(d) Two General Electric CJ610 engines are installed in the Gates Lear Jet Model 24. Later models use the Garrett AiResearch TFE731. (See Fig. 2-19.)

(e) The Northrop T-38 Talon jet trainer. This training version of the highly produced F-5 Freedom Fighter uses two General Electric J85 engines.

(f) Two General Electric CJ610 engines are installed in the Jet Commander, formerly manufactured by the Rockwell Standard Corporation. It is now built by Israel Aircraft Industries Ltd. in a stretched, re-engined, and modified form, and is called the Westwind. [See Fig. 2-19(c).]

(g) The Cessna A-37B is powered by two General Electric J85GE-17 engines instead of the CAE J69. [See Fig. 2-55(d).]

TYPES, VARIATIONS, AND APPLICATIONS 33

In addition to the jet and fan engines, General Electric manufactures the T58 (Fig. 2-28), and the T64 (Fig. 2-29). Both are free-power turbine engines, a major difference being the location of the power takeoff shaft.

The TF34 (Fig. 2-25) is one of General Electric's latest small turbofans driving the Lockheed S-3A and the Fairchild Republic A-10 aircraft.

Still other examples of axial-flow machines are the Detroit Diesel Allison J71 (Fig. 2-7) powering the Douglas B-66, and the Detroit Diesel Allison 501 series or T56 engine (Fig. 2-8) used in the Lockheed Hercules and Electra, Grumman Hawkeye, and the Convair 580 Conversion. Since the 501 is a turboprop, the compressor plus the load of the propeller require the use of many turbine wheels, which is typical of all turboprop/turbofan designs. Detroit Diesel Allison has also designed an axial-flow turboprop engine incorporating a fixed regenerator (Fig. 2-10). The advantages of this cycle are discussed on page 21.

British manufacturers have come up with some interesting variations of the axial-flow engine. For example, the Rolls-Royce Trent (Fig. 2-51) and the Rolls-Royce RB·211 (Fig. 2-52) are three-spool turbofan engines for installation in the Fairchild-Hiller F-228 and Lockheed 1011 respectively. The Rolls-Royce Spey (Fig. 2-50), which powers the DeHavilland Trident, B.A.C. One-Eleven, and Grumman Gulfstream II aircraft, is a multispool turbofan engine with a mixed exhaust (see page 79–81 for a discussion of mixed and non-mixed exhaust systems), and the Rolls-Royce Tyne (Fig. 2-54) is a two-spool turboprop engine with an integral gearbox for use in the Canadair 44. Additionally, Rolls-Royce, in collaboration with S.N.E.C.M.A. of France, builds the Olympus 593 (Fig. 2-49), one of the few afterburning commercial engines, for use in the supersonic Concorde.

The Oryx (Fig. 2-33), manufactured by D. Napier and Son Ltd., is another unusual design of British manufacture. The power produced by the gas generator section of the engine is used to drive another axial-flow compressor. The airflow from both the gas generator and the air pump is mixed together, resulting in an extremely high volume airflow. The engine is specifically designed to drive helicopter rotor blades by a jet reaction at the tips.

The Rolls-Royce Pegasus (Fig. 2-53) is another form of engine designed to produce high-volume airflows. Fan air and primary airflow are both vectored (directed) in an appropriate direction in order to achieve the desired line of thrust. The engine is installed in the V/STOL Hawker Harrier.

AXIAL-CENTRIFUGAL-FLOW COMPRESSOR ENGINES

As a group, the axial-centrifugal-flow engines exhibit the greatest variability and design innovation. The Garrett AiResearch ATF3 is a perfect example of this (Fig. 2-65). All of the various permutations and combinations of compressor design, number of spools, type of combustion chamber, single-shaft versus free-power turbine, location of the power-takeoff shaft, etc., can be found on these engines.

An important producer of axial-centrifugal engines in this country is the Avco Lycoming Corporation. Their T53 and T55 series engines (Fig. 2-12) in their several versions have been designed for wide application in both conventional and rotary wing aircraft. Both engines utilize the same basic concept and arrangement of parts; the main difference is in the number of compressor and free-power turbine stages. The mechanically independent free-power turbine drives a coaxial through-shaft to provide cold, front-end power extraction. A feature of these engines is the reverse-flow combustion chamber design mentioned previously.

Two later engines developed by this company are the LTS/LTP (Fig. 2-11) series of small turbo-shaft/turboprop engines and the ALF502 (Fig. 2-13). At the time of this writing, most turbofan engine fans are either coupled to one of the compressors, or to a group of turbines independent of the gas-generator compressor turbine(s). In either case a compromise has to be made, since the best number of revolutions per minute (rpm) for the fan is, in most cases, lower than the best rpm for the gas-generator compressor (core engine) or any turbine wheel. In the Avco Lycoming ALF502, the fan is geared down, like the propeller on many piston engines, so the low-pressure turbine and high-bypass-ratio fan can each turn at an appropriate rpm.

The Pratt & Whitney Aircraft of Canada Ltd. PT6A engine (Fig. 2-34) also uses a reverse-flow combustion chamber. On this machine, the air enters toward the rear and flows forward, with the power takeoff at the front. It is currently in use in some light twin aircraft, including the Beech King Air and the Piper Aircraft Corp. Cheyenne, in a few Bell helicopters, and in several foreign aircraft. The engine has also been used to power the STP Special at the Indianapolis 500 race.

Detroit Diesel Allison's bid for the small turbine market, the T63 (model 250) (Fig. 2-9) has an axial-centrifugal compressor. It incorporates many unusual design features, for example, it can be disassembled in minutes with ordinary hand tools, it contains a single combustion chamber, and it has an interchangeable gearbox. The axial part of the compressor is only about 4.5 inch (in) [11.4 centimeters (cm)] in diameter, and the engine weighs about 140 pounds (lb) [64 kilograms (kg)] yet produces over 400 hp [298 kw] in some versions. The turboshaft variation of this engine is installed in the Hughes OH-6 Light Observation Helicopter (LOH), the Bell Jet Ranger helicopter, and others. Figure 2-59 shows the world's smallest turbofan, the Williams WR19.

Most small gas turbines utilize the free-power turbine method of driving the load and the Boeing engine is no exception. As shown in Fig. 2-15, air is compressed by a single axial stage, followed by a single centrifugal stage. The compressed air is mixed with fuel and ignited in twin combustors. Hot gases then expand through the single-stage gas producer and power turbines and exhaust through either a single- or double-exhaust nozzle.

GE is now producing an axial-centrifugal engine called the T700 (Fig. 2-32). Because this engine is de-

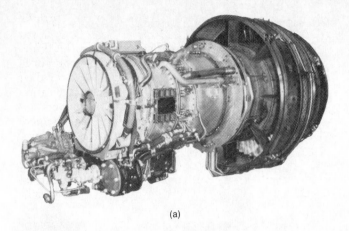

(a)

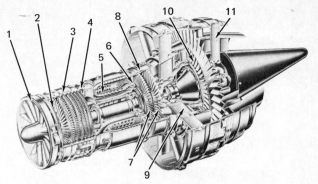

1 FORWARD FRAME
2 EIGHT-STAGE AXIAL-
    FLOW COMPRESSOR
    ROTOR
3 COMPRESSOR CAS-
    ING
4 MAIN FRAME
5 COMBUSTOR
6 TURBINE NOZZLE
7 TWO-STAGE AXIAL-
    FLOW TURBINE
8 TURBINE CASING
9 FAN FRONT FRAME
10 BUCKETS
11 FAN REAR FRAME

(b)

(c)

FIG. 2-21 The General Electric CF700 aft turbofan engine.
(a) External view of the General Electric CF700 turbofan engine.
(b) Cutaway view of the General Electric CF700 engine. Note the independent rear turbine/fan.
(c) The Falcon Fan Jet built by Dassault of France and marketed in the United States by the Falcon Fan Jet Corporation of Teterboro, New Jersey, is powered by two General Electric CF700 engines.

GENERAL ELECTRIC CF700 (*FIG. 2-21*)

This engine is General Electric's aft fan version of their CJ610 engine. The eight-stage axial compressor has a compression ratio of 6.8:1 and a mass airflow of 44 lb/s [20 kg/s] at 16,500 rpm. The compressor also has variable inlet vanes and bleed valves that operate together to improve acceleration characteristics. The engine has an annular combustion chamber and two turbines drive the compressor. The free-power turbine bluckets (a combination of turbine buckets and compressor blades) rotate at the rear at 9000 rpm, have a compression ratio of 1.6:1, and flow 88 lb/s (40 kg/s) of air. Specific fuel consumption is .68 lb/lbt/h [69.3 g/N/h]. The engine weighs 725 lb [329 kg] and produces 4200 lbt [18,682 N].

signed to be installed in the Sikorsky Utility Tactical Transport Aircraft (UTTAS) UH60A, and the Hughes Army Attack Helicopter (AAH) AH64, an integral inlet particle separator is located at the forward end.

An engine that shows great promise, and one which combines many of the design innovations discussed at the beginning of the section on the axial-centrifugal compressor, is the Garrett AiResearch TFE731 (Fig. 2-19). The engine is a medium-bypass, two-spool, geared-front-fan engine coupled, through a planetary gearbox, to the low-pressure axial spool. The centrifugal compressor high-pressure spool is driven by a single turbine. Reverse-flow combustion chambers are also used.

Once again, British designers and manufacturers have produced an unusual axial-centrifugal flow engine. The Bristol Proteus (Fig. 2-16) incorporates a reverse-flow axial-centrifugal compressor and a two-stage free-power turbine driving the propeller output shaft through a series of reduction gears. The engine is used in the Britannia aircraft.

MIXED-FLOW COMPRESSOR ENGINES

This compressor does not fall into any of the three main categories. The mixed-flow design is similar in appearance to the single-entry centrifugal compressor, but the blade arrangement provides a different type of airflow. The compressor receives its air axially as do many other types, but it discharges this air at some angle between the straight-through flow of the axial compressor and the radial flow of the centrifugal compressor. The Fairchild J44 engine (Fig. 2-17) uses this design. (Text continues on page 78.)

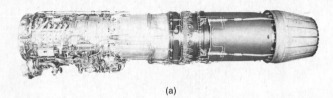

(a)

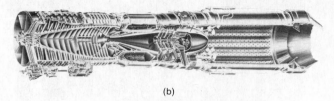

(b)

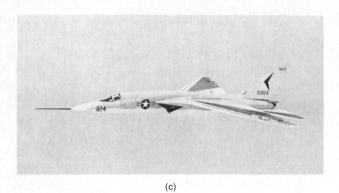

(c)

(d)

(e)

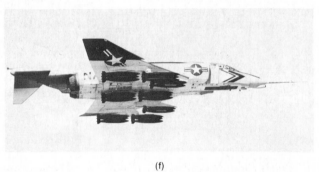

(f)

(g)

FIG. 2-22 The General Electric J79 engine.

(a) External view of one model of the General Electric J79 turbojet engine.

(b) Cutaway view of another model of the G.E. J79 engine.

(c) The North American Aviation RA5C incorporates two General Electric J79 engines.

(d) The Lockheed F104A Starfighter with one G.E. J79-19 engine installed.

(e) The Mach 2.2 B-58 Hustler, manufactured by General Dynamics, Fort Worth Division, and equipped with four G.E. J79 engines, is now phased out of the U.S.A.F. inventory.

(f) The McDonnell Douglas F-4B Phantom II with two G.E. J79 engines is produced in large quantities. Many variations are being used here and abroad.

(g) The Israel Aircraft Industries Kfir C-2 powered by one General Electric J79.

GENERAL ELECTRIC J79 (*FIG. 2-22*)

One of the most highly produced engines in the world is the General Electric J79. It has a 17-stage axial-flow compressor with variable inlet guide vanes and variable stators on the first six stages. Engine operating parameters are approximate values because of the number of different models produced. Compression ratio is 13:1, mass airflow is 170 lb/s [77 kg/s], and rpm is 7680. The combustion section is of the can-annular design with 10 flame tubes. A three-stage turbine drives the compressor. Specific fuel consumption ranges from .84 to 1.96 lb/lbt/h [85.6 to 199.8 g/N/h]. The engine weighs 3670 lb [1665 kg], and produces from 10,900 to 17,900 lbt [48,483 to 79,619 N], depending upon the model, and whether the afterburner is in operation. The J79-17C is one of the low-smoke long-life models recently introduced. New main fuel nozzles, main fuel control, main ignitor plugs, combustor assembly, and first-stage turbine nozzle are combined to provide the benefits of this new model.

GENERAL ELECTRIC CJ805-3 (*FIG. 2-23*)

The General Electric CJ805-3 engine is the commercial turbojet version of the J79. Specifications for the engine are the same as for the CJ805-23 model, except the engine weighs 2815 lb [1277 kg], produces 11,200 lbt [49,818 N], has a specific fuel consumption of .81 lb/lbt/h [82.6 g/N/h], and has no aft fan. (See Fig. 2-24 for further specifications.)

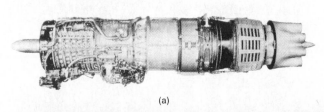

(a)

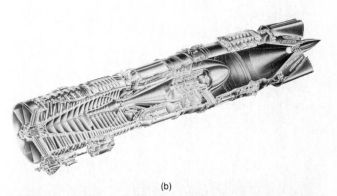

(b)

(c)

FIG. 2-23 The General Electric CJ805, commercial version of the General Electric J79 turbojet engine. (See Fig. 2-22.)

(a) External view of the General Electric CJ805-3 turbojet engine.

(b) Cutaway view of the General Electric CJ805-3 turbojet engine. Note the thrust reverser and sound suppressor.

(c) The Convair 880 is powered by four General Electric CJ805-3 turbojet engines, a commercial derivative of the General Electric J79, but without the afterburner.

(a)

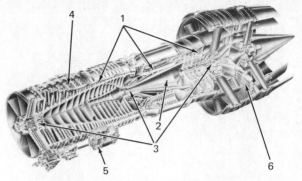

5

GENERAL ELECTRIC CJ805-23 (*FIG. 2-24*)

This General Electric engine is the aft turbofan version of the J79. It has a 17-stage axial-flow compressor, with the inlet guide vanes and the first 6 stages of stator vanes being variable. The compressor has a compression ratio of 13:1 and a mass airflow of 171 lb/s [77.6 kg/s] at 7310 rpm. The can-annular combustion chamber has ten flame tubes. A three-stage turbine drives the compressor, and a single-stage turbine blade-fan blade combination (bluckets) is located to the rear. The fan rotates at 5560 rpm, has a pressure ratio of 1.65:1, and flows 251 lb/s [114 kg/s] of air. Specific fuel consumption is .53 lb/lbt/h [54 g/N/h]. The engine weighs 3760 lb [1706 kg] and produces 16,100 lbt [71,613 N].

1 COMPRESSOR, COM-
BUSTION, AND TUR-
BINE SECTIONS
2 CONICAL COMPRES-
SOR/TURBINE SHAFT
3 THREE MAIN BEAR-
INGS
4 VARIABLE STATORS

5 HYDRO-MECHANI-
CAL CONTROL SYS-
TEMS
6 TURBOFAN: COMBI-
NATION TURBINE
BUCKETS AT ROOT
AND FAN BLADES AT
TIP

(b)

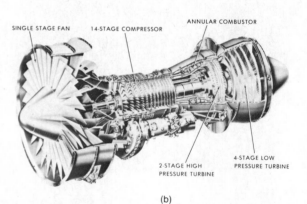

(c)

FIG. 2-24 The General Electric CJ805-23, the aft-fan, free-power version of the General Electric CJ805-3 turbojet engine.
(a) External view of the General Electric CJ805-23 turbofan engine.
(b) Cutaway view of the General Electric CJ805-23 turbofan engine.
(c) The General Dynamics Corporation Convair Division 990 powered by four General Electric CJ805-23 turbofan engines.

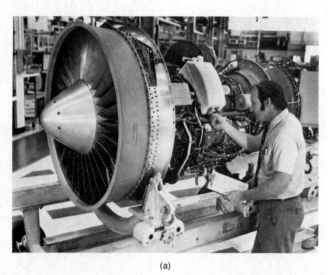

(a)

(b)

FIG. 2-25 The General Electric TF34 dual-spool, high-bypass-ratio turbofan engine.
(a) External view of the G.E. TF34 engine.
(b) Cutaway view of the G.E. TF34 engine. (Continued on following page.)

(c)

(d)

GENERAL ELECTRIC TF34
(*FIG. 2-25*)

The General Electric TF34 is a two-spool, high-bypass-ratio (6.2:1) turbofan engine. The single-stage fan, driven by a four-stage low-pressure turbine, has a pressure ratio of 1.5:1 and bypasses 291 lb/s [132 kg/s] of air at 7365 rpm. The axial-compressor is driven by a single turbine at 17,900 rpm and flows 47 lb/s [21.3 kg/s] of air at a pressure ratio of 14:1. Overall pressure ratio is 21:1 (14 times 1.5). As with most of the second generation turbofan engines (Figs. 2-13, 2-19, 2-26, 2-27, 2-35, 2-44, and 2-52), the TF34 has no fixed inlet struts or guide vanes. Specific fuel consumption is .36 lb/lbt/h [36.7 g/N/h]. The engine weighs 1450 lb [658 kg] and produces about 9200 lbt [40,922 N].

FIG. 2-25 (continued)
(c) Two G.E. TF34–400A engines are used in the Lockheed S-3A. Notice the difference in construction of the engine nacelles in Fig. 2-25(c) and (d).
(d) The Fairchild Republic A-10 with two G.E. TF34–100 engines.

GENERAL ELECTRIC TF39
(*FIG. 2-26*)

The TF39 engine is one of General Electric's larger designs. It is a two-shaft, high-bypass-ratio (8:1), front-turbofan engine. The fan has 1½ stages and no inlet guide vanes. It diverts 1333 lb/s [605 kg/s] of air and has a compression ratio of 2:1 at 3380 rpm. The compressor is a 16-stage axial-flow design, with the first 7 stages having variable stator vanes. Mass airflow through the compressor is 167 lb/s [75.8 kg/s], and total compression ratio is 25:1 at 9513 rpm. The engine has an annular combustion chamber. The first two high-pressure turbines drive the compressor, and the last six turbine stages drive the fan. Air-cooled blades permit a turbine inlet temperature of 2300°F [1260°C] as opposed to 1800°F (980°C) for most engines. Specific fuel consumption is .6 lb/lbt/h [61.2 g/N/h]. The engine weighs 7400 lb [3357 kg] and produces 41,000 lbt [182,368 N].

(a)

FIG. 2-26 The General Electric high-bypass-ratio TF39. This engine shares a common core [see part (c) of this Figure] with the General Electric CF6 series engines. (See Fig. 2-27.)
(a) External view of the General Electric TF39 turbofan engine. (Continued on following page.)

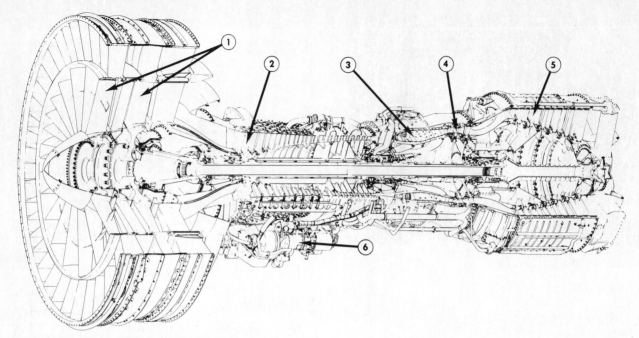

1. FAN: The 1½ stage high-bypass fan, driven by the low-pressure turbine is instrumental in achieving fuel economy and is optimized for the altitude cruise point. Both stages are titanium. They have part span supports for extra stability, vibration control and reduction of blade deflection under load. The first (½) stage fan supercharges the inner flowpath of the second stage which in turn provides core engine supercharging and the bypass flow through the front plug nozzle formed by the core engine cowl. The fan has successfully passed qualification tests at 200 percent of maximum expected stress.

2. COMPRESSOR: The high-pressure (core) compressor has 16 stages. The inlet guide vanes and first six-stage stator vanes are variable and are scheduled to provide an optimum engine cycle, rapid acceleration, and excellent stall margin. Bleed air for the aircraft is drawn from the inner tip of the eighth-stage stator vane to take advantage of the centrifugal action of the compressor in minimizing contamination. Materials are titanium and stainless steels; chosen for reliability, long life, and corrosion resistance.

3. COMBUSTOR: Annular combustor incorporates vortex-inducing swirl cups at each of 30 fuel nozzles. Demonstrated 98.5 percent efficiency and airstart capability well beyond the required envelope.

4. HIGH-PRESSURE TURBINE: The two-stage high-pressure (core) turbine incorporates film and convection cooling in the first-stage nozzle vanes and blades and convection cooling in the second stage. The film cooling system discharges air from holes in the leading edge of the blades, which flows back over the airfoil forming an insulating layer. Actual metal temperatures are comparable to earlier, uncooled systems.

5. LOW-PRESSURE TURBINE: The six-stage low-pressure (fan) turbine is a high-aspect ratio, tip-shrouded, uncooled turbine. Constant diameter is dictated by installation aerodynamic considerations. Case is externally cooled for clearance control and installation compatibility.

6. ENGINE ACCESSORIES

(b)

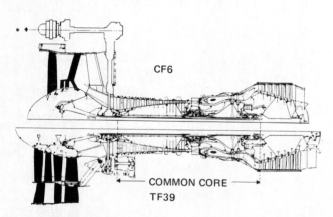

* ACCESSORY GEARBOX IS OUTSIDE FAN DUCT AT BOTTOM OF CF6 ENGINE

(c)

(d)

FIG. 2-26 (continued)

(b) Trimetric view of the General Electric TF39 turbofan engine showing configuration and components.

(c) A sectioned view showing the CF6/TF39 comparison.

(d) The Lockheed Georgia C-5A Galaxy has four General Electric TF39 engines.

(a)

(b)

(c)

(d)

(e)

(f)

FIG. 2-27 The General Electric CF6 series engine was developed from the TF39 engine, which was used in the Lockheed Georgia C-5A Galaxy. It uses a common core with the TF39. (See Fig. 2-26.)

(a) General external view of the G.E. CF6-6 showing the very large diameter fan.

(b) Cutaway view of the G.E. CF6-6 high-bypass-ratio turbofan.

(c) Cutaway view of the G.E. CF6-50. [Compare this version with the CF6-6 model shown in part (b) of this figure.]

(d) The McDonnell Douglas DC-10 is powered by three G.E. CF6-50C engines. Other versions of this airplane may be equipped with CF6-60 engines, or Pratt & Whitney JT9D engines.

(e) The Airbus A300 wide-bodied twin jet is manufactured by Airbus Industrie of France and has two G.E. CF6-50 engines installed.

(f) The Boeing YC-14 has a wing equipped with variable-camber leading-edge flaps. Engine exhaust air from the two G.E. CF6-50s is blown over the upper part of the wing to provide superior STOL performance. This aircraft and the McDonnell Douglas YC-15 [see Fig. 2-43(h).] are in competition for the U.S.A.F. Advanced Medium Short Takeoff and Landing (AMST) transport. The CF6-50 engine is also used in the Boeing 747 Airborne Command Post.

GENERAL ELECTRIC CF6 SERIES (*FIG. 2-27*)

Two engines in the General Electric CF6 Series are the CF6-6 and the growth version, the CF6-50. The major changes between the engines are the different thrusts and turbine-inlet-temperature limits and the introduction of two additional booster stages behind the single-stage low-pressure compressor of the CF6-6. Specifications are given here for the CF6-50A engine. This engine is a two-spool turbofan with a single-stage high-bypass-ratio (4.44:1) fan, a three-stage intermediate booster compressor, a 14-stage high pressure compressor (driven by two turbine wheels), and an annular combustion chamber. The fan diverts 1178 lb/s [534 kg/s] of air at a pressure ratio of 1.69:1 and turns at 3810 rpm. The high-pressure compressor flows 270 lb/s [122 kg/s] of air at 10,275 rpm and has a pressure ratio of 17.2:1 for an overall compressor ratio of 29:1. Specific fuel consumption is .39 lb/lbt/h [39.8 g/N/h]. Weight is 8225 lb [3731 kg], and the engine produces 49,000 lbt [217,952 N]. Growth versions will produce thrusts over 54,000 lbt [240,192 N].

GENERAL ELECTRIC T58 (*FIG. 2-28*)

The General Electric T58 is a free-power axial-flow turboshaft engine. The compressor has 10 stages, with variable inlet guide vanes and variable stators on the first three rows. The compressor has a compression ratio of 8.3:1 and flows approximately 13 lb/s [5.9 kg/s] of air at 26,300 gas producer rpm. The combustion chamber is of the annular design. Two turbines drive the compressor, and one turbine drives the load through the rear at 19,500 rpm. Specific fuel consumption is .64 lb/shp/h [290 g/shp/h]. The engine weighs approximately 350 lb [159 kg] and produces approximately 1300 to 1800 shp, depending upon the model.

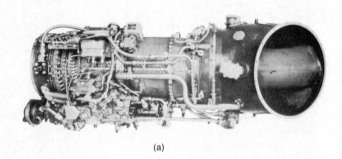

(a)

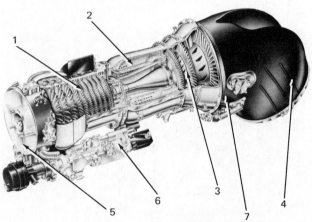

(b)

1 TEN-STAGE, AXIAL-FLOW COMPRESSOR WITH ONE-PIECE STEEL CONSTRUCTION FOR LAST EIGHT STAGES OF ROTOR HUB
2 SHORT, SMALL-DIAMETER ANNULAR COMBUSTOR
3 TWO-STAGE, AXIAL-FLOW GAS GENERATOR TURBINE
4 EXHAUST POSITION, ADJUSTABLE 90° LEFT

OR RIGHT [OPTIONAL TORQUE-SENSING-SPEED DECREASER GEAR (NOT SHOWN) PROVIDES FORE AND AFT POWER TAKEOFF]
5 ANTI-ICED INLET STRUTS AND INLET GUIDE VANES
6 HYDRO-MECHANICAL CONTROL
7 SINGLE-STAGE, AXIAL-FLOW FREE-POWER TURBINE

FIG. 2-28 The General Electric CT58 free-power turbine turboshaft engine.
(a) External view of the General Electric CT58 turboshaft engine.
(b) Cutaway view of the General Electric CT58 engine. (Continued on following page.)

(c)

(d)

(e)

(f)

(g)

(h)

FIG. 2-28 (continued)

(c) The Bell model UH-1F with one General Electric CT58 engine.

(d) The Kamen UH-2 Seasprite has one General Electric CT58 engine.

(e) Two General Electric CT58 engines power the Boeing Vertol Division CH-46 Sea Knight.

(f) The Sikorsky S-62 is driven by one General Electric CT58 engine.

(g) The Sikorsky S-61 (HH-3E) with two General Electric CT58 engines.

(h) The Boeing Vertol Division 107-11 commercial airliner has two General Electric CT58 engines.

(a)

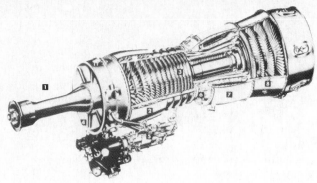

(c)

1 TORQUE SENSOR ON POWER SHAFT (EXCEPT T64-16)
2 HIGH-PRESSURE-RATIO COMPRESSOR
3 BALANCED MOMENT WEIGHT BLADES AND BUCKETS
4 ANTI-ICED FRONT FRAME AND INLET GUIDE VANES
5 FUEL CONTROL, PUMPS, FILTERS, AND ACCESSORY PADS
6 EXTERNAL NOZZLES AND IGNITORS
7 ANNULAR COMBUSTOR
8 TWO-STAGE GAS-GENERATOR TURBINE AND TWO-STAGE FREE-POWER TURBINE

(b)

(d)

(e)

(f)

FIG. 2-29 The General Electric T-64 turboprop/turboshaft engine.
(a) External view of the General Electric T-64 turboshaft/turboprop.
(b) Cutaway view of the General Electric T-64 engine.
(c) The Sikorsky S-65 (CH53A) with two General Electric T-64 turboshaft engines.
(d) Later version of the S-65 (HH53B) is powered by two General Electric T-64 engines.
(e) Vought Systems Division, LTV Aerospace Corp. experimental XC-142, a tilt-wing triservice V/STOL with four General Electric T-64 engines.
(f) The DeHavilland Canada DHC-5 Buffalo with two G.E. T-64-820-1 engines.

GENERAL ELECTRIC T64 (*FIG. 2-29*)

The General Electric T64 is a turboprop/turboshaft engine incorporating a two-stage gas generator turbine and a two-stage free-power turbine. The compressor has 14 stages with the inlet guide vanes and the first 4 stages being variable. The combustion chamber is of the annular variety. Mass airflow at 18,000 rpm is 26 lb/s [11.7 kg/s], and the compression ratio is approximately 13:1, depending upon the model. Specific fuel consumption is about .5 lb/eshp/h [227 g/eshp/h]. Weight with the propeller reduction gearbox is 1130 lb [513 kg]. The shaft version weighs approximately 700 lb [318 kg]. Equivalent shaft horsepower is 2850, or 2770 shp plus 210 lbt [934 N] at 15,600 power turbine rpm for the turboprop, to 3435 shp at 13,600 power turbine rpm for the shaft engine. Power takeoff is at the front.

GENERAL ELECTRIC F101 (*FIG. 2-30*)

The General Electric F101 is a two-spool augmented (afterburning) turbofan in the 30,000 lbt [133,440 N] class. It can produce twice the thrust of the latest augmented J79 (Fig. 2-22) at approximately the same weight and bulk, but with a reduced overall length and with reduced specific fuel consumption. The two-stage fan has a pressure ratio of 2:1 and, together with the compressor, flows 350 lb/s [159 kg/s] of air. Overall compression ratio is 27:1. Turbine inlet temperatures are extremely high, over 2500°F [1371°C], which requires new cooling techniques and materials. The fuel is electronically trimmed with signals received from an infrared pyrometer (see Chap. 8) which averages the high-pressure turbine blade temperature. Weight of the engine is about 4000 lb [1814 kg], and thrust, cold (nonafterburning), is about 17,000 lbt [75,616 N]. At the time of this writing, this engine and the GE F404 (Fig. 2-31) are classified under security restrictions.

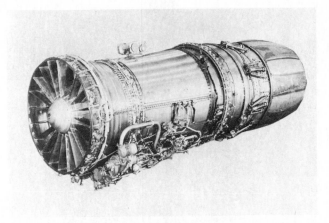

(a)

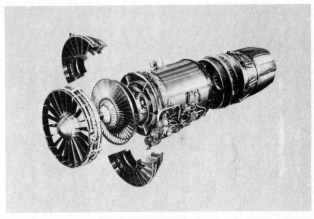

(b)

(c)

FIG. 2-30 The General Electric F101 augmented turbofan engine.
(a) External view of the General Electric F101 engine with a 2:1 bypass ratio.
(b) Exploded view of the General Electric F101 engine.
(c) The Rockwell International B-1 Bomber with four General Electric F101–GE–F100 engines.

GENERAL ELECTRIC F404
(FIG. 2-31)

The General Electric F404 is an advanced technology, low-bypass turbofan (sometimes called a "leaky turbojet"). It is a two-spool engine with each compressor being driven by a single turbine wheel. The low-pressure compressor (fan) is oversize in relation to the core, the surplus air being discharged through a surrounding bypass duct to mix with the core airflow in the afterburner. The engine weighs approximately 2000 lb [907 kg], and produces, with the afterburner on, about 16,000 lbt [71,168 N]. At the time of this writing, this engine and the GE F101 (Fig. 2-30) are classified under security restrictions.

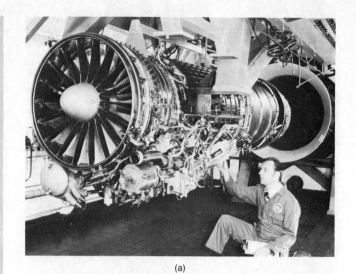

(a)

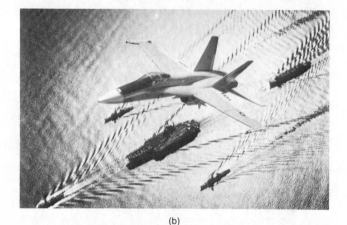

(b)

FIG. 2-31 The General Electric F404 derivative of the General Electric F101 engine.
(a) External view of the low-bypass, augmented F404 turbofan engine.
(b) Two General Electric F404–GE–400 engines power the McDonnell Douglas/Northrop F-18 NACF (Navy Air Combat Fighter). The F-18 is a derivative of the YF-17.

(a)

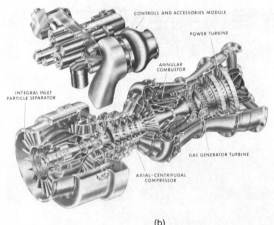

(b)

(c)

FIG. 2-32 The General Electric T700 turboshaft engine.
(a) External view of the General Electric T700 turboshaft engine.
(b) Cutaway view of the G.E. T700 two-spool engine. Notice the integral particle separator.
(c) The Sikorsky UH60A Utility Tactical Transport Aircraft System (UTTAS) driven by two General Electric T700 turboshaft engines. (Continued on following page.)

(d)

FIG. 2-32 (continued)

(d) The Hughes Helicopters Division of Summa Corporation Army Attack Helicopter (AAH) AH-64, with two General Electric T700 engines.

GENERAL ELECTRIC T700
(*FIG. 2-32*)

The General Electric T700 is a free-turbine turboshaft engine with a power takeoff at the front. (See Fig. 2-28 for an example of an engine with the power takeoff at the rear.) A built-in particle separator is designed to remove 95 percent of sand, dust, etc. The extracted matter is discharged by a blower driven from the accessory gearbox. The axial-centrifugal compressor is driven by the first two turbine wheels at 44,720 rpm and flows about 10 lb/s [4.5 kg/s] of air at a pressure ratio of 15:1. The two-stage gas-generator turbine inlet temperature is a high 2010°F [1100 °C], approximately. The two-stage power turbine turns at 20,000 rpm. The engine weighs 400 lb [181 kg] and produces 1536 shp for 30 minutes. Specific fuel consumption during this period is .469 lb/shp/h [213 kg/shp/h].

(a)

FIG. 2-33 The Napier Oryx. A British-designed high-volume-airflow engine. The airflow from the gas generator is joined with airflow from another axial compressor driven by the gas turbine.

(a) External view of the British Napier Oryx.

(b) Sectioned view of the Napier Oryx showing the airflow. (Continued on next page.)

A COMPRESSOR INLETS
B COMPRESSOR TURBINE
C AIRPUMP OUTLET
D THROTTLE VALVE.

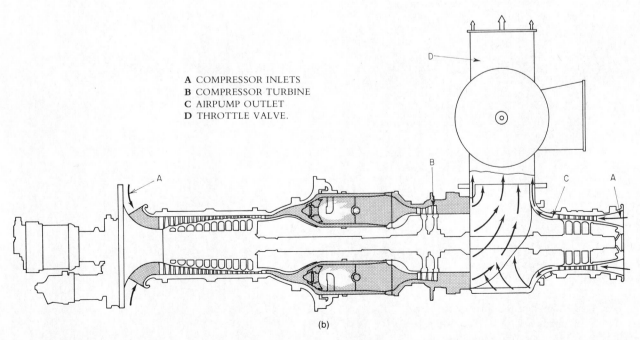

(b)

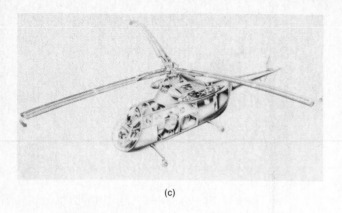

(c)

(c) A proposed Napier Oryx installation using the high-volume airflow ported to the rotor tips.

NAPIER ORYX (*FIG. 2-33*)

Although never produced in quantity, this engine represents an interesting example of the multitude of uses to which the gas turbine can be put. The engine was designed to generate high-volume airflows from the gas generator compressor and the axial compressor load. Air from these two sources were joined together, flowing through a diverter valve, and then to the tips of hollow helicopter rotor blades to provide the thrust necessary to drive the blades. Other unusual uses for high-volume airflow can be seen in Figs. 2-27(f), 2-43(h), and 2-53.

(a)

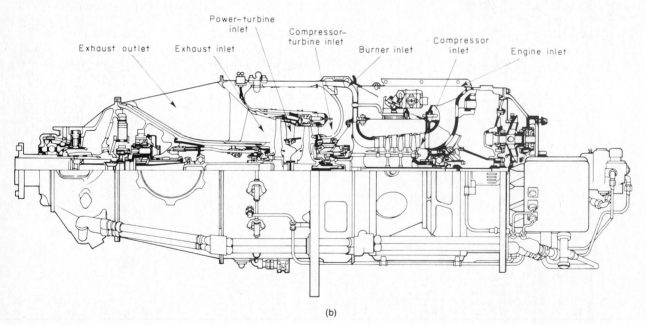

(b)

FIG. 2-34 The Pratt & Whitney Aircraft of Canada Ltd. (PWACL) PT6 series engine is used in a large number of aircraft.
(a) External view of the Pratt & Whitney PT6A-27 (left), and the PT6T-3/T400 Twin Pac (right).
(b) Sectioned view showing the arrangement of the parts of the PWACL PT6 turboprop engine. (Continued on following page.)

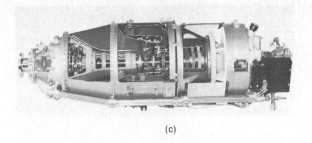

(c)

(d)

(e)

(f)

(g)

(h)

(i)

FIG. 2-34 (continued)

(c) Cutaway view of the PWACL PT6 engine.

(d) The Beechcraft Corporation King Air C-90 with two PWACL PT6A engines.

(e) The Beechcraft Super King Air (C-12A military version) with two PWACL PT6 engines installed.

(f) The American Turbine Beech 18 Conversion uses two PWACL PT6 engines.

(g) Beechcraft/U.S. Navy T-34C Turbo Mentor is powered by the PWACL PT6 engine.

(h) The Bell UH-1N Twin Huey Helicopter with the PWACL T400-CP-400 (PT6T-3) installed. (This model engine is essentially two PT6s joined together.)

(i) The Bell Helicopter 212 Twin is the commercial version of the UH-1N with the PWACL PT6T-3 Twin Pac engine installed. (Continued on following page.)

PRATT & WHITNEY AIRCRAFT OF CANADA PT6A (*FIG. 2-34*)

The Canadian Pratt & Whitney PT6A is a two-shaft turboprop engine. The air enters the three-stage axial compressor from the side of the engine through an inlet screen, then flows forward to a single centrifugal stage. The pressure ratio is 6.7:1, and airflow is 6.8 lb/s [3.1 kg/s] at 37,500 rpm. The reverse-flow combustion chamber surrounds the turbine wheel. One turbine wheel drives the compressors, while the free-power turbine drives the propeller through a 15:1 ratio integral gearbox. Propeller rpm is 2200 maximum. Specific fuel consumption at cruise is .6 lb/eshp/h [272 g/eshp/h]. The engine weighs 300 lb [136 kg] and produces 715 eshp, or 680 shp plus 87.5 lbt [389 N]. Several versions of this engine have higher power settings and slightly different specifications.

(j)

(k)

(l)

(m)

(n)

FIG. 2-34 (continued)
(j) The Piper Aircraft Corporation Cheyenne turboprop has two PT6A-28 engines installed.
(k) Two PWACL PT6 engines are installed in the Israel Aircraft Industries Arava STOL aircraft.
(l) The Pilatus Aircraft Ltd. of Switzerland PC-6 Turbo Porter with the PWACL PT6A-27 engine.
(m) The Pilatus PC-7 Turbo Trainer also uses the PT6A-25 engine.
(n) The DeHavilland Canada DHC-2 MkIII Turbo-Beaver uses the PT6A-6 turboprop engine. (Continued on following page.)

(o)

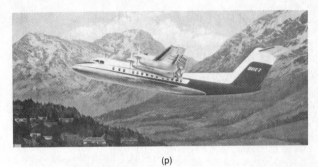

(p)

FIG. 2-34 (continued)
(o) The DeHavilland Canada DHC-6 Twin Otter STOL uses the PWACL PT6 engine.
(p) DeHavilland Canada's latest DHC-7 with four 1120 SHP PT6A-50 engines.

(a)

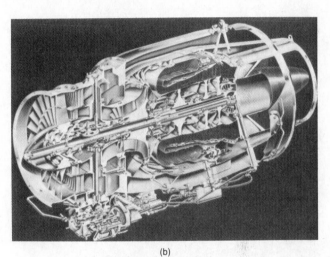

(b)

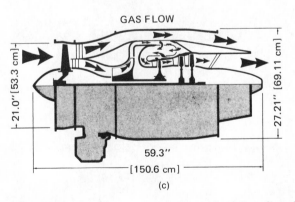

GAS FLOW

21.0" [53.3 cm]

27.21" [69.11 cm]

59.3"
[150.6 cm]

(c)

(d)

FIG. 2-35 The Pratt & Whitney Aircraft of Canada Ltd. JT15D turbofan engine.
(a) External view of the Pratt & Whitney JT15D engine.
(b) Cutaway view of the Pratt & Whitney JT15D engine.
(c) View of the PWACL JT15D showing airflow. Notice the two-spool design and the reverse-flow combustion chamber. (See Chap. 6.)
(d) The Cessna Citation is equipped with two PWACL JT15D engines. Some models use the Garrett AiResearch TFE731 engine. (See Fig. 2-19.)

PRATT & WHITNEY AIRCRAFT OF CANADA JT15D (FIG. 2-35)

The P&WACL JT15D is an advanced technology, two-spool front turbofan engine. The fan is driven by the last two turbines of a three-stage turbine. Total airflow is 75 lb/s [34 kg/s]. Of this amount, 57.5 lb/s [26 kg/s] is secondary airflow for a bypass ratio of 3.3:1. Fan-pressure ratio is 1.5:1, and overall pressure ratio (fan-pressure ratio times the single-stage centrifugal compressor ratio) is almost 10:1. An axial-boost stage is located between the fan and the centrifugal compressor and is driven at the same speed as the centrifugal compressor. Combustion chamber is of the annular reverse-flow type. Specific fuel consumption is .56 lb/lbt/h [57.1 g/N/h]. The engine weighs 557 lb [253 kg] and produces 2500 lbt [11,120 N].

PRATT & WHITNEY AIRCRAFT GROUP J60 (JT12) (FIG. 2-36)

The Pratt & Whitney Aircraft J60 is a single-spool single-shaft axial-flow turbojet engine. The compressor has nine stages and is driven by a two-stage turbine. The can-annular combustion chamber has eight flame tubes. Mass airflow at 16,000 rpm is 50 lb/s [23 kg/s] with a compression ratio of 6.5:1. Specific fuel consumption is .89 lb/lbt/h [90.7 g/N/h]. The engine weighs 468 lb [212 kg] and produces approximately 3000 lbt [13,344 N], depending upon the model. An afterburner may be installed.

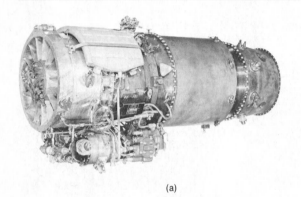

(a)

(b)

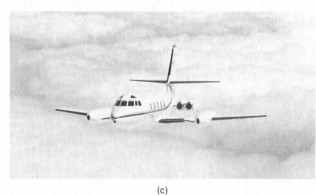

(c)

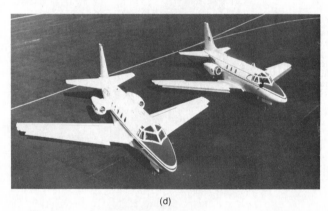

(d)

(e)

FIG. 2-36 The Pratt & Whitney Aircraft Group of United Technologies Corporation J60 (JT12) turbojet engine.
(a) External view of the Pratt & Whitney Aircraft JT12A-6A engine.
(b) Cutaway view of the Pratt & Whitney Aircraft J60 (JT12) engine.
(c) The Lockheed Georgia Division Jet Star (C-140) equipped with four Pratt & Whitney Aircraft JT12 engines. Newer versions of this aircraft are powered by four Garrett AiResearch TFE731 turbofans. (See Fig. 2-19.)
(d) Standard and stretched versions of the North American Sabreliner with two Pratt & Whitney Aircraft JT12 engines.
(e) The North American Aviation T-2B Buckeye equipped with two Pratt & Whitney Aircraft J60 engines.

(a)

(b)

(c)

1 COMPRESSOR INLET OUTER FRONT CONE	10 1ST STAGE TURBINE VANE	18 FREE TURBINE 2ND STAGE BLADE	27 NO. 3 BEARING (ROLLER)
2 COMPRESSOR INLET OUTER REAR CONE	11 1ST STAGE TURBINE BLADE	19 FREE TURBINE AC-CESSORY DRIVE GEARBOX N_2	28 COMBUSTION CHAMBER FUEL DRAIN VALVE
3 NO. 1 BEARING (ROLLER)	12 2ND STAGE TURBINE VANE	20 FREE TURBINE AC-CESSORY DRIVESHAFT	29 COMBUSTION CHAMBER CASE
4 COMPRESSOR INLET VANE	13 2ND STAGE TURBINE BLADE	21 NO. 5 BEARING (ROLLER)	30 MAIN COMPONENT DRIVE TOWER SHAFT
5 COMPRESSOR BLADE (1ST STAGE)	14 FREE TURBINE INLET VANE	22 FREE TURBINE EX-HAUST CASE	31 COMPONENT DRIVES GEARBOX
6 COMPRESSOR VANE (1ST STAGE)	15 FREE TURBINE 1ST STAGE VANE	23 NO. 4 BEARING (BALL)	32 DIFFUSER CASE
7 NO. 2 BEARING (BALL)	16 FREE TURBINE 1ST STAGE BLADE	24 FREE TURBINE CASE	33 COMPRESSOR AIR BLEED VALVE
8 FUEL NOZZLE	17 FREE TURBINE 2ND STAGE VANE	25 FREE TURBINE INLET CASE	34 COMPRESSOR INLET CASE
9 COMBUSTION CHAMBER		26 TURBINE CASE	

FIG. 2-37 The Pratt & Whitney Aircraft Group of United Technologies Corporation JFTD12A (T73-P-700) is a free-power turbine variation of the Pratt & Whitney Aircraft JT12 turbojet engine.
(a) External view of the Pratt & Whitney Aircraft JFTD12A (T73) turboshaft engine.
(b) Cutaway view of the Pratt & Whitney Aircraft JFTD12A (T73) turboshaft engine.
(c) Sectioned view of the Pratt & Whitney Aircraft JFTD12A (T73) showing the major internal parts. (Continued on following page.)

PRATT & WHITNEY AIRCRAFT GROUP JFTD12A (T73-P-700) (*FIG. 2-37*)

The Pratt & Whitney Aircraft JFTD12A engine is a free-power turbine version of the JT12 (J60). It is basically the same as the JT12, except that it has a two-stage free-power turbine. Specific fuel consumption is .7 lb/shp/h [318 g/shp/h]. The engine weighs 980 lb [445 kg], and produces 4800 shp at 9000 power turbine rpm, and 16,700 gas generator turbine rpm.

(d)

FIG. 2-37 (continued)
(d) The Sikorsky Skycrane S-64 (CH54A/B) is equipped with two Pratt & Whitney Aircraft JFTD engines.

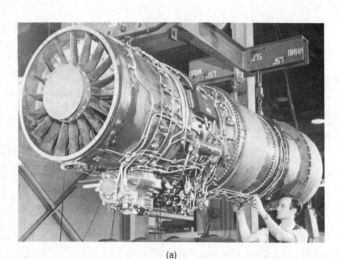

(a)

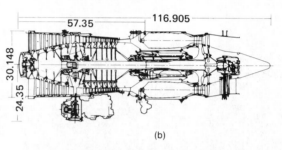

(b)

(d)

(c)

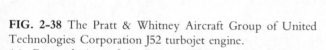

FIG. 2-38 The Pratt & Whitney Aircraft Group of United Technologies Corporation J52 turbojet engine.
(a) External view of the Pratt & Whitney Aircraft J52 turbojet engine.
(b) Sectioned view of the Pratt & Whitney Aircraft J52 two-spool engine.
(c) The Grumman A6 Intruder with two Pratt & Whitney Aircraft J52 engines.
(d) The McDonnell Douglas TA-4F Skyhawk with one Pratt & Whitney Aircraft J52 engine.

PRATT & WHITNEY AIRCRAFT GROUP J52 (JT8) (*FIG. 2-38*)

The Pratt & Whitney Aircraft J52 is an axial-flow two-spool turbojet engine. It has a five-stage low-pressure compressor and a seven-stage high-pressure compressor with a total compression ratio of 12:1. The combustor is of the can-annular design with nine flame tubes. The two-stage turbine is divided so that the first-stage drives the rear, or high-pressure compressor, and the second-stage drives the front- or low-pressure compressor. Mass airflow at 11,000 rpm is 120 lb/s [54.4 kg/s]. Specific fuel consumption is .79 lb/lbt/h [80.5 g/N/h]. The engine produces approximately 9300 lbt [41,366 N] and weighs 2118 lb [961 kg].

(a)

PRATT & WHITNEY AIRCRAFT GROUP J58 (*FIG. 2-39*)

This engine is a single-shaft, supersonic cruise powerplant equipped with an afterburner. It has an eight-stage compressor driven by a two-stage turbine. Six large tubes are used to bypass a part of the compressor airflow to the afterburner. The combustion chamber is a can-annular design with eight flame tubes. No rpm, compression ratio, or mass airflow figures are available. Specific fuel consumption is .8 to 1.9 lb/lbt/h [81.5 to 193.6 g/N/h]. The engine weighs 6500 lb [2948 kg] and produces up to 34,000 lbt [151,232 N] with the afterburner in operation. At the time of this writing, the engine is still classified.

(b)

FIG. 2-39 The Pratt & Whitney Aircraft Group of United Technologies Corporation J58 turbojet engine.
(a) The Pratt & Whitney Aircraft J58 turbojet engine.
(b) The Lockheed YF12A (similar to the SR71) is powered by two Pratt & Whitney Aircraft J58 turbojet engines. The speed of this aircraft is over 2000 mph [3219 km/h] (Mach 3), altitude over 80,000 ft (24,384 m) with skin temperatures ranging from 600°F [316°C] at the nose to 1100°F [593°C] at the tail. It is estimated that the fuselage stretches at least 3 in (7.62 cm) due to thermal expansion.

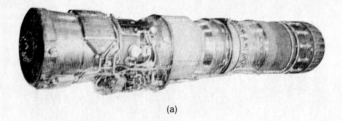

(a)

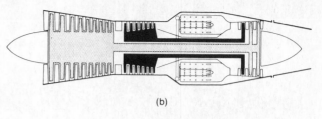

(b)

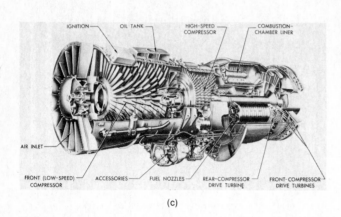

IGNITION   OIL TANK   HIGH-SPEED COMPRESSOR   COMBUSTION-CHAMBER LINER

AIR INLET

FRONT (LOW-SPEED) COMPRESSOR   ACCESSORIES   FUEL NOZZLES   REAR-COMPRESSOR DRIVE TURBINE   FRONT-COMPRESSOR DRIVE TURBINES

(c)

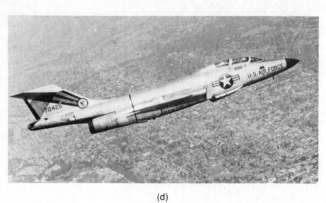

(d)

(e)

(f)

(g)

Fig. 2-40 The Pratt & Whitney Aircraft Group of United Technologies Corporation JT3C (J57) series engine and JT4 (J75) series engine. The J75 is a more powerful version of the J57. The design concept of both of these engines is similar.

(a) External view of the Pratt & Whitney Aircraft JT4 (J75) engine with afterburner.

(b) View of the Pratt & Whitney Aircraft JT3/4 series engine showing the basic two spool design.

(c) Cutaway view of the Pratt & Whitney Aircraft J75 turbojet engine.

(d) The McDonnell Douglas RF101 Voodoo with two Pratt & Whitney Aircraft J57 engines.

(e) The Vought F-8E Crusader is equipped with one Pratt & Whitney Aircraft J57 engine.

(f) The Fairchild Republic F-105 Thunderchief powered by one Pratt & Whitney Aircraft J57 engine.

(g) The Boeing B52G with eight Pratt & Whitney Aircraft J57 engines. (Continued on following page.)

(h)

(i)

FIG. 2-40 (continued)

(h) The Pratt & Whitney Aircraft J57 with afterburner is installed in the North American Aviation F-100 Super Sabre.

(i) The Convair F-106A with one Pratt & Whitney Aircraft J75 engine installed.

(a)

(b)

FIG. 2-41 The Pratt & Whitney Aircraft Group of United Technologies Corporation T34 turboprop is no longer in production, but is included here as an example of a single-spool engine incorporating an integral gearbox.

(a) The Pratt & Whitney Aircraft T34 turboprop engine.

(b) The McDonnell Douglas C-133 Cargomaster with four Pratt & Whitney Aircraft T34 engines.

PRATT & WHITNEY AIRCRAFT GROUP J57/J75 (JT3/JT4) (*FIG. 2-40*)

The J57 is an axial-flow, two-spool turbojet. It has a nine-stage low-pressure compressor driven by the second- and third-stage turbines, and a seven-stage high-pressure compressor driven by the first-stage turbine. The can-annular combustor has eight burner cans with six fuel nozzles in each can. Mass airflow at 9500 rpm (high-pressure compressor) is 180 lb/s [82 kg/s], with a compression ratio of 13:1 total. Specific fuel consumption is .77 lb/lbt/h [78.5 g/N/h] without the afterburner and 2.8 lb/lbt/h [285.4 g/N/h] with the afterburner. Engine weights run from approximately 3800 to 4800 lb [1724 to 2177 kg]. Various engines may be equipped with an afterburner or water-injection system. Thrusts range from approximately 11,000 to 18,000 lbt [48,928 to 80,064 N], depending upon the configuration.

The Pratt & Whitney Aircraft J75 is similar in construction features to the J57. It also is a two-spool axial-flow turbojet engine. The eight-stage low-pressure compressor is driven by the second- and third-stage turbine, and the seven-stage high-pressure compressor is driven by the first-stage turbine. The combustor is of the can-annular design, with six fuel nozzles in each can. It may be equipped with an afterburner or a water injection-system. Mass airflow at 8650 rpm is 265 lb/s [120 kg/s], and the compression ratio is 12:1. The engine weighs 5000 to 6000 lb [2268 to 2722 kg], produces 16,000 to 26,000 lbt [71,168 to 115,648 N], and has a specific fuel consumption of .79 to 2.2 lb/lbt/h [80.5 to 224.2 g/N/h], depending upon configuration.

PRATT & WHITNEY AIRCRAFT GROUP T34 (*FIG. 2-41*)

The Pratt & Whitney Aircraft T34 is a turboprop engine equipped with an integral gearbox. The compressor has 13 stages, with a compression ratio of 6.7:1. The combustor is a can-annular type with eight flame tubes. Mass airflow at 11,000 rpm is 65 lb/s [29 kg/s]. The propeller and compressor are driven by a three-stage turbine. Specific fuel consumption is .7 lb/shp/h [318 g/shp/h], and horsepower equals 6000 to 7000, or 5500 shp plus 1250 lbt [5560 N] to 6500 shp plus 1250 lbt, depending upon the use of water injection and/or configuration. The engine weighs 2870 lb [1302 kg] without the propeller.

(a)

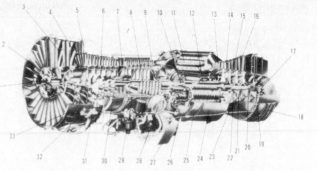

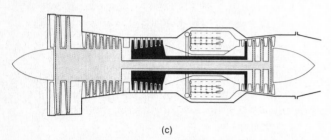

(c)

(d)

(e)

1 COMPRESSOR BLEED CONTROL	18 TURBINE EXHAUST CASE
2 FRONT ACCESSORY SUPPORT	19 FOURTH-STAGE TURBINE DISK AND BLADES
3 FIRST-STAGE BLADES (FAN)	20 THIRD-STAGE TURBINE DISK AND BLADES
4 SECOND-STAGE BLADES (FAN)	21 SECOND-STAGE TURBINE DISK AND BLADES
5 FRONT-COMPRESSOR EXIT VANES	22 FIRST-STAGE TURBINE DISK AND BLADES
6 FRONT-COMPRESSOR ROTOR	23 TURBINE NOZZLE CASE
7 COMPRESSOR INTERMEDIATE INLET VANES	24 COMBUSTION-CHAMBER REAR OUTER CASE
8 REAR-COMPRESSOR ROTOR	25 COMBUSTION-CHAMBER FRONT OUTER CASE
9 COMPRESSOR EXIT GUIDE VANES	26 DIFFUSER CASE
10 FRONT-COMPRESSOR DRIVE TURBINE ROTOR SHAFT	27 ACCESSORY COMPONENTS GEARBOX
11 COMBUSTION CHAMBER	28 COMPRESSOR BLEED VALVE
12 REAR-COMPRESSOR DRIVE TURBINE ROTOR SHAFT	29 COMPRESSOR INTERMEDIATE CASE
13 FIRST-STAGE TURBINE VANES	30 FRONT-COMPRESSOR REAR CASE
14 SECOND-STAGE TURBINE VANES	31 ANTI-ICING AIR VALVE AND ACTUATOR
15 THIRD-STAGE TURBINE VANES	32 FRONT-COMPRESSOR FRONT CASE
16 FOURTH-STAGE TURBINE VANES	33 COMPRESSOR INLET CASE
17 NO. 6 BEARING SUMP AREA	

(b)

FIG. 2-42 The Pratt & Whitney Aircraft Group of United Technologies Corporation JT3D (TF33) turbofan engine.

(a) External view of the Pratt & Whitney Aircraft JT3D (TF33) engine.

(b) Cutaway view of the Pratt & Whitney Aircraft JT3D (TF33) engine showing the location of the major parts.

(c) View of the Pratt & Whitney Aircraft JT3D (TF33) engine. Compare this with Fig. 2-40(b) to see how this manufacturer modified the JT3C turbojet engine to design a turbofan engine.

(d) The Boeing 720 uses four Pratt & Whitney Aircraft JT3D engines.

(e) The Boeing 707 comes equipped with four Pratt & Whitney Aircraft JT3D engines.

(f) The McDonnell Douglas DC-8 with four Pratt & Whitney Aircraft JT3D engines. The stretched version of this aircraft is called the DC-8 Super 61. (Continued on following page.)

(f)

PRATT & WHITNEY AIRCRAFT GROUP TF33 (JT3D) (*FIG. 2-42*)

The Pratt & Whitney Aircraft TF33 is the fan version of the JT3 series engine. It has a two-spool compressor with the six-stage low-speed spool and two-stage fan being driven by the second-, third-, and fourth-stage turbines, and the seven-stage high-speed spool being driven by the first-stage turbine. The can-annular combustor contains eight burner cans, each with six nozzles. Total compression ratio of the compressors is 13:1 to 16:1, with a mass airflow of 180 to 220 lb/s [82 to 100 kg/s] at 9800 rpm. Fan compression ratio is 1.66:1 to 2:1, with a mass airflow of 280 lb/s [127 kg/s] at 6650 rpm. Specific fuel consumption is .53 lb/lbt/h [54 g/N/h]. The engine weighs approximately 4600 lb [2088 kg] and produces 17,000 to 21,000 lbt [75,616 to 93,408 N], depending upon the model.

(g)

(h)

FIG. 2-42 (continued)

(g) The General Dynamics RB-57 Canberra is powered by two Pratt & Whitney Aircraft TF33 engines.

(h) The Lockheed Georgia Division C-141 Starlifter is equipped with four Pratt & Whitney Aircraft TF33 engines.

(i) The later model B-52H is driven by eight Pratt & Whitney Aircraft TF33 engines.

(i)

(a)

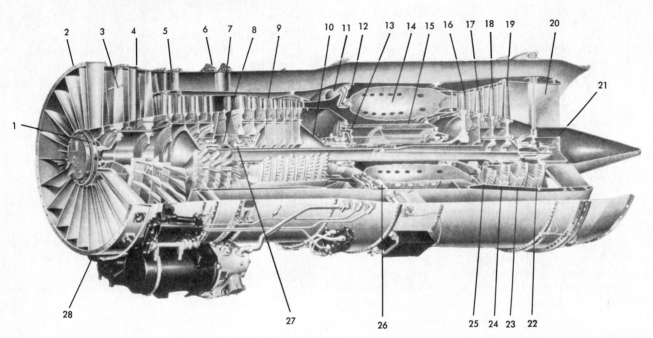

1 ANTI-ICING AIR DIS-CHARGE PORTS	**8** NO. 2 AND 3 BEAR-INGS OIL NOZZLE	CHAMBER INNER CASE
2 FAN INLET CASE	**9** REAR-COMPRESSOR ROTOR	**16** FIRST-STAGE TUR-BINE BLADES
3 FIRST-STAGE FAN BLADES	**10** REAR-COMPRESSOR ROTOR REAR HUB	**17** SECOND-STAGE TURBINE DISK AND BLADES
4 FRONT COMPRES-SOR ROTOR	**11** DIFFUSER CASE AIR MANIFOLD	**18** THIRD-STAGE TUR-BINE DISK AND BLADES
5 FAN DISCHARGE VANES	**12** FUEL NOZZLE	**19** FOURTH-STAGE TURBINE DISK AND BLADES
6 FAN DISCHARGE IN-TERMEDIATE CASE	**13** NO. 4 BEARING OIL NOZZLE	**20** EXHAUST STRUT
7 FAN DISCHARGE IN-TERMEDIATE CASE STRUTS	**14** COMBUSTION CHAMBER	
	15 COMBUSTION-	

21 NO. 6 BEARING HEAT-SHIELD	
22 FOURTH-STAGE TURBINE VANES	
23 THIRD-STAGE TUR-BINE VANES	
24 SECOND-STAGE TURBINE VANES	
25 FIRST-STAGE TUR-BINE VANES	
26 IGNITER PLUG	
27 GEARBOX DRIVE BE-VEL GEAR	
28 NO. 1 BEARING TUBE CONNECTOR	

(b)

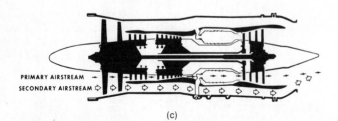

(c)

FIG. 2-43 The Pratt & Whitney Aircraft Group of United Technologies Corporation JT8D series engines.
(a) External view of the Pratt & Whitney Aircraft JT8D-209 turbofan engine.
(b) Cutaway view of the Pratt & Whitney Aircraft JT8D engine.
(c) A schematic view showing primary (core) and secondary (fan) airflow through the Pratt & Whitney Aircraft JT8D turbofan engine. (Continued on following page.)

(d)

(e)

(f)

(g)

THE PRATT & WHITNEY AIRCRAFT GROUP JT8D (*FIG. 2-43*)

Like many of Pratt & Whitney Aircraft engines, the JT8D is a two-spool turbofan. The four-stage low-pressure compressor is connected to the two-stage fan and is driven by the second-, third-, and fourth-stage turbine wheels. The seven-stage high-pressure compressor is driven by the first-stage turbine. A can-annular combustion chamber with nine burner cans is used. Fan, or secondary, mass airflow is 163 lb/s [74 kg/s], with a compression ratio of 1.9:1. Compressor mass airflow is 153 lb/s [69 kg/s], with a compression ratio of 16:1 at 11,450 rpm. The engine is classified as having a low bypass ratio (1:1), and the fan air mixes with the primary air at the exhaust nozzle. Specific fuel consumption is .6 lb/lbt/h [61.2 g/N/h]. The engine weighs 3096 lb [1404 kg] and produces 12,000 to 15,000 lbt [53,376 to 66,720 N] in various configurations.

FIG. 2-43 (continued)

(d) The Boeing 737 with two Pratt & Whitney Aircraft JT8D engines.
(e) The standard McDonnell Douglas DC-9 with two Pratt & Whitney Aircraft JT8D engines.
(f) The series 50, stretched McDonnell Douglas DC-9 aircraft equipped with two Pratt & Whitney Aircraft JT8D-15/17 engines.
(g) The Boeing 727 (standard and stretched versions) uses three Pratt & Whitney Aircraft JT8D engines. Illustrated is the stretched model.
(h) The McDonnell Douglas YC15 Advanced Medium Short Takeoff and Landing Aircraft (AMST) with four Pratt & Whitney Aircraft JT8D-17 engines. [See Fig. 2-27(f).]

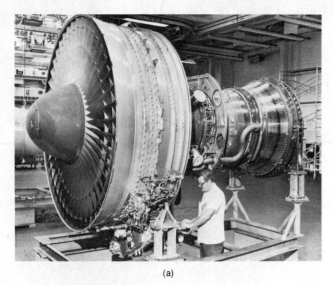

(a)

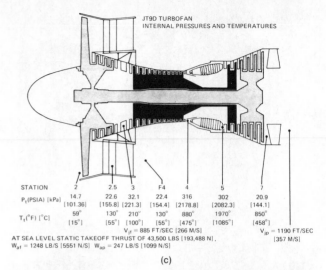

JT9D TURBOFAN
INTERNAL PRESSURES AND TEMPERATURES

STATION	2	2.5	3	F4	4	5	7
P_t(PSIA) [kPa]	14.7 [101.36]	22.6 [155.8]	32.1 [221.3]	22.4 [154.4]	316 [2178.8]	302 [2082.3]	20.9 [144.1]
T_t(°F) [°C]	59° [15°]	130° [55°]	210° [100°]	130° [55°]	880° [475°]	1970° [1085°]	850° [458°]

V_{jf} = 885 FT/SEC [266 M/S] V_{jp} = 1190 FT/SEC [357 M/S]

AT SEA LEVEL STATIC TAKEOFF THRUST OF 43,500 LBS [193,488 N],
W_{af} = 1248 LB/S [5551 N/S] W_{ap} = 247 LB/S [1099 N/S]

(c)

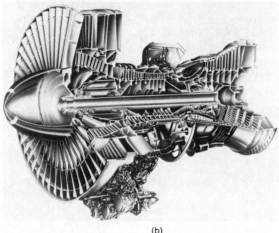

(b)

PRATT & WHITNEY AIRCRAFT GROUP JT9D (*FIG. 2-44*)

The Pratt & Whitney Aircraft high-bypass-ratio (5:1) JT9D is a two-spool turbofan engine. The large single front fan has two midspan supports, diverts 1260 lb/s [571 kg/s] of air, and has no inlet guide vanes. Fan rpm is 3650 with a pressure ratio of 1.6:1. The fan and the 3-stage low-pressure compressor are driven by a 4-stage low-pressure turbine, while the 11-stage high-pressure compressor is driven at 8000 rpm by a 2-stage turbine. Mass airflow through the compressor is 240 lb/s [109 kg/s] with an overall compression ratio of 24.5:1. The combustion chamber is of the annular variety with some modification at the forward end of the inner liner. Turbine inlet temperature is over 2200°F [1200°C] as a result of cooled and coated combustors and blades. Specific fuel consumption is .35 lb/lbt/h [35.7 g/N/h]. The engine weighs approximately 8400 lb [3810 kg] and produces 45,000 lbt [200,160 N]. Growth versions of this engine are intended to produce thrusts of approximately 50,000 lbt [22,240 N].

(d)

(e)

FIG. 2-44 The Pratt & Whitney Aircraft Group of United Technologies Corporation JT9D-59A/-70/-70A engine.
(a) External view of the Pratt & Whitney Aircraft JT9D engine. The fan is 8 ft (2.4 m) in diameter.
(b) Sectioned view of the Pratt & Whitney Aircraft JT9D engine.
(c) A schematic view showing the arrangement of internal parts, temperatures, and pressures at various points, plus air velocity at the core and fan exits of the Pratt & Whitney Aircraft JT9D.
(d) Four Pratt & Whitney Aircraft JT9D turbofan engines are installed in this Boeing 747SP, special cargo version of the Boeing 747 commercial aircraft.
(e) The Boeing 747 with four Pratt & Whitney Aircraft JT9D turbofan engines.

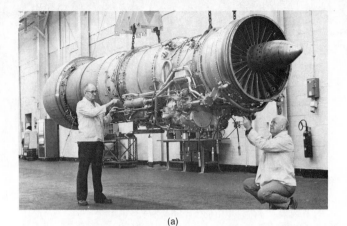

(a)

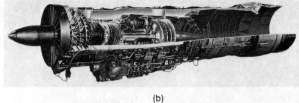

(b)

(c)

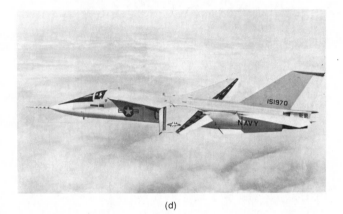

(d)

(e)

(f)

PRATT & WHITNEY AIRCRAFT GROUP TF30 (*FIG. 2-45*)

The Pratt & Whitney Aircraft TF30 is a two-spool turbofan engine. The fan has three stages and is connected to a low-pressure compressor which has six stages. Together they rotate at 9300 rpm. Primary and secondary air both enter the afterburner. The high-pressure compressor has seven stages and rotates at 14,500 rpm. Mass airflow through the fan is 141 lb/s [64 kg/s] with a compression ratio of 2:1. Mass airflow through the compressor is 129 lb/s [58.5 kg/s] with a compression ratio of 17:1. The engine is equipped with a can-annular combustion chamber having eight flame tubes. The low-pressure compressor and fan are driven by the second-, third-, and fourth-stage turbine wheels, and the high-pressure compressor is driven by the first-stage turbine wheel. Specific fuel consumption is .8 to 2.5 lb/lbt/h [81.5 to 254.8 g/N/h] depending upon whether the engine's afterburner is on. Weights range from 2500 to 4100 lb [1134 to 1860 kg] and thrusts from 20,000 to 25,000 lbt [88,960 to 111,200 N] depending upon configuration.

FIG. 2-45 The Pratt & Whitney Aircraft Group of United Technologies Corporation TF30, a turbofan engine equipped with an afterburner.
(a) External view of the Pratt & Whitney Aircraft TF30 augmented turbofan.
(b) Cutaway view of the Pratt & Whitney Aircraft TF30 shows how secondary airflow is handled.
(c) The General Dynamics F-111 with two Pratt & Whitney Aircraft TF30 engines. The wings are partially swept in this view.
(d) The General Dynamics F-111 with the wings extended.
(e) The Vought A-7A Corsair II is powered by one Pratt & Whitney Aircraft TF30 non-afterburning turbofan engine.
(f) The Navy's Grumman F-14A is powered by two uprated Pratt & Whitney Aircraft TF30 engines.

(a)

(b)

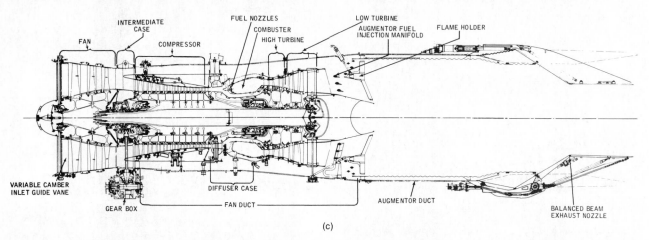

FAN
INTERMEDIATE CASE
COMPRESSOR
FUEL NOZZLES
COMBUSTER
HIGH TURBINE
LOW TURBINE
AUGMENTOR FUEL INJECTION MANIFOLD
FLAME HOLDER
VARIABLE CAMBER INLET GUIDE VANE
GEAR BOX
DIFFUSER CASE
FAN DUCT
AUGMENTOR DUCT
BALANCED BEAM EXHAUST NOZZLE

(c)

(d)

(e)

FIG. 2-46 The Pratt & Whitney Aircraft Group of United Technologies Corporation F100 augmented (afterburning) turbofan engine.

(a) External view of the Pratt & Whitney Aircraft F100 engine.

(b) Cutaway view of the Pratt & Whitney Aircraft F100-PW-100 engine.

(c) Sectioned view of the Pratt & Whitney Aircraft F100-PW-100 two-spool, afterburning turbofan engine.

(d) The McDonnell Douglas F-15 has two Pratt & Whitney Aircraft F100 engines.

(e) The General Dynamics F-16 equipped with one Pratt & Whitney Aircraft F100 engine.

PRATT & WHITNEY AIRCRAFT GROUP F100 (*FIG. 2-46*)

The F100-PW-100 engine is an augmented twin-spool, axial-flow gas turbine engine of modular design. The fan-rotor assembly consists of three fan stages and is driven by a two-stage turbine at 9600 rpm with a low bypass ratio of .7:1. The rear compressor turning at 14,650 rpm is a ten-stage rotor assembly driven by a two-stage turbine. Total airflow is 228 lb/s [103.4 kg/s] and overall pressure ratio is 23:1. Air cooling of the rear-compressor-drive turbine blades and vanes is provided. Variable vanes are located at the fan inlet, rear compressor inlet, and fourth- and fifth-stages of the rear compressor. The engine design includes an annular combustor with 16 fuel nozzles, a mixed flow five-segment augmentor with electric ignition and a lightweight balanced beam variable area, convergent-divergent exhaust nozzle. A rear-compressor driven main gearbox is mounted on the bottom of the engine. This engine has an extremely high (8:1) thrust-to-weight ratio. Specific fuel consumption in the nonafterburning mode is .68 lb/lbt/h [69.3 g/N/h], and 2.55 lb/lbt/h [239.9 g/N/h] with the afterburner on. The engine weighs 3036 lb [1377 kg] and produces 14,375 lbt [63,940 N] dry, and 23,810 lbt [105,907 N] augmented.

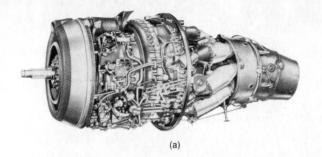
(a)

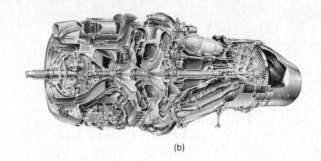
(b)

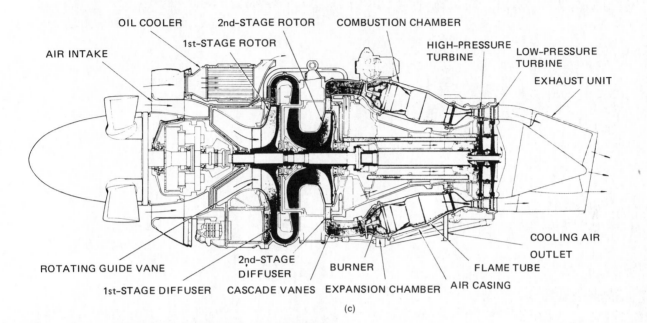

OIL COOLER   2nd-STAGE ROTOR   COMBUSTION CHAMBER

1st-STAGE ROTOR   HIGH-PRESSURE TURBINE   LOW-PRESSURE TURBINE

AIR INTAKE   EXHAUST UNIT

ROTATING GUIDE VANE   2nd-STAGE DIFFUSER   BURNER   COOLING AIR OUTLET

1st-STAGE DIFFUSER   CASCADE VANES   EXPANSION CHAMBER   FLAME TUBE

AIR CASING

(c)

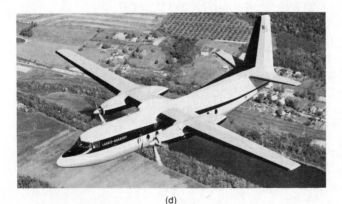
(d)

(e)

(f)

FIG. 2-47 The Rolls-Royce Dart.
(a) External view of the Rolls-Royce Dart.
(b) Cutaway view of the Rolls-Royce Dart R Da7 turboprop engine with three turbine wheels.
(c) Airflow through an early Rolls-Royce engine having two turbine wheels.
(d) The Fairchild Industries F-27 powered by two Rolls-Royce Dart engines.
(e) The Convair Conversion 600 with two Rolls-Royce R Da10 engines.
(f) The Hawker Siddeley 748 uses two R Da7 Mk 535-2 engines generating 2280 ESHP each. (Continued on following page.)

TYPES, VARIATIONS, AND APPLICATIONS 65

ROLLS-ROYCE DART (*FIG. 2-47*)

The Rolls-Royce Dart is another of the most highly produced engines in the world. It has undergone a number of design changes since it was first produced. Basically, the engine is a turboprop with a two-stage centrifugal compressor, two or three turbines and an integral propeller-reduction gearbox. Depending upon the model, the compression ratio is 5.6 to 6.35:1, and mass airflow is from 20 to 27 lb/s [9 to 12 kg/s] at 15,000 rpm (approximately 1400 propeller rpm). There are seven can-type burners set at an angle to the center line of the engine. Specific fuel consumption is approximately .57 lb/eshp/h [259 g/eshp/h]. The engine weighs 1250 lb [567 kg] and produces about 1800 to 3000 eshp. Some of these engines are equipped with water-alcohol injection.

(g)

FIG. 2-47 (continued)
(g) Two Rolls-Royce Dart Mk 529-8X engines are installed in the Grumman Gulfstream.

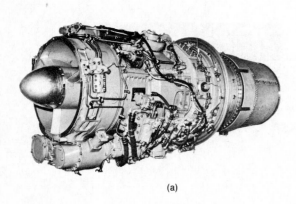

(a)

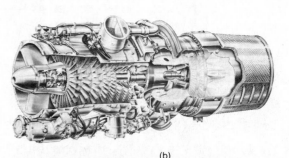

(b)

(d)

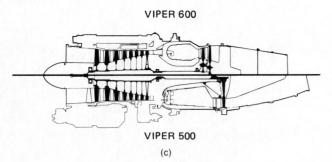

VIPER 600

VIPER 500
(c)

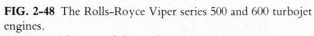

FIG. 2-48 The Rolls-Royce Viper series 500 and 600 turbojet engines.
(a) External view of the Rolls-Royce Viper 600 series turbojet engine.
(b) Cutaway view of the Rolls-Royce 600 series Viper engine.
(c) The Viper 600 has an additional turbine stage, a slightly higher compression ratio, redesigned combustion chamber, and a shorter length, among other differences, than the 500 series.
(d) The BH-125 with two Viper engines installed.

ROLLS-ROYCE VIPER (*FIG. 2-48*)

This is a single-spool eight-stage axial compressor engine with an annular combustion chamber and a two-stage turbine. The compressor has a compression ratio of 5.8:1, flows 58.4 lb/s [26.5 kg/s] and rotates at 13,760 rpm. Turbine inlet temperature is a nominal 1282°F [695°C]. Specific fuel consumption is .9 lb/lbt/h [91.7 g/N/h]. The engine weighs 760 lb [345 kg] and develops 3750 lbt [16,680 N].

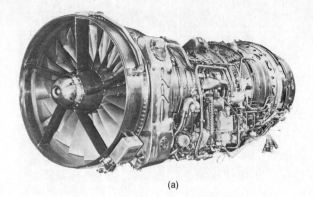

(a)

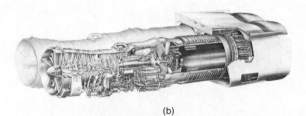

(b)

(c)

ROLLS-ROYCE/S.N.E.C.M.A. OLYMPUS 593 (FIG. 2-49)

This engine is one of the few afterburning commercial engines in service. The seven-stage low-speed spool driven by a single turbine wheel turns at 6500 rpm, and the seven-stage high-speed spool driven by another single turbine turns at 8850 rpm, for an overall pressure ratio of 14:1 and an airflow of 415 lb/s [188 kg/s]. The afterburner section, built by S.N.E.C.M.A., has a variable area exhaust nozzle and provides about 20 percent additional thrust for take-off and transonic acceleration. Specific fuel consumption is .7 to 1.18 lb/lbt/h [71.3 to 120.3 g/N/h] depending upon whether the afterburner is off or on. The engine weighs 5814 lb [2637 kg] and, with the afterburner on, can produce 38,400 lbt [170,803 N].

FIG. 2-49 The Rolls-Royce S.N.E.C.M.A. Olympus 593 Mk 610-14-28, jointly built by British and French companies.
(a) External view of the Olympus 593 engine.
(b) Cutaway view showing the two-spool compressor. The Olympus is one of the few afterburner-equipped commercial engines.
(c) Four Rolls-Royce/S.N.E.C.M.A. Olympus 593 turbojet engines are installed in the B.A.C./Aerospatiale Concorde SST.

(a)

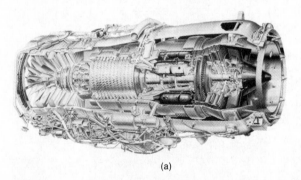

(b)

FIG. 2-50 The Rolls-Royce Spey. (Detroit Diesel Allison Division TF41 turbofan engine.)
(a) Cutaway view of the Rolls-Royce Spey (Mk 505, 506, 555), four-stage low-pressure compressor (fan) engine.
(b) Cutaway view of the Rolls-Royce Spey (Mk 510, 511, 512), five-stage low-pressure compressor (fan) engine. (Continued on following page.)

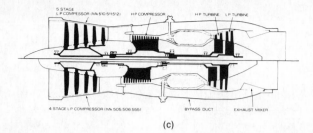

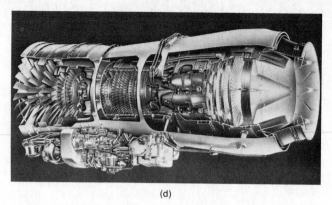

(c)

(d)

(e)

(f)

(g)

(h)

(i)

FIG. 2-50 (continued)

(c) A schematic showing the essential differences between the four- and five-stage low-pressure compressor version of the Rolls-Royce Spey series engines.

(d) Cutaway view of the Detroit Diesel Allison Division TF41, a derivative of the Rolls-Royce Spey series engine.

(e) Two (five-stage) Rolls-Royce Spey engines drive the Grumman Gulfstream II.

(f) The A-7D Corsair II is equipped with one Allison TF41 engine.

(g) The B.A.C 500 One-Eleven series is manufactured by the British Aircraft Corporation, and is powered by two Rolls-Royce Spey Mk 512 (five-stage low-pressure compressor) engines.

(h) Two Rolls-Royce Spey turbofans are installed in the DeHavilland Canada modified Buffalo with an experimental augmentor wing, developed in conjunction with NASA and Boeing. [See Figs. 2-27 and 2-43.]

(i) The Fokker F-28 with two Rolls-Royce Spey Mk 555 engines installed.

ROLLS-ROYCE SPEY (DETROIT DIESEL ALLISON TF41) (*FIG. 2-50*)

The low bypass ratio Spey is a two-shaft turbofan engine having a 4- or 5-stage front fan driven by the 3rd- and 4th-stage turbine, and a 12-stage high pressure compressor driven by the first- and second-turbine stages. The fan compression ratio is approximately 2.7:1 and flows 85 lb/s [38.6 kg/s] of air at 8500 rpm. The high-pressure compressor has a compression ratio of 20:1 overall, and the airflow is 123 lb/s [55.8 kg/s] at approximately 12,600 rpm. The can-annular combustion chamber has ten flame tubes. Specific fuel consumption is .6 to 1.95 lb/lbt/h [61.2 to 198.7 g/N/h] depending upon whether or not the afterburner is used. The engine weighs 2300 to 3600 lb [1043 to 1633 kg] and produces from 10,000 to 21,000 lbt [44,480 to 93,408 N], depending upon the model. The Allison TF41 is an advanced version of this engine having a 3-stage fan, an 11-stage compressor, and other modifications for installation in the Vought A-7D Corsair II. The TF41 was designed and developed jointly by Detroit Diesel Allison Division and Rolls-Royce Ltd.

ROLLS-ROYCE TRENT (*FIG. 2-51*)

The medium bypass ratio (3:1) Rolls-Royce Trent is an interesting three-spool (shaft) turbofan engine. The fan and low-speed and high-speed compressors are all driven by their own individual turbine wheel. The single-stage fan has no inlet guide vanes, and flows 225 lb/s [102 kg/s] of air at 8755 rpm. The low-pressure compressor has four stages and turns at 13,050 rpm, while the high-pressure compressor has five stages and turns at 15,855 rpm. Mass airflow through the compressor is 75 lb/s [34 kg/s], and the total compression ratio is 16:1. An annular combustion chamber is used. Specific fuel consumption is .7 lb/lbt/h [71.3 g/N/h]. The engine weight is 1775 lb [805 kg] and it produces 9980 lbt [44,391 N].

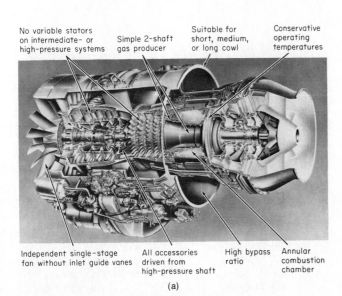

(a)

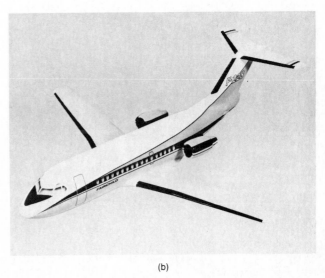

(b)

FIG. 2-51 The Rolls-Royce Trent.
(a) Cutaway view of the Rolls-Royce Trent engine.
(b) The F228 short-haul aircraft manufactured by Fairchild Hiller is designed to use two Rolls-Royce Trent three-spool turbofan engines.

(c)

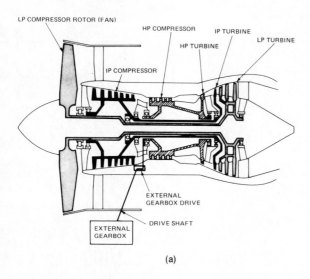

LP COMPRESSOR ROTOR (FAN)

HP COMPRESSOR

IP TURBINE

HP TURBINE

LP TURBINE

IP COMPRESSOR

EXTERNAL GEARBOX DRIVE

DRIVE SHAFT

EXTERNAL GEARBOX

(a)

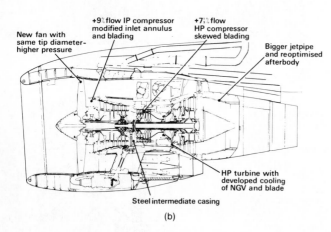

New fan with same tip diameter- higher pressure

+9% flow IP compressor modified inlet annulus and blading

+7½% flow HP compressor skewed blading

Bigger jetpipe and reoptimised afterbody

HP turbine with developed cooling of NGV and blade

Steel intermediate casing

(b)

(d)

FIG. 2-52 The Rolls-Royce RB·211, three-spool (shaft) engine.

(a) Cutaway view with inset of the Rolls-Royce RB·211 engine showing the three-spool concept.

(b) Sectioned view showing the features of one of the newer RB·211 engines.

(c) The very large diameter fan is apparent in this illustration of a Rolls-Royce RB·211 engine during installation into a test cell.

(d) The Lockheed L1011 has three RB·211 engines installed.

ROLLS-ROYCE RB·211 (*FIG. 2-52*)

The Rolls-Royce RB·211 is a high-bypass (5:1), three-shaft (spool) engine similar in concept to the Rolls-Royce Trent (Fig. 2-51). The single-stage midspan-supported front fan turns at 3530 rpm, has an airflow of 1096 lb/s [497 kg/s] and a pressure ratio of 1.6:1. Airflow through the intermediate and high-pressure compressor, each driven by its own turbine wheel, is 274 lb/s [124 kg/s] for a total airflow of 1370 lb/s [621 kg/s] and an overall pressure ratio of 26:1. The combustor is of annular design, and the turbine section blades are air cooled, which allows a turbine inlet temperature of 2300°F [1260°C]. Specific fuel consumption is .62 lb/lbt/h [63.2 g/N/h]. The engine weighs 8108 lb [3678 kg] and produces 53,500 lbt [237,968 N] with a high-speed spool rpm of 9392.

(a)

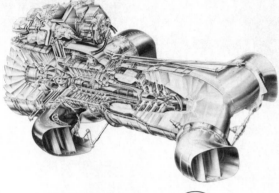

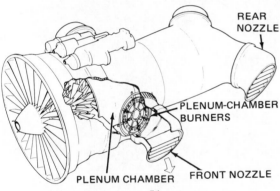

REAR
NOZZLE

PLENUM-CHAMBER
BURNERS

PLENUM CHAMBER   FRONT NOZZLE
(b)

(c)

ROLLS-ROYCE PEGASUS (*FIG. 2-53*)

In this engine, the major portion of the three-stage fan air, 300 lb/s [136 kg/s] at a pressure ratio of 2:1, is diverted to two front vectored-thrust nozzles through a plenum chamber in which fuel may be burned for additional thrust. Two turbines drive the fan and two counterrotating turbines drive the high-pressure compressor which handles 150 lb/s [68 kg/s] of air. Core engine air is directed to a bifurcated duct at the rear whose nozzles move in unison with the front nozzles. Specific fuel consumption is .6 lb/lbt/h [61.2 g/N/h]. Weight of the engine is 2800 lb [1270 kg] and the engine produces 21,500 lbt [95,632 N].

FIG. 2-53 The Rolls-Royce Pegasus Mk 104 vectored-thrust turbofan engine is installed in the British-built Harrier, the only operational jet fighter that takes off and lands vertically.
(a) External view of the Rolls-Royce Pegasus Mk 104. Both primary and secondary (fan) airflow is vectored.
(b) Cutaway view of the Rolls-Royce Pegasus Mk 104 engine.
(c) The Hawker Siddeley Harrier V/STOL, used by the United States Marine Corps. The exhaust gas of the Rolls Royce Pegasus engine is directed straight down for take-off and landing, and toward the rear of the aircraft for normal flight.

(a)

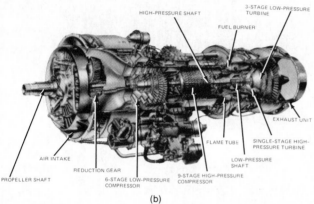

HIGH-PRESSURE SHAFT

FUEL BURNER

3-STAGE LOW-PRESSURE TURBINE

EXHAUST UNIT

SINGLE-STAGE HIGH-PRESSURE TURBINE

FLAME TUBE

LOW-PRESSURE SHAFT

AIR INTAKE

REDUCTION GEAR

PROPELLER SHAFT

6-STAGE LOW-PRESSURE COMPRESSOR

9-STAGE HIGH-PRESSURE COMPRESSOR

(b)

ROLLS-ROYCE TYNE (*FIG. 2-54*)

The Rolls-Royce Tyne is a two-spool axial-flow turboprop engine with an integral propeller reduction gearbox turning the propeller at 975 rpm. The six-stage low-pressure compressor is driven by the second-, third-, and fourth-stage turbines, while the nine-stage high-pressure section is powered by the first-stage turbine. The compression ratio is 13.5:1, and the mass airflow is 46.5 lb/s [21 kg/s] at 15,250 rpm. The can-annular combustor has 10 flame tubes. Specific fuel consumption is .39 lb/eshp/h [177 g/eshp/h] in cruise. Weight of the engine is 2177 lb [987 kg] and it produces 5500 eshp (5095 shp plus 1010 lbt [4492 N]).

(c)

FIG. 2-54 The Rolls-Royce Tyne turboprop engine.
(a) External view of the Rolls-Royce Tyne turboprop engine.
(b) Cutaway view of the Rolls-Royce Tyne turboprop engine.
(c) The Canadair CL-44 powered by the Rolls-Royce Tyne engine.

(a)

FIG. 2-55 The Teledyne CAE J69 series 25 engine.
(a) The Teledyne CAE J69-T-25A, external view. (Continued on following page.)

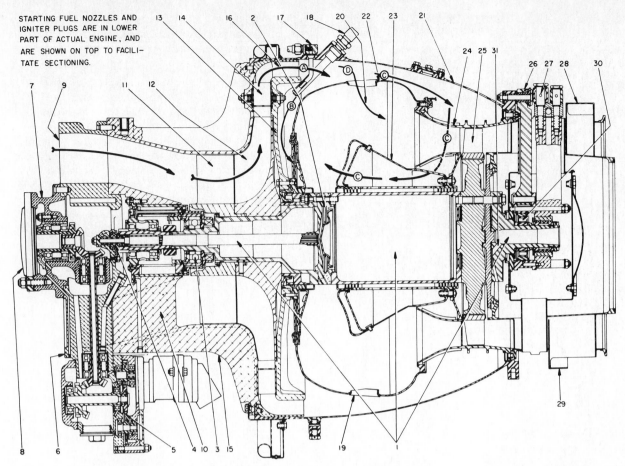

STARTING FUEL NOZZLES AND IGNITER PLUGS ARE IN LOWER PART OF ACTUAL ENGINE, AND ARE SHOWN ON TOP TO FACILITATE SECTIONING.

1 TURBINE SHAFT ASSEMBLY	9 AIR INLET	19 AIR-INLET TUBE	28 STREAMLINE STRUT
2 FUEL DISTRIBUTOR	10 COMPRESSOR-HOUSING STRUT	20 COMBUSTION CHAMBER	29 EXHAUST DIFFUSER
3 FRONT BALL BEARING	11 INDUCER ROTOR	21 TURBINE HOUSING	30 REAR ROLLER BEARING
4 FUEL SEAL	12 COMPRESSOR ROTOR	22 OUTER-COMBUSTOR SHELL	31 TUBULAR AIR PASSAGE
5 ACCESSORY GEAR TRAIN	13 COMPRESSOR COVER	23 INNER-COMBUSTOR SHELL	
6 ACCESSORY CASE	14 RADIAL DIFFUSER	24 TURBINE-INLET NOZZLE	
7 STARTER-GENERATOR DRIVE	15 COMPRESSOR HOUSING	25 TURBINE ROTOR	A INLET AIR
8 STARTER-GENERATOR REPLACEMENT COVER	16 AXIAL DIFFUSER	26 REAR-BEARING HOUSING SUPPORT	B PRIMARY AIR
	17 STARTING-FUEL NOZZLE	27 OIL PASSAGE	C PRIMARY AIR, COOLING
	18 IGNITER PLUG		D SECONDARY AIR

(b)

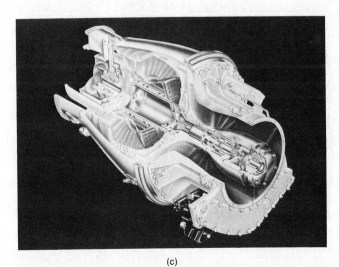

(c)

FIG. 2-55 (continued)

(b) Sectioned view of the Teledyne CAE J69 showing airflow.

(c) Cutaway view of the Teledyne CAE J69-T-25A engine. (Continued on following page.)

TYPES, VARIATIONS, AND APPLICATIONS 73

(d)

FIG. 2-55 (continued)
(d) The Cessna T-37B is powered by two Teledyne CAE J69-T-25 engines.

TELEDYNE CAE J69 (SERIES 25) (*FIG. 2-55*)

The Teledyne CAE J69 is a centrifugal-flow turbojet engine. The single-stage compressor flows 20 lb/s [9 kg/s] of air, and has a compression ratio of 3.8:1 at 21,730 rpm. The engine has an annular side-entry combustion chamber, and a single-stage turbine. Specific fuel consumption is 1.11 lb/lbt/h [113.1 g/N/h]. The engine weighs 364 lb [165 kg] and produces 1025 lbt [4559 N]. Variations include auxiliary power engines with oversized compressors and engines incorporating an axial stage and/or a fan. Thrust and/or power ratings for these engines will vary accordingly. See Figs. 2-56 and 2-57.

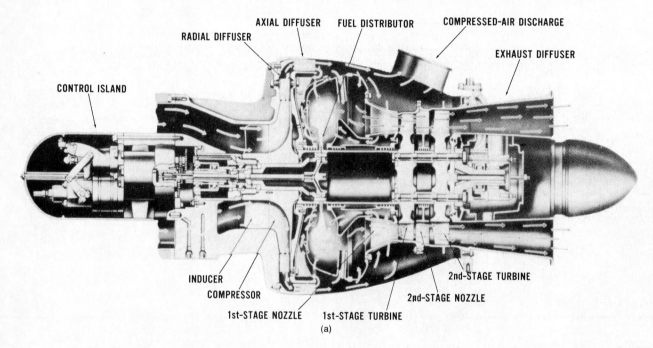

(a)

FIG. 2-56 Teledyne CAE auxiliary-power units based on the J69 Series engine, but with oversized compressors to supply high-volume airflow.
(a) This model is used in the highly produced MA1A starting cart. (See Chap. 17.) (Continued on following page.)

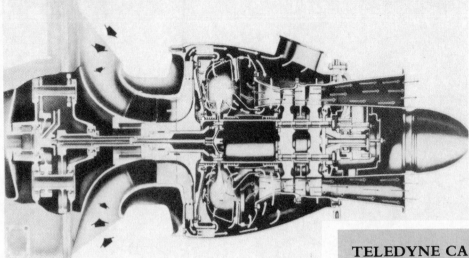

(b)

FIG. 2-56 (continued)
(b) This Teledyne CAE engine can supply shaft power in addition to high-volume airflow.

TELEDYNE CAE AUXILIARY POWER UNITS (*FIG. 2-56*)

As a group, these engines which were originally developed from the French Turbomeca engine designs, are based on the J69 series of gas turbines. They are used as both airborne and ground power units to supply high-volume airflows for starting engines equipped with air turbine starters, or for supplying shaft power to drive electrical generators. The specifications for the TC-106A unit used in a starting cart are as follows: bleed airflow is 90 lb/s [41 kg/s] at a pressure of 45 psia [310 kPa] minimum; rpm is 35,000 and weight of the bare engine is about 240 lb [109 kg]; and the complete trailer unit, including the engine, is 1500 lb [680 kg].

(a)

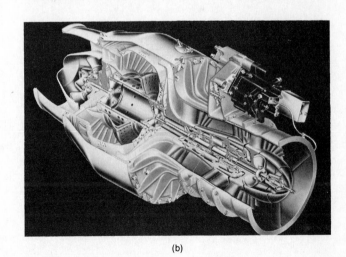

(b)

FIG. 2-57 The Teledyne CAE Series of missile and drone turbine engines. Not included are the J69-T-41A, YJ69-T-406, and the J402-CA-400. All have many of the same construction features as the J69-T-29, but have different performance and specifications. Parts (f) and (g) of this figure represent a radically different approach to engine design from Teledyne CAE's previous concepts.
(a) The Teledyne CAE J69-T-29 (CAE356-7A) external view.
(b) The Teledyne CAE J69-T-29 (CAE356-7A) cutaway view. (Continued on following page.)

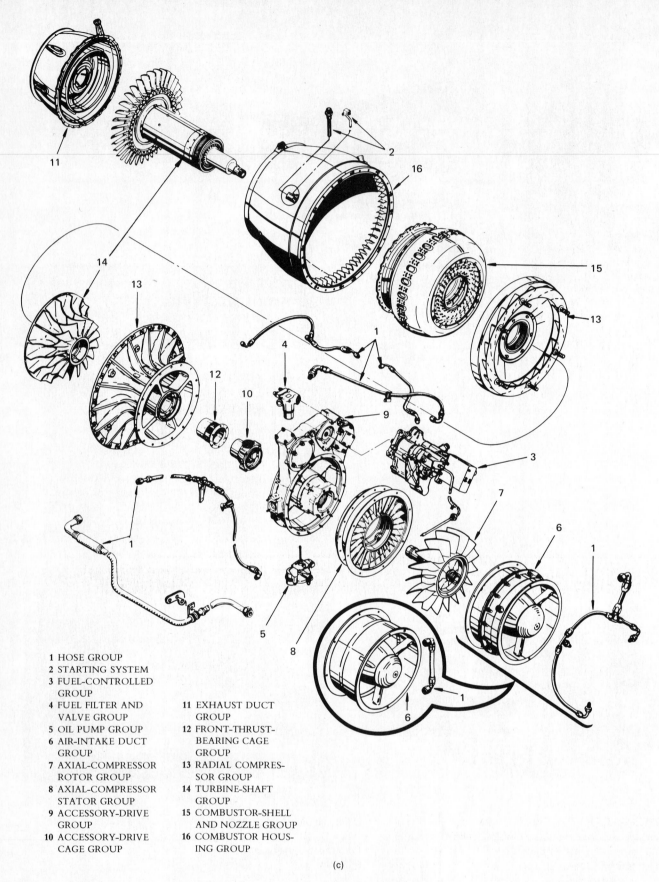

1 HOSE GROUP
2 STARTING SYSTEM
3 FUEL-CONTROLLED
   GROUP
4 FUEL FILTER AND
   VALVE GROUP
5 OIL PUMP GROUP
6 AIR-INTAKE DUCT
   GROUP
7 AXIAL-COMPRESSOR
   ROTOR GROUP
8 AXIAL-COMPRESSOR
   STATOR GROUP
9 ACCESSORY-DRIVE
   GROUP
10 ACCESSORY-DRIVE
   CAGE GROUP

11 EXHAUST DUCT
   GROUP
12 FRONT-THRUST-
   BEARING CAGE
   GROUP
13 RADIAL COMPRES-
   SOR GROUP
14 TURBINE-SHAFT
   GROUP
15 COMBUSTOR-SHELL
   AND NOZZLE GROUP
16 COMBUSTOR HOUS-
   ING GROUP

(c)

FIG. 2-57 (continued)

(c) The Teledyne CAE J69-T-29 (CAE356-7A) sectioned view showing principle parts. (Continued on following page.)

TELEDYNE CAE MISSILE AND DRONE ENGINES (*FIG. 2-57*)

These three Teledyne CAE engines (J69-T-29, J100-CA-100, and the model 490) have all been developed from French Turbomeca designs by Teledyne CAE as powerplants for remote pilot vehicles (RPVs) and other unmanned aircraft. The specifications given here apply only to the J69-T-29 (Teledyne CAE model 356-7A). The J69-T-29 is a single-spool engine with a one-stage axial and one-stage centrifugal compressor that together have a pressure ratio of 5.5:1, an airflow of 28.6 lb/s [13 kg/s] and turn at 22,000 rpm. The side-entry combustion chamber is typical of Teledyne CAE designs. Specific fuel consumption is 1.08 lb/lbt/h [110.1 g/N/h]. The engine weighs 340 lb [154 kg] and produces 1700 lbt [7562 N].

(d)

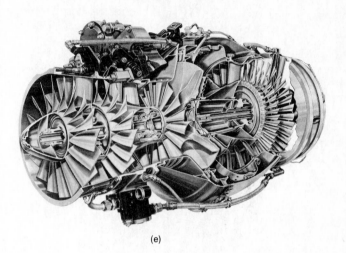

(e)

(f)

FIG. 2-57 (continued)

(d) The Teledyne CAE J100-CA-100 (CAE356-28A) utilizes a two-stage transonic axial plus a single-stage centrifugal compressor. Mass airflow is 44.9 lb/s [20.4 kg/s] with a compression ratio of 6.3:1.

(e) Cutaway view of the Teledyne CAE J100-CA-100 (CAE356-28A). This engine produces 2700 lbt (5449 N), and operates at altitudes in excess of 75,000 ft (22,860 m).

(f) External view of the Teledyne CAE model 490-4 two-spool turbofan based on the Turbomeca-S.N.E.C.M.A. Larzac engine.

(g) Cutaway view of the Teledyne CAE model 490-4 turbofan engine.

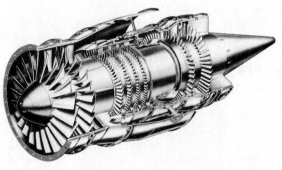

(g)

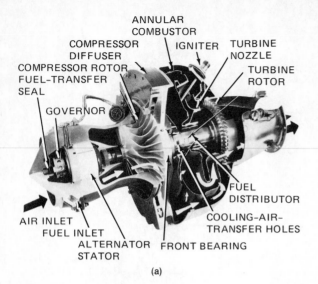

ANNULAR
COMBUSTOR
COMPRESSOR   IGNITER   TURBINE
DIFFUSER   NOZZLE
COMPRESSOR ROTOR   TURBINE
FUEL-TRANSFER   ROTOR
SEAL
GOVERNOR

                                 FUEL
                                 DISTRIBUTOR
                                 COOLING-AIR-
AIR INLET   TRANSFER HOLES
  FUEL INLET
    ALTERNATOR   FRONT BEARING
    STATOR

(a)

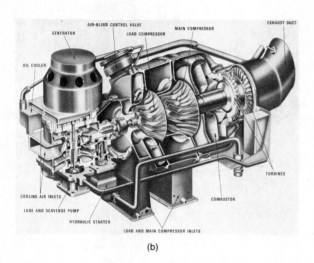

(b)

FIG. 2-58 The Williams Research Corporation WR24-6 and WR27-1 centrifugal-flow gas turbine engines.
(a) Cutaway view showing components of the Williams Research Corporation WR24-6 turbojet engine for use in a target drone.
(b) Cutaway view of the main components of the Williams Research Corporation WR27-1 aircraft auxiliary power unit.

WILLIAMS RESEARCH WR24 AND WR27 (FIG. 2-58)

These two engines are shown as examples of small centrifugal-flow engines used for jet thrust (WR24) and for shaft power (WR27). The WR24-6 turbine is 11 inches [27.9 cm] in diameter and 19 in [48 cm] long. It weighs 30 lb [13.6 kg] and produces 125 lbt [556 N]. Air enters the inlet and passes through the single-stage radial compressor. From the compressor diffuser, it passes into the annular combustion chamber. Fuel enters the inlet housing and is transferred into the rotating governor assembly by the fuel-transfer seal. The fuel then continues through the center of the shaft to the fuel distributors which provide a uniformly distributed fuel fog into the combustor. The products of combustion are cooled by the addition of secondary air through holes near the exit of the combustor. The hot gases then pass through the turbine nozzle and turbine rotor and out the exhaust. The compressor has a 4.2:1 pressure ratio. Turbine inlet temperature is 1750°F [955°C] with an airflow rate of 2.2 lb/s [1 kg/s]. The specific fuel consumption is 1.2 lb/lbt/h [1200 g/N/h] at a rated speed of 60,000 rpm. The WR24-7 version of this engine is similar but incorporates an axial compressor stage and produces 176 lbt [783 N]. The WR27 features a twin compressor configuration. The first compressor rotor (left) provides air to the aircraft systems and the second, higher airflow compressor rotor furnishes air for the fixed-shaft turbine combustor. The two axial turbine stages drive both compressors and the accessory gearbox. The advanced WR27-1 is in U.S. Navy fleet usage in the Lockheed S-3A antisubmarine aircraft. It provides compressed air to start the main TF-34 (Fig. 2-25) engines, as well as electrical power for the electronic equipment and compressed air for the environmental control system. All these functions can be performed both on the ground and in the air. The APU also provides emergency electrical power for aircraft control purposes while airborne.

CHARACTERISTICS AND APPLICATIONS OF THE TURBOJET, TURBOPROP, AND TURBOFAN ENGINES

THE TURBOJET ENGINE

Chapter 3, which deals with engine theory, points out that a turbojet derives its thrust by highly accelerating a small mass of air, all of which goes through the engine. Since a high "jet" velocity is required to obtain an acceptable amount of thrust, the turbine of a turbojet is designed to extract only enough power from the hot gas stream to drive the compressor and accessories. All of the propulsive force produced by a jet engine is derived from the imbalance of forces within the engine itself (Fig. 2-60).

THE TURBOPROP ENGINE

Propulsion in a turboprop engine is accomplished by the conversion of the majority of the gas-steam energy into mechanical power to drive the compressor, accessories, and the propeller load. Only a small amount (approximately 10 percent) of "jet" thrust is available

WILLIAMS RESEARCH WR19 (FIG. 2-59)

The Williams WR19 is a mixed flow (fan air joins core air before exiting) twin-spool fan jet with a bypass ratio of 1:1. It measures only one foot [30.5 cm] in diameter, and two feet [61 cm] in length. The turbine engine weighs 67 lb [30.4 kg] and produces 430 lbt [1913 N]. The fan and low-speed compressor are driven by the second- and third-stage turbines, while the centrifugal high-speed compressor is driven by the first-stage turbine. Specific fuel consumption is .7 lb/lbt/h [71.3 g/N/h].

FIG. 2-59 The Williams Research Corporation WR19 is 1 ft (0.3048 m) in diameter and 2 ft (0.61 m) in length, making it the world's smallest turbofan engine. (No application as of this writing.)
(a) The Williams Research Corporation WR19, external view.
(b) Sectioned view of the WR19 showing the two-spool design.

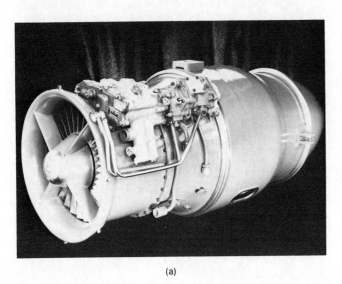

(a)

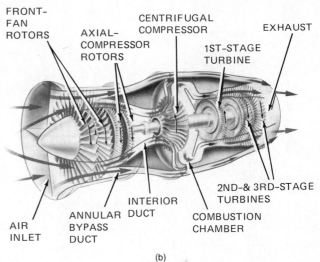

FRONT-FAN ROTORS

AXIAL-COMPRESSOR ROTORS

CENTRIFUGAL COMPRESSOR

1ST-STAGE TURBINE

EXHAUST

AIR INLET

ANNULAR BYPASS DUCT

INTERIOR DUCT

COMBUSTION CHAMBER

2ND- & 3RD-STAGE TURBINES

(b)

in the relatively low-pressure, low-velocity gas stream created by the additional turbine stages needed to drive the extra load of the propeller.

THE TURBOFAN ENGINE

The turbofan engine has a duct-enclosed fan mounted at the front or rear of the engine and driven either mechanically geared down or at the same speed as the compressor, or by an independent turbine located to the rear of the compressor drive turbine. (Refer to Figs. 2-13, 2-19, 2-21, 2-24, 2-26, 2-27, 2-35, 2-42, 2-43, 2-44, 2-51, and 2-52 to see some of these variations.) Figures 2-42 and 2-43 also illustrate two methods of handling the fan air. Either the fan air can exit separately from the primary engine air, or it can be ducted back to mix with the primary engine's air at the rear. If the fan air is ducted to the rear, the total fan pressure must be higher than the static gas pressure in the primary engine's exhaust, or air will not flow. By the same token, the static fan discharge pressure must be less than the total pressure in the primary engine's exhaust, or the

turbine will not be able to extract the energy required to drive the compressor and fan. By closing down the area of flow of the fan duct, the static pressure can be reduced and the dynamic pressure increased. (See Chap. 3 for a discussion of static, dynamic, and total pressure.)

The efficiency of the fan engine is increased over that of the pure jet by converting more of the fuel energy into pressure energy rather than the kinetic (dynamic) energy of a high-velocity exhaust gas stream. As shown in Chap. 3, pressure times the area (PA) equals a force. The fan produces this additional force or thrust without increasing fuel flow. As in the turboprop, primary engine exhaust gas velocities and pressures are low because of the extra turbine stages needed to drive the fan, and as a result this makes the turbofan engine much quieter. (See Chap. 8 on noise.) One fundamental difference between the turbofan and turboprop engine is that the airflow through the fan is controlled by design so that the air velocity relative to the fan blades is unaffected by the aircraft's speed. This eliminates the loss in operational efficiency at high airspeeds which limits the maximum airspeed of propeller-driven aircraft.

The first generation of turbofan designs, such as the

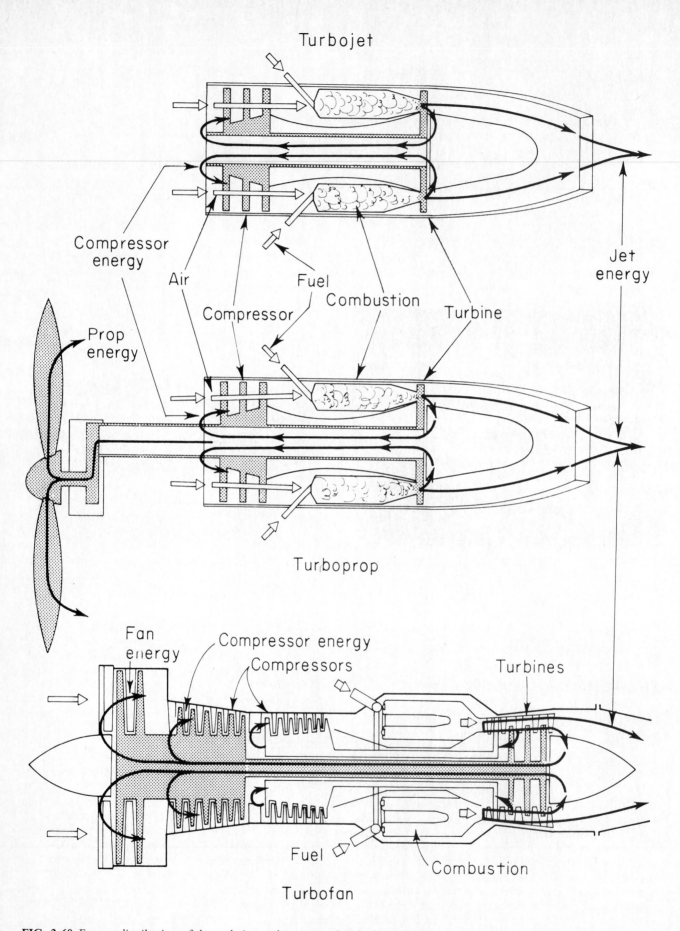

Turbojet

Compressor
energy

Air

Compressor

Fuel

Combustion

Turbine

Jet
energy

Prop
energy

Turboprop

Fan
energy

Compressor energy

Compressors

Turbines

Fuel

Combustion

Turbofan

FIG. 2-60 Energy distribution of the turbojet, turboprop, and turbofan engines.

Pratt & Whitney Aircraft JT3D engine series, had a bypass ratio of approximately 1:1; that is, about 50 percent of the air went through the core of the engine as primary airflow, and about 50 percent went through the fan as secondary airflow. Second generation turbofans like the General Electric CF6 (Fig. 2-27), the Pratt & Whitney Aircraft JT9D (Fig. 2-44), and the Rolls-Royce RB·211 (Fig. 2-52) have bypass ratios on the order of 5:1. Thus, the fan is now providing a much greater percentage of the total thrust produced by the engine.

The emphasis on the use and development of the turbofan engine in recent years is due largely to the development of the transonic blade. The large-diameter fan would require a much lower rpm to keep the blade tips below the speed of sound (see Chap. 3) and this would not be conducive to good gas turbine design.

Fan engines show a definite superiority over the pure jet engines at speeds below Mach 1 (Fig. 2-61), the speed of present-day commercial aircraft. The increased frontal area of the fan presents a problem for high-speed aircraft which, of course, require small frontal areas. At high speeds, the increased drag offered by the fan more than offsets the greater net thrust produced. The disadvantage of the fan for high-speed aircraft can be offset at least partially by burning fuel in the fan discharge air. This would expand the gas, and in order to keep the fan discharge air at the same pressure, the area of the fan jet nozzle is increased. This action results in an increase in gross thrust due to an increase in pressure times an area (PA), and an increase in thrust specific fuel consumption. (See Chap. 3.)

COMPARISON AND EVALUATION OF TURBOJET, TURBOPROP, AND TURBOFAN ENGINES

By converting the shaft horsepower of the turboprop into pounds of thrust, and the fuel consumption per horsepower into fuel consumption per pound of thrust, a comparison between turbojet, turboprop, and turbofan can be made. Assuming that the engines are equivalent as to compressor ratio and internal temperatures and that the engines are installed in equal-sized aircraft best suited to the particular type of engine being used, Fig. 2-61 shows how the various engines compare as to thrust and thrust specific fuel consumption versus airspeed. As the graphs indicate, each engine type has its advantages and limitations. Summaries of these characteristics and uses follow.

Turbojet Characteristics and Uses

1. Low thrust at low forward speeds.
2. Relatively high thrust specific fuel consumption (TSFC) at low altitudes and airspeeds. This disadvantage decreases as altitude and airspeed increase.
3. Long takeoff roll required.
4. Small frontal area results in reduced ground-clearance problems.

5. Lightest specific weight (weight per pound of thrust produced).
6. Ability to take advantage of high ram-pressure ratios.

These characteristics would indicate that the turbojet engine would be best for high-speed, high-altitude, long-distance flights.

Turboprop Characteristics and Uses

1. High propulsive efficiency at low airspeeds, which falls off rapidly as airspeed increases. See p. 107. This results in shorter takeoff rolls. The engine is able to develop very high thrust at low airspeeds because the propeller can accelerate large quantities of air at zero forward velocity of the airplane. A discussion of propulsive efficiency follows in the next chapter.
2. More complicated and heavier than a turbojet.
3. Lowest TSFC.
4. Large frontal area of propeller and engine combination necessitates longer landing gears for low-wing airplanes, but does not necessarily increase parasitic drag.
5. Efficient reverse thrust possible.

These characteristics show that turboprop engines are superior for lifting heavy loads off short and medium runways. Turboprops are currently limited in speeds to approximately 500 mph [805 km/h], since propeller efficiencies fall off rapidly with increasing airspeeds because of shock-wave formations (see pp. 107–109). However research by the Hamilton Standard division of United Technologies Corporation and others is being done to overcome, or extend, this limitation.

Turbofan Characteristics and Uses

1. Increased thrust at forward speeds similar to a turboprop results in a relatively short takeoff. But unlike the turboprop, the turbofan thrust is not penalized with increasing airspeed, up to approximately Mach 1 with current fan designs.
2. Weight falls between the turbojet and turboprop.
3. Ground clearances are less than turboprop, but not as good as turbojet.
4. TSFC and specific weight fall between turbojet and turboprop (Figs. 2-62 and 2-63). This results in increased operating economy and aircraft range over the turbojet.
5. Considerable noise level reduction of 10 to 20 percent over the turbojet reduces acoustic fatigue in surrounding aircraft parts and is less objectionable to people on the ground. No noise supressor is needed. On newer fan engines such as the Pratt & Whitney Aircraft JT9D and General Electric CF6 shown in Figs. 2-44, 2-27, and others, the inlet guide vanes have been eliminated in an attempt to reduce the fan noise, which is considered to be a large problem for high-bypass-ratio fan engines. The noise level is reduced by the elimination of the discrete frequencies that are generated by the fan blades cutting through the wakes behind the vanes. Other fan-noise-reducing features are also incorporated. (See Chap. 8.)

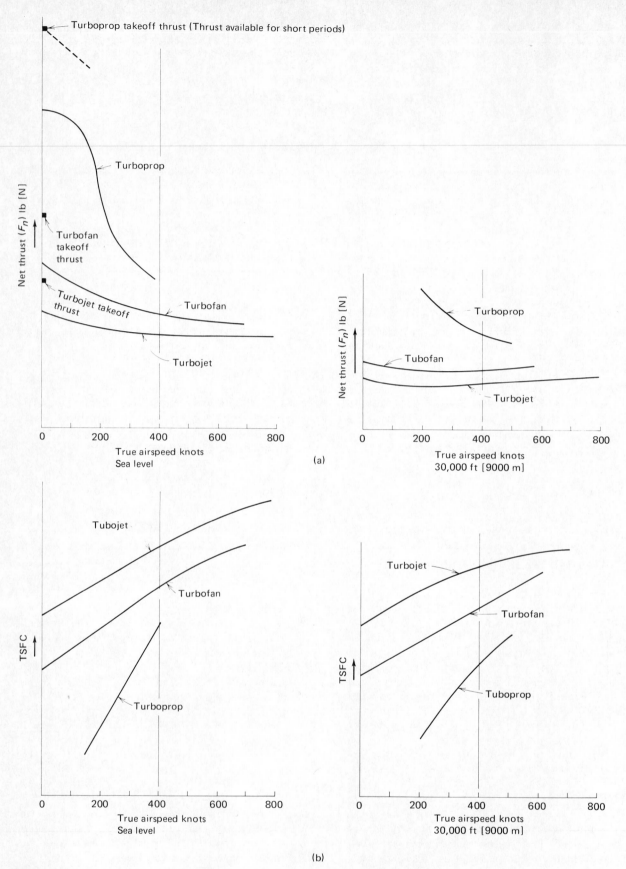

FIG. 2-61 Operating parameters of the turbojet, turboprop, and turbofan engines.
(a) Thrust compared to airspeed at sea level and at 30,000 ft (9144 m).
(b) Thrust specific fuel consumption compared with airspeed at sea level and at 30,000 ft (9144 m).

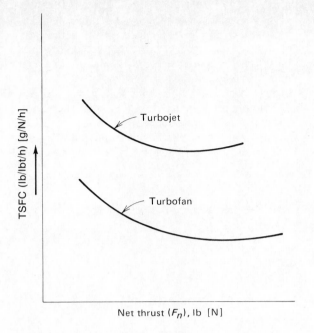

FIG. 2-62 Comparison of thrust specific fuel consumption with thrust for turbojet and turbofan engines shows the superiority of the turbofan.

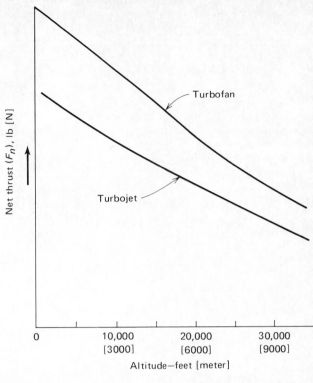

FIG. 2-63 The turbofan engine maintains its superior thrust rating over the turbojet engine at all altitudes.

6. Superior to the turbojet in "hot day" performance (Fig. 2-64).
7. Two thrust reversers are required if the fan air and primary engine air exit through separate nozzles. The advantage of the separate fan nozzle is the short fan duct with corresponding low duct loss.

NOTE: Hamilton Standard is experimenting with a small-diameter, multibladed, wide-chord propeller (propfan) said to be more efficient than the high-bypass-ratio turbofan engine, and with a 20 percent reduction in TSFC.

From the above characteristics it is obvious that the fan engine is suitable for long-range, relatively high-speed flight, and has a definite place in the prolific gas turbine family.

GENERAL TRENDS IN THE FUTURE DEVELOPMENT OF THE GAS TURBINE ENGINE

1. There will be higher compressor airflows, pressure ratios, and efficiencies with fewer compressor stages, fewer parts, and lower cost. Variable-pitch fan or compressor blades will provide reverse thrust for braking.
2. Engines will have a lower specific fuel consumption resulting from component design improvements and other changes.

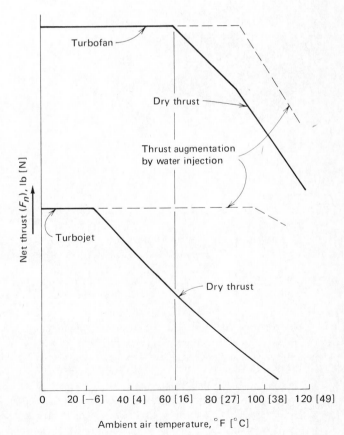

FIG. 2-64 Hot-day performance of the turbojet and turbofan engines at sea level.

3. Increased turbine efficiencies will result in fewer stages to do the necessary work, less weight, lower cost, and decreased cooling air requirements.
4. Increased turbine temperatures will result from better metals and improved cooling techniques (see Chap. 10).
5. There will be less use of magnesium, aluminum, and iron alloys and more of nickel and cobalt-base alloys plus increased use of composite materials.
6. Engines will burn fuel more cleanly and will run more quietly, thus making this type of powerplant less hostile to the environment.
7. More airborne and ground engine-condition-monitoring equipment will be used, such as vibration and oil analyzers, and radiometer sensors to measure turbine blade temperature while the engine is running. There will be increased use of inspection through built-in borescope ports and radiographic techniques, plus many other pressure, temperature, and rpm devices to monitor the engine's health (see Chap. 18).
8. New manufacturing techniques will be used, including diffusion bonding, unusual welding methods, electrochemical devices, and new coating methods, to form and work the new metalic and nonmetalic materials that will be found in the newer gas turbine generation (see Chap. 10).

Many of these trends are discussed in further detail in the chapters that follow.

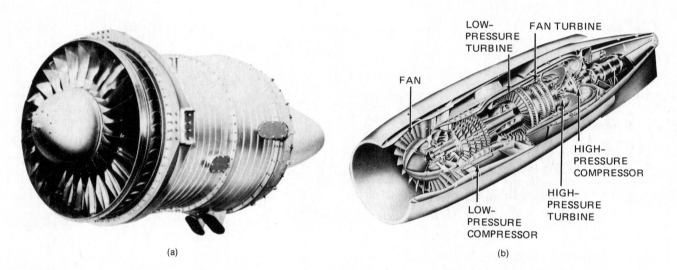

(a) (b)

FIG. 2-65 The Garrett AiResearch ATF3 turbofan engine.
(a) External view of the Garrett AiResearch ATF3.
(b) The gas path, which can be traced in this cutaway view of the ATF3, starts at the pod diffuser and flows into the single-stage fan which turns at 8900 rpm. This fan is driven by a three-stage axial turbine located between the high- and low-pressure compressor drive turbines. The five-stage, axial-flow, low-pressure compressor is driven by a two-stage axial turbine whose rpm is 14,600. After leaving the low-pressure compressor, the airflow is split into eight ducts and turned 180° to enter a centrifugal high-pressure compressor stage rotating at 34,700 rpm. The airflow then enters a reverse-flow, annular combustion chamber, where fuel is atomized through eight fuel nozzles, and the fuel-air mixture burned to attain a turbine inlet temperature of approximately 1600° F (871° C) for cruise operation. A single-stage axial turbine drives the high-pressure compressor. The gases are then expanded through the fan and low-pressure turbines. The turbine exhaust gases are split and turned 113° in eight ducts which fit between the eight ducts connecting the low- and high-pressure compressors. The turbine exhaust gases partially mix with the fan-duct airflow and are exhausted through a common nozzle. Overall pressure ratio is 17:1, and airflow is 140 lb/s (63.6 kg/s) with a bypass ratio of 3. (Continued on following page.)

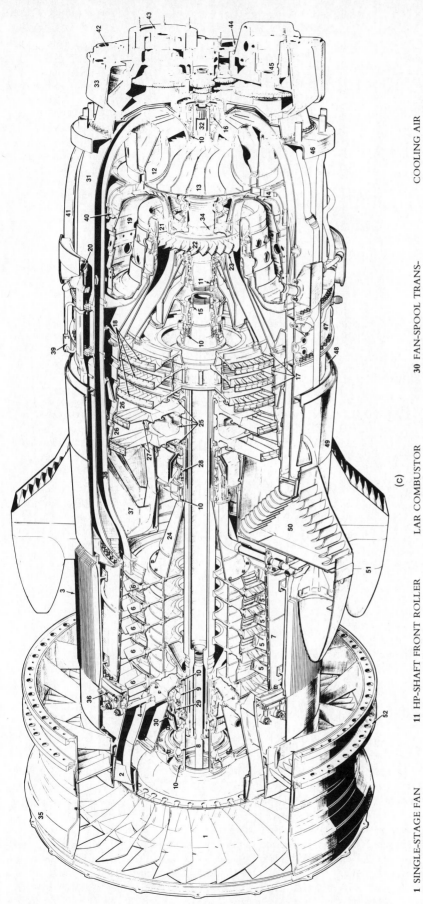

1 SINGLE-STAGE FAN
2 FAN STATORS
3 OIL COOLER
4 COMPRESSOR INLET STA-
  TOR (VARIABLE)
5 LP COMPRESSOR STA-
  TORS
6 FIVE-STAGE LP AXIAL
  COMPRESSOR
7 ALUMINUM-COATED
  COMPRESSOR FIXED STA-
  TOR RING—ALL STAGES
8 FAN-SHAFT THRUST RACE
9 THRUST BALL-RACE LPC
10 AIR-LABYRINTH SEALS
  (BUFFER AIR)

11 HP-SHAFT FRONT ROLLER
  BEARING
12 HP COMPRESSOR FACE
13 HP CENTRIFUGAL COM-
  PRESSOR
14 HP COMPRESSOR DIF-
  FUSER
15 FAN-SPOOL ROLLER
  BEARING
16 HP-SPOOL THRUST BEAR-
  ING
17 FAN-TURBINE STATOR
18 THREE-STAGE FAN TUR-
  BINE
19 REVERSE-FLOW ANNU-

20 LAR COMBUSTOR
21 FUEL MANIFOLD
22 HP TURBINE STATORS
23 SINGLE-STAGE HP AXIAL
  TURBINE WHEEL
24 AIR-COOLED HP TURBINE
  BLADES
25 LP ROTOR SHAFT
26 LP-SHAFT CURVIC COU-
  PLING
27 TWO-STAGE LP TURBINE
28 LP-SPOOL REAR ROLLER
  BEARING
29 LP-SPOOL-SPEED TRANS-
  DUCER GEAR

30 FAN-SPOOL TRANS-
  DUCER GEAR
31 HP CASE
32 ACCESSORY-DRIVE COU-
  PLING SHAFT
33 ACCESSORY GEARBOX
34 HP-TURBINE CURVIC
  COUPLING
35 FAN-DUCT INLET
36 IGV ACTUATION RING
37 SPLITTER
38 CROSSOVER DUCT
39 T₈ HARNESS AND JUNC-
  TION BOX
40 DESWIRL VANES
41 ACCESSORY-GEARBOX

42 GEARBOX VENT
43 PMG
44 FUEL-PUMP PAD
45 OIL-PUMP DRIVE
46 AFT FIRESHIELD
47 MIDFRAME AND MAIN
  MOUNT
48 FORWARD FIRESHIELD
49 AFT FAIRING
50 EXHAUST CASCADES
51 CASCADE DAGMAR
52 FORWARD FRAME AND
  MOUNT

(c)

FIG. 2-65 (continued)
(c) Sectioned view of the Garrett ATF3-6 engine. (Continued on following page.)

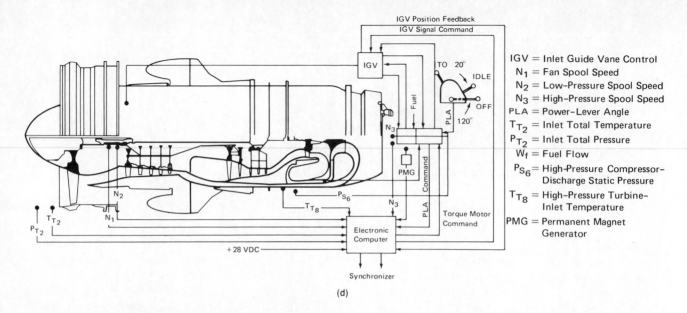

IGV Position Feedback
IGV Signal Command

IGV = Inlet Guide Vane Control
N_1 = Fan Spool Speed
N_2 = Low-Pressure Spool Speed
N_3 = High-Pressure Spool Speed
PLA = Power–Lever Angle
T_{T_2} = Inlet Total Temperature
P_{T_2} = Inlet Total Pressure
W_f = Fuel Flow
P_{S_6} = High-Pressure Compressor-Discharge Static Pressure
T_{T_8} = High-Pressure Turbine-Inlet Temperature
PMG = Permanent Magnet Generator

(d)

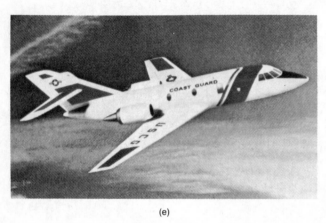

(e)

GARRETT AIRESEARCH ATF3 (*FIG. 2-65*)

This advanced technology three spool engine is listed out of normal sequence, since it is a late production machine from this company. It features modular construction, easy serviceability, low noise and smoke levels, and economical operation among its other attributes. Thrust ranges are from 4000 to 5000 lbt [1814 to 2270 N], specific fuel consumption is approximately .4 to .5 lb/lbt/hr [400 to 500 g/N/h] and the engine weighs 850 to 900 lb [385 to 409 kg].

FIG. 2-65 (continued)

(d) This figure, showing airflow through the engine, also shows the combination electronic and hydromechanical fuel control. (See Chapter 12.)

(e) The Falcon 20G, equipped with two ATF3-6 engines.

REVIEW AND STUDY QUESTIONS

1. What are two methods of classifying gas turbine engines?
2. List some variations of centrifugal-type engines. Name some airplanes in which this type of engine is installed.
3. What are the advantages and disadvantages of the centrifugal compressor?
4. Define "gas generator" and "free-power turbine."
5. What is a regenerator? Of what advantage and disadvantage is this type of device?
6. List some variations of the axial-compressor engine. Name some airplanes in which this type of engine is installed.
7. What are the advantages and disadvantages of the axial compressor?
8. Where is the fan located on the fan-type engine? How is it driven?
9. What are three ways of using a gas turbine engine to power an airplane?
10. What is meant by mixed exhaust and nonmixed exhaust fan engines?
11. Make a table that lists the characteristics and uses of the turbojet, turboprop, and turbofan engines.
12. List the general trend in the future development of the gas turbine engine.

CHAPTER THREE
ENGINE THEORY: TWO PLUS TWO

In order to understand some of the operating fundamentals of the gas turbine engine, it might be well to list the sections of such an engine and then very briefly discuss the series of events that occur.

Basically, a gas turbine engine consists of five major sections: an inlet duct, a compressor, a combustion chamber (or chambers), a turbine wheel (or wheels), and an exhaust duct (Fig. 3-1). In addition to the five major sections, each gas turbine is equipped with an accessory section, a fuel system, a starting system, a cooling system, a lubrication system, and an ignition system. Some engines might also incorporate a water injection system, an afterburner system, a variable-area exhaust nozzle and system, a variable-geometry compressor, a fan, a free-power turbine, a propeller reduction gearbox, and other additional systems and components to improve or change engine operation, performance, and usage.

TYPICAL OPERATION

The front, or inlet, duct is almost entirely open to permit outside air to enter the front of the engine. The compressor works on this incoming air and delivers it to the combustion or burner section with as much as 20 times or more the pressure the air had at the front. In the burner section, fuel, similar to kerosine, is sprayed and mixed with the compressor air. The air-fuel mixture is then ignited by devices similar to spark plugs. When the mixture is lighted, the ignitor can be turned off, as the burning process will continue without further assistance as long as the engine is supplied with the proper fuel/air ratio. The fuel-air mixture burns at a relatively constant pressure with only about 25 percent of the air taking part in the actual combustion process. The balance of the air is mixed with the products of combustion for cooling before the gases enter the turbine wheel. The turbine extracts a major portion of the energy in the gas stream and uses this energy to turn the compressor and accessories. After leaving the turbine, there is still enough pressure remaining to force the hot gases through the exhaust duct and jet nozzle at the rear of the engine at very high speeds. The engine's thrust comes from taking a large mass of air in at the front end and expelling it from the tailpipe at a much higher speed than it had when it entered the compressor. Thrust, then, is equal to *mass flow rate times change in velocity*. In order to appreciate this statement, a review of some basic physics is necessary.

REVIEW OF PHYSICS CONCEPTS

Force

A *force* is defined as a push or a pull which will produce or prevent motion. Gravity, for example, is a force that attracts bodies toward the earth at a rate that will cause the object to increase its velocity by 32.2 feet per second (ft/s) [9.81 meters per second (m/s)] for each second the object is falling. That is, at the end of 2 s, the speed would be 64.4 ft/s [19.62 m/s], at the end of 3 s it would be 96.6 ft/s [29.43 m/s], etc. The figure 32.2 feet per second per second (32.2 ft/s² or 9.81 m/s²) is called the acceleration due to gravity and is represented in formulas by the letter g. This value can also be used to determine the amount of resistance an object of given weight offers to motion. When the weight is divided by the acceleration constant, the quotient is called the mass of the object. (See page 90.)

$$M = \frac{w}{g}$$

Force is also a vector quantity; that is, it has both magnitude and direction. When we speak of 1000 lb [453.6 kg] of force acting upon an object, we cannot know its effect unless we know the direction of the force. Two or more forces acting on a body will produce a resultant force. Vectors and their resultants will be used in Chaps. 5 and 7 to help explain the operation of compressors and turbines.

Work

Mechanical *work* is done when a force acting on a body causes it to move through any distance (Fig. 3-2).

Work = force × distance
$\qquad W = Fd$

It is important to remember that work is accomplished only when an object is moved some distance by an ap-

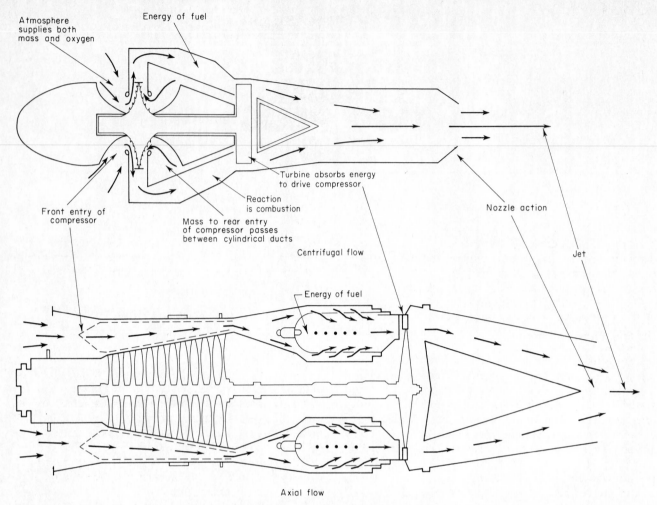

Atmosphere supplies both mass and oxygen

Energy of fuel

Turbine absorbs energy to drive compressor

Reaction is combustion

Front entry of compressor

Mass to rear entry of compressor passes between cylindrical ducts

Centrifugal flow

Nozzle action

Jet

Energy of fuel

Axial flow

FIG. 3-1 Airflow through centrifugal- and axial-flow engines. *(U.S.A.F. AFM 52-2.)*

plied force. For example, if an object that is pushed as hard as possible fails to move, then by the textbook definition, no work has been done.

Force is often expressed in pounds, distance in feet, and work in foot-pounds.

EXAMPLE: A jet engine that is exerting 1000 lb of force [44,480 N] moves an airplane 10 ft [3 m]. How much work is being accomplished?

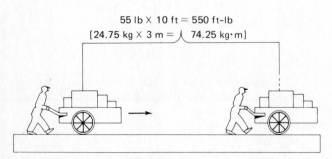

55 lb × 10 ft = 550 ft-lb
[24.75 kg × 3 m = 74.25 kg·m]

FIG. 3-2 Work equals force times distance. *(U.S.A.F. Extension Course Institute and Air University, Course 4301.)*

$$\begin{aligned} \text{Work} &= Fd \\ &= 1000 \times 10 \\ W &= 10{,}000\ \text{ft}\cdot\text{lb}\ [13{,}560\ \text{N}\cdot\text{m}] \end{aligned}$$

Power

Nothing in the definition of work states how fast the work is being done. The rate of doing work is known as *power*.

$$\text{Power} = \frac{\text{force} \times \text{distance}}{\text{time}}$$
$$P = \frac{Fd}{t}$$

As will be learned later when developing a formula for converting thrust to power, power may be expressed in any one of several ways, depending upon the units used for the force, the distance, and the time. Power is often expressed in units of horsepower. One *horsepower* is equal to 33,000 ft·lb/min [4554 kg·m/min] or 550 ft·lb/s [69 kg·m/s]. In other words, a 1-hp motor can raise 33,000 lb a distance of 1 ft in 1 min or 550 lb a distance of 1 ft in 1 s (Fig. 3-3).

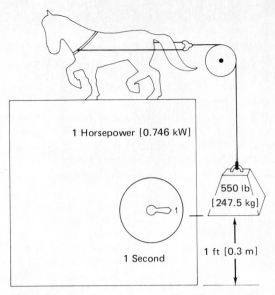

1 Horsepower [0.746 kW]

550 lb [247.5 kg]

1 Second

1 ft [0.3 m]

FIG. 3-3 Most horses are capable of more power than this. (*U.S.A.F. Extension Course Institute and Air University, Course 4301.*)

$$hp = \frac{power\ (ft \cdot lb/min)}{33,000}$$
$$= \frac{Fd/t(min)}{33,000}$$

or

$$hp = \frac{power\ (ft \cdot lb/s)}{550}$$
$$= \frac{Fd/t(s)}{550}$$

EXAMPLE: A 5000-lb [2250·kg] weight is lifted a distance of 10 ft [3 m] in 2 min. How much horsepower is required?

$$P = \frac{Fd}{t}$$
$$= \frac{5000 \times 10}{2}$$
$$= 25,000\ ft \cdot lb/min\ [3450\ kg \cdot m/s]$$

$$hp = \frac{P}{33,000}$$
$$= \frac{25,000}{33,000}$$
$$= 0.75 \quad or \quad \tfrac{3}{4}$$

Energy

Energy is defined as the capacity for doing work. The energy which bodies possess can be classified into two categories: *potential* and *kinetic*. Potential energy may be

due to position, such as water in an elevated storage tank, distortion of an elastic body such as a compressed spring, or a chemical action, for example, from coal.

EXAMPLE: A 20,000-lb [9072 kg] airplane is held 5 ft [1.52 m] off the floor by a jack. How much potential energy does this system possess?

$$PE = WH$$
$$= 20,000 \times 5$$
$$= 100,000\ ft \cdot lb\ [13,830\ kg \cdot m]$$

where PE = potential energy, ft·lb
 W = weight of object, lb
 H = height of object, ft

Kinetic energy is energy of motion. Gases striking the turbine wheel exhibit kinetic energy. If the mass and speed of a body are known, the kinetic energy can be determined from the formula

$$KE = \frac{WV^2}{2g}$$

where W = weight, lb
 V = velocity, ft/s
 g = acceleration due to gravity = 32.2 ft/s² [9.81 m/s²]
 KE = kinetic energy, ft·lb

Notice that the kinetic energy is directly proportional to both the weight and the square of the velocity.

EXAMPLE: An airplane weighing 6440 lb [2921 kg] has a velocity of 205 mph (300 ft/s) [330 km/h (91.6 m/s)]. Find the kinetic energy.

$$KE = \frac{WV^2}{2g}$$
$$= \frac{6440 \times 300^2}{2 \times 32.2}$$
$$= 9,000,000\ ft \cdot lb\ [1,244,700\ kg \cdot m]\ of\ energy$$

Speed

The *speed* of a body in motion is defined as the distance it travels per unit of time.

$$Speed = \frac{distance}{time}$$

Speed units are commonly expressed in miles per hour or feet per second [kilometers per hour or meters per second (m/s)].

Velocity

Velocity can be defined as speed in a given direction. The symbol V is used to represent velocity.

Acceleration

The *acceleration* of a body in motion is defined as the rate of velocity change. The definition is not based on the distance traveled, but on the loss (deceleration) or gain (acceleration) of velocity with time.

$$\text{Acceleration} = \frac{\text{change in motion}}{\text{unit of time}}$$
$$= \frac{\text{final velocity minus initial velocity}}{\text{time}}$$
$$= \frac{V_2 - V_1}{t}$$

where V_1 is the original velocity and V_2 is the final velocity.

Mass

The *mass* of an object is the amount of fundamental matter of which it is composed; it is in a sense the measure of a body's inertia. Mass and weight are often confused because the common method of determining a quantity of matter is by weighing. But, weight is only a measure of the pull of gravity on a quantity of matter. An object that weighs 36 lb [17.2 kg] on earth will weigh 6 lb [2.7 kg] on the moon. Yet the mass is exactly the same. Mass, then, is derived by dividing the weight of the object by the acceleration due to gravity, which, as previously stated is equal to 32.2 ft/s² [9.81 m/s²] on earth.

$$M = \frac{W}{g}$$

Momentum

Mass times velocity or MV defines *momentum*. It is the property of a moving body which determines the length of time required to bring it to rest under the action of a constant force. Large objects with a lot of mass but very little velocity can have as much momentum as low-mass objects with very high velocity. A boat must dock very slowly and carefully because if it touches the dock even gently, it may crush it. On the other hand, a bullet weighs very little but its penetrating power is very high because of its velocity.

NEWTON'S LAWS OF MOTION

The fundamental laws of jet propulsion were demonstrated many years ago by recognized scientists and experimenters. These laws, and the equations derived from them, must be discussed in order to understand the operating principle of the gas turbine engine. Foremost among these scientists was Sir Isaac Newton of England, who derived three laws pertaining to bodies at rest and in motion, and the forces acting on these bodies.

1. *Newton's first law* states that "A body (mass) in a state of rest tends to remain at rest, and a body in motion tends to continue to move at a constant speed, in a straight line unless acted upon by some external force." The portion of the law which states "a body in a state of rest tends to remain at rest" is acceptable from our own experience. But the second part which states "a body in motion tends to remain in motion at a constant speed and in a straight line" is more difficult to accept. For example, the less friction that a bearing offers, the longer a wheel will coast. Therefore, according to this law, if all friction were removed, the wheel would coast forever.

2. *The second law of motion* says that "An unbalance of force on a body tends to produce an acceleration in the direction of the force, and that the acceleration, if any, is directly proportional to the force and inversely proportional to the mass of the body."

$$a = \frac{F}{M}$$

where a = acceleration
F = force
M = mass

A ball thrown with a force that accelerates it at the rate of 50 ft/s² [15.24 m/s²] will need double this force to accelerate the ball to 100 ft/s² [30.48 m/s²]. On the other hand if the mass of the ball is doubled, the rate of acceleration would be halved, or 25 ft/s² [7.62 m/s²]. If each side of the equation is multiplied by M, then

$$F = Ma$$

3. *Newton's third law* states that "For every acting force there is an equal and opposite reacting force." The term *acting* force means the force exerted by one body on another, while the *reacting* force means the force the second body exerts on the first. These forces always occur in pairs but never cancel each other because, although equal in magnitude, they always act on different objects. Examples of the third law are to be found in everyday life (Fig. 3-4). When a person jumps from a boat, it is pushed backwards with the same force that pushes the person forward. It should be noted that the person gains the same amount of momentum as the boat received, but in the opposite direction.

The equation for momentum equals mass times velocity. Since the momentum of both the person and the boat must be equal, then

$$M_1 V_1 = M_2 V_2$$

EXAMPLE: A man weighing 150 lb [68.04 kg] jumps from his boat to shore at a velocity of 2 ft/s [0.61 m/s]. If the boat weighs 75 lb [34.02 kg], what will be its velocity?

$$M_1 V_1 = M_2 V_2$$
$$\frac{150}{32.2} \times 2 = \frac{75 \, V_2}{32.2}$$
$$V_2 = 4 \text{ ft/s } [1.22 \text{ m/s}]$$

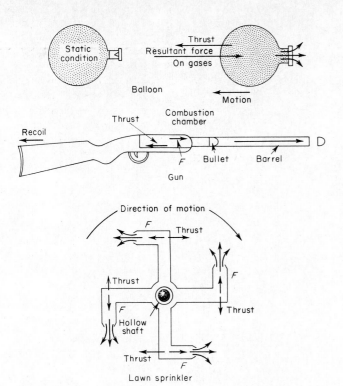

FIG. 3-4 Applications of Newton's Third Law of Motion. (*U.S.A.F. Extension Course Institute and Air University, Course 4301.*)

Everyone knows that when a balloon is blown up and released, it will travel at a fairly high speed for a few seconds. The *gas turbine engine* operates *like a toy balloon,* and the operation of both can be explained using Newton's third law of motion.

When the balloon is inflated, the inside air pressure, which is stretching the skin, is greater than the outside pressure, and if the stem is tied closed, the inside air pushes equally in all directions and the balloon will not move (Fig. 3-5). If the balloon is placed in a vacuum and the stem is released, the escaping air obviously has nothing to push against. Yet the balloon will move in a direction away from the stem just as it does in a normal atmosphere, proving that *it is not the escaping air pushing against anything outside that makes the balloon move.*

Releasing the stem removes a section of the skin on that side of the balloon against which the air has been pushing. On the side directly opposite the stem, however, the air continues to push on an equal area of skin, and it is the uncanceled push on this area that causes the balloon to move in the direction away from the stem.

The balloon's flight is short because the pressure within the skin is lost quickly. This handicap could be overcome by pumping air into the balloon with a bicycle pump so the pressure and airflow are maintained.

To transform this apparatus into a self-contained jet engine, the hand pump is replaced with a compressor. And if the compressor is turned at high speed, a large quantity of air is passed through the balloon while holding a high pressure inside. For energy to turn the compressor, a burner is placed in the airstream. Burning the fuel raises the air temperature rapidly, and the air vol-

ume is greatly increased. Since the compressor pressure blocks the forward flow, the air can only move rearward on the less restricted path leading to the exit. By placing a turbine in the path of the heated air, some of this energy is used to spin the turbine, which in turn, spins the compressor by means of a connecting shaft. The remaining energy is expended in expelling the hot gases through the stem of the balloon, which is in effect a jet nozzle. The transformation is now complete and the balloon "jet engine" can operate as long as there is fuel to burn.

The *acting force* that Newton's third law refers to is the acceleration of the escaping air from the rear of the balloon. The reaction to this acceleration is a force in the opposite direction. In addition, the amount of force acting on the balloon is the product of the mass of air being accelerated times the acceleration of that air. Since the forces always occur in pairs, it can be said that if it takes a certain force to accelerate a mass rearward, the reaction to this force is thrust in the opposite direction.

$$\overleftarrow{\text{Force}} = \overrightarrow{\text{thrust}}$$
$$\text{Action} = \text{reaction}$$

Thrust computation

Using Newton's second law of motion permits the solution of the simple problem that follows.

EXAMPLE: How much force would be necessary to accelerate an object weighing 161 lb [73.03 kg] at the rate of 10 ft/s² [3.05 m/s²]?

$$F = Ma$$

where F = force
$M = \dfrac{\text{weight of the object}}{\text{acceleration of gravity}}$
a = change in velocity

$$F = \frac{W}{g} \times a$$
$$= \frac{161}{32.2} \times 10$$
$$= 50 \text{ lb } [22.68 \text{ kg}]$$

The same formula applies to the jet engine.

EXAMPLE: A large jet engine handles 100 lb [45.36 kg] of air per second. The velocity of this air at the jet nozzle is 659 mph (approximately 966 ft/s [1060.5 m/h (294 m/s)]. What is the thrust of the engine?

$$F = Ma$$
$$= \frac{W_a}{g} \times a$$
$$= \frac{100}{32.2} \times 966$$
$$= 3000 \text{ lb } [1360.8 \text{ kg}] = \text{thrust}$$

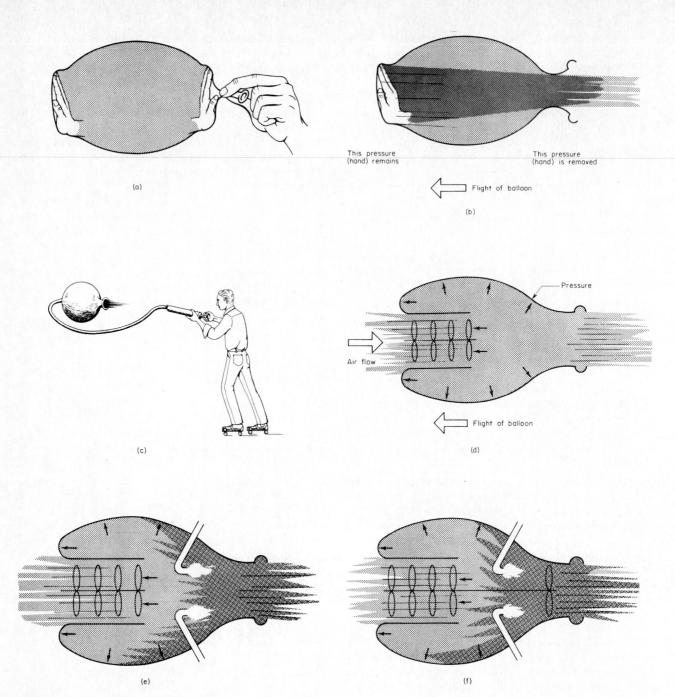

FIG. 3-5 Balloon analogy of the jet engine. *(Pratt & Whitney Aircraft.)*

(a) Pressures are equal in all directions.

(b) An unbalance of force is created when the stem is opened.

(c) Maintaining pressure in the balloon.

(d) Replacing the hand pump with a compressor.

(e) Raising the air temperature and increasing the volume.

(f) The turbine extracts some of the energy in the air to turn the compressor.

From the preceding example, if a 3000-lb force or action is required to produce the 966 ft/s velocity change of 100 lb/s airflow, then an equal but opposite 3000 lb of reaction or thrust will be felt in the structure of the engine.

Gross and Net Thrust

When the *gross thrust* is computed, the velocity of the air coming into the engine due to the velocity of the airplane is disregarded and, as shown in the previous problem, the velocity of the gas leaving the engine is used as the acceleration factor. True acceleration of the gas is the *difference* in the velocity between the incoming and outgoing air, and this difference is used in computing *net thrust*. The loss in thrust involved in taking the air in at the front of the engine is known as the *ram drag*. Net thrust is then gross thrust minus ram drag.

$$F_n = F_g - F_r$$

Engine inlet air velocity times the mass of airflow is the *initial momentum*. The faster the airplane goes, the greater the initial momentum and the less the engine can change this momentum.

$MV_2 - MV_1$ = the acceleration of gases through the engine

where MV_2 = gross thrust or momentum of exhaust gases

MV_1 = ram drag or momentum of incoming air due to airplane speed

EXAMPLE: Using the same engine and values shown in the previous problem, but with the airplane moving at a speed of 220 mph, the net thrust would be

Net thrust = Ma

$$= \frac{W_a}{g} (V_2 - V_1)$$

where V_2 = velocity of air at the jet nozzle, ft/s
V_1 = velocity of the airplane, ft/s (220 mph = 332 ft/s = ram drag) [354.05 km/h (98.1 m/s)]

Net thrust $= \dfrac{100}{32.2} (966 - 322)$

$= 2000$ lb [907.2 kg]

Completing the Jet Engine Equation

Since fuel flow adds some mass to the air flowing through the engine, the same formula must be applied to the weight of the fuel as was applied to the weight of the air, and this must be added to the basic thrust equation

$$F_n = \frac{W_a}{g} (V_2 - V_1) + \frac{W_f}{g} (V_f)$$

where W_f = weight of fuel
V_f = velocity of fuel

Notice that because the fuel is carried along with the engine it will never have any initial velocity relative to

the engine. Some formulas do not consider the fuel flow effect when computing thrust because the weight of air leakage through the engine is approximately equivalent to the weight of the fuel added.

This formula was complete until the development of the "choked" nozzle. When a nozzle is choked, the pressure is such that the gases are traveling through it at the speed of sound and cannot be further accelerated. Any increase in internal engine pressure will pass out through the nozzle still in the form of pressure. Even though this pressure energy cannot be turned into velocity energy, it is not lost. The pressure inside the nozzle is pushing in all directions, but when the neck is open, the air can no longer push in the direction of the nozzle. The pressure in the other direction continues undiminished, and as a result the pressure of the gases will push the engine forward (Fig. 3-6). Any ambient air (air outside the nozzle) is in the way and will cancel out part of the forward thrust. The completed formula for a turbojet engine with a choked nozzle is

$$F_n = \frac{W_a}{g} (V_2 - V_1) + \frac{W_f}{g} (V_f) + A_j(P_j - P_{am})$$

where A_j = area of jet nozzle
P_j = static pressure of jet nozzle
P_{am} = ambient static pressure

EXAMPLE: The following conditions are known about an operating aircraft gas turbine engine.

Barometric pressure	= 29.24 inches of mercury (inHg) = 14.4 pounds per square inch (psi) [99 kilopascals (kPa)]
Aircraft speed	= 310 mph (460 ft/s) [498.88 km/h (140.21 m/s)]
Compressor airflow	= 96 lb/s [43.55 kg/s]
Exhaust nozzle area	= 2 square feet (ft²) [0.19 m²]
Exhaust nozzle pressure	= 80 inHg (39.3 psi) [270.9 kPa]
Exhaust gas velocity	= 1000 mph (1460 ft/s) [1609.3 km/h (445.01 m/s)]
Fuel flow	= 5760 lb/h (1.6 lb/s) [2612.7 kg/h (0.73 kg/s)]

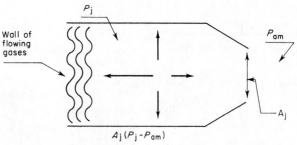

FIG. 3-6 The choked nozzle.

If the acceleration due to gravity = 32 ft/s², then F_n is most nearly equal to

$$F_n = \frac{W_a}{g}(V_2 - V_1) + \frac{W_f}{g}(V_f) + A_j(P_j - P_{am})$$

$$= \frac{96}{32}(1460 - 460) + \frac{1.6}{32}(1460) + 2(5659 - 2074)$$

$$= 3(1000) \qquad\quad + 0.05(1460) + 2(3585)$$

$$= 3000 \qquad\qquad + \quad 73 \quad + \quad 7170$$

$$F_n = 10{,}243 \text{ lb } [4651 \text{ kg}]$$

Thrust Distribution (*FIG. 3-7*)

The net thrust of an engine is a result of pressure and momentum changes within the engine. Some of these changes produce forward forces while others produce rearward forces. Whenever there is an increase in total pressure (by compression) or by a change from kinetic energy to pressure energy, as in the diffuser, forward forces are produced. Conversely, rearward forces or thrust losses result when pressure energy decreases or is converted into kinetic energy, as in the nozzle. The rated net thrust of any engine is determined by how much the forward thrust forces exceed the rearward thrust forces.

If the areas, pressures acting across these areas, velocities, and mass flows are known at any point in the engine, the forces acting at the point can be calculated. For any point in the engine, the force would be the sum of

$$F_n = \frac{W_a}{g}(V_2 - V_1) \quad \text{or} \quad \text{mass} \times \text{acceleration} = Ma$$

$$\text{plus}$$

$$F_n = A_j(P_j - P_{am}) \quad \text{or} \quad \text{pressure} \times \text{area} \quad = PA$$

The completed formula would then read

$$F_n = Ma + PA$$

Using the following values for an engine at rest:

Weight of air	= 160 lb/s [72.58 kg/s]
Inlet velocity	= 0 ft/s
Exhaust gas velocity	= 2000 ft/s [609.6 m/s]
Area of exhaust nozzle	= 330 square inches (in²) [212,916 square millimeters (mm²)]
Pressure at exhaust nozzle	= 6 psi gage [41.4 kPa gage]
Ambient pressure	= 0 psi gage
Acceleration due to gravity	= 32 ft/s² [9.81 m/s²]

the thrust of this engine will be, neglecting fuel flows and losses,

$$F_n = \frac{W_a}{g}(V_2 - V_1) + A_j(P_j - P_{am})$$

$$= {}^{160}/_{32}(2000 - 0) + 330(6 - 0)$$

$$= 11{,}980 \text{ lb } [53{,}287 \text{ N}]$$

The various forward and rear loads on the engine are determined by using the pressure times the area (*PA*) plus the mass times the acceleration (*Ma*) at given points in the engine.

Compressor Outlet

Airflow = 160 lb/s [72.58 kg/s]
Velocity = 400 ft/s [121.92 m/s]
Pressure = 95 psi gage [655 kPa]
Area   = 180 in² [116,136 mm²]

NOTE: The pressure and velocity at the face of the compressor is zero. To compute the forces acting on the compressor, it is necessary to consider only outlet conditions.

$$F_{n,comp} = Ma + PA$$
$$= {}^{160}/_{32}(400) + (95 \times 180)$$
$$= 19{,}100 \text{ lb } [8663.76 \text{ kg}] \text{ of forward thrust}$$

Diffuser Outlet

Airflow = 160 lb/s [72.58 kg/s]
Velocity = 350 ft/s [106.68 m/s]

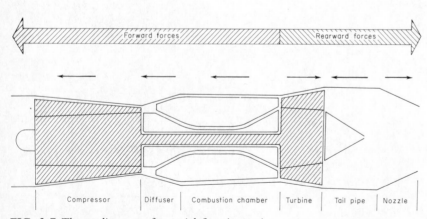

FIG. 3-7 Thrust diagram of an axial-flow jet engine.

Pressure = 100 psi gage [689.5 kPa]
Area   = 200 in² [129,040 mm²]

NOTE: Since the condition at the inlet of the diffuser is the same as that at the outlet of the compressor, that is, 19,100 lb [8663.76 kg], it is necessary to subtract this from the force value found for the diffuser outlet.

$$F_{n,\text{diff}} = Ma + PA - 19,100$$
$$= {}^{160}/_{32}(350) + (100 \times 200) - 19,100$$
$$= 21,750 - 19,100$$
$$= 2650 \text{ lb } [1202.04 \text{ kg}] \text{ of forward thrust}$$

Combustion Chamber Outlet (Burner)

Airflow = 160 lb/s [72.58 kg/s] (neglecting fuel flow)
Velocity = 1250 ft/s [381.00 m/s]
Pressure = 95 psi gage [655 kPa]
Area   = 500 in² [322,600 mm²]

NOTE: The condition at the inlet of the combustion chamber is the same as that at the outlet of the diffuser, that is, 21,750 lb [9865.80 kg].

$$F_{n,\text{burner}} = Ma + PA - 21,750$$
$$= {}^{160}/_{32}(1250) + (95 \times 500) - 21,750$$
$$= 53,750 - 21,750$$
$$= 32,000 \text{ lb } [14,515 \text{ kg}] \text{ of forward thrust}$$

Turbine Outlet

Airflow = 160 lb/s [72.58 kg/s]
Velocity = 700 ft/s [213.36 m/s]
Pressure = 20 psi gage [137.9 kPa]
Area   = 550 in² [354,860 mm²]

NOTE: The condition at the inlet of the turbine is the same as that at the outlet of the combustion chamber, that is, 53,750 lb [24,381 kg], therefore

$$F_{n,\text{turbine}} = Ma + PA - 53,750$$
$$= {}^{160}/_{32}(700) + (20 \times 550) - 53,750$$
$$= 14,500 - 53,750$$
$$= -39,250 \text{ lb } [-17,803.80 \text{ kg}] \text{ of rearward thrust}$$

Exhaust Duct Outlet

Airflow = 160 lb/s [72.58 kg/s]
Velocity = 650 ft/s [198.12 m/s]
Pressure = 25 psi gage [172.4 kPa gage]
Area   = 600 in² [387,120 mm²]

NOTE: The condition at the inlet of the exhaust duct is the same as that at the outlet of the turbine, that is, 14,500 lb [6577 kg], therefore

$$F_{n,\text{exh. duct}} = Ma + PA - 14,500$$
$$= {}^{160}/_{32}(650) + (25 \times 600) - 14,500$$
$$= 18,250 - 14,500$$
$$= 3750 \text{ lb } [1701 \text{ kg}] \text{ of forward thrust}$$

Exhaust Nozzle Outlet

Airflow = 160 lb/s [72.58 kg/s]
Velocity = 2000 ft/s [609.60 m/s]
Pressure = 6 psi gage [41.4 kPa gage]
Area   = 330 in² [212,916 mm²]

NOTE: The condition at the inlet of the exhaust nozzle is the same as that at the outlet of the exhaust duct, that is, 18,250 lb [8278.2 kg], therefore

$$F_{n,\text{nozzle}} = Ma + PA - 18,250$$
$$= {}^{160}/_{32}(2000) + (6 \times 330) - 18,250$$
$$= 11,980 - 18,250$$
$$= -6270 \text{ lb } [-2844.1 \text{ kg}] \text{ of rearward thrust}$$

The sum of the forward and rearward forces is:

		Forward	Rearward
Compressor	=	19,100	
Diffuser	=	2650	
Combustion chambers	=	32,000	
Turbine	=		−39,250
Exhaust duct	=	3750	
Exhaust nozzle	=		− 6270
		57,500	−45,520
		−45,520	
		11,980	

Thrust computed for the complete engine (from page 94) equals 11,980. Thrust computed for the individual sections of the engine equals 11,980.

Fitting the engine with an afterburner will have two large effects on engine operating conditions.

Airflow = 160 lb/s [72.58 kg/s] (neglecting fuel flow)
Velocity = 2500 ft/s [762.00 m/s]
Pressure = 6 psi gage [41.4 kPa gage]
Area   = 450 in² [290,340 mm²]

$$F_{n,\text{nozzle}} = Ma + PA - 18,250$$
$$= {}^{160}/_{32}(2500) + (6 \times 450) - 18,250$$
$$= 15,200 - 18,250$$
$$= -3050 \text{ lb } [-1383.5 \text{ kg}] \text{ of rearward thrust}$$

The amount of rearward thrust for the nonafterburning engine is −6270 lb, and for the afterburning engine it is −3050 lb, a difference of 3220 lb. If 3220 lb is added to the thrust of the nonafterburning engine, the total thrust will be

$$11,980 + 3220 = 15,200 \text{ lb } [6894.7 \text{ kg}]$$

The thrust for the entire engine under afterburning conditions is

$$F_n = \frac{W_a}{g}(V_2 - V_1) + A_j(P_j - P_{am})$$

$$= {}^{160}/_{32}(2500 - 0) + 450(6 - 0)$$
$$= 15,200 \text{ lb } [6894.7 \text{ kg}]$$

Thrust Compared with Horsepower

Thrust and horsepower cannot be directly compared because, by definition, power is a force applied through a distance in a given period of time. All of the power produced by a jet engine is consumed internally to turn the compressor and drive the various engine accessories. Therefore, the jet engine does not develop any horsepower in the normally accepted sense, but supplies only one of the terms in the horsepower formula. The other term is actually provided by the vehicle in which the engine is installed. To determine the thrust horsepower of the jet engine, the following formula is used:

$$thp = \frac{(W_a/g)(V_2 - V_1) + A_j(P_j - P_{am}) \times \text{velocity of a/c (ft/s)}}{550}$$

This can be simplified to

$$thp = \frac{\text{net thrust} \times \text{velocity of plane (ft/s)}}{550}$$

or

$$thp = \frac{F_n V}{550}$$

Since airplane speeds are often given in miles per hour, it may be desirable to compute the thrust horsepower using mile-pounds per hour. If such is the case

$$thp = \frac{\text{net thrust} \times \text{velocity of plane (mph)}}{375}$$

The denominator in these formulas is arrived at in the following manner.

1 hp	= 550 (ft) (lb)/s
550 (ft) (lb)/s × 60	= 33,000 (ft) (lb)/min
33,000 (ft) (lb)/min × 60	= 1,980,000 (ft) (lb)/h
$\dfrac{1,980,000 \text{ (ft) (lb)/h}}{5280}$	= 375 (mi) (lb)/h
375 (mi) (lb)/h	= 1 hp

If an airplane is flying at a velocity of 375 mph and developing 4000 lb of thrust, the thrust horsepower will be:

$$thp = \frac{F_n V_p}{375}$$

where F_n = net thrust, lb
V_p = airplane velocity, mph

$$thp = \frac{4000 \times 375}{375}$$
$$= 4000$$

From this it can be seen that at 375 mph each pound of thrust will be converted to one horsepower, and that

for each speed of the airplane there will be a different thp. At 750 mph this 4000-lb-thrust jet engine will produce 8000 thp.

Factors Affecting Thrust

The jet engine is much more sensitive to operating variables than is the reciprocating engine. Such variables can be divided into two groups: those that change because of design or operating characteristics of the engine, and those that change because of the medium in which the engine must operate.

In the first category are such factors as:

1. Engine rpm (weight of air)
2. Size of nozzle area
3. Weight of fuel flow
4. Amount of air bled from the compressor
5. Turbine inlet temperature
6. Use of water injection

Nondesign factors include:

7. Speed of the aircraft (ram pressure rise)
8. Temperature of the air ⎫
9. Pressure of the air ⎬ density effect
10. Amount of humidity ⎭

For the present, only factors 1, 7, 8, 9, and 10 are discussed. The effect of the other variables on engine operation are covered elsewhere in the book in the appropriate sections.

RPM EFFECT (Fig. 3-8) Engine speed in revolutions per minute has a very great effect on the thrust developed by a jet engine. Figure 3-8 shows that very little thrust is developed at low rpm as compared with the thrust developed at high engine rpm, and that a given rpm change has more effect on thrust at higher engine speeds than at lower engine speeds. The weight of air pumped by a compressor is a function of its rpm. Recalling the formula

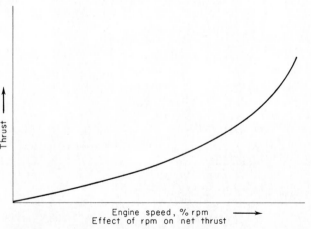

FIG. 3-8 A nonlinear relationship exists between engine rpm and thrust.

$$F_n = \frac{W_a}{g}(V_2 - V_1)$$

it is evident that increasing the weight of air being pumped will result in an increase in F_n or thrust. As we shall see when we get to the section on compressors, engine speed may not be indiscriminately varied, but must be controlled within very close limits. (See Chap. 5.)

SPEED EFFECT The formula

$$F_n = \frac{W_a}{g}(V_2 - V_1)$$

shows that any increase in the forward velocity of the airplane will result in a decrease in thrust. The faster the airplane goes, the greater will be the initial momentum of the air in relation to the engine (V_1 increasing). But the jet nozzle velocity is generally fixed by the speed of sound. Obviously, the $V_2 - V_1$ difference or momentum change will become smaller as airplane speed increases (Fig. 3-9). This loss of thrust will be partially offset by the increase in the W_a due to ram (Fig. 3-10). Not as much thrust is recovered due to the ram pressure rise as would seem to be indicated on first examination. At high airplane speeds there is a considerable temperature rise in addition to the rise in pressure (Fig. 3-11). The actual weight of airflow increase into the engine will be directly proportional to the rise in pressure and inversely proportional to the square root of the rise in temperature. (See Chap. 5.)

$$W_a \frac{\delta_t}{\sqrt{\theta_t}}$$

where δ_t = total pressure
θ_t = total temperature

Losses may also occur in the duct during high speeds as a result of air friction and shock-wave formation. (See Chap. 4.)

TEMPERATURE EFFECT The gas turbine engine is very sensitive to variations in the temperature of the air (Fig. 3-12). All engines are rated with the air at a standard temperature of 59° Fahrenheit (F) [15° Celcius (C)], although some manufacturers will "flat rate" their engines to a higher temperature, that is, the engine is guaranteed to produce a minimum specific thrust at a temperature above 59°F [15°C]. Careful power lever manipulation is required at lower temperatures. In any case, if the engine operates in air temperatures hotter than standard, there will be less thrust produced. Conversely, engine operation in air temperatures colder than standard day conditions will produce a greater than rated thrust.

A rise in temperature will cause the speed of the molecules to increase so that they run into each other harder and move further apart. When they are further apart, a given number of molecules will occupy more space.

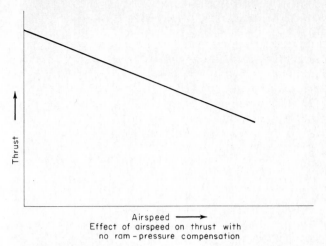

Effect of airspeed on thrust with no ram-pressure compensation

FIG. 3–9 Thrust loss is due to $V_2 - V_1$ difference becoming smaller.

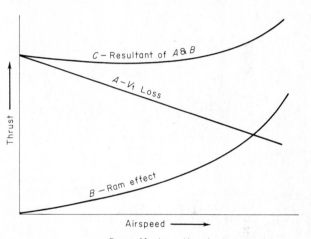

Ram effect on thrust

FIG. 3-10 Combining thrust loss due to $V_2 - V_1$ difference decrease with thrust gain due to ram-pressure rise.

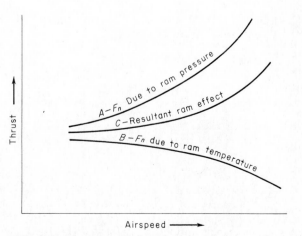

FIG. 3-11 The ram-effect result comes from two factors.

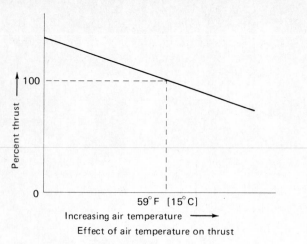

FIG. 3-12 Thrust may vary as much as 20 percent from the specified rating on cold or hot days.

And when a given number of molecules occupy more space, fewer can get into the engine inlet area. This results in a decrease in W_a into the engine with a corresponding decrease in thrust.

PRESSURE EFFECTS An increase in pressure results when there are more molecules per cubic foot. When this situation occurs, there are more molecules available to enter the engine inlet area and as a result, an increase in W_a occurs through the engine (Fig. 3-13). How pressure changes affect thrust will become clearer when we examine the gas turbine cycle later in this chapter.

DENSITY EFFECT Density is defined as the number of molecules per cubic foot, and is affected by both pressure and temperature. When the pressure goes up, the density goes up, and when the temperature goes up, the density goes down. This may be expressed mathematically as

$$\text{Density ratio} = K\frac{P}{T}$$

where K = a constant
P = pressure, inHg.
T = temperature, degrees Rankine (°R)

or density is directly proportional to pressure and inversely proportional to temperature times a constant. A constant of 17.32 is necessary in order to make the density ratio equal 1 under standard conditions of temperature (518.7°R) and pressure (29.92 inHg).

$$\begin{aligned}\text{Density ratio} &= K\frac{P}{T} \\ &= 17.32\frac{29.92}{518.69} \\ &= 1\end{aligned}$$

Density changes are most noticeable, of course, with changes in altitude. The effect of altitude change on thrust is really a function of density. Figure 3-14 shows the result of combining Figs. 3-12 and 3-13. The higher the altitude the less pressure there is. This should result in a decrease in thrust as shown in Fig. 3-13. But the higher the altitude, the colder it gets, and this should result in an increase in thrust as shown in Fig. 3-12. However, the pressure drops off faster than the temperature, so that there is actually a drop in thrust with increased altitude.

At about 36,000 ft [10,973 m], essentially the beginning of the tropopause, the temperature stops falling and remains constant while the pressure continues to fall (Fig. 3-15). As a result, the thrust will drop off more rapidly above 36,000 ft because the thrust loss due to the air pressure drop will no longer be partially offset by the thrust gain due to temperature drop. This makes 36,000 ft the optimum altitude for long-range cruising, because at this altitude, even though the engine's thrust is reduced, the relationship between the thrust produced and the diminished drag on the airplane is most favorable. Most commercial and business jets are certified to a much higher altitude as listed in the type certificate data sheets (TCDS).

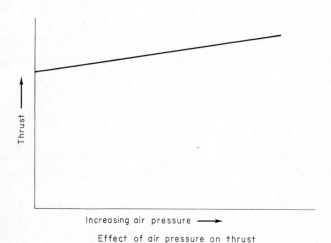

FIG. 3-13 Air pressure drops as altitude is gained.

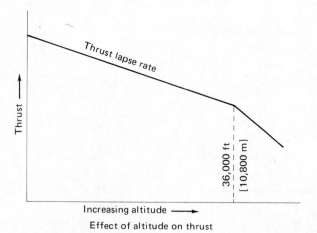

FIG. 3-14 Combining thrust loss due to pressure decrease with altitude and thrust gain due to temperature decrease with altitude.

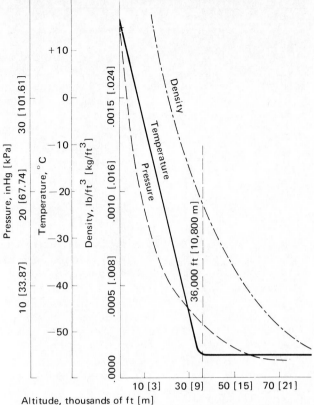

FIG. 3-15 At 36,000 ft (10973 m), temperature stops falling.

pressure-temperature-density variation with altitude increase

HUMIDITY EFFECT While humidity has a fairly large effect on the reciprocating engine, its effect on the gas turbine engine is negligible. Since water vapor weighs only five-eighths as much as dry air, increasing humidity will decrease the weight per unit volume, therefore, the lower density equals less mass at the same rpm. Since a carburetor is essentially a volume-measuring device, it will not sense this decrease in the *weight* of the air and, as a result, will continue to supply the same amount of fuel to the engine, causing the fuel-air mixture ratio to become too rich and the engine to lose power.

On the other hand, a jet engine operates with an excess of air needed for combustion. (Refer to Chap. 6.) Any air needed for the combustion process will come from the cooling air supply. In addition, the fuel control does not measure the volume of air directly, but rather meters fuel flow as a function of pressures, temperatures, and rpm (Fig. 3-16). (Refer to Chap. 12.)

Pressure, Temperature, and Velocity Changes

At the beginning of this chapter, a general description of the series of events that take place in a typical gas turbine was briefly given. A somewhat more specific examination of airflow changes will help to better understand the thrust phenomenon. Refer to Fig. 3-17(a) and (b), both of which illustrate pressure, temperature, and velocity changes through typical gas turbine engines.

PRESSURE CHANGES Air usually enters the front of the compressor at a pressure that is less than ambient, indicating that there is considerable suction at the inlet to the engine. This somewhat negative pressure at the engine inlet may be partly or completely overcome by ram pressure as the airplane speed increases. From this point on, there is a considerable pressure rise through the successive compression stages, with the rate of rise increasing in the later stages of compression. The exit area of the exhaust nozzle, the exit area of the turbine nozzle, and rate of fuel flow all can determine the compression ratio of the compressor. (Refer to Chap. 5.) A final pressure rise is accomplished in the divergent section of the diffuser. From the diffuser, the air passes through the combustion section where a slight pressure loss is experienced. The combustion-chamber pressure must be lower than the compressor discharge pressure during all phases of engine operation in order to establish a direction of airflow toward the rear of the engine and allow the gases to expand as combustion occurs. A sharp drop in pressure occurs as the air is accelerated between the converging passages of the turbine nozzle. The pressure continues to drop across the turbine wheel as some of the pressure energy in the hot gas is converted to a rotational force by the wheel. If the engine is equipped with more than one turbine stage, a pressure reduction occurs across each turbine wheel. Pressure changes after the turbine depend upon the type of exhaust nozzle used, and whether the nozzle is operating in a choked (gas velocity at the speed of sound) or non-choked condition. When the gases leave the exhaust nozzle, the pressure continues to drop to ambient.

TEMPERATURE CHANGES Air entering the compressor at sea level on a standard day is at a temperature of 59°F [15°C]. Due to compression, the temperature through the compressor gradually climbs to a point that is determined by the number of compressor stages and its aerodynamic efficiency. (Refer to Chap. 5.) On some large commercial engines, the temperature at the front of the combustion section is approximately 800°F [427°C]. As the air enters the combustion chambers, fuel is added and the temperature is raised to about 3500°F [1927°C] in the hottest part of the flame. Since this temperature is above the melting point of most metals, the combustion chamber and surrounding parts of the engine are protected by a cooling film of air which is established through proper design of the combustion chamber. (Refer to Chap. 6.) Because of this cooling film, the air entering the turbine section is considerably cooler. The acceleration of air through the turbine section further reduces the temperature. If the engine is operating without the use of an afterburner, there is a slight temperature drop through the exhaust pipe. If the engine is operating with the use of the afterburner, there will be a sharp temperature rise in the exhaust pipe.

VELOCITY CHANGES Since a jet engine derives its thrust mainly from the reaction to the action on a stream of air as it flows through the jet engine, the pressure and temperature changes just discussed are important only because they must be present to accom-

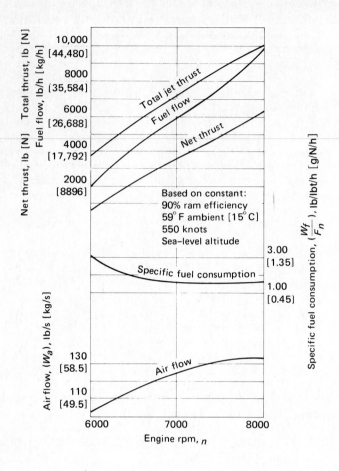

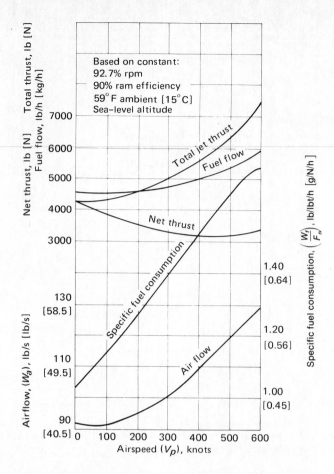

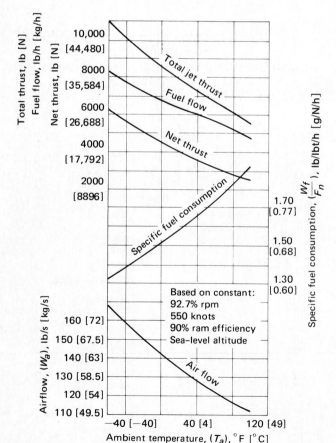

FIG. 3-16. The effects of engine rpm, airspeed, altitude (at a constant temperature), ram efficiency, and temperature on fuel flow, specific fuel consumption, airflow, and thrust for a General Electric J79 engine.

plish the action part of the action-reaction process. What is really desired is a jet of air flowing out of the engine at a speed faster than the speed with which it entered.

The velocity of the air at the front of the compressor must be less than sonic for most present-day compressors. In order to achieve this goal, the design of the inlet duct of the airplane is of paramount importance. (Refer to Chap. 4.) If the ambient air velocity is zero, the air velocity increases as it is drawn into the compressor. On the other hand, if the airplane speed is supersonic, the air's velocity is slowed in the duct. Airflow velocity through the majority of compressors is essentially constant, and in some compressors may decrease slightly. A fairly large drop in airspeed occurs in the enlarging diffuser passage. The turning point where flow velocity starts to increase is in the combustion chamber as the air is forced around the forward end of the combustion chamber inner liner and through the holes along the sides. A further increase occurs at the rear of the combustion chamber as the hot gases expand and are forced through the slightly smaller area of the transition liner. An extremely sharp rise in velocity, with a corresponding loss of pressure, occurs as the air passes through the

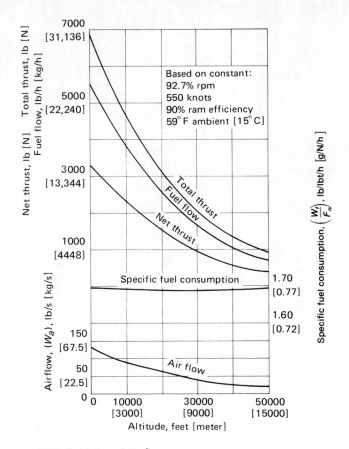

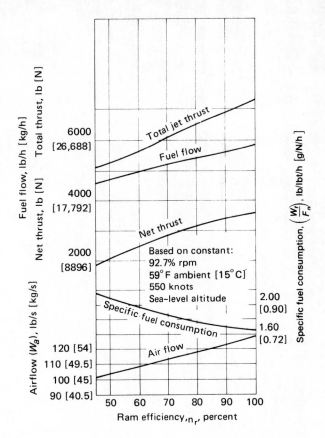

FIG. 3-16 (continued)

converging partitions of the turbine nozzle. This exchange of pressure for velocity is very desirable, since the turbine is designed to operate largely on a velocity drop. As previously explained, the velocity increase is accompanied by a temperature and pressure drop. A large portion of the velocity increase through the nozzle is absorbed by the turbine wheel and applied to drive the compressor and engine accessories. Velocity changes from this point on depend upon the design of the engine. As shown in Fig. 3-17(a), if the engine is not using the afterburner, the velocity is reduced as the air enters the afterburner section because it is a diverging area. As the air is discharged through the orifice formed by the exhaust nozzle, the velocity increases sharply. If the engine is running with the afterburner in operation, the rise in temperature caused by the burning of the afterburner fuel will cause a tremendous velocity increase. In most cases, use of the afterburner produces an increase in exhaust velocity which is approximately equal to the reduction in velocity through the turbine wheel. Notice that the only changes which occur with the use of the afterburner are those of the temperature and velocity in the exhaust pipe. Pressure, temperature, and velocity changes in the basic engine remain the same because the variable-area exhaust nozzle used with afterburner-equipped engines is designed to open to a new position that will maintain the same turbine discharge temperature and pressure that existed when operating at full power without the afterburner. Figure 3-17(b) shows that there is little velocity change after the turbine section of a turboprop. In both engines there is always energy in the form of temperature, pressure, and velocity remaining in the exhaust gases after they leave the turbine, but this energy level is much lower in the turboprop because the turbine extracts more from the gases in order to drive the propeller [Fig. 3-18(a) and (b)]. This is also true for fan-equipped engines. Of course the jet effect is reduced a proportionate amount. In addition, part of the energy is lost because the exhaust gases have not cooled to the same temperature as the air which entered the engine. This problem will be examined in the discussion on engine cycles that follows.

The Gas Turbine Cycle (*FIG. 3-19*)

A cycle is a process that begins with certain conditions and ends with those same conditions. Both reciprocating and jet engines operate on cycles which can be illustrated graphically. As shown in Fig. 3-19, reciprocating and jet engines are similar in many respects. Both types of power plants are called air-breathing engines. Both work by providing rearward acceleration to a mass of air. In the case of the reciprocating engine, the propeller imparts a relatively small acceleration to a large mass of air, while the gas turbine imparts a relatively large acceleration to a small mass of air

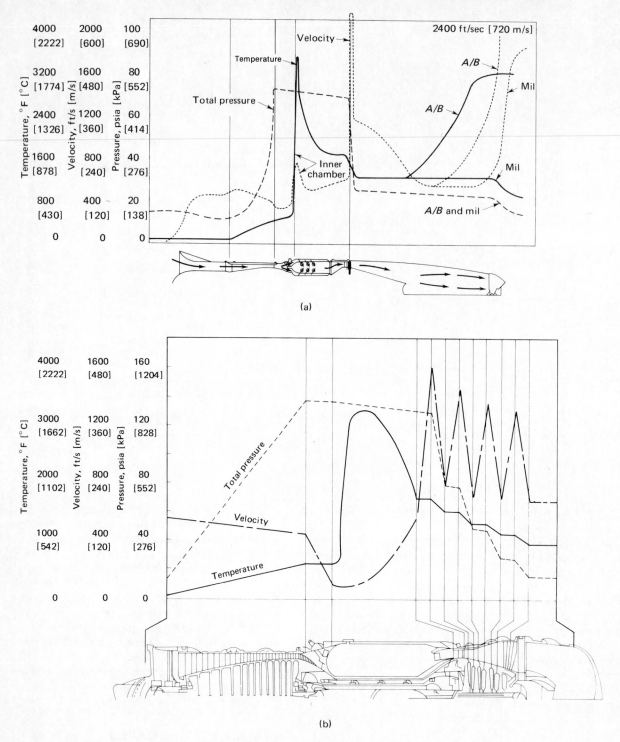

FIG. 3-17 (a) Temperature, pressure, and velocity diagram for a typical turbojet engine with and without afterburner operation. *(Wright Corporation.)*
(b) Temperature, pressure, and velocity diagram for a typical turboprop engine.
(Detroit Diesel Allison Division.)

(Fig. 3–20). Although both convert the energy in expanding gas into thrust, the reciprocating engines does this by changing the energy of combustion into mechanical energy which is used to turn a propeller, while the gas turbine engine produces and uses the propulsive force directly. The same series of events, i.e., intake, compression, power, and exhaust, occur in both, the difference being that in the jet engine all of these events are happening simultaneously, whereas in the reciprocating engine, each event must follow the preceding one. It should also be noted that in the turbine engine each operation in the cycle is not only performed con-

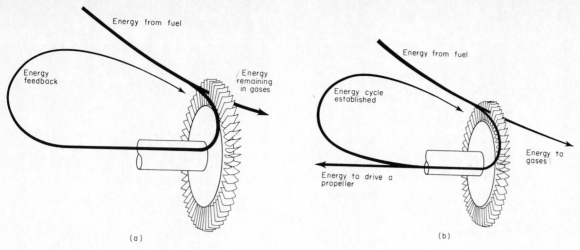

FIG. 3-18 More energy is extracted by the turboprop turbine section than in a turbojet. *(U.S.A.F. Extension Course Institute and Air University, Course 4301.)*
(a) Turbojet turbine.
(b) Turboprop turbine.

tinuously, but also is performed by a separate component designed for its particular function, whereas in the reciprocating engine, all the functions are performed in one section of the engine.

The Otto Cycle (the Constant-Volume Engine) (FIG. 3-21)

The reciprocating engine operates on what is commonly termed a closed cycle. In the series of events shown in Fig. 3-21, air is drawn in on the intake stroke at point 1, where compression by this piston raises the temperature and decreases the volume of the gases. Near the end of the compression stroke, point 2, ignition occurs, which greatly increases the temperature of the mixture. The term "constant volume" is derived from the fact that from point 2 to 3 there is no appreciable change of volume while the mixture is burning. Point 3 to 4 represents the expansion stroke with a loss of temperature and pressure, and a corresponding increase in volume. It might be noted that this is the only stroke of the four from which power may be extracted. When the exhaust valve opens near the end of the power stroke (point 4 to 1), the gases lose their remaining pressure and temperature and the closed cycle starts all over again.

The Brayton Cycle (the Constant-Pressure Engine)

Since all of the events are going on continuously, it can be said that the gas turbine engine works on what is commonly called an open cycle. As in the reciprocating engine, air is drawn in and compressed (point 1 to 2) with a corresponding rise in pressure and temperature, and a decrease in volume. Point 2 to 3 represents the change caused by the fuel mixture burning in the combustion chamber at an essentially constant pressure, but with a very large increase in volume. This increase

in volume shows up as an increase in velocity because the engine area does not change much in this section. From the burner, the gases expand through the turbine wheel causing an increase in volume and a decrease in temperature and pressure (point 3 to 4). This process continues from point 4 to 5 through the exhaust nozzle.

In comparing the two engines, it is interesting to note that the reciprocating engine obtains its work output by employing very high pressures (as much as 1000 psi [6895 kPa]) in the cylinder during combustion. With these high pressures, a larger amount of work can be obtained from a given quantity of fuel, thus raising the *thermal efficiency* (the relationship between the potential heat energy in the fuel and the actual energy output of the engine) of this type of engine. On the other hand, a jet engine's thermal efficiency is limited by the ability of the compressor to build up high pressures without an excessive temperature rise. The shaded area shown in Fig. 3-21 is called the area of useful work, and any increase in this area indicates more energy available for useful output, thus thrust. But increasing the compression ratio would also increase the compressed air temperature. Since most gas turbine engines are already operating at close to maximum temperature limits, this would result in a mandatory decrease in fuel flow (Fig. 3-22), thus making it extremely difficult to increase compression ratios without designing more efficient compressors, i.e., compressors able to pump air with a minimum temperature rise. Ideally, we would like to be able to burn as much fuel as possible in the jet engine in order to raise the gas temperature and increase the area of useful output.

One last illustration will help to understand how the gas turbine works. Note that Fig. 3-23 is similar to the pressure-temperature graph for the gas turbine engine shown in Fig. 3-21.

During steady-state operation, the work done by the turbine must be almost exactly the same as the work needed by the compressor. Although the temperature

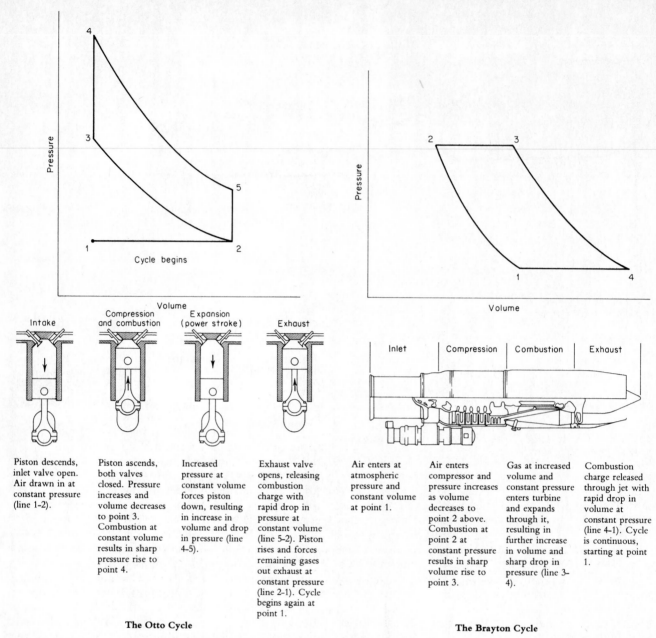

The Otto Cycle

Piston descends, inlet valve open. Air drawn in at constant pressure (line 1-2).

Piston ascends, both valves closed. Pressure increases and volume decreases to point 3. Combustion at constant volume results in sharp pressure rise to point 4.

Increased pressure at constant volume forces piston down, resulting in increase in volume and drop in pressure (line 4-5).

Exhaust valve opens, releasing combustion charge with rapid drop in pressure at constant volume (line 5-2). Piston rises and forces remaining gases out exhaust at constant pressure (line 2-1). Cycle begins again at point 1.

The Brayton Cycle

Air enters at atmospheric pressure and constant volume at point 1.

Air enters compressor and pressure increases as volume decreases to point 2 above. Combustion at point 2 at constant pressure results in sharp volume rise to point 3.

Gas at increased volume and constant pressure enters turbine and expands through it, resulting in further increase in volume and sharp drop in pressure (line 3-4).

Combustion charge released through jet with rapid drop in volume at constant pressure (line 4-1). Cycle is continuous, starting at point 1.

FIG. 3-19 Pressure-volume changes in the reciprocating and gas turbine engines.

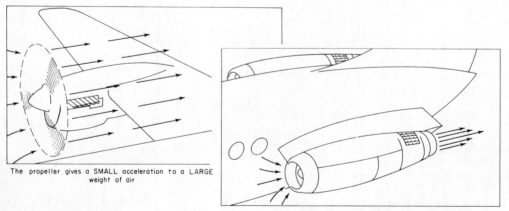

The propeller gives a SMALL acceleration to a LARGE weight of air

The turbojet engine gives a LARGE acceleration to a SMALL weight of air

FIG. 3-20 Propulsive efficiency is high for a propeller, and low for a jet.

drop available to obtain work from the very hot gases passing through the turbines is very much greater than the actual temperature rise through the compressor, the mass (or weight) flow available to the turbine is exactly the same as the mass or weight handled by the compressor. (This assumes that air bleed loss is equal in weight to the fuel added. Refer to page 93.) Therefore, a temperature drop takes place through the turbine that is approximately equivalent to the temperature rise which occurred in the compressor. However, Fig. 3-23 shows that although the compressor temperature rise x equals the turbine temperature drop y, the pressure change needed is much less at the higher temperature.

NOTE: The molecular speed is directly related to the square root of the temperature of the gas. If the temperature is doubled, the molecular speed (pressure) is squared. (See pages 126–127.)

This difference (point D to E) is why pressure is left to produce thrust in the turbojet. In the turboprop, more energy is taken out by the turbines to drive the propeller, with a correspondingly small amount of pressure left over to induce a jet thrust (point F to G). The energy added in the form of fuel has been *more than enough* to drive the compressor. The energy remaining produces the thrust or power for useful work.

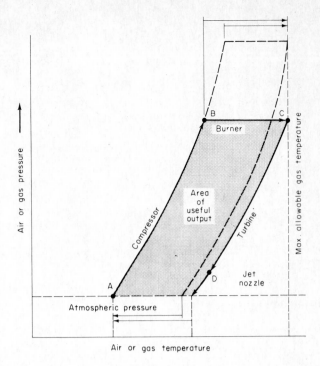

FIG. 3-22 Effect of increasing compressor pressure; pressure-temperature engine cycle diagram. *(Pratt & Whitney Aircraft.)*

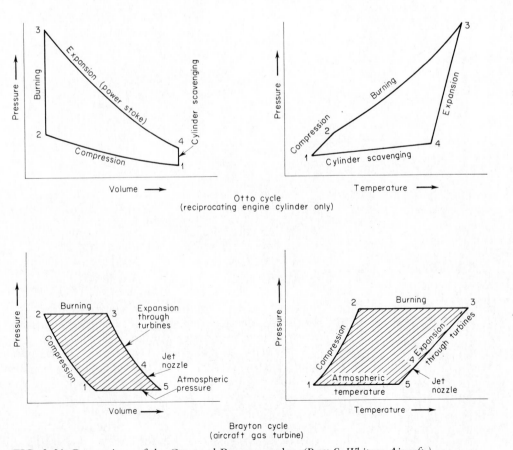

FIG. 3-21 Comparison of the Otto and Brayton cycles. *(Pratt & Whitney Aircraft.)*

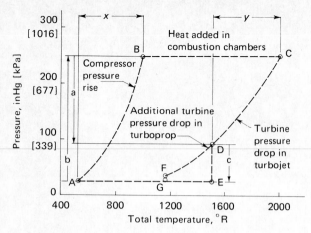

FIG. 3-23 Compression and expansion curves. *(Pratt & Whitney Aircraft.)*

Efficiencies

Engine thermal efficiency is defined as the engine's energy ouput divided by the fuel's energy input. One of the principal measures of jet engine efficiency is called *specific fuel consumption* SFC, or *thrust specific fuel consumption* TSFC. Specific fuel consumption is the ratio between the fuel flow (in pounds per hour) and the thrust of the engine (in pounds).

$$SFC = \frac{W_f}{F_n}$$

EXAMPLE: A 9100-lb-thrust [40,950-N] engine consumes 700 gallons (gal) [2649.5 liters (L)] of fuel per hour. What is the SFC? (One gallon of fuel equals 6.5 lb.)

$$SFC = \frac{W_f}{F_n}$$
$$= \frac{4550}{9100}$$
$$= 0.5 \text{ lb } [0.227 \text{ kg}] \text{ of fuel per hour per pound of thrust}$$

Obviously the more thrust we can obtain per pound of fuel, the more efficient the engine is. Specific fuel consumption enables valid comparisons to be made between engines because the fuel consumption is reduced to a common denominator. For example: If one engine produces 5000 lb [22,240 N] of thrust and consumes 2500 lb [1134 kg] of fuel per hour, while another engine produces 10,000 lb [44,480 N] of thrust and burns 4000 lb [1814 kg] of fuel per hour, which engine is the most efficient?

$$SFC = \frac{W_f}{F_n} \qquad SFC = \frac{W_f}{F_n}$$
$$= \frac{2500}{5000} \qquad = \frac{4000}{10,000}$$
$$= 0.5 \text{ lb } [0.227 \text{ kg}] \text{ of} \qquad = 0.4 \text{ lb } [0.181 \text{ kg}] \text{ of}$$
$$\text{fuel per hour per} \qquad \text{fuel per hour per}$$
$$\text{pound of thrust} \qquad \text{pound of thrust}$$

The 10,000-lb-thrust engine produces twice as much thrust as the 5000-lb-thrust engine, but does not consume twice the fuel.

The term *equivalent specific fuel consumption* ESFC is used to compare fuel consumption between turboprop engines.

$$ESFC = \frac{W_f}{ESHP}$$

The ESHP, or *equivalent shaft horsepower*, is found by the following formula when the aircraft is not moving:

$$ESHP = SHP + \frac{F_n}{2.5} \quad \text{approximately}$$

NOTE: Under static conditions one shaft horsepower equals approximately 2.5 pounds of thrust.

If the airplane is moving, the equivalent shaft horsepower can be found as follows:

$$ESHP = SHP + \frac{F_n V}{550}$$

where F_n is in pounds and V is in feet per second.

EXAMPLE: What is the ESFC of a turboprop engine that consumes 1500 lb of fuel per hour and produces 500 lb of thrust and 2800 hp under static conditions? (Note: Under static conditions, 1 hp = 2.5 lb of thrust.)

$$ESHP = SHP + \frac{F_n}{2.5}$$
$$= 2800 + \frac{500}{2.5}$$
$$= 2800 + 200$$
$$ESHP = 3000 \text{ [2238 kW]}$$

then

$$ESFC = \frac{W_f}{ESHP}$$
$$= \frac{1500}{3000}$$
$$ESFC = 0.5 \text{ lb/ESHP/h } [0.225 \text{ kg/kW/h}]$$

Another way of determining the engine's thermal efficiency is to compare the potential energy stored in a fuel with the amount of power the engine is producing. Assuming that all of the heat energy could be liberated in a pound of gas turbine fuel, the potential power output of the engine can be calculated as follows.

A pound of typical gas turbine fuel contains 18,400 British thermal units (Btu) [19,412,000 joules (J)] of heat energy. Since 1 Btu is the equivalent of 778 ft·lb [1 J = 0.102 kg·m] (see page 128 for an explanation of the Btu and its relationship to work), the fuel will contain

$18,400 \times 778 = 14,315,200$ ft·lb [1,979,792 kg·m] potential work

Dividing the number of foot-pounds of potential work in the fuel by 33,000, the number of foot-pounds per minute in 1 hp (pages 88–89 for the definition and calculation of power), will give the horsepower equivalent of 1 lb of the fuel if it were completely burned in 1 min.

$$hp = \frac{14,315,200}{33,000}$$
$$= 433.79 \text{ potential horsepower [323.48 kW]}$$

The actual power output of the engine as compared to the potential horsepower in the fuel is a measure of the engine's thermal efficiency and can be determined by dividing the engine's actual power output by the fuel's potential power input. They are never equal since all of the potential energy in fuel cannot be liberated, and no engine is capable of taking advantage of all the heat energy available. Typical thermal efficiencies range from 20 percent to 30 percent with a large portion of the heat energy lost to the atmosphere through the exhaust nozzle.

There are many factors that affect the overall efficiency of an engine. Some of these factors are:

1. Component efficiency—Since none of the engine components are perfect, there will always be a certain amount of inefficiency present. For instance, if the compressor is 85 percent efficient, the combustion chamber 95 percent, and turbine 90 percent, the engine will have an overall efficiency of 72.7 percent (85 percent × 95 percent × 90 percent).
2. Type of fuel used—Some fuels have a higher potential heat value than others.
3. Engine operating factors—Already mentioned are rpm, use of water injection or afterburning, density of the air, flight speed, and fuel flow.

While the conversion of fuel energy into kinetic energy determines the thermal or internal efficiency, the efficient conversion of the kinetic energy to propulsive work determines the propulsive or external efficiency. This depends on the amount of kinetic energy wasted by the propelling mechanism which in turn depends on the mass airflow multiplied by the square of its velocity. From this we can see that the high-velocity, relatively-low-weight jet exhaust wastes considerably more energy than the propeller with its low-velocity, high-weight airflow (Fig. 3-24). This condition changes, however, as the aircraft speed increases, because although the jet stream continues to flow at high velocity from the engine, its velocity relative to the surrounding atmosphere is lessened by the forward speed of the aircraft. Thus the energy wasted is reduced. One hundred percent propulsive efficiency would occur if the airplane speed equalled the exhaust velocity.

On some fan engines the fan air is mixed with the exhaust gases, thereby reducing the exit velocity and consequently increasing the propulsive efficiency of the system. This in turn is reflected in lower SFC.

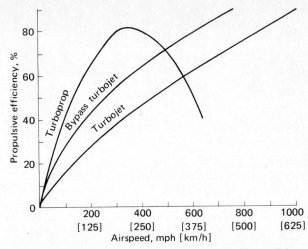

FIG. 3-24 Comparison of the propulsive efficiency of the turbojet, turbofan (bypass), and turboprop with changing airspeed.

Air Behavior at Low and High Velocities

The very nature of the jet engine with its high-speed flight characteristics requires a discussion of the behavior of moving air. Even though a great deal can be written on this subject, coverage here must necessarily be brief.

At low airflows, air is normally treated as an incompressible fluid similar to water because, at these low airspeeds, the air can undergo change in pressure with relatively little change in density. Since this condition of low-speed airflow is analogous to the flow of water or any other incompressible fluid, Bernoulli's theorem and other relationships developed for an incompressible fluid may be used.

BERNOULLI'S THEOREM Most simply stated, Bernoulli's theorem says that when a gas or fluid is flowing through a restricted area, as in a nozzle or venturi, its speed will increase and its temperature and pressure will decrease. If the area is increased as in a diffuser, the reverse is true. The total energy in a flowing gas is made up of static and dynamic temperatures, and static and dynamic pressures. A nozzle or a diffuser does not change the total energy level, but rather changes one form of energy to another (Fig. 3-25). For example, a nozzle will increase the flow, or dynamic pressure, at the expense of the static pressure. If the gas is moving through the pipe at so many pounds per second, the air must continue to flow at the same rate through the nozzle. The only way it can do this is to speed up. A diffuser will do the opposite. Thus by varying the area of the pipe, velocity can be changed into pressure, and pressure into velocity. The jet engine is just such a pipe, with changing areas where the air pressure and velocity are constantly being changed to achieve desired results.

MACH NUMBER At high speeds the pressure changes that take place are very large, resulting in significant changes in air density. The air no longer acts as an incompressible fluid. *Compressibility effects* due to

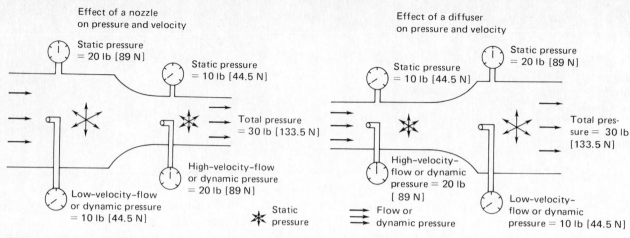

FIG. 3-25 Pressure and velocity changes at subsonic flow.

high-speed airflow are of great importance in the design of both the airplane and the engine. The compressibility phenomenon and the speed of sound are closely related to each other. The speed of sound is the rate at which small pressure disturbances will be propagated through the air. This propagation speed is solely a function of temperature. (See the Appendix for variation of the speed of sound with temperature.) As an object moves through the air, velocity and pressure changes occur which create pressure disturbances in the airflow surrounding the object. Since these pressure disturbances are propagated through the air at the local speed of sound, if the object is traveling below the speed of sound, these pressure disturbances are propagated ahead of the object (as well as in all other directions) and influence the air immediately in front of the object (Fig. 3-26). On the other hand, if the object is traveling at some speed above the speed of sound, the airflow ahead of the object will not be influenced, since the pressure disturbances cannot be propagated ahead of the object. In other words the object is outspeeding its own pressure waves. Thus, as the speed of the object nears the speed of sound, a compression wave will form and changes in velocity, pressure, and temperature will take place quite sharply and suddenly. The compression or shock wave results from the "piling up" of air molecules as they try to move forward, away from the pressure disturbance as fast as the object is moving forward through the air. The shock waves are very narrow areas of discontinuity where the air velocity slows from supersonic to subsonic.

It becomes apparent that all compressibility effects depend upon the relationship of the object's speed to the local speed of sound. The term used to describe this relationship is the Mach number, so named for the Austrian physicist Ernst Mach. Mach number then is the ratio of the speed of an object to the local speed of sound.

$$M = \frac{V}{c_s}$$

where V = velocity of object
c_s = speed of sound

Since the speed of sound varies with temperature, the M number is a very convenient measure of high-speed flow. For example, at sea level, the speed of sound equals 1117 ft/s [340.5 m/s]. If an airplane's speed is also 1117 ft/s, we would say that the airpane was traveling at Mach 1.0. At 36,000 ft [10,973 m], where the speed of sound is only 968 ft/s [295.0 m/s], M = 1.15.

EXAMPLE: An airplane is traveling at M 2.0 at 30,000 ft [9144 m]. What is the plane's airspeed? (Speed of sound at 30,000 ft = 995 ft/s or 680 mph [303 m/s or 1094 km/h].) How fast is the airplane moving?

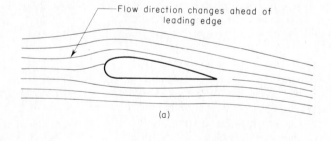

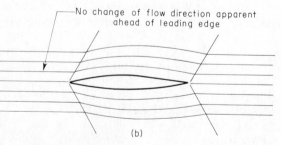

FIG. 3-26 Airflow behavior over an airfoil at subsonic and supersonic speeds.
(a) Typical subsonic flow pattern.
(b) Typical supersonic flow pattern.

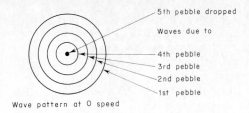

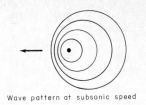

Wave pattern at O speed

Wave pattern at subsonic speed

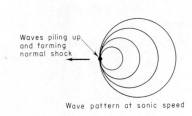

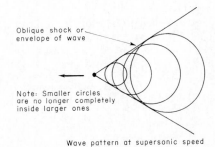

Wave pattern at sonic speed

Oblique shock or envelope of wave

Note: Smaller circles are no longer completely inside larger ones

Wave pattern at supersonic speed

FIG. 3-27 Water wave analogy of shock formation; disturbance caused by pebbles dropped into water at equal time intervals.

$$V = Mc_s \qquad\qquad V = Mc_s$$
$$= 2 \times 995 \quad \text{or} \quad = 2 \times 680$$
$$= 1990 \text{ ft/s [606.6 m/s]} = 1360 \text{ mph [2189 km/h]}$$

Reference was made on page 93 of this section to a choked nozzle, and it was stated that when a nozzle is choked the gases are traveling through it at the speed of sound and cannot be further accelerated. It can now be seen that the velocity of these exhaust gases is much higher than 763 mph [1228 km/h] (the speed of sound at the standard temperature of 59°F [15°C]). In fact, if the exhaust gas temperature is 1040°F [560°C], the exhaust velocity may reach 1896 ft/s [577.9 m/s] or 1293 mph [2081 km/h] before the speed of sound is reached.

In review, a sonic shock wave is the accumulation of sound-wave energy (pressure) developed when a sound moving away from a disturbance (object) is held in a stationary position by the flow of air in the opposite direction. The velocity of airflow across a shock wave will decrease because the air molecules are moving with the sound wave against the air velocity. This decrease in air velocity is accompanied by a pressure rise, because the sound-wave motion, in slowing the air velocity, will convert most of the kinetic energy of velocity into a pressure rise.

There are two types of shock waves, oblique and normal (Fig. 3-27). The oblique shock wave stands off of the moving object at an oblique angle; it occurs at high supersonic velocities, and the velocity drop across this shock is to a *lower supersonic* velocity. The normal shock wave stands perpendicular to the airstream; it occurs at low supersonic speed and its velocity drop is from *supersonic* to *subsonic* behind the normal shock front. In both types, a pressure rise occurs. If the velocity and pressure changes are small, it is called a weak shock; if the velocity drop and pressure rise are high,

it is called a strong or forced shock. Keep in mind that across all shock waves there is a temperature increase. (See Chap. 4.)

REVIEW AND STUDY QUESTIONS

1. Name the five basic sections of the gas turbine engine. What additional sections and/or components may also be present?
2. Briefly describe the series of events that take place in the gas turbine engine.
3. Define the following terms: force, work, power, energy, kinetic energy, speed, velocity, vector, acceleration, mass, and momentum.
4. What is 1 hp equal to in terms of foot-pounds per minute?
5. State Newton's three laws of motion.
6. Using the balloon analogy, explain the reaction principle.
7. What is the relationship between thrust and force?
8. What is the difference between gross thrust and net thrust? How does ram drag enter into this difference?
9. Write the *complete* formula for determining the thrust of a jet engine.
10. What do the words "choked nozzle" mean? How does this effect enter into the determination of thrust?
11. Give the formula for determining the thrust at any given point in the engine.
12. At what major points in the engine is forward thrust being applied? Rearward thrust?
13. State the relationship between thrust and horsepower. At what point are they equal?
14. Make a table to show the effects of increasing and decreasing rpm, airplane speed, temperature, pressure, density, and humidity.

15. How do pressure, temperature, and velocity change through an operating engine? How does the use of the afterburner affect these parameters?
16. Why is the Otto cycle engine called a constant-volume engine? Why is the Brayton cycle engine called a constant-pressure engine?
17. Compare the Otto cycle and Brayton cycle as to differences and similarities.
18. Compare pressure and temperature increases and decreases through the compressor and turbine.
19. Define "specific fuel consumption." Why is this a good way to compare engine efficiencies?
20. List the factors which affect the overall efficiency of the gas turbine engine.
21. Define thermal and propulsive efficiency.
22. What is the relationship between pressure and velocity when a fluid passes through a restriction (venturi) at subsonic velocity? Through a diffuser?
23. What do the terms compressible and incompressible fluid mean? Using these terms, how does air act at subsonic speeds? At supersonic speeds?
24. What influences the speed of sound? Define Mach number.
25. Explain how a shock wave forms.
26. What is the difference between a normal and oblique shock wave?

PART TWO
CONSTRUCTION AND DESIGN

CHAPTER FOUR
INLET DUCTS

Although the inlet duct is made by the aircraft manufacturer, during flight operation, as shown in Chap. 3, it becomes very important to the overall jet engine performance and will greatly influence jet engine thrust output. The faster the airplane goes, the more critical the duct design becomes. Engine thrust can be high only if the inlet duct supplies the engine with the required airflow at the highest possible pressure.

The inlet duct must operate from static ground run up to high aircraft Mach numbers with a high duct efficiency at all altitudes, attitudes, and flight speeds (Fig. 4-1). To compound the problem, the amount of air required by a turbojet engine is approximately 10 times or more than that of a piston engine of comparable size.

Inlet ducts should be as straight and smooth as possible, and should be designed in such a way that the boundary layer air (a layer of still, dead air lying next to the surface) be held to a minimum. The length, shape, and placement of the duct is determined to a great extent by the location of the engine in the aircraft.

Not only must the duct be large enough to supply the proper airflow, but it must be shaped correctly to deliver the air to the front of the compressor with an even pressure distribution. Poor air pressure and velocity distribution at the front of the compressor may result in compressor stall. (See Chap. 5.)

Another primary task a duct must do during flight operation is to convert the kinetic energy of the rapidly moving inlet airstream into a ram pressure rise inside the duct. To do this it must be shaped so that the ram velocity is slowly and smoothly decreasing, while the ram pressure is slowly and smoothly rising.

Inlet ducts are rated in two ways: the duct pressure efficiency ratio and the ram recovery point. The duct pressure efficiency ratio is defined as the ability of the duct to convert the kinetic or dynamic pressure energy at the inlet of the duct into static pressure energy at the inlet of the compressor without a loss in total pressure. (See page 107, Bernoulli's Theorem.) It will have a high value of 98 percent if the friction loss is low and if the pressure rise is accomplished with small losses. The ram recovery point is that aircraft speed at which the ram pressure rise is equal to the friction pressure losses, or that airspeed at which the compressor inlet total pressure is equal to the outside ambient air pressure. A good subsonic duct will have a low ram recovery point (about 160 mph [28 km/h]).

Inlet ducts may be divided into two broad categories.

1. Subsonic ducts
2. Supersonic ducts

Figure 4-2 illustrates the variety of possible inlet duct designs falling within these categories. It is interesting to note that the engine manufacturers rate their engines using a bellmouth inlet (Fig. 4-3). This type of inlet is essentially a bell-shaped funnel having carefully rounded shoulders, which offer practically no air resistance. The duct loss is so small that it is considered zero, and all engine performance data can be gathered without any correction for inlet duct loss being necessary. (See Chap. 19.) Normal duct inefficiencies may cause thrust losses of 5 percent or more because a decrease in duct efficiency of 1 percent will decrease airflow 1 percent, decrease jet velocity ½ percent, and result in 1½ percent thrust loss. The decrease in jet velocity occurs because it is necessary to increase the area of the jet nozzle in order to keep the turbine temperature within limits when duct losses occur.

SUPERSONIC DUCTS

The supersonic inlet duct must operate in three speed zones (Fig. 4-4).

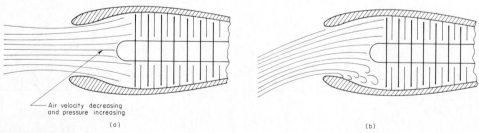

Air velocity decreasing
and pressure increasing

(a)

(b)

FIG. 4-1 (a) Normal and **(b)** distorted flow through an inlet duct. Distorted flow is the result of unusual flight attitudes or ice buildup; the problem is especially critical with short ducts.

Nose Inlet Variations

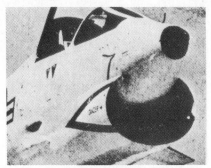

F8U-1, F-84

FJ4, F-86

F-100

Annular Inlet

Navy Demon

Pod Inlet

B-47, B-52, 707, DC-8

Flush-Scoop Inlet

F-80

Wing Inlet Variations

F-101, F9F-8, T-37

F-105, F4D-1

Avro Vulcan

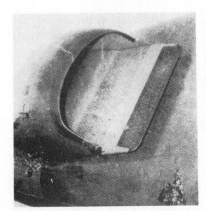

Scoop Inlet Variations
F-104, F-102, F-106, T-33,
F-94, F-89, A4D-1, P6M-1,
F11F-1

FIG. 4-2 Some subsonic and supersonic inlet duct variations.

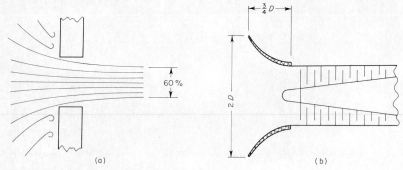

FIG. 4-3 Purpose of the bellmouth inlet.
(a) Low-velocity approach showing vena contracta effect (necking down).
(b) Bellmouth inlet eliminates contraction and allows engine all the air it can handle.

1. Subsonic
2. Transonic
3. Supersonic

Although each of these speed zones needs a slightly different inlet duct design, good overall performance can be achieved by designing to the supersonic shape with some modifications.

The supersonic duct problems start when the aircraft begins to fly at or near the speed of sound. As illustrated in the preceding chapter, at these speeds sonic shock waves are developed which, if not controlled, will give high duct loss in pressure and airflow, and will set up vibrating conditions in the inlet duct called inlet buzz. Buzz is an airflow instability caused by the shock wave rapidly being alternately swallowed and expelled at the inlet of the duct.

Air which enters the compressor section of the engine must usually be slowed to subsonic velocity, and this process should be accomplished with the least possible

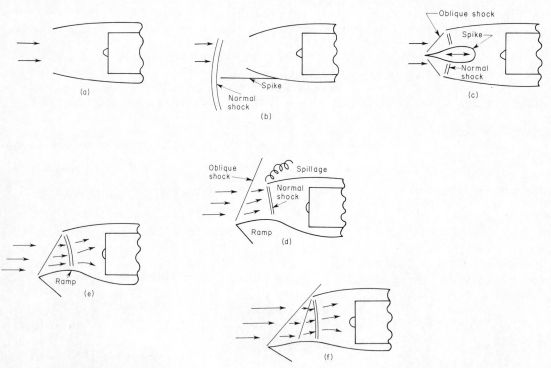

FIG. 4-4 Types of inlet ducts.
(a) Subsonic duct
(b) Transonic duct
(c) Supersonic duct with variable geometry operating at design speed.
(d) Ramp- or wedge-type duct below design speed.
(e) Ramp- or wedge-type duct at design speed.
(f) Ramp- or wedge-type duct establishing multiple oblique shocks.

waste of energy. At supersonic speeds the inlet duct does the job by slowing the air with the weakest possible series or combination of shocks to minimize energy loss and temperature rise.

At transonic speeds (near Mach 1), the inlet duct is usually designed to keep the shock waves out of the duct. This is done by locating the inlet duct behind a spike or probe [Fig. 4-4(b)] so that at airspeeds slightly above Mach 1.0 the spike will establish a normal shock (bow wave) in front of the inlet duct. This normal shock wave will produce a pressure rise and a velocity decrease to subsonic velocities before the air strikes the actual inlet duct. The inlet will then be a subsonic design behind a normal shock front. At low supersonic Mach numbers, the strength of the normal shock wave is not too great, and this type of inlet is quite practical. But at higher Mach numbers, the single normal shock wave is very strong and causes a great reduction in the total pressure recovered by the duct and an excessive air temperature rise inside the duct.

At slightly higher airspeeds the normal bow wave will change into an oblique shock [Fig. 4-4(c) and (d)]. Since the air velocity behind an oblique shock is still supersonic, to keep the supersonic velocities out of the inlet duct, the duct will need to set up a normal shock wave at the duct inlet. The airflow is controlled so that the air velocity at the duct inlet is exactly equal to the speed of sound. At this time the duct pressure rise will be due to

1. An oblique shock pressure rise
2. A normal shock pressure rise
3. A subsonic diverging section pressure rise

As the airspeed is increased, the angle of the oblique shock will be forced back by the higher air velocity until the oblique shock contacts the outer lip of the duct. When this occurs there will be a slight increase in thrust due to an increase in engine inlet pressure and airflow, because the energy contained in the shock front is now enclosed within the duct and delivered to it with less pressure loss. This point is called the *duct recovery point* [Fig. 4-4(e)].

At higher Mach numbers (about 1.4 and above) the inlet duct must set up one or more oblique shocks and a normal shock [Fig. 4-4(f)]. The oblique shocks will slow the supersonic velocities, the normal shock will drop the velocity to subsonic, then the subsonic section will further decrease the velocity before the air enters the compressor. Each decrease in velocity will produce a pressure rise.

VARIABLE-GEOMETRY DUCT

A complication of the supersonic inlet is that the optimum shape is variable with the inlet flow direction and the Mach number. In the higher Mach aircraft the inlet duct geometry is made variable by one of the following:

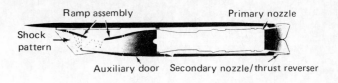

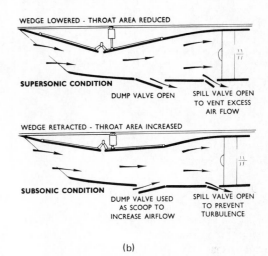

FIG. 4-5 The two-dimensional duct is used on the B.A.C./Aerospatiale Concorde SST. The variable geometry of the nacelle intake and the primary and secondary exhaust nozzles ensures that the engine airflow demand is exactly met and the propulsive efficiency of the power plant is optimized during all flight conditions.

(a) The secondary nozzles or buckets are also used as thrust reversers to provide in-flight retardation and to assist braking.

(b) Intake during supersonic and subsonic flows.

1. Moving the inlet spike in and out so as to maintain the oblique shock on the edge of the outer lip of the duct (axisymmetric duct).
2. Moving the side wall or ramp to a higher angle so as to force a stronger oblique shock front (two-dimensional duct) (Fig. 4-5).
3. Varying the normal shock throat area so as to force a stronger normal shock (expanding center-body).
4. Varying the inlet lip area so as to vary the intake area.

All of these methods have advantages and disadvantages insofar as cost, ease and speed of control, good efficiencies at all flight speeds, and integration into the aircraft aerodynamic and structural design are concerned.

In summation, the supersonic inlet duct must:

1. Have a high duct efficiency and deliver the highest possible ram pressure.
2. Deliver the airflow required, although at low supersonic speed it must dump an excess airflow.

3. Set up and maintain sonic shock waves and keep these shock waves and supersonic velocities out of the engine.
4. Produce the least ram temperature rise, which temperature rise the engine design must take into account. (On some supersonic airplane-engine combinations, compressor inlet temperature may reach 250°F or more from ram air effects.)
5. Deliver ram pressure and airflow that will produce an increase in the net thrust output.

REVIEW AND STUDY QUESTIONS

1. What is the function of the inlet duct? Why is its correct design so important?

2. Define duct pressure efficiency and ram recovery point. What is the importance of these terms in the design of a good inlet duct?
3. What is a bellmouth inlet? When is it used?
4. Into what two broad categories do inlet ducts fall?
5. What are the three speed zones in which the supersonic inlet duct must operate? Give a brief description of the inlet duct airflow under these three conditions.
6. What is the action of inlet duct airflow across a normal shock? Across an oblique shock?
7. Why is it desirable for a high-speed supersonic inlet duct to produce a series of weak oblique shocks?
8. Why must supersonic aircraft be equipped with variable geometry inlet ducts?
9. List some methods of constructing the variable geometry ducts.

CHAPTER FIVE
COMPRESSORS

The role of the compressor in a gas turbine engine is to provide a maximum of high-pressure air which can be heated in the limited volume of the combustion chamber and then expanded through the turbine. The energy that can be released in the combustion chamber is proportional to the mass of air consumed; therefore the compressor is one of the most important components of the gas turbine engine since its efficient operation (maximum compression with minimum temperature rise) is the key to high overall engine performance. The compressor efficiency will determine the power necessary to create the pressure rise of a given airflow and will affect the temperature change which can take place in the combustion chamber. (See Chap. 3.) Present-day compressors have compression ratios approaching 15:1, efficiencies near 90 percent, and airflows up to approximately 350 lb/s [158.8 kg/s]. With the addition of a fan, total pressure ratios of 25:1 and mass airflows of over 1000 lb/s [453.6 kg/s] have been achieved.

TYPES OF COMPRESSORS

All gas turbine engines use one of the following forms of compressor:

1. Centrifugal flow
2. Axial flow

The centrifugal-axial-flow compressor is a combination of the two, with operating characteristics of both.

It will be the purpose of this chapter to examine, in some detail, the construction and operation of each of these compressors.

The Centrifugal-flow Compressor

The centrifugal compressor consists basically of an impeller and a diffuser manifold (Fig. 5-1). Other components such as a compressor manifold may be added to direct the compressed air into the combustion chamber. As the impeller revolves at high speed, air is drawn in at the eye. Centrifugal force provides high acceleration to this air and causes it to move outward from the axis of rotation toward the rim of the rotor where it is ejected at high velocity and high kinetic energy. The pressure rise is produced in part by expansion of the air in the diffuser manifold by conversion of the kinetic energy of motion into static pressure energy (Fig. 5-2).

In Chap. 2 we saw that centrifugal compressors could be manufactured in a variety of designs including single-stage, multiple-stage, and double-sided types. The centrifugal compressor has a number of features to rec-

ommend its use in certain types of gas turbine engines. Chief among its attributes are its simplicity, ruggedness, and low cost. Because of its massive construction, it is much less susceptible to damage from the ingestion of foreign objects. The centrifugal compressor is capable of a relatively high compressor ratio per stage. About 80 percent efficiency may be reached with a compression ratio of 6 or 7 to 1. Above this ratio, efficiency drops off at a rapid rate because of excessively high impeller tip speeds and attending shock-wave formation. This rules out this type of compressor for use in larger engines since high compression ratios are necessary for low fuel consumption (Fig. 5-3). Some centrifugal-flow engines obtain somewhat higher ratios through the use of multistage compressors as shown in Chap 2, Figs. 2-2, 2-18, 2-47, and 2-58(b). Although the tip speed problem is reduced, efficiency is again lost because of the difficulty in turning the air as it passes from one stage to another. Double-entry compressors also help to solve the high-tip-speed problem, but this advantage is partially offset by the complications in engine design necessary to get air to the rear impeller, and by the requirement of a large plenum or air chamber, where the air from the inlet duct is brought to a slower speed for efficient direction change and higher pressures. The plenum chamber acts as a diffuser by which means the rear impeller can receive its air. Examples of airplanes incorporating a plenum chamber are the Air Force T-33 [Fig. 2-4(c)], and the Navy F9F. Both of these airplanes use the Allison J-33 engine (Fig. 2-4).

Because of the problems inherent in this type of design, the centrifugal compressor finds its greatest application on the smaller engines where simplicity, flexibility of operation, and ruggedness are the principal requirements rather than small frontal area, and ability to handle high airflows and pressures with low loss of efficiency.

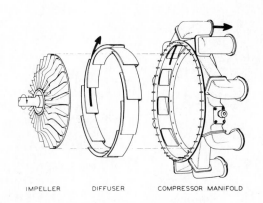

IMPELLER   DIFFUSER   COMPRESSOR MANIFOLD

FIG. 5-1 Typical single-stage centrifugal compressor.

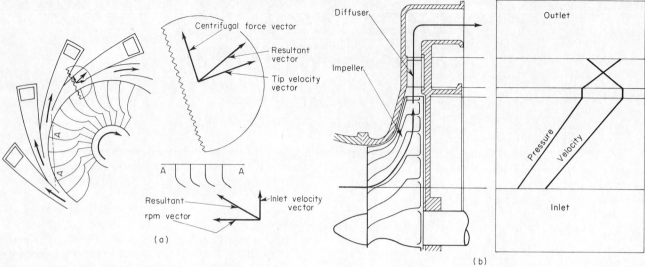

FIG. 5-2 Centrifugal-compressor flow, pressure, and velocity changes.
(a) Airflow through a typical centrifugal compressor.
(b) Pressure and velocity changes through a centrifugal compressor.

The Axial-flow Compressor

The axial-flow compressor is made up of a series of rotating airfoils called rotor blades, and a stationary set of airfoils called stator vanes. As its name implies, the air is being compressed in a direction parallel to the axis of the engine (Fig. 5-4). A row of rotating and stationary blades is called a stage. The entire compressor is made up of a series of alternating rotor and stator vane stages. Some axial-flow designs have two or more compressors or spools which are driven by separate turbines and are therefore free to rotate at different speeds.

Axial compressors have the advantage of being capable of very high compression ratios with relatively high efficiencies (Fig. 5-5). In addition, the small frontal area created by this type of compressor lends itself to installation in high-speed aircraft. Unfortunately, the delicate blading, especially toward the rear, makes this type of air pump especially susceptible to foreign-object damage. Furthermore, the number of compressor blades and stator vanes (which can exceed 1000 in a large jet engine), the close fits required for efficient air pumping,

and the much narrower range of possible operating conditions, make this type of compressor very complex and very expensive to manufacture. Modern manufacturing techniques are bringing down the cost for small axial-flow compressors. For these reasons the axial-flow design finds its greatest application where the demands of efficiency and output predominate considerations of cost, simplicity, flexibility of operation, etc. As we shall see later, most manufacturers utilize several dodges to increase flexibility and to improve the operating characteristics of the axial-flow compressor.

COMPRESSOR THEORY

Any discussion of the operation of the compressor must take into account both aerodynamic and thermodynamic considerations since it is impossible to compress air without incurring a temperature rise.

The temperature rise is a result of the work being done by the rotor blades. In 1843, James Joule demonstrated that the temperature rise in a substance (in this case, air) is always proportional to the energy or work expended. This proposition will be examined in more detail in the section Compressor Thermodynamics.

Compressor Aerodynamics

A vector analysis will help to show airflow, pressure, and velocity changes through a typical axial-flow compressor. Starting with inlet air (Fig. 5-6), notice that the length of arrows (vectors) A and B are the same. This indicates that no change in velocity occurred at this point. The inlet guide vanes only deflect the air to a predetermined angle toward the direction of rotation of the rotor. At points C and D the vectors are of different lengths showing that work is being done upon the air in the form of a velocity loss and a pressure gain. The

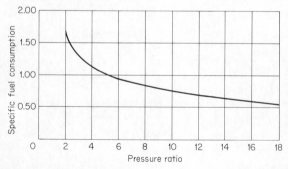

FIG. 5-3 Specific fuel consumption decreases with increasing pressure ratio.

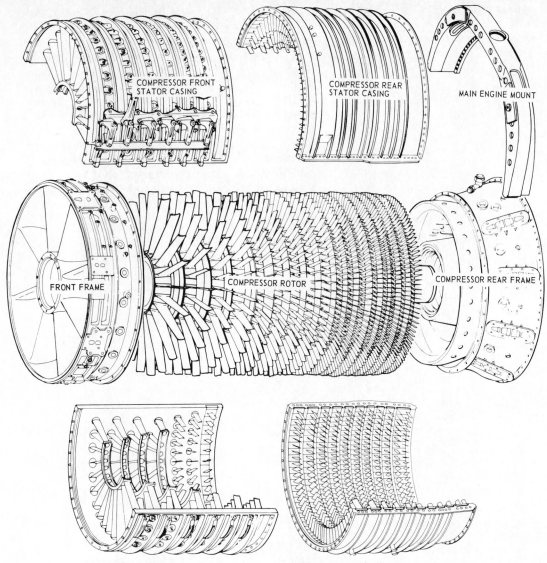

FIG. 5-4 A modern high-performance compressor assembly. (*General Electric*.)

stator entrance (vector E) and the stator discharge (vector F) show another velocity loss and pressure gain exactly like that occurring through the rotor. The discharge air (D) seems to be at an incorrect angle to enter the first-stage stator, but due to the presence of rotary

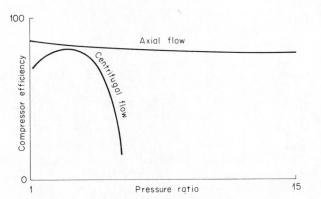

FIG. 5-5 A comparison of centrifugal- and axial-flow compressor efficiencies with increasing pressure ratios.

air motion caused by the turning compressor, the resultants E and G are produced which show the true airflow through the compressor. Notice that these vectors are exactly in line for entrance into the next stage of the compressor. One final aerodynamic point to note is that the stator entrance (vector E) is longer than the stator discharge (vector F) because of the addition of energy to the air by the rotor rotation X. Thus, as each set of blades, rotors, and stators, causes a pressure rise to occur at the expense of its discharge velocity, the air's rotary motion restores the velocity energy at each blade's entrance for it in turn to convert to pressure energy.

You will notice that both the rotor blades and the stator blades are diffusing the airflow. It is much more difficult to obtain an efficient deceleration (diffusion or pressure increase) of airflow than it is to get efficient acceleration, because there is a natural tendency in a diffusion process for the air to break away from the walls of the diverging passage, reverse its direction, and flow back in the direction of the pressure gradient or

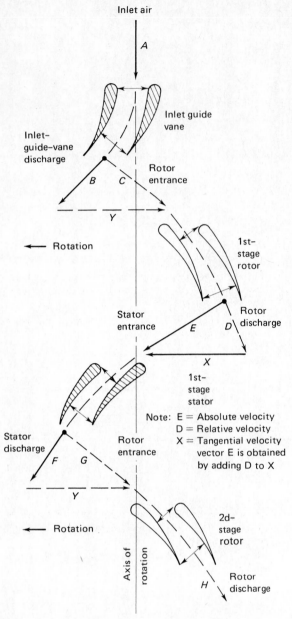

FIG. 5-6 Vector analysis of airflow through an axial-flow compressor (*Pratt & Whitney Aircraft.*)

Note: E = Absolute velocity
D = Relative velocity
X = Tangential velocity
vector E is obtained
by adding D to X

lower pressure (Fig. 5-7). It has been determined that a pressure ratio of approximately 1.2 is all that can be handled by a single compressor stage since higher rates of diffusion and excessive turning angles on the blades result in excessive air instability, hence low efficiency. A desired compression ratio is achieved by simply adding more stages onto the compressor. The amount of pressure rise or compression ratio depends on the mass of air discharged by the compressor, the restrictions to flow imposed by the parts of the engine through which the air must pass, and the operating conditions (pressures) inside the engine compared with the ambient air pressure at the compressor intake. The final pressure is the result of multiplying the pressure rise in each stage.

EXAMPLE: A 13-stage compressor has a pressure ratio across each stage of 1.2 and an ambient inlet pressure of 14.7 psi [101.4 kPa]. What is the final pressure? What is the pressure ratio?

stage 1	stage 2	stage 3
$14.7 \times 1.2 =$	$17.64 \times 1.2 =$	$21.17 \times 1.2 =$

stage 4	stage 5	stage 6
$25.40 \times 1.2 =$	$30.48 \times 1.2 =$	$36.58 \times 1.2 =$

stage 7	stage 8	stage 9
$43.89 \times 1.2 =$	$52.67 \times 1.2 =$	$63.21 \times 1.2 =$

stage 10	stage 11	stage 12
$75.85 \times 1.2 =$	$91.02 \times 1.2 =$	$109.22 \times 1.2 =$

stage 13
$131.07 \times 1.2 = 157.28$

Final pressure = 157.3 psi [1084.5 kPa]
Initial pressure = 14.7 psi [101.4 kPa]

or

$$CR = \frac{157.3}{14.7}$$
$$10.7:1$$

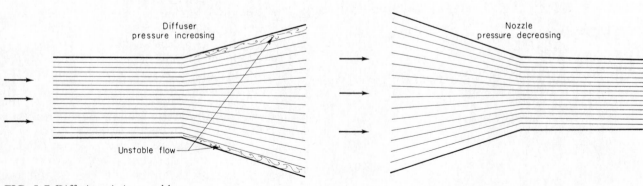

FIG. 5-7 Diffusing air is unstable.

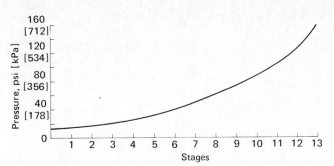

FIG. 5-8 The result of compressor pressure multiplication.

Notice that the pressure rise across the first stage is

17.6 psi	pressure at back of 1st stage
−14.7 psi	pressure at front of 1st stage
2.9 psi	pressure rise across 1st stage

and the pressure rise across the last stage is

157.3 psi	pressure at back of 13th stage
−131.1 psi	pressure at front of 13th stage
26.2 psi	pressure rise across 13th stage

The pressure ratio is the same in both cases but the actual increase in pressure is much greater toward the rear of the compressor (Fig. 5-8) than the front. The compression ratio will increase and decrease with engine speed. It will also be affected by compressor inlet temperature. As the inlet temperature (due to ram) increases, the compression ratio will tend to decrease due to the combined effects of air density decrease and the temperature effects on the angle of attack.

The ideal compression ratio is that which will produce the maximum pressure in the tailpipe. Normally, the higher the compression ratio the higher the tailpipe pressure. However, at high compression ratios and high compressor inlet temperatures, and with turbine inlet temperature limited to a specific value, the compressor discharge temperature would be so high that main fuel flow would have to be limited, and the turbine rotor would have to use more pressure energy in place of combustion energy to drive the compressor. This, then, would result in increased turbine pressure loss in comparison with the compressor pressure increase. (See Chap. 7.)

High-speed flight also has a profound effect on compressor discharge pressure and temperature. High aircraft speeds will cause a ram pressure and ram temperature rise of considerable proportions to exist at the front of the compressor. These high ram pressure ratios and high ram temperature ratios when multiplied by the pressure and temperature rise across the compressor will produce pressures and temperatures in the engine which will be in excess of design limitations. For example, at Mach 3 the pressure ratio due to ram alone will be approximately 30:1. When this is multiplied by a compressor ratio of only 10:1 it will give a total pressure ratio of 300:1. This is far beyond what the combustion chamber can stand. Furthermore, at this high speed, the ram temperature ratio when multiplied by the compressor temperature ratio will give very high temperatures at the compressor discharge, and this in turn will limit the amount of fuel that can be added to obtain a useful temperature rise.

By bypassing some of the excess compressor discharge air around the combustion chambers and turbine, and discharging it into the primary airstream at the jet nozzle, and/or by reducing the number of compressor stages, these high compressor discharge pressures may be used to advantage. A system of "blow-in" doors will be necessary to ensure the proper pressure relationship between the bypassed air and primary air.

A cross-sectional view of the engine (Fig. 5-9) shows

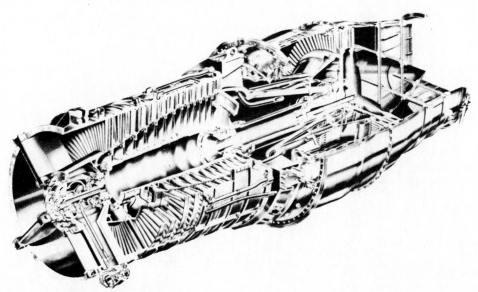

FIG. 5-9 Sectioned view showing a typical compressor taper and constant-diameter case.

that the space formed by the compressor disk rim and the stator casing gradually reduces in area. Figure 3-17 in Chap. 3 shows that the airflow velocity is relatively constant through the compressor. If air is compressed and the volume is not decreased, its velocity will decrease excessively toward the rear stages and the stall area (discussed on pages 123–125), will be approached. Two methods of reducing the volume toward the rear stage are to use a compressor whose case has a constant diameter and whose compressor rotor tapers, or to use a compressor with a tapering case and a constant diameter rotor. Figures 2-40(b) and 2-42(c) show both types used on one engine. Since rpm and airflow are related, the compressor should be turned as fast as possible. Keeping in mind that the blade tips must not exceed the speed of sound in order to avoid shock-wave formation that would destroy pumping efficiency, the tip speed is kept to approximately Mach 0.85.

By using the formula

$$V_{fpm} = \pi D \times \text{rpm}$$

or

$$V_{fps} = \frac{\pi D}{12} \times \frac{\text{rpm}}{60}$$

where V_{fpm} = velocity, ft/min
$\quad\quad V_{fps}$ = velocity, ft/s
$\quad\quad D$ = diameter, if given in feet
$\quad\quad \dfrac{D}{12}$ = diameter, if given in inches
$\quad\quad \dfrac{\text{rpm}}{60}$ = revolutions per second

the tip speed of any rotating wheel, i.e., the compressor or turbine, may be found. For example, if the diameter of a particular compressor is 18 in [45.7 cm] from tip to tip, the compressor rpm is 20,000, and the temperature at the front of the compressor is standard, the tip speed of the blades would be:

$$V_{fps} = \frac{\pi D}{12} \times \frac{\text{rpm}}{60}$$
$$= \frac{3.14 \times 18}{12} \times \frac{20,000}{60}$$
$$= 1570 \text{ [3987.8 cm/s]}$$

Always keeping in mind that the speed of sound is a function of temperature, and that under standard day temperatures the speed of sound is equal to 1117 ft/s [340.5 m/s], we can see that the rpm of this compressor would be much too high, being limited to 1117 × 0.85, or 950 ft/s [289.6 m/s]. But as already indicated, there is a considerable temperature rise at the rear stages of the compressor. This will permit a higher tip speed at the rear stages without exceeding the speed of sound. The Pratt & Whitney two-spool compressor design, among others (Figs. 2-40 and 2-42), takes advantage of this fact. Figure 5-10 shows the advantage of the constant outside case diameter.

To keep blade - tip speed below Mach .85, maximum compressor rpm is limited

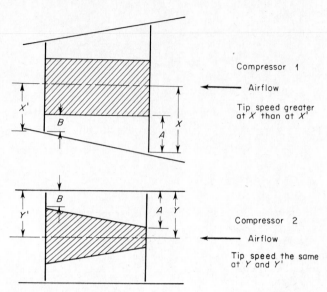

Compressor 1

← Airflow

Tip speed greater at X than at X'

Compressor 2

← Airflow

Tip speed the same at Y and Y'

FIG. 5-10 Comparison of airflow through a constant-outside-diameter and constant-inside-diameter compressor. Conditions: (1) Outlet air temperature for compressors 1 and 2 = 500°F [260°C], (2) A=A and B=B. Critical dimensions for both compressors are the same. Conclusion: A temperature of 500°F would permit a speed of 15,000 rpm, but the speed of compressor 1 is limited by tip speed at X. It is potentially capable of pumping more air.

Other forms of the same formula will allow us to find the maximum rpm or the maximum diameter of a given compressor or turbine.

EXAMPLE: What is the maximum rpm of a compressor whose diameter equals 18 in and whose blade-tip speed is limited to Mach 0.8? Note: The speed of sound equals 1100 ft/s [335.3 m/s] under existing conditions, therefore the tip speed is limited to 1100 × 0.8, or 880 ft/s [268.2 m/s].

$$\text{rpm} = \frac{12 \times 60 \times V_{fps}}{\pi D}$$
$$= \frac{12 \times 60 \times 880}{3.14 \times 18}$$
$$= 11,210$$

EXAMPLE: What is the maximum diameter of a compressor whose rpm equals 10,000 and whose blade-tip speed is limited to Mach 0.7? Note: Again the local speed of sound equals 1100 ft/s.

$$D = \frac{V_{fps} \times 12 \times 60}{\pi \times \text{rpm}}$$
$$= \frac{770 \times 12 \times 60}{3.14 \times 10,000}$$
$$= 17.65 \text{ in [44.83 cm]}$$

As was mentioned previously, the air velocity is fairly constant throughout the compressor and a given air molecule will not wander much more than 180° around the compressor, for although the rotors impart a rotational component to the air, the stators take this rotational component and turn the velocity into pressure.

Compressor Stall

Since an axial-flow compressor consists of a series of alternately rotating and stationary airfoils or wings, the same rules and limitations which apply to an airfoil or wing will apply to the entire compressor. The picture is somewhat more complicated than is the case for a single airfoil, because the blades are close together, and each blade is affected at the leading edge by the passage through the air of the preceding blade. This "cascade" effect can be more readily understood if the airflow through the compressor is viewed as flow through a series of ducts formed by the individual blades, rather than flow over an airfoil that is generating lift. The cascade effect is of prime importance in determining blade design and placement (Fig. 5-11).

The axial compressor is not without its difficulties, and the most vital of these is the stall problem. If for some reason the angle of attack, i.e., the angle at which the airflow strikes the rotor blades, becomes too low, the pressure zones shown in Fig. 5-11 will be of low value and the airflow and compression will be low. (The angle of attack should not be confused with the angle of incidence, which is a fixed angle determined by the manufacturer when the compressor is constructed.) If the angle of attack is high, the pressure zones will be high and airflow and compression ratio will be high. If it is too high the compressor will stall. That is, the airflow over the upper foil surface will become turbulent and destroy the pressure zones. This will, of course, decrease the compression and airflow. The angle of attack will vary with engine rpm, compressor inlet temperature, and compressor discharge or burner pressure. From Fig. 5-12 it can be seen that decreasing the velocity of airflow or increasing engine rotor speed will tend to increase the angle of attack. In general, any action that decreases airflow *relative* to engine speed will in-

crease the angle of attack and increase the tendency to stall. The decrease in airflow may result from the compressor discharge pressure becoming too high, for example, from excessive fuel-flow schedule during acceleration. Or compressor inlet pressure may become too low in respect to the compressor discharge pressure because of high inlet temperatures or distortion of inlet air. Several other causes are possible (Fig. 5-12).

During ground operation of the engine, the prime action that tends to cause a stall is choking. If the engine speed is decreased from the design speed, the compression ratio will decrease with the lower rotor velocities as shown in Fig. 5-12. With a decrease in compression, the volume of air in the rear of the compressor will be greater. This excess volume of air causes a choking action in the rear of the compressor with a decrease in airflow, which in turn decreases the air velocity in the front of the compressor and increases the tendency to stall. If no corrective design action is taken, the front of the compressor will stall at low engine speeds.

Another important cause of stall is high compressor inlet air temperatures. High-speed aircraft may experience an inlet air temperature of 250°F [121°C] or higher because of the ram effect. These high temperatures cause low compression ratios (due to density changes) and will also cause choking in the rear of the compressor. This choking-stall condition is the same as that caused by low compression ratios due to low engine speeds. High compressor inlet temperatures will cause the length of the airflow vector to become longer since the air velocity is directly affected by the square root of any temperature change.

Each stage of a compressor should develop the same pressure ratio as all other stages. But as stated in the two preceding paragraphs, when the engine is slowed down or the compressor inlet temperature climbs, the front stages supply too much air for the rear stages to handle and the rear stages will choke (Fig. 5-13).

There are five basic solutions which can be utilized to correct this front-end, low-speed, high-temperature stall condition.

1. Derate or lower the angles of attack on the front stages so that the high angles at low engine speed are not stall angles.
2. Introduce a bleed valve into the middle or rear of the compressor and use it to bleed air and increase airflow in the front of the compressor at lower engine speeds.
3. Divide the compressor into two sections or rotors (the two-spool rotor) and design the front rotor speed to fall off more than the rear rotor at low speeds so that the low front rotor speed will equal the low choked airflow.
4. Place variable guide vanes at the front of the compressor and variable stator vanes in the front of the first several compressor stages so that the angles of attack can be reset to low angles by moving the variable vanes at low engine speeds. Some advanced engine designs also utilize the variable stator concept on the last several compressor stages.
5. Use a variable-area exhaust nozzle to unload the compressor during acceleration.

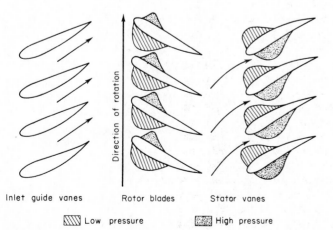

Inlet guide vanes   Rotor blades   Stator vanes

▨ Low pressure   ▧ High pressure

FIG. 5-11 The cascade effect.

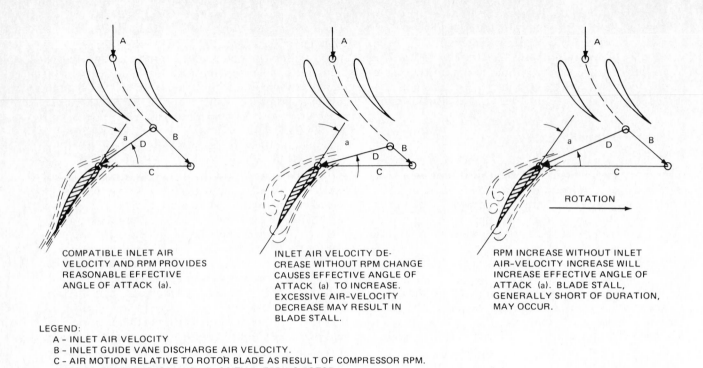

COMPATIBLE INLET AIR
VELOCITY AND RPM PROVIDES
REASONABLE EFFECTIVE
ANGLE OF ATTACK (a).

INLET AIR VELOCITY DE-
CREASE WITHOUT RPM CHANGE
CAUSES EFFECTIVE ANGLE OF
ATTACK (a) TO INCREASE.
EXCESSIVE AIR-VELOCITY
DECREASE MAY RESULT IN
BLADE STALL.

RPM INCREASE WITHOUT INLET
AIR-VELOCITY INCREASE WILL
INCREASE EFFECTIVE ANGLE OF
ATTACK (a). BLADE STALL,
GENERALLY SHORT OF DURATION,
MAY OCCUR.

ROTATION

LEGEND:
A – INLET AIR VELOCITY
B – INLET GUIDE VANE DISCHARGE AIR VELOCITY.
C – AIR MOTION RELATIVE TO ROTOR BLADE AS RESULT OF COMPRESSOR RPM.
D – RESULTANT AIRFLOW AND VELOCITY ENTERING ROTOR.
a – EFFECTIVE ANGLE OF ATTACK.

(a)

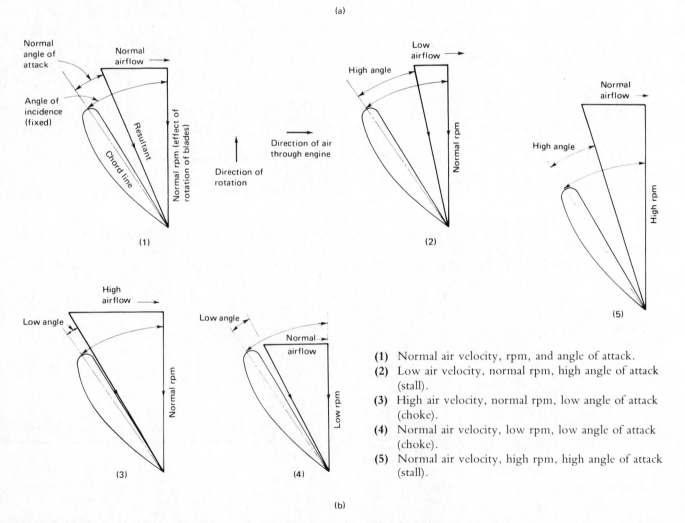

(1) Normal air velocity, rpm, and angle of attack.
(2) Low air velocity, normal rpm, high angle of attack (stall).
(3) High air velocity, normal rpm, low angle of attack (choke).
(4) Normal air velocity, low rpm, low angle of attack (choke).
(5) Normal air velocity, high rpm, high angle of attack (stall).

(b)

FIG. 5-12 The result of changing airflow velocity and rpm on the angle of attack.

NOTE: A combination of these methods may be used.

One last point on compressor aerodynamics needs discussion. Every compressor has a certain ability to maintain a compression ratio for a given mass airflow. Figure 5-14 (a) shows a typical compressor's ability to maintain a compression ratio as mass flow varies. Any point located above the compressor's stall curve represents a compression ratio too high for the existing mass airflow. The normal operating line shows the actual pressure ratio the compressor would be subject to with variations in mass airflow. Any point on this line represents the compression ratio that will exist in a given engine with a specific mass flow. The normal operating line can be raised or lowered by changing the downstream restrictions. Any time the normal operating line crosses the compressor stall curve, the compressor will be subject to compression ratios it cannot maintain. Stall will result. Figure 5-14 (a) illustrates that, while this engine will run unstalled at a compression ratio of 9:1 with a mass airflow of 100 lb/s [45.36 kg/s], it cannot get to this point without encountering the stall zone. All of the listed methods in the preceding paragraph for correcting stall either change the shape of the normal operating line so that it will conform more closely with the compressor stall line or move the compressor stall line to a higher point.

Compressor Thermodynamics

Thermodynamically, the compressor follows natural gas laws which state that a pressure rise is normally accompanied by a temperature rise. In addition, any aerodynamic inefficiencies cause an additional temperature rise and correspondingly less pressure output than the ideal compressor. Since this is not a text on thermodynamics, this phenomena will be investigated only rather briefly.

Stall Indications

Mild stalls may be indicated by abnormal engine noises such as rumble, chugging, choo-choo-ing or, buzzing. Other stall indications may be rapid EGT increase or fluctuation, RPM fluctuation, EPR decrease or fluctuation (see Chap. 19), vibration caused by compressor pulsations, and poor engine response to power-lever movements. More severe stalls can cause very loud bangs and may be accompanied by flame, vapor,

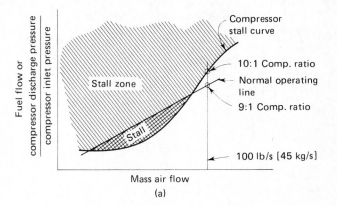

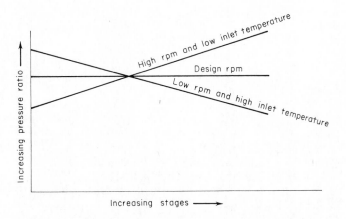

FIG. 5-13 The effect of rpm and temperature on the angle of attack, and the pressure ratio per stage. Off-design RPM and inlet air temperature will cause the pumping characteristics of the individual compressor stages to change. At low RPM and/ or high inlet air temperature, pumping capacity is higher in the forward stages of the axial compressor than in the aft stages. The reverse is true at high RPM and/or low inlet air temperature.

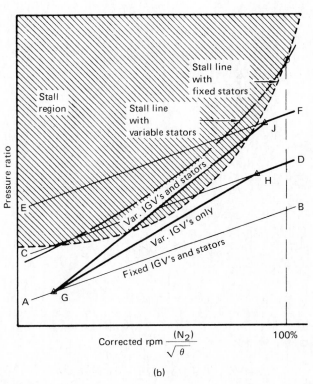

FIG. 5-14 (a) Typical compressor operating curve.
(b) Axial-compressor performance with variable IGVs and variable stators.

or smoke at the engine inlet and/or exhaust. While the explosive-type stall may cause engine failure or malfunction, more commonly, the full effect of a stall, i.e., high EGT, are deferred and are difficult to correlate directly with a specific stall incident. The deferred effects are cumulative and influence engine reliability and durability. Recurring stalls are a malfunction and require maintenance action.

THE BEHAVIOR OF AIR

Air is a mixture of gases, principally oxygen and nitrogen with some water vapor always present. The gases are made up of molecules which are very widely separated in relation to their size. These molecules travel at very high average velocity and cause pressure by hitting the sides of any container.

The force with which any molecule will strike the walls of a container is equal to the mass of the molecule times twice its speed. Two times the speed is used because when a mass has its direction changed, a force results which is equal to the mass times the amount of change, and when a given molecule strikes the wall of the container at a given speed it will rebound with the same speed.

Force = $2mc$

where m = mass of molecule
c = speed of molecule

In addition, the total number of molecules in the container must be considered. Although not all of the molecules strike the container walls, those in the middle form the boundary for the molecules that are hitting the walls, so all of the molecules contribute to the pressure or force in the container.

Force = $2mc$ × number of molecules

Further, each molecule in a container will eventually strike all sides of that container. If a force in one direction is of interest, i.e., pressure on a specific area, the total force must be divided by 3.

Force = $\dfrac{2mc \times \text{number of molecules}}{3}$

One other factor needs to be considered, and that is the number of times each molecule will hit the wall. This is equal to the molecule's speed in feet per second divided by two times the distance the molecule has had to travel to strike the wall. Two times the distance is used because the molecule must travel back to its original point before it can start again.

Force = $\dfrac{2mc \times \text{number of molecules}}{3} \times \dfrac{c}{2d}$

where c = speed of molecule, ft/s
d = molecule distance to wall

Since the mass of the molecule times the number of molecules equals the mass of the whole body, then

$$F = \frac{Mc^2}{3d}$$

where M = mass of entire body.

Pressure times an area equals a force, so

$$PA = \frac{Mc^2}{3d}$$

where P = pressure, psi
A = area, ft^2

By rearranging the formula,

$$P = \frac{Mc^2}{3dA}$$

It is also known that an area times a distance equals a volume.

$$P = \frac{Mc^2}{3V}$$

Mass is equal to w/g and the formula then becomes:

$$P = \frac{wc^2}{3gV}$$

From this formula it can be seen that the internal energy or static pressure will increase as the square of the molecular speed (a function of temperature) and directly in relation to the density. Rearranging the formula again and solving for c^2,

$$c^2 = \frac{3gPV}{w}$$

Pressure × volume = WRT

where W = weight of air
R = gas constant for air (see pages 128–129)
T = absolute temperature, °R

Rearranging this formula gives

$$\frac{PV}{w} = RT$$

Now if

$$c^2 = \frac{3gPV}{w}$$

then

$$c^2 = 3gRT$$

In this formula the 3, g, and R are all constants, and it can be seen that the temperature varies as the square of the molecular speed. The letter K is often used for the combined constant for the 3, g, and R, and this results in the formula:

$$c^2 = KT \quad \text{or} \quad c = \sqrt{KT}$$

which states that the molecular speed is directly related to the square root of the temperature of the gas.

High temperature = high molecular speed
Low temperature = low molecular speed

At absolute zero ($-460°F$) [$-273°C$] the velocity of the gas molecule would theoretically equal zero and the pressure would be zero. On a standard day, and at sea level, the average molecular speed would be 1117 ft/s [340.46 m/s] or 761 mph [1224.68 km/h] (speed of sound) with the pressure equal to 14.7 psi [101.4 kPa]. (See Appendix.)

Air is capable of expanding and contracting, and of absorbing and giving up heat. These effects are inter-related and closely follow known gas laws. For example, adding heat to a fixed volume of air raises its temperature, causes the molecules to move faster and the pressure to increase. Compressing the air delivers energy to the molecules, resulting in higher molecular speed and thus higher temperatures. When the same number of molecules occupies less space, they hit the side of the enclosure with greater frequency, which results in higher pressure. The principal point to remember is that whenever work is done on a gas the gas gets hotter in *proportion* to the amount of work being done on the gas.

BOYLE'S LAW

The three possible properties of air are volume, temperature, and pressure. Since the gas turbine engine is continuously changing these variables, an understanding of the relationship between volume, temperature, and pressure is important. If any two values are known, the third can be determined, assuming that the weight or mass of air has not changed. Boyle's law states that: If the *temperature of a given quantity of gas is kept constant*, absolute pressure is inversely proportional to volume.

$$PV = C \text{ (constant)}$$

if

$$P_1 V_1 = C$$

and

$$P_2 V_2 = C$$

then

$$P_1 V_1 = P_2 V_2 \quad \text{or} \quad \frac{P_1}{P_2} = \frac{V_2}{V_1}$$

where P_1 = initial absolute pressure psi
P_2 = final absolute pressure psi
V_1 = initial volume, ft^3
V_2 = final volume, ft^3

Transposing will give us

$$P_2 = \frac{P_1 V_1}{V_2} \quad \text{or} \quad V_2 = \frac{P_1 V_1}{P_2}$$

CHARLES' LAW (*FIG. 5-15*)

Charles' law states that: if the *pressure on a given quantity of gas is held constant*, the volume is directly proportional to the absolute temperature.

$$\frac{V_1}{V_2} = \frac{T_1}{T_2} \quad \text{or} \quad \frac{T_1}{V_1} = \frac{T_2}{V_2}$$

where V_1 = initial volume, ft^3
V_2 = final volume, ft^3
T_1 = initial absolute temperature, °R
T_2 = final absolute temperature, °R

Transposing will give us

$$V_2 = \frac{V_1 T_2}{T_1} \quad \text{or} \quad T_2 = \frac{T_1 V_2}{V_1}$$

This law may also be written: If the *volume of a given quantity of gas is held constant*, the pressure is directly proportional to the absolute temperature.

$$\frac{P_1}{P_2} = \frac{T_1}{T_2} \quad \text{or} \quad \frac{T_1}{P_1} = \frac{T_2}{P_2}$$

Transposing will give us:

$$P_2 = \frac{P_1 T_2}{T_1} \quad \text{or} \quad T_2 = \frac{T_1 P_2}{P_1}$$

By combining Boyle's and Charles' laws we get the *general gas law* which can be used to find the pressure, volume, or temperature of a gas when any of the other conditions change.

$$\frac{P_1 V_1}{T_1} = \frac{P_2 V_2}{T_2}$$

Transposing will give

$$P_2 = \frac{P_1 V_1 T_2}{V_2 T_1}$$

or

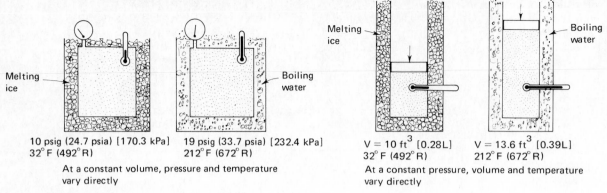

Melting ice

10 psig (24.7 psia) [170.3 kPa]
32° F (492° R)

19 psig (33.7 psia) [232.4 kPa]
212° F (672° R)

Boiling water

At a constant volume, pressure and temperature vary directly

Melting ice

$V = 10$ ft^3 [0.28L]
32° F (492° R)

$V = 13.6$ ft^3 [0.39L]
212° F (672° R)

Boiling water

At a constant pressure, volume and temperature vary directly

FIG. 5-15 Diagram illustrating Charles' law.

$$V_2 = \frac{P_1 V_1 T_2}{P_2 T_1}$$

or

$$T_2 = \frac{P_2 V_2 T_1}{P_1 V_1}$$

SPECIFIC HEAT

The *British thermal unit* is a standard measure of energy and is defined as the amount of energy required to raise the temperature of one pound of water 1°F at a constant volume. If the temperature of 1 lb [0.454 kg] of air is raised the same amount only 0.1715 Btu [180.9325 J] is required to do the job. The numerical comparison of Btu's between water and air is called the *specific heat* at a constant volume (c_v). According to Charles' law, the pressure of a given weight of gas, when heated at a constant volume, varies directly as the absolute temperature. When a gas does not expand, its volume remains the same and no external work is done. All of the heat added to a gas at a constant volume is effective only in raising its temperature. But a gas will require *more than* 0.1715 Btu to raise its temperature 1°F [5/9°C] at a constant pressure than at a constant volume, because at a constant pressure the gas can expand and do useful work. In other words, the heat added must be sufficient to raise the gas temperature 1°F as well as to supply energy equal to the external work done. The specific heat of air at a constant pressure (c_p) is 0.24 Btu [253.20 J]. Now if it takes 0.1715 Btu to raise the gas temperature 1°F at a constant volume, and 0.24 Btu to raise the gas temperature and furnish expansion energy for useful work at a constant pressure, then

$$\begin{array}{ll} 0.2400 & c_p \\ -0.1715 & c_v \\ \hline 0.0685 \ [72.2675 \ J] & \end{array}$$ = amount of Btu that went into the expansion of 1 lb of air when the temperature was raised 1°F.

It has been determined, through experimentation, that 1 Btu = 778 ft·lb [107.58 kg·m]. Multiplying the energy (Btu's) that goes into the expansion of 1 lb of air by 778 will give the work in foot-pounds that the air is capable of doing for each degree of temperature rise per pound of air. This is the mechanical equivalent of heat.

$$\begin{aligned} 1 \ \text{Btu} &= 778 \ (c_p - c_v) \\ &= 778 \ (0.24 - 0.1715) \\ &= 53.3 \ \text{ft} \cdot \text{lb} \ [7.37 \ \text{kg} \cdot \text{m}] \ \text{of mechanical work} \end{aligned}$$
done by the expansion of 1 lb of air at a constant pressure when its temperature is raised 1°F

The number 53.3 is practically constant for all temperatures and pressures of air and is designated by the letter R. Other gases will have other gas constants.

PERFECT-GAS EQUATION

According to Boyle's law, the product of the pressure times the volume of one pound of air at a given temperature equals a constant.

$$PV = \text{constant}$$

According to Charles' law the volume varies as the temperature varies. In addition it is known that the pressure will vary as the temperature. Combining these statements will give:

$$\frac{PV}{T} = \text{const} \qquad \text{or} \qquad R$$

By transposing,

$$PV = RT$$

If the values for R (53.3 for air as found above) and any

two of the quantities P, V, or T is known, the third can be found by using one of the following equations.

$$P = \frac{RT}{V} \qquad V = \frac{RT}{P} \qquad T = \frac{PV}{R} \qquad R = \frac{PV}{T}$$

HORSEPOWER REQUIRED TO DRIVE THE COMPRESSOR

The compressor requires a considerable amount of shaft horsepower to pump air and to give this air a pressure and temperature rise. The horsepower requirements can be determined by finding out how much energy has been put into the air by the compressor. Multiplying the specific heat of air at a constant pressure by the temperature rise across the compressor will give the energy put into each pound of air during the pressurizing process. If this result is then multiplied by the total weight of airflow through the compressor, the total energy put into the air by the compressor can be obtained:

$c_p \Delta T W_a$ or $c_p(T_2 - T_1)W_a$ = Btu/s being put into the air, or total energy of air

Since 1 Btu = 778 ft·lb, and 1 hp equals 550 ft·lb/s, the horsepower requirements for the compressor are

$$\frac{c_p \Delta T W_a 778}{550}$$

NOTE: The specific heat of air at a constant pressure is used any time the volume of a gas is increasing or decreasing because while the c_v and c_p both take into account the energy required to increase molecular speed, only the c_p takes into consideration the change in volume or change in distance the molecules travel.

EXAMPLE: An axial-flow engine is running at 100 percent rpm under standard day conditions. The compressor is pumping 200 lb [90.72 kg] of air per second, and the temperature rise across the compressor is 600°F [333°C]. How much power is needed by this compressor? Disregard any losses and assume 100 percent efficiency.

$$\text{Compressor horsepower requirements} = \frac{c_p \Delta T W_a 778}{550} \text{ or } \frac{c_p(T_2 - T_1) W_a 778}{550}$$

$$= \frac{0.24 \times 600 \times 200 \times 778}{550}$$

$$= 40,739 \qquad \text{at 100 percent}$$

Construction Features

Centrifugal-flow jet engines usually use heat-treated forged aluminum compressors, although cast compressors are being used on small engines of this type. The aluminum or magnesium diffuser is also generally manufactured by casting. In many cases the inducer or guide vanes, which smooth and direct the airflow into the engine and thus minimize the shock in the impeller, are manufactured separately from the impeller or rotor. Rotor vanes may either be all full length as in Fig. 5-16, or some may be half length as shown in Fig. 5-17. A close fit is important between the compressor and its case in order to obtain maximum compressor efficiency. The clearance is usually checked with a feeler gage or with a special fixture. Balancing of the rotor may be accomplished by removing material from specified areas of the compressor or by using balancing weights installed in holes in the hub of the compressor. On some engines in which the compressor and turbine wheel are balanced as a unit, special bolts or nuts having slight variations in weight are used. Compressor support bearings may be either ball or roller, although most manufacturers use at least one ball bearing on the compressor to support both axial and radial loads.

Axial-flow engines have compressors that are constructed of several different materials depending upon the load and temperature under which the unit must operate. The rotor assembly illustrated in Fig. 5-18 consists of stub shafts, disks, blades, ducts, spacers, and

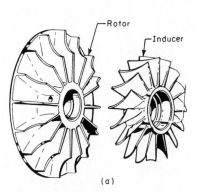

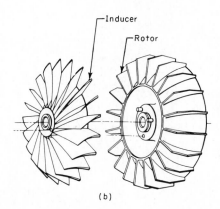

FIG. 5-16 Typical single-stage centrifugal compressors.
(a) Teledyne CAE
(b) Fairchild J44.

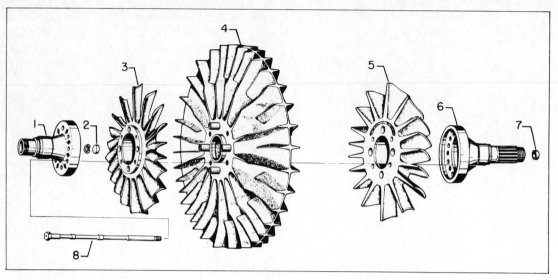

1 IMPELLER FRONT SHAFT
2 OIL PLUG
3 FRONT ROTATING GUIDE
 VANE
4 IMPELLER

5 REAR ROTATING GUIDE
 VANE
6 IMPELLER REAR SHAFT
7 OIL PLUG
8 IMPELLER BOLT

FIG. 5-17 Double-sided centrifugal compressor with half-vanes (Detroit Diesel Allison J33).

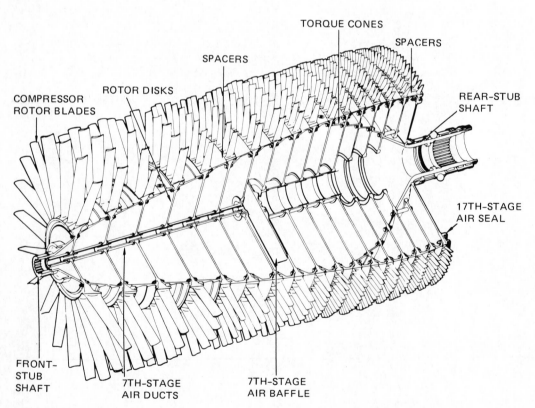

FIG. 5-18 Compressor construction features of the General Electric CJ805-23 (J79) engine.

torque cones. The rotor blades are generally machined from stainless-steel forgings, although some blades may be of titanium in the forward or colder part of the compressor; the remainder of the components are machined, low-alloy steel forgings. The clearance between the rotating blades and the outer case is most important with many manufacturers depending upon a "wear fit" between the blade and the compressor case. Many companies design their blades with knife-edge tips so that the blades will wear away and form their own clearance as they expand from the heat generated from compression of the air. Other companies coat the inner surface of the compressor case with a soft material which can be worn away without damage to the blade. The several stages of the compressor are composed of disks which can be joined together by means of a tie bolt as shown in Fig. 5-19. Serrations or splines prevent the disks from turning in relation to each other. Other manufacturers eliminate the tie bolt and join the stages together at the disk rim (Figs. 5-18 and 5-20). Methods of attaching the blade to the disk also vary between manufacturers, with the majority using some variation of the dovetail to hold the rotor blade in the disk. Various locking methods are used to anchor the blades in place. The blades do not have a tight fit in the disk, but rather are seated by centrifugal force during engine operation. By allowing the blades some movement, vibrational stresses, produced by the high-velocity airstreams from between the blades, are reduced.

On some of the latest engines, for example the Pratt & Whitney TF30-P-100, JT9D and others, airflow has been increased by designing a bulged inner diameter flow path. This configuration provides a greater linear blade velocity increasing the fan-root pressure ratio and work capability of the low-pressure compressor. The canted vanes and blades in the first few stages of the compressor section improve efficiency by reducing boundary layer flow separation which occurs when the vanes and blades are parallel to the plane of rotation. External case dimensions of the engine remain unchanged (Fig. 5-21).

Stator cases also show great variability in design and construction features. Figure 5-22 shows most of the features to be found in a typical compressor case, with an additional feature being the variable stator vanes. The stator vanes may be either of solid or hollow construction and may or may not be connected together at their tips by a shroud. The shrouding serves two purposes. First, it provides support for the longer stator vanes located in the forward stages of the compressor, and second, it provides the absolutely necessary air seal between rotating and stationary parts. Detroit Diesel Allison, General Electric, Avco Lycoming, and others use split compressor cases (Figs. 5-23 and 5-24), while Pratt & Whitney favors a weldment, forming a continuous case. The advantage to the split case lies in the fact that the compressor and stator blades are readily available to inspection. On the other hand, the continuous case offers simplicity and strength, since it requires no horizontal parting surface (Fig. 5-25).

Both the case and the rotor are very highly stressed parts. Since the compressor turns at very high speeds, the disks must be able to withstand very high centrifugal force, and, in addition, the blades must resist bending loads and high temperatures. When the compressor is constructed, each stage is balanced separately and after

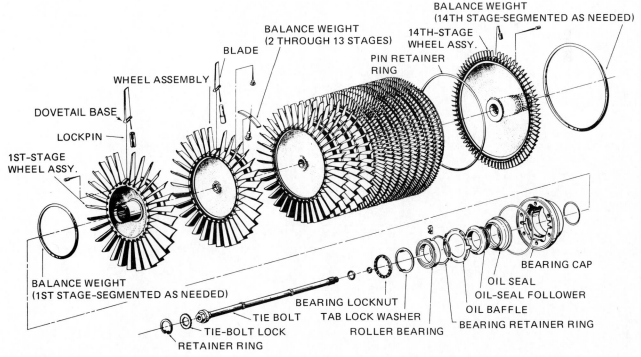

FIG. 5-19 The Detroit Diesel Allison 501-D13 compressor is held together with a tiebolt. Notice the dovetail blade base, balancing weight, and lock pin.

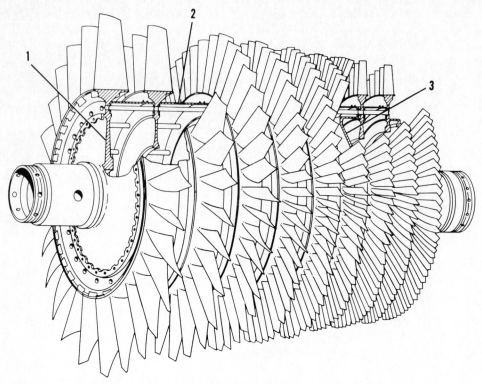

FIG. 5-20 Low-speed (N_1) compressor construction for the Pratt & Whitney Aircraft JT3C. Notice the dovetail blade attachment. (1) Front hub, (2) spacer assembly, (3) rear hub.

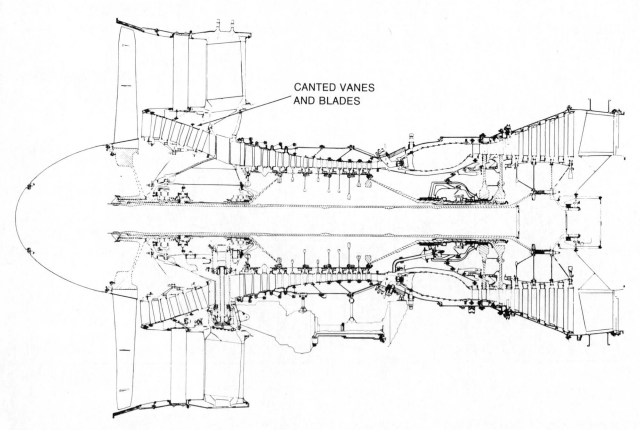

CANTED VANES
AND BLADES

FIG. 5-21 The Pratt & Whitney Aircraft JT9D showing canted vanes and blades.

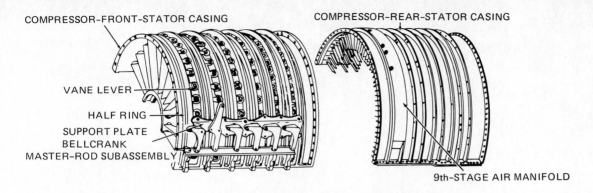

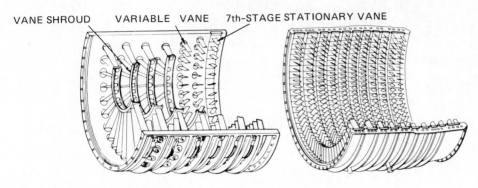

FIG. 5-22 Stator case for the General Electric CJ805-23 (J79) engine.

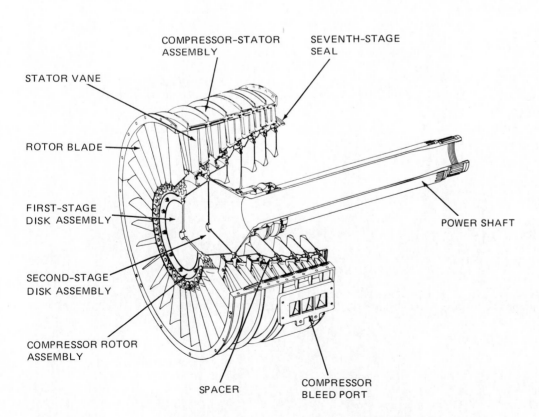

FIG. 5-23 Rotor and stator assembly for the General Electric CJ610 (J85). Note the first-stage blade attachment.

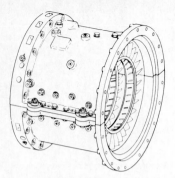

FIG. 5-24 The Avco Lycoming T53 has a split compressor case.

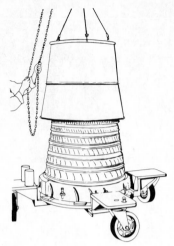

FIG. 5-25 Disassembly of a Pratt & Whitney Aircraft N_1 compressor having a continuous, one-piece case.

assembly, the compressor is balanced as a unit. Figure 5-19 shows the balance method used by Allison on their 501-D13 engine. The compressor case in most instances is one of the principal structural, load-bearing members of the engine, and may be constructed of aluminum as shown in Fig. 5-26 or steel as shown in Figs. 5-22, 5-23, and 5-25.

REVIEW AND STUDY QUESTIONS

1. How would you define an efficient compressor?
2. List some typical operating specifications in terms of compression ratios, airflows, and efficiencies for a large axial-flow compressor.
3. Name the two basic types of gas turbine compressors. Describe the operating principles of each.
4. Using vectors, show the pressure and velocity changes through an axial compressor.
5. What is the practical maximum pressure rise per stage for an axial-flow compressor? For a centrifugal-flow compressor? Why?
6. What is the purpose of the curved section at the front of the centrifugal-flow compressor?
7. Compare the pressure increase per stage at the front and rear of the axial-flow compressor.
8. Determine the compression ratio for a 10-stage axial-flow compressor (1.2 CR and 14.7 psi [101.4 kPa] ambient pressure).
9. What are two methods of reducing the flow area toward the rear of the compressor? Why is it necessary to do this?
10. What formula is used to determine the tip speed of the compressor?
11. Explain the phenomenon of compressor stall. What conditions bring a compressor closer to stall? How may the stall problem be reduced?
12. What is the relationship between temperature and the speed of sound? Between temperature and work done on a gas?
13. State Boyle's law and Charles' law.
14. What is meant by specific heat?
15. Give the formula for determining the power required by the compressor.
16. Describe the construction features and materials of the axial and centrifugal compressors.

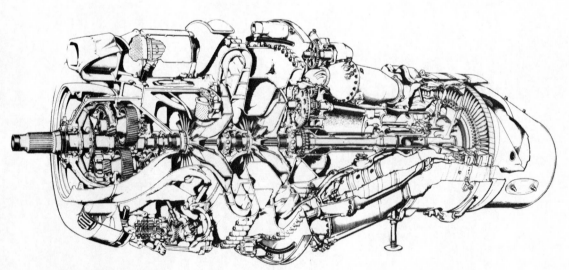

FIG. 5-26 Sectioned Rolls-Royce Dart showing aluminum compressor housing and diffuser construction.

CHAPTER SIX
COMBUSTION
CHAMBERS

The development of burner systems for aircraft gas turbine engines presents a number of challenging problems. These problems involve thermodynamics, fluid mechanics, heat transfer, chemistry, metallurgy, and many other phases of development.

TYPES OF BURNERS

The three basic types of burner systems in use today are the:

1. Can type (Fig. 6-1)
2. Annular type (Fig. 6-2)
   (a) Through-flow annular
   (b) Side-entry annular
   (c) Reverse-flow annular
3. Can-annular type (Fig. 6-3), a combination of the can and annular styles

Can Types

Can-type combustion chamber versions can be seen in Figs. 2-4, 2-5, 2-9, and 2-47.

Annular Types

Typical through-flow annular combustion chambers are shown in Figs. 2-20, 2-21, 2-25, 2-26, 2-27, 2-28, 2-29, 2-32, 2-46, 2-48, 2-49, 2-51, 2-52, and 2-53. Variations of the annular combustion chamber include reverse-flow designs illustrated in Figs. 2-11, 2-12, 2-13, 2-18, 2-19, 2-34, and 2-35, and side-entry annular combustion chambers used in such engines as the Teledyne CAE series engine (Figs. 2-55, 2-56, and 2-57), and the Williams Research Corp. WR series (Figs. 2-58 and 2-59).

Can-annular Types

The can-annular combustion chamber is represented by the engines shown in Figs. 2-7, 2-8, 2-10, 2-22,

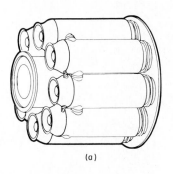

(a)

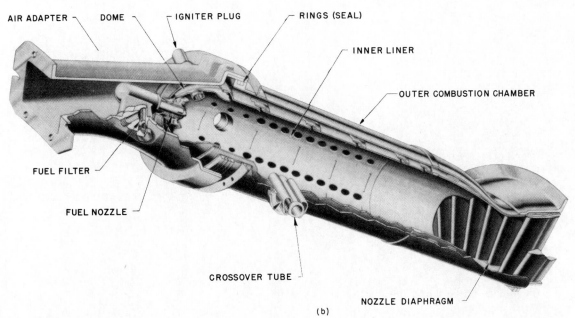

AIR ADAPTER   DOME   IGNITER PLUG   RINGS (SEAL)

INNER LINER

OUTER COMBUSTION CHAMBER

FUEL FILTER

FUEL NOZZLE

CROSSOVER TUBE

NOZZLE DIAPHRAGM

(b)

FIG. 6-1 (a) The can-type burner.
(b) Sectioned view of a typical can-type burner.

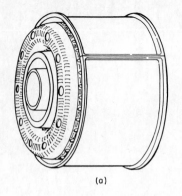

(a)

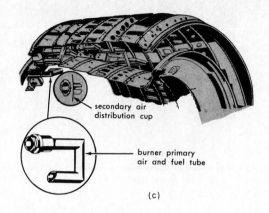

secondary air
distribution cup

burner primary
air and fuel tube

(c)

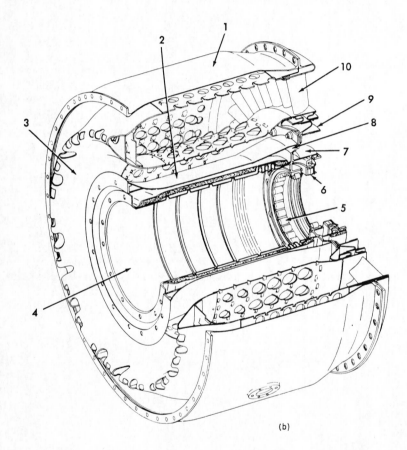

(b)

1 OUTER COMBUSTION
  CASING
2 INNER COMBUSTION
  CASING
3 COMBUSTION LINER
4 SHAFT SHIELD
5 NO. 3 (REAR) BEARING
6 NO. 3 (REAR) CARBON
  SEAL
7 NO. 3 (REAR) SEAL SUP-
  PORT
8 NO. 3 (REAR) BEARING
  SUPPORT
9 TURBINE STATIONARY
  SEAL
10 FIRST-STAGE TURBINE
   NOZZLE

FIG. 6-2 (a) The annular-type burner.
(b) Sectioned view of the annular combustion chamber used on the General Electric
  (J85) engine.
(c) An annular combustion chamber using vaporizing tubes instead of injection
  nozzles.

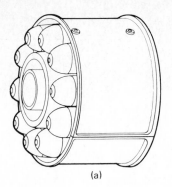

(a)

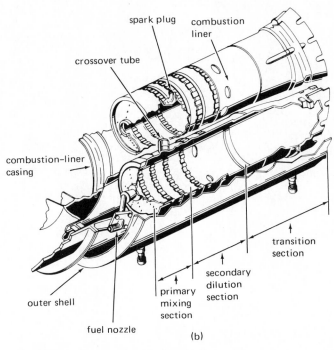

FIG. 6-3 (a) The can-annular-type burner.
(b) Sectioned view of the can-annular burner used on the Detroit Diesel Allison 501-D13 engine. Note how the transition section changes the circular cross section of the liner to an annular shape so that the gases will impinge on the entire nozzle face.

FIG. 6-4 The Pratt & Whitney Aircraft JT9D combustion chamber is basically of annular design but the forward end is divided into individual inner liners permitting fuel and airflow patterns to be more accurately controlled.

2-23, 2-24, 2-36, 2-37, 2-38, 2-40, 2-42, 2-43, 2-45, and 2-50.

NOTE: Certain combustion chamber designs such as that illustrated in Fig. 6-4 do not fall readily into any of the aforementioned various categories.

ADVANTAGES AND DISADVANTAGES OF THE DIFFERENT TYPES OF BURNERS

In the can type, individual burners, or cans, are mounted in a circle around the engine axis, with each one receiving air through its own cylindrical shroud. One of the main disadvantages of can-type burners is that they do not make the best use of available space and this results in a large-diameter engine. On the other hand, the burners are individually removable for inspection and air-fuel patterns are easier to control than in annular designs. The annular burner is essentially a single chamber made up of concentric cylinders mounted coaxially about the engine axis. This arrangement makes more complete use of available space, has low pressure loss, fits well with the axial compressor and turbine, and from a technical viewpoint has the highest efficiency, but has a disadvantage in that structural problems may arise due to the large-diameter, thin-wall cylinder required with this type of chamber. The problem is more severe for larger engines. There is also some disadvantage in that the entire combustor must be removed from the engine for inspection and repairs. The can-annular design also makes good use of available space, but employs a number of individually replaceable cylindrical inner liners that receive air through a common annular housing for good control of fuel and airflow patterns. The can-annular arrangement has the added advantage of greater structural stability and lower pressure loss than that of the can type (Fig. 6-5).

OPERATION OF THE COMBUSTION CHAMBER

Fuel is introduced at the front end of the burner in either a highly atomized spray from specially designed nozzles (see Chap. 12), or in a prevaporized form from devices called vaporizing tubes. Air flows in around the fuel nozzle and through the first row of combustion air holes in the liner. The burner geometry is such that the air near the nozzle stays close to the front wall of the liner for cooling and cleaning purposes, while the air entering through opposing liner holes mixes rapidly with the fuel to form a combustible mixture. Additional air is introduced (Fig. 6-6) through the remaining air holes in the liner. The air entering the forward section of the liner tends to recirculate and move upstream against the fuel spray. During combustion this action permits rapid mixing and prevents flame blowout by forming a low-velocity stabilization zone which acts as a continuous pilot for the rest of the burner. The air entering the downstream part of the liner provides the

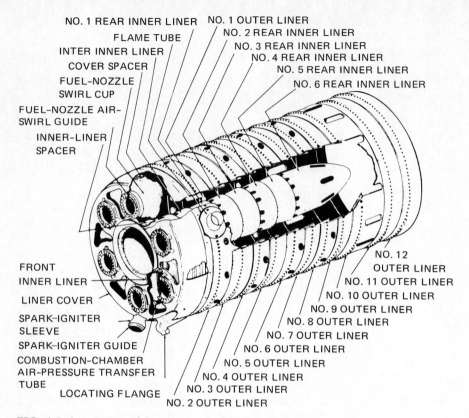

NO. 1 REAR INNER LINER
NO. 1 OUTER LINER
FLAME TUBE
NO. 2 REAR INNER LINER
INTER INNER LINER
NO. 3 REAR INNER LINER
COVER SPACER
NO. 4 REAR INNER LINER
FUEL-NOZZLE
NO. 5 REAR INNER LINER
SWIRL CUP
NO. 6 REAR INNER LINER
FUEL-NOZZLE AIR-
SWIRL GUIDE
INNER-LINER
SPACER

FRONT
INNER LINER
NO. 12
OUTER LINER
LINER COVER
NO. 11 OUTER LINER
SPARK-IGNITER
NO. 10 OUTER LINER
SLEEVE
NO. 9 OUTER LINER
SPARK-IGNITER GUIDE
NO. 8 OUTER LINER
COMBUSTION-CHAMBER
NO. 7 OUTER LINER
AIR-PRESSURE TRANSFER
NO. 6 OUTER LINER
TUBE
NO. 5 OUTER LINER
LOCATING FLANGE
NO. 4 OUTER LINER
NO. 3 OUTER LINER
NO. 2 OUTER LINER

FIG. 6-5 A variation of the can-annular combustion chamber as used on the Pratt & Whitney Aircraft JT3 (J57) series engines, where each inner liner is actually a miniature annualar combustion chamber.

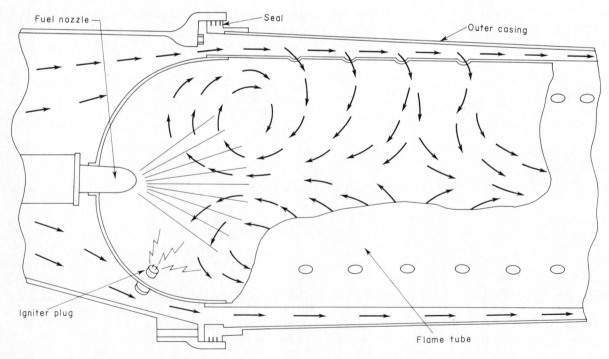

FIG. 6-6 Typical airflow through a burner.

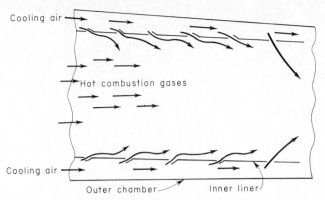

FIG. 6-7 Cooling the burner walls. Large arrows to the rear represent dilution airflow.

correct mixture for combustion, and it creates the intense turbulence that is necessary for mixing the fuel and air and for transferring energy from the burned to the unburned gases.

Since there are usually only two ignitor plugs in an engine, cross ignition tubes are necessary in the can and can-annular types of burners in order that burning may be initiated in the other cans or inner liners. The ignitor plug is usually located in the upstream reverse-flow region of the burner. After ignition, the flame quickly spreads to the primary or combustion zone where there

is approximately the correct proportion of air to completely burn the fuel. If all the air flowing through the engine were mixed with the fuel at this point, the mixture would be outside the combustible limits for the fuels normally used. Therefore only about one-third to one-half is allowed to enter the combustion zone of the burner. About 25 percent of the air actually takes part in the combustion process. The gases that result from combustion have temperatures of 3500°F [1900°C]. Before entering the turbine the gases must be cooled to approximately half this value, which is determined by the design of the turbine and the materials involved. Cooling is done by diluting the hot gases with secondary air that enters through a set of relatively large holes located toward the rear of the liner. The liner walls must also be protected from the high temperatures of combustion. This is usually accomplished by introducing cooling air at several stations along the liner, thereby forming an insulating blanket between the hot gases and the metal walls (Fig. 6-7).

Higher metal temperatures required the development of an advanced burner can. The Finwall® design, a unique cooling concept, reduces cooling air requirements by 50 percent and decreases burner weight by 20 percent. This is accomplished by designing a burner can wall containing many longitudinal passages to substantially increase cooling efficiency. Finwall's construction, similar to a honeycomb, provides a stronger wall at less weight through the use of thinner materials (Fig. 6-8).

CURRENT BURNER CAN

FINWALL® BURNER CAN

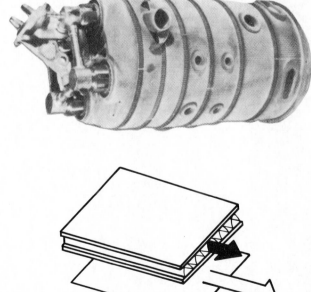

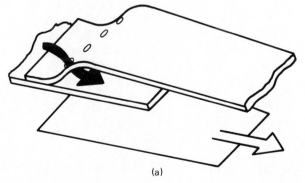

(a)

(b)

FIG. 6-8 **(a)** Conventional, and **(b)** Finwall® burner can construction. (*Pratt & Whitney Aircraft.*)

PERFORMANCE REQUIREMENTS

1. *High combustion efficiency*—This is necessary for long range.
2. *Stable operation*—Freedom from blowout at airflows ranging from idle to maximum power and at pressures representing the aircraft's entire altitude range is essential.
3. *Low-pressure loss*—It is desirable to have as much pressure as possible available in the exhaust nozzle to accelerate the gases rearward. High-pressure losses will reduce thrust and increase specific fuel consumption.
4. *Uniform temperature distribution*—The average temperature of the gases entering the turbine should be as close to the temperature limit of the burner material as possible to obtain maximum engine performance. High local temperatures or hot spots in the gas stream will reduce the allowable average turbine inlet temperature in order to protect the turbine. This will result in a decrease in total gas energy and a corresponding decrease in engine performance (Fig. 6-9).
5. *Easy starting*—Low pressures and high velocities in the burner make starting difficult; therefore, a poorly designed burner will start within only a small range of flight speeds and altitudes, whereas a well-designed burner will permit easier air restarts.
6. *Small size*—A large burner requires a large engine housing with a corresponding increase in the airplane's frontal area and an increase in aerodynamic drag. This will result in a decrease in maximum flight speed. Excessive burner size also results in high engine weight, and for a given aircraft means a lower fuel capacity and payload and shorter range. Modern burners release 500 to 1000 times the heat that a domestic oil burner or heavy industrial furnace of equal unit volume does. Without their high heat release the aircraft gas turbine could not have been made practical.
7. *Low smokey burner*—Smoke is not only annoying on the ground, but may also allow easy tracking of high-flying military aircraft.
8. *Low carbon formation*—Carbon deposits can block critical air passages and disrupt airflow along the liner walls causing high metal temperatures and low burner life.

All of the burner requirements must be satisfied over a wide range of operating conditions. For example, airflows may vary as much as 50:1, fuel flows as much as 30:1, and fuel/air ratios as much as 5:1. Burner pressures may cover a ratio of 100:1, while burner inlet temperatures may vary by more than 700°F [390°C].

EFFECT OF OPERATING VARIABLES ON BURNER PERFORMANCE

The operating variables are:

1. Pressure
2. Inlet air temperature
3. Fuel/air ratio
4. Flow velocity

Combustion Efficiency

As the pressure of the air entering the burner increases, the combustion efficiency rises and levels off to a relatively constant value. The pressure at which this leveling off occurs is usually about 1 atmosphere (atm), but this may vary somewhat with different burner configurations. As the inlet air temperature is increased, combustion efficiency rises until it reaches a value of substantially 100 percent. If the fuel/air ratio is increased, combustion efficiency first rises, then levels off when the mixture in the combustion zone is close to the ideal value, and then decreases as the fuel-air mixture becomes too rich. An increase in fuel/air ratio will result in increased pressure loss because increasing fuel/air ratios cause higher temperatures with a corresponding decrease in gas density. In order to maintain continuous flow, these gases must travel at higher velocities and the energy needed to create these higher velocities must come from an increase in pressure loss. Increasing the flow velocity beyond a certain point reduces combustion efficiency, probably because it reduces the time available for mixing and burning.

Stable Operating Range

The stable operating range of a burner also changes with variations in pressure and flow velocity. As the pressure decreases, the stable operating range becomes narrower until a point is reached below which burning will not take place. As the flow velocity increases, the stable operating range again becomes narrower until a critical velocity is reached, above which combustion will not take place. Increasing the temperature of the incoming charge usually increases the fuel/air ratio range for stable operation. In addition, as the flow velocity is increased, the burner pressure loss will rise. This is mainly due to higher expansion losses as the air flows through the restricting or metering holes in the liner.

Temperature Distribution

The temperature distribution of the burner exit is also affected by changes in the operating variables. Reducing the pressure below a set point tends to upset tempera-

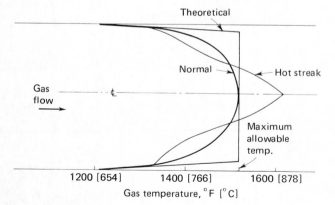

FIG. 6-9 Temperature distribution.

ture uniformity. On the other hand, for a given size burner, more uniform temperatures may be obtained by creating better mixing of the hot and cold gases at the expense of an increase in pressure loss. If the fuel/air ratio and flow velocity are increased, the exit temperatures tend to become less uniform because more heat is released and there is less time for mixing.

Starting

Starting is usually easier with high temperature, high pressure, and low velocity. In addition there is an optimum fuel/air ratio for starting, above or below which ignition of the fuel-air mixture becomes increasingly difficult.

Carbon Deposits

The operating variables have some effect upon the accumulation of carbon deposits in the burner, but their effects may vary with different burner types and configurations. Generally, deposits get worse with increasing temperatures and pressures until a point is reached where they begin to burn off. Increasing the fuel/air ratio has a tendency to increase deposits, probably because the proportion of oxygen in the combustion zone becomes too low to burn the fuel completely. In addition, changes in fuel/air ratios may change the location of carbon deposits within the burner. It should also be noted that the properties of the fuel have a significant effect on carbon accumulation and burner performance, and must be considered in the design of the burner.

Temperature and Cooling Requirements

Changes in the operating variables have a direct bearing on the temperature and cooling requirements of the liner. If the pressure and temperature of the incoming charge are increased, more heat is transferred from the burning gases to the liner, partly by radiation through the insulating blanket of cool air and partly by forced convection, and the lining temperature goes up. If the fuel/air ratio is increased, combustion temperatures become higher, and again the liner temperature goes up, mainly due to increased radiation. On the other hand, an increase in flow velocity outside the liner tends to increase external convection, thereby reducing the temperature of the liner.

INFLUENCE OF DESIGN FACTORS ON BURNER PERFORMANCE

The design factors include:

1. Methods of air distribution
2. Physical dimension of burner
3. Fuel-air operating range
4. Fuel nozzle design

Methods of Air Distribution

Since the quantity of air required for efficient combustion is much less than the total amount pumped through the engine, an important factor in burner design is the correct distribution of air between the combustion zone and the dilution zone. As more of the total airflow is used for combustion, a higher overall fuel/air ratio is needed to maintain maximum efficiency. The manner in which air is introduced into the burner also has a substantial effect on combustion efficiency, therefore the size, number, shape, and location of the air inlet holes has a marked influence on the burner performance.

Physical Dimensions

One method of reducing pressure loss is to increase the diameter or length of the burner. The increase allows more time for the mixing of the hot and cold gases; hence the amount of energy required for mixing which must be supplied by a loss in pressure does not have to be as great. If the burner diameter is made too large, the pressure loss may have to be increased in order to produce adequate mixing and provide sufficient cooling for the added liner surface area. If a greater proportion of air is used to cool the liner, an increase in pressure loss is required since the cooling air filtered in along the liner walls must eventually be mixed with the central high-temperature stream in order to maintain uniform discharge temperatures.

Fuel-air Operating Range

There are several ways in which the fuel/air ratio operating range or blowout limit of the burner can be increased. One is to cut down the flow velocity through the burner by increasing the diameter. Another is to improve fuel atomization and distribution by increasing the pressure drop across the fuel nozzle, or by improving the design of the nozzle-metering elements. Once proper atomization has been established, an increase in pressure drop or further changes in the metering elements will have no appreciable effect. The fuel-air operating range of a burner can also be increased by improving the manner in which combustion air is introduced and distributed. The starting ability of a burner is closely related to its blowout limits, therefore starting can also be improved by increasing the burner diameter, by improving the distribution of combustion air, or by improving the fuel spray pattern, either through a change in fuel nozzle design or an increase in pressure drop across the nozzle. The life of the burner liner depends, to a large extent, upon its operating temperature. This temperature can be lowered by using a larger portion of the total airflow as a convective cooling film and insulating blanket along the liner wall. However, since an increase in liner cooling requires an increase in pressure loss, the quantity of air used for this purpose should be the least amount needed to maintain safe operating temperatures.

Fuel Nozzle Design

Fuel nozzle design plays a major part in burner performance. Not only must the nozzle atomize and dis-

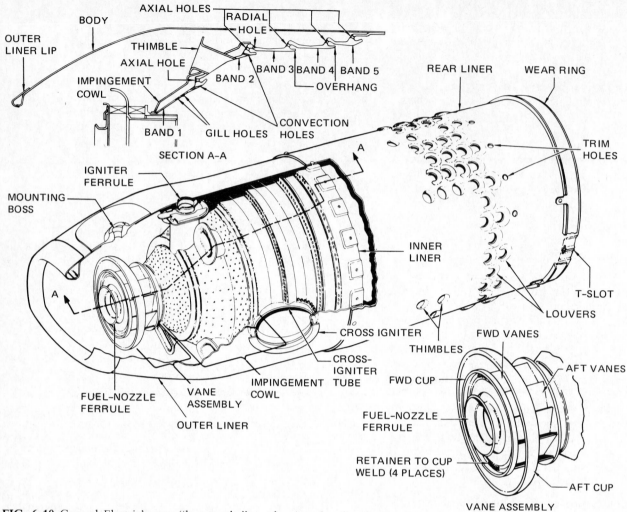

OUTER LINER LIP
BODY
AXIAL HOLES
RADIAL HOLE
THIMBLE
AXIAL HOLE
IMPINGEMENT COWL
BAND 2
BAND 3 BAND 4 BAND 5
OVERHANG
CONVECTION HOLES
BAND 1
GILL HOLES
SECTION A-A

REAR LINER
WEAR RING
TRIM HOLES

IGNITER FERRULE
MOUNTING BOSS
INNER LINER
T-SLOT
LOUVERS
CROSS IGNITER
THIMBLES
CROSS-IGNITER TUBE
FUEL-NOZZLE FERRULE
VANE ASSEMBLY
IMPINGEMENT COWL
OUTER LINER

FWD VANES
FWD CUP
FUEL-NOZZLE FERRULE
AFT VANES
RETAINER TO CUP WELD (4 PLACES)
AFT CUP
VANE ASSEMBLY

FIG. 6-10 General Electric's new "low smoke" combustion chamber for the J79 (see Fig. 2-22).

tribute the fuel, but it must also be able to handle a wide range of fuel flows. For a given fuel system there is a small pressure drop across the nozzle that must be maintained for good atomization, and there is a maximum pressure that a practical fuel pump is able to produce. With a simple swirl-type nozzle like those used in many domestic burners, the range of fuel flows that can be handled within these pressure limitations is usually far short of the engine's requirements. One way of meeting the engine's fuel requirements while still maintaining good atomization is to use a two-stage fuel system. The primary stage functions alone at low fuel flows until the maximum available pressure is reached; then a pressure-sensitive valve is opened and fuel begins to spray through the secondary passages. The total flow in the primary and secondary stages fulfills the engine's fuel requirements without the use of excessive pressures and without a sacrifice in atomization qualities and spray pattern at low fuel flows. (See Chap. 12 on fuel nozzles.)

All of the operating and design variables must be taken into account when the burner is designed and manufactured. The final configuration is, at best, a compromise to achieve the desired operating characteristics, because it is impossible to design and construct a given

burner that will have 100 percent combustion efficiency, zero pressure loss, maximum life, minimum weight, and minimum frontal area, all at the same time.

Combustion chambers are constantly being experimented with in an effort to achieve optimum performance. An example of this effort is General Electric's "low smoke" redesigned combustion chamber for the J79-17C shown in Fig. 6-10.

REVIEW AND STUDY QUESTIONS

1. List the three basic types of combustion chambers. Give the advantages and disadvantages of each type.
2. Describe the gas flow through a typical combustion chamber.
3. How many ignitors are generally used in the combustion section? How is the flame front propagated into the rest of the section?
4. List the requirements for a good combustion chamber.
5. Discuss the effects of the operating variables such as pressure, inlet air temperature, fuel/air ratio, and flow velocity on burner performance.
6. What is the relationship between fuel nozzle design and burner performance?

CHAPTER SEVEN
TURBINES

The function of the turbine is to drive the compressor and accessories, and, in the case of the turboprop, the propeller, by extracting a portion of the pressure and kinetic energy from the high-temperature combustion gases. In a typical jet engine about 75 percent of the power produced internally is used to drive the compressor. What is left is used to produce the necessary thrust. In order to furnish the drive power to compress the air, the turbine must develop as much as 100,000 hp [74,570 kW] or more for the larger jet engines. One blade or bucket of a turbine can extract about 250 hp [186 kW] from the moving gas stream. This is equivalent to the power produced by a typical eight-cylinder automobile engine. It does all of this in a space smaller than the average automobile engine, and with a considerable advantage in weight. See Chaps. 3 and 5 for a detailed discussion of turbine power output. (Although Chap. 5 deals with the power required to drive the compressor, this power is supplied by the turbine. Therefore the same formula which is used to compute compressor power requirements can also be used to determine turbine power output by simply using the temperature drop (ΔT) across the turbine instead of the temperature rise across the compressor.)

EXAMPLE: A small engine is pumping 5 lb [2.27 kg] of air per second. The temperature at the inlet of the turbine is 1700°F [930°C] and at the outlet is 1300°F [700°C]. How much power is the turbine delivering?

$$\begin{aligned} \text{hp} &= \frac{c_p \, \Delta T \, W_a \, 778}{550} \\ &= \frac{0.24 \times 400 \times 5 \times 778}{550} \end{aligned}$$
$$\text{hp} = 679 \; [506 \text{ kW}]$$

TYPES OF TURBINES

With a few exceptions, gas turbine manufacturers have concentrated on the axial-flow turbine (Fig. 7-1), although some manufacturers are building engines with a radial inflow turbine (Fig. 7-2). The radial inflow turbine has the advantage of ruggedness and simplicity, and is relatively inexpensive and easy to manufacture when compared with the axial-flow type. On this type of turbine, inlet gas flows through peripheral nozzles to enter the wheel passages in an inward radial direction. The speeding gas exerts a force on the wheel blades and then exhausts the air in an axial direction to the atmosphere. These turbine wheels used for small engines are well suited for a lower range of specific speeds and work at relatively high efficiency.

The axial-flow turbine is comprised of two main elements consisting of a set of stationary vanes and one or more turbine rotors. The turbine blades themselves are of two basic types, the impulse and the reaction. The modern aircraft gas turbine engine utilizes blades which have both impulse and reaction sections (Fig. 7-3).

The stationary part of the turbine assembly consists of a row of contoured vanes set at an angle to form a series of small nozzles which discharge gases onto the blade of the turbine wheel. For this reason, the stationary vane assembly is usually referred to as the turbine nozzle, and the vanes themselves are called nozzle guide vanes.

FUNCTION OF THE NOZZLE GUIDE VANES (*FIG. 7-4*)

The nozzle guide vanes (diaphragm) have two principal functions. First, they must convert part of the gas heat and pressure energy into dynamic or kinetic energy, so that the gas will strike the turbine blades with some degree of force. Second, the nozzle vanes must turn this gas flow so that it will impinge upon the turbine buckets in the proper direction; that is, the gases must impact on the turbine blade in a direction that will have a large component force in the plane of the rotor. The nozzle does its first job by utilizing the Bernoulli theorem. As through any nozzle, when the flow area is restricted, the gas will accelerate and a large portion of the static pressure in the gas is turned into dynamic pressure. The degree to which this effect will occur depends upon the relationship between the nozzle guide vane inlet and exit areas, which, in turn is closely related to the type of turbine blade used.

The turbine nozzle area is a critical part of engine design. Making the nozzle area too small will restrict the airflow through the engine, raise compressor discharge pressure, and bring the compressor closer to stall. This is especially critical during acceleration when the nozzle will have a tendency to choke (gas flowing at the speed of sound). Many engines are designed to have the nozzle operate in this choked condition. Small exit areas also cause slower accelerations because the compressor will have to work against an increased back pressure. Increasing the nozzle diaphragm area will result in faster engine acceleration, less tendency to stall, but higher specific fuel consumption. The area of the nozzle is adjusted at the factory or during overhaul so that the gas velocity at this point will be at or near the speed of sound. (See Chap. 18, Maintenance and Overhaul Procedures.)

The second function, that of turning the gases so that they strike the turbine blades at the correct angle, is

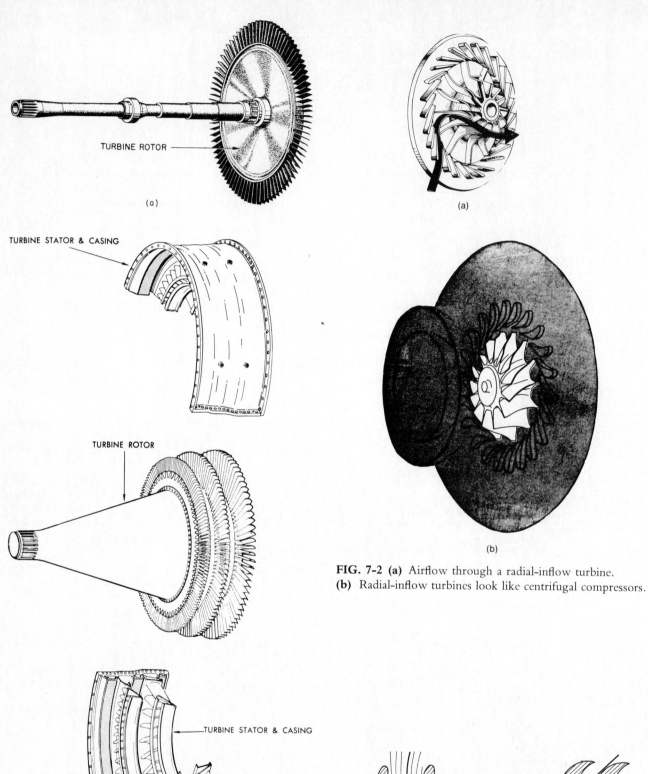

TURBINE ROTOR

(a)

TURBINE STATOR & CASING

TURBINE ROTOR

TURBINE STATOR & CASING

(b)

(a)

(b)

FIG. 7-2 (a) Airflow through a radial-inflow turbine.
(b) Radial-inflow turbines look like centrifugal compressors.

FIG. 7-1 (a) A single-stage axial-flow turbine wheel.
(b) A multistage axial-flow turbine with turbine stators (nozzles). First-stage nozzle is not shown. Blades may be solid or hollow on either nozzle and/or rotor.

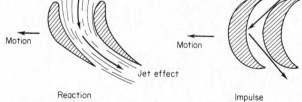

Motion
Jet effect
Reaction

Motion
Impulse

FIG. 7-3 Reaction and impulse turbine blading (tangential velocity vectors omitted).

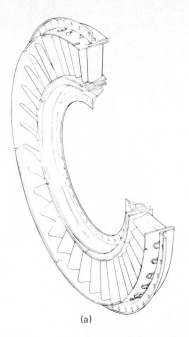

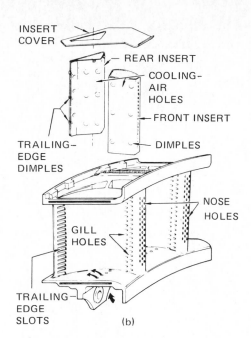

INSERT
COVER

REAR INSERT

COOLING-
AIR
HOLES

FRONT INSERT

DIMPLES

TRAILING-
EDGE
DIMPLES

NOSE
HOLES

GILL
HOLES

TRAILING
EDGE
SLOTS

(a)

(b)

FIG. 7-4 (a) A typical nozzle diaphragm.
(b) In this nozzle diaphragm section both film and impingement cooling techniques
are being used. (See Chap. 10, page 185 for an explanation of these terms.)

accomplished by setting the blades at a specific angle to the axis of the engine. Ideally, this angle should be variable as a function of engine rpm and gas flow velocity, but in practice, the vanes are fixed in one position. It should be noted that the auxiliary power unit (APU) for the DC-10 and several turbine-powered ground vehicles are equipped with variable-angle nozzle vanes.

CONSTRUCTION OF THE NOZZLE

Nozzle vanes may be either cast or forged. Some vanes are made hollow (Fig. 7-5) to allow a degree of

cooling using compressor bleed air. In all cases the nozzle assembly is made of very high temperature, high-strength steel to withstand the direct impact of the hot, high-pressure, high-velocity gas flowing from the combustion chamber.

Several companies are experimenting with transpiration-cooled nozzle and turbine blading in which the air flows through thousands of small holes in a porous airfoil made from a sintered wire mesh material. (Refer to Fig. 10-5.) Since the performance of the gas turbine engine is to a large measure dependent upon the temperature at the inlet of the turbine, increasing the turbine inlet temperature from the present average limit of

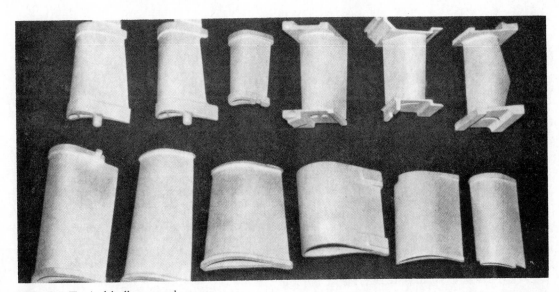

FIG. 7-5 Typical hollow nozzle vanes.

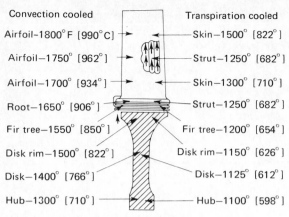

Convection cooled | Transpiration cooled

Airfoil—1800°F [990°C] → | ← Skin—1500° [822°]

Airfoil—1750° [962°] → | ← Strut—1250° [682°]

Airfoil—1700° [934°] → | ← Skin—1300° [710°]

Root—1650° [906°] → | ← Strut—1250° [682°]

Fir tree—1550° [850°] | Fir tree—1200° [654°]

Disk rim—1500° [822°] | Disk rim—1150° [626°]

Disk—1400° [766°] | Disk—1125° [612°]

Hub—1300° [710°] | Hub—1100° [598°]

FIG. 7-6 Typical turbine metal temperatures (°F) [°C].

about 1800°F [982°C] to the 2500°F [1370°C] possible with transpiration cooled blades will result in about a 100 percent increase in specific horsepower. Transpiration cooling may be a promising development in gas turbine design (Fig. 7-6). See Chap. 10, page 179, for a more detailed treatment of material and cooling advancements.

THE IMPULSE TURBINE (*FIG. 7-7*)

A characteristic of an impulse turbine and the nozzle used with it is that gases entering the nozzle diaphragm are expanded to atmospheric pressure. In the ideal impulse turbine, all pressure energy of the gas has been converted into kinetic energy; hence, no further pressure drop can occur across the blades.

Gases enter the nozzle diaphragm in the direction A and leave at a specific velocity indicated by vector V_1 in the turbine gas flow diagram (Fig. 7-7). The speed of rotation of the turbine wheel is represented by the length of vector U.

In order to determine the speed and angle that the entering gases make with the turbine blades, the *relative* velocity must be found. To find the relative velocity it is necessary to subtract one vector from another ($V_1 - U$). Note: This process is just the opposite of a vector addition in which a resultant is found. From the inlet velocity diagram, the relative velocity is found to be V_R. V_R represents the speed and angle of the entering gases as seen by the rotating turbine blades.

Another characteristic of an impulse turbine is that the area of the inlet and exit between the blades is equal. If the area is equal, then the relative velocity V_R at the exit will be equal to the relative velocity V_R at the inlet,

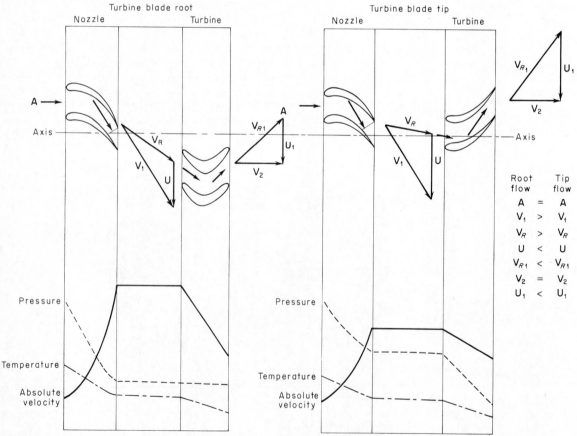

FIG. 7-7 Vector analysis of turbine gas flow. Note: Gases working on turbine = relative velocity, or vector subtraction. Turbine working on gases = resultant velocity or vector addition.

minus friction losses which we will not consider. The descending curve of velocity on the graph is due to the change in direction of V_R as it flows across the blade. A change in velocity may occur in direction and/or magnitude. If the turbine is fixed and unable to rotate, there would be no loss of velocity across the turbine blades. Maximum force would be applied to the blades at this time, but no work will be done because the turbine is not moving.

The speed and direction of the gases at the outlet of the turbine may be determined by a vector addition to find the resultant of V_{R_1} and U. From the gas flow diagram the resultant is found to be V_2. V_2 is less than V_1 due to the rotation of the turbine. The gas gives up some of its kinetic energy to do work on the buckets; that is, the change in momentum of the jet develops a force on the buckets between which the jet passes. The presence of the turbine blade in the path of the gases results in an impulse force being exerted on the gases, which changes the direction of the V_R across the blade. The greater the mass of gas, the faster the gases strike the blades, and the more they are turned, the greater will be their impulse force. Expressed mathematically:

$$\text{Momentum change} = \frac{2W_\Delta V}{g}$$

where $\dfrac{W}{g}$ = mass gas flow, lb/s [kg/s]

$_\Delta V$ = velocity change, ft/s [m/s]

Since the force imparted to the blades is proportional to the momentum change, then

$$\text{Force} = \frac{2W_\Delta V}{g}$$

The impulse force acting on the blade is represented by a vector in Fig. 7-8. From this figure it can be seen that the impulse force does not act directly in the plane of rotation of the turbine wheel, but is resolved into

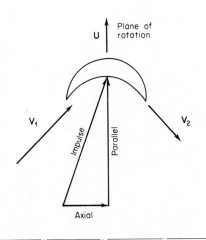

FIG. 7-8 Forces exerted on an impulse blade.

two components. The parallel component acts in the plane of rotation and causes the turbine wheel to rotate. The axial component acts as a thrust along the center line of the shaft and has to be taken up by a thrust bearing.

THE REACTION TURBINE

Again referring to Fig. 7-7, as the gases from the combustion chamber enter the first row of stator blades, they experience a drop in pressure and an increase in velocity through the nozzle, but to a lesser degree than through the nozzle used with the impulse turbine. The gases leave the nozzle at a specific velocity indicated by the vector V_1. The speed of rotation of the turbine wheel is represented by the length of the vector U. From the inlet velocity diagram, the relative velocity is found to be V_R, which is the vector difference of V_1 and U.

On entering the first rotor stage, the gases see the rotor as a convergent passage (outlet area less than inlet area). The change in area produces an increase in the *relative* velocity with an accompanying pressure drop across the blades. The acceleration of the gases generates a reaction force like that produced on a wing. It is from this feature of the reaction turbine that its name is derived. The relative velocity increase is represented by the length of the vector V_{R_1}.

The velocity of the gases at the outlet of the turbine may be determined by the vector addition of the relative velocity of V_{R_1} and the rotational speed U_1 of the turbine wheel. From the outlet velocity diagram, the resultant is found to be V_2. Note: V_2 is less than V_1 which indicates a loss in *absolute* velocity across the blade, but as stated in the preceding paragraph, an increase in *relative* velocity. (The definition of absolute and relative velocity may be viewed as follows: Relative velocity changes assume that the turbine is not turning, whereas absolute velocity changes take into account the rotation of the turbine.)

The presence of the turbine blade in the path of the gases causes a force to be exerted on the gases. The force acting on the gases is represented by the deflecting force vector in Fig. 7-9(a). The deflecting force acts on the gases to change the direction of the relative velocity from V_R to V_{R_1} across the blade. A change in the momentum of the gas flow caused by its change in direction through the rotor blading results in an impulse force also shown in Fig. 7-9(a). Note that a certain amount of impulse force is always present in a reaction-type turbine, but reaction force is not present in an impulse turbine.

The reaction force results from the acceleration of the gases across the blade. The direction in which the reaction force acts may be determined by considering the blade as an airfoil. The reaction force, like lift, may be drawn perpendicular to the relative wind which is represented by V_R in Fig. 7-9(b).

From Fig. 7-9(a) and (b) it can be seen that both impulse and reaction forces are acting on the blade of a reaction turbine. This is why the impulse turbine requires high-velocity gas in order to obtain the maxi-

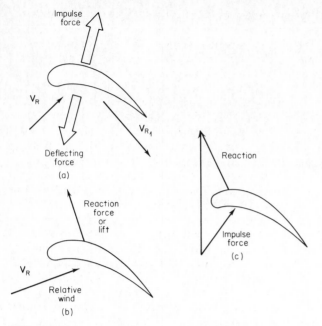

FIG. 7-9 Forces exerted on a reaction blade.

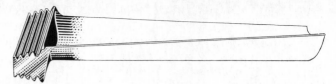

FIG. 7-11 An impulse-reaction blade.

mum rate of momentum change, while the reaction turbine causes its rate of momentum change by the nozzling action of the rotor blading, and therefore does not require excessively high nozzle diaphragm exit velocities. The presence of an impulse and reaction force may then be represented by vectors as shown in Fig. 7-9(c). It can be seen that the two forces combine vectorially into a resultant which acts in the plane of rotation to drive the turbine.

REACTION-IMPULSE TURBINE (FIG. 7-10)

It is important to distribute the power load evenly from the base to the tip of the blade. An uneven distribution of the work load will cause the gases to exit from the blade at different velocities and pressures. Obviously, the blade tips will be traveling faster than

the blade roots (as can be seen by the length of the vector U in Fig. 7-7), because they have a greater distance to travel in their larger circumference. If all the gas velocity possible is made to impinge upon the blade roots, the difference in wheel speed at the roots and the tips will make the relative speed of the gases less at the tips, causing less power to be developed at the tips than at the roots.

To cope with this problem, in actual practice, the turbine blading is a blending of the impulse type at the roots and the reaction type at the tips (Fig. 7-11). Figure 7-10 shows that by making the blade "impulse" at the root and "reaction" at the tip, the blade exit pressure can be held relatively constant. The changing height between the two pressure lines indicates the pressure differential across the blade. From previous discussion it can be seen that the required pressure drop for "reaction" is present at the tip, and gradually changes to the "no pressure loss" condition required for "impulse" at the root. In addition, the higher pressures at the tip will tend to make the gases flow toward the base of the blade which counteracts the centrifugal forces trying to throw the air toward the tip.

Of course, with every change in engine speed and gas-flow velocity, the vector triangle will be shaped considerably differently. The angle of the nozzle and the turbine blades are such that optimum performance is achieved only during a small range of engine rpm.

One can see that if the length of the vector U in Fig. 7-7 is changed (rpm varied), the gases will not enter the turbine in the correct direction and loss of efficiency will occur. In addition, it is desirable to have the gases exit from the turbine with as much of an axial-flow component as possible. Changing rpm will cause V_2 to be angled off the axis and the result will be a swirling motion to the gas with a consequent loss of energy. To counteract the swirling of the gases, straightening vanes are located immediately downstream of the turbine. These vanes also serve the function in many engines of providing one of the main structural components and additionally act as a passageway for oil, air, and other lines (Fig. 7-12).

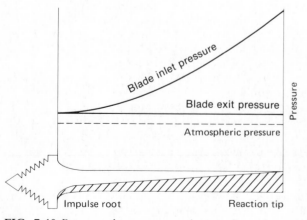

FIG. 7-10 Pressure changes across the impulse and reaction sections of a turbine blade.

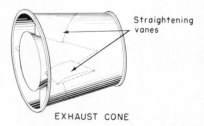

FIG. 7-12 Vanes are used for straightening gas flow and for structural support.

TURBINE CONSTRUCTION

The turbine wheel is one of the most highly stressed parts in the engine. Not only must it operate at temperatures of approximately 1800°F [982°C], but it must do so under severe centrifugal loads imposed by high rotational speeds of over 60,000 rpm for small engines to 8000 rpm for the larger ones. Consequently, the engine speed and turbine inlet temperature must be accurately controlled to keep the turbine within safe operating limits.

The turbine assembly is made of two main parts, the disk and blades. The disk or wheel is a statically and dynamically balanced unit of specially alloyed steel usually containing large percentages of chromium, nickel, and cobalt. After forging, the disk is machined all over and carefully inspected using x-rays, sound waves, and other inspection methods to assure structural integrity. The blades or buckets are attached to the disk by means of a "fir tree" design (Fig. 7-13) to allow for different rates of expansion between the disk and the blade while still holding the blade firmly against centrifugal loads. The blade is kept from moving axially either by rivets, special locking tabs or devices, or another turbine stage.

Some turbine blades are open at the outer perimeter, as shown in Fig. 7-1, whereas in others a shroud is used as in Fig. 7-13. The shroud acts to prevent blade-tip losses and excessive vibration. Distortion under high loads, which tend to twist the blade toward low pitch, is also reduced. The shrouded blade has an aerodynamic advantage in that thinner blade sections can be used and tip losses can be reduced by using a knife edge or labyrinth seal at this point. Shrouding, however, requires that the turbine run cooler or at a reduced rpm because of the extra mass at the tip. On blades that are not shrouded, the tips are cut or recessed to a knife edge to permit a rapid "wearing-in" of the blade tip to the turbine casing with a corresponding increase in turbine efficiency.

Blades are forged from highly alloyed steel and are passed through a carefully controlled series of machining and inspection operations before being certified for use. Many engine manufacturers will stamp a "moment weight" number on the blade to retain rotor balance when replacement is necessary.

The temperature of the blade is usually kept within limits by passing relatively cool air bled from the compressor over the face of the turbine, thus cooling the disk and blade by the process of convection. This method of cooling may become more difficult, as high Mach number flights develop high compressor inlet and outlet temperatures.

A discussion of some newer methods of blade temperature control which include convection, impingement, film and transpiration cooling techniques, and the use of ceramic materials is included in Chap. 10, pages 185–186.

As shown in Fig. 7-1, some gas turbine engines use a single-stage turbine, whereas others employ more than one turbine wheel. Multistage turbines are used where the power required to drive the compressor would necessitate a very large turbine wheel. Multistage wheels are also used for turboprops where the turbine has to extract enough power to drive both the compressor and the propeller. When two or more turbine wheels are used, a nozzle diaphragm is positioned directly in front of each turbine wheel. The operation of the multiple-stage turbine is similar to that of the single stage, except that the succeeding stages operate at lower gas velocities, pressures, and temperatures. Since each turbine stage receives the air at a lower pressure than the preceding stage, more blade area is needed in the rear stages to assure an equitable load distribution between stages. The amount of energy removed from

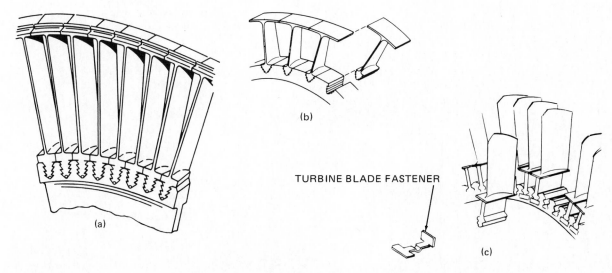

TURBINE BLADE FASTENER

(a)

(b)

(c)

FIG. 7-13 (a) The shroud is formed by the tips of the blades touching each other.
(b) Notice the "fir tree" attachment.
(c) This method of turbine blade construction and attachment isolates the hot blade and allows the disk to run at a cooler temperature.

each stage is proportional to the amount of work done by each stage.

Most multistage turbines are attached to a common shaft. However, some multistage turbine engines have more than one compressor. In this case, some turbine wheels drive one compressor and the remaining turbines drive the other. (See. Fig. 2-42.)

As stated at the beginning of this section, the turbine wheel is subjected to both high speed and high temperature. Because of these extreme conditions, blades can easily deform by growing in length (a condition known as "creep") and by twisting and changing pitch. Since these distortions are accelerated by exceeding engine operating limits, it is important to operate within the temperature and rpm points set by the manufacturer.

REVIEW AND STUDY QUESTIONS

1. What is the function of the turbine?
2. Name two types of turbines. Describe the gas flow in each.
3. What is the function of the nozzle guide vanes? How does the nozzle area influence engine performance?
4. What is meant by impulse and reaction blading? With the use of vectors, show the gas flow through both types of blades.
5. Make a table showing relative and absolute velocity gasflow changes across an impulse and a reaction blade.
6. Describe the physical construction of the axial-flow turbine wheel.
7. What dictates the number of turbine wheels employed in the gas turbine?

CHAPTER EIGHT
EXHAUST
SYSTEMS

EXHAUST DUCTS

The exhaust duct takes the relatively high-pressure (relative to the exhaust nozzle), low-velocity gas leaving the turbine wheel and accelerates this gas flow to sonic or supersonic speeds through the nozzle at its rear. It is desirable in a pure jet engine to convert as much of the pressure energy in the gas into kinetic energy in order to increase the gas momentum and therefore the thrust produced. If a majority of the gas expansion occurs through the turbine section, as for example in a turboprop, the duct does little more than conduct the exhaust stream rearward with a minimum energy loss. However, if the turbine operates against a noticeable back pressure, the nozzle must convert the remaining pressure energy into a high-velocity exhaust. As stated previously, the duct also serves to reduce any swirl in the gas as it leaves the turbine, thereby creating as much of an axial-flow component as possible (Fig. 8-1).

CONSTRUCTION

Basically, the exhaust duct is constructed of two stainless steel cones. The outer cone is usually bolted to the turbine casing with the inner cone supported from the outer cone. The vanes used to support the inner cone straighten the swirling gas flow. Many engines are instrumented at the duct immediately to the rear of the turbine for turbine temperature (Fig. 8-2) and for turbine discharge pressure (see Chap. 18). Although it would be more desirable to measure turbine inlet temperature, most manufacturers prefer to locate the thermocouples at the rear of the turbine since there is no

chance of turbine damage in the event of mechanical failure of the thermocouple; in addition, it is easier to inspect and will have a longer service life. The exhaust gas temperature is an indication of the turbine inlet temperature because the temperature drop across the turbine is calculated when the turbine is designed and the inlet temperature is then indirectly sensed or controlled through the turbine discharge temperature.

The jet nozzle, which on most engines is formed by the converging section or taper at the rear of the outer cone, may be located at the aft end of a tailpipe if the engine is buried in the airplane (Fig. 8-3). The tailpipe will connect the outer cone of the exhaust duct with the nozzle at the rear of the airplane. In order to keep the velocity of the gases low and reduce skin friction losses, tailpipes are made with as large a diameter as possible. The tailpipe is also kept short and constructed with few or no bends in order to reduce pressure-loss effects. If the engine is buried, any bends necessary to duct air in or out of the power plant are made in the inlet duct, since pressure losses are less damaging there than in the exhaust duct.

EXHAUST NOZZLES

Two types of nozzles in use today are the convergent type and the convergent-divergent type. Generally the convergent nozzle will have a fixed area, while the convergent-divergent nozzle area will be variable.

The area of the jet nozzle is critical, since it affects the back pressure on the turbine and hence the rpm, thrust, and exhaust gas temperature. Decreasing the exhaust nozzle area a small amount will sharply increase the exhaust gas temperature, pressure, and velocity, and will also increase thrust. Although rapidly disappearing as a method of nozzle adjustment, on some engines this area is still adjustable, as shown in Fig. 8-3(b), by the insertion of small metal tabs called *mice*. By use of these tabs, the engine can be trimmed to the correct rpm, temperature, and thrust settings.

The Convergent Nozzle

The typical convergent nozzle is designed to maintain a constant internal pressure and still produce sonic velocities at the nozzle exit. In this type of nozzle the gas flow is subsonic as it leaves the turbine. Each individual gas molecule is, in effect, being squeezed by the converging shape and pushed from behind. This three-dimensional squirting action causes the velocity to increase. Since this velocity increase is faster than the

FIG. 8-1 A typical exhaust duct with straightening vanes.

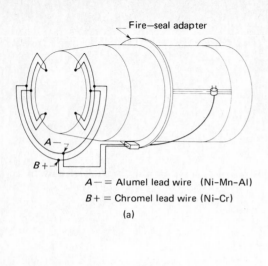

Fire—seal adapter

$A-$ = Alumel lead wire (Ni-Mn-Al)
$B+$ = Chromel lead wire (Ni-Cr)

(a)

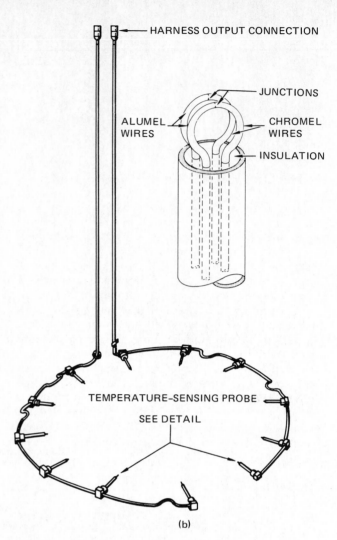

HARNESS OUTPUT CONNECTION

JUNCTIONS

ALUMEL WIRES

CHROMEL WIRES

INSULATION

TEMPERATURE-SENSING PROBE

SEE DETAIL

(b)

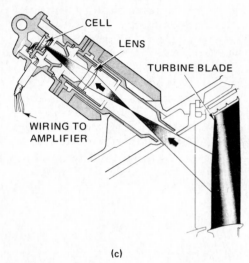

CELL

LENS

TURBINE BLADE

WIRING TO AMPLIFIER

(c)

FIG. 8-2 **(a)** Thermocouples, located in the exhaust duct, are all hooked in parallel so that several may fail without losing the temperature indication. Many thermocouples are used to obtain an average reading.

(b) This system actually has 24 thermocouples for safety. Each sensing probe is a double thermocouple.

(c) A radiation pyrometer is a device used for measuring turbine blade temperature by converting radiated energy into electrical energy. The pyrometer consists of a photovoltaic cell, sensitive to radiation over a band in the infrared region of the spectrum, and a lens system to focus the radiation onto the cell. The pyrometer is positioned on the nozzle casing so that the lens system can be focused, through a sighting tube, directly onto the turbine blades. The radiated energy emitted by the hot blades is converted to electrical energy by the photovoltaic cell and is then transmitted to a combined amplifier/indicating instrument that is then calibrated in degrees Celsius.

volume expansion, a converging area is necessary to maintain the pressure or squirting action. In the convergent nozzle, the gas velocity cannot exceed the speed of sound because, as the gas velocity increases, the ability of the gas pressure to move the molecules from behind becomes less. In fact, the pushing action will drop to zero when the gas moves at the speed of sound. The speed of sound is the speed of a natural pressure wave movement. It is dependent upon the natural internal molecular velocity which is limited by the amount of internal temperature energy of these gas molecules. In other words, the speed of sound, although a pressure wave, is limited by the molecular velocity (or sound-temperature energy).

The Convergent-divergent Nozzle

If the pressure at the entrance to a convergent duct becomes approximately twice that at the exhaust nozzle, the change in velocity through the duct will be enough

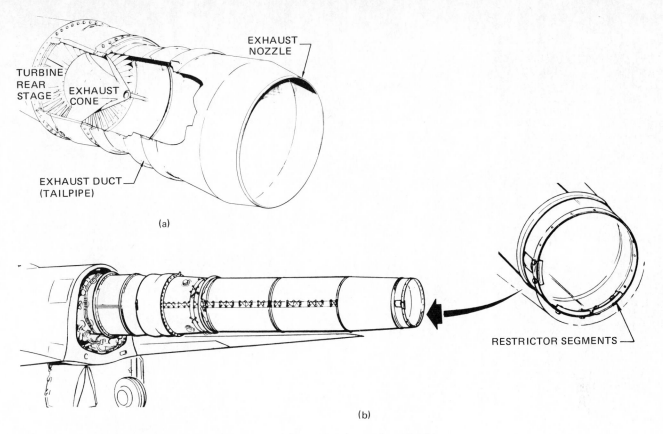

TURBINE
REAR
STAGE

EXHAUST
CONE

EXHAUST
NOZZLE

EXHAUST DUCT
(TAILPIPE)

(a)

RESTRICTOR SEGMENTS

(b)

FIG. 8-3 (a) and **(b)** A buried engine requires a tailpipe or exhaust duct. In **(b)** nozzle adjustments to a fixed area nozzle can be made by using restrictor segments called "mice."

to cause sonic velocity at the nozzle. At high Mach numbers the pressure ratio across the duct will become greater than 2.0, and unless this pressure can be turned into velocity before the gases exit from the nozzle, a loss of efficiency will occur. Since the maximum velocity that a gas can attain in a convergent nozzle is the speed of sound, a convergent-divergent nozzle must be used. In the diverging section, the gas velocities can be increased above the speed of sound. Since the individual gas molecules cannot be pushed by the pressure or molecules behind them, the gas molecules can be accelerated only by increasing the gas volume outward and rearward. The diverging section of the convergent-divergent nozzle allows expansion outward, but also holds in the expansion so that most of the expansion is directed rearward off the side wall of the diverging section. In other words, the diverging action accelerates the airflow to supersonic velocities by controlling the expansion of the gas so that the expansion (which is only partially completed in the converging section) will be rearward and not outward to the side and wasted.

An example of the action that produces an increase in thrust through a diverging nozzle can be shown with the following experiment. If a greased rubber ball is pushed down into a funnel and then released, the ball would shoot out of the funnel (Fig. 8-4). If only the funnel were released, it would move away from the ball. What is happening is that the ball is partially com-

pressed when it is pushed down into the funnel, increasing the pressure of the air inside the ball. When the funnel is released, the air in the ball expands, returning the ball to its normal size and pushing the funnel away. This same type of action occurs in the diverging section of a converging-diverging nozzle. As the gases expand against the side of the duct, they produce a pushing effect even though they are decreasing in pressure.

To sum up then, supersonic, or compressible, flow through a diverging duct will result in increasing ve-

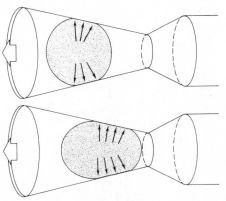

FIG. 8-4 The ball analogy showing thrust increase by means of a divergent nozzle.

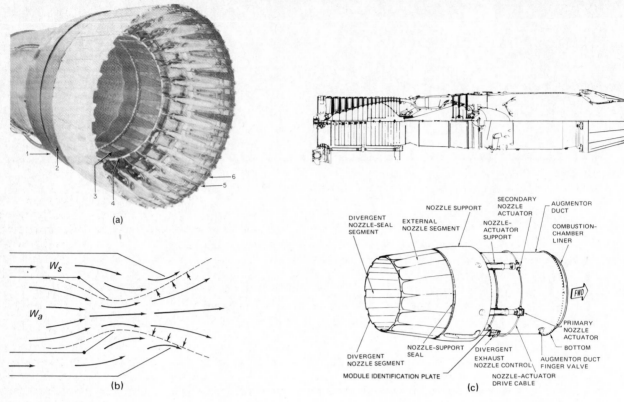

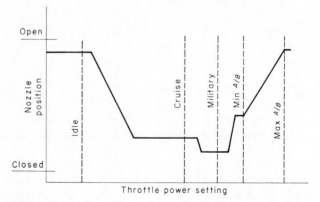

FIG. 8-5 (a) The variable-area exhaust nozzle used on some General Electric J79 engines.

(b) A convergent-divergent nozzle airflow. The diverging section on some C-D nozzles is formed by a wall of air, while on others, the diverging air is guided by the walls of the nozzle itself, as shown in part (c) of this figure.

(c) The diverging section of this later model C-D nozzle is formed by the walls of the nozzle. Compare this to part (b) of this figure. (See Fig. 2-22.)

locity above the speed of sound, decreasing pressure, and decreasing density (increasing volume).

Unfortunately the rate of change in the divergent duct may be too large or small for a given engine operating condition to allow a smooth supersonic flow. That is, the rate of increase in area of the duct is not correct for the increase in the rate of change in volume of the gases. If the rate of increase is too small, maximum gas velocities will not be reached. If the rate of increase is too great, turbulent flow along the nozzle wall will occur. To correct this problem, and for use with an afterburner, all engines capable of supersonic flight incorporate a variable-area exhaust nozzle (Fig. 8-5). Older engines utilized a two-position eyelid or "clamshell" type of variable-area orifice, but modern jet nozzles are infinitely variable aerodynamic converging-diverging ejector types utilizing primary and secondary nozzle flaps to control the main and secondary airflows. As shown in Fig. 8-5(a), the converging section is formed by a series of overlapping primary flaps through which the air is accelerated to sonic velocity. The diverging section may be formed by a secondary airflow instead of a metal wall as shown in Fig. 8-5(b). The ejector action produced by the main airflow shooting out of the primary nozzle or nozzle flap draws the secondary flow into the outer shroud and ejects it out the rear end.

The ejected secondary air which comes from the outside of the engine will be accelerated to a higher velocity and will therefore contribute to jet thrust. The nozzle and shroud flap segments are linked to open and close together.

In addition to forming an aerodynamically correct shape for supersonic flow, modern variable-area exhaust nozzles can be used to advantage to improve engine performance in other ways. For example, by keep-

FIG. 8-6 The nozzle schedule can be designed so that the nozzle is open in both idle and afterburner regimes.

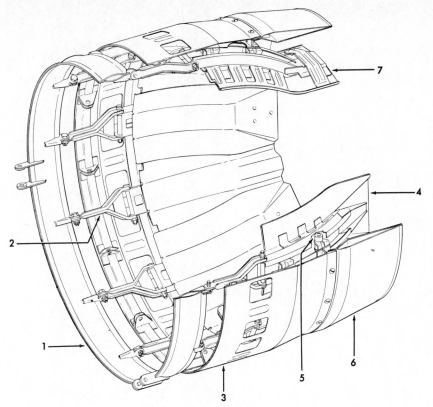

FIG. 8-7 The variable-area exhaust nozzle for the General Electric J85.

1 ACTUATOR RING
2 LINK
3 HOUSING
4 INNER LEAF
5 ROLLER
6 HOUSING EXTENSION
7 OUTER LEAF

ing the nozzle wide open, the engine idle speed can be held high with a minimum of thrust produced (Fig. 8-6). This is beneficial since maximum available thrust can then be made available quickly by merely closing the nozzle, thereby eliminating the necessity of having to accelerate the engine all the way up from a low-idle rpm. High-idle rpm will also provide a higher quantity of bleed air for operation of accessories, and in addition, make "go-arounds" safer. Of course, variable-area nozzles incur a penalty in the form of additional weight and complexity, as indicated in Fig. 8-7, and at the present time are being used only on aircraft operating at sufficiently high Mach numbers to make their use worthwhile.

SOUND SUPPRESSION

THE NOISE PROBLEM

The noise problem created by commercial and military jet takeoffs, landings, and ground operations at airports near residential areas has become a very serious problem within the last several years. Figure 8-8 illustrates the several levels of sound, in decibels, from various sources. The *decibel* (dB) is defined as being approximately the smallest degree of difference of loudness ordinarily detectable by the human ear, the range of which includes about 130 dB.

The pattern of sound from a jet engine makes the noise problem even more bothersome than that coming from other types of engines. For example, the noise from a reciprocating engine rises sharply as the airplane propeller passes an observer on the ground, and then drops off almost as quickly. But as shown in Fig. 8-9, a jet reaches a peak after the aircraft passes, and is at an angle of approximately 45° to the observer. This noise then stays at a relatively high level for a considerable length of time. The noise from a turbojet is also more annoying because it overlaps the ordinary speech frequencies more than the noise from a reciprocating engine and propeller combination (Fig. 8-10).

Since the noise is produced by the high-velocity exhaust gas shearing through the still air (see page 156–162), it follows that if the exhaust velocity is slower and the mixing area wider, the exhaust noise levels can be brought down to the point where a sound suppressor is not necessary. The exhaust-gas velocity of a turbofan is slower than a turbojet of comparable size because more energy must be removed by the turbine to drive the fan. The fan exhaust velocity is relatively low and creates less of a noise problem. Noise levels are also lower in the high-bypass-ratio turbofan engine through the elimination of the inlet guide vanes (see Figs. 2-13, 2-19, 2-25, 2-26, 2-27, 2-35, 2-44, 2-51, and 2-52) and the resulting reduction of the "siren" effect. The noise generated by this effect occurs when the columns of air created by the compressor inlet guide vanes are cut by the rapidly moving compressor blades, generating high-frequency pressure fluctuations. Further noise reductions are achieved by lining the fan shroud with acoustical materials [see Fig. 8-17(b)], thus dampening the

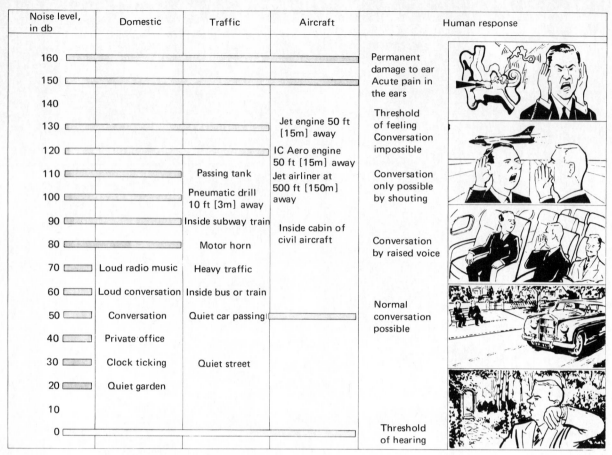

Noise level, in db	Domestic	Traffic	Aircraft	Human response
160				Permanent damage to ear
150				Acute pain in the ears
140				
130			Jet engine 50 ft [15m] away	Threshold of feeling / Conversation impossible
120			IC Aero engine 50 ft [15m] away	
110		Passing tank	Jet airliner at 500 ft [150m] away	Conversation only possible by shouting
100		Pneumatic drill 10 ft [3m] away		
90		Inside subway train	Inside cabin of civil aircraft	
80		Motor horn		Conversation by raised voice
70	Loud radio music	Heavy traffic		
60	Loud conversation	Inside bus or train		Normal conversation possible
50	Conversation	Quiet car passing		
40	Private office			
30	Clock ticking	Quiet street		
20	Quiet garden			
10				
0				Threshold of hearing

FIG. 8-8 Comparison of the level of sound from various sources.

pressure fluctuations, and by spacing the outlet guide vanes farther away from the compressor. For these reasons, fan engines in general do not need sound suppressors.

The function of the noise suppressor is to lower the level of the sound, about 25 to 30 dB, as well as to change its frequency (Fig. 8-11), and to do this with a minimum sacrifice in engine thrust or additional weight.

The two facets of the noise problem, ground operation and airborne operation, lend themselves to two solutions. Noise suppressors can be portable devices for use on the ground by maintenance personnel, or they can be an integral part of the aircraft engine installation. Examples of various types of ground and airborne suppressors can be seen in Figs. 8-12 and 8-13. Of the two, airborne suppressors are more difficult to design because of the weight limitations and the necessity of having the air exit in an axial direction to the engine.

THE SOURCE OF SOUND

Jet engine noise is mainly the result of the turbulence produced when the hot, high-velocity exhaust jet mixes with the cold, low-velocity or static ambient atmosphere around the exhaust. The turbulence increases in proportion to the speed of the exhaust stream and produces noise of varying intensity and frequency until mixing is completed (Fig. 8-14). Since there is little mixing close to the nozzle, the fine grain turbulence produces a relatively high frequency sound in this area. But as the jet stream slows down, more mixing takes place, resulting in a coarser turbulence and correspondingly lower frequency. Therefore the noise produced from a jet engine exhaust is a "white" noise (an analogy to white light) consisting of a mixture of all frequencies with intensity peaks at certain frequencies due to the characteristic note of the turbine wheel, which makes an excellent siren (Fig. 8-15). At low power settings, for example, during landing, noise generated from the compressor blades may become predominant. The noise emanating from the compressor consists of more discrete higher frequencies. High-frequency sounds are attenuated (weakened) more rapidly by obstructions and distance, and are more directional in nature, whereas low-frequency sounds will travel much farther. Therefore compressor noise is less of a problem than noise from the jet exhaust stream.

THEORY OF OPERATION

The high-intensity sound generated by the shearing action of the "solid" jet exhaust in the relatively still air can be reduced and modified by breaking up this "solid" air mass by mechanical means and thus make the mixing

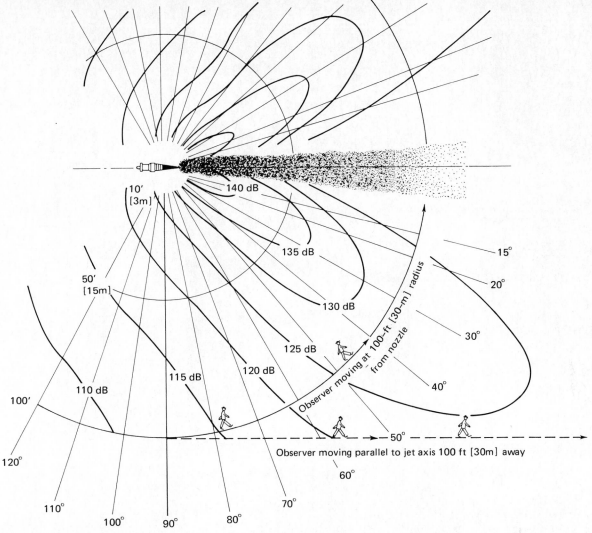

FIG. 8-9 Typical noise field from a jet spreading in still air. The curved lines represent equal sound levels.

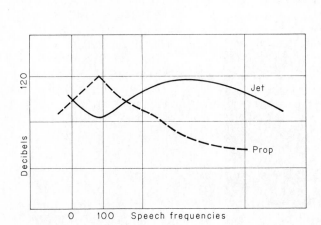

FIG. 8-10 The jet engine produces its maximum noise in the speech frequency.

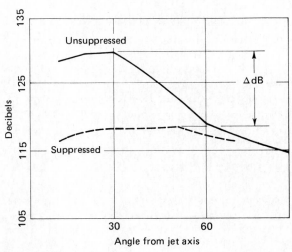

FIG. 8-11 The sound level intensities are reduced by means of a noise suppressor.

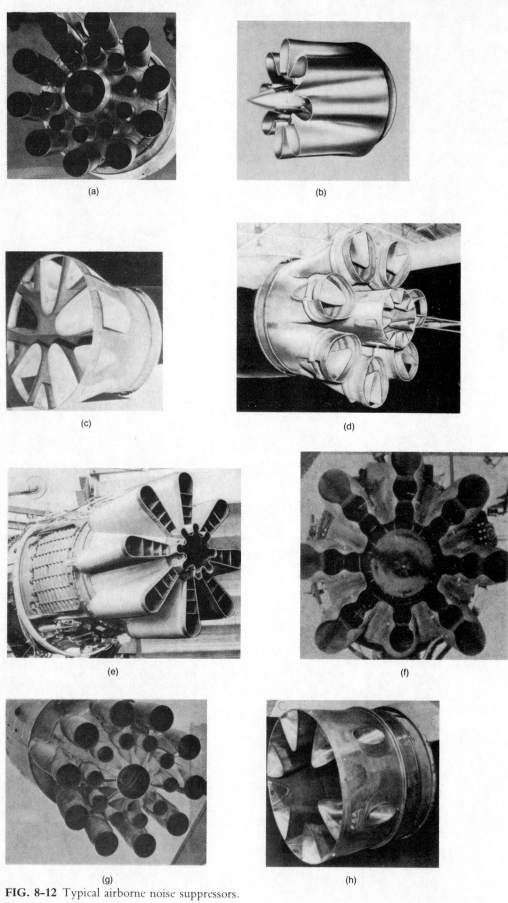

FIG. 8-12 Typical airborne noise suppressors.

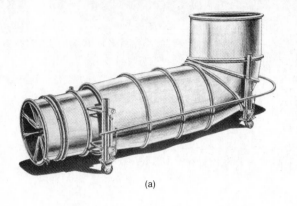

(a)

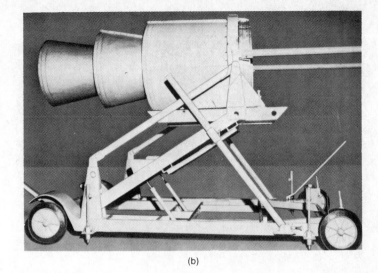

(b)

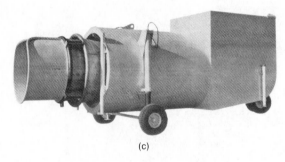

(c)

30'–6'' [930 cm]

15'–6'' [457 cm]

7'–0'' D [213 cm]

TENSION BAR
ATTACHMENT
TO ENGINE

SUPPRESSOR INLET
SHIELD

CABLE
ATTACHMENT

7'–0'' D.
[213 cm]

8'–0''
[244 cm]

6'–0'' D
[182 cm]

4'–2'' [127 cm]
to   to
9'–7'' [292 cm]

ALTERNATE
ATTACHMENT
TO GROUND

PANTOGRAPH
STEERING

IAC THRUST TRAILER ALLOW-
ING VERTICAL, HORIZONTAL,
ANGULAR & ROTATIONAL
MODES OF ADJUSTMENT
EST. WT. 2500 LB [1125 kg]

UNIVERSAL UDAC NOISE
SUPPRESSOR DESIGNED
FOR USE WITH JT-3, JT-4
CJ-805, ROLLS-ROYCE,
CONWAY & AVON ENGINES
EST. WT. 1300 LB [585 kg]

ADD-ON VERTICAL EXHAUST
SILENCER (CAN BE EASILY
ATTACHED OR DETACHED
EST. WT. 6200 LB [2790 kg]

15'–0'' [457 cm]

TOTAL EST. WT. 10,000 LB [4500 kg]

(d)

FIG. 8-13 Various types of ground noise suppressors.
(a) A ground noise suppressor with water ring for cooling. *(Air Logistics Corporation.)*
(b), (c), *and* **(d)** Suppressors made by Industrial Acoustics Company, Inc.
(Continued on the following page.)

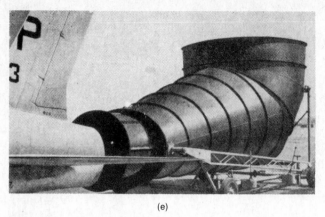

(e)

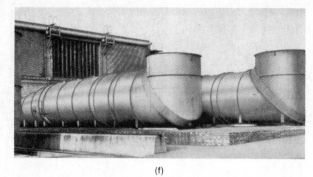

(f)

FIG. 8-13 (cont.)

(e) Another suppressor made by Industrial Acoustics Company, Inc.

(f) Large ground noise suppressors are used with test cells.

process more gentle. Most airborne noise suppressors are of the "corrugated perimeter" design. This type of silencer alters the shape of the exhaust stream as it leaves the engine and permits air to be entrained with the gases which, in turn, encourages more rapid mixing. Examples of the corrugated perimeter can be seen in Fig. 8-12(b), (e), and (f). The *multitube suppressor* [Fig. 8-12(a) and (g)] breaks the primary exhaust stream up into a number of smaller jets, each with its own discharge nozzle which is surrounded by ambient air.

Some suppressor designs [Fig. 8-12(c), (d), and (h)] admit a secondary flow of air into the exhaust through the corrugations. In this manner the mixing process is promoted even more quickly.

Deep corrugations give greater noise reduction. But if the corrugations are too deep, excessive drag will result because the overall diameter may have to be increased in order to maintain the required nozzle area. Also, a nozzle designed to give a large reduction in noise may involve a considerable weight penalty because of the additional strengthening required.

CONSTRUCTION

Noise suppressors are generally built of welded stainless steel sheet stock and are of relatively simple construction. Overhaul life should be about the same as the engine's, but since the suppressor is exposed to the high-temperature gas flow, periodic inspection must be made for cracks and other signs of impending failure.

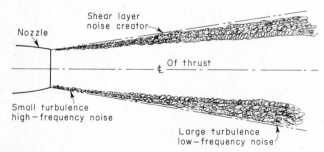

FIG. 8-14 The shear-layer noise source.

PROTECTION AGAINST SOUND

Although the airborne sound suppressor is installed principally for the comfort of those people living in residential areas surrounding the municipal airport, mechanics and others required to work within several hundred feet of operating engines must have some form of ear protection. Not only can noise permanently damage hearing, but at the same time can reduce a person's efficiency by causing fatigue. It also affects both physical and psychological well being and may result in increasing job errors. Susceptibility to temporary or permanent hearing loss varies with the individual and the quality of the noise. Such factors as frequency, intensity, and the time of exposure to the noise all point up the need for protection. For these reasons, earplugs (Fig. 8-16) and/or ear protectors [Fig. 8-17(a)] are required pieces of equipment for all personnel who must be in the proximity of running jet engines (Fig. 8-18).

A general rule of thumb to follow for determining noise levels with the distance from the source is that as

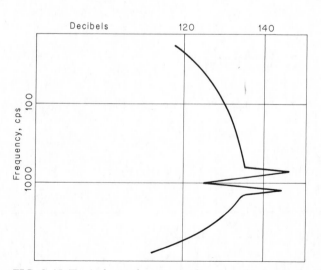

FIG. 8-15 Typical sound spectrum showing a high noise level over a wide frquency range. The two peaks are fundamental notes of the turbine wheel acting as a siren.

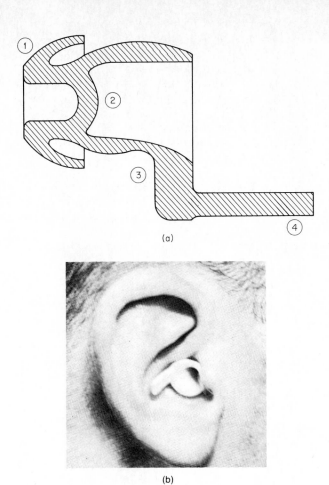

(a)

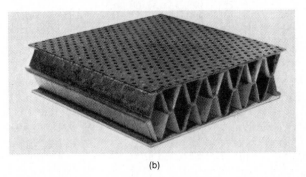

(b)

FIG. 8-16 (a) Ear insert. (1) Conforms to the shape of the ear canal. (2) Diaphragm permits wearer to hear normal conversation. (3) Prevents plug from being inserted too far into the ear. (4) Tab for removal and insertion.
(b) Placement of ear inserts. *(American Optical Co.)*

FIG. 8-17 Two methods of dealing with noise.
(a) Ear protectors with fluid-filled ear cushions deal with sound at the receiving end. *(Wilson Products.)*
(b) Acoustical treatment of the engine (in this case, the fan cowl for the General Electric CF6 engine) provides a good broadband absorption characteristic, and thus deals with sound at the origin.

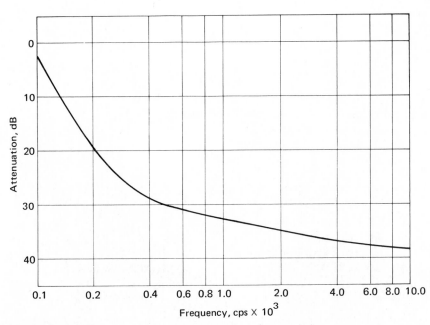

FIG. 8-18 Muff-type ear protector attenuation characteristics.

distance is doubled, the noise level decreases 6 dB, and conversely, as distance is halved, the noise level increases 6 dB. Noise levels of 140 dB are not uncommon close to a running engine. Protection is required from approximately 100 dB on up, but is difficult to declare a fixed safe distance from an operating engine, since the sound levels vary depending upon the angle of the observer in relation to the engine. Refer to Fig. 8-9.

THRUST REVERSERS

A jet-powered aircraft, during its landing run, lacks the braking action afforded by slow-turning propellers, which on larger aircraft are capable of going into reverse pitch, thus giving reverse thrust. The problem is further compounded by the higher landing speeds due to the highly streamlined, low-drag fuselage and the heavier gross weights common to modern jet airplanes. Standard wheel brakes are no longer adequate under these adverse conditions, and larger brakes would incur a severe weight and space penalty and decrease the useful load of the aircraft. In addition, brakes can be very ineffective on wet or icy runways.

Many solutions to this problem have been advanced. One method, used extensively by the Air Force, is a "drag chute," or "parabrake" (Fig. 8-19). The parachute stowed at the rear of the fuselage is deployed upon landing, or in some cases, while the airplane is still airborne to hasten emergency descents. The parabrake does not lend itself to commercial operation, since the chute is easily damaged, must be repacked after each use, and is not absolutely dependable. In addition, once deployed, the pilot has no control over the drag on the aircraft except to release the chute.

Experiments in the use of large nets for stopping aircraft and in the use of arresting gear similar to that used in the Navy for carrier landings, are being performed at some airports. These methods cause fast deceleration which could result in excessive stresses being applied to the airframe and in discomfort to the passengers, and are therefore not suitable for commercial operations at busy airports except for use as an emergency overrun barrier. A simple but effective method of slowing aircraft quickly after landing is to reverse the direction of engine thrust by reversing the exhaust gas stream. Devices in use today allow the pilot to control the degree of reverse thrust. In addition, the modern thrust reverser can be used for emergency descents and to slow the aircraft and vary the rate of sink during approaches while still keeping rpm high to minimize the time needed to accelerate the engine in the event of a "go-around." High-idle rpm will also provide sufficient compressor bleed air for the proper performance of air-driven accessories. The use of the thrust reverser is also being evaluated as a method of improving air combat maneuvering.

TYPES OF THRUST REVERSERS

The two basic types of thrust reversers (Fig. 8-20) are:

1. Postexit or target type
2. Preexit using cascades or blocker/deflector doors

Postexit reversing is accomplished simply by placing an obstruction in the jet exhaust stream about one nozzle diameter to the rear of the engine (Fig. 8-21). The gas stream may be deflected in either a horizontal or vertical direction, depending upon the engine's placement on the airframe.

In the preexit type shown in Figs. 8-22 and 8-23 the gases are turned forward by means of doors which are normally stowed or airfoils which are normally blocked during forward thrust operation (Figs. 8-22 and 8-23). During reverse thrust, doors are moved so that they now block the exhaust gas stream. The gas now exits and is directed in a forward direction through turning vanes or by deflector doors. Figure 8-24 shows that on some aircraft the thrust reverser and the noise suppressor are combined into one integrated unit.

THRUST REVERSER DESIGNS AND SYSTEMS

A good thrust reverser should:

1. Be mechanically strong and constructed of high-temperature metals to take the full force of the high-velocity jet, and at the same time turn this jet stream through a large angle.

FIG. 8-19 Drag parachute deployed to shorten the landing roll of a McDonnell Douglas F4H.

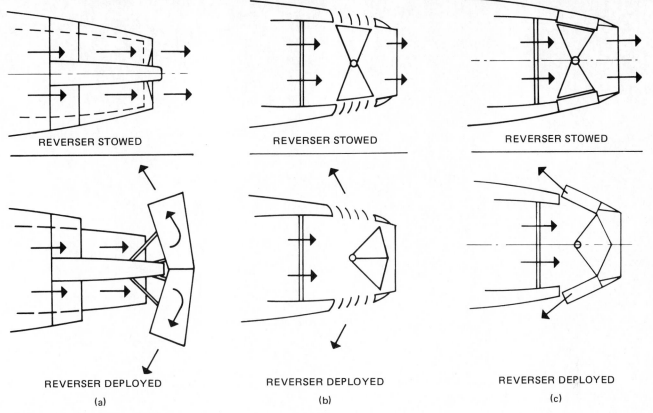

FIG. 8-20 Types of thrust reversers in use.

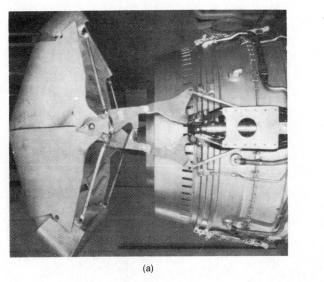

(a)

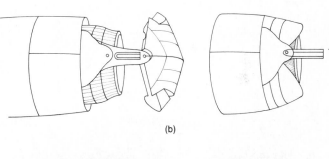

(b)

FIG. 8-21 Photographs and drawings of the target, or post-exit type of thrust reverser used to reverse both primary and secondary (fan) streams on the General Electric CJ805-23 aft-fan engine.

(c)

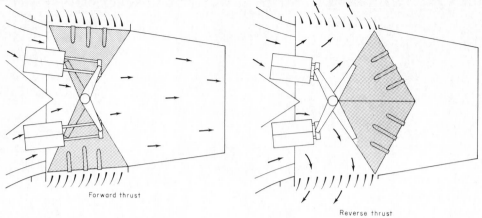

Forward thrust

Reverse thrust

FIG. 8-22 Schematic of the preexit type of thrust reverser.

2. Not affect the basic operation of the engine, whether the reverser is in operation or not.
3. Provide approximately 50 percent of the full forward thrust.
4. Operate with a high standard of fail-safe characteristics.
5. Not increase drag by increasing engine and nacelle frontal area.
6. Cause little increased maintenance problems.
7. Not add an excessive weight penalty.
8. Not cause the reingestion of the gas stream into the compressor nor cause the gas stream to impinge upon the airframe. That is, the discharge pattern must be correctly established by the placement and shape of the target or vane cascade.

9. Allow the pilot complete control of the amount of reverse thrust.
10. Not affect the aerodynamic characteristics of the airplane adversely.

Naturally the designer cannot incorporate all of these requirements to the fullest extent. In order to incorporate some of the features listed above, thrust reverser systems can become somewhat complicated. Typical of the systems in use today are the ones used on the Pratt & Whitney JT3D engine and the General Electric CJ805-23 engine, shown in Figs. 8-25 and 8-26, respectively.

Of the two, the Pratt & Whitney system is the more complicated in that there are actually three separate re-

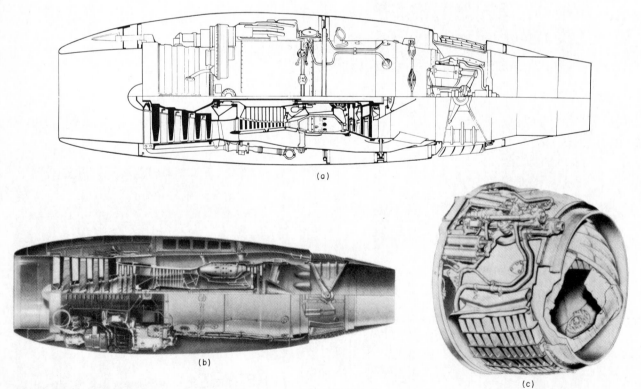

(a)

(b)

(c)

FIG. 8-23 (a), (b), *and* **(c)** Preexit reversers showing the use of "cascade" turning vanes. (Continued on following page.)

164 CONSTRUCTION AND DESIGN

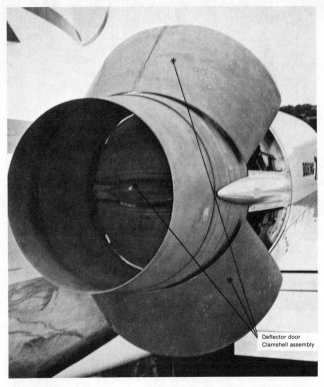

FIG. 8-23 (cont.)

(d) The Boeing 727 and 737 use deflector doors to direct the gases forward, as opposed to cascade vanes.

FIG. 8-24 A combination suppressor and reverser in the General Electric CJ805-3 engine, installed in the Convair 880.

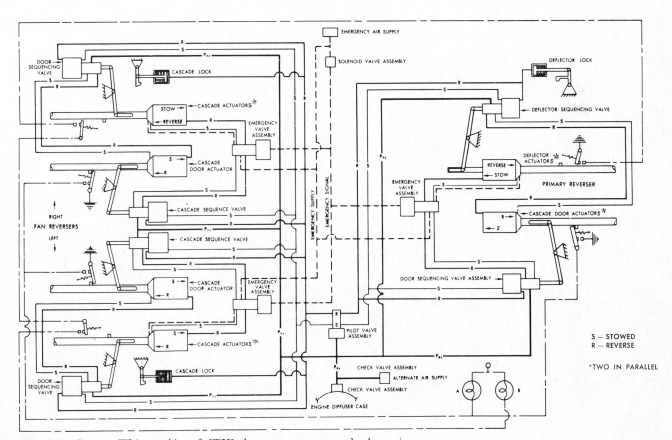

FIG. 8-25 Pratt & Whitney Aircraft JT3D thrust-reverser control schematic.

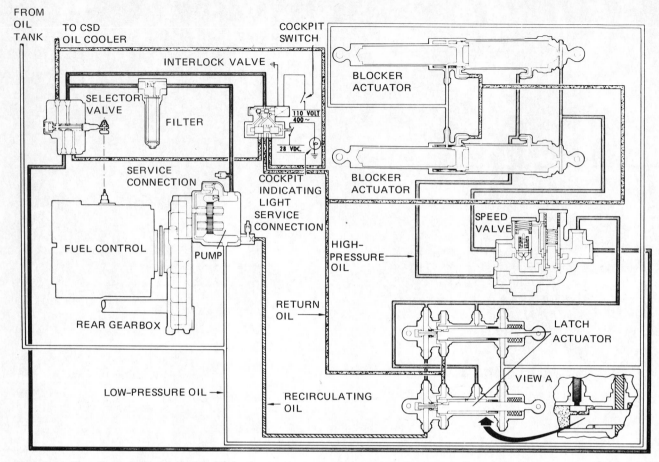

FIG. 8-26 The General Electric CJ805-23 thrust-reverser control schematic.

versers (Fig. 8-27). There is a system for the primary stream and one for each of the bifurcated fan ducts. Actuation of this reverser system is by means of compressor bleed air, while actuation of the General Electric system is by means of oil pressure. In some systems, when the pilot moves the power lever past the idle detent, the reverser system operates, while in others, thrust reverse operation is accomplished with a separate lever. The amount of forward or reverse thrust desired may then be controlled by the position of the power or thrust-reverser lever. It should be noted that forward-fan engines designed with mixed-flow exhaust systems (see Chap. 2) require only one reverser, and that on high-bypass-ratio turbofans it is more important to reverse the flow of the secondary (fan) air than the primary (core) air, since most of the thrust comes from the fan airflow. (See Figs. 8-27, 8-28 and 8-29).

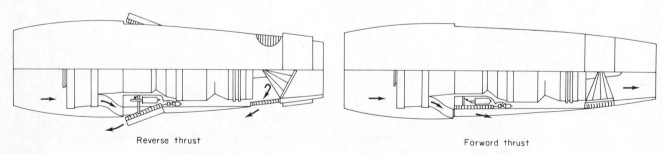

Reverse thrust   Forward thrust

FIG. 8-27 The Pratt & Whitney Aircraft JT3D fan engine requires a separate reverser for primary and secondary airflows, as do several other engines, including the Pratt & Whitney Aircraft JT9D engine shown in Fig. 8-28, and the General Electric CF6 series engine shown in Fig. 8-29.

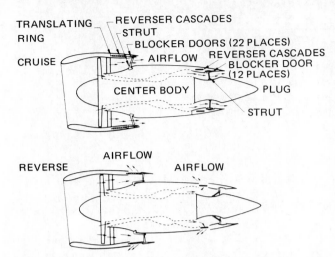

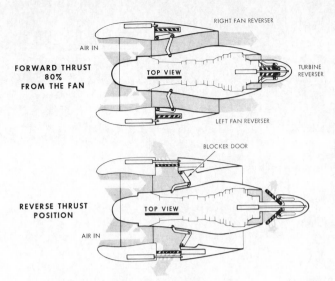

FIG. 8-28 The Boeing 747 nacelle employs a dual annualar nozzle concept. Therefore a separate nozzle and reverser is provided for the fan section discharge and for the turbine section discharge.

FIG. 8-29 The General Electric CF6-6 thrust reverser in the stowed and deployed positions uses both pre- and post-exit types.

REVIEW AND STUDY QUESTIONS

1. What is the function of the exhaust duct? How does the type of engine influence duct design?
2. Discuss the physical construction of the exhaust duct.
3. Where are the thermocouples located? Why?
4. Describe the operating principles of the convergent and convergent-divergent exhaust nozzles.
5. Why is it best to incorporate a variable geometry exhaust nozzle on an engine equipped with an afterburner?
6. Describe the sound pattern provided by a jet engine. By a reciprocating engine.

CHAPTER NINE
METHODS OF
THRUST
AUGMENTATION

WATER INJECTION

In Chap. 3 we learned that one of the more important determiners of the power output of any gas turbine is the weight of airflow passing through the engine. A reduction in atmospheric pressure due to increasing altitude or temperature will therefore cause a reduction in thrust or shaft horsepower.

Power under these circumstances can be restored or even boosted as much as 10 to 30 percent for takeoff by the use of water or water-alcohol injection (Fig. 9-1). Engine power ratings during the period water injection is used are called "wet thrust" ratings as opposed to "dry thrust" ratings when water injection is off.

The alcohol adds to the power by providing an additional source of fuel, but because the alcohol has a low combustion efficiency, being only about half that of gas

turbine fuel, and because the alcohol does not pass through the central part of the combustion chamber where temperatures are high enough to efficiently burn the weak alcohol-air mixture, the power added is small. Water alone would provide more thrust per pound than a water-alcohol mixture due to the high latent heat of vaporization and the overall decrease in temperature. The addition of alcohol has two other effects. If water only is injected, it would reduce the turbine inlet temperature, but with the addition of alcohol, the turbine temperature is restored. Thus the power is restored without having to adjust the fuel flow. The alcohol also serves to lower the freezing point of the water.

The water provides additional thrust in one of two ways, depending upon where the water is added. Some engines have the coolant sprayed directly into the compressor inlet, whereas others have fluid added at the

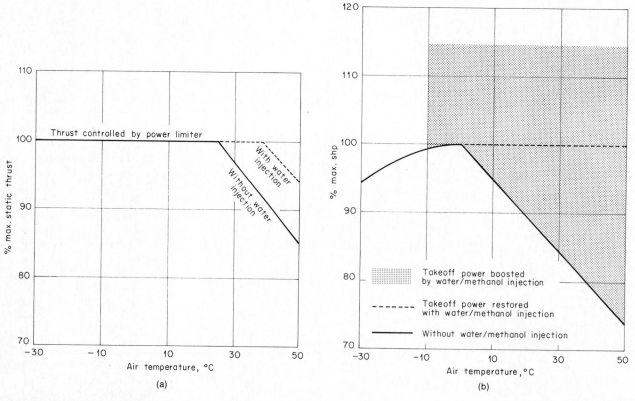

FIG. 9-1 The effect of water injection on a turbojet and turboprop engine.
(a) Turbojet thrust increase with water injection.
(b) Turboprop power with and without water injection.

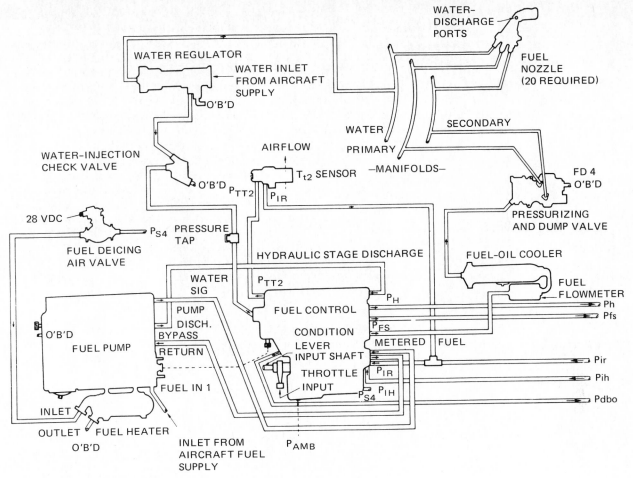

FIG. 9-2 On the Pratt & Whitney Aircraft JT9D, as on most other engines, the fuel and water injection systems are integrated. In this system, water will not be supplied unless the power lever and burner pressure are at appropriate settings. Water will also reset the fuel control for a different fuel flow.

diffuser. Figure 9-2 shows a system where the water is added at the fuel nozzles. The system described at the end of this section injects water at both points. (See Fig. 13-3).

When water is added at the front of the compressor, power augmentation is obtained principally by the vaporizing liquid cooling the air. This increases the density and mass airflow. Furthermore, if water only is used, the cooler, increased airflow to the combustion chamber permits more fuel to be burned before the turbine temperature limits are reached. As shown in Chap. 3, higher turbine temperatures will result in increased thrust.

Water added to the diffuser increases the mass flow through the turbine relative to that through the compressor. This results in a decreased temperature and pressure drop across the turbine which leads to an increased pressure at the exhaust nozzle. Again, the reduction in turbine temperature when water alone is used allows the fuel system to schedule an increased fuel flow, providing additional thrust.

In both cases water is the fluid used because its high heat of vaporization results in a fairly large amount of cooling for a given weight of water flow. Demineral-

ized water is generally used to prevent deposit buildup on compressor blades which will lead to deterioration of thrust requiring more frequent "field cleaning" of the compressor and engine trimming. (See Chap. 19.)

Generally water/air ratios are in the order of 1 to 5 lb [0.45 to 2.25 kg] of water to 100 lb [45 kg] air. For example, Fig. 9-2 shows a schematic of the water injection system used on the B-52G and earlier models. Later models of this airplane use the fan engine and there is no provision for water injection. The water tank holds approximately 1200 gallons (gal) [4542 liters (L)] which is usually exhausted during takeoff. About 110 s are required to consume all of the liquid. Any water not used during takeoff is drained overboard.

OPERATION (FIG. 9-3)

The first step in the operating procedure is to place the system control switch in the ON position. This supplies electrical power to energize the tank-mounted water-injection boost pumps and to light up the low-pressure warning lights on the control panel to indicate that the engines are *not* receiving water. As the tank

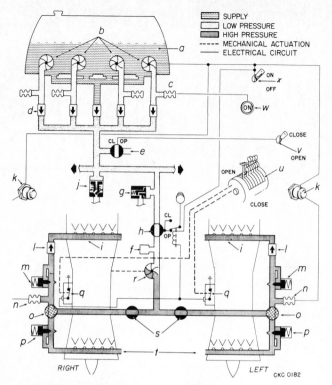

SUPPLY
LOW PRESSURE
HIGH PRESSURE
---- MECHANICAL ACTUATION
—— ELECTRICAL CIRCUIT

RIGHT LEFT CKC 0182

a	WATER-INJECTION TANK		REGULATOR
b	TANK BOOST PUMPS	n	PRESSURE SWITCHES
c	BOOST-PUMP PRESSURE-INDICATING SWITCHES	o	WATER STRAINERS
		p	LOW-PRESSURE WATER REGULATORS
d	BOOST-PUMP OUTLET CHECK VALVES	q	MICROSWITCH ON FUEL CONTROL UNIT
e	TANK DRAIN VALVE	r	ENGINE-DRIVEN WATER PUMP
f	SURGE CHAMBER	s	MANUALLY OPERATED WATER SHUTOFF VALVE
g	DRAIN VALVE		
h	MOTOR-OPERATED WATER SHUTOFF VALVE	t	ENGINE-INLET SPRAY RINGS
i	DIFFUSER-CASE NOZZLE RINGS	u	THROTTLE LEVERS
j	SIPHON-BREAK VALVE	v	DRAIN SWITCH
k	LOW-PRESSURE WARNING LIGHTS	w	TANK-BOOST-PUMP PRESSURE INDICATOR
l	CHECK VALVES	x	SYSTEM CONTROL SWITCH
m	HIGH-PRESSURE WATER		

FIG. 9-3 Water injection system for one model of the B52.

boost pumps deliver water to the system, the tank boost pump indicators on the control panel move from the OFF to the ON position.

When the throttle is advanced to approximately 86 percent rpm (this percentage of rpm varies on different engines), a microswitch in the fuel control unit (Fig. 9-3) is actuated. This supplies power to open the motor-operated shutoff valve, thus permitting water to reach the engine-driven water pump. The water pump boosts the water pressure to 385 to 440 psi [2654 to 3034 kPa], with a rated flow of 160 gal/min [605.6 L/min]. At the water-pressure regulators, the water pump output is divided and sent to the inlet spray ring and to the diffuser case nozzle ring.

Approximately one-third of the water passes through the low-pressure regulator into the inlet spray ring, and two-thirds of the water passes through the high-pres-

sure regulator into the diffuser case nozzle rings. When the water-pressure regulators receive water from the water pump, the low-pressure warning lights go out, indicating that the engines are receiving water. The water pressure is also directed from the water-pressure regulator to the fuel control unit to reposition the control unit maximum speed limit. This will permit an increased fuel flow during the period of water injection.

As stated previously, the water supply is sufficient to allow 110 s of continuous maximum takeoff rated (wet) thrust. When the water pressure is lowered by depletion or by placing the system control switch in the OFF position, the water drain valves open automatically and permit the water *remaining in the system* to drain overboard. The drain switch on the water-injection control panel is placed in the OPEN position, and this permits draining of the *water tank*. To ensure proper drainage of the water tank, the eight amber low-pressure warning lights will remain illuminated until the drain switch is moved to the OPEN position, even though the system control switch is OFF and the system is depleted.

AFTERBURNING (*FIG. 9-4*)

Afterburning or reheating is one method of periodically augmenting the basic thrust of the turbojet, and more recently, the turbofan engine, without having to use a larger engine with its concurrent penalties of increased frontal area, weight, and fuel consumption. The afterburner, whose operation is much like a ramjet, increases thrust by adding fuel to the exhaust gases after they have passed through the turbine section. At this point there is still much uncombined oxygen in the exhaust. (See Chap. 6.) The resultant increase in the temperature raises the velocity of the exiting gases and therefore boosts engine thrust. Most afterburners will produce an approximate 50 percent thrust increase, but with a corresponding threefold increase in fuel flow. Since the specific and actual fuel consumption is considerably higher during the time the engine is in afterburning or "hot" operation, as compared to the non-afterburning or "cold" mode of operation, reheating is used only for the time-limited operation of takeoff, climb, and maximum bursts of speed. Afterburning rather than water injection as a method of gaining ad-

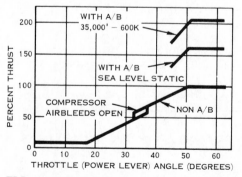

FIG. 9-4 Typical thrust augmentation due to afterburning. (*Pratt & Whitney Aircraft.*)

ditional thrust is used extensively in, but not limited to, fighter aircraft because of the higher thrust augmentation ratios possible.

Late-model engines utilizing an afterburner are the Pratt & Whitney Aircraft TF30 augmented turbofan engine (Fig. 9-5) that is being used in the General Dynamics Corporation "Swing Wing" F-111, Grumman F14, and the Olympus 593 engine used on the Concorde (see Figs. 2-45 and 2-49). Investigations have indicated the desirability of using this reheating method in combination with the turbofan engine by heating the duct or fan air to high temperatures.

All engines which incorporate an afterburner must, of necessity, also be equipped with a variable-area exhaust nozzle in order to provide for proper operation under afterburning and nonafterburning conditions. The nozzle is closed during nonafterburning operation, but when afterburning is selected, the nozzle is automatically opened to provide an exit area suitable for the increased volume of the gas stream. This prevents any increase in back pressure from occurring which would slow the airflow through the engine and affect the compressor's stall characteristics. A well-designed afterburner and variable-area exhaust nozzle will not influence the operation of the basic turbojet engine.

Specific requirements for a reheat augmentation device are:

1. *Large temperature rise*—The afterburner does not have the physical and temperature limits of the turbine. The temperature rise is limited mainly by the amount of air that is available.
2. *Low dry loss*—The engine does suffer a very slight penalty in thrust during "cold" operation due principally to the restriction caused by the flame holders and fuel spray bars.
3. *Wide temperature modulation*—This is necessary to obtain "degrees" of afterburning for better control of thrust.

CONSTRUCTION

The typical afterburner consists of the following components (Fig. 9-6):

1. Engine- or turbine-driven afterburner fuel pump
2. Afterburner fuel control
3. Pressurizing valve—if multistage operation is possible
4. Spray nozzles or spraybars
5. Torch ignitor and/or ignition system
6. Flame holders
7. Variable-area exhaust nozzle
8. Connections (mechanical and pressure) from main fuel control, throttle, and engine
9. Screech liner

A detailed examination of an actual afterburner system is made at the end of this chapter, and in Chap. 21.

OPERATION

The gases enter the afterburner at the approximate temperature, pressure, and velocity of 1022°F, 40 psi, and 2000 ft/s [550°C, 276 kPa, and 610 m/s], respectively, and leave at about 2912°F, 40 psi, and 3000 ft/s [1600°C, 276 kPa, and 914 m/s], respectively. These values can vary widely with different engines, nozzle configurations, and operating conditions. The duct area to the rear of the turbine is larger than a normal exhaust duct would be in order to obtain a reduced-velocity gas stream, and thus reduce gas friction losses. This reduced velocity is still too high for stable combustion to take place, since the flame propagation rate of kerosene is only a few feet per second. It becomes necessary to use a form of flame stabilizer or holder located downstream of the fuel spraybars to provide a region in which tur-

FIVE-ZONE, FULLY VARIABLE AFTERBURNER AUGMENTATION SYSTEM

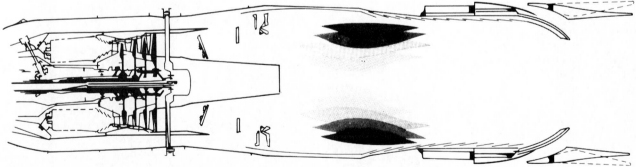

FIG. 9-5 One of the newer types of afterburners is used on the Pratt & Whitney Aircraft TF30-P-100 engine. This system utilizes a multizone afterburner fuel system which provides smooth transient thrust increases from the minimum afterburner thrust level to maximum. The five-zone, fully variable afterburner augmentation system utilizes a 4 joule (See Chap. 16) electrical ignition design in place of either hot streak or torch ignition, thus reducing pressure excursions during initial light-off by 30 to 40 percent. Notice the translating primary iris nozzle combined with an aerodynamically actuated blow-in ejector. This arrangement provides an increase in aircraft subsonic operating range through a reduction in base drag. Drag is reduced by the smaller "boat-tail" angle of the iris nozzle.

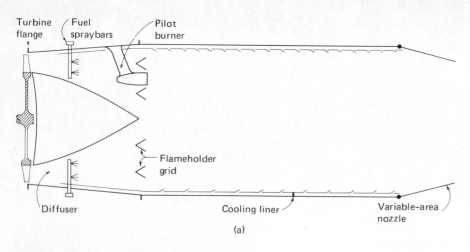

Turbine flange
Fuel spraybars
Pilot burner
Flameholder grid
Diffuser
Cooling liner
Variable-area nozzle

(a)

(b)

FIG. 9-6 (a) Simple afterburner schematic.
(b) The Pratt & Whitney Aircraft TF30-P-100 afterburner showing the fuel manifold and flameholder.

bulent eddies are formed, and where the local gas velocity is further reduced. Fuel is fed into the afterburner through a series of nozzles or spraybars. In some engines the afterburner is either on or off, while in others, degrees of afterburning are available. Ignition occurs in one of several ways.

1. *Hot streak ignition*—In this system an extra quantity of fuel is injected into one of the combustion chambers. The resulting streak of hot gases ignites the afterburner fuel.
2. *Torch ignition*—A "pilot light" located in the area of the spraybars is fed fuel and ignited with its own ignition system. The system works continuously during afterburner operation.

These systems are utilized because spontaneous ignition of the afterburner fuel cannot be depended upon, especially at high altitudes where the atmospheric pressure is low.

A screech or antihowl liner fits into the inner wall of the duct. The liner is generally corrugated and perfo-rated with thousands of small holes. The liner prevents extreme high frequency and amplitude pressure fluctuations resulting from combustion instability or the unsteady release of heat energy. Screech results in excessive noise, vibration, heat transfer rates, and temperatures which will cause rapid physical destruction of the afterburner components. The screech liner tends to absorb and dampen these pressure fluctuations.

The flame holder mentioned above usually takes the form of several concentric rings with a V cross-sectional shape.

THRUST INCREASE (*FIG. 9-7*)

For a constant pressure ratio, the amount of thrust increase, in terms of percent, due to afterburning, is directly related to the ratio of the exhaust gas temperature before and after the afterburner. For example, if the gas temperatures before and after the afterburner are 1140°F (1600°R) [615°C (888°Kelvin, °K)] and 3040°F (3500°R) [1670°C (1940°K)], respectively,

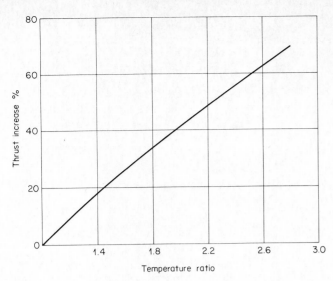

FIG. 9-7 Thrust increase versus temperature ratio increase.

$$\text{Temperature ratio} = \frac{3500}{1600} = 2.19$$

But since the velocity of the jet stream increases as the square root of the temperature, then

$$\sqrt{2.19} = 1.48$$

or a jet-stream velocity and thrust increase of 48 percent at sea level static conditions.

$$
\begin{aligned}
\text{Percent increase} &= \left(\sqrt{\frac{3500}{1600}} - 1 \right)(100) \\
&= (\sqrt{2.19} - 1) \quad (100) \\
&= (1.48 - 1) \quad (100) \\
&= (0.48) \quad (100) \\
\text{Percent increase} &= 48
\end{aligned}
$$

NOTE: The generalized formula for finding the percentage difference between two numbers is:

$$\Delta \text{ percent} = \left(\frac{x}{y} - 1 \right)(100)$$

or

$$\Delta \text{ percent} = \left(\frac{x - y}{y} \right)(100)$$

On a "net thrust" basis the advantage increases directly with increases in airplane speed. For example, this same engine-airplane combination with the same cycle temperature would realize a net percentage augmentation of 85 percent at Mach 1, and 130 percent at Mach 2.

There are also small effects on "wet" thrust due to changes in total F/F weight across the nozzle, in total pressure, and in the specific heat of air as the temperature increases.

J57 AFTERBURNER SYSTEM *(FIG. 9-8)*

This afterburner is composed of the afterburner diffuser and the afterburner duct and exhaust nozzle assembly. The exhaust nozzle assembly is variable and is operated by pneumatic actuating cylinders moved by compressor bleed air, which is metered by the exhaust nozzle control valve. During normal engine operation, the cylinders hold the nozzle iris or flaps in the CLOSED position. When afterburning occurs, the cylinders open the nozzle to permit the less restricted passage of exhaust gases.

Afterburning operation is normally controlled by a switch installed in the throttle quadrant. This switch is actuated when the throttle lever is moved outboard while the engine is operating above the 80 percent range. The switch connects an electrical circuit to a 28-volt (V) direct current (dc) afterburner actuator motor shown in Fig. 9-8, mounted on the fuel transfer valve body (l). This causes the fuel shuttle valve (m) to open the fuel ports, first in the afterburner exhaust nozzle actuator control, and then in the afterburner ignition fuel valve.

The fuel shuttle valve (m) directs fuel from the afterburner stage of the fuel pump (r) into the afterburner fuel control. Metered fuel from the afterburner fuel control enters the manifold (b) and is atomized for burning by the spray nozzles mounted in the afterburner diffuser. A mechanical afterburner fuel shutoff valve (j) is installed on the engine, in conjunction with the throttle linkage, to prevent fuel flow from going to the afterburner until approximately 80 percent of engine power has been reached. Field adjustment of any of the afterburner fuel regulation and control components should not be attempted. The units must be set with the use of proper flow bench facilities.

The order of the fuel flow through the units of the system is as follows:

The engine and afterburner fuel pump assembly consists of two gear-type pumps and one impeller mounted on the same gear shaft. Fuel routed to the fuel pump inlet (p) enters the throat of the impeller, which discharges fuel under boosted pressure through the pump inlet screen(s). The filtered fuel (t) then passes through the inlet side of the afterburner and engine stages of the pump assembly. Relief valves (u) located in the discharge side of the gear pumps relieve excess fuel pressure back to the inlet side of the fuel pumps. A connecting point is provided on the outlet side of the engine fuel pump for connecting the low-fuel-pressure warning system (o). The afterburner fuel pump (r) is a part of the engine-driven fuel pump assembly, which has a gear stage for each system. The inlet of both stages is fed by a common centrifugal boost pump (q). An automatic emergency transfer valve is incorporated in the pump housing for the purpose of diverting fuel flow from the afterburner stage of the pump to the main fuel system if the main fuel pump fails.

The afterburner fuel shuttle valve and actuator unit (m) is incorporated as a part of the fuel-pump transfer valve assembly to control the flow of fuel to the afterburner fuel system. The actuator, a 28-V dc motor that

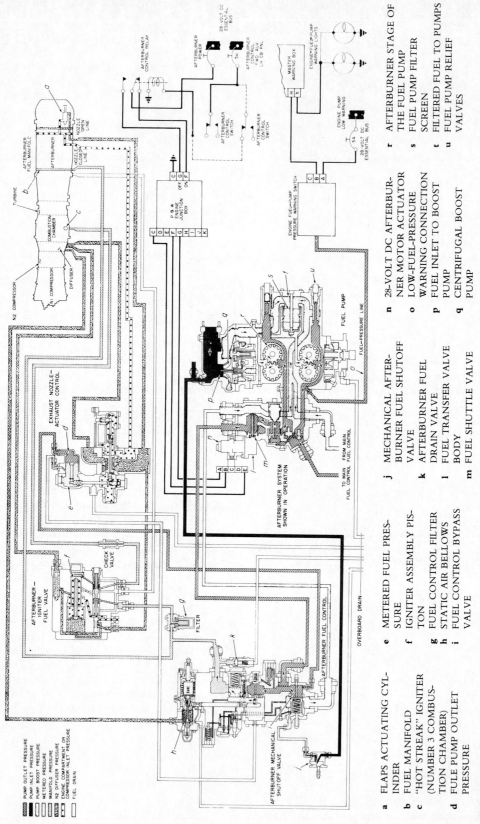

a FLAPS ACTUATING CYL-
INDER
b FUEL MANIFOLD
c "HOT STREAK" IGNITER
(NUMBER 3 COMBUS-
TION CHAMBER)
d FULE PUMP OUTLET
PRESSURE

e METERED FUEL PRES-
SURE
f IGNITER ASSEMBLY PIS-
TON
g FUEL CONTROL FILTER
h STATIC AIR BELLOWS
i FUEL CONTROL BYPASS
VALVE

j MECHANICAL AFTER-
BURNER FUEL SHUTOFF
VALVE
k AFTERBURNER FUEL
DRAIN VALVE
l FUEL TRANSFER VALVE
BODY
m FUEL SHUTTLE VALVE

n 28-VOLT DC AFTERBUR-
NER MOTOR ACTUATOR
o LOW-FUEL-PRESSURE
WARNING CONNECTION
p FUEL INLET TO BOOST
PUMP
q CENTRIFUGAL BOOST
PUMP

r AFTERBURNER STAGE OF
THE FUEL PUMP
s FUEL PUMP FILTER
SCREEN
t FILTERED FUEL TO PUMPS
u FUEL PUMP RELIEF
VALVES

FIG. 9-8 Pratt & Whitney Aircraft J57 afterburner schematic.

opens and closes the valve, is controlled by the after-burner switch located in the cockpit throttle quadrant.

The afterburner fuel control is installed on the right-hand side of the engine at the engine wasp-waist section. The control is provided to meter fuel for use during afterburner operation. Fuel metering is accomplished by an internal mechanical linkage that adjusts the metering valve opening. The internal mechanical linkage is actuated by a static air bellows (h) which extends and retracts with variations of N_2 compressor discharge pressure.

Fuel metering is also affected by spring-loaded fuel pressure valves within the control. Control inlet fuel is routed to the antispring side of the fuel control bypass valve (i). This valve maintains a constant-pressure head across the metering valve. Most of the fuel is then routed through the control filter (g), but some passes through a damping restriction to the spring side of the valve. This permits the spring load on the valve to determine the pressure differential between metered and unmetered fuel.

Ignition for the afterburning system is provided by a "hot streak" type of system. The igniter assembly contains a cylinder and piston (f), which discharges a quantity of fuel through the igniter injection nozzle into the number 3 burner can (combustion chamber, c). The fuel ignites in the burner and travels aft to ignite fuel being supplied by the afterburner discharge manifold (b). The igniter control operates on a combination of fuel pressure, spring tension, and N_2 compressor discharge pressure. At the time of afterburner actuation, the fuel piston (f) is actuated and injects fuel into the burner. The piston then remains in the actuated position until afterburning is terminated and then returns to its preoperational position.

Some of the J57 engine afterburning systems are equipped with an afterburner recirculating igniter control, a modification of the "hot streak" afterburner igniter system. The modification results in continuous fuel circulation through the igniter fuel chamber, thereby reducing igniter coking. Because the fuel chamber is full at all times, a continuous leakage of fuel will occur from the chamber rather than only during afterburning as on the earlier configuration of the igniter. A check valve is installed in the exhaust nozzle actuator control to the recirculating igniter tube assembly, and another check valve is installed in the diffuser case to the recirculating igniter tube assembly. The two valves prevent static fuel from leaking to the bottom of the diffuser case and the exhaust nozzle actuator control assembly.

The afterburner exhaust nozzle actuator control is a spring- and fuel-pressure-actuated component that directs air pressure to the actuating cylinder to open or close the afterburner exhaust nozzle. Metered fuel pressure (e) and fuel pump outlet pressure (d) are the two pressures used to actuate this component. The control assembly is equipped with a relay valve, which is spring-loaded to the exhaust nozzle CLOSED position. At the time of afterburner actuation, afterburner fuel pressure is directed to the exhaust nozzle open end of the relay valve. This repositions the valve and routes air pressure to open the exhaust nozzle. The control assembly is equipped with two flapper valves which vent exhaust air from the nozzle actuator pneumatic system.

Multiple exhaust nozzle actuating cylinders (a) are installed around the engine tailpipe to actuate the iris-type or flap-type exhaust nozzle shutters. The shutters (or flaps) are installed on the aft end of the afterburner to increase or decrease the exhaust nozzle opening. The nozzle must be closed when the afterburner is not in operation to prevent excessive loss of engine thrust. The nozzle must be open during afterburner operation to prevent excessive engine pressures and temperatures.

The actuating cylinders are operated by 16th-stage air pressure from the engine and are directed by the exhaust nozzle actuator control. Operation of the actuator is begun when the pilot moves the throttle into afterburner range, completing the electrical control circuit to the afterburner fuel shutoff valve. The afterburner fuel control governs fuel pressure and then actuates the nozzle actuator control valve, which in turn directs air pressure to the nozzle actuators. The nozzle actuators are protected from excessive heat radiation by a metal insulating blanket installed between the actuators and the afterburner duct.

An afterburner drain valve (k) is installed in the line between the afterburner regulator and the fuel-injection manifold. This valve is closed by fuel pressure during afterburner operation and is opened by spring pressure to drain the manifold when the afterburner is not in use.

THE GENERAL ELECTRIC J85 AFTERBURNER SYSTEM (FIG. 9-9)

In this particular system, the afterburner fuel and nozzle control is designed to maintain the proper fuel flow to the engine's afterburner section as a function of power lever angle and compressor discharge pressure; and the proper exhaust nozzle opening as a function of power lever angle, compressor inlet temperature, and turbine discharge temperature. Late model engines incorporate air-cooled variable exhaust nozzle (VEN) actuators to improve the service life of this part. The system consists of the afterburner fuel pump and shutoff valve, the afterburner fuel and nozzle control, the fuel manifold drain valve, the main and pilot burner spraybars, and the turbine-discharge-temperature sensing system.

Operation (FIG. 9-10)

Fuel enters the afterburner pump through the pump shutoff valve. The valve opens when the power lever is placed in the afterburner ON position, the engine speed is 100 percent, and the main fuel acceleration valve is closed. The pump is an engine-driven centrifugal type, and can handle fuel flows in excess of 10,000 lb/h [4536 kg/h].

From the pump, the fuel flows to the afterburner fuel

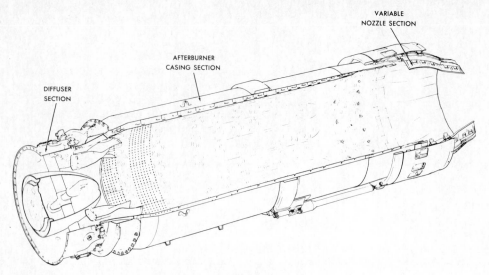

DIFFUSER
SECTION

AFTERBURNER
CASING SECTION

VARIABLE
NOZZLE SECTION

FIG. 9-9 Sectioned view of the General Electric J85 afterburner.

and nozzle control, which is mounted directly on the pump. It is a hydromechanical device consisting of three main parts: the fuel-metering section, the computer section, and the afterburner nozzle control section.

The fuel-metering section meters the fuel flow required during afterburner operation to the pilot burner and main spray bars as determined from information received from the computing section.

The computer section positions the main and pilot burner valves of the fuel-metering section as a function of compressor discharge pressure and power lever angle. The power lever angle input is in turn limited by signals from the nozzle and the turbine discharge temperature system.

The afterburner nozzle control section schedules the afterburner nozzle area, as directed by the power lever, and the turbine discharge temperature. The leaves of the variable-area exhaust nozzle are positioned by three mechanical screwjack actuators powered by the nozzle actuator control via flexible drive cables.

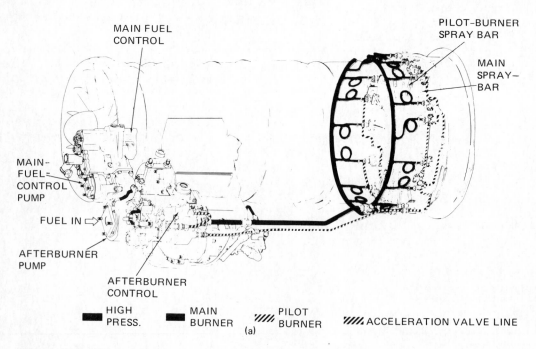

MAIN FUEL
CONTROL

PILOT-BURNER
SPRAY BAR

MAIN
SPRAY-
BAR

MAIN-
FUEL-
CONTROL
PUMP

FUEL IN ⇨

AFTERBURNER
PUMP

AFTERBURNER
CONTROL

■ HIGH PRESS.   ■ MAIN BURNER   ▨ PILOT BURNER   ▨ ACCELERATION VALVE LINE

(a)

FIG. 9-10 The General Electric J85 afterburner and nozzle system.
(a) Afterburner and fuel system components.
(b) Afterburner fuel system.
(c) Afterburner and nozzle control.
(Continued on following page.)

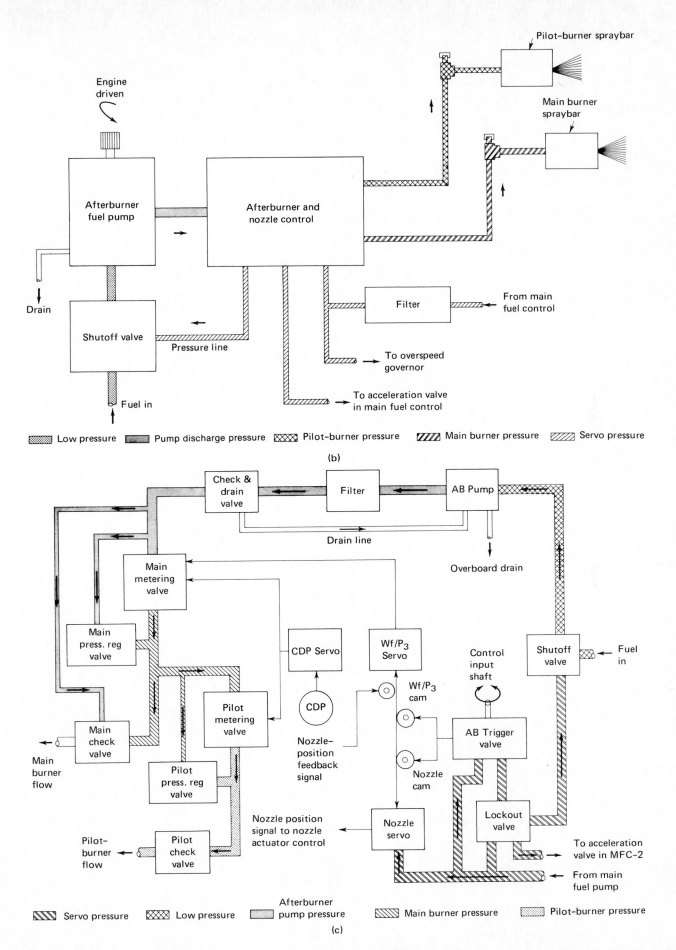

Low pressure ▨ Pump discharge pressure ▨ Pilot–burner pressure ▨ Main burner pressure ▨ Servo pressure

(b)

Servo pressure ▨ Low pressure ▨ Afterburner pump pressure ▨ Main burner pressure ▨ Pilot-burner pressure

(c)

REVIEW AND STUDY QUESTIONS

1. Name two methods of thrust augmentation.
2. Why is alcohol added to water in water injection systems? What effect does the alcohol have on the operation of the engine?
3. Tell how water injection increases thrust. Afterburning.
4. Where can water be injected?
5. How much thrust increase does water injection give; afterburning? Why?
6. Describe a typical water injection system.
7. Discuss the principle behind the afterburner.
8. Why are afterburners used mainly for military aircraft as a method of boosting thrust?
9. Why is it necessary to have a variable-area geometry exhaust on an afterburner-equipped engine?
10. What are the requirements of a good afterburner?
11. Describe a typical afterburner system. What type of ignition systems are used?

CHAPTER TEN
MATERIALS
AND
METHODS OF
CONSTRUCTION

GAS TURBINE MATERIALS

High-temperature, high-strength materials and unique methods of manufacture have made the gas turbine engine a practical reality in a few decades. To a large measure, the performance of turbojet and turboprop engines is dependent upon the temperature at the inlet to the turbine. It can be shown that increasing the turbine inlet temperature from the present limit (for most engines in high production) of approximately 1700 to 2500°F [927 to 1370°C] will result in a specific thrust increase of approximately 130 percent along with a corresponding decrease in specific fuel consumption. For this reason, obviously, high cycle temperatures are desirable. Just as obvious is the fact that not all materials can withstand the hostile operating conditions found in parts of the gas turbine engine.

Commonly Used Terms

Some of the more commonly used terms in the field of metallurgy and metalworking are listed below.

1. *Strength*
   (a) *Creep strength*—Defined as the ability of a metal to resist slow deformation due to stress, but less than the stress level needed to reach the yield point. Creep strength is usually stated in terms of time, temperature, and load.
   (b) *Yield strength*—This point is reached when the metal exhibits a permanent set under load.
   (c) *Rupture strength*—That point where the metal will break under a continual load applied for periods of 100 and 1000 h. Metals are usually tested at several temperatures.
   (d) *Ultimate tensile strength*—The load under which the metal will break in a short time.
2. *Ductility*—The ability of a metal to deform without breaking.
3. *Coefficient of expansion*—A measure of how much a metal will expand or grow with the application of heat.
4. *Thermal conductivity*—The measure of the ability of a metal to transmit heat.
5. *Corrosion and oxidation resistance*—An important factor which indicates how well a metal can resist the corrosive effects of the hot exhaust stream.
6. *Melting point*—The temperature at which the metal becomes a liquid.
7. *Critical temperature*—As a metal is cooled, it passes through distinct temperature points where its internal structure and physical properties are altered. The rate of cooling will greatly influence the ultimate properties of the metal.
8. *Heat treatability*—This is a measure of how the metal's basic structure will vary under an operation, or series of operations, involving heating and cooling of the metal while it is in a solid state. Ferritic, austenitic, and martensitic steels all vary as to their heat treatability. (All of these terms have to do with the physical and chemical properties of metal.)
9. *Thermal shock resistance*—The ability of a metal to withstand extreme changes in temperature in short periods of time.
10. *Hardness*—An important characteristic in that it influences ease of manufacture and therefore cost.

Metal working terms include:

1. *Casting*—A process whereby metal, in a molten state, solidifies in a mold.
2. *Forging*—This is a process of plastic deformation under a pressure that may be slowly or quickly applied.
3. *Electrochemical machining (ECM)*—ECM is accomplished by controlled high-speed deplating using a shaped tool (cathode), an electricity-conducting solution, and the workpiece (anode).
4. *Machining*—Any process whereby metal is formed by cutting, hot or cold rolling, pinching, punching, grinding, or by means of laser beams.
5. *Extrusion*—Metal is pushed through a die to form various cross-sectional shapes.
6. *Welding*—A process of fusing two pieces of metal together by locally melting part of the material through the use of arc welders, plasmas, lasers, or electron beams.
7. *Pressing*—Metals are blended, pressed, sintered (a process of fusing the powder particles together through heat), and coined out of prealloyed powders.
8. *Protective finishes and surface treatments*—This includes plating by means of electrical and chemical processes, by use of ceramic coatings, or painting. Surface treatments for increased wear may take the form of nitrid-

Fe Base Alloys	C	Cr	Ni	Co	Mo	Ti	Cb	Other
Chromoloy	0.20	1	—	—	1	—	—	0.1 V
Lapelloy	0.30	12	—	—	3	—	—	0.3 V
17-7PH	0.07	17	7	—	—	—	—	1 A1
321 SS	0.05	18	10	—	—	0.4	—	—
A286	0.05	15	26	—	1	2	—	0.2 A1
Incoloy T	0.08	20	32	—	—	1	—	—
Timken 16-25-6	0.10	16	25	—	6	—	—	0.15 N
N-155	0.30	20	20	20	3	—	1	2 W
15-7Mo	0.09	15	7	—	2.5	—	—	1.0 A1
19-9DL	0.32	18.5	9	—	1.4	0.25	—	1.35 W
B5 F5	0.45	.95	—	—	0.55	—	—	0.30 V
M308	0.08	13.75	32.5	—	4.1	2.0	—	6.5 W
V57	0.06	15	25.5	—	1.25	3.0	—	0.25 A1

Ni Base Alloys	C	Cr	Fe	Co	Mo	Ti	Al	Other
Inconel	0.05	14	7	—	—	—	—	—
Inconel W	0.05	14	7	—	—	2.5	0.6	—
Inconel X	0.05	15	7	—	—	2.5	0.7	1 Cb
Inco 702	0.02	16	—	—	—	0.5	3.7	—
Inco (Cast)	0.2	12	5	—	4.5	0.75	6.0	2 Si + Mn 2 Cb + Ta
Hastelloy B	0.10	1	5	—	28	—	—	—
Hastelloy X	0.15	22	23	—	9	—	—	—
M-252	0.10	19	2	10	10	2.5	0.8	—
Hastelloy R-235	0.15	16	10	—	6	3	2	—
Udimet 500	0.15	17.5	4 max	16	4	3	3	1.5 Si + Mn
Astroloy	0.055	15.0	0.20	15.0	5.0	3.5	4.4	—
Cosmoloy F	0.04	15.0	0.20	—	3.75	3.45	4.75	2.25 W
Hastelloy C	0.08	15.5	5.5	2.5	16.0	—	—	3.75 W
René 41	0.12	19.0	5.0	11.0	9.75	3.15	1.5	—
U 700	0.10	15.0	4.0	18.5	5.0	3.25	4.25	—
Waspalloy	0.10	19.5	2.00	13.5	4.25	3.0	1.25	—

Co Base Alloys	C	Cr	Ni	Fe	Mo	W	Cb	Other
HS 21	0.25	27	3	1	5	—	—	—
HS 25 (L-605)	0.12	20	10	1	—	15	—	—
HS 31 (X-40)	0.40	25	10	1	—	8	—	—
HE1049	0.40	26	10	2	—	15	—	0.4 B
S-816	0.40	20	20	3	4	4	4	—
V-36	0.30	25	20	3	4	3	2	—

Al Base Alloys	Cu	Si	Mg	Mn
355	1.5	5	0.5	—
14 S	4.5	1	1	1

Mg Base Alloys	Al	Zr	Zn	Th
Dow C	9	—	2	—
HK 31	—	—	0.7	3
HZ 32	—	0.75	2.1	3.25

Ti Base Alloys	Al	Mn	Fe	Cr	Mo	V	Other
C 130 AM	4	4	—	—	—	—	★
Ti 140A	—	—	2	2	2	—	★
C 110 M	—	8	—	—	—	—	★
6 Al-4 V	6	—	—	—	—	4	★
7Al-4Mo-Ti	7	—	—	—	4	—	★
A 110 AT	4.0–6.0						2.0–4.5 Sn ★

★For all alloys C = 0.2 max; O_2 = 0.25 max; N = 0.1 max; H_2 = 0.015 max.

FIG. 10-1 Compositions of some jet engine alloys.

ing, cyaniding, carburizing, diffusion coating, and flame plating.

9. *Shot peening*—A plastic flow or stretching of a metal's surface by a rain of round metallic shot thrown at high velocity.

10. *Heat treatment*—A process to impart specific physical properties to a metal alloy. It includes normalizing, annealing, stress relieving, tempering, and hardening.

11. *Inspection*—Strictly speaking, not a part of the metal working process, inspection is nevertheless integrally associated with it. Inspection methods include magnetic particle and dye penetrant inspection, x-ray inspection, dimensional and visual inspection, and inspection by devices utilizing sound, light, and air. Some of these inspection procedures will be discussed in Chap. 18.

Heat Ranges of Metals (FIG. 10-1)

The operating conditions within a gas turbine engine vary considerably, and metals differ in their ability to satisfactorily meet these conditions.

ALUMINUM ALLOYS Aluminum and its alloys are used in temperature ranges up to 500°F [260°C]. With low density and good strength-to-weight ratios, aluminum forgings and castings are used extensively for centrifugal compressor wheels and housings, air inlet sections, accessory sections, and for the accessories themselves.

TITANIUM ALLOYS Titanium and its alloys are used for axial-flow compressor wheels, blades, and other forged components in many large, high-performance engines. Titanium combines high strength with low density and is suitable for applications up to 1000°F [538°C].

STEEL ALLOYS This group includes high-chromium and high-nickel iron-base alloys in addition to low-alloy steels. Because of their relatively low material cost, ease of fabrication, and good mechanical properties, the low-alloy steels are commonly used for both rotating and static engine components such as compressor rotor blades, wheels, spacers, stator vanes, and structural members. Low-alloy steels can be heat-treated and can be used in temperatures up to 1000°F. High nickel-chromium-iron base alloys can be used up to 1250°F [677°C].

NICKEL-BASE ALLOYS The nickel-base alloys constitute some of the best metals for use between 1200°F and 1800°F [649°C and 982°C]. Most contain little or no iron. They develop their high-temperature strength by age hardening and are characterized by long-time creep-rupture strength, and high ultimate and yield strength combined with good ductility. Many of these materials, originally developed for turbine bucket applications, are also being used in turbine wheels, shafts, spacers, and other parts. Their use is somewhat restricted because of their cost and because of their requirement for critical materials.

COBALT-BASE ALLOYS Cobalt-base alloys form another important group of high-temperature, high-strength, and high-corrosion-resistance metals. Again, as a group, they contain little or no iron. These alloys are used in afterburners and other parts of the engine subjected to very high temperatures.

Chemical Elements Used in Alloys

The number of alloying materials is large. Some of the commonly used elements are listed in Table 10-1.

The percentages of the various elements used, partially determines the physical and chemical characteristics of the alloy and its suitability to a particular application (Fig. 10-2). Tempering and other processes determine the rest. Three characteristics that must be considered are:

1. High-temperature strength
2. Resistance to oxidation and corrosion
3. Resistance to thermal shock

High-temperature Strength (FIG. 10-3)

The most highly stressed parts of the gas turbine engine are the turbine blades and disks. Centrifugal forces tending to break the disk vary as the square of the speed. For example, the centrifugal force on a disk rotating at 20,000 rpm will be 4 times that at 10,000 rpm. Blades weighing only 2 ounces (oz) [6.2 grams (g)] may exert loads of over 4000 lb [1814 kg] at maximum rpm. The blades must also resist the high bending loads applied by the moving gas stream to produce the thousands of horsepower needed to drive the compressor. There is also a severe temperature gradient (difference) between the central portion of the disk and its periphery of several hundred degrees centigrade. Many metals which would be quite satisfactory at room temperatures will lose much of their strength at the elevated temperatures encountered in the engine's hot section. The ultimate tensile strength of a metal at one temperature is not necessarily indicative of its ultimate tensile strength at a higher temperature. For example, at 1000°F [538°C] Inconel X has an ultimate tensile strength of approximately 160,000 psi [1,103,200 kPa] and S 816 at the same temperature has an ultimate tensile strength of 135,000 psi [930,825 kPa]. At 1500°F [816°C] their positions are reversed. Inconel X has an ultimate tensile strength of 55,000 psi [379,225 kPa], while S 816 has an ultimate tensile strength of 75,000 psi [517,125 kPa]. The creep strength, which is closely associated with ultimate tensile strength, is probably one of the most important considerations in the selection of a suitable metal for turbine blades (Fig. 10-4). Engine vibration and fatigue resistance will also have some influence on the selection and useful life of both disks and blades.

Although there are many materials which will withstand the high temperatures encountered in the modern gas turbine engine (for example, carbon, columbium, molybdenum, rhenium, tantalum, and tungsten, all have melting points above 4000°F [2200°C]), the ability to withstand high temperatures while maintaining a reasonable tensile strength is not the only consideration.

TABLE 10-1

Element	Chemical symbol	Element	Chemical symbol
Aluminum	Al	Molybdenum	Mo
Boron	B	Nitrogen	N
Carbon	C	Nickel	Ni
Chromium	Cr	Silicon	Si
Cobalt	Co	Tantalum	Ta
Columbium	Cb	Titanium	Ti
Copper	Cu	Tungsten	W
Iron	Fe	Vanadium	V
Manganese	Mn	Zirconium	Zr

	DESIGNATION	C	Mn	Si	Cr	Ni	Co	Fe	Mo	W	Others	GENERAL USE
HIGH STRENGTH HIGH TEMPERATURE ALLOYS	21 Alloy (AMS 5385B)	.20 .30	1.0 Max.	1.0 Max.	25.0 29.0	1.75 3.75	Bal.	3.0 Max.	5.0 6.0		B—.007 Max.	Good high temperature strength and shock resistance. Oxidation resistance to 2100°F
	X-40 31 Alloy (AMS 5382B)	.45 .55	1.0 Max.	1.0 Max.	24.5 26.5	9.5 11.5	Bal.	2.0 Max.		7.0 8.0	P—.04 Max. S—.04 Max.	Maximum high temperature strength. Oxidation resistance to 2100°F
	Hastelloy "C" (AMS 5388)	.15 Max.	1.0 Max.	1.0 Max.	15.5 17.5	Bal.	2.5 Max.	4.5 7.0	16.0 18.0	3.75 5.25	V—.2—.6	Intermediate high temp strength. Oxidation resistance to 2100°F. Resistance to thermal shock
	N-155 (AMS 5376B)	.20 Max.	1.0 2.0	1.0 Max.	20.0 22.0	19.0 21.0	18.5 21.0	Bal.	2.5 3.5	2.0 3.0	Cb+Ta—.75 —1.25 N—.1—.2 P—.04 Max. S—.03 Max.	Good strength at intermediate temperature. Oxidation resistance to 2000°
	309 Mod. H.R. Crown Max.	.15 .30	1.0 Max.	.75 2.00	22.0 25.0	11.0 14.0		Bal.		2.5 3.5	P—.04 Max. S—.04 Max.	Good strength at intermediate temp. with low alloy content
	Inconel X	.08 Max.	.3 1.0	.50 Max.	14.0 16.0	70.0 Min.		5.0 9.0			Cb—.7—1.0 Ti—2.25—2.75 Al—.4—1.0 S—.01 Max.	Maximum elevated temp strength properties in the heat treated condition
AUSTENITIC STAINLESS STEELS	Type 302 (AMS 5358)	.25 Max.	2.0 Max.	1.0 Max.	17.0 19.0	8.0 10.0		Bal.	.50 Max.		P—.04 Max. S—.03 Max. Cu—.5 Max.	Good corrosion resistance. Oxidation resistance to 1600°F
	Type 310 (AMS 5366A)	.18 Max.	2.0 Max.	.50 1.50	23.0 26.0	19.0 22.0		Bal.	.50 Max.		P—.04 Max. S—.03 Max. Cu—.5 Max.	Excellent oxidation resistance to 2000°F. Moderate high temp. strength
	Type 316 (AMS 5360)	.15 Max.	2.0 Max.	.75 Max.	16.0 18.0	12.0 14.0		Bal.	1.50 2.25		P—.04 Max. S—.03 Max. Cu—.5 Max.	Maximum corrosion resistance and moderate high temperature strength to 1600°F
	Type 303	.20 Max.	1.5 Max.	2.0 Max.	18.0 21.0	9.0 12.0		Bal.	.40 .80		P—.05 Max. S—.2—.4	Free machining grade for corrosion service
MARTENITIC STAINLESS STEELS	Type 410 (AMS 5350C)	.05 .15	1.0 Max.	1.0 Max.	11.5 13.5	.5 Max.		Bal.	.50 Max.		P—.04 Max. S—.03 Max. Cu—.5 Max.	Moderate corrosion and heat resistance, service to 1200°F. Maximum damping capacity
	Type 431	.12 .20	1.0 Max.	1.0 Max.	15.0 17.0	1.5 3.0		Bal.	.50 Max.		P—.04 Max. S—.04 Max. Cu—.5 Max.	Moderate corrosion and heat resistance, service to 1500°F. Maximum tensile strength
CARBON AND LOW ALLOY STEELS	1020	.18 .23	.30 .60	.75 Max.				Bal.			S—.05 Max. P—.04 Max.	Carburized parts or structural shapes requiring good ductility
	1035	.32 .38	.60 .90	.75 Max.				Bal.			S—.05 Max. P—.04 Max.	Stressed structural shapes
	1095	.90 1.05	.30 .50	.75 Max.				Bal.			S—.05 Max. P—.04 Max.	Tools and wear resistant parts
	4140	.38 .43	.75 1.00	.75 Max.	.80 1.10			Bal.	.15 .25		S—.04 Max. P—.04 Max.	Highly stressed parts and service to 900°F
	4340	.38 .43	.60 .80	.75 Max.	.70 .90	1.65 2.00		Bal.	.20 .30		S—.04 Max. P—.04 Max.	Highly stressed parts. Good impact resistance with high strength. Service to 1000°F
	8620	.18 .23	.70 .90	.75 Max.	.40 .60	.40 .70		Bal.	.15 .25		S—.04 Max. P—.04 Max.	Highly stressed carburized parts

FIG. 10-2 Several representative alloys and their properties.

CAST-ABILITY	CONDITION	TEST TEMP. °F	Y.S. 2% OFFSET psi	T.S. psi	ELONG. %	R.A. %	HARD-NESS	STRESS RUPTURE DATA 100 HR. LIFE STRESS psi	1000 HR. LIFE STRESS psi	CREEP DATA 1% 10,000 HR. STRESS psi	FATIGUE STRESS 10^8 CY. STRESS psi
Excellent	As Cast	70	80,000	100,000	8	10	28R_c				35,000
	As Cast	1200	35,000	70,000	15	40		50,000	45,000		44,000
	As Cast	1500	20,000	65,000	15	30		21,000	15,000	7,000	33,000
	As Cast	1800		33,000	35	50		9,000	6,000		
	Aged (1)	70	100,000	120,000	2	3					
	Aged (1)	1200	70,000	90,000	2	5					
Excellent	As Cast	70	80,000	110,000	8	10	30R_c				
	As Cast	1200	40,000	75,000	15	25		55,000	46,000		56,000
	As Cast	1500	25,000	60,000	15	20		28,000	22,000	12,000	46,000
	As Cast	1800		30,000	30	40		10,000	8,000		
	Aged (1)	70	110,000	125,000	2	3					
	Aged (1)	1200	60,000	80,000	5	6					
Good	As Cast	70	50,000	80,000	5	5	20R_c				
	As Cast	1200		60,000	15	15		45,000	30,000		
	As Cast	1500		50,000	18	15		18,000	14,000		
	As Cast	1800		20,000	20	50		5,000	2,000		
Excellent	As Cast	70	55,000	95,000	25	15	85R_b				
	As Cast	1200	32,000	60,000	25	25		45,000	35,000		
	As Cast	1500	20,000	50,000	15	20		18.000	14,000	7,000	
	As Cast	1800		20,000	25	50		5,000	2,000		
Excellent	As Cast	70	40,000	80,000	15	25	80R_b				
	As Cast	1200	25,000	50,000	10	10		25,000	15,000		
	As Cast	1500	20,000	30,000	25	30		12,000	5,000		
Poor	Ht. Tr. (2)	70	85,000	105,000	4	6	29R_c				
	Ht. Tr. (2)	1200	70,000	100,000	8	10		75,000	60,000		
	Ht. Tr. (2)	1500	50,000	55,000	20	30		28,000	18,000		
Good	As Cast	70	30,000	80,000	40	50	80R_b				
	Ann'ld.(2)	70	25,000	70,000	50	70	76R_b				
	As Cast	1200	12,000	40,000	30	40		18,000	13,000		
	As Cast	1500	10,000	20,000	30	40					
Excellent	As Cast	70	25,000	60,000	40	60	70R_b				
	Ann'ld. (3)	70	25,000	60,000	50	60					
	As Cast	1200	20,000	50,000	20	25		24,000	18,000		
	As Cast	1500	15,000	25,000	10	15					
Good	As Cast	70	30,000	75,000	40	50	77R_b				
	Ann'ld. (3)	70	30,000	70,000	50	60	75R_b				
	As Cast	1200	20,000	55,000	30	50		26,000	20,000		
	As Cast	1500	12,000	25,000	20	40					
Good	As Cast	70	30,000	75,000	35	35	80R_b				
	Ann'ld. (3)	70	35,000	80,000	40	40	81R_b				
Fair	Pro. Ann. (4)	70	90,000	120,000	10	20	25R_c				
	H.T. (4&5)	70	110,000	140,000	10	30	30R_c				
	H.T. (4&5)	70	85,000	110,000	20	50	24R_c				
Fair	Pro. Ann. (4)	70	90,000	120,000	15	20	26R_c				
	H.T. (4&6)	70	130,000	180,000	10	15	42R_c				
	H.T. (4&5)	70	100,000	140,000	15	30	30R_c				
Good	Ann'ld. (7)	70	35,000	70,000	25	40	77R_b				
	H.T. (8)	70	100,000	125,000	15	40	30R_c				
	H.T. (8)	70	80,000	95,000	20	50	95R_b				
Excellent	Ann. (7)	70	45,000	85,000	20	35	84R_b				
	H.T. (8)	70	130,000	150,000	5	15	38R_c				
	H.T. (8)	70	95,000	110,000	10	30	28R_c				
Excellent	Ann. (7)	70	65,000	125,000	5	5	25R_c				
	H.T. (8)	70									
Excellent	Ann. (7)	70	50,000	100,000	10	20	92R_b				
	H.T. (8)	70	160,000	180,000	5	10	40R_c				
	H.T. (8)	70	110,000	140,000	10	20	33R_c				
Excellent	Ann. (7)	70	60,000	100,000	20	45	93R_b				
	H.T. (8)	70	170,000	200,000	5	20	43R_c				
	H.T. (8)	70	130,000	150,000	15	30	35R_c				
Good	Ann. (7)	70	40,000	85,000	20	40	85R_b				
	H.T. (8)	70	120,000	150,000	10	30	33R_c				
	H.T. (8)	70	90,000	110.000	20	50	22R_c				

(1) 50 hr. 1350°F

(2) 3 hr. 2100°F, air cooled, 24 hr. 1550°F plus 20 hr. 1300°F

(3) Oil or water quench from 1900 to 2100°F

(4) 4 hr. 1200°F to 1400°F, air cool

(5) 1 hr. 1800°F, oil quench, draw 1 hr. 1100 to 1300°F

(6) 1 hr. 1875°F, oil quench, draw 3 hrs. 675°F or 2 hrs. 1100°F

(7) 2 hr. 1600 to 1650°F, furnace cool

(8) Water or oil quenched from 1600 to 1650°F and drawn to hardness indicated

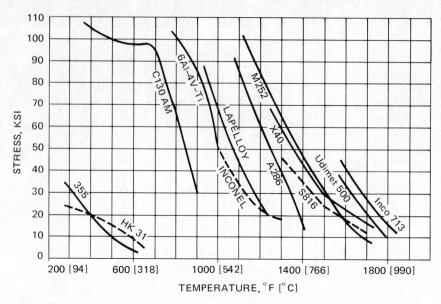

FIG. 10-3 100-hour stress-rupture strengths of some turbine engine alloys.

Such factors as critical temperature, rupture strength, thermal conductivity, coefficient of expansion, yield strength, ultimate tensile strength, corrosion resistance, workability, and cost must all be taken into account when selecting any particular metal.

Resistance to Oxidation and Corrosion

Corrosion and oxidation are results of electrical and chemical reactions with other materials. The hot exhaust gas stream encountered in the engine speeds up this reaction. While all metals will corrode or oxidize, the degree of oxidation is determined by the base alloy and the properties of the oxide coating formed. If the oxide coating is porous, or has a coefficient of expansion different from that of the base metal, the base metal will

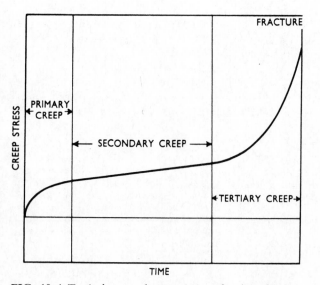

FIG. 10-4 Typical creep characteristics of turbine blades.

be continually exposed to the oxidizing atmosphere. One solution to the problem of oxidation at elevated temperatures has been the development and use of ceramic coatings. One product called Solaramic coating, manufactured by Solar, a division of International Harvester Company located in San Diego, California, is a ready-to-use ceramic slurry that can be thinned with water and applied to a part by spraying, brushing, or dipping. After drying, the Solaramic material will change to a white powder which in turn is transformed to a ceramic coating when baked at 950°F [510°C]. Ceramic-coated afterburner liners and combustion chambers are in use today. The ceramic coating has two basic functions:

1. Sealing the base metal surface against corrosion, oxidation, and carbonizing
2. Insulating the base metal against high temperatures

These coatings are not without disadvantages, in that they are more susceptible to thermal shock, they must have the same coefficient of expansion as the base metal, they are brittle, and they have low tensile strength which, of course, restricts their use in the engine. Some work which shows promise is being done with various metal-ceramic combinations called *Cermets* or *Ceramels*. Ceramic materials being used include aluminum, beryllium, thorium, and zirconium oxides to name a few.

Thermal Shock Resistance

Many materials which would otherwise be quite suitable must be rejected because of their poor thermal shock characteristics. Several engine failures have been attributed to thermal shock on the turbine disk. Ceramic coatings in particular are vulnerable to this form of stress. Improved fuel controls, starting techniques, and engine design have lessened this problem.

Convective, Film, and Impingement Cooling (FIG. 10-5)

The effort to achieve higher turbine inlet temperatures, and therefore higher thermal efficiencies, has been approached from two directions. The first has been the development and use of high-temperature materials, both metals and ceramics. The second avenue of approach has been to cool the highly stressed turbine components. One method of cooling the nozzle guide vanes and turbine blades on gas turbine engines is to pass compressor bleed air through the hollow blades to cool them by convective heat transfer.

A newer procedure called film cooling also uses compressor bleed air which is made to flow along the outside surface of both vanes and blades, thus forming an insulating blanket of cooler air between the metal and the hot gas stream. The layer of air also reduces temperature gradients and thermal stress. Advanced manufacturing techniques such as shaped-tube electrolytic machining (STEM) and Electro-Stream (trademark of General Electric) drilling have made the production of the necessary small holes in the superhard turbine material possible (Fig. 10-6). (See Fig. 10-19 for some other nontraditional machining techniques.)

Some engines also use the air bled from the compressor to cool the front and rear face of the turbine disks.

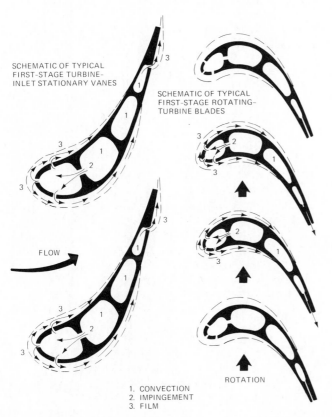

SCHEMATIC OF TYPICAL FIRST-STAGE TURBINE-INLET STATIONARY VANES

SCHEMATIC OF TYPICAL FIRST-STAGE ROTATING-TURBINE BLADES

FLOW

ROTATION

1. CONVECTION
2. IMPINGEMENT
3. FILM

FIG. 10-5 Types of air-cooling techniques used with turbine vanes and blades.

Transpiration Cooling (FIG. 10-7)

This is a novel and efficient method of allowing the turbine blades and other parts within the hot section to operate at much higher turbine inlet temperatures. In this type of cooled blade the air passes through thousands of holes in a porous airfoil made from a sintered wire mesh material. Since the sintered wire mesh is not strong enough by itself, an internal strut is provided as the main structural support carrying all airfoil and centrifugal loads. Fabrication techniques involve rolling layers of woven wire mesh and then sintering these layers to form a porous metal sheet, which is then rolled into an airfoil shape.

Porous materials, for example, Poroloy made by the Bendix Corporation, have been tested for use in combustion chambers and for afterburner liners. A similar material called Rigimesh has also been used in rocket engines to help keep the fuel nozzles cool. Many manufacturers are experimenting with other types of porous materials for use in blades in an attempt to obtain higher turbine inlet temperatures.

Ceramics

Experiments are being performed using ceramic materials in many of the engine's hot section parts, such as the combustor, nozzle diaphragm, turbine blades, and turbine disks. Materials being looked at are hot-pressed and/or bonded silicon nitride or silicon carbide.

Advances in material development and new cooling techniques have allowed engines such as the General Electric F101 (See Fig. 2-30) and others to be designed that have operating turbine inlet temperatures of 2500°F [1371°C] and higher, with a resulting 100 percent increase in specific weight (thrust-to-weight ratio), and with a lower specific fuel consumption in comparison with previous engines.

Other Materials

Relatively new types of materials called composites are coming to the foreground for use in both airframes and engines. In these products, graphite, glass, or boron filaments are embedded in an epoxy-resin matrix or base substance. Other types of filaments and matrices are being tried to meet the demands of higher temperature and/or stress. The chief advantage of the composite material is its very favorable strength-to-weight ratio, which can lead to the lightening of many structural parts. For example, a lighter fan blade will allow a lighter fan disk, which will in turn permit a lightening of other parts all the way down the line. Composite materials may be used in conjunction with other load-bearing materials to provide a support function. Typical of this type of structure are fan blades made with a steel spar and base and with an airfoil composite shell. Closely associated with the future use of composite materials is the development of new manufacturing techniques to produce these materials.

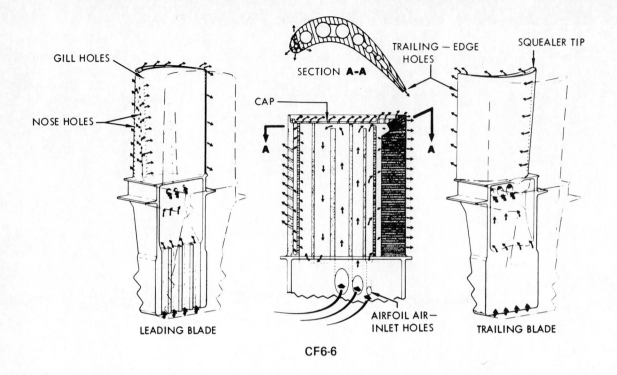

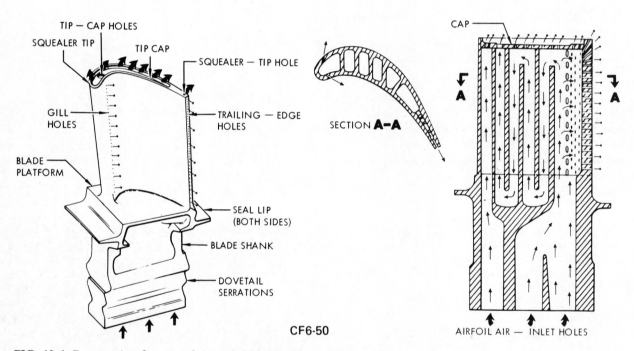

FIG. 10-6 Construction features of air-cooled blades. *(General Electric.)*

MANUFACTURING TECHNIQUES

The variety of manufacturing techniques is large, and is dependent upon a number of factors such as the material from which the part is made, the duties the part must perform, and the cost of the process. As a result, basic parts of the engine are produced by several casting and forging processes, literally dozens of machine operations, and fabrication procedures using a variety of metal-joining methods.

Casting

Several engine parts are cast in aluminum, magnesium, or steel alloys. These parts include intake and compressor housings, accessory cases, and blading to name a few. Casting methods differ. They include:

1. Sand casting
2. Spin casting
3. Lost-wax or investment casting

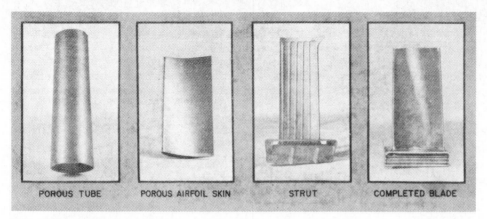

FIG. 10-7 Steps in the manufacture of a transpiration-cooled blade.

4. Resin-shell mold casting
5. Slip casting
6. Mercasting

Sand casting (Fig. 10-8) uses a wood or metal pattern around which a clay-free sand has been packed to form the mold. The mold is then split, the pattern removed, the mold reassembled, and any cores that are necessary added. Molten metal at a precise temperature is poured into the mold and allowed to cool. The mold is removed and various heat treatments may be performed to obtain the desired physical characteristics. The casting may be spun while being poured. Spin casting results in a denser, more sound casting. Spinning is normally performed on small ring sections. Cooling of the metal radially inward results in fewer stresses. Other casting techniques result in greater tensile strength by causing the normally random grain structure of the casting to become oriented in one direction like the grain of wood. (Fig. 10-9)

Basically the investment casting process (Fig. 10-10) involves the use of heat-disposable wax or plastic pat-

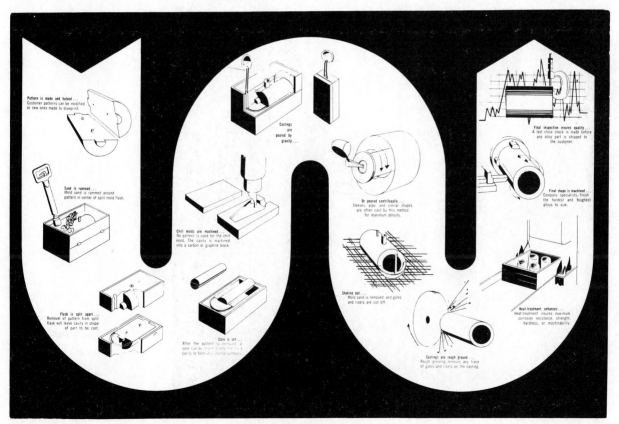

FIG. 10-8 Typical sand-casting procedures. (*Union Carbide Corp.*)

terns which are surrounded with a refractory material to form a monolithic mold. Patterns are removed from the mold in ovens, and molten metal is poured into the hot mold. Sometimes this pouring is done in a vacuum furnace. After cooling, the mold material is quite fragile and is easily removed from the castings. Because the finished product duplicates the pattern exactly, the production of patterns is a critical factor. They are made by injecting molten wax or plastic into metal dies. The finished castings have an exceptionally smooth surface finish and require very little further machining. Incidentally, this process is not new. It was used by the ancient Greeks and Egyptians to cast lightweight statues, intricate bowls, and pitchers.

Resin-shell mold casting (Fig. 10-11) is a high production method similar to investment casting except that the tolerances are not held as closely. In many ways it rivals sand casting in economy.

Slip casting (Fig. 10-12), borrowed from the ceramics industry, is used to form super-heat-resistant materials. Often it is the only way certain materials can be shaped. Metal ceramics, silicon nitride, and refractory metals cast this way can be used in temperatures over 2200°F [1200°C].

The Mercast process is a precision-casting technique. It is essentially the same kind of method as the lost-wax precision-investment process, except that frozen mercury is used as a pattern instead of wax. Liquid mercury

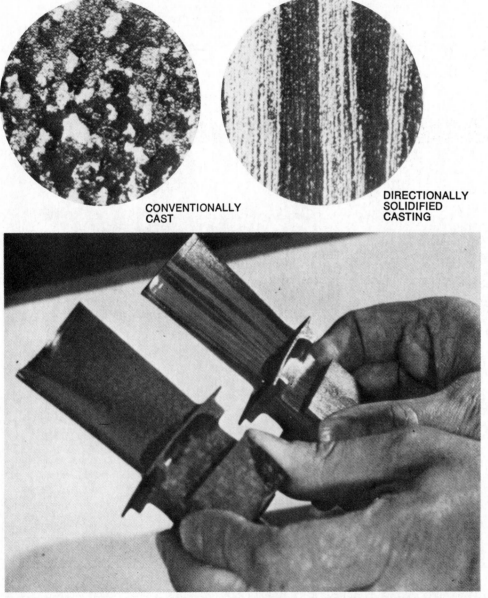

CONVENTIONALLY CAST

DIRECTIONALLY SOLIDIFIED CASTING

FIG. 10-9 Conventional and directionally solidified cast turbine blades. *(Pratt & Whitney Aircraft.)*

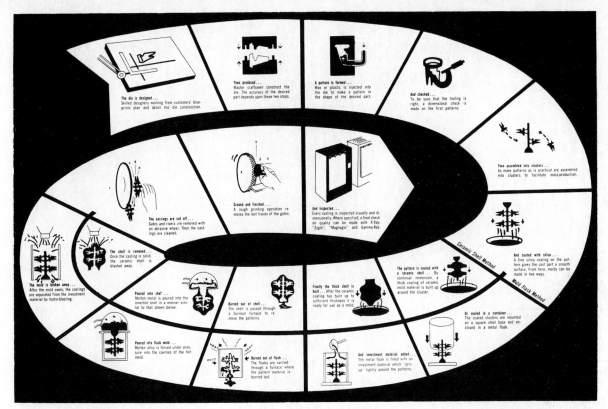

FIG. 10-10 Two methods of investment casting. *(Union Carbide Corp.)*

is poured into a master mold, where it is frozen at temperatures below −40°F [−40°C]. Then it is removed and coated with a cold refractory slurry to a thickness of ⅛ in. [3.175 mm] or more. The refractory shell is dried at low temperature; then the shell and mercury are brought to room temperature, and the mercury is melted out. The refractory shell is fired to give it strength and then is used as the mold for a usual casting process. Complicated parts can be made by use of the Mercast process, and very close tolerances and excellent surface finish can be obtained. The cost however, is higher than that of some other methods.

Forging *(FIG. 10-13)*

Disks, drive shafts, rings, gears, vanes, blades, and numerous other parts of the gas turbine engine are manufactured by forging. This process allows the development of a grain structure and results in a fine-grain, more ductile, strong, dense product (Fig. 10-14). Forging can be accomplished by rapid hammering or slow pressing. The choice of technique depends mainly upon the resistance of metal to rapid deformation. The workpiece is generally heated to improve plasticity and reduce forging forces and will often pass through several different dies before the final shape is obtained (Fig. 10-15). All ductile materials can be forged, but their forgeability varies considerably. At the forging temperature, forgeability generally depends upon the melting point,

ductility, yield strength, crystallographic structure, recovery from forging stresses, surface reactivity, die friction, and cost.

Some parts are rolled or swaged, which essentially simulates the forging process. By using this method, a well-defined grain structure is established, which increases tensile strength considerably (Fig. 10-16).

Prior to forging some turbine blades, the end of the forging blank (usually a rod) is upset by heating, or the shank is swaged to develop natural "flow lines" in the root and shank section of the blade.

Machining *(FIGS. 10-17 AND 10-18)*

In addition to the hammers, presses, and other tools mentioned above, the inventory of machinery for manufacturing gas turbine parts includes all of the common varieties such as lathes, mills, broaches, grinders, shapers and planers, polishers and buffers, drills, saws, shears, filers, threaders, contour machines of all kinds, and a host of other devices to cut and form metal. Many of these devices utilize a numerical tape control or other automatic control devices to reduce human error and produce a more uniform, less expensive product.

Some nontraditional machining techniques for removing metal from superhard and supertough alloys, and from other materials whose complex shapes preclude machining with conventional metal-cutting tools include chemical milling, electrochemical machining

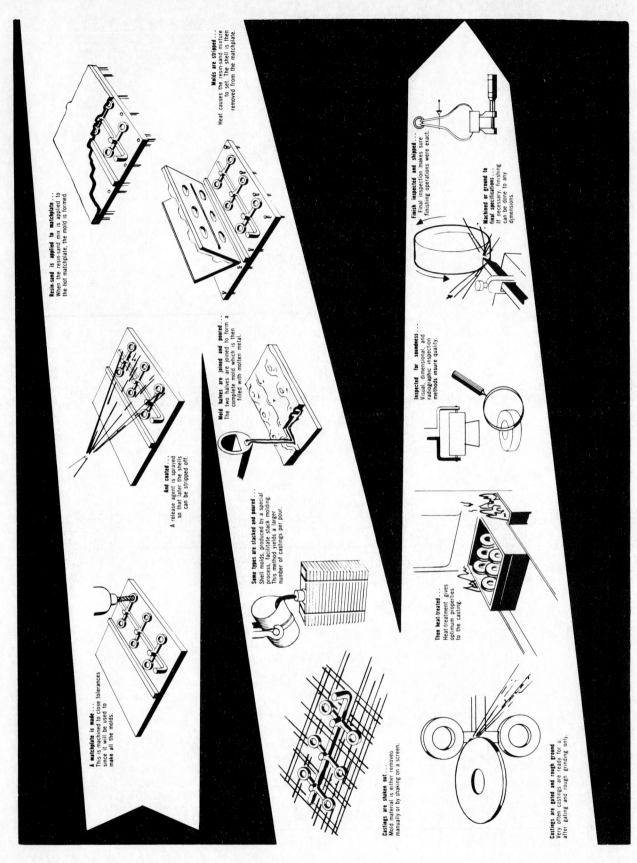

A matchplate is made ...
This is machined to close tolerances since it will be used to make all the molds.

Resin-sand is applied to matchplate ...
When the resin-sand mix is applied to the hot matchplate, the mold is formed.

Molds are stripped ...
Heat causes the resin-sand mixture to set. The shell is then removed from the matchplate.

And coated ...
A release agent is sprayed so that later the shells can be stripped off.

Mold halves are joined and poured ...
The two halves are joined to form a complete mold which is then filled with molten metal.

Some types are stacked and poured ...
Shell molds, produced by a special process, facilitate stack molding. This method yields a larger number of castings per pour.

Castings are shaken out ...
Mold material is either removed manually or by shaking on a screen.

Castings are gated and rough ground ...
Very often castings are ready for u. after gating and rough grinding only.

Then heat-treated ...
Heat-treatment gives optimum properties to the casting.

Inspected for soundness ...
Visual, dimensional, and radiographic inspection methods insure quality.

Machined or ground to final specifications ...
If necessary, finishing can be done to any dimensions.

Finish inspected and shipped ...
Final inspection makes sure finishing operations were exact.

FIG. 10-11 Resin-shell mold casting procedure. (*Union Carbide Corp.*)

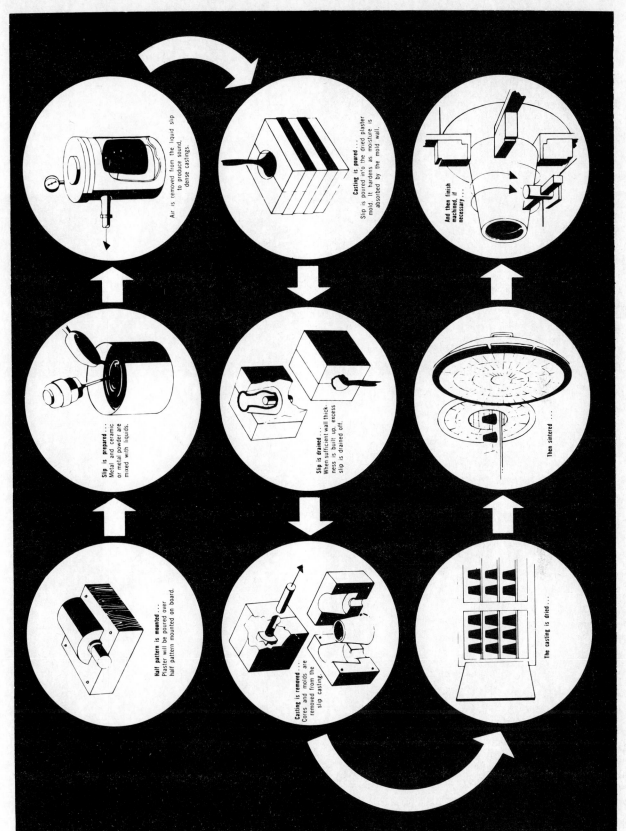

FIG. 10-12 Slip-casting procedure. (*Union Carbide Corp.*)

FIG. 10-13 A 35,000-lb [7200-kg] drop-hammer forge.

FIG. 10-14 Grain flow is developed through forging for additional strength.

FIG. 10-15 Typical forging steps.

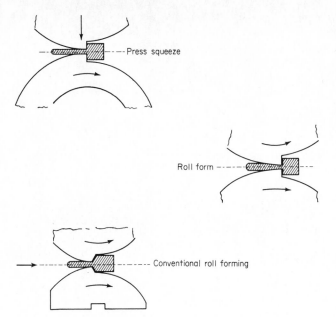

FIG. 10-16 Forming small compressor blades. Upper drawings show how the blade is first squeezed by the press action, and then rolled to form the foil section. The bottom drawing points up the problem of controlled forming at the foil root with conventional rolling techniques.

FIG. 10-17 Broaching a fir-tree root on a turbine blade.

(ECM), electric discharge machining (EDM), electron-beam machining, and laser-beam machining. Other nonconventional machining includes everything from abrasive jet cutting to ultrasonic machining.

Chemical milling involves the removal of metal by dissolving it in a suitable chemical. Those areas that are not to be dissolved away are masked with nonreactive materials. The process can be used on most metals, including aluminum, magnesium, titanium, steels, and superalloys for surface sculpturing. Both sides of the workpiece can be chemically milled simultaneously. In

addition, the process can be used to machine very thin sheets.

ECM is basically a chemical deplating process in which metal, removed from a positively charged workpiece using high-amperage–low-voltage dc, is flushed away by a highly pressurized electrolyte before it can plate out on the cathode tool. The cathode tool is made to produce the desired shape in the workpiece, and both must be electrically conductive. The work proceeds while the cathode and workpiece are both submerged in an electrolyte such as sodium chloride. A variation

FIG. 10-18 Milling several compressor blades at one time.

ELECTROCHEMICAL MACHINING
ECM

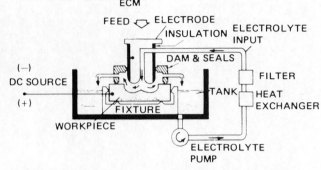

ECM IS THE REMOVAL OF CONDUCTING MATERIAL BY THE ANODIC DISSOLUTION OF A POSITIVE WORKPIECE SEPARATED FROM A SHAPED NEGATIVE ELECTRODE TOOL BY A MOVING CONDUCTIVE ELECTROLYTE

ELECTRICAL DISCHARGE MACHINING
EDM

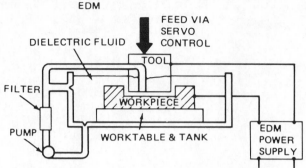

EDM IS THE REMOVAL OF A CONDUCTIVE MATERIAL BY THE RAPID REPETITIVE SPARK DISCHARGE BETWEEN A TOOL AND A WORKPIECE SEPARATED BY A FLOWING DIELECTRIC FLUID

ABRASIVE JET MACHINING
AJM

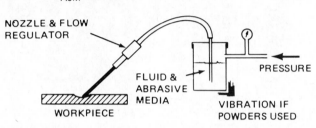

AJM IS THE REMOVAL OF MATERIAL THRU THE ACTION OF A FOCUSED STREAM OF FLUID, GENERALLY CONTAINING ABRASIVE PARTICLES.

CHEMICAL MACHINING
CHM

CHM IS THE CONTROLLED DISSOLUTION OF MATERIAL BY CONTACT WITH STRONG CHEMICAL REAGENTS.

ELECTROCHEMICAL GRINDING
ECG

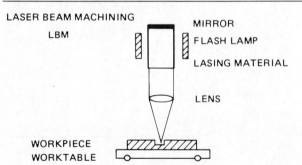

ECG IS THE ANODIC DISSOLUTION OF A POSITIVE WORKPIECE UNDER A CONDUCTIVE ROTATING ABRASIVE WHEEL WITH A MOVING CONDUCTIVE ELECTROLYTE

ELECTRON BEAM MACHINING
EBM

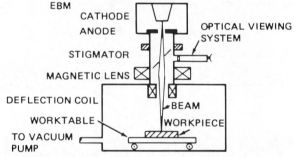

EBM REMOVES MATERIAL WITH A HIGH-VELOCITY FOCUSED STREAM OF ELECTRONS THAT MELTS & VAPORIZES THE WORKPIECE AT THE POINT OF IMPINGEMENT

LASER BEAM MACHINING
LBM

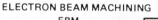

LBM REMOVES MATERIAL WITH A HIGHLY FOCUSED MONOFREQUENCY COLLIMATED BEAM OF LIGHT THAT SUBLIMES MATERIAL AT THE POINT OF IMPINGEMENT

ULTRASONIC MACHINING
USM

USM IS THE ABRASIVE CUTTING OF MATERIAL BY HIGH-FREQUENCY REPETITIVE IMPACT BETWEEN A TOOL & A WORKPIECE WITH AN ABRASIVE SLURRY IN BETWEEN-GENERALLY ABOVE AUDIBLE RANGE

FIG. 10-19 Some nontraditional machining processes.

and extension of electrochemical machining is electro-stream drilling. In this process a negatively charged electrolyte, usually an acid, drills holes in a workpiece that has been positively charged. Holes as small as 0.005 in. [0.127 mm] in diameter and 0.5 in. [12.7 mm] deep in superalloys can be drilled in this manner.

In EDM, high voltages are used to produce a high electrical potential between two conductive surfaces, the workpiece and electrode tool, both of which are immersed in a dielectric fluid. Material is removed from both the electrode and the workpiece by a series of very short electric discharges or sparks between the two, and is swept away by the dielectric fluid. More material is removed from the workpiece than from the tool by proper selection of the two materials. This process can be used to shape complex parts to very close tolerances from refractory metals and alloys that were formerly impossible to machine. The use of electric discharge machining is limited in that it is slower than electrochemical machining, tool replacement can become expensive, and the surface of the workpiece is damaged as a result of the sparks. On the other hand, the EDM process is less expensive than the ECM process.

Electron-beam machining and laser-beam machining are being used experimentally, and may find future use in the production of gas turbines and other aerospace components. Many of the nontraditional machining techniques are illustrated in Fig. 10-19.

Fabrication

Welding is used extensively to fabricate and repair many engine parts. Fabricated sheet steel is used for combustion chambers, exhaust ducts, compressor casings, thrust reversers, silencers, etc. Common methods include resistance and inert-gas (usually argon) welding. Uncommon methods utilize plasmas (see page 197) and lasers (*Light Amplification by Stimulated Emission of Radiation*). Electric resistance welding is used to make spot, stitch (overlapping spots), and continuous seam welds (Fig. 10-20). Inert-gas welding employs a non-consumable electrode (tungsten-thorium alloy) surrounded by some inert gas such as argon or helium (Fig. 10-21). The gas prevents an adverse reaction with the oxygen present in the normal atmosphere. The inert gas can be applied in the immediate area of the arc, or in the case of production runs, the workpiece and/or the entire welding machine can be enclosed in a thin plastic balloon, sometimes as large as a room. The entire plastic bubble is filled with and supported by the inert gas. The operator stands on the outside and works through specially designed armholes. After welding, many parts must be stress-relieved. Where temperature or working loads are not large, brazing or silver soldering may be used to join such parts as fittings and tube assemblies.

Electron-beam welding (Fig. 10-22) is showing great promise as a method of fabricating parts from heretofore difficult-to-weld or unweldable materials. Electron-beam welding uses a stream of focused electrons traveling at speeds approaching 60 percent the speed of light. Even though the mass of electrons which form the beam is small, they are traveling at such speeds that they contain a great amount of kinetic energy. When the beam strikes the workpiece, the kinetic energy is transformed into heat energy. The welding usually takes place in a vacuum, although nonvacuum techniques can be used. Deep, narrow welds with a very narrow heat-affected zone in the base metal, the ability to weld materials as thin as 0.00025 in. [0.00635 mm]

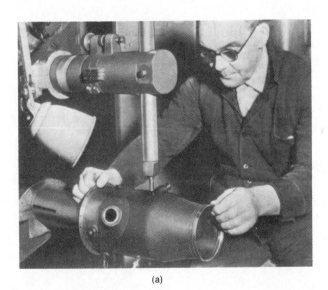

(a)

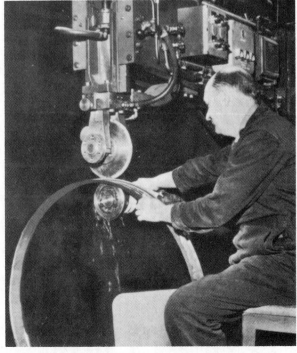

(b)

FIG. 10-20 (a) Spot welding machine.
(b) Continuous-seam welding machine.

FIG. 10-21 A Heliarc welding setup.

FIG. 10-23 An inertia welder. *(General Electric.)*

and as thick as 4 in. [101.6 mm] of stainless steel, and the ability to weld many different types of materials, make this welding process a valuable one in the gas turbine manufacturing area.

Another new welding method is called friction or inertia welding (Fig. 10-23). In this process the parts are joined through the friction generated when they are rubbed together. Strictly speaking, the joint is not a weld. It is more closely related to forming by hot forging, and the "welded" joint is actually bonded in a solid state. This results in a quality joint of great strength.

Finishing

The basic material, the properties desired in the finished product, and the kind of protection desired will determine the type of surface and internal treatment received. The variety is considerable and includes the following.

CHEMICAL TREATMENT Chrome pickling is the most commonly used of all chemical treatments of magnesium. The part is dipped in a solution of sodium dichromate, nitric acid, and water.

ELECTROCHEMICAL TREATMENT Anodizing is a common surface treatment for aluminum alloys whereby the surface aluminum is oxidized to an adherent film of aluminum oxide.

PAINTING A thin preservative resin varnish coating is used to protect internal steel, aluminum, and magnesium parts. The characteristic color of this shiny, transparent coating is usually green or blue-green. A graphite powder may be mixed with the varnish to act as an antigallant. Gray, black, or aluminum enamel or epoxy paint is also used extensively as a protective finish.

SHOT PEENING This procedure can increase the life of a part many times. It is essentially a plastic flow or stretching of a metal's surface by a rain of round metallic shot thrown at high velocity by either mechanical or pneumatic means. The 0.005- to 0.035-in. [0.127- to 0.889-mm] stretched layer is placed in a state of compression with the stress concentration uniformly distributed over the entire surface. Glass beads are sometimes used as the shot for cleaning purposes.

PLATING A great number of plating materials and procedures are used. Plating materials involving the use of chemical or electrochemical solutions include cadmium, chromium, silver, nickel, tin, and others. The exact procedure is determined by the plating and base metal.

Aluminizing is another plating method whereby pure molten aluminum is sprayed onto the aluminum alloy base material to form a protective coating against oxidation and corrosion.

The Coating Service of Union Carbide Corporation has developed and is producing machines for applying extremely wear-resistant and other specialized coatings

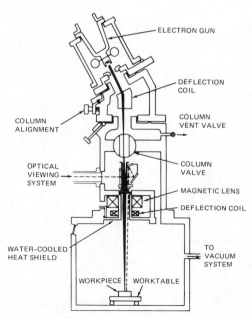

FIG. 10-22 The Hamilton Standard electron-beam welder.

ELECTRON GUN

DEFLECTION COIL

COLUMN VENT VALVE

COLUMN ALIGNMENT

OPTICAL VIEWING SYSTEM

COLUMN VALVE

MAGNETIC LENS

DEFLECTION COIL

WATER-COOLED HEAT SHIELD

TO VACUUM SYSTEM

WORKPIECE   WORKTABLE

to gas turbine parts (Fig. 10-24), tools, and other machines. The different coatings are applied by either of two methods—*the detonation gun* or the *plasma gun*. In the detonation gun (Fig. 10-25), measured quantities of oxygen, acetylene, and suspended particles of the coating material are fed into the chamber of the gun. Four times a second, a spark ignites the mixture and creates a detonation which hurls the coating particles, heated to a plastic state by the 6000°F [3316°C] temperature in the gun, out of the barrel at a speed of 2500 ft/s [762 m/s]. The part to be plated is kept below 300°F [149°C] by auxiliary cooling streams. The high-level sound of 150 dB necessitates housing the gun in a double-walled sound-insulated construction. Operation is controlled from outside this enclosure.

The plasma gun or *torch* (Fig. 10-26) produces and controls a high-velocity inert-gas stream that can be maintained at temperatures above 20,000°F [11,000°C]. Unlike the D-gun process, no combustion takes place. The high-temperature plasma is formed by ionizing argon gas in the extreme heat of an electric arc. Gas molecules absorb heat energy from the arc, split into atoms, and then further decompose into electrically charged particles called ions. The hot gas stream can melt, without decomposition, any known material. When the molten particles, which are introduced in powdered form, strike the part being coated, a permanent welded bond is formed. While the D-gun is a patented Union Carbide machine, other manufacturers make and distribute a variety of plasma plating and cutting devices.

HEAT TREATMENTS All the following procedures alter the mechanical properties of steel to suit the end use. They include:

1. *Normalizing*—The steel is heated to a temperature above the critical range and allowed to cool slowly. Normalizing promotes uniformity of structure and alters mechanical properties.
2. *Annealing*—Consists of heating to a point at or near the critical range, then cooling at a predetermined rate. It is used to develop softness, improve machinability, reduce stress, improve or restore ductility, and modify other properties.
3. *Stress relieving*—The metal is heated throughout to a point below the critical range and slowly cooled. The

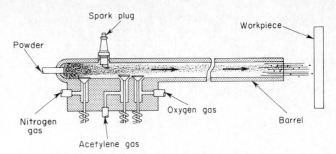

FIG. 10-25 The Union Carbide detonation gun.

object of this treatment is to restore elastic properties or reduce stresses that may have been induced by machining, cold working, or welding.

4. *Hardening*—Involves heating the metal to a temperature above the critical range and then quenching. The cooling rate will determine hardness.
5. *Tempering*—The steel is usually too brittle for use after quenching. Tempering restores some of the ductility and toughness of steel at the sacrifice of hardness or strength. The process is accomplished by heating the hardened steel to a specific point below the critical temperature.

Nonmetallic Materials

Teflon, nylon, carbon, rubber, Bakelite, and a host of plastic materials are used in the gas turbine engine mainly as sealing and insulation materials. For example, nylon and Teflon are used to insulate and protect the shielded electrical wiring located on the outside of the engine. Teflon is also used on the J79 for the seal on the variable-stator-vane actuators. Carbon is used largely inside the engine in the form of carbon rubbing oil seals. Some of these "face" carbon rubbing seals must be flat to within two helium light bands, or approximately 23 millionths of an inch [0.5842 micrometer (μm)]. Rubber and rubberized fabric materials make up the sealing edge of the fire seal which divides the hot and cold sections of the engine when mounted in the nacelle. Synthetic rubber is used extensively throughout the engine in the form of O-ring or other-shaped seals.

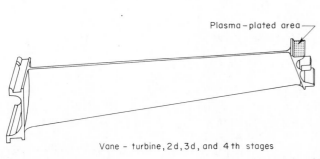

Vane - turbine, 2d, 3d, and 4th stages

FIG. 10-24 One of many gas turbine engine plasma-plated parts. This part is plated with chrome carbide.

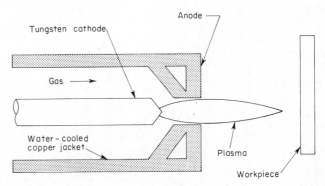

FIG. 10-26 A schematic of the plasma torch.

CONCLUSION

Materials and methods of construction for the gas turbine engine are many, and are rapidly changing. This chapter has dealt with some of the more commonly used substances employed in the manufacture of this type of engine. Advances in design, chemistry, and metallurgy are sure to bring some changes in this area.

REVIEW AND STUDY QUESTIONS

1. What is the relationship between the science of metallurgy and the gas turbine engine?
2. What is the present turbine inlet temperature limit?
3. Define the following terms: strength, ductility, thermal conductivity, corrosion, critical temperature, heat treatability, thermal shock resistance, and hardness.
4. List the methods of working and forming metal.
5. How may the surface or internal properties of a metal be altered?
6. Name some inspection methods.
7. List five alloys that are used in the construction of the gas turbine engine. Describe the properties of each that make it desirable for use in specific locations in the engine.
8. List some of the major metallic elements that are used in gas turbine alloys.
9. What are three characteristics that must be considered in the selection of any metal for use in the gas turbine engine? Discuss these characteristics.
10. What is meant by the term transpiration cooling?
11. List six methods of casting. Give a short description of each method and its advantages and disadvantages.
12. What are the advantages of forging? Give some variations of the forging process.
13. Name as many of the machines as you can that are used in the manufacturing of parts for the gas turbine engine.
14. Discuss the welding processes used in the fabrication of the gas turbine engine.
15. What are some factors which will determine the finishing processes used on a gas turbine part?
16. List the finishing processes and briefly discuss each.

PART THREE
SYSTEMS AND ACCESSORIES

CHAPTER ELEVEN
FUELS

During the early years of gas turbine development it was popularly believed that this type of engine could use almost any material that would burn—and in theory this is true. Of course we now know that, practically speaking, this is not correct. As a matter of fact, because of the high rate (Table 11-1) of consumption, the effect of any impurities in the fuel will be greatly magnified. This, coupled with wide temperature and pressure variations, results in the modern gas turbine engine having a fussy appetite that calls for fuels with carefully controlled characteristics if the engine is to function efficiently and with a reasonable service life under all operating conditions.

TABLE 11-1
Rate of fuel consumption for some piston and turbine engines

Engine Model	U.S. Gal per Hour of Operation
Piston:	
Pratt & Whitney R-1830	45
Pratt & Whitney R-2800	100
Wright R-3350 Turbo Compound	125
Turbine:	
Rolls-Royce Dart	100
Allison 501	175
Rolls-Royce Avon	400
Rolls-Royce Conway	550
Pratt & Whitney JT3	450
Pratt & Whitney JT4	550
General Electric CF6	2800

FUEL SOURCES

Jet fuel is manufactured or refined from crude oil petroleum, which consists of a great number of hydrocarbons, or compounds of the elements hydrogen and carbon. A pound of a typical gas turbine fuel might be composed of approximately 16 percent hydrogen atoms, 84 percent carbon atoms, and a small amount of impurities, such as sulfur, nitrogen, water, and sediment and other particulate matter. (See the Appendix for gallons–weight conversions for a typical jet fuel.) There are literally thousands of combinations that these carbon and hydrogen atoms may form. General families include the paraffins, naphthenes, olefins, and aromatics. Other components in the crude include asphalts, resins, organic acid material, and sulfur compounds. Crude oil from the eastern sections of the country has a relatively large amount of paraffinic hydrocarbons, while crudes taken from the west and gulf coasts contain larger proportions of naphthenic and aromatic hydrocarbons. Crude oil may vary greatly in color, appearance, and in physical characteristics.

JET FUEL FROM CRUDE OIL

The Physical Process (*Fig. 11-1*)

The first step in the process of turning crude oil into gas turbine fuel is to allow the physical impurities such as mud, water, and salts to settle out. Next, the crude must be separated into its various fractions. This process uses heat and is called *fractional distillation*. Each of the various fractions or families of hydrocarbons is of a different size and has a different boiling temperature. The lighter fractions boil at lower temperatures, and as the molecule size increases, the boiling point becomes higher. The asphalts and resins are very large in size and do not boil at all at the temperatures used in the distillation process.

Distillation takes place in a "bubble tower" [Fig. 11-1(b)]. The crude oil is first pumped through a furnace in a rapid continuous flow, where it is heated quickly to a relatively high temperature. Almost all of the crude oil is vaporized simultaneously. The vapors from the furnace are then piped to tall condensers which are the fractionating or bubble towers previously mentioned. As the vapors rise in the towers, the highest boiling point material becomes liquid first, intermediate boiling point materials liquefy next, and lowest boiling point material last. The towers are equipped with a number of trays placed at different levels to catch the liquid that condenses as the vapors are rising. The condensed liquid from each tray is drawn off, separating the crude into as many portions as there are trays. In general practice there are only about seven trays for drawing off distilled fractions of the crude. Each of the fractions will have a range of boiling points, with the fractions and their ranges dependent upon the crude and the number of trays in the fractionating column.

A typical midcontinent crude might yield the following approximate percentages of the various fractions.

Gas	=	3 percent
Gasoline and naphtha	=	18 percent
Kerosene	=	15 percent
Gas-oil	=	39 percent
Lube oil	=	7 percent
Residual material	=	18 percent

The percentages will vary depending upon the source and type of the crude.

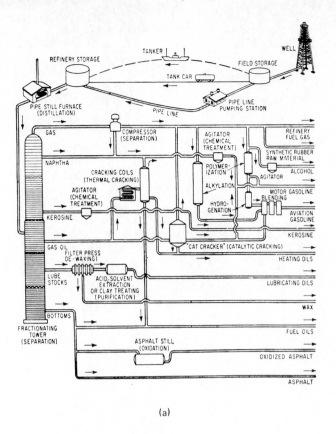

(a)

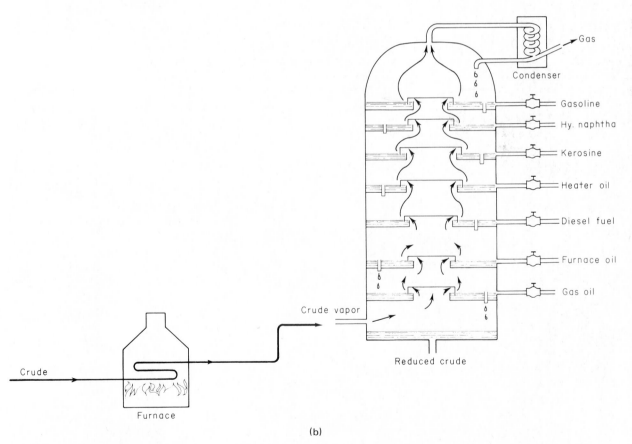

(b)

FIG. 11-1 (a) Petroleum flow chart: from the well to the refinery (*Humble Oil and Refining Co.*)
(b) Detail of the bubble tower used for fractional distillation

The Chemical Process

From the purely physical processes of separation, the products of distillation are further refined to:

1. Remove undersirable components such as sulfur and gums and resins by means of sulfuric acid and other chemicals, or by a selective solvent extraction process.
2. Split heavy molecules or combine lighter ones to obtain more of a particular fraction of the crude. Thermal and catalytic cracking are two methods of splitting large molecules into smaller ones. While in the process of polymerization, two lighter molecules are combined into one large molecule.
3. Impart to a fraction certain desirable properties with the inclusion of "additives." Such additives might include chemical compounds for inhibiting microbial growths. Other additives reduce the tendency for the ever-present water in fuel to form ice crystals at the low temperatures encountered at high altitudes. Lubricating oils also contain many additives.

Step two represents processes performed mainly to develop more gasoline from a barrel of crude.

DEVELOPMENT OF JET FUELS (*TABLE 11-2*)

Various grades of jet fuel have evolved during the development of jet engines in an effort to ensure both satisfactory performance and adequate supply. The JP series have been used by the military and its behavior is outlined in Specification MIL-J-5624. For commercial use the American Society for Testing and Materials (ASTM) has specification D-1655, which covers Jet A, A-1, and B fuels.

JP-1 Fuel

Grade JP-1 fuel was the original low-freezing-point kerosene-type fuel. In the United States, kerosene is required to have a minimum flash point of 120°F [48.9°C] and to have an endpoint in the ASTM distillation test of not more than 572°F [300°C]. (See the following section on fuel tests for a definition of terms used in this section.) Its characteristics were low vapor pressure, good lubricating qualities, and high energy content per unit volume. It was thought to be a safer fuel than gasoline because of its higher flash point. Ker-

---

TABLE 11-2
Comparison of the JP series fuels

	5616	—	5624A	5624A
Specification MIL-F-				
Nominal Grade	JP-1	JP-2	JP-3	JP-4
Flash point, min °F	110.0	—	—	—
Reid vapor press., psi min	—	—	5.0	2.0
Reid vapor press., psi max	—	2.0	7.0	3.0
Initial boiling point, min °F	—	150.0	—	—
10% evaporated point, max °F	410.0	—	—	250.0
90% evaporated point, max °F	490.0	—	400.0	—
Endpoint, max °F	572.0	500.0	600.0	550.0
Color	White	White	White	—
Saybolt color, max	+12.0	+12.0	—	—
Freezing point, max °F	−76.0	−76.0	−76.0	−76.0
Sulfur, % by wt max	0.20	0.20	0.40	0.40
Mercaptans, % by wt max	—	—	0.005	0.005
Inhibitors permitted*	No	No	Yes	Yes
Existent gum, mg/100 ml max	5.0	5.0	10.0	10.0
Accelerated gum (7.0 hr), mg/100 ml max	—	8.0	—	—
Accelerated gum (16.0 hr), mg/100 ml max	8.0	—	20.0	20.0
Corrosion, copper strip	None	None	None	None
Water tolerance permitted	None	None	None	None
Specific gravity (60/60) max†	0.850	0.850	0.802	0.8017
Specific gravity (60/60) min†	—	—	0.728	0.7507
Viscosity at −40°F, centistokes max	10.0	10.0	—	—
Viscosity at 100°F, centistokes min	—	0.95	—	—
Aromatics, % by vol max	20.0	20.0	25.0	25.0
Bromine no. max	3.0	3.0	30.0	30.0
Heating value (lower or net), Btu/lb min	—	—	18,400	18,400

* Inhibitors may be added to the extent required (max 1.0 lb approved inhibitor for each 5000 U.S. gal of finished fuel) to prevent formation of excessive gum during the oxygen bomb test. Several inhibitors are approved for use.

† $\dfrac{\text{Density of fuel at 60°F}}{\text{Density of water at 60°F}}$

osene-type fuels like JP-1 proved to have many disadvantages. Cold-weather starts were quite difficult (in part due to poor ignition and early, less sophisticated, fuel controls), and at high altitudes, kerosene was prone to cause engine flame-outs, and air starts were nearly impossible. Kerosene has a tendency to hold both water and solids in suspension, making filtration and ice formation a problem. In addition, the potential supply of kerosene is more limited than gasoline since more gasoline can be produced from a barrel of crude than kerosene.

JP-2 Fuel

This fuel was an experimental blend of gasoline and kerosene. A large percentage of the blend was kerosene, and therefore it did not appreciably save enough crude oil to warrant its widespread adoption.

JP-3 Fuel

Grade JP-3 fuel with a Reid vapor pressure of 5 to 7 psi [35.5 to 48.3 kPa], a flash point of about −40°F [−40°C], and an endpoint of 550°F [287.8°C], superseded Grade JP-1. The fuel was a blend of 65 to 70 percent gasoline, 30 to 35 percent kerosene, and had handling characteristics very similar to gasoline. Cold-weather starting was improved, as was the chance of an air restart at high altitude. Its chief disadvantages were high vapor locking tendencies and high fuel losses through the aircraft's fuel tank vents during high rates of climb because of both evaporation of the lighter fractions and entrainment of the liquid fuel with the escaping vapor. JP-3 also had poor lubricating characteristics because of the high gasoline content.

JP-4 Fuel

One of the most commonly used fuels for both civil and military jet engines is JP-4. It is a wide-cut blend of kerosene with some naphtha fractions, and gasoline, but has a much lower Reid vapor pressure of 2 to 3 psi [13.8 to 20.7 kPa] and a flash point of about −35°F [−37.2°C]. Its distillation range is 200 to 550°F [93.3 to 287.8°C], and its freezing point is −76°F [−60°C]. The lower vapor pressure reduces fuel tank losses and vapor lock tendencies. The absence of the lighter ends or fractions not only reduces vapor pressure, but reduces the combustion performance during cold-weather and high-altitude starting.

JP-5 Fuel

JP-5 fuel was developed as a heavy kerosene to be blended with gasoline to produce a fuel similar to JP-4 for use on aircraft carriers. The gasoline is carried on board the ship for use in reciprocating engine aircraft. JP-5 has a high flash point of 140°F [60°C], a very low volatility, a distillation range of 350 to 550°F [176.7 to 287.8°C], and a freezing point of −55°F [−50°C] maximum. Because of this low volatility, it can be stored safely in the skin tanks of the ship rather than in the high-priority, protected space in the center of the ship that is required by Av-Gas. The mixed fuel requires only one protected service tank. Several engines are now designed to use straight JP-5. Although cold-weather starts are marginal, the altitude restarting problem appears to have greatly diminished because of the development of high-energy ignition systems.

JP-6/JP-7 Fuel

These fuel specifications were developed by the Air Force for use in supersonic aircraft. Their low freezing point of −65°F [−53.9°C] makes them suitable for use in cold climates and high altitudes. JP-6 is designated as a wide-cut kerosene, with a distillation range of 250 to 550°F [121.1 to 287.8°C]. JP-7 is the fuel used in the Lockheed SR71.

Jet A and A-1 Fuel (*TABLE 11-3*)

The most commonly used commercial fuels are Jet A and Jet A-1. Both are kerosene-type fuels, and both are alike except that Jet A has a freezing point below −40°F [−40°C], and Jet A-1 has a freezing point below −58°F [−50°C]. Another kerosene specification used by British manufacturers is D. Eng. R-D-2482.

Jet B Fuel

Jet B fuel and JP-4 are basically alike. They are wide-boiling-range fuels covering the heavy gasoline-kerosene range and are sometimes called gasoline-type fuels. They have an initial boiling point considerably below that of kerosene. They also have a lower specific gravity. Specifications for current fuels are listed in Table 11-3.

FUEL TESTS

In order to determine the physical and chemical characteristics of a fuel, a number of tests are performed. Most of these tests have been devised by the ASTM, composed of a group of people representing the oil companies, airline operators, and engine manufacturers. The ASTM has also published several fuel specifications defining properties of fuels suitable for commercial gas turbine use. Among these are the *Specification A*, describing a kerosene-type fuel similar to JP-5, and *Specification B*, describing a gasoline-type fuel like JP-4 (both are listed in the preceding section). The Detroit Diesel Allison Division of General Motors, and the Pratt & Whitney Aircraft Group of United Technologies Corporation both have written their own fuel specifications as a guide for airplane operators to follow when purchasing gas turbine fuel. These specifications will produce a fuel similar in volatility characteristics to both JP-4 and JP-5.

TABLE 11-3
Aircraft turbine fuel specifications

TEST	Kerosene-Type Fuels			JP-4 Type Fuels	
	ASTM D-1655 Jet A	D.Eng.R.D. 2482	MIL-J-5624E JP-5	ASTM D-1655 Jet B	MIL-J-5624F JP-4
Gravity, °API	39–51	40–51	36–48	45–57	45–57
Specific gravity	0.8299–0.7753	0.8251–0.7753	0.8448–0.7883	0.8017–0.7507	0.8017–0.7507
Viscosity, centistokes					
0°F	—	6 max	—		
−30°F	15 max	—	16.5 max	—	—
Flash point, °F	110–150	100 min	140 min	—	—
Freezing point, °F	−40 max	−40 max	−55 max	−60 max	−76 max
Pour point, °F	−40 max	—	—		
Color, 18-in Lovibond	—	4 max	—	—	—
Distillation, °F					
10% Evap.	400 max	—	400 max	—	
20%	—	392 (rec.) max	—	290 max	290 max
50%	450 max	—	—	370 max	370 max
90%	—	—	—	470 max	470 max
Endpoint	550 max	572 max	550 max	—	—
Reid vapor pressure, lb	—	—	—	3 max	2–3
Total sulfur, %	0.3 max	0.2 max	0.4 max	0.3 max	0.4 max
Mercaptan sulfur, %	0.003 max	0.005 max	0.001 max	0.003 max	0.001 max
Aromatics, % vol	20 max	20 max	25 max	20 max	25 max
Olefins, %	—	5 max	5 max	5 max	5 max
Net heating value, Btu/lb	18,400 min	18,300 min	18,300 min	18,400 min	18,400 min
Aniline-gravity constant	—	4500 min	4500 min	5250 min	5250 min
Combustion properties					
(1) Luminometer no. min	45	—	—	50	—
or					
(2) Smoke point, min	25	—	19	—	—
or					
(3) Smoke point, min	20	—	—	—	—
Burning test, 16 h	Pass	—	—	—	—
or					
(4) Smoke point, min	20	—	—	—	—
Naphthalenes, max %	3	—	—	—	—
or					
(5) Smoke volatility index, min	—	—	—	54	52
Burning test, 16 h	Pass	—	—	—	—
Copper-strip corrosion					
3 h at 122°F	1 max	—	1 max	—	—
2 h at 212°F	—	1 max	—	1 max	1 max
Water reaction, ml	2 max	1 max	1 max	1 max	1 max
Separometer	—	—	85	—	95
Existent gum, mg/100 ml	7 max	3 max	7 max	7 max	7 max
Accelerated gum, mg/100 ml	14 max	6 max	14 max	14 max	14 max
Total acidity	—	Nil max	—	—	—
Thermal stability					
Preheater tube deposits, 300°F	Less than code 3	—	Less than code 3	Less than code 3	Less than code 3
Filter pressure drop, 400°F, max	12	—	13	12	13
Additives	See note 3				

NOTES

1 The above specifications are considered only a summary. In case of question the detailed specification must be consulted.

2 ASTM Jet A-1 is identical to Jet A except that the freezing point is −58°F max and the pour point is eliminated.

3 In general ASTM specifications permit approved oxidation and corrosion inhibitors and metal deactivators. However, the quantities and types must be declared and agreed to by the consumer. Military specifications permit the inclusion of oxidation inhibitors. MIL-J-5624E, Grade JP-4, requires the addition of anti-icing additive and corrosion inhibitor.

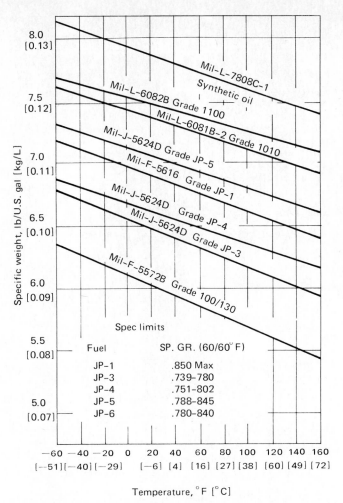

FIG. 11-2 Temperature effects on density of aviation fuels and oils.

Description of Fuel Tests (*FIG. 11-2*)

SPECIFIC GRAVITY This is the ratio of the weight of a substance (fuel in this case) to an equal volume of water at 60°F [15.6°C]. Most often the specific gravity is given in terms of degrees A.P.I. This is a scale arbitrarily chosen by the American Petroleum Institute in which the specific gravity of pure water is taken as 10. Liquids lighter than water have values greater than 10, and those liquids heavier than water have a value smaller than 10.

$$\text{Degrees A.P.I.} = \frac{141.5}{\text{sp. gr. at } 60°F} - 131.5$$

For example: JP-4 has a minimum specific gravity in degrees A.P.I. of 57.

$$57 = \frac{141.5}{x} - 131.5$$

$$x = 0.7507 \text{ sp. gr.}$$

Specific gravity is an important factor since fuel is metered and sold by volume. To arrive at the correct volume, gravity must be known. Notice that the formula indicates that the specific gravity is affected by temperature (Fig. 11-3).

ANILINE-GRAVITY CONSTANT This is the product of two fuel properties which has been related to the net heating value. By measuring fuel aniline point and gravity and using standard conversion tables, it is unnecessary to burn a fuel sample to obtain the heating value.

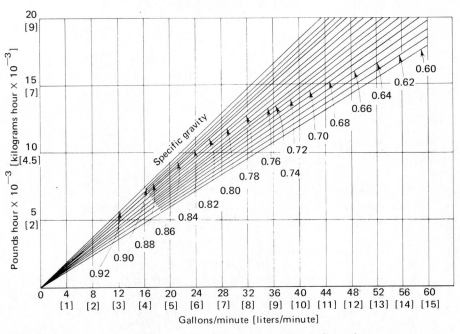

FIG. 11-3 Fuel conversion chart: gallons per minute to pounds per hour.

SMOKE POINT This is the height of a flame in millimeters when the fuel is burned in a special lamp. This height is measured when the flame just starts to smoke.

LUMINOMETER This is a test in which the radiation from a special lamp is measured by a photocell and is compared with the radiation from reference fuels. Increasing luminometer numbers indicate decreasing radiation and therefore better combustion characteristics.

SMOKE VOLATILITY INDEX This is a mathematical combination of smoke point and fuel distillation.

COPPER STRIP CORROSION This is a test to measure the corrosivity of fuel toward copper. Copper appearance is rated numerically with increasing numbers indicating increasing corrosion.

WATER REACTION This is a test to check the separation characteristics of fuel and water. To be accurate it must be conducted under strictest laboratory conditions.

SEPAROMETER This is a test in which a fuel-water emulsion is pumped through a miniature filter separator. Water removal is rated by measuring the haziness of the filtered fuel with a photocell.

EXISTENT GUM This is the amount of nonvolatile material present in fuel. Such material is usually formed by the interaction of fuel and air.

ACCELERATED GUM This is the amount of nonvolatile material formed when a fuel is put into contact with pure oxygen at high pressure and elevated temperature.

BURNING TEST, 16 H Kerosene is burned in a standard lamp for 16 h. Changes in flame shape, density, and color of deposit as well as wick condition are reported.

TOTAL ACIDITY This is the acidity of fuel.

VISCOSITY This is measured in centistokes, and gives an indication of the fuel's ability to flow at different temperatures (Fig. 11-4).

REID VAPOR PRESSURE This is the approximate vapor pressure exerted by a fuel when heated to 100°F [37.8°C]. This value is important in that it indicates the tendency of fuel to "vapor lock."

A.P.I. GRAVITY This is an indication of liquid density as measured by the buoyancy of special hydrometers. Increasing A.P.I. numbers indicate decreasing liquid density. Water has an A.P.I. gravity of 10/0 at 60°F [15.6°C].

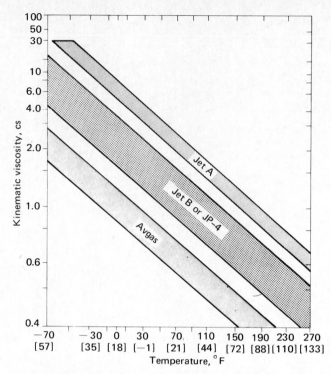

FIG. 11-4 Viscosity range of aviation fuels. The viscosity change influences the rate of flow through a given size filter.

FLASH POINT This is the liquid temperature at which vapors from the heated liquid can first be ignited by a flame under closely controlled conditions.

FREEZING POINT This is the temperature at which solids such as wax crystals separate from a fuel upon cooling.

POUR POINT This is the lowest temperature at which a fuel or oil will pour from a special test tube, cooled at a specified rate. Pour point is measured at 5°F [2.8°C] intervals.

COLOR This is the color of a fuel compared with a numbered standard color. Jet fuel grades vary from water white to light yellow.

DISTILLATION These are the temperatures at which various portions of fuel boils under closely controlled conditions. Since any turbine fuel is a mixture of many hydrocarbons which all have differing boiling points, a turbine fuel does not have a single boiling point, but boils over a range of temperatures. The distillation range is an approximation of a fuel's boiling range.

NET HEATING VALUE This is the amount of heat liberated when a pound or gallon of fuel is burned completely. A correction is included for the heat removed to condense the water formed in this burning. This value, which is usually expressed in Btu, influences the range of a particular aircraft, for where the limiting factor is the capacity of the aircraft tanks, the calorific value per unit volume should be as high as possible, thus enabling more energy, and hence more aircraft range, to be obtained from a given volume of fuel. When the useful payload is the limiting factor, the calorific value per unit of weight should be as high as possible, because more energy can then be obtained from a minimum weight of fuel. Other factors, which affect the choice of heat per unit of volume or weight, such as the type of aircraft, the duration of flight, and the required balance between fuel weight and payload must also be taken into consideration (Table 11-4).

TOTAL SULFUR This is the total amount of sulfur in a fuel.

MERCAPTAN SULFUR These are special sulfur compounds which smell unpleasantly and are corrosive to certain materials.

AROMATICS These are certain unsaturated hydrocarbon compounds. These have a higher ratio of carbon to hydrogen than other hydrocarbons.

OLEFINS These are unsaturated hydrocarbons formed by cracking processes. Such compounds are not normally found in crude oils.

NAPHTHALENES These are certain highly unsaturated hydrocarbon compounds felt to have poor combustion characteristics.

THERMAL STABILITY This is the tendency of a fuel at high temperatures to form deposits on a heater tube and to form material which will plug a fuel line filter. Both properties are measured in apparatus which holds the heater tube and the filter at high temperatures.

FUEL HANDLING AND STORAGE

The amount of fuel consumed by many gas turbine engines makes the delivery of clean, dry fuel essential to proper engine performance. This makes handling and storage of the fuel of prime importance to the operator of gas turbine engines. Elaborate methods of filtration for the removal of solids, which may take the form of rust, scale, sand or dust, pump wear, rubber or elastomers, and lint or other fibrous materials are employed.

Water, in gas turbine fuels, is a particularly troublesome problem in both aircraft and fuel storage tanks because of the fuel's affinity to water. This has led to malfunctioning of fuel controls, ice plugging of fuel filters, and freezing of fuel boost and transfer pumps.

TABLE 11-4
Comparison of net heating values by unit weight and unit volume*

Gravity		Density	Net Heat of Combustion at Constant Pressure, Q_p		
Degrees A.P.I. at 60°F	Specific at 60°/60°F	lb/gal	cal/g	Btu/lb	Btu/gal
40	0.8251	6.879	10,280	18,510	127,300
41	0.8203	6.839	10,300	18,530	126,700
42	0.8155	6.799	10,310	18,560	126,200
43	0.8109	6.760	10,320	18,580	125,600
44	0.8063	6.722	10,330	18,600	125,000
45	0.8017	6.684	10,340	18,620	124,400
46	0.7972	6.646	10,360	18,640	123,900
47	0.7927	6.609	10,370	18,660	123,300
48	0.7883	6.572	10,380	18,680	122,800
49	0.7839	6.536	10,390	18,700	122,200
50	0.7796	6.500	10,400	18,720	121,700
51	0.7753	6.464	10,410	18,740	121,100
52	0.7711	6.429	10,420	18,760	120,600
53	0.7669	6.394	10,430	18,780	120,100
54	0.7628	6.360	10,440	18,800	119,500
55	0.7587	6.326	10,450	18,810	119,000
56	0.7547	6.292	10,460	18,830	118,500
57	0.7507	6.258	10,470	18,850	118,000
58	0.7467	6.225	10,480	18,870	117,500
59	0.7428	6.193	10,490	18,880	116,900
60	0.7389	6.160	10,500	18,900	116,400

* Heat energy per pound of fuel decreases with increasing molecular weight or specific gravity but heat energy per gallon of fuel increases under the same conditions.

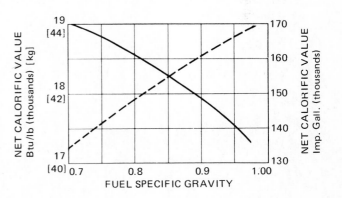

In addition, certain microbial and fungal growths thrive on the interface provided by the water environment and the hydrocarbon food supply necessary for their existence. The growth of the "bug" population results in a buildup of bacterial slime which can clog small metering orifices and penetrate the coatings and sealants used in fuel tanks, and thus expose the aluminum surfaces to corrosion.

The Phillips Petroleum Company has developed a fuel additive PFA 55MB which is now being added to all JP-4 turbine fuels. Many commercial fuel suppliers are also using this additive. This substance provides excellent biocidal and anti-icing protection. Dispensers are available for adding PFA 55MB (Prist) at those airports that do not sell treated fuel.

Water in fuel takes two forms.

1. *Dissolved water*—Dissolved water is similar to humidity in the atmosphere, and is a function of the temperature and the type of fuel (Fig. 11-5). The dissolved water content may be as much as 1 pint [0.473 L] per 1000 gal [3785 L] of fuel. The concentration may be checked with the use of a hydrokit water-detector system.
2. *Free water*—Free water may take two forms,
   (a) *Entrained water*—when the water in a fuel is suspended in minute globules. The water may not be perceptible to the naked eye, but it may cause the fuel to take on a hazy appearance.
   (b) *Water slugs*—visible droplets or pools of water.

The amount of free water is usually checked with litmus paper, with water-detecting paste, or by drawing off a small amount of fuel and observing its clarity.

Water problems are solved in a number of ways.

1. *Settling*—Most entrained water will eventually settle out of the fuel, but it takes a much longer time (approximately 4 times as long) for the water to settle out of turbine fuel than out of Av-Gas due to its more viscous nature and higher specific gravity (Fig. 11-6).
2. *Coalescing tanks*—These are packed with Fiberglas or other cellulose material which causes the finely divided water particles to join together, creating large globules which then settle to a sump where they can be drained off (Fig. 11-7).

In addition to its slow settling time, and its affinity to water, the fuel vapor-air mixture above the surface of JP-4 and Jet B fuel in a storage tank or fuel cell is nearly always in the explosive range through the wide temperature extremes of 80 to −10°F [26.7 to −23.3°C], whereas Av-Gas in storage forms a mixture that is too rich to burn, and kerosene forms a mixture that is too lean to burn.

Precautions While Handling Jet Fuels

Because of the reasons indicated above, special precautions must be taken when handling or storing jet fuels.

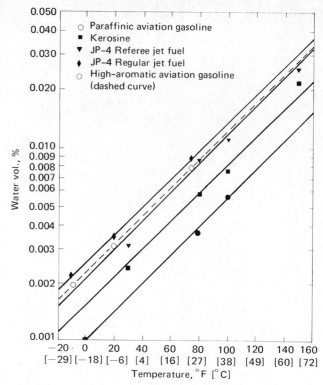

FIG. 11-5 Solubility of water in aviation fuels.

1. Static electrical charges accumulate very rapidly with high fuel flow rates, high gravity fuels, and wider boiling range fuels. Flow rates must be restricted to a specific maximum, depending upon the hose diameter. Grounding or bonding is an absolute essential.
2. Observe all "No Smoking" requirements.
3. Since jet fuels tend to soften asphalt and do not evap-

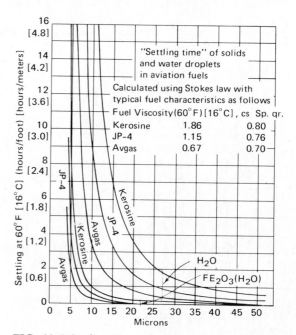

FIG. 11-6 Settling time of solids and water droplets in aviation fuels.

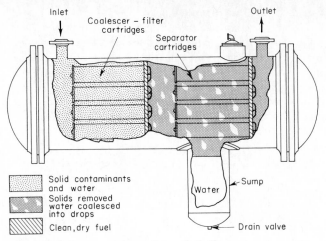

FIG. 11-7 Cutaway view of a typical filter-separator.

Legend:
Solid contaminants and water
Solids removed water coalesced into drops
Clean, dry fuel

Labels: Inlet, Coalescer – filter cartridges, Separator cartridges, Outlet, Water, Sump, Drain valve

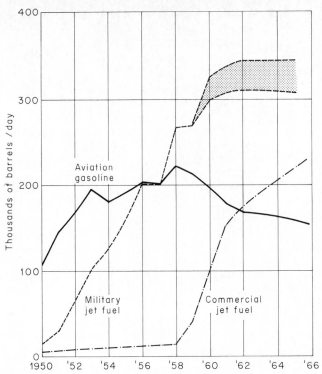

FIG. 11-8 U.S.A. aviation fuel requirements. Military fuel demands are uncertain. Projected commercial jet fuel requirements to 1980 is well over 1.5 million barrels/day.

orate readily, spillage should be avoided. Remove small amounts of jet fuel with a commercial absorbing agent. Wash large spills with copious amounts of water. Caution should be observed to see that the fuel is not washed into a sanitary or storm sewer system.

4. Approved fire extinguishing equipment should be readily available.

5. Jet fuels should not be used for cleaning purposes. Excessive inhalation and skin contact should be avoided. The skin should be washed thoroughly with soap and water, and clothing should be removed and laundered as soon as possible after contact. Because jet fuels are less volatile than gasoline, they do not evaporate as readily and therefore are more difficult to remove from clothing.

Future Developments *(FIG. 11-8)*

In recent years there have been large-scale efforts by the government and industry to develop more powerful and more suitable fuels for supersonic and hypersonic aircraft. Trends in this direction include research into the following areas (Fig. 11-9):

1. Jet fuels with a thermal stability up to 700°F [371.1°C] and a heat of combustion of 18,400 Btu/lb [2103.2 cal]. Thermal stability is important because at Mach 2, skin temperatures reach 194°F [90°C] and at Mach 3, 482°F [250°C]. Under these conditions, trace components in a fuel concentrate into deposits which plate out on critical surfaces and block fine orifices in the fuel system.

2. Jet fuels that are capable of absorbing the aerodynamic heat generated by aircraft operating in the Mach 3 to Mach 6 range. An interesting development in this area is the endothermic fuel. Basically, the idea is to use a chemical that decomposes into a good fuel mixture as a result of a heat-absorbing (endothermic) reaction. This reaction not only boosts the heat-sink capacity, but also increases the amount of energy that can be extracted from any given fuel. *Reason:* The heat taken

up during the reaction has to be released when the fuel is burned. Thus the fuel, which initially enters the system as a liquid, constantly soaks up heat as it cools the engine walls and changes from the liquid to the vapor phase. It continues to absorb heat and then undergoes the endothermic reaction that decomposes the chemical or alters its structure.

3. High-density jet fuels that have a high-energy content per unit volume for use on volume-limited aircraft such as fighter airplanes and missiles.

4. Low volatility fuels because of high engine temperatures and high altitudes.

5. Jet fuels with adequate low-temperature characteristics

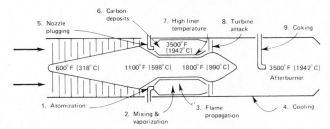

FIG. 11-9 Engine operating conditions compared with the fuel properties which affect those conditions. Fuel properties: (1) viscosity, surface tension; (2) volatility, vapor pressure; (3) kinetics; (4) specific heat; (5) thermal stability (6) heavy aromatics; (7) luminosity; (8) sulfur metals; (9) thermal cracking.

(viscosity and freezing point) for cold weather starting and subsonic operation.

6. Jet fuels with low water content to prevent the formation of ice resulting from the low temperatures encountered at high altitudes.
7. Gelled fuels to reduce the danger of fire in the event of an abnormal landing.

A radically different approach to the problem of finding suitable fuels for supersonic engines would involve the use of Liquefied Natural Gas (LNG). The advantage claimed for such a product is greater thermal stability, allowing cleaner burning at Mach 3 temperatures without forming varnishes and other deposits that could foul injectors. In addition, LNG would burn with a lower radiant heat output, resulting in lower metal temperature of the combustion chamber component. Another advantage claimed for this fuel is its high hydrogen/carbon ratio and therefore higher heat content per unit weight.

Because it is a cryogenic fluid with a boiling point of $-260°F$ $[-162.2°C]$, LNG offers substantially more heat-sink capacity than present fuels. This greater cooling capacity could be used to cool the turbine cooling air and thus raise cycle efficiency or increase engine life.

Liquefied natural gas also has some drawbacks. Its extremely low temperature and volatility could cause handling and storage problems. The low density of LNG—roughly half that of kerosene—means lower heat content per unit volume. In addition, completely new servicing and distribution methods would be required. Cost and availability would have to be determined.

Some other considerations involved in developing new fuels are those of economy, compatibility with engine and airframe materials, availability, and storageability. In addition, safety in storage might necessitate fuel tank inerting or filling the space on top of the fuel with an inert gas such as nitrogen, and thus precluding the likelihood of explosion.

REVIEW AND STUDY QUESTIONS

1. What is the source of gas turbine fuels?
2. List the steps in the refining process.
3. What is the military specification number with which gas turbine fuels must comply; the civil specification number?
4. List the JP series of fuels. Give a brief description of each.
5. Describe the two generally used fuels for civil aircraft.
6. Make a table of some major fuel tests. Tell why each test is performed.
7. Of what significance is the net heating value of a fuel?
8. Why is water such a problem in relation to gas turbine fuel? In what form may water exist in turbine fuel? How is the water problem solved?
9. What precautions should be taken while handling turbine fuels?
10. List some possible future developments for turbine fuels.

CHAPTER TWELVE
FUEL SYSTEMS
AND
COMPONENTS

FUEL CONTROLS

Introduction

Depending upon the type of engine and the performance expected of it, fuel controls may range in complexity from simple valves to automatic computing controls containing hundreds of intricate, highly machined parts.

Strictly speaking, a pilot of a gas-turbine-powered airplane does not directly control the engine. The pilot's relation to the power plant corresponds to that of the bridge officer on a ship who obtains engine response by relaying orders to an engineer below deck who, in turn, actually moves the throttle of the engine (Fig. 12-1). But before moving the throttle, the engineer monitors certain operating factors which would not be apparent to the captain, such as pressures, temperatures, and rpm. The engineer then refers to a chart and computes a fuel flow or throttle movement rate which will not allow the engine to exceed its operating limits.

Types of Controls

Modern fuel controls can be divided into two basic groups, hydromechanical and electronic, and they may sense some or all of the following engine operating variables:

1. Pilot's demands
2. Compressor inlet temperature
3. Compressor discharge pressure
4. Burner pressure
5. Compressor inlet pressure
6. Rpm
7. Turbine temperature

The more sophisticated controls will sense even more operating parameters.

There are as many variations in controls as there are fuel control manufacturers. Although each type of fuel control has its particular advantage, most controls in use today are of the hydromechanical type. Regardless of type, all controls accomplish the same things, although some may sense more of the aforementioned variables than others. At best, fuel controls are extremely complex devices composed of speed governors, servo systems and feedback loops, valves, metering systems, and various sensing mechanisms. The electronic fuel controls contain thermocouples, amplifiers, relays, electrical servo systems, switches, solid-state devices, and solenoids.

The discussion of fuel-control theory will limit itself mainly to the hydromechanical type, but included in this section (Fig. 12-2) is a schematic showing the general functions of the components used in an integrated electronic fuel control system. A much later combination hydromechanical and electronic fuel control is shown in Fig. 12-3. Also included in this section is the electronic fuel trimming mechanism used on the Detroit Diesel Allison 501-D13 engine.

Theory of Operation

The simplest conceivable control would consist of a plain metering valve to regulate fuel flow to the engine. This type of control could be installed on an engine used for thrust or the generation of gas (Fig. 12-4).

Some refinements might include a:

1. Pump to pressurize the fuel
2. Shutoff valve to stop fuel flow
3. Relief valve to protect the control when the shutoff valve is closed
4. Minimum fuel control adjustment to prevent complete stoppage of fuel by the metering valve

A flow of fluid may be metered by keeping the pressure drop or difference across the metering valve a constant value while varying the valve orifice, or the valve orifice may be kept a constant size and the pressure difference varied. Most modern fuel controls meter fuel by using the first method, so an additional refinement would consist of a device to maintain a constant pressure drop across the metering valve regardless of the

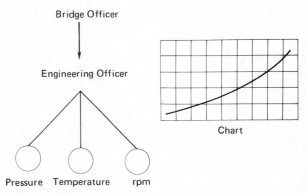

FIG. 12-1 Boat power-plant control analogy to a fuel control.

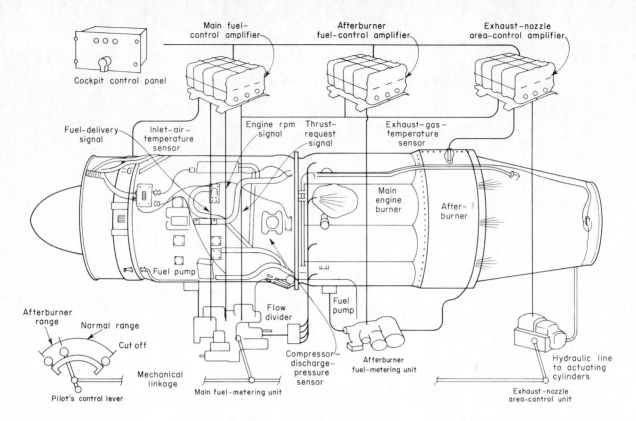

Cockpit control panel This panel provides various indicator lights and control switches including a starter switch. A selector lever allows the pilot to transfer to hydromechanical control for comparison or as an emergency measure.

Pilot's control lever The pilot selects engine thrust by the position of this lever. Regardless of how far or how rapidly the lever is moved, automatic features ensure maximum rates of engine acceleration or deceleration, but within safe engine operating limits.

Main fuel-control amplifier The "thrust request" from the pilot's control lever is signaled electrically to this amplifier; also it receives sensor signals covering various engine operating conditions. From these a control signal is integrated and sent out to the metering unit.

Main fuel-metering unit The fuel-metering valve of this unit is electrically controlled from the main fuel-control amplifier. In this way, fuel is metered to the engine spray nozzles in response to the integrated fuel demands, and within safe operating limits.

FIG. 12-2 An early integrated electronic fuel control system.

Afterburner fuel-control amplifier When the requested engine thrust calls for afterburner operation, this amplifier signals the afterburner fuel-metering unit, causing afterburner lightup. Following that it regulates fuel metering to the afterburner for the additional thrust.

Afterburner fuel-metering unit This fuel-metering unit controls lightup and fuel flow to the afterburner spray nozzles. It is electrically regulated from the afterburner fuel-control amplifier. The amplifier also senses and integrates suitable engine parameters.

Exhaust-nozzle area control amplifier This amplifier receives a signal from a thermocouple which senses the temperature of the exhaust gases. This signal is electronically compared with other engine operating conditions. A resulting working signal directs the nozzle area control.

Exhaust-nozzle area control unit This is a hydraulic unit which is under electrical control from the amplifier. It serves to position the exhaust-nozzle area mechanism through hydraulic actuating cylinders. Nozzle area is varied for optimum operating conditions.

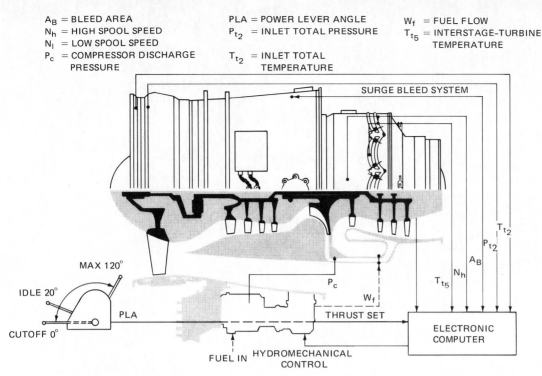

FIG. 12-3 The Garrett AiResearch TFE731 fuel control.

pressure level on either side of the valve, or the valve opening.

The fuel components discussed are important. It would take a very careful operator to run the engine without one or more of these refinements. Another component which is only slightly less essential is an acceleration limiter. Since these engines are internally air cooled, much of the air pumped by the compressor is used to cool the combustion gases to the point where they can run the turbine without melting the blades. In order to accelerate the engine, the fuel flow is increased, but only to the point where the limiting temperature is reached. As the engine accelerates and increases the air-flow, more fuel can be added. This function can be performed by the operator, but if it must be done often, it can probably be done better and more cheaply by an automatic device. If turbine inlet temperature were the only engine limitation, a control sensing this tempera-ture could be used, although such controls are generally complex and expensive. In most cases it is also necessary to avoid the compressor surge and stall lines (Fig. 12-5). (Refer to Chap. 5.) Since a good incipient stall de-tector has not yet been developed, it is necessary to schedule the accelerating fuel in accordance with some engine parameter or combination of engine parameters. When the shortest possible acceleration time is impor-tant, the control becomes rather involved. This is the primary reason why aircraft gas turbine controls are so complicated. For smaller engines with less critical com-pressors, or for applications where the cost and sim-plicity of the control are more important than optimum performance, a simpler control is used, giving equally

effective engine protection, but with a longer accelera-tion time. Compressor discharge pressure or burner pressure is commonly used as the sensed variable for these simple controls, since each varies with both engine speed and inlet air conditions, and thus gives a fair in-dication of the amount of fuel which can be burned safely.

The amount of fuel required to run the engine at rated rpm varies with the inlet air conditions. For example, it requires less fuel to run the engine on a hot day than on a cold day. In order to relieve the operator of the necessity of resetting the power lever, the final refine-ment, a speed governor, is added to the simple fuel control. A speed governor becomes necessary when:

1. The turbojet is used in an airplane where air inlet con-ditions change drastically.
2. The simple machine delivers hot gases to another wheel, or is subject to variation in back pressure from any other source.
3. Any shaft power is taken from the machine so that the fuel flow required is a function of load as well as speed.

A simple governor consists of flyweights balanced by a spring. When the engine is running unloaded, at rated speed, the metering valve is open only far enough to supply the small fuel flow required. If a load is applied to the engine, the speed will decrease. This causes the flyweights to move in under the force of the spring, and the fuel valve to open wider and admit more fuel. With the additional fuel, the engine picks up speed again, and as the set speed is reached, the flyweights

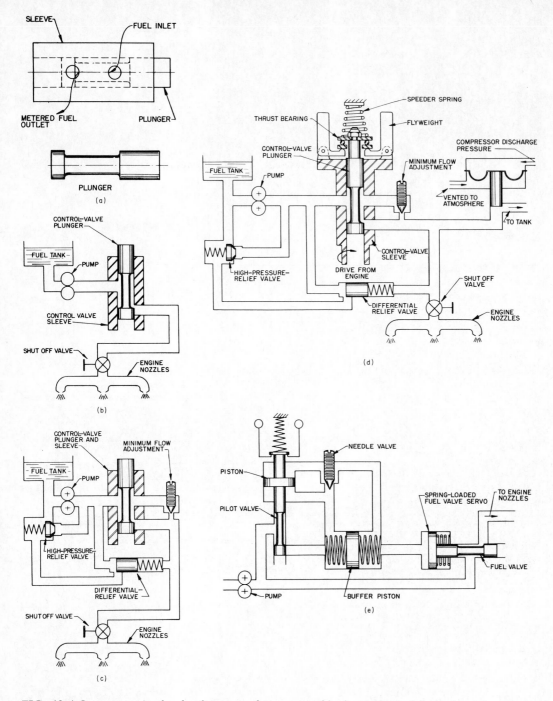

FIG. 12-4 Some steps in the development of one type of hydromechanical fuel control.

(a) Basic fuel control.

(b) Adding a shutoff.

(c) Adding a high-pressure relief valve, a differential relief valve, and a minimum-flow valve.

(d) Adding a governor and an acceleration scheduling valve.

(e) Adding a droop control. (*Woodward Governor Company.*)

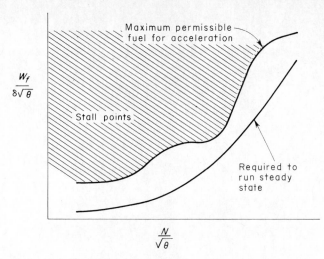

FIG. 12-5 Typical acceleration limit curve.

move the fuel valve in the closing direction until the proper steady-state fuel flow is reached. Note, however, that since the engine now requires more fuel to carry the load, the fuel valve must be farther open than it was at no load. This causes the flyweights to run slightly farther in, so that the spring is relaxed somewhat and exerts a little less force. This means that the system will come to equilibrium at a speed slightly lower than the unloaded speed. Thus, as load is progressively added to the engine, the speed will progressively decrease. If speed versus load is plotted, a drooping line would result (Fig. 12-6). Therefore, there is always this characteristic "speed droop," sometimes called *regulation* or *proportional control*. The droop characteristic is common to all mechanical governors. It is a good thing from the standpoint that it makes the engine-governor system stable, but it does mean that the engine is either running below rated speed when loaded or above rated speed when unloaded.

The problem of droop can be reduced in various ways.

1. Use a weaker spring which requires a lower force necessary for valve movement.

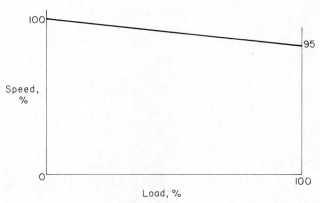

FIG. 12-6 The effect of speed droop.

2. Make the metering port very wide radially so that a very small movement of the valve is required.
3. Increase the pressure differential across the metering valve to force the required fuel through a smaller opening of the valve.

All of these methods have limitations. The problem is solved in several different ways, some of which are discussed in the detailed examination of the following four fuel controls.

Many modern fuel controls will also incorporate:

1. Servo systems to boost weak input signals and thus make the control more sensitive.
2. Devices to prevent "undershooting" and "overshooting" by returning the metering valve to its desired position before the governor alone can do the job.
3. Auxiliary functions such as inlet guide vane positioning and nozzle, afterburner, and thrust reverser signals. Several other auxiliary functions are discussed on the following pages.

FOUR FUEL CONTROLS

The principle of operation of several fuel controls used on current jet engines is examined in this section. It would be difficult to include information on all of the fuel controls now used. The controls selected for discussion are those commonly used and/or those illustrating certain fuel-metering principles. Included are the following controls:

1. DP-F2 manufactured by the Bendix Energy Controls Division and used on the UACL PT6.
2. AP-B3 manufactured by Bendix and used on the 501-D13 (T56) with electronic fuel trimming.
3. 1307 manufactured by Woodward and used on the CJ805 (J79).
4. JFC25-7 manufactured by Hamilton Standard and used on the JT4. (This control with some change is similar to the JFC25-16 being used on the JT3D fan engine.)

BENDIX DP-F2 FUEL CONTROL (FIG. 12-7)

General Description

The model DP-F2 gas turbine fuel control is mounted on the engine-driven fuel pump and is driven at a speed proportional to gas-producer turbine speed N_g. The control determines the proper fuel schedule for the engine to provide the power required as established by the throttle lever. This is accomplished by controlling the speed of the gas-producing turbine, N_g. Engine power output is directly dependent upon gas-producer turbine speed. The fuel control governs N_g, thereby actually

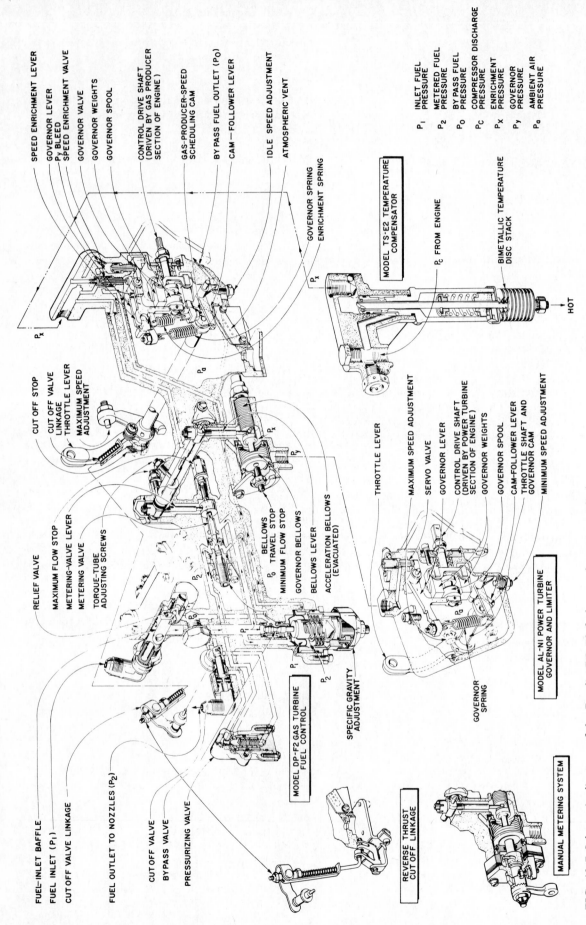

SPEED ENRICHMENT LEVER
GOVERNOR LEVER
Py BLEED
SPEED ENRICHMENT VALVE
GOVERNOR VALVE
GOVERNOR WEIGHTS
GOVERNOR SPOOL
CONTROL DRIVE SHAFT (DRIVEN BY GAS PRODUCER SECTION OF ENGINE)
GAS-PRODUCER-SPEED SCHEDULING CAM
BY PASS FUEL OUTLET (Po)
CAM—FOLLOWER LEVER
IDLE SPEED ADJUSTMENT
ATMOSPHERIC VENT
GOVERNOR SPRING
ENRICHMENT SPRING

P_1 INLET FUEL PRESSURE
P_2 METERED FUEL PRESSURE
P_o BY PASS FUEL PRESSURE
P_c COMPRESSOR DISCHARGE PRESSURE
P_x ENRICHMENT PRESSURE
P_y GOVERNOR PRESSURE
P_a AMBIENT AIR PRESSURE

MODEL TS-E2 TEMPERATURE COMPENSATOR
P_c FROM ENGINE
BIMETALLIC TEMPERATURE DISC STACK
HOT

CUT OFF STOP
CUT OFF VALVE LINKAGE
THROTTLE LEVER
MAXIMUM SPEED ADJUSTMENT

THROTTLE LEVER
MAXIMUM SPEED ADJUSTMENT
SERVO VALVE
GOVERNOR LEVER
CONTROL DRIVE SHAFT (DRIVEN BY POWER TURBINE SECTION OF ENGINE)
GOVERNOR WEIGHTS
GOVERNOR SPOOL
CAM-FOLLOWER LEVER
THROTTLE SHAFT AND GOVERNOR CAM
MINIMUM SPEED ADJUSTMENT

RELIEF VALVE
MAXIMUM FLOW STOP
METERING-VALVE LEVER
METERING VALVE
TORQUE-TUBE ADJUSTING SCREWS

BELLOWS TRAVEL STOP
MINIMUM FLOW STOP
GOVERNOR BELLOWS
BELLOWS LEVER
ACCELERATION BELLOWS (EVACUATED)

GOVERNOR SPRING
MODEL AL-N1 POWER TURBINE GOVERNOR AND LIMITER

FUEL-INLET BAFFLE
FUEL INLET (P1)
CUT OFF VALVE LINKAGE
FUEL OUTLET TO NOZZLES (P2)

CUT OFF VALVE
BYPASS VALVE
PRESSURIZING VALVE

MODEL DP-F2 GAS TURBINE FUEL CONTROL

SPECIFIC GRAVITY ADJUSTMENT

REVERSE THRUST CUT OFF LINKAGE

MANUAL METERING SYSTEM

FIG. 12-7 Schematic diagram of the Bendix DP-F2 fuel control system with the AL-N1 power turbine governor and limiter, and the TS-E2 temperature compensator. (Bendix Energy Controls Division.)

governing the power output of the engine. Control of N_g is accomplished by regulating the amount of fuel supplied to the engine burner chamber.

The model TS-E2 temperature compensator alters the acceleration fuel schedule of the fuel control to compensate for variations in compressor inlet air temperature. Engine characteristics vary with changes in inlet air temperature, and the acceleration fuel schedule must, in turn, be altered to prevent compressor stall and/or excessive turbine temperatures.

On some engines there is no model AL-N1 power turbine governor. The fuel topping governor function is performed by the propeller governor. A pneumatic line P_y from the drive body adapter of the DP-F2 fuel control connects with the propeller overspeed governor.

An engine-furnished starting fuel control incorporates the cutoff and pressurizing valves. A link between the starting control lever and the lever on the DP-F2 fuel control is employed to provide high idle when this position of the starting control lever is selected.

Variations of this basic fuel control are used on the Detroit Diesel Allison Division model 250 series engine, some models of the Garrett AiResearch TPE331, and several other engines.

Fuel Section

The fuel control is supplied with fuel at pump pressure P_1. Fuel flow is established by a metering valve and bypass valve system. Fuel at P_1 pressure is applied to the entrance of the metering valve. The fuel pressure immediately after the metering valve is called metered fuel pressure P_2.

The bypass valve maintains an essentially constant fuel pressure differential (P_1-P_2) across the metering valve. The orifice area of the metering valve will change to meet specific engine requirements. Fuel pump output in excess of these requirements will be returned to the pump inlet. This returned fuel is referred to hereafter as P_0.

The bypass valve consists of a sliding valve working in a ported sleeve. The valve is actuated by means of a diaphragm and spring. In operation, the spring force is balanced by the P_1-P_2 pressure differential working on the diaphragm. The bypass valve will always be in a position to maintain the P_1-P_2 differential and bypass fuel in excess of engine requirements.

A relief valve is incorporated parallel to the bypass valve to prevent a buildup of excessive P_1 pressure in the control body. The valve is spring-loaded closed and remains closed until the inlet fuel pressure P_1 overcomes the spring force and opens the valve. As soon as the inlet pressure is reduced, the valve again closes.

The metering valve consists of a contoured needle working in a sleeve. The metering valve regulates the flow of fuel by changing the orifice area. Fuel flow is a function of metering valve position only, since the bypass valve maintains an essentially constant differential fuel pressure across this orifice regardless of variations in inlet or discharge fuel pressures.

The pressurizing valve is located between the metering valve and the cutoff valve. Its function is to maintain adequate pressures within the control to assure correct fuel metering.

The cutoff valve provides a positive means of stopping fuel flow to the engine. During normal operation, this valve is fully open and offers no restriction to the flow of fuel to the nozzles.

An external adjustment is provided on the bypass valve spring cover which was initially intended to compensate for the difference in specific gravity of various fuels. It is sometimes used to match accelerations between engines on multiengine installations. Compensation for variations in specific gravity resulting from changes in fuel temperature is accomplished by the bimetallic disks under the bypass valve spring.

Throttle Input, Speed Governor, and Enrichment Section

Figure 12-8 illustrates details of the governor and enrichment levers. Views A and B identify the individual levers and their relationship to each other. Views C and D show these levers in operation. The following text is coordinated with Fig. 12-8.

The throttle input shaft incorporates a cam which depresses an internal lever when the throttle is opened. A spring connects this cam follower lever to the governor lever. The governor lever is pivoted and has an insert which operates against an orifice to form the governor valve. The enrichment lever pivots at the same point as the governor lever. It has two extensions which straddle a portion of the governor lever so that after a slight movement a gap will be closed and then both levers must move together. The enrichment lever actuates a fluted pin which operates against the enrichment "hat" valve. Another smaller spring connects the enrichment lever to the governor lever. A roller on the arm of the enrichment lever contacts the end of the governor spool.

The speed scheduling cam applies tension to the governor spring through the intermediate lever which applies a force to close the governor valve. The enrichment spring between the enrichment and governor levers provides a force to open the enrichment valve.

As the drive shaft revolves, it rotates a table on which the governor weights are mounted. Small levers on the inside of the weights contact the governor spool. As gas-producer turbine speed N_g increases, centrifugal force causes the weights to apply increasing force against the spool. This tends to move the spool outward on the shaft against the enrichment lever. As governor weight force overcomes opposing spring force, the governor valve is opened and the enrichment valve is closed.

The enrichment valve will start to close whenever N_g increases enough to cause the weight force to overcome the force of the smaller spring. If N_g continues to increase, the enrichment lever will continue to move until it contacts the governor lever as shown in view C, at which time the enrichment valve will be fully closed. The governor valve will open if N_g increases sufficiently

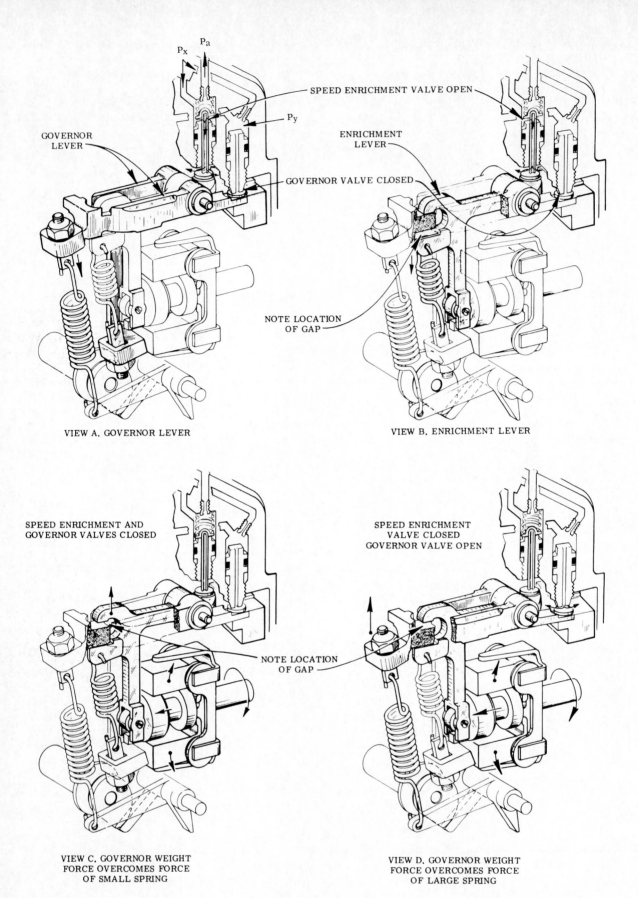

FIG. 12-8 Operation of the drive body assembly. (*Bendix Energy Controls Division.*)

to cause the weight force to overcome the force of the larger spring. At this point the governor valve will be open and the enrichment valve closed, as shown in view D.

The main body incorporates a vent port which vents the inner body cavity to atmospheric pressure P_a. Modified compressor discharge pressure, P_x and P_y, will be bled off to P_a when the respective enrichment and governor valves are open.

Bellows Section

The bellows assembly consists of an evacuated (acceleration) bellows and a governor bellows connected by a common rod. The end of the acceleration bellows opposite the rod is attached to the body casting. The acceleration bellows provides an absolute pressure reference. The governor bellows is secured in the body cavity and its function is similar to that of a diaphragm.

Movement of the bellows is transmitted to the metering valve by the cross shaft and associated levers. The cross shaft moves within a torque tube which is attached to the cross shaft near the bellows lever. The tube is secured in the body casting at the opposite end by means of an adjustment bushing. Therefore, any rotational movement of the cross shaft will result in an increase or decrease in the force of the torque tube. The torque tube forms the seal between the air and fuel sections of the control.

The torque tube is positioned during assembly to provide a force in a direction tending to close the metering valve. The bellows act against this force to open the metering valve.

P_y pressure is applied to the outside of the governor bellows. P_x pressure is applied to the inside of the governor bellows and to the outside of the acceleration bellows.

Figure 12-9 illustrates the forces applied to the bellows and their function. For explanation purposes, the governor bellows is illustrated as a diaphragm.

P_y pressure is applied to one side of the "diaphragm" and P_x is applied to the opposite side. P_x is also applied to the evacuated bellows attached to the diaphragm.

The force of P_x applied against the evacuated bellows is cancelled by application of the same pressure on an equal area of the diaphragm, as the forces act in opposite directions.

All pressure forces applied to the bellows section can be resolved into forces acting on the diaphragm only. These forces are P_y pressure acting on the entire surface of one side, the internal pressure of the evacuated bellows acting on a portion of the opposite side (within the area of pressure cancellation), and P_x acting on the remainder of that side. Any change in P_y will have more effect on the diaphragm than an equal change in P_x, because of the difference in effective area.

P_x and P_y vary with changing engine operating conditions as well as inlet air temperature. When both pressures increase simultaneously, as during acceleration, the bellows cause the metering valve to move in an opening direction.

When P_y decreases as the desired N_g is approached

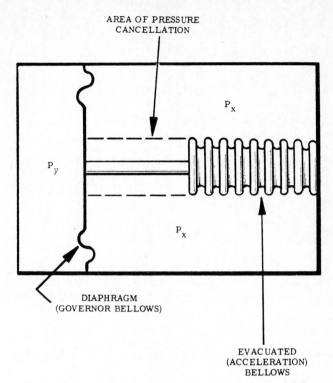

AREA OF PRESSURE CANCELLATION

P_x

P_y

P_x

DIAPHRAGM (GOVERNOR BELLOWS)

EVACUATED (ACCELERATION) BELLOWS

FIG. 12-9 Functional diagram of the bellows section. (*Bendix Energy Controls Division.*)

(for governing after acceleration), the bellows will travel to reduce the opening of the metering valve.

When both pressures decrease simultaneously, the bellows will travel to reduce the metering valve opening because a change in P_y is more effective than the same change in P_x. This occurs during deceleration and moves the metering valve to its minimum flow stop.

Model AL-N1 Power Turbine Governor

The model AL-N1 power turbine governor is mounted on the reduction gear case of the engine, is driven at a speed proportional to power turbine speed N_f, and provides power turbine overspeed protection.

The function of the AL-N1 governor is to limit the maximum speed N_f of the power turbine as, during normal operation, N_f is controlled by the propeller governor. However, in the event of a system malfunction, the AL-N1 governor will prevent N_f from exceeding 105 percent. This is accomplished by reducing the fuel flow of the DP-F2 fuel control.

The governor employs a drive body similar to the drive body of the fuel control, with the main difference being the elimination of the speed enrichment mechanism. The cover incorporates vent holes which maintain the inner cavity of the governor at atmospheric pressure P_a. During normal operation, the governor throttle lever is positioned against the maximum speed stop and locked in this position.

P_y pressure from the DP-F2 control is applied to the AL-N1 governor valve. If a power turbine overspeed occurs, the governor weight force overcomes the spring force which opens the valve to bleed off P_y pressure.

This, in turn, reduces P_y pressure on the governor bellows in the fuel control, and results in a reduction of fuel flow and consequently, N_g.

As the overspeed condition is corrected, the governor weight force diminishes and the spring force again overcomes the reduced weight force. This action closes the valve restoring control of P_y pressure and engine fuel flow to the DP-F2 fuel control.

Model TS-E2 Temperature Compensator

The model TS-E2 temperature compensator is mounted on the compressor case with the bimetallic discs extending into the inlet air stream. Compressor discharge pressure P_c is applied to the compensator. This pressure source is used to provide a P_x pressure signal to the DP-F2 fuel control.

The TS-E2 compensator changes the P_x pressure to provide an acceleration schedule biased by inlet temperature to prevent compressor stall or excessive turbine temperatures.

Operation of the Complete Fuel Control System

STARTING THE ENGINE The gas-producing turbine is cranked with the starter until sufficient speed is obtained for light-off. The pilot's throttle lever is then moved to the IDLE position to provide fuel.

At the time of light-off, the fuel control metering valve is in a low-flow position. As the engine accelerates, the compressor discharge pressure P_c increases, causing an increase in P_x pressure. P_x and P_y increase simultaneously since $P_x = P_y$ during engine acceleration. The increase in pressure sensed by the bellows causes the metering valve to move in an opening direction.

As N_g approaches idle, the centrifugal force of the drive body weights begins to overcome the governor spring force and opens the governor valve. This creates a $P_x - P_y$ differential which causes the metering valve to move in a closing direction until the required-to-run idle fuel flow is obtained.

Any variation in engine speed from the selected (IDLE) speed will be sensed by the governor weights and will result in increased or decreased weight force. This change in weight force will cause movement of the governor valve which will then be reflected by a change in fuel flow necessary to reestablish the proper speed.

ACCELERATION As the throttle lever is advanced above idle, the speed scheduling cam is repositioned, moving the cam-follower lever to increase the governor spring force. The governor spring then overcomes the weight force and moves the lever closing the governor valve, P_x and P_y immediately increase and cause the metering valve to move in an opening direction. Acceleration is then a function of increasing P_x ($P_x = P_y$).

With the increase in fuel flow, the gas-producer turbine will accelerate. When N_g reaches a predetermined point (approximately 70 to 75 percent), weight force overcomes the enrichment spring and starts to close the enrichment valve. When the enrichment valve starts to close, P_y and P_x pressures increase, causing an increase in the movement rate of the governor bellows and metering valve, thus providing speed enrichment to the acceleration fuel schedule. Continued movement of the enrichment lever will cause the valve to close and enrichment will then be discontinued.

Meanwhile, as N_g increases and the exhaust gas velocity increases, the propeller governor increases the pitch of the propeller to prevent N_f from overspeeding and to apply the increased power as additional thrust.

Acceleration is completed when the centrifugal force of the weights again overcomes the governor spring and opens the governor valve.

Once the acceleration cycle has been completed, any variation in engine speed from the selected speed will be sensed by the governor weights and will result in increased or decreased weight force. This change in weight force will cause the governor valve to either open or close, which will then be reflected by the change in fuel flow necessary to reestablish the proper speed. When the fuel control is governing, the valve will be maintained in a regulating or "floating" position.

Altitude compensation is automatic with this fuel control system since the acceleration bellows is evacuated and affords an absolute pressure reference. Compressor discharge pressure is a measurement of engine speed and air density. P_x is proportional to compressor discharge pressure, so it will decrease with a decrease in air density. This is sensed by the acceleration bellows which acts to reduce fuel flow.

DECELERATION When the throttle is retarded, the speed scheduling cam is rotated to a lower point on the cam rise. This reduces the governor spring force and allows the governor valve to move in an opening direction. The resulting drop in P_y moves the metering valve in a closing direction until it contacts the minimum flow stop. This stop assures sufficient fuel to the engine to prevent flame-out.

The engine will continue to decelerate until the governor weight force decreases to balance the governor spring force at the new governing position.

STOPPING THE ENGINE The engine is stopped by placing the throttle lever in the CUTOFF position. This action moves the DP-F2 cutoff valve to its seated position, stopping all fuel flow to the engine.

BENDIX AP-B3 FUEL CONTROL (FIG. 12-10)

General Description

The fuel control is a hydromechanical metering device used on the Allison 501-D13 (T56) engine, and accomplishes the following:

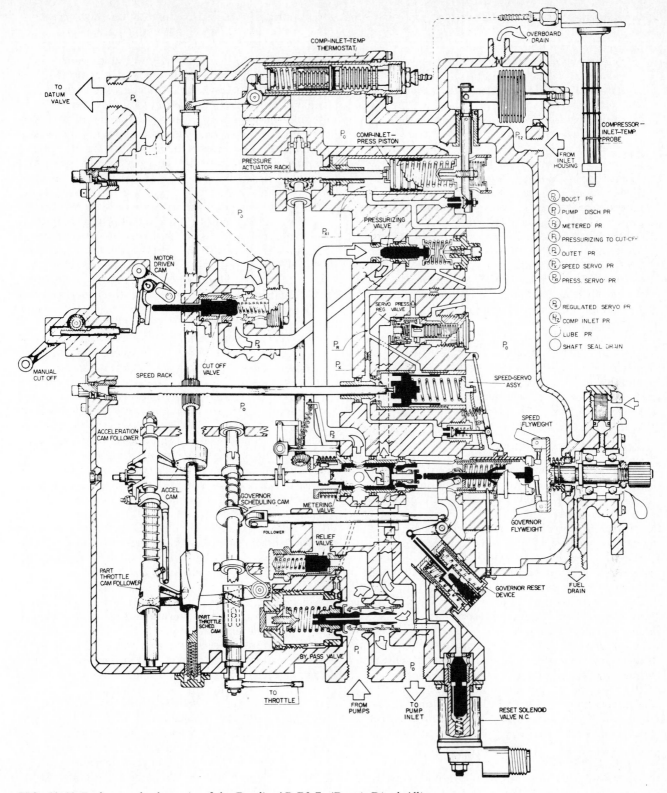

FIG. 12-10 Fuel control schematic of the Bendix AP-B3-7. (*Detroit Diesel Allison Service School.*)

1. Supplies a controlled fuel flow to initiate an engine fire-up.
2. Supplies a controlled fuel flow during acceleration from fire-up to the stabilized starting rpm (either low-speed taxi or high-speed taxi) to assist the 5th- and 10th-stage air-bleed system in the prevention of compressor surge.
3. Meters fuel flow in accordance with variations in air density caused by compressor inlet air temperature or pressure changes.
4. Permits pilot to vary fuel flow to the engine by movement of the throttle.
5. Meters approximately 20 percent more fuel than is required to operate the engine based upon rpm, air density, and throttle setting. This provides the temperature datum valve with a definite amount of fuel to trim.
6. Provides the means of completely stopping fuel flow to the engine at shutdown.
7. Limits the maximum possible fuel flow.
8. Provides overspeed protection for the engine.
9. Provides the means of selecting either low-speed or high-speed taxi operation.
10. Controls power available in maximum reverse.

The fuel control is only one part of the total fuel-metering system and operates in conjunction with a temperature datum valve and temperature amplifier. (see pages 226–227.)

Principles of Operation

The bypass valve assembly bypasses the excess fuel delivered to the fuel control, and establishes a pressure differential of fuel pump discharge pressure $P1$ minus metered fuel pressure $P2$ across the metering valve. This differential ($P1 - P2$) will remain practically constant during all operation. The position of the bypass valve is determined by the differential forces acting on the valve's flexible diaphragm. When the bypass valve is stabilized, the opening force of $P1$ equals the closing force of $P2$ plus spring force. Therefore, $P1 - P2$ equals spring force. The length of the spring is a function of bypass valve position, and spring force is a function of spring length. Thus, when the bypass valve moves to assume a new position, the spring length will vary, causing the spring force to change slightly.

The relief valve establishes the maximum fuel pressure within the fuel control, and thus establishes the maximum possible fuel flow from the fuel control. It is set to open when $P1$ exceeds $P0$ (low-pressure filtered fuel pressure) by a preset amount.

The metering valve meters all fuel flow to the temperature datum valve in accordance with variations in engine rpm, throttle setting, and compressor air inlet temperature and pressure. The metering valve also provides protection from overspeed by reducing the fuel flow when a certain overspeed rpm is exceeded. The size of the metering-valve orifice may be changed either by rotation or linear movement of the valve. The metering valve is rotated due to changes in compressor air inlet pressure. Changes in rpm, throttle settings, and compressor air inlet temperature actuate a cam assembly which results in linear movement of the metering valve. The metering valve linear opening force is a spring. The cam assembly establishes the linear position, and thus the orifice size of the metering valve during normal operation. The governor spring force is established by the governor setting lever which is controlled either by a cam and cam follower positioned by the throttle, or by the solenoid-operated governor reset mechanism. When the solenoid is energized, the governor reset mechanism positions the governor setting lever such that the governor spring is set for 10,000 (+300/−100) rpm low-speed taxi operation. When the solenoid of the governor reset solenoid assembly is deenergized, the reset mechanism positions the governor setting lever such that the governor spring is set for the overspeed rpm of the taxi and flight ranges. The solenoid of the reset solenoid assembly is controlled by the cockpit low-speed taxi switch. The force of the governor weights serves as a closing force for the metering valve. During low-speed taxi operation, the linear position of the metering valve is established by a balance of two forces—governor spring force and governor weight force. The cam assembly does not position the metering valve in low-speed taxi operation. During high-speed taxi and flight operation, the governor spring force is greater than the governor weight force. Thus, the governor spring tends to move the metering valve fully open. However, the maximum linear opening of the metering valve will be established by the cam assembly. In the event of over-speeding, the governor weight force increases with rpm. When governor weight force overcomes the governor spring force, the metering valve moves to decrease its linear opening, and thus reduce fuel flow from the fuel control. This limits the engine speed at a definite speed above 13,820 rpm. During an overspeed condition, the cam assembly has no control over the linear position of the metering valve because the governor weight force moves the metering valve away from the cam assembly. When the overspeed is corrected, the governor weight force decreases, and the governor spring begins moving the metering valve open. Then the cam assembly again determines the maximum linear opening of the metering valve.

The inlet pressure actuator assembly senses compressor air inlet pressure changes by means of a pressure probe in the left horizontal air inlet housing strut. The inlet pressure actuator initiates an action which causes the metering valve to rotate to provide a corrected fuel flow required by any air-pressure variation. A partially evacuated bellows, sensitive to air-pressure changes, repositions the pressure actuator servo valve by means of a lever action whenever compressor air inlet pressure changes. The position of the pressure actuator servo valve establishes servo pressure Px' which, in turn, establishes the position of the pressure piston and the pressure actuator rack. When the air pressure changes, the pressure actuator servo valve causes Px' to change. This moves the pressure piston and pressure actuator rack, and results in the pressure actuator servo valve being moved to a stabilized position. When the pressure actuator rack moves, the metering valve drive gear causes the metering valve to rotate by means of a bevel gear. Any change in air pressure results in the rotational movement of the metering valve.

The speed servo control assembly senses rpm changes, and initiates an action which causes the metering valve to move linearly and monitor fuel flow to prevent compressor surges during engine accelerations. The position of the speed servo valve establishes servo pressure Px which, in turn, establishes the position of a speed piston and speed rack. When rpm increases, the speed weights actuate linkage to move the speed servo valve toward a closed position. This causes Px to increase. Thus, the speed piston moves the speed rack and linkage to stabilize the speed servo valve. Movement of the speed rack rotates the speed and temperature shaft which has two cams—an acceleration cam and a part throttle cam. Each of these cams has a follower—the acceleration cam follower and the part throttle cam follower. The acceleration and part throttle cams are designed such that, during rpm changes, the acceleration cam positions the acceleration cam follower; and when there is no rpm change, the part throttle cam positions the part throttle cam follower, its shaft, and the acceleration cam follower. The acceleration cam follower's position establishes the metering valve's linear position, and thus the flow of metered fuel from the fuel control. Any change in rpm results in linear movement of the metering valve.

The temperature compensation section senses compressor air inlet temperature changes by means of a probe inserted through the air inlet housing beneath the left horizontal strut. The temperature compensation section initiates an action which causes the metering valve to move linearly to provide a corrected fuel flow required by an air temperature variation. The probe and a bellows in this section are filled with alcohol. Any air temperature variation causes the alcohol's volume to change. Thus, the length of the bellows is dependent upon the sensed air temperature. When air temperature changes, the temperature compensation section causes the speed and temperature shaft, with the two cams on it, to move linearly. Either the acceleration or the part throttle cam, whichever is in control of the acceleration cam follower at the time of the air temperature change, will reposition the acceleration cam follower. This causes the metering valve to move linearly to provide a corrected fuel flow. A return spring always retains the speed and temperature shaft in contact with the bellcrank of the temperature compensation section. Another bellows in this section is used to prevent any change in $P0$ fuel temperature or pressure from moving the speed and temperature shaft. The fuel control only compensates for variations in air temperature, and never for fuel temperature changes.

The part throttle scheduling cam is positioned by the throttle. When the throttle is moved, the part throttle scheduling cam moves its follower. The part throttle scheduling cam follower moves the part throttle cam follower linearly on its shaft. This changes the relative position of the part throttle cam follower in relation to the part throttle cam. The contour of the part throttle cam causes the part throttle cam follower to pivot slightly. The shaft of the part throttle cam follower then moves the acceleration cam follower. The acceleration cam follower moves the metering valve linearly to schedule fuel flow as required by the throttle movement.

The servo pressure valve assembly establishes and maintains the regulated servo pressure PR at a predetermined value above $P0$. PR is used by the inlet pressure actuator assembly and the speed servo control assembly, along with servo valves to establish Px and Px' required by the assemblies. The position of the servo pressure valve is determined by PR, the closing force, and $P0$ plus a spring force, the opening force. The pressurizing valve causes metered pressure $P2$ to build up to a predetermined value before it opens to allow metered fuel pressure $P3$ to flow to the cutoff valve assembly. This causes $P1$ to build up before fuel can flow from the fuel control, and results in quicker stabilization of the fuel-control components during the initial phases of an engine start. The pressurizing valve is not designed to have any metering effect on the fuel, but there is a small decrease in pressure across the pressurizing valve. This is the reason for indicating a $P2$ and $P3$ metered fuel pressure. The opening force on the pressurizing valve is $P2$, and the closing force is $P0$ plus spring force.

The cutoff valve assembly provides the means by which fuel flow to the engine is started or stopped. Electrical actuation of the cutoff valve is desirable for automatic engine starts and normal shutdowns. Since the possibility of an electrical failure exists, mechanical actuation of the cutoff valve is required for emergency shutdowns. Therefore, the cutoff valve may be moved to the closed position, either electrically or mechanically. The cutoff valve must be permitted to open both electrically and mechanically during an engine start.

The only time the cutoff valve will be mechanically held closed is when the emergency shutdown handle is pulled to the emergency position. Pulling the emergency handle causes the normal cutoff lever to move such that a bellcrank and lever move a plunger to compress a spring within the cutoff valve. When the force of this spring exceeds the force of an opposing spring, the cutoff valve moves against its seat to stop fuel flow. Energizing the cutoff valve actuator motor causes the cutoff cam to move the bellcrank, lever, and plunger to compress the spring and close the valve. When the cutoff valve is closed, $P3$ is ported to $P0$. The cutoff valve is not designed to have any metering action, but a small decrease in pressure does occur across the cutoff valve. This is the reason for indicating a $P3$ and $P4$ metered pressure.

Engine lubricating oil is used to lubricate the drive bearings of the fuel control. This oil is supplied and scavenged by the power section's lubricating system. Internal components of the fuel control are lubricated by the fuel which flows through the fuel control.

Operating Characteristics

The operating characteristics of the engine with respect to fuel flows furnished by the AP-B3 control are illustrated in Fig. 12-11. Disregarding for the moment

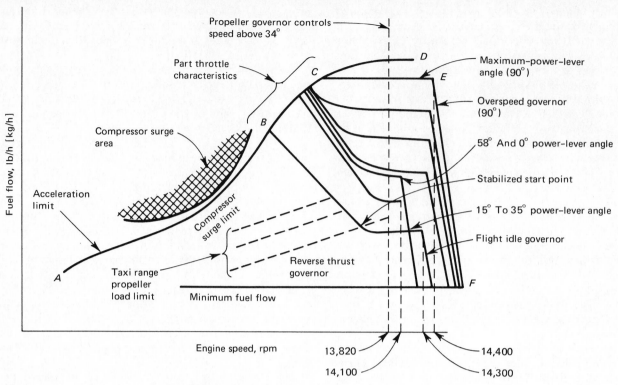

FIG. 12-11 Fuel curve at sea level. (*Bendix Energy Controls Division.*)

any altitude or temperature corrections, this diagram may be assumed to represent engine operation at some average constant engine air inlet condition. The acceleration curve (*ABCD*) represents the fuel flow required at different speeds to develop maximum allowable turbine inlet temperatures for engine acceleration, except for those limitations imposed by the necessity of circumventing the compressor surge area. The minimum fuel-flow curve represents the minimum desired fuel flow to prevent loss of fire in the engine burners (flame-out). At part throttle the governor slopes, as illustrated, serve two purposes.

1. To control engine speed outside the limits of the propeller governor setting, namely: start, flight idle on the ground, taxi, and reverse thrust.
2. To provide a measure of protection in overspeed conditions by reducing fuel flow and turbine temperature.

Fuel is metered in the control through one controlled orifice area. All the input variables are mechanically integrated and result in the creation of a specific orifice area. The fuel head across the orifice is maintained constant within close limits.

In Fig. 12-11 the acceleration curve represents acceleration during an engine start. The compressor surge area is avoided by action of the acceleration cam contour. Compressor inlet pressure and temperature change also contribute to the shape of this curve, and the combination of acceleration cam contour and compressor inlet pressure effect would continue to produce the acceleration curve *BCD*; however, during start, fuel flow is trimmed at *B* by the action of the part throttle cam,

causing the curve to assume the lower level until the stabilized start point is reached, as illustrated in Fig. 12-11.

During an acceleration beyond point *B* to a maximum throttle opening of 90°, the curve is limited at point *C* by the part throttle scheduling cam (face cam) and the part throttle setting cam. The combination of the effect of both of these cams produces the curve *CE*. Temperature change will shift the plotted curves upward or downward as required to avoid a corresponding shift of the compressor surge area on a temperature basis, always avoiding the surge area but at the same time maintaining maximum permissible fuel flows and corresponding engine efficiency by following the upward shift of the surge area. The higher the temperature, the more the curve would shift away from the presently indicated surge area. At lower temperatures the fuel curve would follow the receding surge area within predetermined limits.

However, before engine speed reaches point *E*, propeller governor action (separate from the fuel control system) maintains a predetermined maximum engine speed at any throttle setting above a tentatively established position of 34°. If for any reason the propeller governor were ineffective or sluggish in its operation, maximum fuel flow from the control would be limited at point *E* at 90° throttle lever position, thus defining maximum engine speed by operation of the centrifugal governor weights. When the governor weight force overcomes the governor spring force, the metering valve is moved toward a closed position, thus limiting maximum engine speed.

Each of the represented curves preceding point *C* in

Fig. 12-11 are part throttle curves determined by degrees of throttle opening below 90°; these curves are defined by cam contours and the effect of compressor inlet pressure. In each case the desirable engine speed is maintained by the propeller governor with the exception of points below a tentative point of approximately 34° throttle opening. The overspeed governor cutoff curve *EF* represents return to the minimum fuel flow from a 90° throttle opening. Other curves originating at the governor break point (dotted line extending below point *E*) and connecting with the ends of the part throttle characteristics curves would terminate parallel to *EF*.

Figure 12-12 illustrates the result of governor scheduling cam operation. The cam rise begins at a tentative point of approximately 20° of throttle opening and reaches its maximum speed setting at approximately 37.5° of throttle opening. Beyond this point the governor speed is maintained relatively constant.

Below approximately 20° the governor break point is determined by the governor spring force versus the governor weight force. Taxi range propeller load limits illustrate fuel requirements and engine speeds during ground operation (forward or reverse thrust). Reverse-thrust operation is at reduced speed on part throttle curve back-slopes; forward thrust is indicated at normal operating speeds, as illustrated in Fig. 12-11.

Referring to Fig. 12-13, which is a detailed enlargement of the part throttle characteristics portion of Fig. 12-11, the approximate fuel curves for various throttle angle positions are illustrated. Note that the 0° and 58° curves are similar, thus providing increased fuel flow for reverse thrust operation. Also, as lower fuel flows are selected (for instance, the idle and land band), the curves become somewhat sharper. Governing, provided by the propeller governor, ceases below approximately 34° of throttle opening. The significance of the slightly increasing increments of dip or hook in the lower fuel curves is to provide proportional increments of engine speed control for fixed pitch propeller operation, approaching the region of idle and start, by reduced fuel flow.

At altitude, the fuel curve will be similar to that illustrated in Fig. 12-11, except that fuel flow will be lessened as altitude increases, thus flattening out the curve.

It must be remembered that the operation of the AP-

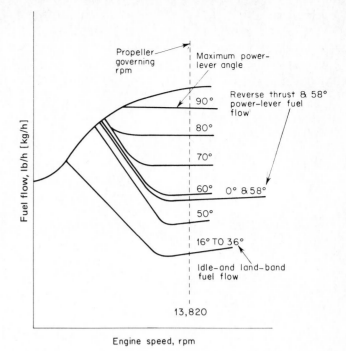

FIG. 12-13 Throttle angle effect. (*Bendix Energy Controls Division.*)

B3 gas turbine fuel control is allied with other components of the engine fuel system, and that the fuel flows provided by this control are not (as a result) necessarily the same fuel flows delivered to the engine burners. An example of this is illustrated in Fig. 12-14, wherein a relative comparison of these values is available at sea-level standard conditions. A basic knowledge of the complete fuel system is helpful in order to fully understand the function performed by this fuel control. (See the following section, Temperature Datum Valve.)

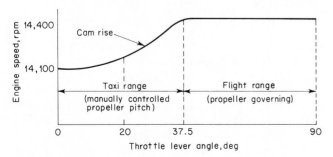

FIG. 12-12 Governor cam effect. (*Bendix Energy Controls Division.*)

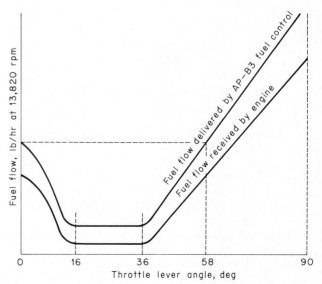

FIG. 12-14 Fuel control and engine nozzle fuel flow comparison at sea level. (*Bendix Energy Controls Division.*)

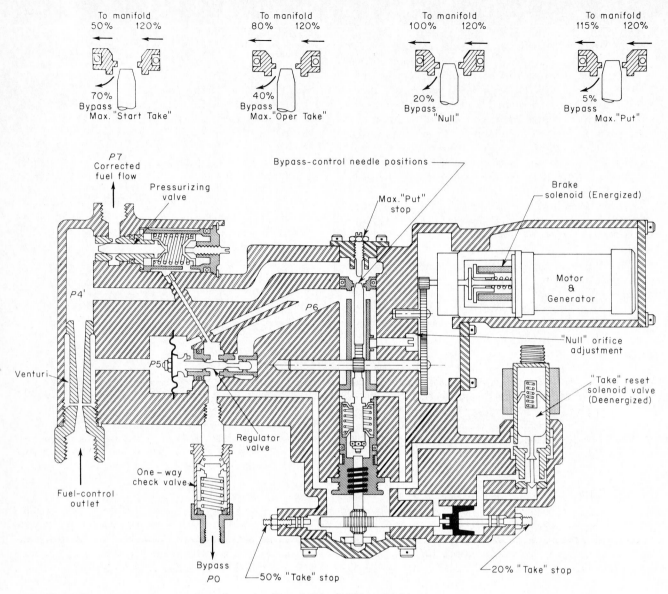

To manifold
50%   120%
70%
Bypass
Max. "Start Take"

To manifold
80%   120%
40%
Bypass
Max. "Oper Take"

To manifold
100%   120%
20%
Bypass
"Null"

To manifold
115%   120%
5%
Bypass
Max. "Put"

P7
Corrected
fuel flow

Pressurizing
valve

Bypass-control needle positions

Max. "Put"
stop

Brake
solenoid (Energized)

P4'

P6

Motor
&
Generator

Venturi

P5

"Null" orifice
adjustment

"Take" reset
solenoid valve
(Deenergized)

Regulator
valve

One-way
check valve

Fuel-control
outlet

Bypass
P0

50% "Take" stop

20% "Take" stop

FIG. 12-15 Temperature datum valve schematic. (*Detroit Diesel Allison Division.*)

Temperature Datum Valve *(FIG. 12-15)*

The temperature datum valve is a part of the electronic fuel trimming system of the 501-D13 engine, and is located between the fuel control and the fuel manifold. It receives 120 percent of the engine fuel requirements from the fuel control. The extra 20 percent of fuel enables the fuel trimming system to adjust fuel flow to compensate for variations in density and Btu content of the fuel, manufacturing tolerances in the components of the fuel system, and turbine inlet temperature limitations (TIT). The amount of fuel bypassed by the temperature datum valve is controlled by the temperature datum control.

In describing the operation of the temperature datum valve, certain terms are used to indicate conditions of trimming or bypassing. *Null* is that condition during which the electronic trim system makes no correction to fuel flow, and the extra 20 percent of fuel delivered to the temperature datum valve is bypassed. *Take* is that condition during which more than 20 percent of the fuel is bypassed in order to prevent excessive temperature during acceleration, and to compensate for "rich" fuel schedules and high-Btu-content fuel. *Put* is that condition during which less than 20 percent of fuel is bypassed in order to compensate for "lean" fuel schedules and low-Btu-content fuel.

Components of the electronic fuel trimming system are:

1. Temperature datum valve (mounted on bottom of compressor housing)
2. Temperature datum control (engine furnished, but aircraft mounted)
3. Relay box (engine furnished, but aircraft mounted)
4. Coordinator control (mounted on fuel control)
5. Temperature trim light (in cockpit)
6. Temperature trim switch (in cockpit)
7. Temperature datum control switch (in cockpit)

The electronic fuel trimming system has two ranges of operation, *temperature limiting* and *temperature control*. Temperature-limiting operation is desirable during engine starts and accelerations where the "rich" mixtures, required for acceleration, could result in excessive turbine inlet temperatures. If the turbine inlet temperatures become excessive during temperature limiting, the temperature datum valve must take (bypass more) fuel. When rpm is constant, temperature-control operation is desirable. Therefore, turbine inlet temperature is scheduled. In order that the turbine inlet temperature remains as scheduled, it may be necessary for the temperature datum valve to take or put fuel.

The electronic fuel trimming system is in the temperature-limiting range of operation if one or more of the following conditions exist.

1. Engine rpm is less than 13,000 (temperature limit is 871°C).
2. Throttle setting is less than 65° (temperature limit is 977°C if rpm is above 13,000).
3. Temperature trim switch is in LOCKED (temperature limit is 977°C if rpm is above 13,000).

The temperature limit of 871°C is required when the bleed air valves on the fifth and tenth stages of the compressor are open. When these bleed valves are closed, a temperature limit of 977°C is possible.

The electronic fuel trimming system is in the temperature-control range of operation *only* if all three of the following conditions exist.

1. Engine rpm greater than 13,000
2. Throttle setting greater than 65°
3. Temperature trim switch is in CONTROLLED

The temperature datum control compares two input signals.

1. Temperature signal (from 18 thermocouples wired in parallel at the inlet of the turbine)
2. Reference signal (from one of three potentiometers, depending upon engine operation)

As a result of comparing these two signals, the temperature datum control may complete a circuit to the temperature datum valve motor. Energizing this motor will move the bypass control needle in the temperature datum valve either to put or take as required to establish the selected turbine inlet temperature, or limit the turbine inlet temperature.

During the starting cycle (engine rpm below 13,000), a "start" potentiometer (adjusted to 871°C) in the temperature datum control provides the reference signal. When rpm exceeds 13,000, the speed-sensitive control initiates an action which causes the normal potentiometer (adjusted to 977°C) in the temperature datum control to provide the reference signal. If TIT exceeds the referenced temperature limit of either 871 or 977°C, a take signal is sent to the temperature datum valve motor which moves the bypass control needle to increase the amount of fuel bypassed. Bypassing more fuel results

in a reduced fuel flow to the engine which limits the TIT to prevent excessive temperatures. When operating in the temperature-control range, the reference signal to the temperature datum control will be provided from the variable potentiometer in the coordinator. The intensity of this signal is controlled by throttle position. The voltage difference of the reference signal and the temperature signal, as compared by the temperature datum control, will determine the signal to be sent to the temperature datum valve motor. This will result in the bypass control needle being repositioned to change the amount of fuel being bypassed, and thus altering fuel flow to the engine as required to permit the temperature signal (turbine inlet temperature) to equal (balance) the reference signal.

Prior to a landing approach and with the throttle above 65°, the pilot may elect to *lock in* a fuel correction. This is done by placing the temperature trim switch in LOCKED when the turbine inlet temperature is stabilized. The solenoid of the temperature datum valve is deenergized when the fuel correction is locked in. This allows a spring to move the temperature datum valve brake to the APPLIED position, resulting in the bypass control needle being locked in a corrected fuel-flow position, which will provide for a fixed percentage correction of metered fuel flow during the approach and landing. It must be understood that locking in a fuel correction never locks in a specific volume of fuel flow to the engine, but does lock in a fixed percentage of any fuel metered by the datum valve. Locking in a fuel correction will permit a more accurate control of horsepower output during the approach and landing.

Fuel, delivered to the temperature datum valve from the fuel control, must flow through the venturi. The velocity of the fuel increases as it approaches the throat of the venturi, and decreases as it flows away from the venturi throat. As a result of the velocity changes, the static pressure $P5$ at the venturi throat is lower than the static pressure $P4'$ at the venturi outlet. Venturi outlet ($P4'$) fuel is delivered to the pressurizing valve and the bypass-control needle.

The pressurizing valve is set to open whenever venturi outlet ($P4'$) fuel pressure exceeds bypass ($P0$) pressure by approximately 50 psi [344.8 kPa]. The setting of the pressurizing valve is established by a spring. When the pressurizing valve opens during an engine start, fuel flow to the engine begins. At engine shutdown, the pressurizing valve prevents drainage of fuel out of the temperature datum valve. Restriction to the flow of fuel through the pressurizing valve and fuel nozzles is constant when the pressurizing valve is open, and the fuel nozzles' metering valves are fully open. Thus, venturi outlet ($P4'$) pressure will determine the volume of fuel which will flow to the fuel nozzles through the pressurizing valve.

The volume of fuel which will flow through an orifice is a function of restriction (orifice size) and the pressure differential across the orifice. Thus, the amount of fuel which will pass through the orifice established by the bypass-control needle is a function of the difference in pressure between venturi outlet ($P4'$) and metered bypass ($P6$) pressure. The orifice, established by the bypass-control needle, can be made larger or smaller by moving the needle out of or into the orifice opening. Movement of the bypass-control needle is accomplished by means of the motor generator, which is controlled by the temperature datum control (amplifier). The rpm of the motor generator is lessened by the reduction gear and a spur gear train so that the bypass-control needle may be moved very slowly when the motor rotates. The motor is reversible, and its direction of rotation is determined by the signal sent to the motor by the temperature datum control. The output of the generator, which is driven by the motor, is a function of motor rpm, and "tells" the temperature datum control how fast the fuel correction is being made by the bypass needle. The brake solenoid controls the position of a brake which acts on the shaft of the motor. The brake is spring-loaded to the LOCKED position, and is "released" when the brake solenoid is energized. When the brake is locked, the bypass-control needle cannot move because the motor shaft and reduction gearing cannot rotate. A two-way compression spring is compressed whenever the bypass control needle is moved by the motor. This spring will return the bypass-control needle to NULL if the brake is released and the motor has no torque.

The maximum "put" stop establishes how far the motor can move the bypass-control needle into the orifice. As this orifice decreases in size, its restriction to the flow of venturi outlet ($P4'$) fuel through it increases. This results in a smaller percentage of fuel flow through the bypass-control needle, and a greater percentage of fuel flow through the pressurizing valve and fuel nozzles. The 20 percent take stop and 50 percent take stop are used to establish how far the motor can move the bypass control needle out of its orifice. As the orifice size increases, its restriction to the flow of venturi outlet ($P4'$) fuel through it decreases. Thus, a greater percentage of fuel flows through the bypass-control needle, and a smaller percentage flows through the pressurizing valve and fuel nozzles. The take reset solenoid valve is used to establish the take mechanism (rack and piston) against either the 20 percent stop or 50 percent stop. When the take-reset solenoid is energized, metered bypass ($P6$) fuel is ported on both sides of the take mechanism piston, and bypass ($P0$) is on the end of the rack. The differential areas and forces act to position the take mechanism rack against the 50 percent take stop. When the solenoid is deenergized, one side of the take mechanism piston is ported to metered bypass ($P6$) fuel, and the other side is ported to bypass ($P0$) fuel. Since metered bypass ($P6$) fuel is the higher pressure, the take mechanism moves, and the piston contacts the 20 percent take stop. When the mechanism moves from the 50 percent stop to the 20 percent stop, the rack rotates a pinion. This rotation moves the pinion nearer to the end of the bypass-control needle due to worm gear action. The take reset solenoid valve is deenergized at 13,000 rpm. Therefore, maximum possible take is 20 percent at all rpm in excess of 13,000. The null orifice adjustment has an eccentric projection on one end which fits into an opening of the sleeve surrounding the bypass control needle. When the null orifice adjustment is

turned, the sleeve moves in relation to the bypass control needle, and the orifice size established by the bypass control needle is varied slightly. The restriction of this orifice is thus varied to cause a slight change in the percentage of venturi outlet ($P4'$) fuel which will flow through the bypass control needle. A null orifice adjustment should be made if the peak starting turbine inlet temperatures are too high or too low.

The regulator valve is a double-ported valve that is secured to a flexible diaphragm. It is double-ported to prevent the flow through the valve from having an effect upon the valve's position. The position of the regulator valve is determined by a balance of forces acting on the diaphragm. One side of the diaphragm is ported to venturi throat pressure $P5$, while the other side is ported to metered bypass pressure $P6$. When the regulator valve is stabilized, venturi throat pressure $P5$ is equal to metered bypass pressure $P6$. If either of these pressures exceeds the other, the regulator valve moves to again establish their equality.

The one-way check valve will allow the temperature datum valve to bypass fuel with a common bypass line attached to the fuel control. The one-way check valve prevents fuel, bypassed by the fuel control, from flowing into the temperature datum valve. Thus, during cranking up to 2200 rpm, any fuel bypassed by the fuel control cannot reverse flow into the temperature datum valve and through the pressurizing valve to the fuel nozzles.

The entire system is illustrated in Fig. 13-4.

HAMILTON STANDARD MODEL JFC25 FUEL CONTROL

General Description

The JFC25 fuel control is designed to control the engine in either forward- or reverse-thrust operation under all operating conditions. Two control levers are provided. One to control the engine during forward or reverse operation, and the other to effect engine shutdown and starting by closing and opening a fuel shutoff valve. Provision has been made for future incorporation of an emergency system in the control.

The fuel control accurately governs the engine steady-state selected speed, acceleration, deceleration, maximum burner pressure, and thrust reversal, and provides a number of auxiliary signals. The speed-governing system is of the proportional or droop type (Figs. 12-16 and 12-17).

The JFC25 may be considered as consisting of a metering and a computing system. The metering system selects the rate of fuel flow to be applied to the engine burners in accordance with the amount of thrust demanded by the pilot, but subject to engine operating limitations as scheduled by the computing system as a result of its monitoring various operating parameters. The computing section senses and combines various engine operational parameters to control the output of the metering section of the fuel control during all re-

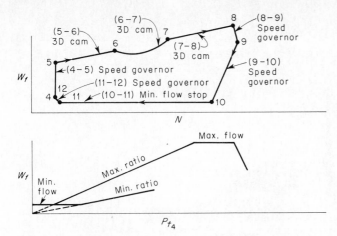

FIG. 12-16 Characteristic fuel-flow curves. (*Hamilton Standard Division, United Technologies Corp.*)

gimes of engine and aircraft operation. This fuel control, with some modification, is similar to the ones used on the JT3 and JT4 turbojet engines.

Principles of Operation (Typical Control) (*FIG. 12-18*)

The metering system operates in the following manner. High-pressure fuel is supplied to the control inlet from the engine-driven pump. This fuel initially encounters the filtration system which consists of a coarse filter and a fine filter. The coarse filter protects the metering section against damage by fuel contaminants.

This filter will open when clogged to permit continued operation with unfiltered fuel. The fine filter protects the computing section against damage by fuel contaminants. This filter is self-cleaning, whereby trapped particles are removed and passed through the metering system. The fuel next encounters the *pressure-regulating valve*. This valve is designed to maintain a constant pressure differential across the throttle valve. All high-pressure fuel in excess of that required to maintain this pressure differential is bypassed to pump interstage pressure by the pressure-regulating valve. The valve is servo controlled, whereby the actual pressure drop is compared with a selected pressure drop and any error is hydraulically compensated. The error resets the pressure-regulating-valve spring on this valve so that sufficient high-pressure fuel is bypassed to maintain the selected pressure drop. The high-pressure fuel, as altered by the pressure-regulating valve, then passes through the metering valve. This valve consists of a contoured plunger which is positioned by the computing section of the control within a sharp-edged orifice. By virtue of the constant pressure drop maintained across the valve, the fuel flow is proportional to the position of the plunger. An adjustable stop is provided to limit the motion of this plunger in the decrease-fuel direction to permit selection of the proper minimum fuel flow.

The final component to act upon the metered flow prior to its entry into the fuel manifold is the minimum pressure and shutoff valve. This valve is designed to shut off the flow of metered fuel to the engine when the pilot moves the shutoff lever to the OFF position. This generates a high-pressure signal by means of the

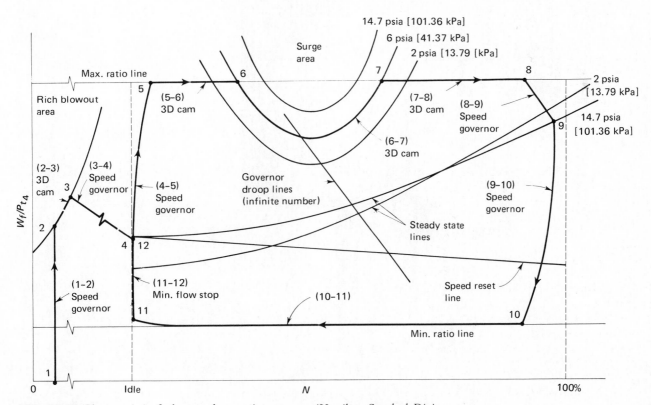

FIG. 12-17 Characteristic fuel-control operating curves. (*Hamilton Standard Division, United Technologies Corp.*)

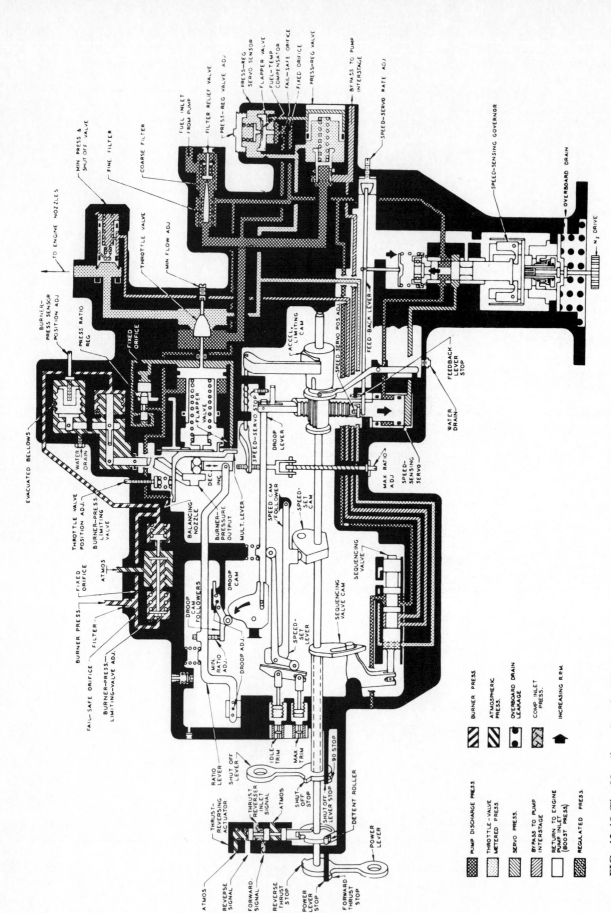

TO ENGINE NOZZLES

MIN PRESS & SHUT OFF VALVE
FINE FILTER
COARSE FILTER
FUEL INLET FROM PUMP
FILTER RELIEF VALVE
PRESS-REG VALVE ADJ
PRESS-REG SERVO SENSOR
FLAPPER VALVE
FUEL-TEMP COMPENSATOR
FAIL-SAFE ORIFICE
FIXED ORIFICE
PRESS-REG VALVE
BY PASS TO PUMP INTERSTAGE
SPEED-SERVO RATE ADJ

SPEED-SENSING GOVERNOR

OVERBOARD DRAIN
N₂ DRIVE

BURNER-PRESS SENSOR POSITION ADJ
PRESS RATIO REG
FIXED ORIFICE
THROTTLE VALVE
MIN FLOW ADJ

SPEED SERVO POS ADJ
ACCEL LIMITING CAM
FEED-BACK LEVER
FEEDBACK LEVER STOP
WATER DRAIN

EVACUATED BELLOWS
THROTTLE VALVE POSITION ADJ
BURNER-PRESS LIMITING-VALVE ADJ
WATER DRAIN

SPEED-SERVO STOP
FLAPPER VALVE
DROOP LEVER
MAX RATIO ADJ
SPEED-SENSING SERVO

BURNER PRESS
FIXED ORIFICE
FILTER
ATMOS
BALANCING NOZZLE
BURNER-PRESSURE OUTPUT
MULT. LEVER
SPEED CAM FOLLOWER
SPEED-SET CAM
SEQUENCING VALVE

FAIL-SAFE ORIFICE
BURNER-PRESS LIMITING-VALVE ADJ
DROOP CAM FOLLOWERS
DROOP CAM
MIN RATIO ADJ
DROOP ADJ
SPEED-SET LEVER
SEQUENCING VALVE CAM

DEC INC

RATIO LEVER
SHUT OFF LEVER
ATMOS
REVERSE SIGNAL
THRUST-REVERSING ACTUATOR
FORWARD SIGNAL
REVERSE THRUST STOP
THRUST REVERSER INLET SIGNAL
SHUT OFF STOP
MAX TRIM
IDLE TRIM
SHUT OFF LEVER STOP
DETENT ROLLER
90 STOP
POWER LEVER STOP
POWER LEVER
FORWARD THRUST STOP

BURNER PRESS.
ATMOSPHERIC PRESS.
OVERBOARD DRAIN LEAKAGE
COMP INLET PRESS.
INCREASING R.P.M.

PUMP DISCHARGE PRESS.
THROTTLE-VALVE METERED PRESS.
SERVO PRESS.
BYPASS TO PUMP INTERSTAGE
RETURN TO ENGINE PUMP INLET (BOOST PRESS)
REGULATED PRESS.

FIG. 12-18 The Hamilton Standard JFC25. Similar controls are used on several versions of the JT3 and JT4 series engines. (*Hamilton Standard Division, United Technologies Corp. from J. V. Casamassa and R. D. Bent, Jet Aircraft Power Systems, 3d ed., McGraw-Hill, New York, 1965.*)

throttle-operated pilot valve, which acts on the spring side of the shutoff valve, forcing it down against a seat, thus shutting off the flow of fuel to the engine. When the shutoff lever is moved into the ON position, the high-pressure signal is replaced by pump interstage pressure, and when metered fuel pressure has increased sufficiently to overcome the spring force, the valve opens and fuel flow to the engine is initiated. Thereafter, the valve will provide a minimum operating pressure within the fuel control, ensuring that adequate pressure is always available for operation of the servos and valves at minimum flow conditions.

The computing system positions the throttle valve to control the steady-state engine speed, acceleration, and deceleration. This is accomplished by using the ratio W_f/P_B (the ratio of metered fuel flow to engine burner pressure) as a control parameter. This parameter has been pioneered by Hamilton Standard and is presently used on all Hamilton Standard main fuel controls.

The positioning of the throttle valve by means of the W_f/P_B parameter is achieved through a multiplying system whereby the W_f/P_B signal for acceleration, deceleration, or steady-state speed control is multiplied by a signal proportional to P_B to provide the required fuel flow.

P_B is sensed as follows: A metallic bellows is internally exposed to P_B and the resulting force is opposed by an evacuated bellows of equal size. The net force, which is proportional to absolute burner pressure, is transmitted through a lever system to a set of rollers whose position is proportional to W_f/P_B. These rollers ride between the above lever and a second lever. Thus the force proportional to P_B is transmitted through the rollers to the second lever. Any change in the roller position (W_f/P_B) or the P_B signals causes the upsetting of the equilibrium of this lever which results in the opening or closing of a variable or metering orifice which is supplied with regulated high-pressure fuel through a fixed-bleed orifice. The opening or closing of the metering orifice modulates the fuel pressure between the two orifices, and this pressure is used to control the position of a piston attached to the throttle valve plunger. The motion of this piston compresses or relaxes a spring which will return the second lever to its equilibrium position.

Deceleration control near the end is accomplished by placing a minimum stop on the W_f/P_B signal. This provides a linear relationship between W_f and P_B which results in blowout-free decelerations. This stop is adjustable.

Acceleration control is accomplished by placing a maximum stop on the W_f/P_B signal. This stop is positioned by a three-dimensional cam which is rotated by a signal proportional to engine speed and translated by a signal proportional to engine inlet pressure. The three-dimensional cam is so contoured as to define a schedule of W_f/P_B versus engine speed for each value of engine inlet pressure. This combination will permit engine accelerations which avoid the overtemperature and surge limits of the engine without compromising engine acceleration time.

Engine speed is sensed by an engine-driven flyweight governor of the centrifugal type which controls the movement of the speed servo piston through a pilot valve. When speed changes, the flyweight force varies and the pilot valve is positioned to meter high-pressure fuel to the speed servo piston or to allow it to drain. The motion of this piston repositions the pilot valve until the speed-sensing system returns to equilibrium. The position of the servo piston is indicative of actual engine speed. This piston incorporates a rack which meshes with a gear segment on the acceleration three-dimensional cam to provide the speed signal for acceleration limiting. The piston is also utilized to control selected engine speed as described below.

The compressor inlet pressure-sensing servo translates the speed set cam in proportion to expansion or contraction of the evacuated bellows for speed reset at altitude.

Decreased compressor inlet pressure will expand the sensing bellows allowing servo pressure to bleed at the flapper valve. Servo pressure decreasing, high pressure will move the servo to the left, translating the cam to the right, causing the cam follower to rise in conformity to the contour of the cam, therefore decreasing the W_f/P_B ratio. The acceleration cam is simultaneously translated.

As the servo moves to the left, spring pressure is diminished, tending to close the flapper valve, thereby allowing sufficient servo pressure to accumulate so that a new position of equilibrium is established.

With increased CIP (compressor inlet pressure), the flapper valve will tend to close further, thereby causing servo pressure to overcome pump discharge pressure acting on a smaller area, resulting in a reverse action to that effected by a decreased CIP.

Engine speed control is accomplished by comparing the actual speed as indicated by the speed servo piston to the desired speed as selected by the pilot through a power-lever-operated three-dimensional cam. This cam defines schedules of W_f/P_B versus power lever angle (selected speed) for each value of CIP. The deviation of desired speed from the actual speed, or the speed error, effects a change in the W_f/P_B signal in the multiplying linkage. This causes an increase or decrease in fuel flow to effect the necessary speed correction. The speed-setting three-dimensional cam is rotated by the pilot's power lever and translated by a signal proportional to CIP. The CIP signal is obtained from the CIP servo system previously described. The bias of selected engine speed by CIP is required to compensate for the speed governing error due to CIP variations which is normally encountered with proportional speed controls. Many models of the series 25 fuel control do not incorporate the CIP sensor. In addition, some models may include provision for increasing fuel flow during periods of water injection.

The engine thrust reversal mechanism consists of a cam-operated pilot valve which meters compressor discharge pressure bleed air to the thrust reversal actuators when the power lever is retarded to the REVERSE position. The speed-setting cam incorporates a reverse power regime to schedule higher engine speeds as the power lever is retarded beyond idle into reverse. The cam includes a flat or idle portion prior to the initiation of thrust reversal.

Auxiliary functions include a throttle-operated pilot valve which provides the shutoff signal to the shutoff and minimum pressure valve, and a pressure signal to the manifold drain valve. The valve is positioned by a shutoff-lever-operated cam so that the signals are generated at the desired lever positions. The pilot valve also provides a windmill bypass feature when the shutoff valve is closed. This feature bleeds the throttle valve downstream pressure to increase the throttle-valve pressure drop. This allows the pressure-regulating valve to continue to operate normally. Thus damage to the pumps due to excessive pressure is prevented during engine windmilling.

WOODWARD TYPE 1307 FUEL CONTROL (*FIG. 12-19*)

General Description

The Woodward main fuel control is used on the J79 and CJ805 series turbojet engines, and provides the following engine control functions:

1. Maintains engine speed (rpm) according to the throttle schedule.
2. Schedules maximum and minimum fuel rate limits.
3. Limits maximum compressor discharge pressure limit ($P3$) by limiting fuel flow.
4. Reduces maximum speed (rpm) limit in the low engine inlet temperature range.
5. Provides a pilot-actuated reset of minimum fuel, when required, for high-altitude starting.
6. Resets the flight idle speed as a function of compressor inlet air temperature.
7. Controls the position of the inlet guide vanes by providing control fuel for the inlet guide vane actuators.
8. Provides the exhaust-nozzle area control with an interlock signal when the engine is being accelerated at maximum fuel limit (J79 only).
9. Provides the afterburner pump with a pressure signal as a function of power-lever position and engine speed (J79 only).

To enable the control to perform the above objectives, there must be inputs in addition to the supply of fuel. The inputs include the compressor inlet temperature (CIT), the compressor discharge pressure ($P3$), and the speed of the main shaft (rpm). These three inputs, referred to as parameters, and a power-lever (throttle) setting determine the outputs of the control. The compressor inlet temperature is sensed by the compressor inlet temperature sensor which is a separate piece of equipment. The outputs of the control include a controlled fuel supply to the combustion chamber of the engine, a control fuel pressure signal for the inlet guide vane actuators, an interlock pressure signal to the exhaust-nozzle area control, and an afterburner pressure signal to the afterburner fuel control.

NOTE. Sections dealing with afterburning and nozzle area control apply to military J79 engines only.

The airframe boost pump supplies fuel to the main fuel pump from the fuel tank. The boost pump is a low-pressure pump with a discharge pressure $P0$ that does not exceed 65 psi absolute (psia) [448.2 kPa absolute]. Working with the airframe boost pump as a unit is another low-pressure pump, the engine boost pump. These two pumps supply the main fuel pump, which, in turn, supplies the main fuel control with a maximum fuel flow of about 20,000 lb/h [9072 kg/h] at a pressure ($P1$) that ranges from 150 to about 900 psi gage (psig) [6205.5 kPa gage]. Fuel is bypassed from the regulator to the engine boost pump discharge line at case pressure Pb. A line also bypasses fuel from the regulator to the airframe boost pump discharge line at $P0$ pressure.

Principles of Operation

NOTE: All numbers refer to Fig. 12-19.

Maximum fuel is supplied to the control inlet port (107) from the main fuel pump discharge at pressure $P1$. The flow of fuel is divided at the metering valve, sending fuel to the engine at one pressure ($P2$) and a bypass flow to the pump inlet at a second pressure (Pb).

Fuel inlet pressure $P1$ and outlet pressure $P2$ are applied to opposite sides of sensing land on the differential pilot valve plunger (110). Inlet pressure $P1$ on the bottom side of the land is opposed by $P2$ plus the force of the pressure regulator reference springs (111). The controlling action of the valve plunger regulates pressure $P4$ until the bypass flow causes the force produced by $P1$—$P2$ to be equal to the force of the spring. As $P2$ increases, or $P1$ decreases, the valve plunger is forced downward by $P2$, venting $P1$ to $P4$. This action forces the bypass valve plunger (109) to reduce the opening, increasing $P1$ enough to restore the $P1$—$P2$ differential and recenter the differential pilot valve plunger. When $P1$ increases, or $P2$ decreases, the valve plunger is forced upward by $P1$, venting $P4$ to Pb. This action allows $P1$ pressure to force the bypass valve plunger to the left, thus opening the port and increasing bypass flow. This decreases $P1$, restoring the $P1$—$P2$ differential. Therefore a constant differential pressure is maintained across the fuel valve plunger (108), resulting in a fuel-flow rate that is a function of valve position alone, and is substantially independent of pump delivery rate or pressure level in the system. Adjustment of the specific-gravity cam (112) sets the preload on the pressure regulator reference spring (111), allowing manual compensation for differences in the density of various fuels. The $P1$—$P2$ differential increases as the spring preload is increased.

Fuel at $P1$ is supplied through a filter to the pressure-regulating valve (105), which regulates the servo pressure, Pc, for the pilot valves and servos. Fuel at Pc is opposed by the pressure-regulating-valve spring (106) and case reference pressure, which is the same as bypass pressure Pb. The combined action of the pressure and spring positions the valve plunger so that $Pc—Pb$ is essentially constant.

Flow of fuel to the engine is determined by the position of the fuel valve plunger (108), which is normally positioned by the fuel valve servo piston (30), which is controlled by the speed-control plunger (67) of the governor pilot valve assembly. The two fuel-limit pistons (27 and 28) act as scheduled limit stops, so the plunger action cannot cause the rate of fuel flow to exceed scheduled maximum or minimum limits.

When the fuel rate is within the scheduled limits, the fuel limit pistons are vented to Pb, exerting no force on the fuel-valve plunger (108). When this condition exists, the fuel-valve plunger is positioned by the speed control plunger (67) acting to regulate pressure on the fuel-valve servo piston (30), which in turn acts to balance a counterforce produced by Pc on the return servo piston (26). Since the fuel-valve servo piston has a surface area twice that of the return servo piston, the speed-control plunger regulates the fuel-valve plunger position by varying the pressure on the fuel-valve servo piston between the limits of Pb and Pc.

When the speed governor is in transient condition, the following action takes place. The fuel-valve servo piston (30) is supplied with fuel at Pc at an underspeed signal or with fuel at Pb at an overspeed signal through the action of the governor pilot valve assembly. The flyweights of the governor ballhead assembly (66) act on the speed-control plunger. Their centrifugal force is translated to axial force at the toes of the flyweights; this force is opposed by the force of the governor reference spring (64). The position of the speed-setting cam (76) determines the compression of this spring and the speed that the engine must attain so that the flyweights will balance the force of the spring. The speed-setting cam is adjusted by rotation of the power-lever shaft (75). Uniform governor operation is accomplished through a compensating system consisting of a buffer-valve piston (32) floating between two buffer springs (33), a compensating land on the speed-control plunger, and a variable-compensation plunger (31). When the speed of the engine falls below its set value, the governor reference spring overcomes the reduced centrifugal force of the flyweights, and the speed control plunger moves downward. This downward movement uncovers the port at the upper end of the plunger, permitting fuel at Pc to enter the passage leading to the fuel valve servo piston (30) by displacing the buffer-valve piston. The change in pressure forces the servo piston downward, rotating the fuel servo lever (29) clockwise, opening the fuel-valve plunger (108). Displacement of the buffer-valve piston compresses one of the buffer springs (33) causing a pressure differential across the buffer-valve piston (32) which is transmitted to the compensating land of the speed-control plunger. The greater pressure on the lower side of this land acts to supplement the force of the flyweights, causing the speed-control plunger to close before the required speed has been attained. As fuel leaks across the variable-com-

pensation plunger, this false speed signal is dissipated and the buffer-valve piston recenters; at the same rate the engine speed returns to normal. Action resulting from engine overspeed is similar but in the reverse direction. Increased centrifugal force of the flyweights, due to increased engine speed, overcomes the force of the governor reference spring, resulting in an upward movement of the speed-control plunger. This plunger movement opens the regulating port to Pb, thereby causing the return servo piston to move the fuel valve in the closed direction to decrease the flow of fuel. While this action is taking place, the decreased pressure on the left-hand end of the buffer-valve piston allows it to move to the right. The resulting pressure differential, caused by the buffer-valve piston compressing the spring on the right-hand end of the piston, acting on the compensating land of the speed-control plunger, recenters the plunger before the required speed is attained. As fuel leaks across the variable-compensation plunger, the pressure differential due to this action is dissipated, at the same rate as the engine speed and flyweight force, to normal. The buffer-valve piston is designed to bypass fuel after a given displacement, in order to accommodate large flows of fuel, either to or from the fuel-valve servo piston, resulting from abrupt changes in fuel requirements. The bypass ports of the buffer-valve piston are offset so that the piston must be displaced to a greater degree during a reduction in the flow of fuel, therefore reducing the undershoot when the speed is suddenly reduced to idle. To keep the buffer-valve piston displaced while the engine is decelerating, the undershoot-valve plunger (23) maintains a minimum pressure on the fuel-valve servo piston. This pressure balance is accomplished by the end of the plunger being exposed to the same pressure as the servo piston. When this pressure falls below a specified minimum, the undershoot-valve spring (24) overcomes the upward force of the plunger, forcing it downward and allowing fuel at Pc to bleed through an orifice (25) of the plunger to the servo piston. This action maintains pressure in the line at a level sufficient to keep the buffer-valve piston displaced. As the pressure increases, the undershoot-valve plunger closes and the Pc supply is cut off. An orifice in the valve plunger drains valve leakage to Pb.

The maximum fuel limit, or the acceleration fuel limit, is determined by three inputs received by the regulator from the engine. The three inputs, compressor discharge pressure ($P3$), regulator speed, and compressor air inlet temperature, work together according to the relation: maximum fuel limit $= f(P3) \times f$(speed and compressor inlet temperature). This relation is established by a mechanical-hydraulic computer in the following manner.

The $P3$ transducer changes $P3$ pressure into an angular displacement of the $P3$ cam (95). Fuel at $P1$ flows through an orifice (14) at pressure $P5$ into the $P3$ bellows assembly (6). $P5$ pressure is opposed by $P3$ and the $P3$ sensor adjustment spring (2). The opposing forces regulate the flow through the port of the $P3$ sensor valve seat assembly (7) which acts as the $P5$ regulating valve to maintain a constant pressure differential, $P5 - P3$.

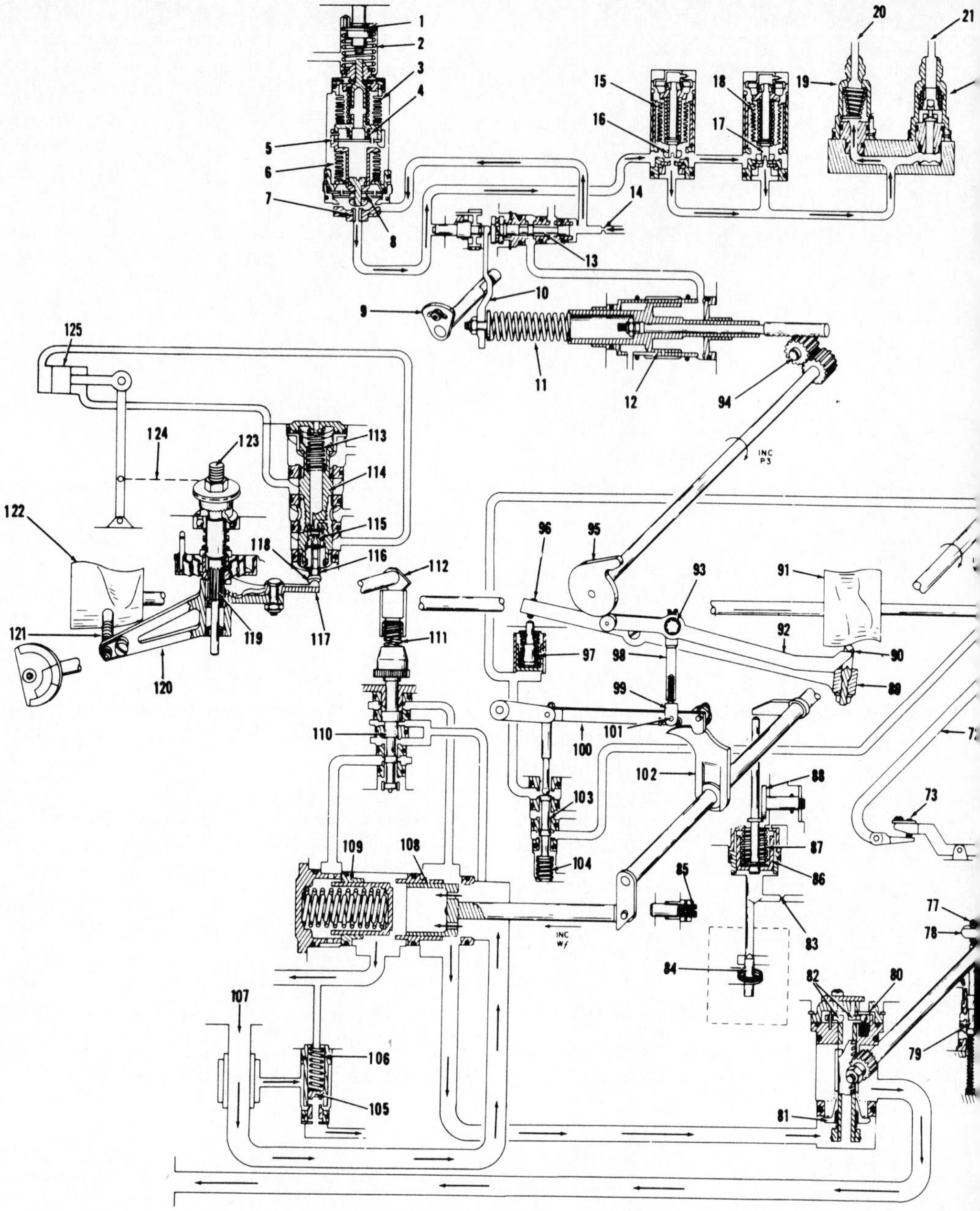

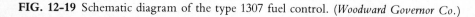

1 P3 LEVEL-ADJUSTMENT SCREW
2 P3 SENSOR-ADJUSTMENT SPRING
3 P3 AUXILIARY-BELLOWS ASSEMBLY
4 P3 AUXILIARY-BELLOWS TRANSFER SPRING
5 P3 BELLOWS-ORIFICE DIAPHRAGM
6 P3 BELLOWS ASSEMBLY
7 P3 SENSOR-VALVE SEAT ASSEMBLY
8 P3 SENSOR-DAMPER PISTON
9 P3 FEEDBACK-FULCRUM ADJUSTMENT
10 P3 FEEDBACK LEVER
11 P3 FEEDBACK SPRING
12 P3 SERVO PISTON
13 P3 PILOT-VALVE PLUNGER
14 ORIFICE
15 P3 REFERENCE BELLOWS ASSEMBLY
16 STEM OF P3 REFERENCE VALVE
17 STEM OF EXTERNAL P3 REFERENCE VALVE
18 EXTERNAL P3 REFERENCE BELLOWS ASSEMBLY
19 P3 REFERENCE-PRESSURE CHECK-VALVE ASSEMBLY
20 DRAIN TO AIRFRAME BOOST PUMP
21 TO OVERBOARD DRAIN (ATMOSPHERIC PRESSURE)
22 P3 REFERENCE PRESSURE- RELIEF-VALVE ASSEMBLY
23 UNDERSHOOT VALVE PLUNGER
24 UNDERSHOOT VALVE SPRING
25 ORIFICE
26 RETURN SERVO PISTON
27 FUEL LIMIT PISTON
28 FUEL LIMIT PISTON
29 FUEL-SERVO LEVER
30 FUEL-VALVE SERVO PISTON
31 VARIABLE-COMPENSATION PLUNGER
32 BUFFER-VALVE PISTON
33 BUFFER SPRING
34 COMPRESSOR-INLET-TEMPERATURE SENSOR
35 COMPRESSOR-INLET-TEMPERATURE FEEDBACK LEVER
36 COMPRESSOR-INLET-TEMPERATURE PILOT-VALVE PLUNGER
37 COMPRESSOR-INLET-TEMPERATURE PILOT-VALVE SLEEVE
38 NOZZLE LOCK-VALVE PLUNGER
39 ON-OFF SPEED HIGH-POINT ADJUSTING SCREW

40 ON-OFF SPEED LOW-POINT ADJUSTMENT SCREW
41 ON-OFF SPEED ADJUSTING PISTON
42 PILOT-VALVE PLUNGER
43 TACHOMETER SERVO PISTON
44 ON-OFF SPEED LEVER
45 RACK OF THE TACHOMETER-SERVO PISTON ROD
46 GEAR
47 SPEED FEEDBACK CAM

48 TACHOMETER FEED-BACK-FOLLOWER LEVER
49 TACHOMETER LINKAGE-FULCRUM ADJUSTMENT SCREW
50 TACHOMETER SPEEDER-SPRING ADJUSTING SCREW
51 TACHOMETER FEEDBACK LEVER
52 TACHOMETER REFERENCE SPRING
53 TACHOMETER PILOT-VALVE PLUNGER

54 FLYWEIGHTS OF THE TACHOMETER-BALL-HEAD ASSEMBLY
55 BIMETAL STRIP ASSEMBLY
56 IDLE-SPEED ADJUSTMENT
57 HIGH-TEMPERATURE SPEED RESET LEVER SCREW
58 COMPRESSOR-INLET-TEMPERATURE SERVO PISTON
59 THREE-DIMENSIONAL

FIG. 12-19 Schematic diagram of the type 1307 fuel control. (*Woodward Governor Co.*)

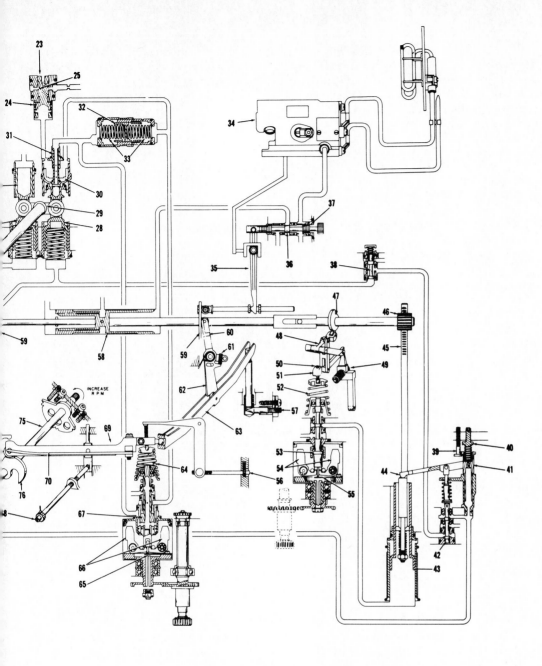

The $P3$ auxiliary bellows assembly (3) normally has no differential pressure across it and is provided as an emergency substitute for the $P3$ bellows. Should the $P3$ bellows fail, $P5$ would in effect be vented to the $P3$ bellows orifice diaphragm (5). The pressure drop across the orifice causes the diaphragm to overcome the force of the $P3$ auxiliary bellows transfer spring (4), and move upward, closing the $P3$ connection to the $P3$ bellows (6). This results in continued normal operation through the $P3$ auxiliary bellows assembly (3). Pressure $P5$ on the right end of the $P3$ pilot-valve plunger (13) is balanced by the force of the $P3$ feedback spring (11) and $P3$ reference pressure on the left end of the plunger. The $P3$ pilot valve plunger regulates the supply of fuel to or from the $P3$ servo piston (12). The servo piston is opposed by Pc supplied to the left end of the piston and has half the surface area. As $P3$ increases, $P5$ also increases, applying more force on the right end of the $P3$ pilot valve plunger, moving it to the left. This movement vents Pc to the right end of the $P3$ servo piston, forcing it to the left. As the piston moves to the left, the $P3$ feedback spring is compressed, the $P3$ feedback lever (10) transmits this force to the pilot valve plunger, moving it to the right, closing the port, and stopping the Pc supply to the servo piston. As $P3$ decreases, the reverse action takes place, the pilot valve moves to the right, venting the servo piston to Pb (case pressure), allowing the servo to move to the right. This movement decreases the tension on the $P3$ feedback spring allowing it to recenter the pilot valve. Movement of the $P3$ servo piston (12) is transmitted to the $P3$ cam (95) through the $P3$ servo rack and the $P3$ servo idler gear (94), resulting in a rotation of the cam in proportion to $P3$. The cam is machined to produce a follower output proportional to the $\log f(P3)$ and moves one end of the cam summing lever (92) accordingly. Setting the $P3$ feedback fulcrum adjustment (9) sets the ratio of $P3$ servo piston travel to $P3$. The evacuated $P3$ reference bellows assembly (15) regulates the flow through the stem of the $P3$ reference valve (16) to produce a constant $P3$ reference pressure on the left end of the $P3$ pilot-valve plunger. This prevents random fluctuations in fuel pressure in the case from affecting calibration. Fuel passing the stem of the $P3$ reference valve drops to airframe boost pressure $P0$ and passes to the drain to airframe boost pump (20) through the $P3$ reference pressure check valve assembly (19). Should there be a surge of $P0$ between the airframe boost pump and the check valve assembly, the check valve assembly would close. Under these conditions, $P0$ pressure would rise between the stem of the $P3$ reference valve and the check valve until this slightly higher pressure would open the $P3$ reference pressure relief-valve assembly (22) directing a small amount of fuel to the overboard drain (21). The external $P3$ reference bellows assembly (18) will remain inactive and the stem of the external $P3$ reference valve (17) will remain closed until there is a rupture or failure of the internal $P3$ reference bellows assembly (15). Such a failure of the internal bellows will cause the spring inside the bellows to close the valve stem, allowing the $P3$ reference pressure to rise. At the new higher pressure the external bellows will function, allowing fuel to pass the stem of the external $P3$ reference valve to the airframe boost pump or to the overboard drain.

The control speed and compressor inlet temperature section works as follows. Rotation of the corrected fuel cam (91) and the inlet guide vane cam (122) is a function of speed and is accomplished by the tachometer ballhead assembly and an associated servo system. As engine speed increases, the centrifugal force of the flyweights of the tachometer ball head assembly (54) overcomes the opposing force of the tachometer reference spring (52), resulting in an upward movement of the tachometer pilot-valve plunger (53). This movement of the plunger vents Pc to the lower surface of the tachometer servo piston (43). Since the area on the lower surface of the piston is twice the area on the upper surface of the piston, the servo piston moves upward. The rack of the tachometer servo piston rod (45) rotates the corrected fuel cam and the inlet guide vane cam. The speed feedback cam (47) also rotates with the movement of the piston rod. This cam, in turn, depresses the tachometer feedback follower lever (48) and the tachometer feedback lever (51). This downward movement compresses the reference spring, and the tachometer pilot-valve plunger is recentered, stopping the supply of Pc to the servo piston. When speed decreases, the reaction is the opposite; as the lower surface of the servo piston is vented to Pb, Pc on the upper surface forces the piston downward, and the pilot-valve plunger is recentered. Movement of the servo piston (43) in relation to the speed feedback cam is set through the adjustment of the tachometer speeder-spring adjusting screw (50) and the tachometer linkage-fulcrum adjustment screw (49). The speeder-spring adjusting screw adjusts the preload on the reference spring and has a greater effect at low speeds. Adjustment of the linkage fulcrum adjustment screw determines the feedback ratio. The corrected fuel cam and the inlet guide vane cam are positioned axially in relation to the compressor inlet temperature. Movement is initiated by the compressor-inlet-temperature sensor (34) and transferred to the two cams by the compressor-inlet-temperature servo piston (58), the compressor-inlet-temperature pilot-valve plunger (36), and the compressor-inlet-temperature feedback lever (35). For each temperature input there is a unique horizontal position for the cams, resulting in a unique schedule of follower positions versus speed. The corrected fuel cam for each axial position produces a function of the controlled fuel supply over the compressor discharge pressure ($P3$) equal to a function of the speed as specified for the particular input temperature.

The cam summing lever (92) combines the final inputs from the corrected fuel cam (91) and the $P3$ cam (95), and transfers this movement through the fuel summing link (98) and cam summing link fork (99) to the fuel limit lever (100). This movement is proportional to log maximum fuel limit/$P3$ + log $P3$, which is equivalent to log maximum fuel limit/$P3$ × $P3$ or log maximum fuel limit. The porting of the fuel-valve plunger (108) is accompanied by movement of the fuel supply fuel cam (102) so that cam follower displacement is proportional to log fuel supply. Log maximum fuel limit and log fuel supply are compared and their difference results in a displacement of the fuel limit plunger (103). If the maximum fuel limit is greater than the fuel supply, the fuel limit lever (100) positions the fuel limit plunger so that the fuel limit piston (28) is vented to *Pb*. This action permits the speed-control plunger (67) to regulate the fuel supply by controlling the pressure on the fuel-valve servo piston (30). When the fuel supply reaches the maximum fuel limit, the fuel cam (102) has rotated clockwise, moving the left end of the fuel limit lever downward, forcing the fuel limit plunger to close the lower port. Further movement of the fuel valve plunger is stopped since the return servo piston (26) and the fuel limit piston effectively overcome the force produced by the fuel-valve servo piston. The accelerating fuel limit is thereafter determined as follows: The fuel limit lever (100) pivots on the summing link fork (99) and positions the fuel limit plunger (103). This controls the fuel limit piston (28) until the governor speed matches the speed setting. The speed-control plunger (67) then rises, reducing the pressure on the fuel-valve servo piston to regulate fuel supply and maintain the set speed.

Deceleration fuel limit is established by the fuel limit plunger (103) moving upward, as a result of the rotation of the fuel cam (102) caused by a change in the speed or the speed setting. The plunger covers the upper port leading to the deceleration fuel limit piston (27). As the fuel-valve plunger (108) continues to close, pressure is exerted on the decamming piston (97), moving it upward. When the piston stops, pressure builds up against the fuel limit piston (27), preventing further closing of the fuel-valve plunger until the computer output indicates a lower fuel limit. During the upward movement of the decamming piston, it rotates the cam summing idler lever (89) counterclockwise by contacting the deceleration adjustment screw (96). This movement moves the corrected fuel cam follower (90) away from the corrected fuel cam (91) to a position predetermined by the setting of the adjustment screw. The result is a deceleration fuel schedule equal to a constant fuel supply/*P3*, which corresponds to the deceleration fuel varying in direct proportion to *P3*.

When the fuel flow reaches the specified normal low limit, the fuel-valve plunger (108) is stopped mechanically by the minimum fuel reset adjustment piston (86). The reset piston is held against the minimum fuel reset adjustment cam (88) as long as the solenoid reset valve (84) remains closed. Switching action by the pilot opens the reset valve, venting the reset piston to overboard drain. The solenoid reset valve is not part of the fuel regulator. Fuel at *Pc* is metered to the reset piston through an orifice (83) with less capacity than the reset valve; therefore the minimum fuel reset spring (87) overcomes the force of the piston when the valve is open. A lower minimum fuel limit is then established by the minimum fuel adjustment screw (85), limiting the travel of the fuel-valve plunger.

An auxiliary function of this control is the inlet guide vane mechanism. The inlet guide vane cam (122) is mounted in tandem on the same shaft as the corrected fuel cam. It produces a follower displacement proportional to a function of the engine speed and the com-

pressor inlet temperature, expressed as a function of the speed and compressor inlet temperature. Positioning of the inlet guide vane actuators is accomplished by a flow of fuel controlled by the inlet guide vane plunger (114). The position of the inlet guide vane feedback gear segment (117) compares the computed position with the feedback position. Movement of the actuator will persist until the feedback position corresponds to the schedule computed position. This will occur when the inlet guide vane plunger is at a neutral position. A change in speed or compressor inlet temperature repositions the inlet guide vane cam so the inlet guide vane cam stylus (121) may contact a different radius of the cam. If the stylus is displaced to a larger radius, it rotates the inlet guide vane cam follower lever (120) and the feedback gear segment in a counterclockwise direction. This movement forces the plunger assembly upward, venting *P1* to the right surface of the inlet guide vane servo (125) and the left surface to the *Pb*. The inlet guide vane servo (125) is not part of the control. The servo moves toward the rod end, pulling the feedback cable (124) in the same direction. The cable turns the inlet guide vane shaft assembly (123) counterclockwise and the feedback gear segment clockwise, allowing the inlet guide vane plunger spring (113) to recenter the plunger assembly. This action stops the servo at a position corresponding to the radius on the cam. If the spring fails to recenter the plunger, the inlet guide vane valve stem spring (116) moves the power actuator plunger (118) and the inlet guide vane valve disk (115) downward. This vents *P1* to the top end of the inlet guide vane plunger, which forces the plunger down.

When speed reaches a specified maximum in relationship to compressor inlet temperature, a further decrease in compressor inlet temperature will start reducing the set maximum speed. The three-dimensional cam end plate (59), actuated by the compressor inlet temperature servo piston (58), contacts the maximum speed reset crank (72). This causes clockwise movement of the maximum speed reset lever (74). The speed-control lever (69) moves in a decrease-speed direction, overriding the speed setting cam (76). This is done by the speed-control lever leaf spring (70) yielding, allowing the speed-control lever to shift downward in relation to the speed-control lever bearing roller (71). The point that compressor inlet temperature will start limiting maximum speed is set by the adjustment of the maximum speed reset lever screw (73).

When speed decreases to a specified minimum in relationship to compressor inlet temperature, a further increase in compressor inlet temperature will start increasing the set idle speed. The three-dimensional cam end plate (59), actuated by the compressor inlet temperature servo piston, contacts the high-temperature speed reset arm (60) turning it and the high-temperature speed reset cam (62) in a clockwise direction. The cam forces the high-temperature speed reset lever (63) to rotate in a counterclockwise direction. This depresses the governor reference spring (64), resulting in a higher speed setting. The temperature at which the idle speed setting begins to increase is set by the adjustment of the high-temperature speed reset adjusting screw (61). The maximum idle limit is adjusted by the high-temperature speed reset lever screw (57).

The speed signals function as follows. If the engine has not exceeded the specified speed value, the pilot-valve plunger (42) will supply P1 to the nozzle lock-valve plunger (38). If the specified speed value is exceeded, the ON-OFF speed lever (44) will raise the pilot-valve plunger, venting the nozzle lock plunger to Pb. The same action will supply P1 to the afterburner valve plunger (79). As P1 is supplied to the afterburner valve plunger, P1 also acts on the ON-OFF speed adjusting piston (41), raising it. This changes the fulcrum of the ON-OFF speed lever from the ON-OFF speed high-point adjusting screw (39) to the ON-OFF speed low-point adjustment screw (40). Under this condition the pilot-valve plunger will continue to supply P1 to the afterburner valve plunger and the ON-OFF speed adjusting piston until the speed falls below a lower specified value. When the speed falls below the lower specified value, the pilot valve plunger moves downward, venting the afterburner valve plunger to Pb. This action allows the ON-OFF speed-adjusting piston to drop the fulcrum of the ON-OFF speed lever to the specified speed-value adjusting screw. With the pilot-valve plunger down, P1 is again vented to the nozzle lock-valve plunger.

The nozzle lock signal, another auxiliary function, operates as follows. The nozzle lock-valve plunger (38), supplied with P1 from the pilot-valve plunger, is also subject to the same pressure applied to the acceleration fuel limit piston (28). When fuel is being scheduled at the maximum acceleration rate, the pressure applied to the fuel limit piston acts on the bottom of the nozzle lock plunger, forcing it upward. This action vents the P1 pressure to the exhaust-nozzle area control.

The power-lever shaft (75) rotates to provide four scheduled inputs received by the main fuel control.

1. *Speed setting*—The position of the speed setting cam (76) determines the load on the governor reference spring (64) by positioning the speed control lever (69). The military speed adjustment (68) adjusts the fulcrum of the speed control lever to the military speed setting. The idle speed adjustment (56) acts as a stop for the speed control lever to set the idle speed.
2. *Stopcock operation*—If the power-lever shaft is rotated to an extreme counterclockwise position, the shutdown valve plunger (81) will be forced upward to a closed position. This action stops the flow of fuel to the engine regardless of the position of the fuel-valve plunger (108).
3. *Shutdown bypass valve operation*—When the shutdown valve plunger closes, it forces the shutdown bypass valve lever (82) to open the shutdown bypass valve plunger (80). This reduces P2 by venting it to Pb, allowing the differential pilot valve plunger (110) to continue regulating P1 − P2 and to maintain P1 to operate the inlet guide vane servo while unloading the supply pump.
4. *Afterburner pressure signal*—When the afterburner signal lever (78) rotates with the power-lever shaft in the increase speed direction, it depresses the afterburner

valve plunger. This action opens the port to the afterburner pump at a predetermined power-lever setting, adjusted by the afterburner signal adjustment screw (77).

To compensate for the effect of changes in fuel temperature, bimetal strip assemblies (55 and 65) are used in parallel with the reference springs on the speed control plunger (67) and the tachometer pilot-valve plunger (53).

FUEL PUMPS

Fuel pumps for gas turbine engines generally employ one or two gear-type pumping elements. Some pumps also incorporate an integral centrifugal boost stage. If the pump contains two gear stages, they may be connected in series as on the JT3C and D engine, in parallel as on the CJ805 engine, or either way as is the case with the 501-D13.

The Chandler Evans Corporation is one of the principal manufacturers of fuel pumps, which, like fuel controls, are produced in a wide variety of designs. Described and pictured in this section are several representative examples of this unit as used on the JT4, JT3C and D, CJ805-23, and 501-D13.

SINGLE-GEAR ELEMENT WITH CENTRIFUGAL BOOST (JT4) *(FIG. 12-20)*

Fuel enters the pump at the fuel INLET port, flows across the centrifugal boost element, out the INTRCLR OUT (intercooler out) port to the fuel-deicing heat exchanger, returning to the pump through the INTRCLR RET port. The fuel then passes through the inlet screen assembly, the main gear stage, and out the DISCHARGE port to the engine main fuel control. Fuel not required by the main fuel control is returned to the pump through the BYPASS port, located schematically between the impeller and the inlet screen assembly. Fuel leakage from the main fuel control is returned to the pump via the low-pressure return port.

Fuel entering the pump at the INLET port is boosted by the centrifugal pumping element (gear-driven centrifugal impeller) prior to entering the single positive-displacement gear-type pumping element. The pressure rise across the centrifugal boost element (boost discharge pressure minus inlet pressure) is a function of impeller rpm and fuel flow and reaches a maximum of approximately 70 psi [482.6 kPa] at 3710 pump rpm. Fuel pressure is further increased across the gear-type pumping element, with gear stage discharge pressure, which is controlled by main fuel control pressure regulation, reaching a maximum of 765 psig [5275 kPa gage] at rated conditions.

Additional pump components and their functions are as follows:

1. *High-pressure relief valve*—The piston-type spring-loaded valve is designed to limit the pressure rise across

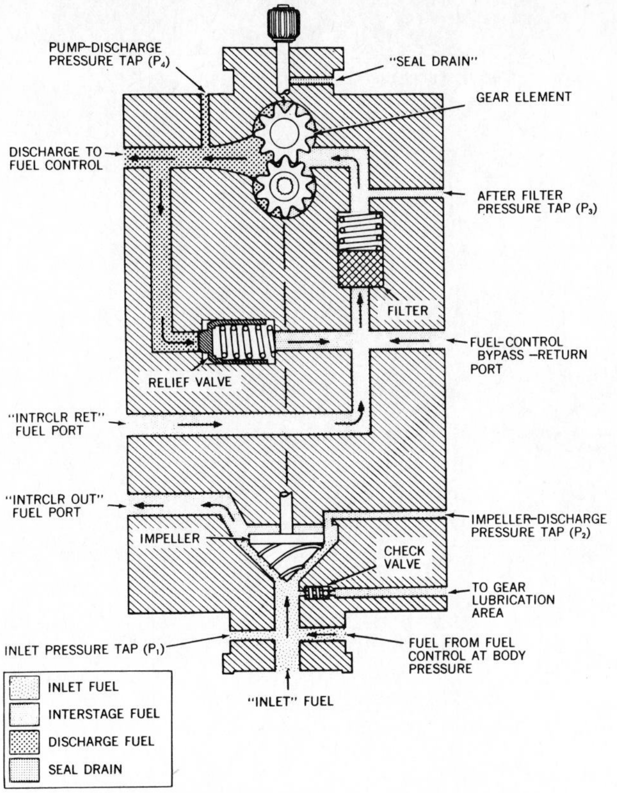

PUMP-DISCHARGE PRESSURE TAP (P_4)
"SEAL DRAIN"
GEAR ELEMENT
DISCHARGE TO FUEL CONTROL
AFTER FILTER PRESSURE TAP (P_3)
FILTER
FUEL-CONTROL BYPASS -RETURN PORT
RELIEF VALVE
"INTRCLR RET" FUEL PORT
"INTRCLR OUT" FUEL PORT
IMPELLER-DISCHARGE PRESSURE TAP (P_2)
IMPELLER
CHECK VALVE
TO GEAR LUBRICATION AREA
INLET PRESSURE TAP (P_1)
FUEL FROM FUEL CONTROL AT BODY PRESSURE
"INLET" FUEL
INLET FUEL
INTERSTAGE FUEL
DISCHARGE FUEL
SEAL DRAIN

FIG. 12-20 Single-gear-element pump with centrifugal boost stage. (*Chandler Evans Corp.*)

the main gear stage (discharge minus after-filter pressure) and begins relieving at approximately 825 psi [5688 kPa] rise, bypassing the full output of the pump internally to the inlet side of the inlet screen assembly without exceeding a pressure rise of 900 psi [6206 kPa].

2. *Slippage check valve*—The spring-loaded ball check valve is designed to ensure positive pump lubrication pressures at high altitude in the event of pump operation with negative inlet pressures, such as might be experienced with failure of the aircraft tank-mounted boost pumps. The pressure differential across this valve ranges from 10 to 19 psi [69 to 131 kPa].
3. *Self-relieving inlet screen assembly*—The inlet screen assembly, fabricated from 40- by 40-in. mesh stainless steel wire with a perforated outer stainless steel shell reinforcement, is designed to limit the pressure drop across the screen element in the presence of ice or contaminant, to a maximum of 10 psi.
4. Several pressure measuring taps are provided in the pump for use on the flow bench.

DOUBLE-GEAR ELEMENTS (SERIES) WITH NO CENTRIFUGAL STAGE (JT3C AND JT3D) *(FIG. 12-21)*

This pump includes the following basic components: inlet fuel filter with self-relieving valve, two positive-displacement gear-type pumping elements, two relief valves, one check valve, one control valve, and one drive shaft equipped with a rotary seal.

Two positive-displacement gear-type pump elements (boost stage and main stage) operate in series to supply fuel to the engine fuel control. The boost stage acts as a pressure boost for the main stage, which supplies the fuel to the engine fuel control. Control of internal fuel pressures and main-stage discharge pressure is maintained by a group of three valves. The pump is installed on the engine and functions as follows.

Power to drive the pump is supplied by the engine through a mounting pad which accommodates the main drive shaft spline. Exterior plumbing brings fuel to:

1. Inlet port, FUEL IN
2. Fuel from the engine fuel regulator at case pressure to the bypass return port, BY PS RET
3. Engine fuel controller main bypass fuel to the main-stage return port, PRIM RET
4. Fuel from the boost-stage discharge port, SEC OUT through the engine fuel deicer system to the boost stage return port, SEC RET
5. Fuel from the main stage discharge port, PRIM OUT to the engine fuel controller

Under normal operating conditions, fuel flows through the pump from the FUEL IN port and the inlet fuel filter element and self-relieving valve to the boost stage, out the SEC OUT port to the SEC RET port, to the main stage and out of the PRIM OUT port to the engine fuel controller. Fuel not required by the engine is returned to the pump by the fuel control through the PRIM RET and BY PS RET ports.

MAIN-STAGE PRESSURE-RELIEF VALVE A [Fig. 12-21(a)] This valve controls the maximum value of pump discharge pressure, and is set to open when main stage discharge pressure reaches approximately 1050 psi [7240 kPa]. When the valve is open, fuel flow is bypassed internally to the inlet side of the main stage pumping element. The *A* valve is normally closed during operation.

MAIN-STAGE INLET CHECK VALVE B If the boost element fails, this valve will open, providing a fuel supply to the main element. The *B* valve is normally closed during operation.

BOOST-STAGE PRESSURE-REGULATING VALVE D This regulating valve controls the pressure of the fuel delivered to the engine fuel deicer system, and therefore the pressure of fuel delivered to the main stage inlet. The valve is set to open at between 45 and 65 psi [310 and 448 kPa] above pump inlet pressure. When the valve is open, fuel is recirculated internally to the pump inlet.

DOUBLE-GEAR ELEMENTS (PARALLEL) WITH CENTRIFUGAL BOOST (CJ805-23) *(FIG. 12-22)*

In this pump fuel first enters at the centrifugal-type boost element. The boost element, which is driven at a greater speed than the high-pressure elements, in-

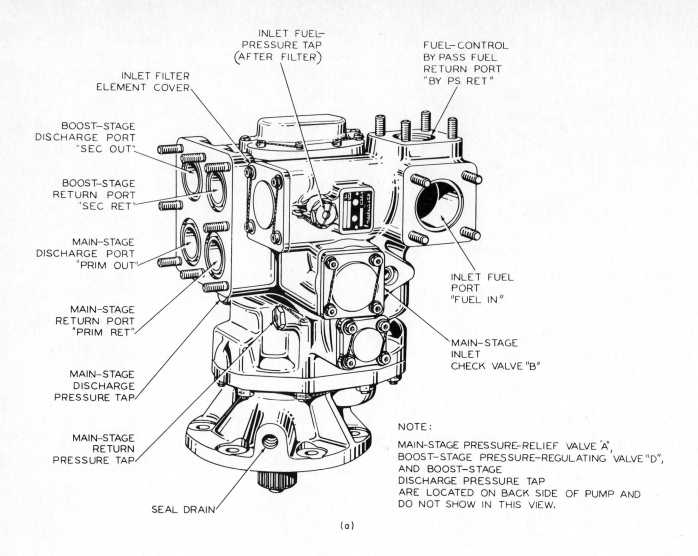

INLET FUEL-
PRESSURE TAP
(AFTER FILTER)

FUEL-CONTROL
BY PASS FUEL
RETURN PORT
"BY PS RET"

INLET FILTER
ELEMENT COVER

BOOST—STAGE
DISCHARGE PORT
"SEC OUT"

BOOST—STAGE
RETURN PORT
"SEC RET"

MAIN—STAGE
DISCHARGE PORT
"PRIM OUT"

INLET FUEL
PORT
"FUEL IN"

MAIN—STAGE
RETURN PORT
"PRIM RET"

MAIN—STAGE
INLET
CHECK VALVE "B"

MAIN—STAGE
DISCHARGE
PRESSURE TAP

MAIN—STAGE
RETURN
PRESSURE TAP

NOTE:

MAIN-STAGE PRESSURE-RELIEF VALVE "A",
BOOST—STAGE PRESSURE-REGULATING VALVE "D",
AND BOOST—STAGE
DISCHARGE PRESSURE TAP
ARE LOCATED ON BACK SIDE OF PUMP AND
DO NOT SHOW IN THIS VIEW.

SEAL DRAIN

(a)

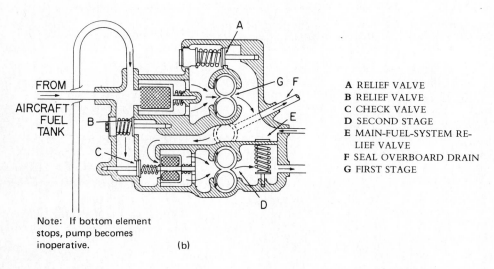

FROM

AIRCRAFT
FUEL
TANK

A RELIEF VALVE
B RELIEF VALVE
C CHECK VALVE
D SECOND STAGE
E MAIN-FUEL-SYSTEM RE-
   LIEF VALVE
F SEAL OVERBOARD DRAIN
G FIRST STAGE

Note: If bottom element
stops, pump becomes
inoperative.   (b)

FIG. 12-21 (a) This pump is composed of two gear elements connected to flow fuel in series. There is no centrifugal stage. Note: The letter references in the text refer to Fig. 12-21(a).

(b) Schematic of Chandler Evans pump used on the Pratt & Whitney Aircraft JT3 series engines. NOTE: If bottom element stops, pump becomes inoperative.

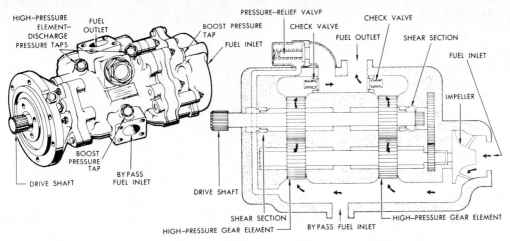

FIG. 12-22 Parallel arrangement of a double-gear element pump with booster stage. (*General Electric Co.*)

creases the pressure of the fuel 15 to 45 psi [103 to 310 kPa], depending upon engine speed. The fuel is discharged from the boost element to the two gear-type positive-displacement, high-pressure elements. Each of these elements discharges fuel through a check valve to a common discharge port. At a discharge pressure of 850 psig [5861 kPa gage], the high-pressure elements deliver approximately 51 gal/min [193 L/min].

Shear sections are incorporated in the drive systems of each element. Thus, if one element fails, the remaining element remains operative, and the check valves prevent recirculation through the inoperative element. One element can supply sufficient fuel for moderate aircraft speeds.

A relief valve in the discharge port of the pump opens at approximately 900 psi [6206 kPa] and is capable of bypassing the total flow at 960 psi [6619 kPa]. This permits fuel at pump discharge pressure to be recirculated as a protection against "dead-heading" the pump. The bypass fuel is routed to the inlet side of the two high-pressure elements.

The pump always supplies more fuel than is needed in the system. The fuel control determines the amount of fuel required for engine operation and bypasses the remainder back to the pump. This bypass flow is routed to the intake side of the high-pressure elements.

The fuel pump is mounted at the seven o'clock position on the rear face of the transfer gearbox and requires a power input of 48 hp [35.8 kW] maximum. Pump speed is 3960 rpm at 100 percent engine speed. Four ports on the pump are for attaching pressure-measuring instruments. Two ports, located between the boost and gear elements, may be used to measure pressure at the inlet of the gear elements (engine boost). The two other ports, located downstream of each gear element, may be used to measure discharge pressure of each element. A port on the pump mounting flange provides for drainage of the interstage shaft seal area.

DOUBLE-GEAR ELEMENTS (SERIES OR PARALLEL) WITH CENTRIFUGAL BOOST STAGE (501-D13) *(FIG. 12-23)*

This fuel pump is engine-driven and is attached to the rear of the accessory case. It incorporates two gear-type pressure elements supplied by one centrifugal boost pump. The design of the pump is such that the capacity of the primary gear-type element is 10 percent greater than that of the secondary element. This feature allows series operation with the primary element taking the full load without the need for a bleed valve bypassing the secondary element during normal operation. The fuel pump operates in conjunction with the high-pressure fuel filter which is mounted on the bottom of the fuel pump.

During an engine start the pump elements are in parallel operation, and the paralleling valve in the fuel filter is energized closed. The pressure switch in the fuel filter is closed, causing a cockpit-mounted warning light to illuminate and indicate that the secondary element is operating properly (2200 to 9000 rpm). If the primary element fails (indicated by a warning light being lit) while the engine is running, the secondary element provides sufficient fuel flow and pressure to operate the power unit in flight. During a start, if a primary element has failed, fuel flow may not be sufficient for a satisfactory start. The primary-fuel-pump failure warning light is cockpit mounted. It goes on when the primary element of the fuel pump has failed. The light also goes on during the engine starting cycle when the fuel pump elements are operating in parallel (2200 to 9000 rpm).

The fuel pump and filter assembly consists of two castings which may be separated. The upper portion is the fuel pump which contains the boost pump and the two pump elements. The lower portion is the filter assembly and contains the removable high-pressure filter, the check valves, the paralleling valve, and the pressure switch.

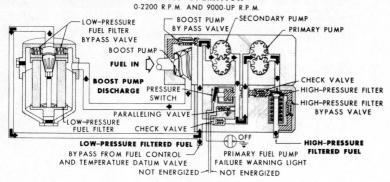

SERIES OPERATION
0-2200 R.P.M. AND 9000-UP R.P.M.

LOW-PRESSURE
FUEL FILTER
BYPASS VALVE

BOOST PUMP
BY PASS VALVE — SECONDARY PUMP
PRIMARY PUMP

BOOST PUMP

FUEL IN

BOOST PUMP
DISCHARGE

PRESSURE
SWITCH

PARALLELING VALVE

LOW-PRESSURE
FUEL FILTER

CHECK VALVE

CHECK VALVE
HIGH-PRESSURE FILTER
HIGH-PRESSURE FILTER
BYPASS VALVE

LOW-PRESSURE FILTERED FUEL
BYPASS FROM FUEL CONTROL
AND TEMPERATURE DATUM VALVE
NOT ENERGIZED

OFF

PRIMARY FUEL PUMP
FAILURE WARNING LIGHT
NOT ENERGIZED

**HIGH-PRESSURE
FILTERED FUEL**

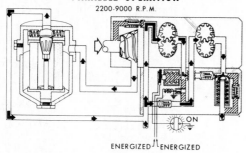

PARALLEL OPERATION
2200-9000 R.P.M.

ON

ENERGIZED — ENERGIZED

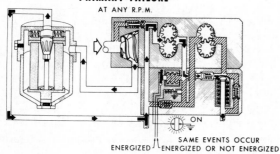

PRIMARY FAILURE
AT ANY R.P.M.

ON

SAME EVENTS OCCUR
ENERGIZED — ENERGIZED OR NOT ENERGIZED

FIG. 12-23 Fuel pump and high-pressure filter flow schematic. (*Detroit Diesel Allison Division of General Motors Corp.*)

FUEL NOZZLES

On most gas turbine engines, fuel is introduced into the combustion chamber through a fuel nozzle whose function is to create a highly atomized, accurately shaped spray of fuel suitable for rapid mixing and combustion with the primary airstream under varying conditions of fuel and airflow (Fig. 12-24). Most engines use either the single (simplex) or the dual (duplex) nozzle.

THE SIMPLEX NOZZLE

Figure 12-25 illustrates a typical simplex nozzle. This nozzle, as its name implies, has the advantage of being simpler in design than the duplex nozzle. Its chief disadvantage lies in the fact that it is unable to provide a satisfactory spray pattern with the large changes in fuel pressures and airflows encountered in bigger engines.

THE DUPLEX NOZZLE

At starting and low rpm, and at low airflow, the spray angle needs to be fairly wide in order to increase the chances of ignition and to provide good mixing of fuel and air, but at higher rpm and airflow, a narrow pattern is required to keep the flame of combustion away from the walls of the combustion chamber (Fig. 12-26). The small fuel flow that is used in idling is broken up into a fine spray by being forced through a small outlet formed by the primary holes. The secondary holes are larger, but still provide a fine spray at higher rpm because of the higher fuel pressure. The chief advantage then, of the duplex nozzle, is its ability to provide good fuel atomization and proper spray pattern at all rates of fuel delivery and airflow without the necessity of utilizing abnormally high fuel pressures.

In order for the duplex nozzle to function, there must be a device to separate the fuel into low (primary) and high (secondary) pressure supplies. This flow divider

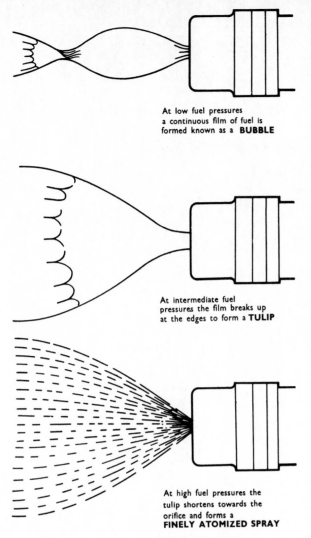

FIG. 12-24 Various stages of fuel atomization.

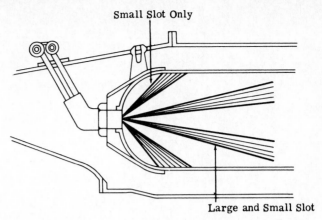

FIG. 12-26 The spray angle changes when fuel flows in the primary or primary and secondary manifolds.

may be incorporated in each nozzle, as is the case with the single-entry duplex type, as shown in Fig. 12-27(a) and (b), or a single-flow divider may be used with the entire system (Fig. 12-28).

Single-entry duplex nozzles incorporating an internal flow divider require only a single fuel manifold (Fig. 12-29), while, as shown in Fig. 12-28, dual-entry fuel nozzles require a double fuel manifold. Some dual fuel manifolds may not be apparent as such. For example, the Pratt & Whitney JT3 and JT4 series engines utilize a concentric manifold system (Fig. 12-30).

The flow divider, whether self-contained in each noz-

zle, or installed in the manifold, is usually a spring-loaded (Fig. 12-31) valve set to open at a specific fuel pressure. When the pressure is below this value, the flow divider directs fuel to the primary manifold and/or nozzle orifice. Pressures above this value cause the valve to open and fuel is allowed to flow in both manifolds and/or nozzle orifices.

Most modern nozzles have their passages drilled at an angle so that the fuel is discharged with a swirling motion in order to provide low axial air velocity and high flame speed. In addition, an air shroud surrounding the nozzle cools the nozzle tip and improves combustion by retarding the accumulation of carbon deposits on the face. The shroud also provides some air for combustion and helps to contain the flame in the center of the liner (Fig. 12-32).

Extreme care must be exercised when cleaning, repairing, or handling the nozzles, since even fingerprints on the metering parts may produce a fuel flow which is out of tolerance.

Some models of the Lycoming T-53 and T-55 and others use devices called vaporizing tubes (Fig. 12-33) instead of injector nozzles. The vaporizing tube is essentially a U-shaped pipe whose exit faces upstream to the compressor airflow. Excellent mixing of the fuel and air results from this arrangement.

OTHER FUEL SYSTEM COMPONENTS

FUEL FILTERS

All gas turbine engines will have several fuel filters installed at various points along the system. It is common practice to use at least one filter before the fuel pump and one on the high-pressure side of the pump. (See Chap. 13 on the complete fuel system for the location of the filters in some commonly used engines.) In most cases the filter will incorporate a relief valve set to open at a specified pressure differential in order to provide a bypass for fuel when filter contamination becomes excessive.

FIG. 12-25 The simple or simplex nozzle.

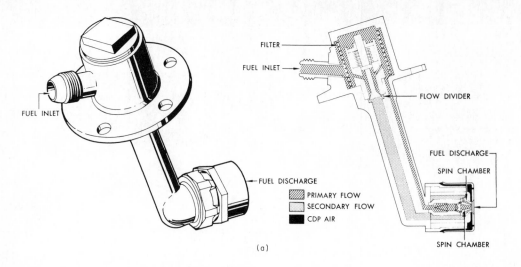

(a)

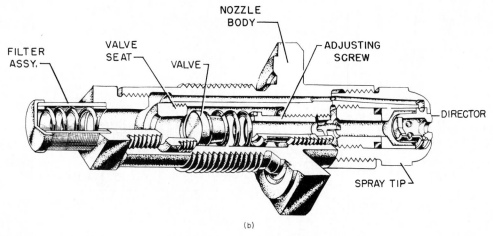

(b)

FIG. 12-27 (a) The CJ805 (J79) single-entry duplex nozzle. The spray angle of this nozzle is 80° to 100° with primary flow, and 80° to 90° with combined primary and secondary flows. The angle varies inversely with pressure.
(b) Another form of single-entry duplex nozzle.

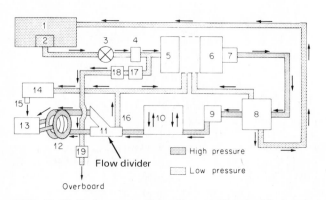

1 A/C FUEL CELL
2 A/C BOOST PUMP
3 A/C FUEL SHUT OFF
4 A/C LOW PRESSURE FIL-
TER
5 BOOST FUEL PUMP
6 MAIN FUEL PUMP
7 HIGH-PRESSURE FILTER
8 PE 3A FUEL REGULATOR
9 FLOW METER
10 OIL COOLER
11 FLOW DIVIDER
12 FUEL MANIFOLDS
13 FUEL NOZZLE
14 COMBUSTION STARTER
15 MANUAL DRAIN
16 SEAL DRAIN
17 FUEL-PRIMING SOLENOID
18 CHECK VALVE
19 DRIP DRAIN VALVE

FIG. 12-28 Fuel system of the General Electric J73 showing the flow divider and the required double fuel manifold.

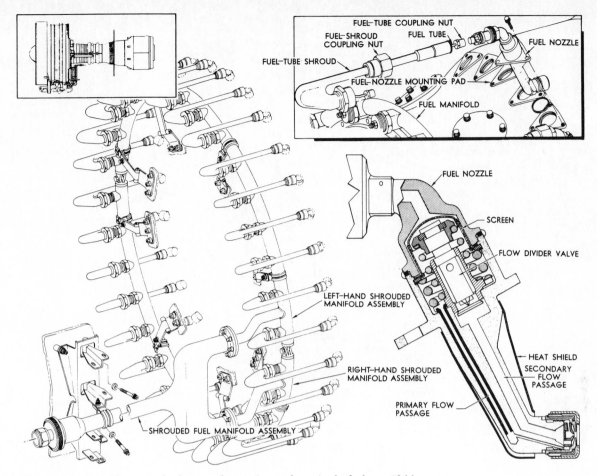

FIG. 12-29 A single-entry duplex nozzle requires only a single fuel manifold.

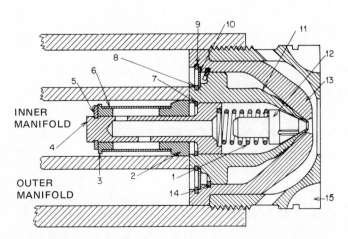

1 SPRING
2 LOWER PRIMARY FER-
    RULE
3 UPPER PRIMARY FERRULE
4 SPRING SEAT
5 PRIMARY-STRAINER SNAP
    RING
6 PRIMARY STRAINER
7 SPRING SEAT SNAP RING
8 INNER SECONDARY-
    STRAINER SNAP RING
9 OUTER SECONDARY-
    STRAINER SNAP RING
10 SECONDARY STRAINER
11 NOZZLE BODY INSERT
12 PRIMARY PLUG
13 NOZZLE BODY
14 INSERT SNAP RING
15 RETAINING NUT

FIG. 12-30 A double-entry nozzle with a concentric fuel manifold used on the Pratt & Whitney Aircraft JT3D engine.

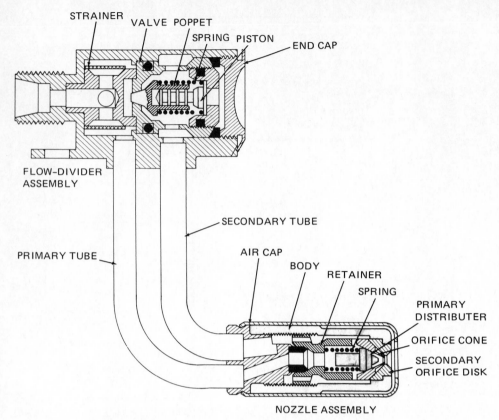

FIG. 12-31 The flow divider and nozzle are an integral unit on the General Electric J85 (CJ610). The flow divider is located outside the hot combustion area.

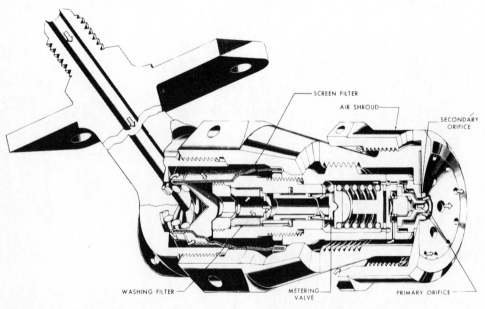

FIG. 12-32 The Detroit Diesel Allison 501-D13 engine uses a single-entry duplex nozzle incorporating an air shroud.

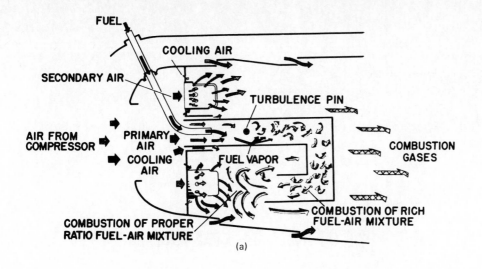

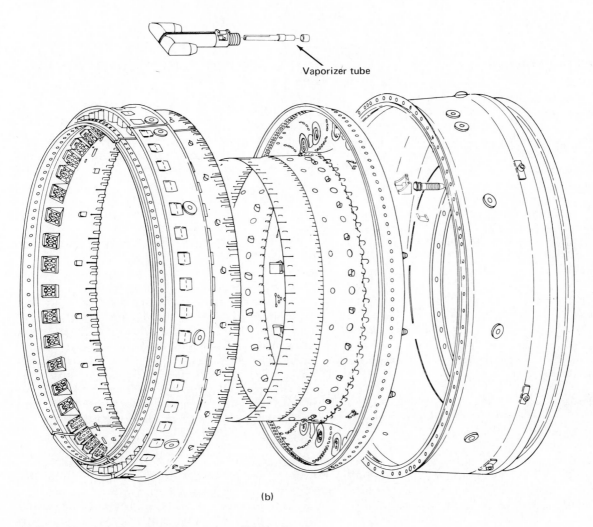

Vaporizer tube

(b)

FIG. 12-33 (a) Forms of vaporizer tubes. (*Wright Corp.*)
(b) T-type vaporizer tube. (*Avco Lycoming Corp.*)

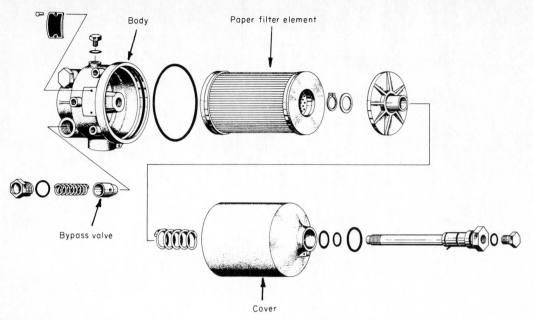

FIG. 12-34 A typical low-pressure filter with a paper element.

TYPES OF FILTERS

1. *Paper cartridge type* (Fig. 12-34)—This type of filter is usually used on the low-pressure side of the pump. It incorporates a replaceable paper filter element which is capable of filtering out particles larger than 50 to 100 microns (μ). Its function is to protect the fuel pump from being damaged due to fuel contamination. One micron equals 0.000039 inch, or 25,400 microns equal 1 in. To obtain an idea of how small a micron is, the following comparisons can be made:
Human hair = 70 μ or 0.00273 in
Smallest dirt eye can see = 50 μ or 0.00195 in
White blood cell = 20 μ or 0.00078 in
Aircraft filter = 5 μ or 0.000195 in
Red blood cell = 4 μ or 0.000156 in
Test stand filter = 3 μ or 0.000117 in.
2. (a) *Screen type* [Fig. 12-35(a)]—Generally used as a low-pressure fuel filter, some of these filter screens are constructed of sinter-bonded stainless steel wire cloth and are capable of filtering out particles larger than 40μ.
  (b) *Convoluted screen* [Fig. 12-35(b)]—Another form of the basic screen-type filter but with increased filtering area due to the convolutions (folding or pleating of the screen material.)
  (c) *Screen disk type* [Fig. 12-35(c)]—Located on the outlet side of the pump, this filter is composed of a stack of removable fine wire mesh screen disks which must be disassembled and cleaned periodically in an approved solvent.

In addition to the main-line filters, most fuel systems will incorporate several other filtering elements, locating them in the fuel tank, fuel control, fuel nozzles, and in any other place deemed desirable by the designer. (See Figs. 13-1 to 13-12 for the location of the filtering elements in some typical fuel systems.)

PRESSURIZING AND DRAIN (DUMP) VALVES *(FIG. 12-36)*

The purpose of the pressurizing and drain valve on the CJ805 and the CJ610 engines (see Figs. 13-1 and 13-5) is to prevent flow to the fuel nozzles until sufficient pressure is attained in the main fuel control to operate the servo assemblies which are used to compute the fuel-flow schedules. It also drains the fuel manifold at engine shutdown to prevent post-shutdown fires, but keeps the upstream portion of the system primed to permit faster starts.

The pressurizing and dump valve used on the Pratt & Whitney Aircraft JT3 engines (Fig. 12-37) has a somewhat different function. In addition to the drainage or dumping function, this unit also serves as a flow divider. At the beginning of an engine "start" the fuel control supplies a pressure signal to the pressurizing and dump valve, causing the valve to close the manifold drain and open a passage for fuel flow to the engine. On engine shutdown, fuel flow is cut off immediately by a valve in the fuel control. The pressure signal drops and the dump valve will open, allowing fuel to drain from the manifold. The flow divider allows fuel to flow to the primary and secondary manifolds depending on fuel pressure. (See Fig. 13-3.)

The Allison 501-D13 engine incorporates an electrically actuated drain valve (which Allison calls a drip valve), activated automatically at a specific rpm (Fig. 12-38) to assure a clean cutoff and prevent fuel from dripping into the combustion chambers at low fuel pressures. In this way gum and carbon deposits are prevented, and the fire hazard is reduced. The General Electric CJ610 fuel manifold drain valve performs the same function, but works when fuel pressure drops below a specified minimum. (See Figs. 13-1 and 13-4.)

Some manufacturers, such as Detroit Diesel Allison and Pratt & Whitney Aircraft install a pressure-operated

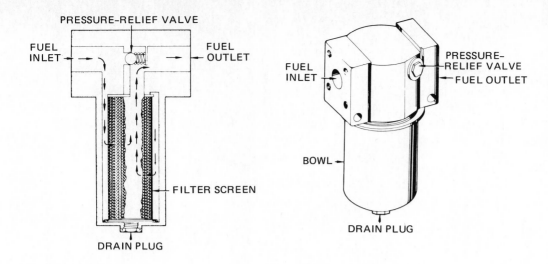

(a)

(b)

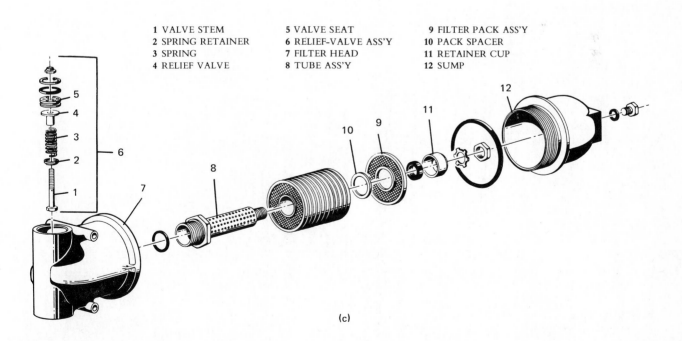

1 VALVE STEM	5 VALVE SEAT	9 FILTER PACK ASS'Y
2 SPRING RETAINER	6 RELIEF-VALVE ASS'Y	10 PACK SPACER
3 SPRING	7 FILTER HEAD	11 RETAINER CUP
4 RELIEF VALVE	8 TUBE ASS'Y	12 SUMP

(c)

FIG. 12-35 **(a)** A typical screen-type filter.
(b) A convoluted screen element.
(c) A typical mesh-screen disk-type filter.

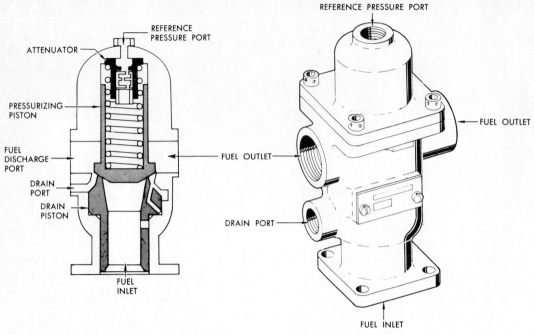

FIG. 12-36 Pressurizing and drain valve for the General Electric CJ805 (J79) engine.

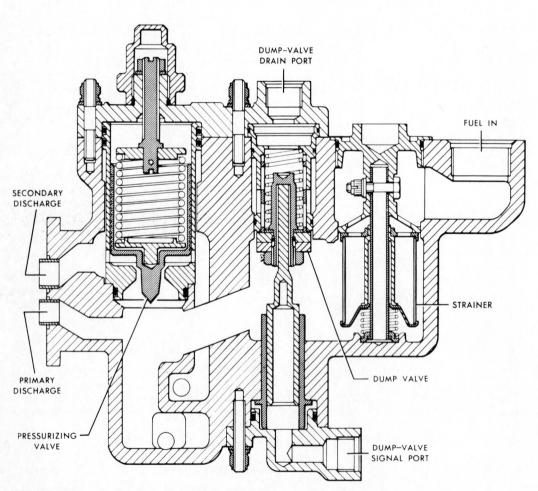

FIG. 12-37 Pratt & Whitney Aircraft pressurizing and dump valve schematic.

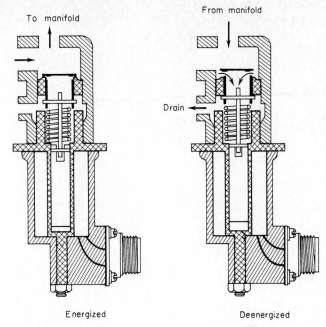

To manifold

From manifold

Drain

Energized

Deenergized

FIG. 12-38 Drip valve used on the Detroit Diesel Allison 501-D13 engine.

valve in the combustion-chamber section. When the pressure in the burners drops below a specified minimum, usually a few pounds per square inch, this valve will open and drain any residual fuel remaining after a false start or normal shutdown. (Refer to Fig. 13-3 to see where this drain valve fits into the system.)

FLOW METERS

Most fuel systems will incorporate a flow meter as a measure of fuel consumption in pounds per hour. While the flow meter is usually an airframe-supplied accessory, it is included here to show its location in the system, which is generally after the fuel control. (Figures 13-1, 13-2, and 13-5 show the placement of this unit.)

FUEL-OIL COOLERS

Some engines use fuel as a cooling medium for the oil. (Refer to Figs. 13-1 and 13-5. For a more detailed look at fuel-oil coolers refer to Fig. 15-8.)

FUEL HEATERS

The General Electric CJ805-23, CJ610, some models of the Pratt & Whitney JT3 engines, and others may incorporate an additional unit to reduce the possibility of ice crystals forming in the fuel. The fuel heater consists basically of an air-to-liquid heat exchanger and a thermostatically controlled valve to regulate airflow. The thermostatic valve is responsive to the temperature of the outgoing liquid. The liquid is turbojet engine fuel and the air is compressor bleed air supplied by the engine. (See Fig. 13-5.)

The need for the fuel heater has been reduced because of the increasing use of PFA 55MB, an anti-icing inhibitor and biocidal agent. See Chap. 11, page 208.

REVIEW AND STUDY QUESTIONS

1. What is the relationship between the operator, fuel control, and engine in the control of power?
2. List the basic inputs to a fuel control.
3. Name two groups (types) of fuel controls.
4. Discuss the essential requirements of any fuel control.
5. What is the purpose of the servo system used in many modern fuel controls?
6. Why is an acceleration limit system needed on the fuel control?
7. What are two ways to meter fuel?
8. Explain briefly the operation of the following fuel controls: DP-F2, AP-B3, 1307, and JFC25.
9. What type of fuel pump is used on most turbine engines?
10. Describe the operation of a typical double-element fuel pump with integral booster impeller.
11. Explain what is meant by "series" and "parallel" pump operation.
12. Why is it necessary to use duplex fuel nozzles on many engines?
13. What is a flow divider? Where may it be located?
14. What device is used in place of the fuel nozzle? Tell how it works.
15. Name three types of fuel filters.
16. What is the purpose of the pressurizing and drain valve?
17. What is the purpose of the fuel heater used on some engines?

CHAPTER THIRTEEN
TYPICAL FUEL SYSTEMS

The fuel system must supply clean, accurately metered fuel to the combustion chambers. Although the systems will differ in many respects, they all have certain elements in common. For example, all fuel systems must have a fuel pump, fuel control, pressurizing valve or its equivalent, fuel manifold, and fuel nozzles (or vaporizers). How these specific units do their jobs differs radically from one engine to another. Some systems incorporate features that are not, strictly speaking, necessary to the metering of fuel. Examples of this can be found in the use of fuel as a cooling medium for the oil, and in the use of fuel to operate variable inlet guide vanes, stators, and compressor bleed mechanisms.

It is the purpose of this section to illustrate typical fuel systems so that the reader will obtain some idea of fuel flows and the location of the several parts that constitute the system. Because of space limitations, Figs. 13-9, 13-10, 13-11, and 13-12 are not accompanied by a detailed system explanation, but are included here as examples of other fuel systems used on modern engines.

THE GENERAL ELECTRIC CJ610 FUEL SYSTEM *(FIG. 13-1)*

This fuel and control system meters fuel to the engine and provides actuation pressure for the variable-inlet-geometry (VG) system. It consists of the following engine-mounted components:

Fuel pump
Fuel control
Overspeed governor
Fuel-oil cooler
Fuel pressurizing valve
Fuel manifolds
Fuel manifold drain valve
Fuel nozzles (with integral flow divider)
Actuator assembly (VG)
Bleed valves
Fuel flowmeter (airframe-furnished equipment)

Control System

The control system schedules the rate of engine fuel flow for variations in air density, air temperature, and engine speed.

Fuel Pump

The fuel pump comprises a single-element positive-displacement pump, centrifugal boost pump, filter, and bypass circuit with a pressure-relief valve. The pump supplies fuel to the fuel control and is mounted on and driven by the accessory gearbox.

Fuel Control

The fuel control is mounted on and driven by the fuel pump. The control incorporates a hydromechanical computer section and fuel-regulating section to operate the control servos. Parameters of engine speed, power-lever setting, compressor inlet temperature, and compressor discharge pressure are used in the computer section to schedule the operation of the fuel-metering valve and the VG servo valve. The fuel-regulating section meters fuel to the engine under all operating conditions.

Overspeed Governor

The isochronous overspeed governor is mounted on and driven by the accessory gearbox. Fuel is supplied to the governor bypass section from the fuel control, and to the governing section from the fuel pump. Overspeed governing is controlled by bypassing the fuel, when it is in excess of engine maximum limiting speed requirements, to the fuel pump inlet port.

Fuel-oil Cooler

For a description of the fuel-oil cooler, refer to Chap. 15.

Fuel Pressurizing Valve

The pressurizing valve is mounted on the fuel-oil cooler and connects to the fuel manifolds, manifold drain valve, and fuel pump interstage reference pressure line. During starting, boost pressure and spring force close the pressurizing valve to prevent low-pressure fuel flow to the fuel nozzles, and to allow the fuel control to build up sufficient pressure to operate the control servos and (VG) actuators. The control pressure then opens the pressurizing valve and closes the manifold drain valve. Fuel is then distributed to the fuel nozzles at sufficient pressure for satisfactory atomization.

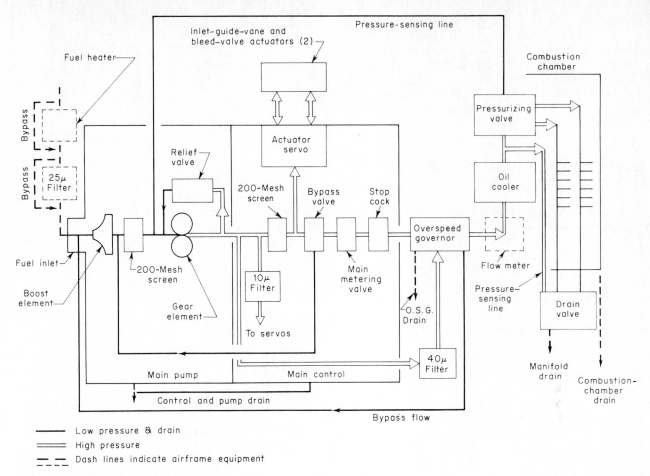

FIG. 13-1 General Electric CJ610 fuel system schematic.

Fuel Manifolds

Two fuel manifold tubes are located around the main-frame casing. Each manifold tube connects to six fuel nozzles. Fuel is supplied from the pressurizing valve, through the manifold tubes, to the fuel nozzles.

Fuel Manifold Drain Valve

The fuel manifold drain valve drains the fuel manifolds at engine shutdown to prevent residual fuel from dribbling out the fuel nozzles, thus creating a fire hazard. It also prevents the formation of gum and carbon deposits in the manifold and nozzles. The valve consists of a piston, which is spring-loaded, to open the manifold drain passage at shutdown, and a fuel filter with a bypass valve that opens if the filter becomes clogged. During engine operation, the pressurizing valve actuates to close the manifold drain passage of the valve and admit fuel to the fuel manifolds.

Fuel Nozzles

Twelve fuel nozzles, mounted on the main frame, spray atomized fuel into the combustion chamber. The fuel nozzle incorporates a flow divider, a primary and secondary flow passage, and an air-shrouded spin-chamber-type orifice. During starting, low-pressure

fuel in the primary passage sprays a mixture adequate for ignition. As the engine accelerates, increased fuel pressure opens the flow divider and additional fuel flows into the secondary passage to the spin chamber where it merges with the primary passage fuel flow. The air shroud sweeps air across the nozzle orifice to prevent carbon formation. (See Fig. 12-31.)

Actuator Assembly (VG)

Two variable-geometry actuators, mounted on the compressor casing, position the inlet guide vanes and interstage bleed valves. They are linear-travel, piston-type actuators, hydraulically actuated by high-pressure fuel from a servo valve in the fuel control. The actuator piston rods are connected to bellcranks which position the inlet guide vanes and interstage bleed valves. A feedback cable is connected from the bellcrank assembly to the fuel control and supplies the fuel-control servo valve with a position signal.

Bleed Valves

Two bleed valves are mounted on each side of the compressor stator casing. During transient engine speeds, the valves bleed air from the third, fourth, and fifth stages of the compressor according to a bleed schedule, which is a function of compressor speed and inlet air temperature, prescribed by the fuel control. The valves are actuated by the fuel control and two VG actuators through a bellcrank-linkage arrangement. A synchronizing cable synchronizes the bleed-valve positions and, in case of malfunction in either VG actuator, transmits the motion of the functioning VG actuator to the other.

THE PRATT & WHITNEY AIRCRAFT JT3D FUEL SYSTEM (FIG. 13-2)

This engine fuel system is designed to satisfy the requirements peculiar to a turbofan engine. The fuel must be pressurized, filtered, metered, distributed, and atomized before it can finally be ignited and burned in the combustion section of the engine. To improve system reliability, provisions are also made for heating the fuel to prevent fuel system icing.

The operation of the fuel system in the engine and the functions of its various components may be described by tracing the passage of fuel from the inlet connection of the fuel pump to its ultimate discharge through the fuel nozzles in the engine combustion chambers.

The system consists of the following components:

Engine-driven fuel pump
Fuel heater
Heater air shutoff valve
Fuel filter
Fuel-control unit
Pressurizing and dump valve
Fuel manifolds and spacer
Fuel nozzles

Fuel Pump (FIG. 12-21)

The engine-driven fuel pump is a two-stage, pressure-loaded, gear-type pump incorporating the following:

Combination 40 and 80 mesh, self-relieving inlet filter
Gear-type secondary (boost) stage
Secondary stage pressure-regulating valve
Secondary stage bypass valve
Gear-type primary (main) stage
Primary stage pressure-relief valve

Fuel Heater

The fuel heater utilizes compressor discharge air to heat the fuel and thus prevent fuel system icing. Fuel heater operation is selected from the cockpit by electrically opening or closing the air shutoff valve in the heater air supply line.

Fuel Filter

The fuel filter contains a 40-μ paper-type element designed to protect the fuel control unit from ice crystals and other contaminants. A pressure switch mounted on the filter housing provides an indication of filter con-

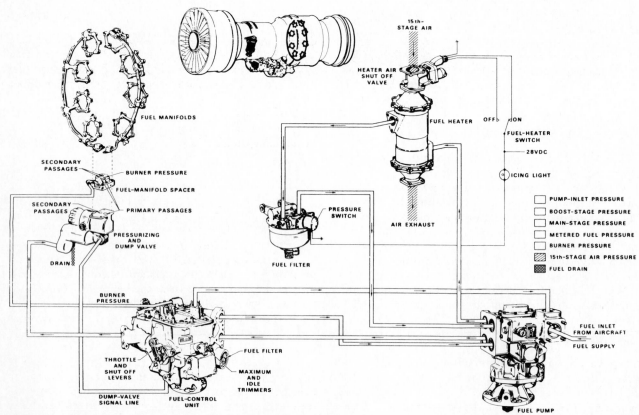

FIG. 13-2 JT3D engine fuel system. (*Pratt & Whitney Aircraft Group.*)

tamination by illuminating a light in the cockpit. To prevent a clogged filter element from interrupting fuel flow through the system, a bypass valve is incorporated in the filter housing.

Fuel Control *(FIG. 12-18)*

The fuel-control unit meters fuel to the fuel nozzles. It is designed to satisfy fuel-flow requirements for starting, acceleration, deceleration, and stabilized (steady-state) operation. To enable the control to determine the proper fuel-flow rate for any operating condition, three separate senses, or inputs, are provided to the control. They are:

Throttle position
N_2 compressor speed
P_b (burner pressure)

A hydraulically actuated fuel shutoff valve in the control is positioned in response to movement of the fuel cutoff lever on the throttle quadrant. Included on the fuel-control body are two trim adjustments, idle trim and maximum trim. These adjustments are used to adjust fuel flow to achieve specific N_2 rpm and engine thrust values.

Fuel Pressurizing and Dump Valve *(FIG. 12-37)*

The fuel pressurizing and dump valve actually contains two valves within one housing. The dump valve is spring-loaded open and was originally designed to drain the primary fuel manifold at engine shutdown. It is closed by a fuel pressure signal from the fuel-control unit. The other valve in the housing, the pressurizing valve, is spring-loaded closed. This valve limits starting fuel flow to the primary manifold. As the engine is accelerated to higher power settings, the pressurizing valve is forced open by the steadily increasing fuel pressure. When the valve opens, fuel will be allowed to flow into the secondary manifold. This flow will supplement the flow in the primary manifold to satisfy engine fuel requirements at higher power settings.

Fuel Manifold Spacer

The fuel manifold spacer, or adapter, interconnects the pressurizing and dump valve and the fuel manifolds. It provides both primary and secondary passages for the two manifold halves and a connector for the fuel-control P_b signal.

Fuel Manifolds

The fuel manifolds distribute fuel to the fuel nozzles. The manifolds are coaxial and are split into two sections. The secondary manifold is inside the primary manifold.

Fuel Nozzles *(FIG. 12-30)*

The fuel nozzles are dual-orifice type and are utilized to atomize the fuel. They are located in clusters of six at the forward end of each of the eight combustion chambers.

Fuel-oil Cooler *(optional)*

On some JT3 series engines, a fuel-oil cooler is used. This unit will be discussed under the lubrication system since its prime function is to help cool the oil in conjunction with the airframe-supplied air-oil cooler. It consists of a cylindrical oil chamber surrounded by a jacket through which the fuel passes. Heat from the oil is transferred to the fuel via radiation.

Water-injection System *(optional)* *(FIG. 13-3)*

Because of the functional interrelationship of the water-injection system with the engine fuel system, they should be discussed together. A detailed description of one form of this system is covered in Chap. 9. It must be mentioned, however, that, in conjunction with the fuel control, a switch is operated which passes electrical power for opening or closing the airframe-supplied water-injection shutoff valve. Also, a sensing line from the water-injection control is attached to the fuel control for resetting the fuel control's maximum speed adjustment to a higher setting during water injection.

THE DETROIT DIESEL ALLISON 501-D13 FUEL SYSTEM *(FIG. 13-4)*

The system consists of the following components:

Engine-driven fuel-pump assembly
Paralleling valve
High-pressure fuel-filter assembly
Low-pressure fuel-filter assembly
Pressure switch
Fuel control
Primer valve
Temperature datum valve
Fuel manifold
Fuel nozzles
Drip valve

Fuel Pump

Fuel is supplied by the aircraft fuel system to the inlet of the engine-driven fuel-pump assembly. The fuel-pump assembly consists of a boost pump and two spur gear-type pumps. The gear-type pumps may be placed in either series or parallel by an electrically operated paralleling valve located in the high-pressure fuel-filter assembly. The boost pump output is delivered to the low-pressure fuel-filter assembly which filters the fuel and delivers it to the high-pressure fuel-filter assembly, where it is directed to the inlets of the two gear-type pumps. The output of the two gear pumps is filtered by the high-pressure fuel filter. A pressure switch, in the high-pressure fuel-filter assembly, completes an

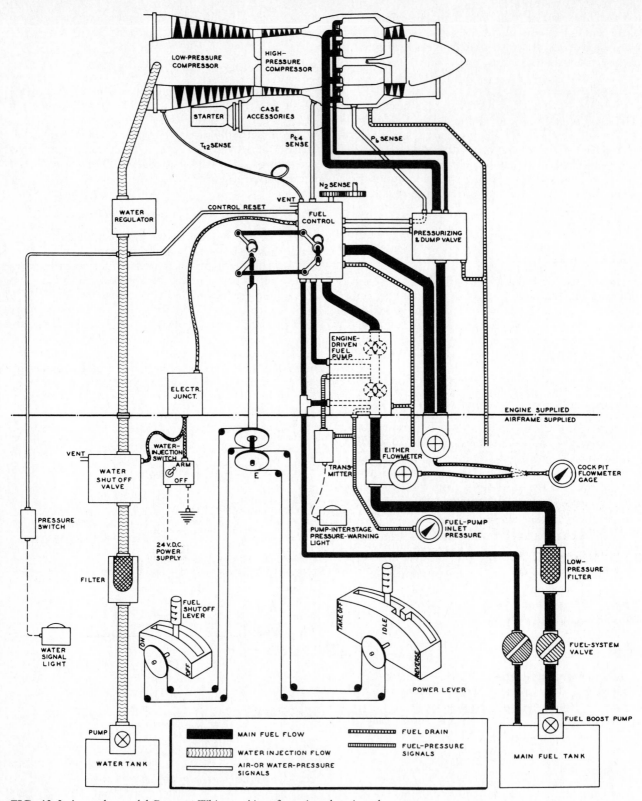

FIG. 13-3 An early-model Pratt & Whitney Aircraft engine showing the water–injection system. (See Chapter 9.)

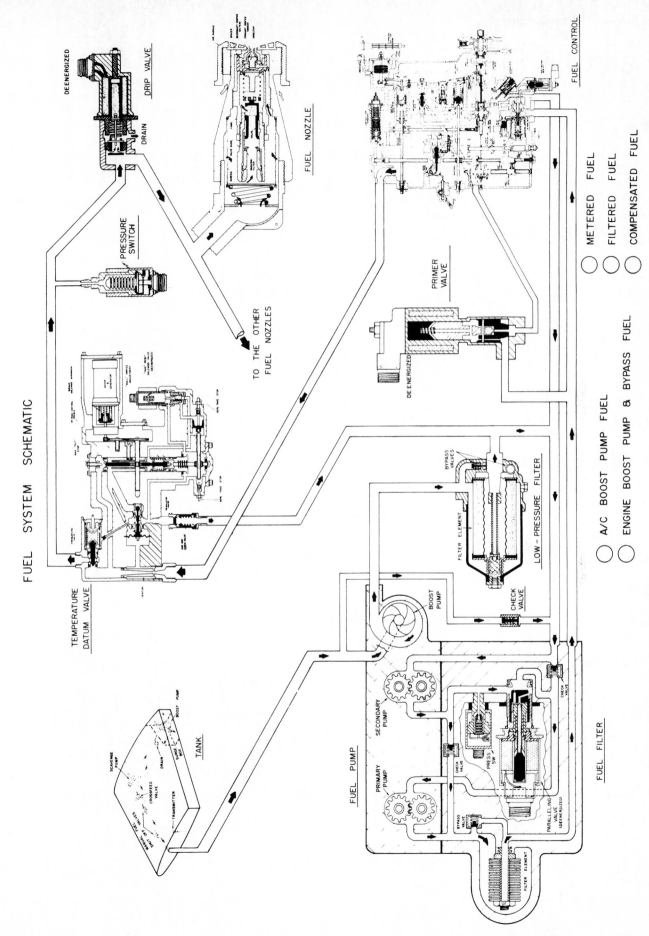

FIG. 13-4 501-D13 fuel system schematic. (*Detroit Diesel Allison Div.*)

FUEL SYSTEM SCHEMATIC

DEENERGIZED

DRIP VALVE

DRAIN

FUEL NOZZLE

FUEL CONTROL

PRESSURE SWITCH

TO THE OTHER FUEL NOZZLES

PRIMER VALVE

DE ENERGIZED

METERED FUEL

FILTERED FUEL

COMPENSATED FUEL

A/C BOOST PUMP FUEL

ENGINE BOOST PUMP & BYPASS FUEL

TEMPERATURE DATUM VALVE

BYPASS VALVES

FILTER ELEMENT

LOW - PRESSURE FILTER

CHECK VALVE

BOOST PUMP

SECONDARY PUMP

PRIMARY PUMP

CHECK VALVE

CHECK VALVE

PRESS SW

BYPASS VALVE

PARALLELING VALVE (DEENERGIZED)

FUEL PUMP

FUEL FILTER

FILTER ELEMENT

TANK

SCAVENGE PUMP

DRAIN

CROSSFEED VALVE

SURGE BOX

BOOST PUMP

TRANSMITTER

electrical circuit to the fuel-pump light in the cockpit to give a warning of a primary gear pump failure. (See Fig. 12-23.)

Fuel, leaving the high-pressure fuel filter, may take two paths. One path enters the fuel control and flows through the fuel-metering section. Here the fuel volume is corrected to 120 percent of engine demand. This correction is for rpm, throttle, and air-density variations. The second path enters the fuel control through the primer valve and bypasses the metering section. The latter path is utilized only during the initial phase of the starting cycle when the use of the primer system is selected by a manually positioned cockpit primer switch.

Fuel Control *(FIG. 12-10)*

The fuel control delivers metered fuel to the temperature datum valve which provides further correction to the fuel flow. The temperature datum valve is part of the electronic fuel-trimming system, and the fuel-flow correction made by the temperature datum valve is established by the temperature datum control (not pictured). The electronic fuel-trimming system compensates for variations in fuel density and Btu content. The temperature datum valve receives more fuel from the fuel control than it delivers to the fuel manifold, and is always bypassing fuel. The amount of fuel bypassed is determined by the position of a bypass-control needle which varies in response to an electrical signal from the temperature datum control (amplifier). The amplifier determines this electrical signal by comparing a desired turbine inlet temperature signal to the actual turbine inlet temperature signal provided by a parallel circuit of 18 thermocouples located in the turbine inlet. (See Fig. 12-15.)

Fuel Manifold

Fuel flow from the temperature datum valve is delivered to the fuel manifold through an aircraft-furnished flowmeter. The fuel manifold distributes the fuel to six fuel nozzles which atomize and inject the fuel into the forward end of the six combustion liners. A drip valve, located at the lowest point of the fuel manifold, is used to drain the fuel manifold at engine shutdown. During the starting cycle, a solenoid is energized to close the drip valve, and fuel pressure holds the drip valve closed during normal operation. At shutdown, a spring opens the drip valve.

Fuel, bypassed by the fuel control and temperature datum valve, is returned to the fuel-pump assembly by way of the high-pressure fuel-filter assembly. Any fuel leakage past the seals of the fuel-pump assembly and fuel control is drained overboard through a common manifold.

Primer System

During an engine start, it is desirable to fill the fuel manifold rapidly so that an initial high pressure to the fuel nozzles will allow the nozzles to better atomize the fuel. This ensures a better light-off during engine starts.

The secondary and primary fuel pumps are placed in parallel during a start to ensure sufficient fuel flow to fill the fuel manifold rapidly. If a starting attempt is not successful, additional fuel can be delivered to the fuel manifold on the next starting attempt by using the primer system. The primer system must be "armed" by the cockpit primer switch. If the primer system is armed, the primer valve will open at 2200 rpm due to speed-sensitive control and ignition relay operation. When the pressure in the fuel manifold exceeds approximately 50 psi, a pressure switch, connected to the fuel manifold, opens an electrical circuit which will cause the primer valve to close. When the primer valve is open, fuel will flow through the primer valve to the upstream side of the fuel control cutoff valve. Functionally, the primer valve is in parallel with the metering section of the fuel control.

THE GENERAL ELECTRIC CJ805-23 FUEL SYSTEM *(FIG. 13-5)*

This fuel system consists of the following components:

Fuel filter
Fuel heater
Fuel pump
Fuel nozzle
Pressurizing and drain valve
Inlet guide vane actuator
Fuel control
Inlet guide vane mechanical feedback assembly
CIT sensor
Variable stator reset mechanism

Fuel Pump

The pump consists of three elements, two gear-type and one boost-type, arranged in tandem. All three pumping elements are driven by coaxial shafts and incorporate individual shear sections so that failure of one pumping element will not adversely effect the operation of the remaining two pumping elements. A gear train is also provided in the pump to drive the boost pump at a higher speed than the gear pumping elements. In addition to the pumping elements, the housing also contains a check valve, located at the outlet of both gear elements, which serves to prevent counterflow through a sheared gear pump element. A pressure-relief valve is also incorporated in the housing and is set to open at 900 to 1000 psig [6206 to 6895 kPa] discharge pressure. Four ports are provided in the pump. These are inlet, boost bleed, bypass, and outlet. Three pressure taps are incorporated in the pump to measure pressure. (See Fig. 12-22).

Fuel Heater

The fuel heater is an air-to-liquid heat exchanger, incorporating a thermostatically controlled valve to reg-

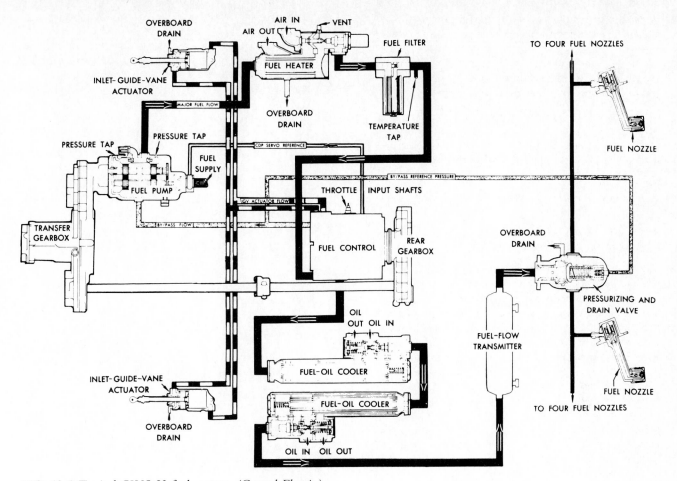

FIG. 13-5 Typical CJ805-23 fuel system. (*General Electric.*)

ulate airflow. The thermostatic valve is responsive to the temperature of the outgoing liquid. The liquid is turbojet engine fuel and the air is compressor bleed air supplied by the engine.

Fuel Filter

The fuel filter, which is in the fuel line between the fuel heater and the fuel control, protects the fuel controller from contaminants in the fuel. Fuel enters the filter and surrounds the filter screen. The fuel passes through the screen into an inner chamber, then flows out the discharge port. If the filter becomes clogged, fuel is bypassed through the relief valve. The filter screen is constructed of sinter-bonded, stainless steel wire cloth and filters out 98 percent of all particles larger than 43 μ. At the base of the filter is a drain port to facilitate draining the filter prior to removal for cleaning or replacement.

Fuel Control

The fuel control maintains engine rpm according to the throttle schedule, schedules maximum and minimum fuel rate limits, reduces maximum rpm limit in the low engine inlet temperature region, controls the position of the inlet guide vanes (IGV) by providing

control fuel for the inlet guide vane actuators, and provides fuel shutoff that is separate from the throttle schedule. To enable the control to perform the above objectives, there must be inputs in addition to the supply of fuel. The inputs include the air inlet temperature (CIT), the compressor discharge pressure $P3$, and the speed of the main shaft. These three inputs, referred to as parameters, and a power-lever (throttle) setting determine the outputs of the control. The air inlet temperature is sensed by the compressor inlet temperature sensor. The outputs of the control include a controlled fuel supply (W_f) to the combustion chamber of the engine and control fuel pressure signal for the inlet guide vane actuators.

Compressor-inlet-temperature Sensor

The compressor-inlet-temperature (CIT) sensor is a temperature sensor and transducer unit for the fuel regulator. It responds to the air inlet temperature of CJ805 turbojet engines and positions the temperature transducer differential servo of the main regulator.

Variable Stator Reset Mechanism

The variable stator reset mechanism is a feedback bias mechanism used to alter the schedule of the inlet guide vanes and variable stator vanes of the engine during certain operating conditions.

Inlet Guide Vane Actuator and Inlet Guide Vane Mechanism Feedback Assembly

The vane actuator assembly is a valve-operated fuel piston-type actuator with a fixed cooling flow orifice across the piston. This actuator controls the position of the inlet guide vanes of the engine through a fuel signal from the fuel control. The mechanical feedback assembly informs the fuel control where the guide vanes are positioned.

Pressurizing and Drain Valve

This valve is a pressure-operated valve designed to prevent flow to the engine fuel nozzles until sufficient pressure is attained in the fuel control. The valve operates to drain the fuel manifold at engine shutdown, but will maintain pressure between the fuel control and the fuel nozzles. There are five ports on the valve: reference pressure, two fuel outlets, fuel inlet, and drain port. (See Fig. 12-36.)

Fuel Nozzle

The fuel nozzle is a fuel-metering device. It produces a conical fuel spray of fine droplets, uniform density, and uniform thickness over its entire range of operating pressures. [See Fig. 12-27(a)].

THE AVCO LYCOMING T53 FUEL SYSTEM *(FIG. 13-6)*

The fuel system consists of the following components:

Fuel regulator
Overspeed governor
Starting fuel solenoid shutoff valve
Main and starting fuel manifolds
Igniter nozzles
Fuel vaporizers [See Fig. 12-33(b)]
Combustion-chamber drain valve

System Operation

Fuel enters the engine fuel regulator and after metering goes to the starting and main discharge ports. The

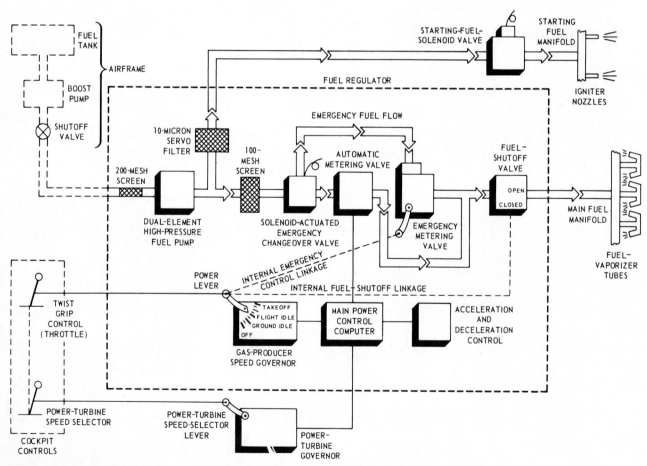

FIG. 13-6 Fuel system for the Avco Lycoming T53 incorporating separate controls for the gas-producer and power turbines of the engine.

starting fuel flows to a solenoid shutoff valve, wired in conjunction with the ignition system. Energizing the ignition system activates the solenoid valve, allowing starting fuel to enter the starting fuel manifold and combustion chamber through five igniter nozzles. Two igniter plugs initiate the flame. Main fuel is delivered from the fuel regulator to the main fuel system when the engine rpm is great enough to deliver minimum fuel pressure. After combustion occurs and ignition is de-energized, the solenoid valve shuts off the flow of starting fuel. The igniter nozzles are self-purging and remove excess fuel automatically. Main fuel flow is maintained as the engine flame is propagated. An electrical cable is connected to the starting fuel solenoid shutoff valve and to the engine fuel regulator. After engine shutdown, the pressure-actuated combustion-chamber drain valve opens automatically and drains unburned fuel from the combustion chamber. The engine is designed to use JP-4 fuel.

Fuel Control

The fuel control itself consists of a fuel regulator for the gas-producer section and an overspeed governor for the power turbine section. An integral dual fuel pump and an emergency control system are incorporated into the fuel regulator. For emergency fuel system operation, a special emergency valve is connected to the power-lever control.

The fuel-regulator is of the hydromechanical type incorporating an all-speed flyball governor for acceleration and deceleration control and a droop-type governor for a steady-state speed control.

Inputs to the regulator consist of a speed selector lever, compressor inlet pressure and temperature, and gas-producer speed. The compressor pressure and temperature sensors act to limit fuel flow to prevent the turbine inlet temperature from exceeding the limits under all operating conditions. The inlet pressure sensor biases the fuel flow at the main metering valve through a multiplying linkage. The inlet temperature sensor biases the fuel flow for acceleration, deceleration, and maximum permissible steady-state speed through the rotation of a three-dimensional cam which is translated as a function of gas-producer speed. The control in sensing these parameters monitors fuel flow, preventing temperature limits from being exceeded.

The overspeed governor is a flyball droop-type governor. This governor acts as a topping device by limiting fuel flow in the event the power turbine tends to exceed the power turbine rpm selected.

THE TELEDYNE CAE J69 FUEL SYSTEM (FIG. 13-7)

This system contains the following units:

Fuel pump
Fuel filter
Fuel control consisting of a starting fuel system and a main fuel system

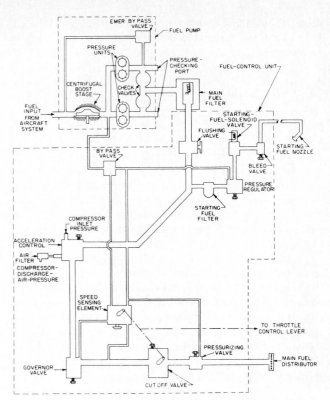

FIG. 13-7 The Teledyne CAE J69 fuel system utilizes a rotating main fuel distributor supplied with fuel through the shaft, and a stationary starting fuel nozzle.

Fuel Pump

The engine fuel system starts with the fuel pump. This pump, driven off the accessory gear train, has a centrifugal booster stage which is intended to provide boost pressure if the boost provisions in the aircraft system should fail. It also reduces vapor effects by raising total boost pressure. The centrifugal booster stage feeds two gear-pump pressure sections operating in parallel. Either section will provide full pressure and flow for the engine and each section is independent of the other.

Fuel Filter

From the main fuel pump, fuel is carried by a hose to the fuel filter built into the fuel control unit. This filter incorporates two separate filtering elements with provisions to bypass the fuel if the elements should become clogged. A manually operated flushing valve permits closing off the rest of the fuel system when reverse flushing of the filter is accomplished. This valve is at the filter outlet.

Fuel Control

Within the fuel control there are two separate fuel paths. From the flushing valve outlet, starting fuel is led through a starting fuel filter to a pressure regulator, then to the starting fuel solenoid valve. From this valve, starting fuel passes through adjustable bleed valves to the external piping that leads the fuel to the starting fuel nozzles.

The main fuel path feeds to the acceleration control, to the governor valve, then to the cutoff valve from which flow is to the pressurizing valve and thence into the engine fuel tube. The acceleration control is designed to influence fuel input during acceleration and also to compensate for changes of altitude or other ambient air conditions. The governor valve influences flow to hold the speed called for by the throttle lever setting. The governor valve is servo operated and responds to pressure signals developed in the "speed sensing" element. The latter also sends pressure signals to the bypass valve. The function of the bypass valve is to maintain a design pressure differential across the metering elements (which are the acceleration control and the governor valve). This pressure differential is maintained by bypassing fuel back to the fuel-pump inlet. Since the design pressure differential must change with speed, the bypass valve is made responsive to a signal from the speed-sensing element. The pressurizing valve is designed to open only above a minimum pressure and so prevents "dribble" of fuel or drainage of the control unit when the engine comes to a stop.

The fuel control also contains check valves, "trim" provisions, and passages for return of fuel bleed-off or seepage.

The fuel control is the key element affecting engine control. Provided the proper volume and pressure of fuel are fed into the fuel control unit, it regulates and meters engine fuel input to cover all operating conditions automatically. During starting the separate starting fuel path sets up fuel flow to the starting fuel nozzles. The fuel-control starting fuel solenoid valve opens this path and closes it in response to signals from a control element not supplied with the engine. Then the acceleration control sets up fuel flow to the main fuel distributor such as to speed the engine up from starting speed to idle without surge or overtemperature. As the engine reaches the speed set by the throttle lever position, the governor valve will come into action to hold the speed as set. After the engine has reached idle, the acceleration control is designed to control fuel input for all changing conditions for all operation from idle up to full speed without allowing surge or overtemperature. It compensates for acceleration, for change of altitude, and for other changes of ambient air characteristics. The governor valve, in all cases, acts to hold engine speed to the value set by throttle lever position.

THE PRATT & WHITNEY AIRCRAFT JT9D FUEL SYSTEM (*FIG. 13-8*)

The final fuel system to be discussed, the JT9D engine fuel system, consists of the:

Fuel pump (containing the filter)
Fuel control
Flow meter

Fuel-oil heat exchanger
Pressurizing and drain valve
Deicing and indicating system
Distribution system
Fuel nozzles

Fuel Pump

The engine-driven fuel pump is located on the forward face of the main gearbox. The fuel pump is, basically, a three-stage pump. It consists of a boost stage, main stage, and hydraulic stage. The pump housing also contains a fuel filter and a fuel deicing system.

A new feature of this fuel pump is its hydraulic stage. The pump supplies boosted fuel pressure to act as the hydraulic agent for engine variable-stator-vane control. The variable stator vanes are a part of engine airflow control.

With the fuel shutoff valve open, fuel flows first to the fuel pump boost stage. The boost stage increases fuel pressure and pumps the fuel through an external fuel heater, back through the fuel filter to the main and hydraulic stages of the pump. Notice that the fuel normally flows through the heater, even when the heater is not being used. Both the heater and the fuel filter are equipped with bypass valves. Clogging of either unit will not result in engine fuel starvation. A bypass valve is also incorporated into the boost surge pump circuit. If the boost stage fails, the bypass valve will open, allowing the main and hydraulic stages to continue normal operation. Output of the main stage will be sufficient for cruise power, and maybe even takeoff power, under this condition.

Fuel Filter

The fuel filter is integral with the fuel pump—that is, the filter housing attaches directly to the fuel pump. A drain plug is provided which allows purging of the fuel lines with boost-pump pressure. The filter has a bypass that allows passage of fuel if the filter becomes clogged.

The filter differential pressure switch is mounted on the fuel-filter housing. It monitors the icing condition of the filter by sensing an increase in differential pressure across the filter. This pressure can be created by ice, or by an accumulation of foreign matter in the filter element. In either event, the switch closes and turns on the amber filter-icing light on the flight engineer's panel. The icing light indicates that the filter is clogging up which normally means that the fuel requires heating. When fuel icing conditions are present, 15th-stage compressor air is used for deicing.

Fuel Heater

The fuel heater is located with the fuel pump on the right side of the engine. It is connected between the boost and main stages of the pump. The heater consists of a core of air tubes and a series of baffles. Fifteenth-stage air passes through the core air tubes, and fuel is baffled around the tubes. The airplane is equipped with a fuel-temperature-measuring system.

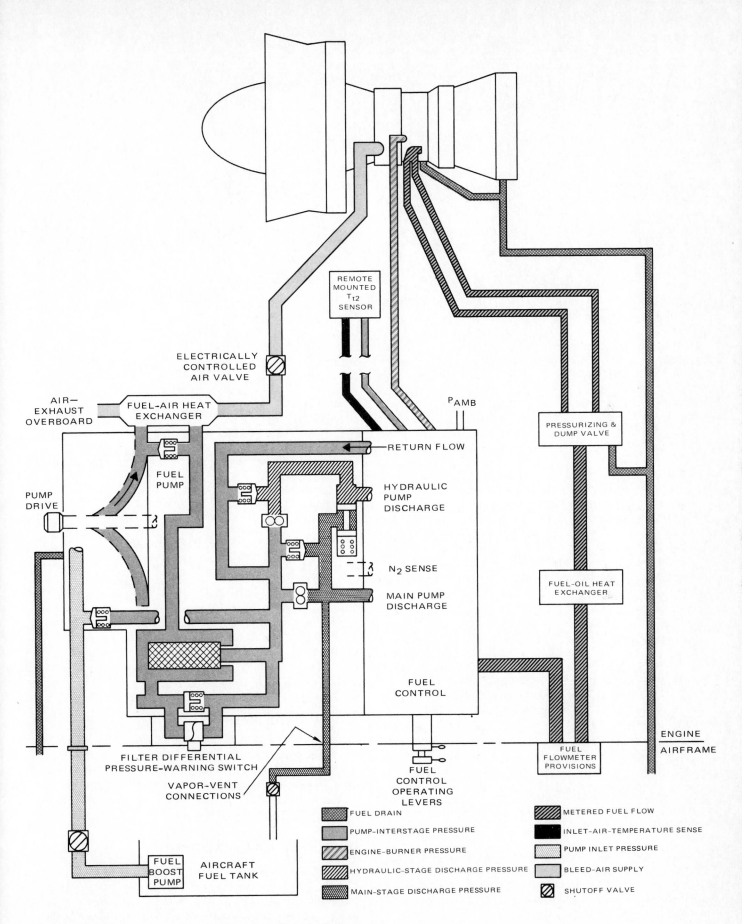

REMOTE
MOUNTED
T_{t2}
SENSOR

ELECTRICALLY
CONTROLLED
AIR VALVE

P_{AMB}

AIR—
EXHAUST
OVERBOARD

FUEL-AIR HEAT
EXCHANGER

PRESSURIZING &
DUMP VALVE

RETURN FLOW

FUEL
PUMP

HYDRAULIC
PUMP
DISCHARGE

PUMP
DRIVE

N_2 SENSE

MAIN PUMP
DISCHARGE

FUEL-OIL HEAT
EXCHANGER

FUEL
CONTROL

ENGINE
AIRFRAME

FILTER DIFFERENTIAL
PRESSURE-WARNING SWITCH

VAPOR-VENT
CONNECTIONS

FUEL
CONTROL
OPERATING
LEVERS

FUEL
FLOWMETER
PROVISIONS

FUEL DRAIN

PUMP-INTERSTAGE PRESSURE

ENGINE-BURNER PRESSURE

HYDRAULIC-STAGE DISCHARGE PRESSURE

MAIN-STAGE DISCHARGE PRESSURE

METERED FUEL FLOW

INLET-AIR-TEMPERATURE SENSE

PUMP INLET PRESSURE

BLEED-AIR SUPPLY

SHUTOFF VALVE

FUEL
BOOST
PUMP

AIRCRAFT
FUEL TANK

FIG. 13-8 The Pratt & Whitney Aircraft JT9D basic fuel system.

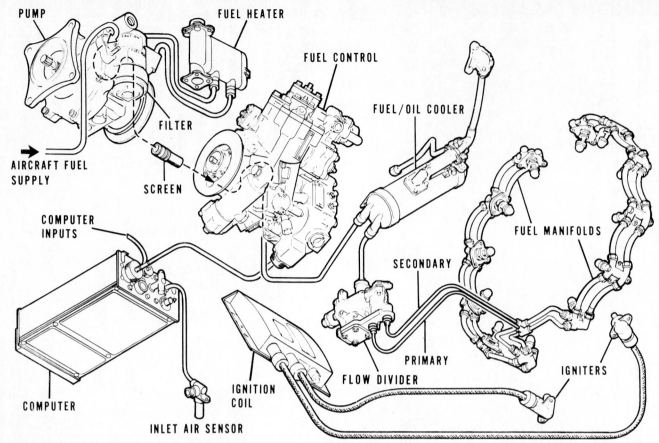

FIG. 13-9 A simplified drawing of the Garrett AiResearch TFE731 fuel system. Fuel under slight pressure is provided from the aircraft tanks to the high-pressure pump where its pressure is increased. It flows through a screen to the fuel control which meters it in response to signals received from the computer. It then passes through the fuel-oil cooler to the flow divider which directs it initially through the primary manifold and nozzles to the combustion chamber, where it is lit by the igniters. As the engine accelerates after lightup, the rising fuel flow causes the flow divider to permit flow through the secondary manifold and nozzles as well as the primaries.

Fuel Control

The fuel control is mounted piggyback style to the engine fuel pump. It is a hydromechanical control that in addition to its fuel-control function, is used with the engine vane control to regulate the thrust output of the engine. Inputs consist of:

Engine speed N_2
Ambient pressure P_{amb}
Burner pressure P_{s4}
Inlet total temperature T_{t2}

Fuel-flow Meter

The fuel-flow meter is actually the transmitter portion of the fuel-flow indicating system. The transmitter is located on the right side of the engine, just below the fuel-oil cooler. The fuel-flow transmitter measures the fuel flow rate and converts it to electrical pulse signals.

The pulse signals are processed in the fuel-flow electronics modules in the main equipment center and sent to the indicators.

Fuel-oil Heat Exchanger

From the flow meter, the metered fuel flows through the fuel-oil cooler. The fuel-oil cooler is the standard heat-exchanger type. Its primary purpose is to use engine fuel as a cooling agent to reduce the temperature of engine oil. The fuel-oil cooler is located on the right side of the engine, just above the fuel-flow transmitter.

Pressurizing and Drain Valve

From the fuel-oil cooler the fuel flows to the pressurizing and dump valve, or P&D valve. During engine operation the P&D valve supplies fuel only to the primary nozzles until demand is sufficient to require both

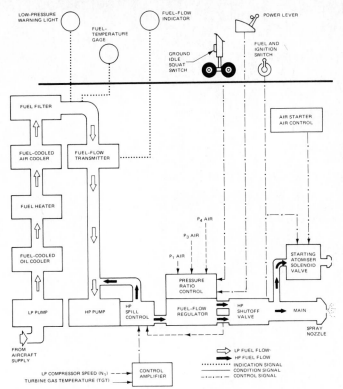

FIG. 13-10 Rolls-Royce RB·211 engine fuel distribution system. This engine fuel system is completely self-contained and delivers fuel at the proper pressure and flow to satisfy thrust levels selected at the throttle quadrant. The system automatically distributes the fuel to the annular combustion chamber as a finely atomized spray suitable for efficient and stable combustion, at a rate consistent with engine requirements under all operating conditions. The automatic control function of the fuel system also enables climb and cruise ratings to be maintained for a fixed power-level setting irrespective of ambient temperature up to International Standard Atmosphere (ISA) + 15° C. Fuel received from the aircraft supply system is distributed by a centrifugal-type low-pressure (LP) fuel pump through an oil cooler, fuel heater, air cooler, fuel filter, and fuel-flow transmitter to a gear-type high-pressure (HP) fuel pump. The HP pump distributes fuel through a fuel-flow regulator, a HP shutoff valve, and the fuel manifolds, to the fuel spray nozzles.

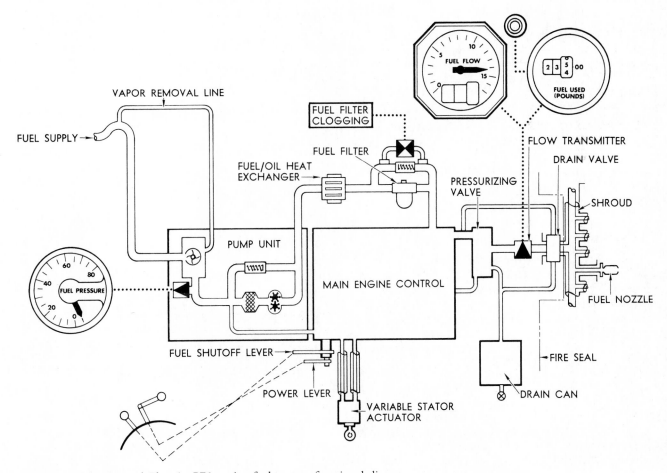

FIG. 13-11 The General Electric CF6 engine fuel system functional diagram.

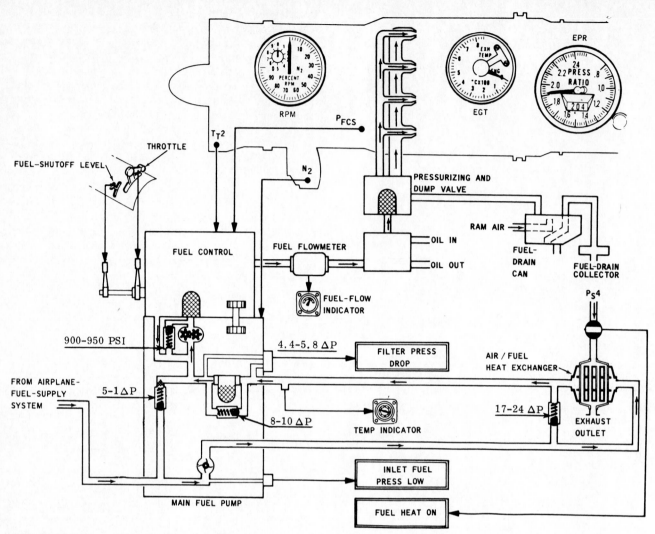

FIG. 13-12 Pratt & Whitney Aircraft JT8D fuel system schematic.

primary and secondary nozzles. The dump valve section drains the fuel from the fuel manifold into the drain tank at engine shutdown.

Fuel Nozzles

From the P&D valve the metered fuel flows through two manifolds to the 20 fuel nozzles. There it is sprayed under fuel-control-unit metered pressure into the annular combustion chamber. The 20 duplex fuel nozzles are located around the forward end of the combustion chamber. They are mounted to fuel nozzle supports on the diffuser case. They are referred to as duplex nozzles because the primary and secondary nozzles are in the

same housing. If water injection is used, the water is injected at the fuel nozzle. (See Fig. 9-2.)

NOTE: As previously mentioned, because of space limitations Figs. 13-9, 13-10, 13-11, and 13-12 are not accompanied by a detailed system explanation, but they are included here as examples of other fuel systems used on modern airplanes.

REVIEW AND STUDY QUESTIONS

1. What is the purpose of any fuel system?
2. Very briefly describe the fuel systems used on the: CJ610, JT3D, 501-D13, CJ805-23, T53, J69, and JT9D.

CHAPTER FOURTEEN
LUBRICATING OILS

GAS TURBINE OILS

Lubricants now must perform under environmental and mechanical conditions much more severe than a few years ago. Early aircraft gas turbine engines operated on light mineral oils, and some engines requiring such oils are still in service. But the low-temperature requirements imposed by high-altitude flight, coupled with the higher engine operating temperatures, are not met by existing petroleum-based oils. Inasmuch as a mineral oil is generally not capable of giving satisfactory performance at both very low and very high temperatures, most modern turbojet and turboprop engines are lubricated with synthetic oils. Synthetic oils are also used in some accessories on the engine such as starters and constant-speed drives to preclude using the wrong oil in these units. These oils can also be found in modern airplane hydraulic systems and in some instruments.

The characteristics of both the natural petroleum oil and the synthetic oils have been outlined in Military Specifications MIL-O-6081 for the natural oil and MIL-L-7808 and MIL-L-23699 for the synthetic oils.

MIL-O-6081

MIL-O-6081 is a narrow-cut (see the section on fuels) light mineral oil containing additives to enhance oxidation resistance and improve viscosity-temperature properties. It generally has a low pour point, low viscosity at low temperatures, reasonable stability in the presence of heat, and is noncorrosive to metals commonly used in engines. It is used in applications where the bearing temperatures are about 300°F [148.9°C] or less. At elevated temperatures this oil suffers large evaporation losses, inadequate viscosity, and causes large coking deposits. The lubricant is processed from crude oil obtained from various parts of the world. The crude oil can be broadly separated into two groups—paraffinic oils and naphthenic oils. The division is based on the way the hydrogen and carbon atoms are linked together. The paraffinic oils are relatively stable at high temperatures, have a high viscosity index (see the following section for a definition of terms), and contain a high percentage of dissolved wax. Naphthenic oils are less stable at elevated temperatures, but have little or no wax, and therefore tend to remain liquid at low temperatures. The viscosity index of naphthenic oils is poor. Most natural petroleum jet engine oils employ a mixed base stock.

MIL-L-7808 (Type I)

This oil is a widespread synthetic lubricant used in the United States. The specifications for both the natural and synthetic oils list −65°F [−53.9°C] for starting requirements, but the synthetic lubricant is rated for temperatures up to 400°F [204.4°C]. Although there are many synthetic lubricants on the market, the one most commonly used is classified as a dibasic-acid ester. It can be made by using animal tallow or vegetable oils (castor bean) as the raw material in a reaction with alcohol, or from petroleum hydrocarbon synthesis. The exact identities of the compounds used in the construction of these oils is kept under proprietary secrecy. Since the processing required for a synthetic oil is complex, its current price of approximately $16.00 per gallon as compared with $4.00 per gallon for natural oil can be readily understood. Oils meeting the MIL-L-7808 specification are sometimes called type I oils (Table 14-1).

TABLE 14-1
Synthetic lubricant MIL-L-7808 specification

Viscosity, centistokes:	
210°F	3.0 min
100°F	11.0 min
−65°F	13,000 max
Viscosity change, −65°F, % at 3 h	±6.0 max
Pour point, °F	−75 max
Flash point, °F	400 min
Neutralization no.	—
SOD lead corrosion, 325°F, mg/in² in 1 h	6.0 max
450°F corrosion, mg/in²:	
Copper	3.0 max
Silver	3.0 max
347°F corrosion and oxidation stability, mg/cm²:	
Copper	±0.4
Magnesium	±0.2
Iron	±0.2
Aluminum	±0.2
Silver	±0.2
% viscosity change, 100°F	−5 to +15
Neutralization no. increase	2.0 max
Evaporation loss, 400°F, %	35 max
"H" rubber swell, %	12 to 35
Panel coke, 600°F, mg	80 max
Deposition no.	5.0 max
Foam test	Pass
72-h low-temperature stability, −65°F, centistokes	17,000 max
Compatibility with MIL-L-7808 oils	Pass
Ryder gear test, relative rating, %	68 min

FIG. 14-1 The high thermal stability of type II oils results in reduced deposits.
(a) Type I oil.
(b) Type II oil.

MIL-L-23699 (Type II)

Several companies have developed a type II lubricant meeting the Military Specification No. MIL-L-23699. It uses a new synthetic base and new additive combinations to cope with the more severe operating conditions of the second and third generations of jet engines (Fig. 14-1) and is being widely adopted by military and civilian operators. The new oil's chief advantages are as follows:

1. High viscosity and viscosity index
2. High load-carrying characteristics
3. Better high-temperature-oxidation stability
4. Better thermal stability

MIL-L-7808 may be mixed with MIL-L-23699 since they are required by specification to be compatible with each other, but this practice should be avoided since the MIL-L-7808 oil will tend to degrade the MIL-L-23699 oil to the MIL-L-7808 level and nullify the new oil's benefits as listed above.

CHARACTERISTICS OF LUBRICATING OILS

Lubricants for jet engines must exhibit certain physical and performance properties in order to perform satisfactorily. The following is a list of tests performed on gas turbine oils to determine their physical and performance properties.

Physical Properties

1. *Viscosity index*—This refers to the effect of temperature on viscosity. All petroleum products thin with a temperature increase and thicken with a temperature decrease. A high viscosity index number indicates a comparatively low rate of change.
2. *Viscosity*—This is the measure of the ability of an oil to flow at a specific temperature.
3. *Pour point*—This refers to the effect of low temperatures on the pourability of the oil.
4. *Flash point*—This is the lowest temperature at which the oil gives off vapors that will ignite when a small flame is periodically passed over the surface of the oil.
5. *Fire point*—This is the lowest temperature at which an oil ignites and continues to burn for at least 5 s.
6. *Volatility*—This is the measure of the ease with which a liquid is converted to a vaporous state.
7. *Acidity*—This test indicates the corrosive tendencies of the oil.

Performance Factors

1. *Oil foaming*—This is the measure of the ability of the oil to separate from entrained air.
2. *Rubber swell*—This is the measure of how much the oil will cause swelling in a particular rubber compound.
3. *Oxidation and thermal stability*—This is a measure of how well an oil can resist the formation of hard carbon and sludge at high temperatures.
4. *Corrosivity to metals*—This is a test to determine the

corrosivity of the oil by its effect on a small strip of polished copper. Other metals may also be used.

5. *Gear or pressure tests*—This shows the ability of the oil to carry a load.
6. *Carbon residue or coking tests*—This measures the amount of carbon residue remaining in an oil after subjecting it to extreme heating in the absence of air.
7. *Engine tests*—This demonstrates the characteristics of the oil in an actual engine.

Additional tests such as the water emulsion test, compatibility test, storage stability test, interfacial tension test, and several others may be performed to determine several other physical and performance properties of an oil.

In many cases suitable chemical substances are mixed with the oil to impart desirable characteristics. These additives include such materials as detergents, rust preventatives, dyes, anticorrosives, antioxidants, foam inhibitors, viscosity index improvers, pour point depressants, and a host of other additives for improving performance and imparting new properties to the lubricant. Much of the research in lubricants is concentrated in this area.

REQUIREMENTS OF A GAS TURBINE LUBRICANT

As mentioned elsewhere in this book, jet engine temperatures may vary from −60 to over 400°F [−51.1 to 204.4°C]. Since oil must be fluid enough at the low-temperature extreme to permit rapid starting and prompt flow of oil to the parts to be lubricated, the jet engine oils must have a fairly low viscosity and pour point. On the other hand, the viscosity index must be as high as possible or the oil will become too thin to support the bearing and gear loads when the engine comes up to operating temperatures.

The flash point, fire point, oxidation resistance, thermal stability, and volatility of an oil are also very important in view of the high operating temperatures in the hot section of the engine and the high altitude, low ambient pressure in which the engine normally operates. Temperatures of the hotter bearings of some gas turbine engines reach from 400 to 500°F [204.4 to 260°C] or higher during operation. The relatively few gallons of oil in the system are circulated at a high rate from the tank through the coolers, to bearings and gears, and then back to the tank. Bulk oil temperatures in some engines may run to slightly less than 300°F [148.9°C], with some oil being heated locally to the much higher temperatures of bearing surfaces. These extreme conditions coupled with the fact that scavenge oil is very thoroughly mixed with air used to pressurize the bearing sumps (see Chap. 15 on oil systems), promotes thermal decomposition, oxidation, and volatilization of the lubricating medium. Results of these harmful processes include the formation of sludge, corrosive materials, and other deposits. They also increase viscosity and oil consumption. In addition, excess deposits can increase bearing friction and temperatures,

clog filters and oil jets, interfere with oil flow, and cause increased seal wear. Sludge deposits may coat tube surfaces in oil coolers and prevent normal removal of heat from the oil.

Resistance to foaming is also a very important property of an oil. In the preceding paragraph it was noted that a large quantity of air is put into the system by the scavenge pumps and bearing sumps. This air-oil mixture is carried to the oil supply tank or special air-oil separator, where, with good oil, rapid separation occurs and excess air can be vented off harmlessly. On the other hand, an oil unsuitable in this respect will foam, and much of the air-oil mixture will be vented overboard. A substantial amount of oil can be lost this way. Furthermore, an air-oil mixture supplied to the bearings will not remove heat nor lubricate as efficiently as a solid oil flow.

HANDLING SYNTHETIC LUBRICANTS

Synthetic oils are not as storage-stable as conventional petroleum oils. Temperature extremes should be avoided, oil stock should be used as soon as possible, and partial stock withdrawals should be avoided. In general, most commercial engine operators either limit or entirely prohibit mixing different brands of oil, although the oil specifications require that every oil shall be compatible with previously approved oils. Military services routinely mix oil brands with no gross ill effects noted.

Synthetic lubricants have a deleterious effect on some types of paints, electrical insulation, and elastomer materials used in seals, although in some cases a slight

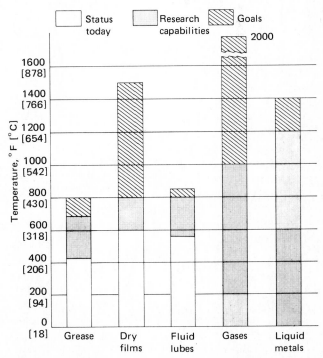

FIG. 14-2 Potentials of high-temperature lubricants.

swelling of rubber seals is desirable to prevent leakage of oil. It has been found that some people show a skin sensitivity to this type of lubricant, but in general, synthetic oils may be classed in the same category as mineral oils, both in liquid and vapor states, with regard to toxicity. As in the case of mineral oil, ingestion and prolonged skin contact are to be avoided.

FUTURE DEVELOPMENTS

Higher temperatures will necessitate oils with improved thermal stability. Experiments are under way looking into the possibility of using liquid metals, solid lubricants, and gases or vapors containing various additives. Figure 14-2 shows the lubrication potential of some other lubricating mediums.

The effort to find a suitable lubricant for high- and low-temperature applications may come full circle with the adoption of a superrefined petroleum oil which shows promise of having all of the properties of the best synthetics, but at a lower cost.

REVIEW AND STUDY QUESTIONS

1. Name two basic types of lubricating oils now in use for gas turbine engines. Describe the properties of each.
2. Make a table of the characteristics of lubricating oils.
3. Why are additives placed in the oil?
4. Discuss the requirements for an oil to be used in gas turbine engines.
5. What precautions are necessary when handling synthetic lubricants?
6. What can be expected in the future in relation to lubricants for gas turbines?

CHAPTER FIFTEEN
LUBRICATING
SYSTEMS

Although the oil system of the modern gas turbine engine is quite varied in design and plumbing, most have units which are called upon to perform similar functions. In a great majority of cases a pressure pump or system furnishes oil to the several parts of the engine to be lubricated and cooled, after which, the oil having done its job, a scavenging system returns the oil to the tank for reuse. It is interesting to note, in relation to the cooling function of the oil, that the problem of overheating is more severe after the engine has stopped than while it is running. Oil flow which would normally have cooled the bearings has stopped, and the heat stored in the turbine wheel will raise the bearing temperature to a much higher degree than that reached during operation. Most systems include a heat exchanger to cool the oil. Many are designed with pressurized sumps, the function of which will be discussed more fully in the following pages. Some systems incorporate a pressurized oil tank which ensures a constant head pressure to the pressure lubrication pump to prevent pump cavitation at high altitude.

Oil consumption in a gas turbine engine is rather low compared with that of a reciprocating engine of equal power. For example, the large J79 jet engine has a maximum oil consumption of 2.0 lb [0.907 kg] [approximately 1 quart (qt) or 1 L] per hour as compared to the R3350 engine which may consume as much as 50 lb [22.7 kg] (approximately 26 qt [24.5 L]) per hour. Oil consumption on the turbine engine is primarily a function of the efficiency of the seals. However, oil can be lost through internal leakage, and on some engines by malfunction of the pressurizing or venting system. Oil sealing is very important in a jet engine because any wetting of the blades or vanes by oil vapor will encourage the accumulation of dust and dirt. A dirty blade or vane represents high friction to airflow, thus decreasing engine efficiency, and resulting in a noticeable decrease in thrust or increase in fuel consumption. Since oil consumption is so low, oil tanks can be made relatively small with a corresponding decrease in weight and storage problems. Tanks may have capacities ranging from ½ to 8 gal [1.89 to 64 L]. System pressures may vary from 15 psig [103.4 kPa gage] at idle, to 200 psig [1379 kPa gage] during cold starts. Normal operating pressures and bulk temperatures are about 50 to 100 psig [344.7 to 689.4 kPa gage] and 200°F [93.3°C] respectively.

In general, the parts to be lubricated and cooled include the main bearings and accessory drive gears, and in the turboprop, the propeller gearing. This represents a gain in gas turbine engine lubrication simplicity over the complex oil system of the reciprocating engine. The main rotating unit can be carried by only a few bearings, while in a piston power plant there are hundreds more moving parts to be lubricated. On some turbine engines the oil may also be used to operate the servo mechanism of some fuel controls, to control the position of the variable-area exhaust nozzle vanes, and to operate the thrust reverser.

Because each bearing in the engine receives its oil from a metered or calibrated orifice, the system is generally known as the *calibrated* type. With a few exceptions the lubricating system used on the modern jet engine is of the *dry-sump* variety. In this design the bulk of the oil is carried in an airframe or engine-supplied separate oil tank, as opposed to the *wet-sump* system in which the oil is carried in the engine itself. Although our discussion will limit itself to dry-sump systems, an example of the wet-sump design can be seen in the now obsolete Detroit Diesel Allison J33 engine, Fig. 15-13. In this engine, the oil reservoir is an integral part of the accessory drive gear case.

OIL-SYSTEM COMPONENTS

The oil-system components used on gas turbine engines are as follows:

Tank(s)
Pressure pump(s)
Scavenger pumps
Filters
Oil coolers
Relief valves
Breathers and pressurizing components
Pressure and temperature gages
Temperature regulating valves
Oil jet nozzles
Fittings, valves, and plumbing
Seals

It should be noted that not all of the units will necessarily be found in the oil system of any one engine, but a majority of the parts listed will be found in most engines.

Oil Tanks *(FIG. 15-1)*

Tanks can be either an airframe or engine manufacturer-supplied unit. Usually constructed of welded sheet aluminum or steel, it provides a storage place for the oil, and in most engines is pressurized to ensure a constant supply of oil to the pressure pump. The tank can

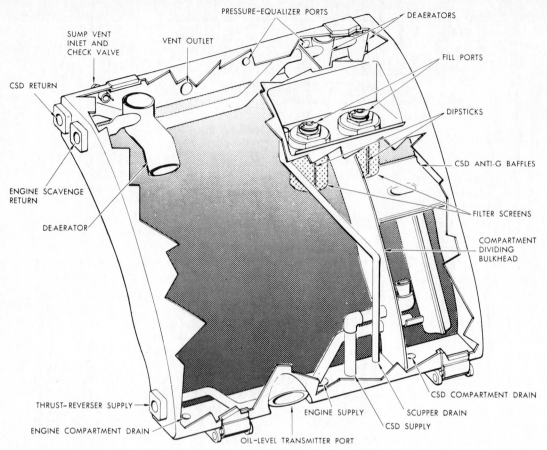

FIG. 15-1 This tank, used on the General Electric CJ805 engine, incorporates a separate oil supply for the constant-speed drive (CSD). The CSD is needed to enable the alternator to produce a constant electrical frequency regardless of engine rpm within the limits of the drive. Note: Electronic constant frequency controlling devices are being tested to eliminate the need for the CSD.

contain a venting system, a deaerator to separate entrained air from the oil, an oil level transmitter and/or dipstick, a rigid or flexible oil pickup, coarse mesh screens, and various oil and air inlets and outlets.

Pressure Pumps (FIG. 15-2)

Both the gear- and gerotor-type pumps are used in the lubricating system of the turbine engine. The gear-type pump consists of a driving and driven gear. The rotation of the pump, which is driven from the engine accessory section, causes the oil to pass around the outside of the gears in pockets formed by the gear teeth and the pump casing. The pressure developed is proportional to engine rpm up to the time the relief valve opens, after which any further increase in engine speed will not result in an oil-pressure increase. Although Fig. 15-2(a) shows the relief valve in the pump housing, it may be located elsewhere in the pressure system for both types of pumps. Figure 15-14 and others show the relief valve in the system instead of the pump.

The gerotor pump has two moving parts, an inner toothed element meshing with an outer toothed element. The inner element has one less tooth than the

outer, and the "missing tooth" provides a chamber to move the fluid from the intake to the discharge port. Both elements are mounted eccentrically to each other on the same shaft.

Scavenger Pumps

These pumps are similar to the pressure pumps but are of much larger total capacity. An engine is generally provided with several scavenger pumps to drain oil from various parts of the engine. Often one or two of the scavenger elements are incorporated in the same housing as the pressure pump (Fig. 15-3). Different capacities can be provided for each system despite the common driving shaft speed, by varying the diameter or thickness of the gears to vary the volume of the tooth chamber. A vane-type pump may sometimes be used.

Filters (FIGS. 15-4 to 15-7)

The three basic types of oil filters for the jet engine are the cartridge, screen, and screen-disk types shown in Figs. 15-4, 15-5, and 15-6, respectively. The cartridge filter must be replaced periodically, while the other two

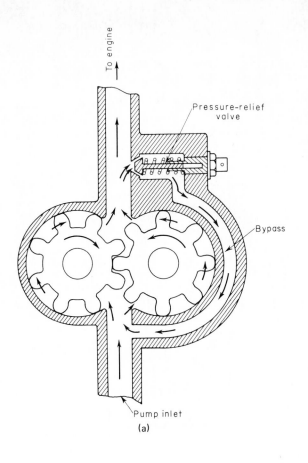

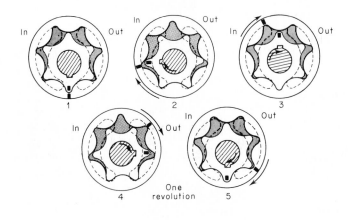

FIG. 15-2 The two basic types of oil pumps used on gas turbine engines.
(a) Gear oil pump
(b) Gerotor oil pump

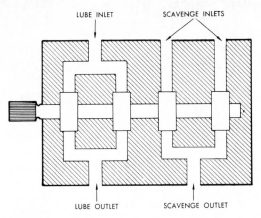

FIG. 15-3 Schematic of a double-element pressure and scavenger lube pump in a common housing.

can be cleaned and reused. In the screen-disk filter shown in Fig. 15-7, there are a series of circular screen-type filters, with each filter being composed of two layers of mesh so as to form a chamber between the mesh layers. The filters are mounted on a common tube and arranged in a manner to provide a space between each circular element. Lube oil passes through the circular mesh elements and into the chamber between the two layers of mesh. This chamber is ported to the center of a common tube which directs oil out of the filter.

All of the various types of filters will incorporate a bypass or relief valve, either as an integral part of the filter, or in the oil passages of the system, to allow a flow of oil in the event of filter blockage. When the pressure differential reaches a specified value (about 15 to 20 psi [103 to 138 kPa]), the valve will open and allow oil to bypass the filter. Some filters incorporate

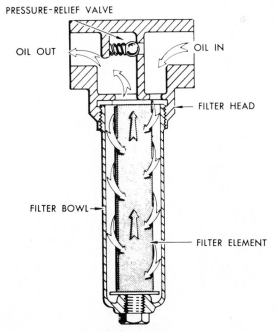

FIG. 15-4 Cartridge-type oil filter. The valve will open if the element becomes clogged.

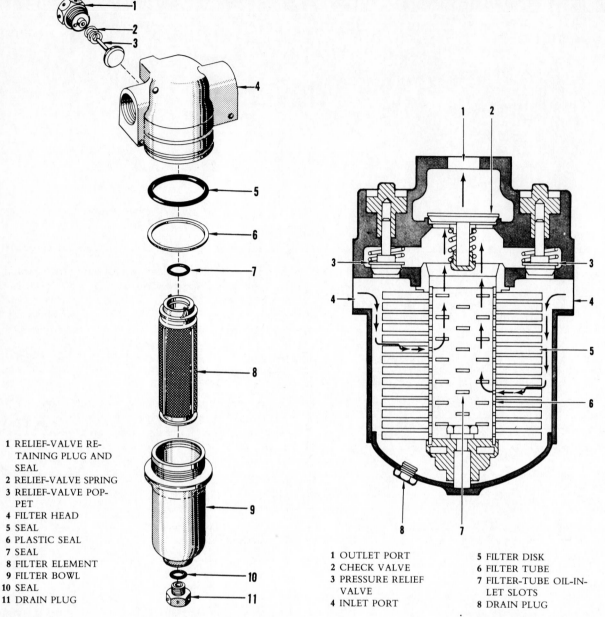

1 RELIEF-VALVE RE-
  TAINING PLUG AND
  SEAL
2 RELIEF-VALVE SPRING
3 RELIEF-VALVE POP-
  PET
4 FILTER HEAD
5 SEAL
6 PLASTIC SEAL
7 SEAL
8 FILTER ELEMENT
9 FILTER BOWL
10 SEAL
11 DRAIN PLUG

FIG. 15-5 A screen-type filter with a bypass (relief) valve.

1 OUTLET PORT   5 FILTER DISK
2 CHECK VALVE   6 FILTER TUBE
3 PRESSURE RELIEF   7 FILTER-TUBE OIL-IN-
  VALVE   LET SLOTS
4 INLET PORT   8 DRAIN PLUG

FIG. 15-7 Oil flow through a typical screen disk filter.

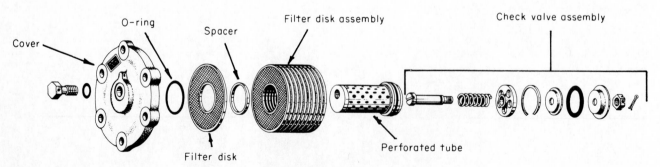

FIG. 15-6 In this disk-type filter, installed in the 501-D13 (see Fig. 15-11), a check valve prevents oil from flowing from the tank into the accessories section when the engine is stopped. The bypass valve is located elsewhere in the system.

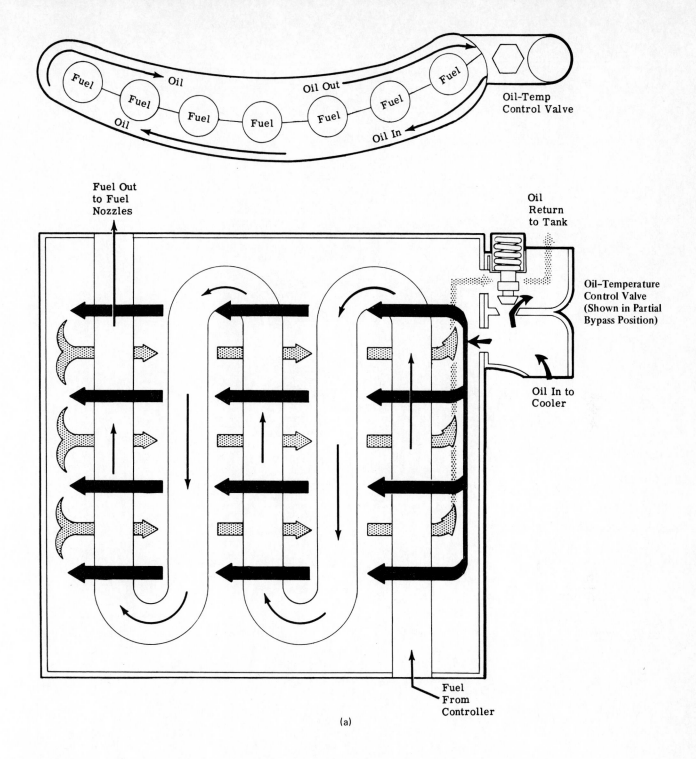

(a)

a check valve (Figs. 15-6 and 15-7) which will prevent either reverse flow, or flow through the system when the engine is stopped. Filtering characteristics vary, but most filters will stop particles of approximately 50 μ.

Oil Coolers (FIG. 15-8)

The oil cooler is used to reduce the temperature of the oil by transmitting heat from the oil to another fluid. The fluid is usually fuel, although air-oil coolers have been used. Since the fuel flow through the cooler is much greater than the oil flow, the fuel is able to absorb

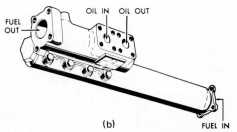

(b)

FIG. 15-8 (a) A diagram of a typical fuel-oil cooler. (b) An actual fuel-oil cooler.

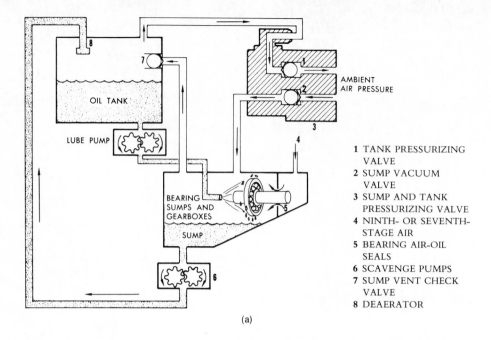

1 TANK PRESSURIZING
VALVE
2 SUMP VACUUM
VALVE
3 SUMP AND TANK
PRESSURIZING VALVE
4 NINTH- OR SEVENTH-
STAGE AIR
5 BEARING AIR-OIL
SEALS
6 SCAVENGE PUMPS
7 SUMP VENT CHECK
VALVE
8 DEAERATOR

(a)

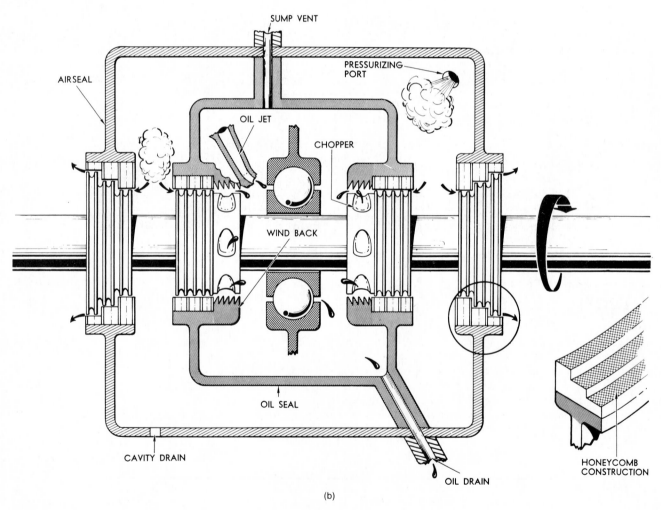

(b)

FIG. 15-9 (a) A simplified diagram of a tank and sump pressurizing system.
(b) The sump sealing system used on the General Electric CF6 high-bypass-ratio turbofan engine.

a considerable amount of heat from the oil, thus reducing the size of the cooler greatly, as shown in Fig. 15-8(b), as well as the weight. Thermostatic or pressure-sensitive valves control the temperature of the oil by determining whether the oil shall pass through or bypass the cooler.

Breathers and Pressurizing Systems (FIG. 15-9 and also refer to FIG. 21-29)

In many modern engines internal oil leakage is kept to a minimum by pressurizing the bearing sump areas with air that is bled off the compressor. The airflow into the sumps minimizes oil leakage across the seals in the reverse direction. The system operates as follows.

The oil scavenge pumps exceed the capacity of the lubrication pressure pump(s) and are therefore capable of handling considerably more oil than actually exists in the bearing sumps and gearboxes. Because the pumps are a positive-displacement type, they make up for the lack of oil by pumping air from the sumps. Thus, large quantities of air are delivered to the oil tank. Sump and tank pressures are maintained close to each other by a line which connects the two. If the sump pressure exceeds the tank pressure, the sump vent check valve opens, allowing the excess sump air to enter the oil tank. The valve allows flows only into the tank, so oil or tank vapors cannot back up into the sump areas. Tank pressure is maintained a small amount above ambient.

Functioning of the scavenge pumps and sump vent check valve results in a relatively low air pressure in the sumps and gearboxes. These low internal sump pressures allow air to flow across the oil seals into the sumps. This airflow minimizes lube oil leakage across the seals. For this reason, it is necessary to maintain sump pressures low enough to ensure seal air leakage into the sumps. But under some conditions the ability of the scavenge pumps to pump air forms a pressure low enough to cavitate the pumps or cause the collapse of the sump, while under other conditions too much air can enter the sump because of excessive quantities of air entering through worn seals.

If the seal leakage is not sufficient to maintain proper internal pressure, check valves in the sump and tank pressurizing valves open and allow ambient air to enter the system. If the seal leakage exceeds that required to maintain proper internal sump and gearbox pressure, air flows from the sumps, through the sump vent check valve, the oil tank, the tank and sump pressurizing valves, and then to the atmosphere. Tank pressure is always maintained a few pounds above ambient pressure by the sump and tank pressurizing valve.

Seals (FIG. 15-10)

Dynamic (running) seals used in gas turbine engines can basically be divided into two groups:

1. *Rubbing or contact seals*—Two varieties are face [Fig. 15-10(a)] and circumferential [Fig. 15-10(b)] types and are

constructed of metals, carbon, elastomers, and rubbers or combinations of these materials.

2. *Nonrubbing labyrinth or clearance seals* [FIG. 15-10(c) and (d)].

In both cases the type of seal and the material used is determined mainly by the range of pressures, temperatures, and speeds over which the seal must operate, the requirements of a reasonable service life, the media to be sealed, and the amount of leakage that can be tolerated.

Rubbing or contact seals are used in applications where a minimum amount of leakage is allowed and a high degree of sealing required. For example, they are used to seal accessory drive shafts where the shaft exits from the accessory gear case, and for variable-stator vane bearings in the compressor case. Carbon rubbing seals are often used for, but not limited to, sealing the main internal bearing areas, especially in the engine's hot section.

Nonrubbing clearance- or labyrinth-type seals are, as the name implies, devices through which a specific amount of leakage can take place because there is no actual contact between the rotating and stationary part of the seal. The unit consists essentially of one or more thin strips of metal attached to a housing through which the shaft rotates. This arrangement may occasionally be reversed with the thin metal strips attached to the rotating shaft. By establishing the correct pressure differential across the seal the designed amount of leakage can occur in the desired direction. (See Fig. 15-9.)

TYPICAL OIL SYSTEMS

The Allison 501-D13 (FIG. 15-11)

This sytem incorporates a low-pressure, independent dry sump oil system which includes one combination main pressure and scavenge oil pump assembly, three separate scavenge pumps, a pressure-regulating valve, an oil filter and check valve, a filter bypass valve, and a scavenge pressure-relief valve. Oil is supplied to the power unit by an aircraft-furnished tank and cooled by an aircraft-furnished cooler.

The main oil pump is located on the center of the front face of the accessory drive housing cover. Oil, supplied to the pressure pump from the aircraft-furnished tank, is pumped through a disk-type filter and check valve, through drilled and cored passages and internal and external lines to those parts of the power unit which require lubrication. A pressure-regulating valve located in the main oil pump regulates the oil pressure delivered by the pump. A filter check valve is included in the system to prevent oil from seeping into the power unit after shutdown.

Scavenge oil is returned to the aircraft tank by the rear turbine scavenge pump, the front turbine scavenge pump, the diffuser scavenge pump, and the main scavenge pump. Scavenge oil is carried by drilled passages, and internal and external lines.

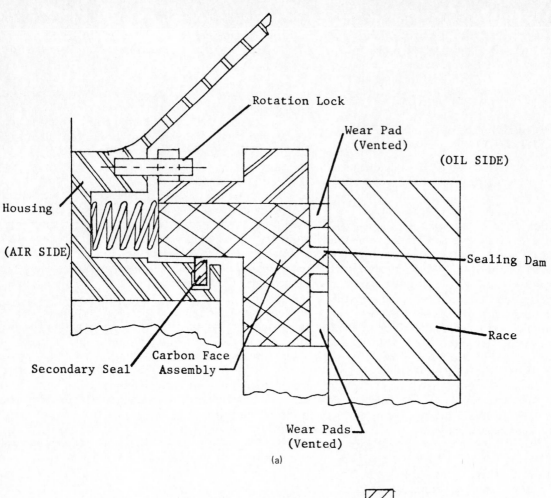

(a)

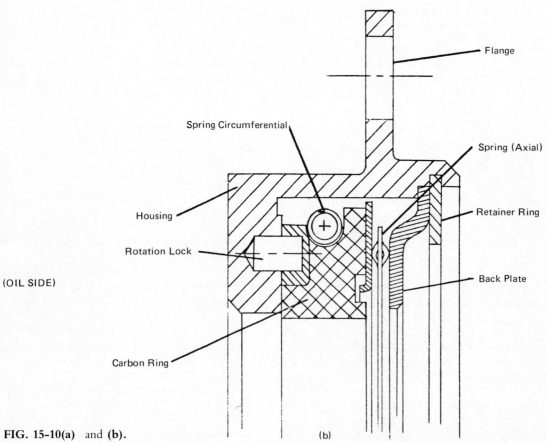

FIG. 15-10(a) and **(b).**

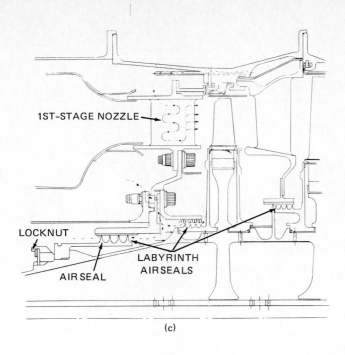

(c)

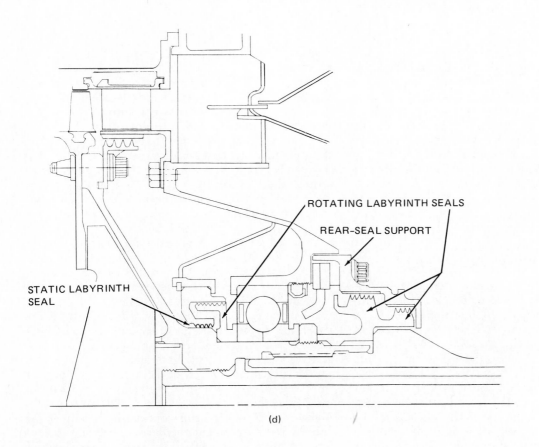

(d)

FIG. 15-10 Types of carbon rubbing and labyrinth seals.
(a) A typical "face rubbing" carbon seal.
(b) A typical circumferential bore or "edge rubbing" carbon seal.
(c) Labyrinth air seals used in the turbine area of the General Electric T58 engine.
(d) Labyrinth oil seals used in the no. 2 bearing area of the General Electric T58 engine.

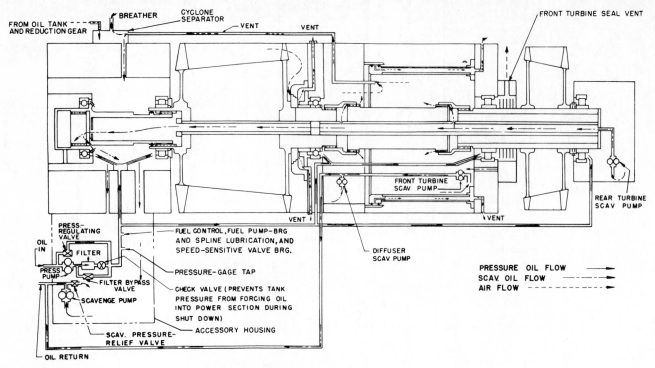

FROM OIL TANK
AND REDUCTION GEAR

BREATHER

CYCLONE
SEPARATOR

VENT

VENT

PRESS-
REGULATING
VALVE

OIL
IN

FILTER

PRESS
PUMP

FILTER BYPASS
VALVE

SCAVENGE PUMP

SCAV. PRESSURE-
RELIEF VALVE

OIL RETURN

FUEL CONTROL, FUEL PUMP-BRG
AND SPLINE LUBRICATION, AND
SPEED-SENSITIVE VALVE BRG.

PRESSURE-GAGE TAP

CHECK VALVE (PREVENTS TANK
PRESSURE FROM FORCING OIL
INTO POWER SECTION DURING
SHUT DOWN)

ACCESSORY HOUSING

VENT

DIFFUSER
SCAV. PUMP

FRONT TURBINE
SCAV. PUMP

VENT

REAR TURBINE
SCAV. PUMP

PRESSURE OIL FLOW ⟶
SCAV. OIL FLOW ------⟶
AIR FLOW -·-·-·-·-⟶

FIG. 15-11 The Detroit Diesel Allison 501-D13 lubrication system.

Two magnetic drain plugs are provided on the accessory drive housing; one at the bottom of the housing and one in the scavenge oil outlet connection.

The power section breather is located on top of the air inlet housing. It is a cyclonic separator of air and oil, in that oil-saturated air coming through the vent line from the diffuser enters the breather and is rotated by the spiral passageway in the breather. Movement of the air through this passageway imparts a centrifugal force on the air that separates the air and oil. The air is vented overboard, and the oil flows through passages to the scavenge oil sump in the accessory drive housing. (See Chap. 23.)

The General Electric CJ610 *(FIG. 15-12)*

The engine has a pressurized closed-circuit lube system designed to furnish oil to parts which require lubrication during engine operation. After oil has been supplied to these parts it flows to the sumps; here it is recovered and recirculated throughout the system. All system components are engine-furnished and engine-mounted except the oil pressure transducer, which is airframe-furnished equipment but also engine-mounted. The main components of the lube system are located beneath the compressor section. They include: a lube and scavenge pump mounted on the rear right-hand pad of the accessory gearbox; an oil tank, secured to the lube and scavenge pump; an oil cooler, mounted on the oil tank; and an oil filter, contained within the lube and scavenge pump casing; a scavenge oil temperature tap, located at the gearbox on the left-hand scavenge tube; and the oil pressure transducer (airframe-supplied).

The pressure element of the lube pump pumps oil from the tank, through the oil cooler, through the oil filter, and into the accessory gearbox. Part of the oil services the gearbox; the remainder flows through the gearbox to two OUT ports on the left-side of the gearbox. Oil flows from one OUT port through an external line to the number 1 bearing. It flows from the second OUT port through an external line which is branched to service the transfer gearbox. The main line connects to the main frame. This line services, through internal tubing, the internal bearings, seals, and gears of the engine. The five scavenge elements of the lube and scavenge pump pick up the oil which collects in the engine sumps and return it to the oil tank.

A closed vent circuit provides for pressurization of all parts of the lube system, including the oil tank, sumps, and gearboxes. This ensures pressurization to the inlets of the pump service element and scavenge elements. The sumps are pressurized by seal-leakage air that enters the sumps. Vent lines and check valves between the engine sumps, gearboxes, and oil tank maintain a balanced pressure between the sumps, gearboxes, and oil tank. These vent lines and check valves also prevent overpressurization of the sumps by venting air (in excess of what the scavenge elements can handle) directly to the oil tank. A pressure-relief valve on the outboard side of the oil tank controls the pressure in the entire system by venting overboard any air in excess of 2.5 (±0.5) psig [17.2 (±3) kPa gage]. A small orifice through the center of this valve provides for depressurization of the system at engine shutdown.

The lube and scavenge pump is mounted on three studs at the right rear pad of the accessory gearbox. It is a positive-displacement pump consisting of six guided-vane-type elements mounted along a common

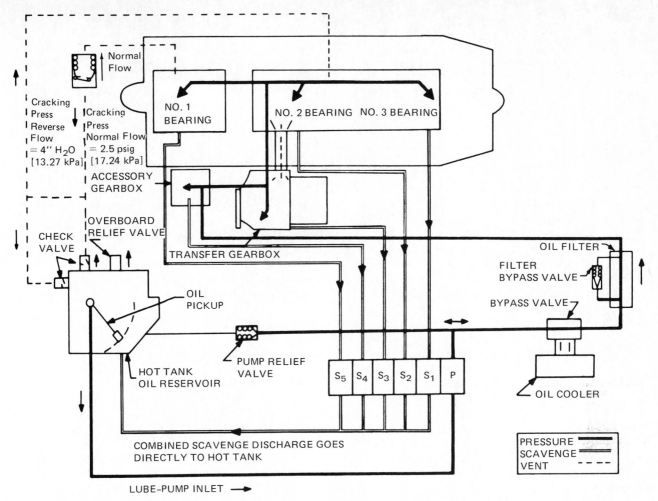

FIG. 15-12 The General Electric CJ610 oil system. The weighted oil pickup prevents the pump from drawing air. Note the "hot tank" oil system with the oil cooler on the pressure side of the system.

drive shaft and contained within the same housing. Five of the elements scavenge oil from the engine sumps; one element is the pressure element which supplies oil to the engine. The operating elements extend aft into the oil tank from the pump flange on which the oil tank is mounted. Scavenge oil passes from the sumps, through the accessory gearbox, through one of five scavenge ports on the face of the pump mounting flange, through the sump scavenge element, and into the oil-tank dwell chamber through a common discharge port. Oil from the tank enters the pressure element through a pendulum-type swivel pickup tube that extends out from the aft right-hand side of the pump body. The oil is pumped by the pressure element directly to the oil cooler, through the oil filter, and into the gearbox; from here it is then distributed throughout the system.

A pressure-relief valve, mounted in parallel with the lube discharge passage, is included to prevent damage that might result from excessive oil pressure due to cold starting or restriction of normal oil flow. The relief valve is factory-adjusted to open at a differential pressure of 90 (±5) psi [620 (±34) kPa] across the valve. If

this pressure differential is exceeded, the relief valve opens and oil from the pressure element is discharged back into the oil tank.

The rear pad of the pump is a drive capable of driving the tachometer-generator.

Lube pump discharge pressure is transmitted from a pressure tap located on the lube pump housing, to the cockpit lube pump pressure indicator (airframe-furnished equipment). Pressure readings indicate lube filter condition and lube pump operation.

Scavenge oil temperature is transmitted from a temperature tap in the left-hand scavenge tube to the cockpit scavenge oil temperature indicator (airframe-furnished equipment). Temperature readings indicate the operating conditions of the lube pump, lube filters, oil cooler, and engine bearings.

The oil filter assembly is mounted within the lube pump housing at the bottom, and is accessible for removal. It is a full-flow, in-line-type filter with a screen element of corrugated corrosion-resistant steel. The screen filters out contaminants over 40 μ in size. A filter bypass valve is included in the core of the filter element. If a pressure difference between oil entering the filter and oil leaving the filter exceeds 20 to 24 psi [138 to 166 kPa], the valve opens to permit a direct flow of oil through the unit without filtration.

The fabricated-steel oil tank is mounted on the rear flange of the lube and scavenge pump. Included within the tank is a separate air chamber, a dwell chamber, and a system of vent tubes. The filler port is located on the rear face of the tank, and the oil level in the tank is indicated by a dipstick graduated in pints to be added. A remote-fill line (airframe-furnished) connection is available on the rear face of the tank. When it is used, a vent line is required from the oil tank to the remote filler assembly. Oil that overflows during filling is collected in a scupper and may be drained overboard through the scupper drain port. The oil tank, when FULL is read on the dipstick, has a total capacity of 4.0 qt [3.78 L] of which 3.0 qt [2.84 L] are usable. There is adequate space allowed for expansion. The bottom of the tank is formed into a concave well which provides a recess for mounting the tachometer-generator unit.

The oil cooler is a shell and tube heat exchanger mounted on the front face of the oil tank at the right-hand side. Fuel flows through the tubing and absorbs heat from the hot engine oil flowing over the tube bundle. Oil enters and leaves the cooler through ports located in the housing. A pressure bypass valve is designed to bypass oil around the cooler in response to overpressure. If the cooler clogs, the pressure valve control bypasses the oil when the pressure differential across the valve exceeds 20 (±4) psi [138 (±28) kPa].

The Allison J33 (FIG. 15-13)

Although obsolete, the Detroit Diesel Allison Division J33 engine is shown in Fig. 15-13 as an example of a "wet-sump" oil system.

The General Electric CJ805-23 Aft Fan (FIG. 15-14)

In this system the main lube and scavenge pump supplies high-pressure oil from the engine-mounted tank to the areas requiring positive lubrication, and four scavenge pumps return the oil to the tank. Filters remove foreign material from the oil, and coolers prevent the oil from rising to destructive temperatures. Air valves in the system maintain the correct pressure balance. The system supplies oil to lubricate the five rotor-support bearings and the gears and bearings in the gearboxes. It scavenges, filters, and cools the used oil to prepare it for recirculation through the system, and regulates air pressure in the system to maintain a positive head of oil pressure at the inlet to the lube and scavenge pumps. It also establishes a pressure differential across the bearing seals, thus controlling oil consumption. The components required to perform the above tasks are divided into three functional subsystems: the oil supply, the scavenge, and the sump and tank pressurization subsystems.

The oil tank is a two-compartment tank. Oil in one compartment is for the oil supply subsystem. Oil in the other compartment is used for the hydraulic fluid used in the thrust reverser and constant-speed drive systems. One element of the lube and scavenge pump receives oil at tank pressure and discharges oil at higher pressures to the lube oil filter. The two other elements scavenge oil from the numbers 3 and 4 bearing sumps. This oil is discharged to the scavenge oil filter. The oil lube filter prevents the oil jet nozzles from clogging by filtering the oil flowing from the lube pump. The oil-pressure tap lube distribution manifold contains two calibrated orifices which work with a pressure-relief valve to protect the oil-pressure transducer from extreme pressure surges. The pressure-relief valve limits the maximum

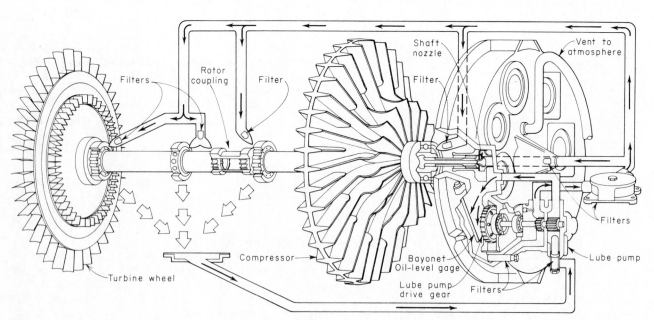

FIG. 15-13 The Detroit Diesel Allison Division J33 engine's "wet-sump" oil system.

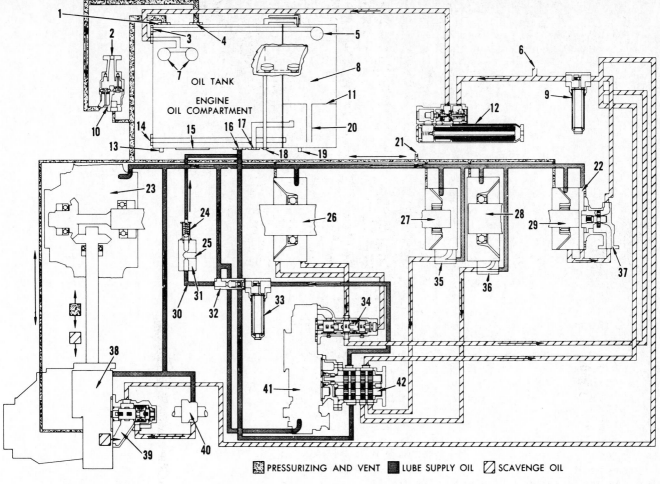

OIL TANK

ENGINE
OIL COMPARTMENT

PRESSURIZING AND VENT ■ LUBE SUPPLY OIL ◻ SCAVENGE OIL

1 SUMP-VENT INLET-PRES- SURE VALVE	10 SUMP AND TANK PRES- SURIZING VALVE	21 SUMP PRESSURE REFER- ENCE TAP	33 SUPPLY OIL FILTER
2 TO OVERBOARD VENT	11 ANTI-G BAFFLES	22 FAN-SPEED TACHOMETER	34 REAR-GEARBOX SCAV- ENGE PUMP
3 CSD AND THRUST-RE- VERSER RETURN	12 FUEL-OIL COOLER	23 NO. 1 BEARING AND FRONT GEARBOX	35 EDUCTOR
4 TANK PRESSURIZING VENT	13 ENGINE OIL DRAIN	24 PRESSURE-RELIEF VALVE	36 EDUCTOR
5 DEAERATOR	14 THRUST-REVERSER SUPPLY	25 OIL-PRESSURE TAP	37 NO. 5 BEARING SCAV- ENGE PUMP
6 TEMPERATURE REFER- ENCE TAP	15 OIL LEVEL	26 NO. 2 BEARING	38 TRANSFER GEARBOX
7 DEAERATORS	16 OIL SUPPLY	27 NO. 3 BEARING	39 TRANSFER-GEARBOX SCAVENGE-PUMP
8 CSD AND THRUST RE- VERSER COMPARTMENT	17 CSD SUPPLY	28 NO. 4 BEARING	40 DAMPER BEARING
9 SCAVENGE-OIL FILTER	18 SCUPPER DRAIN	29 NO. 5 BEARING	41 REAR GEARBOX
	19 CSD AND THRUST-RE- VERSER COMPARTMENT DRAIN	30 BYPASS FLOW	42 LUBE AND SCAVENGE PUMP
	20 DOWNCOMER TUBE	31 ORIFICE BLOCK	
		32 CHECK VALVE	

FIG. 15-14 The General Electric CJ805-23 lubrication system.

pressure sensed by the transducer, and bypasses extreme oil pressure surges from the lube pump discharge back to the lube pump inlet. Supplied by the airframe manufacturer, the oil-pressure transducer senses oil pressure and generates an electrical signal for the cockpit indicator. The last parts in the oil supply subsystem are the oil jet nozzles which spray lubricating oil over the engine bearings, gears, and seals.

The scavenge subsystem begins with the transfer-gearbox scavenge pump which scavenges used oil from the number 1 bearing front gearbox, damper bearing, and transfer gearbox and discharges oil to the scavenge oil filter. The rear-gearbox scavenge pump scavenges

oil from the number 2 bearing and rear gearbox, and discharges this oil to the scavenge oil filter. The number 5 bearing scavenge pump scavenges used oil from the number 5 bearing and aft fan tachometer-generator and discharges this oil to the scavenge oil filter. The used oil received from the four scavenge pumps is filtered by the scavenge oil filter and delivered to the engine fuel-oil cooler. The engine fuel-oil cooler cools used oil received from the scavenge-oil filter by using engine fuel as the coolant. Oil discharged from the cooler returns to the oil tank. An optional fuel-oil cooler supplied by the engine manufacturer cools the return oil from the constant-speed drive. Engine fuel is the coolant. Another optional air-oil cooler, supplied by the airframe manufacturer, may also be used to cool used oil returning from the constant-speed drive.

The final subsystem contains the sump and tank pressurizing valve which regulates pressure in the bearing sumps, oil tank, gearboxes, and connecting pipes. Also incorporated in this subsystem is the sump vent check valve which vents sump air pressure to the tank, yet prevents reverse oil flow.

The Teledyne CAE J69 *(FIG. 15-15)*

Oil from the engine oil tank (not supplied with the engine) is led to the main oil pump where the pressure section develops main oil pressure. The output of the pressure pump is led through an antileak valve to the main oil filter. This filter system incorporates a pressure-regulating arrangement as well as bypass provisions to pass oil beyond the filter element if it should become clogged. From the oil filter output, oil for the rear bearing is carried by an external hose. Another external hose picks up return oil from the rear bearing to carry this oil to the rear-bearing scavenge section of the oil pump. The rear-bearing housing incorporates a vent passage as well as passages to feed the oil to and from the rear bearing. At the front of the engine, oil is led

from the main oil filter output through passages to the front bearings and front-end gears. Oil is also fed to the accessory gear train that fans out across the lower part of the compressor housing. Oil from the front bearings and upper gears drains down to the accessory case from which one scavenge section of the oil pump pulls return oil. All scavenge sections of the oil pump lead return oil back to the engine oil tank. The front-end section is vented by a passage to the top of the upper gear housing. (See Chap. 24.)

The Avco Lycoming T53 *(FIG. 15-16)*

Engine lubricating oil supplied from the aircraft oil tank enters the oil pump located on the accessory gearbox. The two-element, gear-type oil pump is driven by a single, splined drive shaft. One element is used to supply main lubricating oil pressure, the other element is used to return scavenge oil to the aircraft oil tank. A pressure-relief valve in the oil pump is adjusted to deliver between 60 and 80 psi [414 and 552 kPa] oil pressure. This setting is rated for a maximum inlet oil temperature of 200°F [93°C] and an oil flow of 3300 lb/h [1497 kg/h] at sea level and 3000 lb/h [1361 kg/h] at 25,000 ft. [7620 m]. Pump pressure is directly proportional to compressor rotor speed at pressures below the

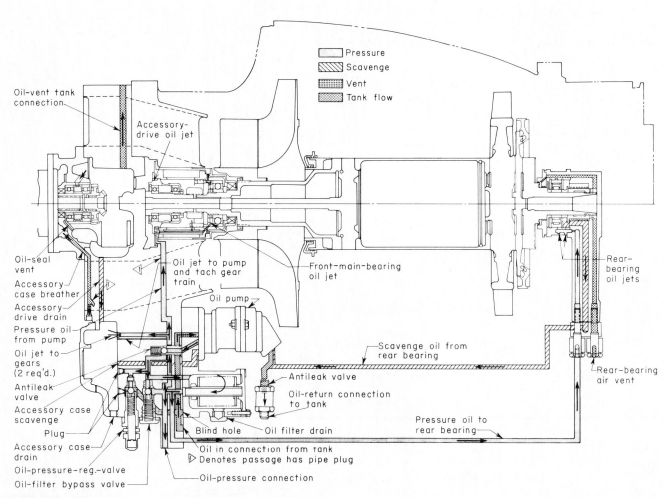

FIG. 15-15 The Teledyne CAE J69 oil system schematic

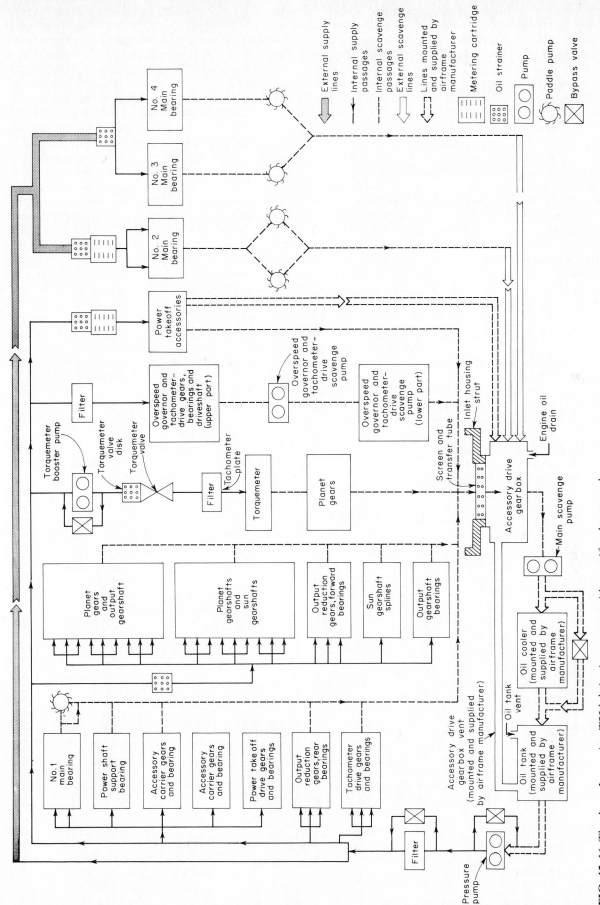

FIG. 15-16 The Avco Lycoming T53 lubrication system with centrifugal-type paddle pumps.

relief-valve setting. From the oil pump, the engine oil flows through internal passages to the oil filter. The oil filter is a wafer-disk type. A bypass valve, set at a differential pressure of between 15 and 20 psi [103 and 138 kPa], allows oil flow to bypass the filter elements and to supply emergency lubrication to the engine in the event the filter becomes clogged. Filtered oil is directed into two flow paths. One path flows internally through passages in the accessory drive gearbox to the inlet housing to lubricate the output reduction carrier and gear assembly, torquemeter, overspeed governor, and tachometer drive support and gear assembly, accessory drive carrier assembly, sun gearshaft, and number 1 main bearing. The second path flows externally and leads to the numbers 2, 3, and 4 main bearings. Scavenge oil is drained by paddle pumps through various internal passages and external lines to the accessory drive gearbox where the main scavenge pump picks up the oil and sends it first to the oil cooler, and then to the tank. A more detailed examination of this system and its components is given in Chap. 22.

The Pratt & Whitney Aircraft JT3D *(FIG. 15-17)*

The JT3D engine lubrication system is a self-contained, high-pressure design consisting of a pressure system which supplies oil to the main engine bearings and to the accessory drives, and a scavenge system which scavenges the bearing compartments and accessory drives. The oil is cooled by passing through a fuel-oil cooler. A breather system, interconnecting the individual bearing compartments and the oil tank, completes the engine lubricating system. The engine requires a synthetic lubricant.

The engine oil tank on most models is mounted in the upper right quadrant of the intermediate case, by being fastened with two straps attached to brackets on the front and rear flanges of the intermediate case. On some models the oil tank is supplied by the airframe manufacturer. Strips of resilient material serving as vibration isolators are installed between the tank and the engine, and the tank and the straps. The tank has a capacity of 6.0 gal [22.7 L] with a minimum usable quantity under all operating attitudes of 3.25 gal. [12.3 L]. Internally, the tank incorporates a flow deaerator, which is so located that the outlet is submerged even at low tank levels to prevent reaeration of the oil. Various holes in the tank permit the tank to breathe and the oil to enter and leave. There are also other holes for draining, cleaning, and inspection.

Oil is supplied to the inlet of the spur gear pressure pump. This pump is a duplex unit having a single gear stage for the pressure and scavenge section, separated by a center body. The pump forces the oil through the oil filter into an adapter and through external tubing, to the engine components requiring lubrication and cooling. Proper distribution of the total oil flow among the various locations is maintained by metering orifices and clearances. Oil-pressure differential is controlled by an oil-pressure relief valve.

An oil filter assembly, equipped with a bypass relief valve assembly and located forward of the accessory and component drive gearbox, assures a clean supply of oil to the lubrication system. The valve permits the oil to bypass the filter element in the event the screens become clogged. The filter assembly is composed of a series of screens in disk form, separated alternately by stamped inlet and outlet spacers, assembled around an inner filter element. The filter assembly is easily accessible for removal for disassembly and cleaning. Connections are provided for an oil-pressure transmitter and a differential pressure switch to activate with the buildup of screen contaminants.

The scavenge oil system contains five gear-type pumps located throughout the engine. They are located in the front accessory section for scavenging the no. 1 bearing compartment and front accessory section; the accessory and component drive gearbox for scavenging the nos. 2, 2½, and 3 bearing compartments through a common adapter; then through external tubing to the accessory and component drive gearbox; the dual scavenge pump in the diffuser case for scavenging the nos. 4, 4½, and 5 bearing compartments through the oil tube heat shields; and the turbine rear pump for scavenging the no. 6 bearing compartment. (Figure 20-27 shows the location of the various bearings.) The nos. 1 and 6 bearing scavenge oil flows through external tubing to the gearbox to join the scavenge oil from the nos. 4, 4½, and 5 bearing areas on its way to the tank. The oil handling capacity of the combined scavenge-oil pumps is approximately 3 times the quantity output of the pressure oil pump. All scavenge oil is routed through the fuel-oil cooler and to the oil tank by means of external tubing. The scavenge oil contains a considerable amount of entrapped air which must be vented overboard. This entrapped air is handled and dissipated by the engine breather system through the breather centrifuge.

Each of the separate scavenge areas is vented through external tubing and inner passages to a breather chamber formed by the compressor intermediate case annulus and then to a cavity in the accessory and component drive gearbox. The common overboard vent is from this cavity through a rotary breather which prevents the majority of oil particles from being carried overboard in the breather airflow. (See Chap. 20.)

The Pratt & Whitney Aircraft F100-PW-100 *(FIG. 15-18)*

INTRODUCTION The F100-PW-100 oil system is integral with the engine, and is composed of three major subsystems: the oil pressure, oil scavenge, and oil breather systems. They combine to satisfy the system requirement of providing the bearings with filtered oil at the proper pressures and temperatures.

OIL PRESSURE SYSTEM The oil pressure system is a "nonregulated pressure" system since the engine oil pressure is determined by the speed of the rear compressor rotor. Because oil viscosity increases as oil temperature decreases, there is a need to limit engine oil pressure during cold weather starting. This is accom-

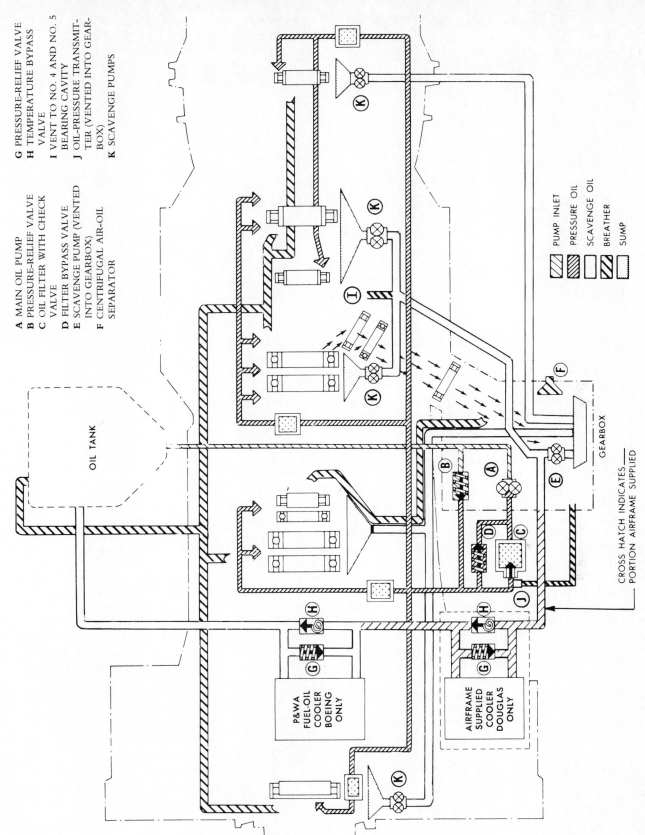

A MAIN OIL PUMP
B PRESSURE-RELIEF VALVE
C OIL FILTER WITH CHECK VALVE
D FILTER BYPASS VALVE
E SCAVENGE PUMP (VENTED INTO GEARBOX)
F CENTRIFUGAL AIR-OIL SEPARATOR
G PRESSURE-RELIEF VALVE
H TEMPERATURE BYPASS VALVE
I VENT TO NO. 4 AND NO. 5 BEARING CAVITY
J OIL-PRESSURE TRANSMITTER (VENTED INTO GEARBOX)
K SCAVENGE PUMPS

PUMP INLET
PRESSURE OIL
SCAVENGE OIL
BREATHER
SUMP

OIL TANK

GEARBOX

CROSS HATCH INDICATES PORTION AIRFRAME SUPPLIED

P&WA FUEL-OIL COOLER BOEING ONLY

AIRFRAME SUPPLIED COOLER DOUGLAS ONLY

FIG. 15-17 The Pratt & Whitney Aircraft JT3D oil system.

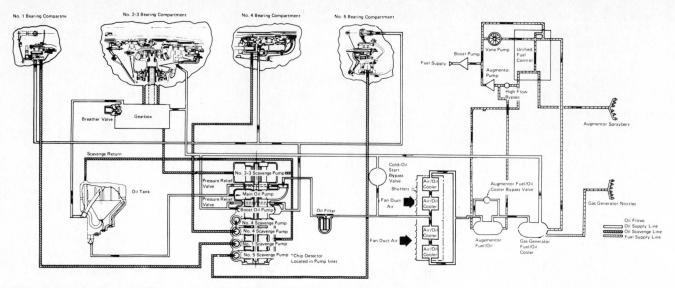

FIG. 15-18 The Pratt & Whitney Aircraft F100-PW-100 lubrication/fuel system.

plished by a pressure relief valve at the main-stage oil pressure pump, which limits the pump discharge pressure to a preselected value. Oil is gravity-supplied from the oil tank to the main pressure stage of the oil pump assembly, and is then directed through a full-flow, no-bypass oil filter. During normal engine starting and operation, the filtered oil flows through four fan air-oil coolers, and through the main fuel-oil cooler. When the engine is operating in augmentation, an augmenter fuel flow control valve shuttles and causes the oil to also flow through the augmenter fuel/oil cooler. When the engine is not in augmentation, this cooler is bypassed. During cold-weather starting, a bypass valve located in the oil filter housing permits the filtered oil to bypass the oil coolers and flow directly to the bearings. After the oil flows through the oil coolers, part of it is sent directly to the no. 2 and no. 3 bearing compartment, and the engine gearbox. The rest of the oil is sent to the oil boost pump for distribution to the no. 1, no. 4, and no. 5 bearing compartments. They are "capped compartments" and the oil boost pump ensures that the oil to them is at high enough pressure to provide proper lubrication. The oil jets of the system are protected from clogging by in-line, screen-type filters. These are frequently called "last chance" filters. Taps are provided for oil pressure and temperature transmitters to sense these values before the oil reaches the bearing compartments.

OIL SCAVENGE SYSTEM The function of the scavenge oil system is to collect the oil from the bearing compartments and return it to the oil tank. A single pump scavenges the engine gearbox and the no. 2 and no. 3 bearing compartment oil, which drains into the gearbox via the towershaft cavity. The no. 4 bearing compartment requires two pumps to ensure proper oil scavenging under all flight conditions. The no. 1 bearing compartment and the no. 5 bearing compartment are each scavenged by their own respective pumps. All scavenge pumps are connected to a common oil-tank-

return line. As the scavenged oil enters the oil tank, it flows through a stationary deaerator. For system inspection, five magnetic chip detectors are located in the scavenge system to collect chips from the engine gearbox and bearing compartments.

OIL BREATHER SYSTEM The no. 1, no. 4, and no. 5 bearing compartments are referred to as "capped compartments" because they vent breather air through their scavenge lines. For this reason, the compartment breather pressures are higher than engine gearbox breather pressure and vary as a function of flight conditions. The no. 2 and no. 3 bearing compartment breather pressure vents to the engine gearbox via the towershaft cavity. Oil tank breather pressure is vented to the engine gearbox by an external line. The gearbox breather passes through a deaerator impeller and is then vented to atmosphere through a breather pressurizing valve.

OIL TANK Because the oil returns to the tank uncooled, it is known as a "hot tank" system. Oil specification is MIL-L-7808G (type I). The oil tank features a:

Spectographic oil-analysis port
Sight gage
Deaerator (internal)
Tank drain and remote-fill provision
Overflow port

The oil tank and system capacities are as follows:

Tank maximum capacity (3.7 gal) [14 L].
Usable oil (2.5 gal) [9.46 L]. (Usable oil is a quantity equal to 10 times the maximum hourly oil consumption.)
Unusable oil (0.4 gal) [1.5 L]. (Unusable oil is the minimum amount needed to provide oil, containing no more than 10 percent by volume entrained air, to the engine.)
Expansion space (0.8 gallons) [3 L].
Engine oil wetdown estimate (4.5 to 5.0 gal) [17.0 to 18.9 L].

OIL PUMPS The oil pump assembly is composed of six stages of positive displacement, gear-type pumps that are mounted on the front of the engine gearbox. Two of the pumps (main pressure and boost pressure) function in the oil-pressure system. Both pumps incorporate pressure-relief valves, which start to open at 175 psia [1207 kPa absolute] and are full open at 225 psia [1551 kPa absolute]. The no. 1, no. 4 (two each), and no. 5 pumps serve to scavenge their respective bearing compartments.

The gearbox scavenge pump is a positive displacement, gear-type pump located in the engine gearbox and scavenges the no. 2 and no. 3 bearing compartment and the engine gearbox.

OIL FILTER The main oil filter is a 70-μ, metal wire mesh, full-flow, non-bypass-type filter. The conventional oil filter bypass valve has been eliminated to ensure delivery of only clean filtered oil to the engine. The system features a high-capacity oil filter and a feature to indicate filter clogging. This design allows engine operation with a partially clogged oil filter. A visual indicator (red button) is incorporated in the oil filter to indicate filter clogging. The indicator is activated when the differential pressure across the filter element exceeds 35 psid [241 kPa differential] for oil temperatures greater than 180°F (82.2°C). The filter capacity has been designed to ensure that the flow of filtered oil is sufficient to sustain the engine, even with a partly clogged filter. This design allows normal engine operation until corrective maintenance is performed.

A cold-oil bypass valve is located in the filter housing downstream of the filter. If the oil-cooler pressure drop exceeds 75 psid [517 kPa differential], the valve opens, allowing the oil to bypass the oil coolers. Two shutoff valves, located in the filter housing, prevent oil from draining out of the system when the oil filter is removed. These valves are unseated when the filter element and bowl are installed in the filter housing.

OIL COOLERS The engine oil system is provided with six oil coolers. The four fan air-oil coolers are of the plate-fin design and are located in the fan duct. They are in series and use fan air as the coolant. The main fuel-oil cooler is of the tube, baffle, and shell design and uses gas generator fuel as the coolant. The augmenter fuel-oil cooler is also of the tube, baffle, and shell design, but uses augmenter fuel as the coolant.

BREATHER PRESSURIZING VALVE The breather pressurizing valve is mounted on the engine gearbox and is of the aneroid bellows-spring poppet valve type.

From sea level to 30,000–35,000 ft [9144–10,668 m], the bellows holds the poppet valve off its seat at sea level, and positions the poppet valve closer to its seat as a function of increased altitude. This action maintains a breather pressure equal to ambient pressure. At approximately 30,000 ft altitude, the poppet valve has reached the closed position. Above 30,000–35,000 feet, the poppet valve will start to move off its seat at 1.5 psid [10.3 kPa differential] and will be full open at 2.0 psid [13.8 kPa differential]. A slip connection of the poppet valve allows this action to occur with no interference of the aneroid bellows.

OIL-SYSTEM OPERATION VALUES

Main oil-pressure range: 20 to 80 psig [138 to 552 kPa gage] at 200°F [93.3°C]

Main oil pressure minimum at idle: 20 psig [138 kPa gage]

Main oil pressure maximum one minute: 300 psig [2068 kPa gage] at −65°F [−53.9°C]

Boost oil pressure: 40 to 80 psid [276 to 552 kPa] referenced to no. 4 and no. 5 scavenge pump inlet pressure at 200°F [93.3°C]

Oil temperature normal: 150 to 300°F [121 to 150°C] intermediate thrust

Oil temperature maximum: 315°F [163°C] maximum steady state

Oil temperature maximum transient: 365°F [185°C] one minute or less

Breather pressure: 6 inHg [20.3 kPa] at steady state, 15 inHg [50.6 kPa] at transient

Oil consumption: 0.2 gal/h [0.76 L/h] average during service use for the first overhaul period

Pratt & Whitney Aircraft of Canada Ltd. JT15D Lubrication System *(FIG. 15-19)*

GENERAL The lubrication system is designed to supply clean lubricating oil, at a constant pressure, to the engine bearings and all accessory drive gears and bearings. The oil flow lubricates and cools the bearings and carries foreign matter to the oil filter where it is retained. Calibrated oil nozzles on the main engine bearings ensure that an optimum oil flow is maintained under all operating conditions. The three-element oil pump assembly is mounted on and driven from the accessory gearbox. Pressure oil is routed through an external tube to the oil filter housing. From the oil filter housing, oil is transferred through an internal tube to the accessory gearbox to lubricate its bearings, and two external transfer tubes, branching from a single oil-filter outlet tube, duct oil to the engine bearings.

The oil tank is an integral part of the intermediate case and is sealed at the rear by a cover which provides transfer tube pickup locations and internal passageways for pressure oil to the no. 3 bearing and scavenge oil from the nos. 3 and 3½ bearings. The JT15D-1 tank has a total capacity of 2.39 U.S. gal [9.04 L], of which 1.5 U.S. gal [5.68 L] are usable oil. This capacity provides an expansion space of approximately 1.00 U.S. gal [3.78 L]. The JT15D-1A total tank capacity is 2.14 U.S. gal [8.10 L] and usable oil is 1.25 U.S. gal [4.73 L] resulting in approximately the same 1.00 U.S. gallon expansion space.

The oil tank is provided with an oil filler neck, dipstick, and cap assembly which can be mounted to either side of the intermediate case front flange. The oil level in the tank is equal to the level in the filler neck and is indicated by the dipstick.

PRESSURE OIL SYSTEM Oil drawn from the tank by the pressure oil pump element is ducted through

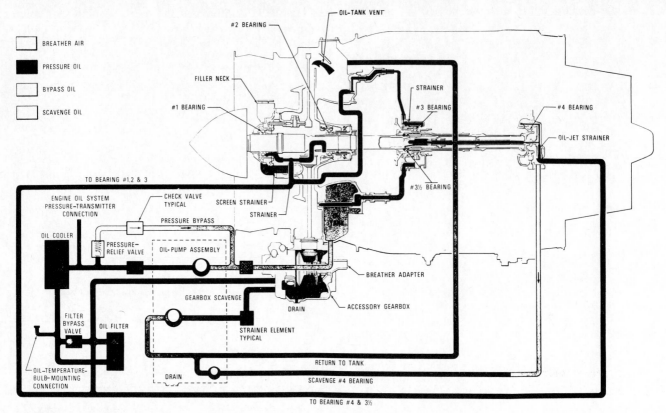

FIG. 15-19 The Pratt & Whitney Aircraft of Canada Ltd. JT15D oil lubrication system schematic.

a check valve to the pressure-relief-valve inlet of the oil filter assembly. The oil is then passed through the oil cooler, which is mounted on the oil filter housing and the filter element, which, in the event of clogging, is bypassed by a valve. Oil pressure in excess of 73 (±6) psi [503 (±41) kPa] at the oil filter outlet opens the pressure relief valve and some of the oil is bypassed and ducted externally through a second check valve to the oil pump pressure inlet.

An external transfer tube routes oil to a boss located in the five o'clock position at the rear of the engine, and an internal transfer tube takes the oil to the no. 4 bearing housing. In the no. 4 bearing housing, part of the oil is passed through a calibrated lubrication nozzle that sprays no. 4 bearing, and the rest of the oil is passed through the intershaft oil transfer tube to no. 3½ bearing. From the transfer tube, oil is centrifuged through two drillings in the low turbine shaft to a center annular groove in the inner surface of the no. 3½ bearing inner race. The groove channels oil to six axial grooves having alternate front and rear radial drillings through the race that spray oil into the opposite sides of the bearing cage.

A second external transfer tube routes oil to a boss in the four o'clock position on the intermediate case, to provide lubrication for the nos. 1, 2, and 3 bearings and the bevel and spur gears located in the intermediate case.

PRESSURE AND SCAVENGE OIL PUMP ASSEMBLY Pressure oil is circulated from the inte-

gral oil tank through the engine lubricating system by one of the three gear-driven rotor-type pump elements of the oil pump assembly. The two other pump elements operate in parallel to pump scavenge oil from the accessory gearbox and the no. 4 bearing housing to the oil tank via the top left mount pad. The pump housing incorporates a drain plug for draining the oil tank which must be accomplished before the pump assembly is removed.

CHECK VALVES The two check valves in the system prevent gravity oil flow when the engine is not running and also allow oil system components, such as filter and external transfer tubes downstream from the check valves, to be removed for servicing without draining the oil tank.

OIL COOLER The JT15D-1 engine oil cooler is essentially an oil-to-fuel heat exchanger. The cooler is considered adequate to handle all the cooling requirements of the engine, which has two hot bearing areas: nos. 3 and 3½, which constitute one area, and no. 4. For this reason, it is not necessary to install an oil cooler in the airframe.

The oil cooler consists of a core assembly of 85 beaded tubes through which fuel flows. The tubes project through circular end support plates and are enclosed within a cylindrical shell which extends beyond the core at each end. Baffles, pierced to accommodate the tubes, and from which a segment has been cut, are assembled at intervals along the core, the cutaway of alternate baffles lying on diametrically opposite sides within the shell. Diametrically opposed holes at opposite ends of the shell communicate with external manifolds which run longitudinally to the midpoint of the shell. The whole assembly is fabricated of stainless steel sheet and brazed into an integral unit. Axially drilled and internally threaded plugs welded into the projecting ends of the shell provide fuel inlet and outlet fittings; oil enters and leaves the unit through passages in the mounting which communicate with the external manifolds.

Fuel entering the cooler passes through the core tubes from inlet to outlet. Oil for the inlet manifold enters the shell at the fuel outlet end outside the core tubes and flows, in the opposite direction to the fuel flow, to the exit manifold. The baffles ensure that oil traverses the core tubes repeatedly in its passage in order to obtain maximum heat transfer.

OIL FILTER The 40-μ filter element, which may be cleaned and reused, is housed in the oil-filter-housing assembly and retained within a cover. Oil passes from the outside of the filter to the center and then out through the housing at the top. In the event that the oil filter becomes blocked, the bypass valve in the housing will open, allowing unfiltered oil to pass through to the engine. A plug at the bottom of the cover allows the filter assembly to be drained before removal.

OIL-FILTER-HOUSING ASSEMBLY The oil-filter-housing assembly comprises the following: two oil check valves, a pressure-relief valve assembly, an oil filter and an oil-filter bypass valve. Bosses at the top of the housing provide for the installation of an oil temperature bulb and a pressure transmitter. The oil cooler is mounted on the side of the housing.

SCAVENGE OIL SYSTEM The function of the scavenge oil system is to return used oil to the oil tank. This is achieved by allowing the oil from nos. 1, 2, 3, and 3½ bearings to drain into the accessory gearbox, aided by the airflow from the bearing compartment labyrinth seals. The no. 4 bearing scavenge oil is pumped by a separate pump element in the oil pump assembly.

The scavenge oil returned to the accessory gearbox collects in a sump at the bottom of the housing. Sump oil is pumped out by a separate and larger scavenge pump element. This pump element returns both the no. 4 bearing and gearbox scavenge oil to the oil tank. Scavenge oil is returned to the oil tank through an external transfer tube on the left-hand side of the engine which connects to a boss in the 12 o'clock position on the intermediate case; from this boss, the oil flows directly to the tank.

BREATHER SYSTEM Air from the engine bearing compartments and the accessory gearbox is extracted from the air-oil mist, and vented overboard through the action of an aluminum-alloy, impeller-type centrifugal breather. The breather is mounted on the main shaft assembly of the gearbox. The pressure difference between the air in the gearbox and the ambient atmosphere causes the air-oil mist in the gearbox to flow radially inward through the impeller. As the mist passes through the impeller, the oil particles adhere to the vanes and are thrown radially outward by centrifugal force. The relatively oil-free air passes through the hollow main shaft to a breather adapter, mounted at the rear on the gearbox cover. An airframe-supplied overboard vent line must be connected to the gearbox breather adapter.

ACCESSORIES DRIVE SPLINES LUBRICATION The accessory gearbox hydraulic pump and fuel-pump drive splines are oil-mist lubricated. Continuous wet-spline drive is provided by means of two diametrically opposite holes in each gearshaft picking up oil mist in the gearbox.

Pratt & Whitney Aircraft JT8D Lubrication System (FIG. 15-20)

GENERAL The JT8D has what is referred to as a "hot tank" system. This term refers to the technique of returning hot scavenge oil directly from the bearing compartments to the deaerator located in the oil tank. In a "cold tank" system, the scavenge oil is passed through the oil cooler prior to being returned to the oil tank. The advantage of the hot tank system is more efficient removal of entrapped air.

PRESSURE SYSTEM The oil is gravity-fed from the tank to the main oil pump via a transfer tube and a cored passage in the accessory gearbox. Pump discharge pressure is then directed to the main oil filter through another cored passage. A bypass valve located in the main oil filter provides oil to the system if the main filter becomes obstructed. External pressure taps are provided to sense oil pressure before and after the filter. This permits in-flight monitoring of the main oil filter via a differential pressure switch and flight-deck annunciator light.

Oil from the main oil filter, regulated to provide operating pressure after the fuel-oil cooler, is directed to the fuel-oil cooler through a passage and an external line. Oil at the desired system pressure and temperature exits from the fuel-oil cooler, and is delivered to the engine bearing compartments and accessory gearbox. A system pressure sense line located on the discharge side of the fuel-oil cooler provides an input of system working pressure to the regulating valve. The oil-pressure regulating valve is located in a cored passage which interconnects the main-oil-pump discharge pressure to the pump inlet. If system working pressure should decay, as the result of an obstructed oil cooler core or partial obstruction of the main oil filter, the regulating valve will be biased by a decrease in sense pressure. Any decrease in sense pressure causes the regulating valve to close proportionally, thus increasing pump output pressure sufficiently to return system working pressure to normal.

The surface area of the fuel-oil cooler element is adequate to provide sufficient cooling when fuel flow is

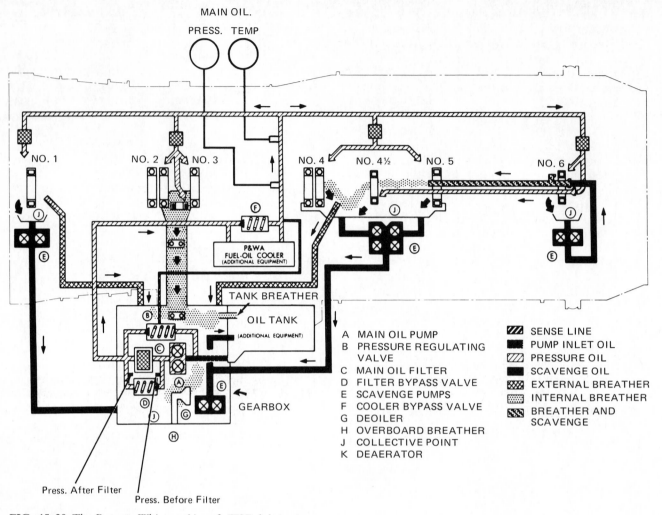

MAIN OIL.

PRESS. TEMP

NO. 1   NO. 2   NO. 3   NO. 4   NO. 4½   NO. 5   NO. 6

P&WA
FUEL-OIL COOLER
(ADDITIONAL EQUIPMENT)

TANK BREATHER

OIL TANK

(ADDITIONAL EQUIPMENT)

GEARBOX

A   MAIN OIL PUMP
B   PRESSURE REGULATING
    VALVE
C   MAIN OIL FILTER
D   FILTER BYPASS VALVE
E   SCAVENGE PUMPS
F   COOLER BYPASS VALVE
G   DEOILER
H   OVERBOARD BREATHER
J   COLLECTIVE POINT
K   DEAERATOR

SENSE LINE
PUMP INLET OIL
PRESSURE OIL
SCAVENGE OIL
EXTERNAL BREATHER
INTERNAL BREATHER
BREATHER AND
SCAVENGE

Press. After Filter

Press. Before Filter

FIG. 15-20 The Pratt & Whitney Aircraft JT8D lubrication system.

in the mid-to-high thrust range. Thus, the requirement for thermostatic control of oil temperature is eliminated. At prolonged idle settings, however, an increase in oil temperature is sometimes noted. This is the result of reduced fuel flow and reduced capacity to dissipate heat from the engine oil. The higher oil temperatures associated with prolonged idling can be controlled by periodically advancing the power lever to increase fuel flow so that excessive heat can be adequately rejected by the oil system through the cooler.

A bypass valve is incorporated in the fuel-oil cooler to assure sufficient oil flow if the cooler core should become obstructed. Oil discharged from the cooler is delivered to the engine bearing compartments through a network of external and internal stainless steel tubing.

SCAVENGE SYSTEM After the oil has lubricated and cooled the main engine and accessory gearbox bearings, it is returned to the oil tank by the scavenge system.

The main collection points for scavenge oil are located in the nos. 1, 4, 5, and 6 bearing compartments and the accessory gearbox. Located in each of these compartments is a gear-type pump which returns scav-

enge oil to the oil tank. Scavenge oil from the no. 1 bearing compartment is returned directly to the gearbox. Nos. 2 and 3 bearings scavenge to the gearbox via gravity and breather flow through the towershaft housing. Gearbox lube oil and scavenge oil from the nos. 1, 2, and 3 bearings is then returned to the oil tank via the gearbox scavenge pump.

Scavenge oil from the no. 6 bearing area is pumped to the no. 4½ bearing area through transfer tubes located inside the low-pressure-compressor drive turbine shaft.

Centrifugal force causes the oil to be ejected from the no. 4½ bearing nut through the high-pressure turbine shaft scavenge holes to the no. 4 and no. 5 bearing compartment.

The combined scavenge oil from the nos. 4, 4½, 5, and 6 bearings is then returned directly to the oil tank from the scavenge pump located in the nos. 4 and 5 bearing collection point.

Breather System

To assure proper oil flow, and to maintain satisfactory scavenge pump performance, the pressure in the

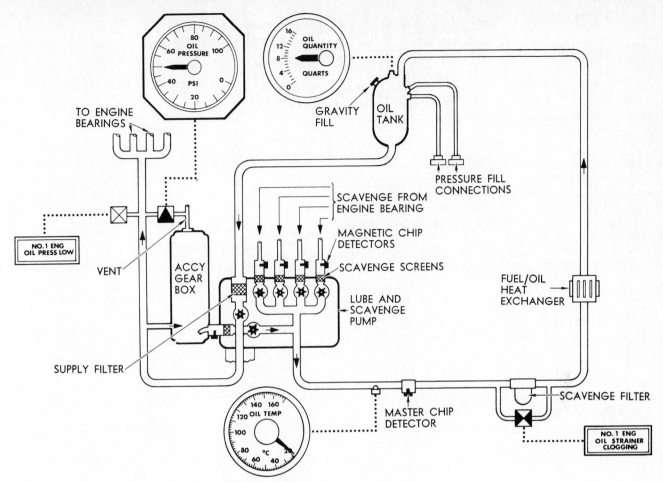

FIG. 15-21 The General Electric CF6 engine oil system functional diagram. Note: No text material accompanies this diagram. It is included here as another example of a modern high-bypass-ratio turbofan lubrication system.

bearing cavities is controlled by the breather system. The breather air from all of the main bearings is vented to the accessory gearbox as follows:

No. 1 bearing breather air is vented to the accessory gearbox via external tubing.

Nos. 2 and 3 bearings are vented internally to the accessory gearbox through the towershaft housing.

Nos. 4½ and 6 bearings breathe through the scavenge system into the number 4 and 5 bearing collection point.

The combined breather air from nos. 4, 4½, 5, and 6 bearings is vented to the accessory gearbox through an external line.

A deoiler located in the accessory gearbox serves to remove oil particles in the breather air before it is discharged into the airframe waste tube.

REVIEW AND STUDY QUESTIONS

1. Compare the oil system requirements of the reciprocating and gas turbine engines.
2. Other than lubricating, what jobs can the oil do?
3. List the several components contained in a typical gas turbine lubricating system. Discuss each unit in some detail.
4. Very briefly describe the lubricating system of the 501-D13, CJ610, CJ805-23, J69, T53, JT3D, F100-PW-100, JT15D, and the JT8D engines.

CHAPTER SIXTEEN
IGNITION
SYSTEMS

Jet engine ignition systems fall into two general classifications: first, the induction type (now obsolete), producing high-tension sparks by conventional induction coils, and, second, the capacitor type, or those that cause ignition by means of high-energy and very-high-temperature sparks produced by a condenser discharge. A third kind of ignition system not widely adopted, but incorporated on the Pratt-Whitney Aircraft PT6A, utilizes a glow plug.

REQUIREMENTS FOR THE GAS TURBINE IGNITION SYSTEM

The advent of various types of jet engines toward the end of World War II created an entirely new set of problems for the manufacturers of ignition equipment.

In conventional reciprocating engines, accurately timed sparks occur between the spark-plug electrode when the fuel-air mixture has been subjected to a pressure of about 5 to 10 atm. Furthermore, the mixture has been heated by rapid compression and remains somewhat turbulent, although it is ignited when the piston velocity is nearly zero. Under these conditions ignition is relatively easy.

The nearly ideal fuel-air ratios and essentially stable conditions within the combustion chamber have been replaced by a very cold and considerably overlean (too much air in relation to fuel) fuel-air mixture that rushes past the igniter plugs at a high velocity. This causes difficulty because, in order to start a fire, a mixture, in spite of its low temperature and excessive air content, must be brought to a high temperature in the brief instant that it is adjacent to the igniter plugs.

In addition, spark-plug fouling is a major problem. Since jet engine combustion is a self-sustaining process, most ignition systems are required to operate only during the starting cycle. The spark plug or igniter is not able to keep itself clean by continuous arcing across its gap, as is the case with reciprocating engine spark plugs.

The lower volatility of jet fuels, coupled with the extremely high altitudes and correspondingly low temperatures in which the gas turbine engine operates, makes the conditions for an in-flight relight in the event of a flame-out even more difficult.

Continuously operating ignition systems are being installed on several of the newer high-bypass-ratio turbofan engines in order to ensure an immediate relight in case of flame-out due to any number of flight and/or environmental situations. Also being looked at is a second type of flame-out insurance consisting of a pressure-sensitive switch which rapidly senses flame-out

through a pressure decay in the combustion chambers. This automatically reactivates the engine's standard ignition system, eliminating the need for continuous ignition flame-out protection.

EARLY INDUCTION-TYPE IGNITION SYSTEMS (*FIG. 16-1*)

Early jet engine ignition systems evolved using the tried principles that were developed for the reciprocating engine. Some of these early systems employed a vibrator and transformer combination somewhat similar to the booster coils used for starting purposes on reciprocating engines. Other units substituted a small electric motor driving a cam to provide the necessary pulsating magnetic field to the primary coil of the transformer. Several variations appeared, all using the same basic principle of high-voltage induction by means of a transformer in order to reach the necessary voltage capable of causing an arc across the wide-gap jet igniter plug. A typical unit of this kind is illustrated in Fig. 16-1. An interesting variation of this transformer type of ignition system is the opposite-polarity system used on some models of the General Electric J47 (Fig. 16-2). In this circuit two electrodes extend into the combustion

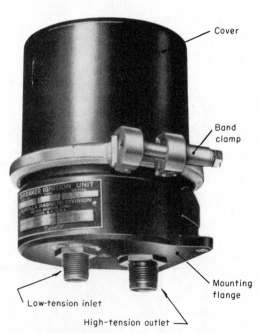

FIG. 16-1 Early type of induction ignition system. (*Bendix Electrical Components Division.*)

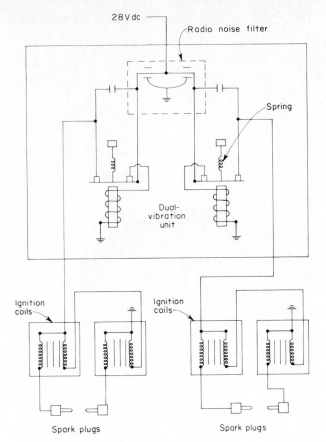

FIG. 16-2 The opposite-polarity ignition system, an early form of gas turbine ignition system.

chamber. Each electrode alternately becomes highly positively and negatively charged, thus causing a very high potential difference to exist across the electrodes.

MODERN CAPACITOR-TYPE IGNITION SYSTEMS

In modern engines it is necessary to have not only a high voltage to jump a wide-gap igniter plug, but also a spark of high heat intensity for the reasons mentioned in the section dealing with ignition-system requirements. The high-energy, capacitor-type ignition system has been universally accepted for gas turbine engines, because it provides both high voltage and an exceptionally hot spark which covers a large area. Excellent chances of igniting the fuel-air mixture are assured at reasonably high altitudes.

The term "high energy" has been used throughout this section to describe the capacitor type of ignition system. Strictly speaking, the amount of energy produced is very small. The intense spark is obtained by expending a small amount of electric energy in a very small amount of time.

Energy is the capacity for doing work. It can be expressed as the product of the electrical power (watt) and time. Gas turbine ignition systems are rated in *joules*. The joule is also an expression of electric energy, being

equal to the amount of energy expended in one second by an electric current of one ampere through a resistance of one ohm. The relationship between these terms can be expressed by the formula:

$$W = \frac{J}{t}$$

where W = watts (power)
J = joules
t = time, s

All other factors being equal, the temperature of the spark is determined by the power level reached. It can be seen from the formula that a high-temperature spark can result from increasing the energy level J, or by shortening the duration t of the spark. Increasing the energy level will result in a heavier, more bulky ignition unit, since the energy delivered to the spark plug is only about 30 to 40 percent of the total energy stored in the capacitor. Higher erosion rates on the igniter plug electrodes would also occur because of the heavy current flowing for such a comparatively long time. Furthermore, much of the spark would be wasted, since ignition takes place in a matter of microseconds. On the other hand, since heat is lost to the igniter plug electrodes, and since the fuel-air mixture is never completely gaseous, the duration of the spark cannot be too short.

An example of the relationship between watts and time is shown in the following table for a 4-joule (J) ignition unit (4 J appearing at the plug.)

Time, Seconds	Power, Watts
1	4
0.01 (hundredths)	400
0.001 (thousandths)	4,000
0.0001 (ten-thousandths)	40,000
0.00001 (hundred-thousandths)	400,000
0.000001 (millionths)	4,000,000

In an actual capacitor discharge ignition system, most of the total energy available to the igniter plug is dissipated in 10 to 100 μs (0.000010 to 0.000100 s), so the system described above would actually deliver 80,000 W if the spark duration was 50 μs.

$$W = \frac{J}{t}$$
$$= \frac{4}{0.000050}$$
$$= 80,000$$

Because of this high power, to prevent receiving a *lethal electrical shock* from capacitors, avoid contact, directly or through uninsulated tools, with leads, connections, and components, until capacitors have been grounded and are known to be fully discharged. All capacitor ignition boxes are labeled with an appropriate warning to this effect.

To review, the spark temperature (a function of the watts value) is the most important characteristic of any ignition system, but all three factors—watts, energy, and time—must be considered before the effectiveness of any ignition system can be determined.

TWO TYPES OF HIGH-ENERGY IGNITION SYSTEMS

Just as ignition systems for jet engines are divided into induction and capacitor-discharge types, the capacitor discharge type can be further divided into two basic categories.

1. High-voltage capacitor ignition system with dc or ac input
2. Low-voltage capacitor ignition system with dc or ac input

High-voltage Capacitor System—DC Input (More than 5000 V to the Plug) (*FIG. 16-3*)

As the cycle of operation begins, the power source delivers 28-V-dc input to the system at the exciter. Dual ignition is provided on the engine by twin circuits throughout the exciter as shown in Fig. 16-4, or by two separately mounted exciters, depending on engine configuration, as shown in Fig. 16-5. In either case, each triggering circuit is connected to a spark igniter. The operation described here takes place in each individual circuit and is essentially the same in both units, with the exception of the mechanical features of the armature. Figure 16-6 shows a unit with two exciters and a separate ignition transformer. The operation is the same as the devices shown in Figs. 16-4 and 16-5.

The following description of a typical dc input ignition unit, refers to Fig. 16-5. In this system the direct current supply, after passing through a radio noise filter (1) to prevent high-frequency feedback into the aircraft electrical system, is fed to the primary of the transformer, which is an integral part of the vibrator assembly (2). From the primary this current is passed through a pair of breaker contacts, normally closed, to ground. A primary capacitor is connected across these contacts to damp excessive arcing.

The action of the vibrator is produced by the transformer, which has a laminated core. With the contacts in a closed position, the flow of current through the coil produces a magnetic field. This field exerts a force against the armature, which is mounted on a pivot.

The armature is pulled downward, and after a certain degree of travel to acquire momentum, strikes the end of the contact spring. With further movement the contacts are separated, the flow of current stops, and the magnetic field collapses. With the cessation of magnetic force against the underside of the armature, its movement slows to a halt, and it is positively returned to its original position, first by the tension of the contact spring, and finally by the pull of the permanent magnet mounted above it. The spring meanwhile returns the lower contact to a closed position, and the vibrating cycle resumes.

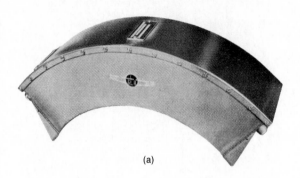

(a)

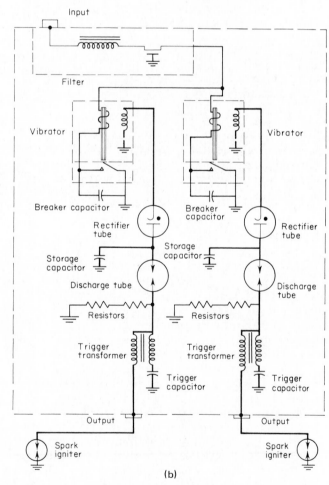

(b)

FIG. 16-4 (a) General Laboratories Associates ignition unit used on the Detroit Diesel Allison 501-D13 engine.
(b) Capacitor-discharge electronic ignition exciter; a dc input is used on this unit. (*General Laboratories Associates, Inc.*)

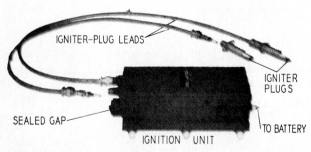

FIG. 16-3 Bendix dual high-voltage system with dc input.

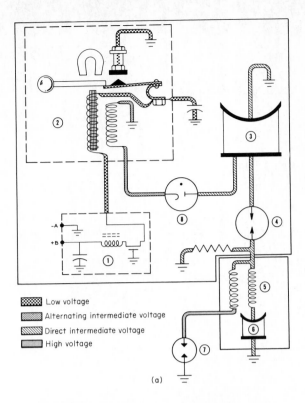

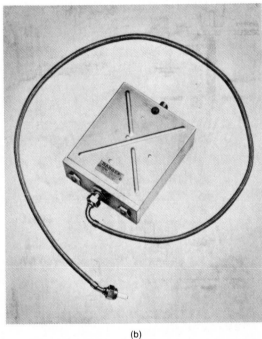

Low voltage
Alternating intermediate voltage
Direct intermediate voltage
High voltage

(a)

(b)

FIG. 16-5 (a) The electrical circuit shown in this schematic is typical of the type used on the Pratt & Whitney Aircraft JT3 and JT4 series engines. It is one of two separate units with a dc input and a high-voltage output.
(b) External appearance of the GLA high-energy unit.
(General Laboratories Associates, Inc.)

The collapse of the magnetic field in the transformer causes a high alternating voltage to be induced in the secondary. This produces successive pulses flowing into the storage capacitor (3) through the gas-charged rectifier tube (8), which limits the flow to a single direc-

tion. With repeated pulses the capacitor assumes a greater and greater charge, at a constantly increasing voltage.

When this intermediate voltage reaches the predetermined level for which the spark gap (4) has been set, the gap breaks down, allowing a portion of the accumulated charge to flow through the primary of the triggering transformer (5) and the trigger capacitor (6) connected in series with it.

The surge of the current through the primary induces a very high voltage in the secondary of the triggering transformer, which is connected to the spark igniter (7). This voltage is sufficient to ionize the gap, and produces a trigger spark of approximately 5000 V.

When the gap at the spark igniter is thus made conductive, the storage capacitor discharges the remainder of its accumulated energy through it, together with the charge from the trigger capacitor. This results in a capacitive spark of very high energy, capable of vaporizing globules of fuel and overcoming carbonaceous deposits.

The bleeder resistor is provided in the discharge circuit to dissipate the residual charge on the trigger capacitor between the completion of one discharge at the spark igniter and the beginning of the next cycle.

The spark rate will vary depending on the value of the input voltage. At lower voltage values, more time will be required to raise the intermediate voltage on the storage capacitor to the level necessary to break down the spark gap. Since that level remains constant, however, being established by the physical properties of the gap, the full normal store of energy will always be accumulated by the storage capacitor before discharge.

Typical specifications for this system are as follows:

Input voltage:	Normal: 24 V dc
	Operating limits: 14 to 30 V dc
Spark rate:	4 to 8 per second at each plug, depending on input voltage
Designed to fire:	2 igniter plugs
Accumulated energy:	3 J
Duty cycle:	2 min on, 3 min off,
	2 min on, 23 min off

Two igniter plugs are mounted in the combustion section outer case. The spark igniters are generally located in two diametrically opposite combustion liners. The igniters receive the electrical output from the ignition exciter unit and discharge the electric energy during engine starting to ignite the fuel-air mixture in the combustion liners.

Figure 16-7 shows a typical high-voltage, high-energy, capacitor-type ignition system using a motor-driven cam to operate the breaker points instead of a vibrator, and a motor-driven single-lobe cam instead of a sealed spark gap tube.

High-voltage Capacitor System—AC Input (FIG. 16-8)

As shown in Fig. 16-8, power is supplied to the input connector of the unit from the 115-V, 400-cycle-per-second (hertz, Hz) source in the aircraft, and is first led

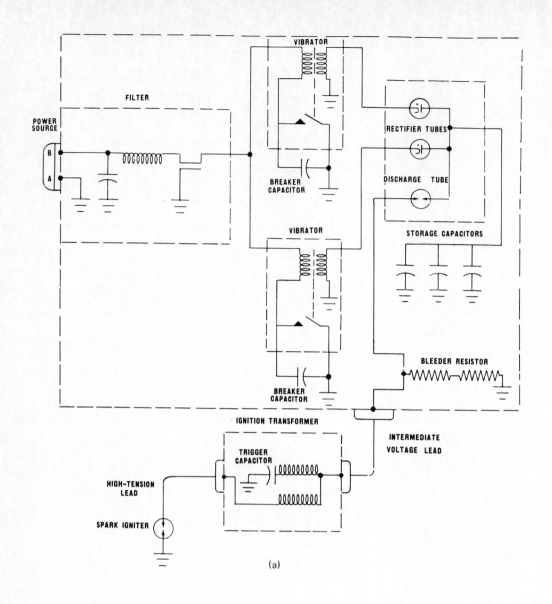

(a)

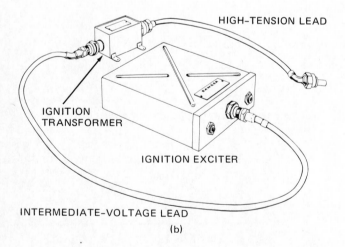

(b)

FIG. 16-6 (a) High-energy ignition system with a separate transformer and dc input.
(b) External appearance of the ignition system with the separate transformer.

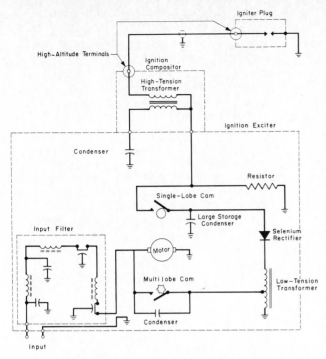

FIG. 16-7 A high-energy system with cam-operated breaker points. In most modern systems, all mechanical parts have been replaced by electronic solid-state devices.

through a filter which serves to block conducted noise voltage from feeding back into the airplane electrical system. From the filter, the circuit is completed through the primary of the power transformer to ground.

In the secondary of the power transformer an alternating voltage is generated at a level of approximately 1700 V. During the first half-cycle this follows a circuit through the doubler capacitor and rectifier A to ground, leaving the capacitor charged. During the second half-cycle, when the polarity reverses, this circuit is blocked by rectifier A; the flow of this pulse is then through ground to the storage capacitor, through rectifier B, the resistor, and the doubler capacitor back to the power transformer.

With each pulse the storage capacitor thus assumes a greater and greater charge, which by virtue of the action of the doubler capacitor approaches a voltage approximately twice that generated in the power transformer. When this voltage reaches the predetermined level for which the spark gap in the discharge tube X (the control gap) has been calibrated, this gap breaks down, allowing a portion of the accumulated charge to flow through the primary of the high-tension transformer and the trigger capacitor in series with it. This surge of current induces a very high voltage in the secondary of the high-tension transformer, sufficient to ionize the gap in discharge tube Y. The storage capacitor

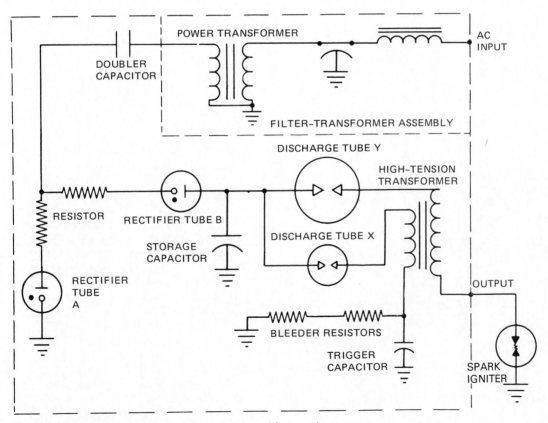

FIG. 16-8 A high-energy, high-voltage system with an ac input.

immediately discharges the remainder of its accumulated energy through the spark igniter. This produces a capacitive spark of very high energy.

The bleeder resistors are provided to dissipate the residual charge on the trigger capacitor between the completion of one discharge at the spark igniter and the beginning of the next cycle.

Typical specifications for this system are as follows:

Input voltage: Normal: 115-V 400-Hz ac
                    Operating limits: 90 to 120 V
Spark rate: Normal: 1.50 to 2.75 per second
                    Operating limits: 0.75 to 5.00 per second
Designed to ignite: One spark igniter
Accumulated energy: 14 to 17 J
Duty cycle: 2 min on, 3 min off,
                    2 min on, 23 min off

Low-voltage Capacitor System-DC Input (Less Than 1000 V to the Plug) *(FIG. 16-9)*

The basis of operation upon which the low-voltage, high-energy ignition system is built is the *self-ionizing* feature of the igniter plug. In the high-voltage system a double spark is produced, the first part consisting of a high-voltage component to ionize (make conductive) the gap between the igniter plug electrodes in order that the second high-energy low-voltage portion may follow. The low-voltage, high-energy spark is similar to the above except that ionization is effected by the self-ionizing igniter plug discussed above.

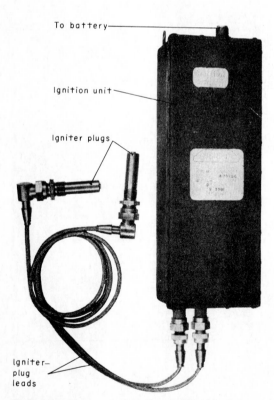

To battery

Ignition unit

Igniter plugs

Igniter-plug leads

FIG. 16-9 Two complete Bendix Electrical Components Division high-energy, low-voltage systems in one case.

The explanation refers to Fig. 16-10, which shows one of two separate and independent low-tension, high-energy ignition units used on some models of the General Electric J79 engine.

The main ignition unit changes the amplitude and the frequency characteristics of aircraft power into pulsating dc. To do this the components in the ignition unit are grouped in stages to filter, amplify, rectify, and store an electric charge.

A pi-type filter ($C1$, $L1$, $C2$), located in the input stage, grounds out radio interference entering or leaving the unit. This prevents the ignition unit from disrupting the operation of other aircraft electronic equipment and stabilizes the output of the unit itself.

At radio frequencies the choke coil ($L1$) blocks current flow in either direction. Capacitors $C1$ and $C2$ act as short circuits to ground. Radio-frequency noise pulses, approaching the filter from either direction are blocked by the coil and shunted to ground through the capacitors.

The choke coil ($L1$) passes aircraft dc easily, and the capacitors ($C1$ and $C2$) now act as blocking devices to prevent grounding out the current. Current flows through the filter to the primary of the step-up transformer ($L2$).

The opposition to current flow varies because of the change in reactance (resistance) of the choke and capacitor when frequency changes. At radio frequencies, inductive reactance (resistance) of the coil is high and capacitive reactance of the capacitors is low.

When power is applied to the unit, current flows from ground through the normally closed contacts of the vibrator, through the transformer primary winding ($L2$), the radio-frequency filter, and back to the power source.

Current through the primary winding causes the normally closed contacts of the vibrator to open, momentarily halting any further current flow. This action changes the dc to a pulsating dc in the primary. The pulsating dc induces a high-voltage ac across the transformer secondary.

Capacitor $C3$, wired across the vibrator contacts, can be referred to as a buffer capacitor. It protects or buffers the contacts against a voltage arc which might occur during normal operation. Such an arc develops carbon deposits and pit marks on the contacts, and reduces the service life of the vibrator. The ac voltage developed across the transformer secondary winding ($L3$) is next applied across the half-wave rectifier formed by two diode gas-rectifier tubes ($V1$, $V2$). The rectifier circuit converts the ac voltage to a pulsating dc. The rectifying action of this circuit depends on the cathode to plate polarity of diode tube $V1$. When the top of the secondary winding ($L3$) is positive, the plate of $V1$ becomes positive with respect to its cathode. Diode $V1$ is ionized by this potential and starts to conduct. At this point diode $V2$ is ionized and also conducts. On the opposite half-cycle of the ac voltage across the transformer secondary ($L3$), the voltage at the top of the winding is negative. The plate of diode $V1$ becomes negative with respect to its cathode, and neither rectifier conducts.

The output of the half-wave rectifier is a pulsating

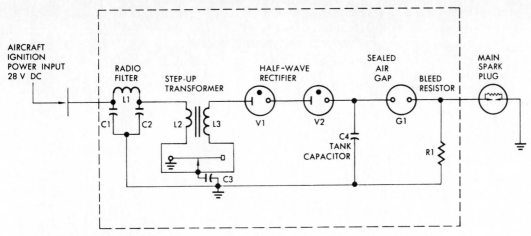

FIG. 16-10 High-energy, low-voltage system with a vibrator. This system is used on the General Electric J79.

direct current that flows from the tubes, down the transformer secondary winding (*L*3) to ground, and up from ground to the lower side of the tank capacitor (*C*4). The current supplying the diodes is derived from electrons leaving the top side of *C*4 because of capacitor reaction to the charge building up on the lower side.

As the rectified current flows to the tank capacitor (*C*4), a charge of energy is built up across *C*4. Each time the tank capacitor is ready to discharge, its voltage has reached the ionizing potential of the sealed air gap.

The function of the sealed air gap *G*1 is similar to that of an automatic switch. While the gap is deionized, the switch is open and no ignition voltage can appear across the spark plug. Once ionized, the gap allows the tank capacitor voltage to ionize the spark-plug gap. With both gaps ionized, the tank capacitor has a complete current path and discharges instantly across the spark-plug electrodes.

A bleed resistor (*R*1) is wired across the output circuit to act as a dummy load in the event the ignition unit is energized while the spark plug is disconnected. This eliminates the possibility of damaging the ignition unit.

As stated previously, the spark plugs used in this ignition system are the shunted-gap type, which are self-ionizing and designed for low-tension (relatively low voltage) applications. See Figs. 16-14(b) and 16-15.

Although the spark plug fires at relatively low volt-

age, a high-temperature spark is obtained from the speed at which the energy is discharged across the gap. The spark is of very short duration (40 μs), but momentarily expends a great amount of power. Tank capacitor discharge current from the main ignition unit surges to the spark-plug electrodes, building a potential between the center electrode and ground electrode. The semiconducting material shunts the electrodes. When the potential between electrodes reaches approximately 800 V, it forces enough current through the semiconductor to ionize the air gap between the electrodes. The full tank capacitor current arcs instantly across the ionized gap, emitting a high-energy spark. Figure 16-11 shows a typical low-tension system with ac input.

Combination or Dual-duty Ignition System used on the JT3D Engine *(FIG. 16-12)*

This ignition system includes one intermittent-duty exciter, one continuous-duty exciter, one intermediate voltage lead, and two high-tension leads. It is designed to fire two spark igniters during ground starts by means of the 20-J intermittent-duty exciter, or one spark igniter during flight by means of the 4-J continuous-duty exciter. This functional description covers the operation of the complete system.

When intermittent operation is to be employed, dc

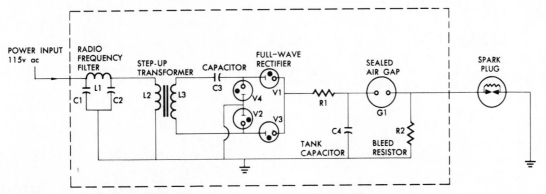

FIG. 16-11 High-energy, low-voltage system without a vibrator. This system is used on some models of the General Electric J79.

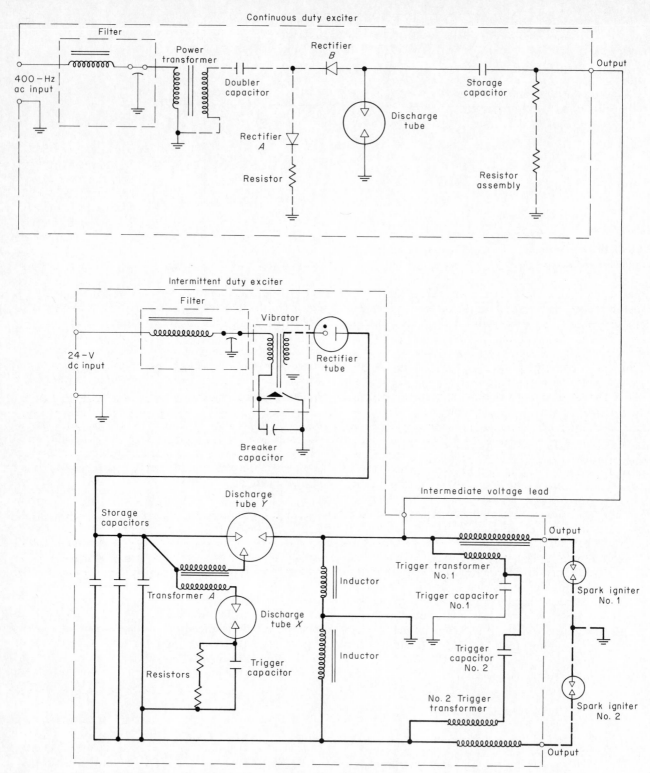

FIG. 16-12 Dual system used on the Pratt & Whitney Aircraft JT3D engine.

power is supplied to the input of the intermittent-duty exciter from the 24-V aircraft electrical system. It is first passed through a radio noise filter to prevent high-frequency feedback.

From the filter, input voltage is fed to the primary of the transformer in the vibrator, which is an integral part of the vibrator assembly. From the primary, a current thus flows through a pair of breaker contacts, normally closed, to ground. A capacitor is connected across these contacts to damp excessive arcing.

With the contacts closed, the flow of current through the coil produces a magnetic field. The force exerted by this field pulls the armature free from the permanent magnet above it. Rapid acceleration builds up kinetic energy in the armature for a brief period before it strikes the contact spring. This opens the contact points quickly, the flow of current stops, and the magnetic field collapses. The armature is returned by the tension of the contact spring, and is positively held in its original position by the permanent magnet. The spring having meanwhile closed the contacts, the vibrating cycle recommences.

Each collapse of the magnetic field induces a high voltage in the secondary of the transformer. This produces successive pulses flowing through the gas-charged rectifier, which limits the flow to a single direction, into the storage capacitors, which thus assume a greater and greater charge, at a constantly increasing voltage.

When this intermediate voltage reaches the predetermined level for which discharge tube X has been calibrated, this tube breaks down. A small portion of the accumulated charge, flowing through the primary of transformer A, induces a high voltage in the secondary. This voltage triggers the threepoint, discriminating discharge tube Y, which breaks down and permits a surge of current to flow from the storage capacitors through the primary of the trigger transformers and trigger capacitor no. 2.

The very high voltage thus induced in the secondary of the trigger transformers is sufficient to ionize the gaps at the spark igniters, producing a trigger spark.

The remainder of the energy in the storage capacitors is immediately discharged, following a path through the secondary of the trigger transformer and the high-tension lead to spark igniter no. 1, through ground to spark igniter no. 2, and back through the other high-tension lead and trigger transformer secondary to the storage capacitors.

The inductance in the inductors is high enough that the current shunted through them is not significant, but after completion of the spark cycle they provide a return path to bleed off any residual charge on the trigger capacitors.

If one spark igniter is shorted, the operation is the same, producing only one spark. If the circuit to one spark igniter is open, the operation is the same, producing only one spark; the path from the operating spark igniter returns through ground and the inductor on the opposite side of the exciter circuit to the storage capacitors.

When continuous operation is to be employed, power is supplied to the input of the continuous-duty exciter from the 115-V, 400-Hz ac source in the aircraft.

It is first led through a filter which serves to block conducted noise voltage from feeding back into the aircraft electrical system. From the filter the circuit is completed through the primary of the power transformer to ground.

In the secondary of the power transformer an alternating voltage is generated at a level of approximately 1500 V. During the first half-cycle this follows a circuit through the doubler capacitor and rectifier A to ground, leaving the capacitor charged. During the second half-cycle, when the polarity reverses, this circuit is blocked by rectifier A; the flow of this pulse is then through ground and the resistors to the storage capacitor, through rectifier B and the doubler capacitor back to the power transformer.

With each pulse the storage capacitor thus assumes a greater and greater charge, which by virtue of the action of the doubler capacitor approaches a voltage approximately twice that generated in the power transformer. When this voltage reaches the predetermined level for which the spark gap in the discharge tube has been calibrated, this gap breaks down, and the accumulated charge on the storage capacitor reaches the output terminal of this exciter.

From the output terminal it is carried to the intermittent-duty exciter by the intermediate voltage lead. Being prevented from reaching the storage capacitors in this unit by the discriminating discharge tube Y, a portion of the charge flows through the primary of trigger transformer no. 1 and associated trigger capacitor no. 1.

This surge of current induces a very high voltage in the secondary of the trigger transformer, sufficient to ionize the gap at spark igniter no. 1.

The remainder of the charge is immediately dissipated as a spark at the spark igniter, the return circuit being completed through ground to the continuous-duty exciter.

The inductor in the intermittent-duty exciter serves to bleed off any residual charge on trigger capacitor no. 1 between spark cycles.

JET-ENGINE IGNITERS

Jet-engine igniters come in many sizes and shapes depending upon the type of duty they will be subjected to. The electrodes of the plugs used with high-energy ignition systems must be able to accommodate a current of much higher energy than the electrodes of conventional spark plugs are capable of handling. Although the high-energy current causes more rapid igniter-electrode erosion than that encountered in reciprocating-engine spark plugs, it is not of large consequence because of the relatively short time that a turbine engine ignition system is in operation. It does, however, constitute one of the reasons for not operating the gas turbine ignition system any longer than necessary. Igniter plug gaps are

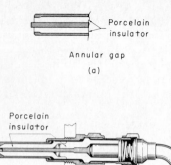

Porcelain insulator

Annular gap

(a)

Porcelain insulator

Constrained gap

(b)

FIG. 16-13 Two types of igniter gaps.
(a) Annular gap.
(b) Constrained gap.

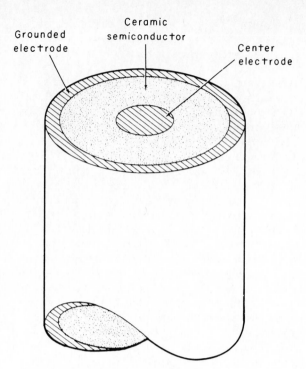

FIG. 16-15 The tip of an igniter plug used with low-voltage systems. (*General Electric.*)

large in comparison with those of conventional spark plugs, because the operating pressure at which the plug is fired is much lower than that of a reciprocating engine.

Most igniter plugs are of the annular-gap type shown in Fig. 16-13(a), although constrained gaps as shown in Fig. 16-13(b) are used in some engines. Normally, the annular-gap plug projects slightly into the combustion chamber liner in order to provide an effective spark. The spark of the constrained-gap plug does not closely follow the face of the plug; instead it tends to jump in an arc which carries it beyond the face of the chamber

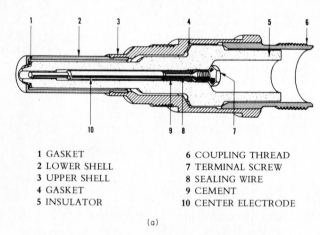

1 GASKET
2 LOWER SHELL
3 UPPER SHELL
4 GASKET
5 INSULATOR
6 COUPLING THREAD
7 TERMINAL SCREW
8 SEALING WIRE
9 CEMENT
10 CENTER ELECTRODE

(a)

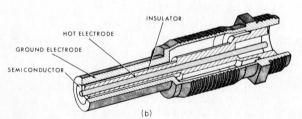

(b)

FIG. 16-14 (a) Typical igniter plug used on the Pratt & Whitney Aircraft JT3 series engines.
(b) Igniter plug for the General Electric J85.

liner. The constrained-gap plug need not project into the liner, with the result that the electrode operates at a cooler temperature than that of the annular-gap plug. Figures 16-14, 16-15, 16-16, 16-17, and 16-18 illustrate some of the varieties of gas turbine igniters.

The turbojet ignition system, designed for severe altitude conditions common to the military form of operation is rarely, if ever, taxed to its full capabilities by transport use. Flame-out is much less common than it was, and flight relight is not normally required of the ignition system. Ignition problems in general are of a minor nature in comparison to the constant attention required by the piston engine system. Airborne ignition analysis equipment is unnecessary. Spark igniter plug replacement is greatly minimized. Only two plugs per engine are used, compared with the three dozen used in some reciprocating engines.

The trends that are taking place in the gas turbine ignition area are:

1. Use of ac power inputs, thus eliminating the vibrator, a major source of trouble.
2. Use of solid-state rectifiers.
3. Use of two discharge tubes permits the level of the stored energy per spark to be more consistent throughout the life of the exciter.
4. Sealed units.
5. Longer time between overhauls.
6. The advent of short-range jets has increased the ratio of ignition "on" time to engine operation. This has led to the development of dual systems, one of which was described in this section.

| HIGH-VOLTAGE AIR GAP | HIGH-VOLTAGE SURFACE GAP | HIGH-VOLTAGE AIR-SURFACE GAP | LOW-VOLTAGE SHUNTED-SURFACE GAP | HIGH-VOLTAGE SHUNTED-SURFACE GAP |

FIG. 16-16 Types of igniter tips.

REVIEW AND STUDY QUESTIONS

1. List the requirements for a gas turbine ignition system. How does this compare to a reciprocating engine's requirements?
2. Describe a typical induction-type ignition system. Describe the variations of this system.
3. Discuss the general operating principal behind the modern gas turbine ignition system.
4. Make a list of the input and output variations possible with capacitor-discharge ignition systems.
5. What is the relationship between joules, watts, and time? Of what significance is this in the design of a gas turbine ignition system?
6. Briefly describe the operation of the following capacitor-discharge gas turbine ignition systems: high-voltage dc; high-voltage ac; low-voltage dc; low-voltage ac.
7. Discuss the trends that are taking place in the gas turbine ignition field.

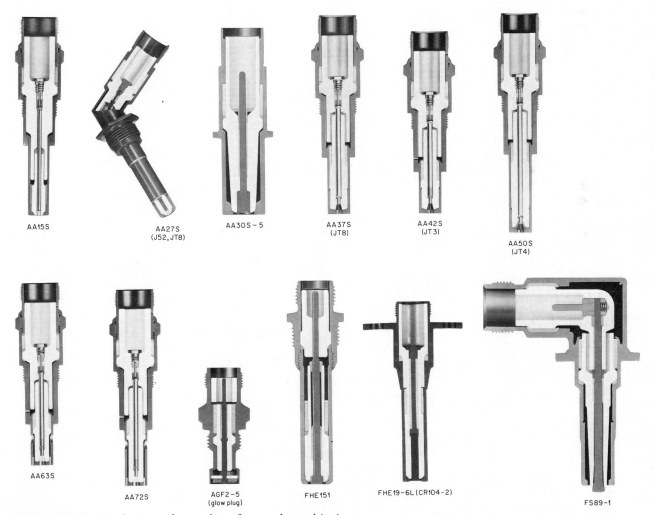

AA15S
AA27S (J52, JT8)
AA30S-5
AA37S (JT8)
AA42S (JT3)
AA50S (JT4)
AA63S
AA72S
AGF2-5 (glow plug)
FHE151
FHE19-6L (CR104-2)
FS89-1

FIG. 16-17 Sectional views of a number of currently used igniters.

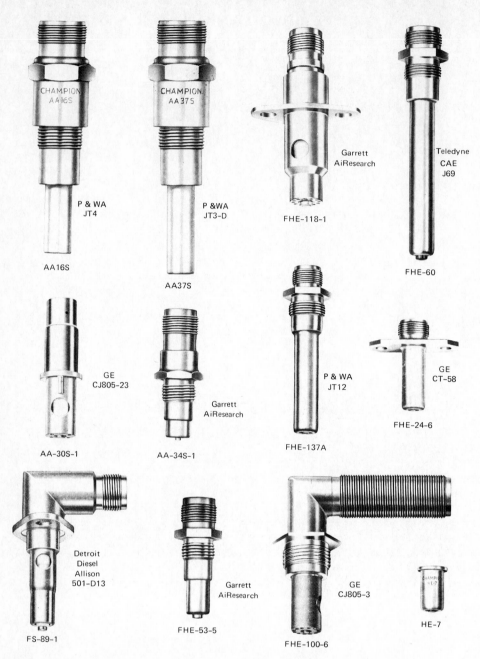

FIG. 16-18 A great variety of igniter plug types are used today. An application of each is indicated next to the igniter. (*Champion Spark Plug Co.*)

CHAPTER SEVENTEEN
STARTING SYSTEMS

The purpose of any starter system is to accelerate the engine to the point where the turbine is producing enough power to continue the engine's acceleration. This is called the *self-accelerating* speed. The proliferation of gas turbine starter types seems to indicate that no one starter shows a definite superiority, for all situations, over other types. The choice of a starting system depends upon several factors.

1. *Length of starting cycle*—For military equipment, starting time may be of primary importance. In addition, the speed with which the starter can accelerate the engine to idle speed will influence not only peak exhaust gas temperatures, but also the length of time the engine spends at these high starting temperatures (Fig. 17-1). Unlike the reciprocating engine starter, the gas turbine starter must continue to accelerate the engine even after "light-off." Slower than normal accelerations or starters which "drop out" too soon may cause "hot" or "hung" starts. (See Chap. 19.) A hung start is a situation where the engine accelerates to some intermediate rpm below idle, and stays there. Hot starts are, of course, what the name implies: a start where turbine or exhaust gas temperature limits are exceeded.

2. *Availability of starting power*—Even small gas turbine engines require large amounts of either electric or pressure energy. Large engines require correspondingly more. Some starting systems are completely self-contained, while others require power from external sources. Many airplanes carry their own energy source

in the form of a self-contained small auxiliary gas turbine engine which produces electric and/or pressure energy. Power may also be taken from a running engine in multiengine installations. In such a situation, one engine might be started using a starter requiring no external source of power such as a solid propellant, or fuel-air combustion starter. The other engine(s) can then be started in turn with power taken from the running engine. Starting power requirements for gas turbine engines differ from those of reciprocating engines. In the reciprocating engine, the peak load to the starter is applied in the first moments of starter engagement, but because of the increasing compressor load, the load on the turbine starter is actually increasing during engine acceleration prior to light-off.

3. *Design features*—Included in this area are such things as specific weight (pounds of starter weight per foot-pound of torque produced), simplicity, reliability, cost, and maintainability.

The following is a list of the various forms of gas turbine starters.

1. Electric motor starter
2. Electric motor-generator (starter-generator)
3. Pneumatic or air turbine starter
4. Cartridge or solid-propellant starter
5. Fuel-air combustion starter
6. Gas turbine starter (jet fuel starter)
7. Hydraulic motor starter

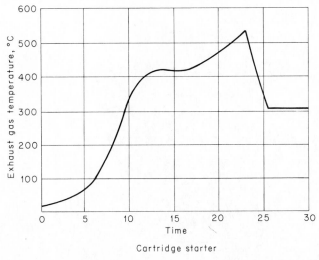

Cartridge starter

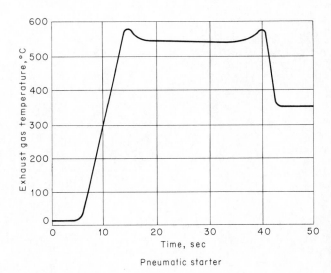

Pneumatic starter

FIG. 17-1 Exhaust-gas temperature vs. starting time for a cartridge and pneumatic starter.

8. Liquid monopropellant starter
9. Air-impingement starter
10. Hand-crank starter

NOTE: Of these, the starter-generator and the air turbine starter are the ones used most often in small and large engines respectively.

ELECTRIC MOTOR STARTER *(FIG. 17-2)*

These starters are 28-V, series-wound electric motors, designed to provide high starting torque. Their use is limited to starting smaller engines because of the very large current drain (over 1000 A for some models) and because they are relatively heavy for the amount of torque they produce. The starter includes an automatic jaw meshing mechanism, a set of reduction gears, and a clutch. The straight electric motor starter as a means for starting gas turbine engines has generally given way to the starter-generator in order to save weight and simplify accessory gear arrangements. A typical electric motor starting system is illustrated in Fig. 17-3. It may have provisions for automatically engaging the ignition units when the starter switch is thrown. If the circuit is so arranged, the ignition system is constructed so that it may be separately energized for air restarts. The starter system may also be equipped with a relay to

"drop" the starter out when a specified rpm has been reached or the starter load reduced. Some systems incorporate a timing switch to permit a gradual voltage buildup as the starter gains speed.

ELECTRIC MOTOR-GENERATOR (STARTER-GENERATOR) *(FIG. 17-4)*

Most small gas turbine engines such as the General Electric CJ610, Pratt & Whitney Aircraft JT12 and PT6, Detroit Diesel Allison T63, Teledyne CAE J69, and the Avco Lycoming T53 utilize a starter-generator. This system has the advantage of being lighter than a separate starter and generator since a common armature is used and it requires no engaging or reduction gear mechanism. The engine accessory section also requires one less gear.

As shown in Fig. 17-5, a splined drive, which usually incorporates a torsional vibration dampener to protect the drive quill against engine torsional vibration, connects the starter-generator to the engine. The unit incorporates two field windings. The series winding is used to develop the low-speed, high starting torque necessary to crank the engine, while the shunt or parallel winding functions when the unit is acting as a generator. There are four external connections (*A*+ or shunt field connection; *B*+ or armature positive; *C* or series

FIG. 17-2 Typical electric motor starter.

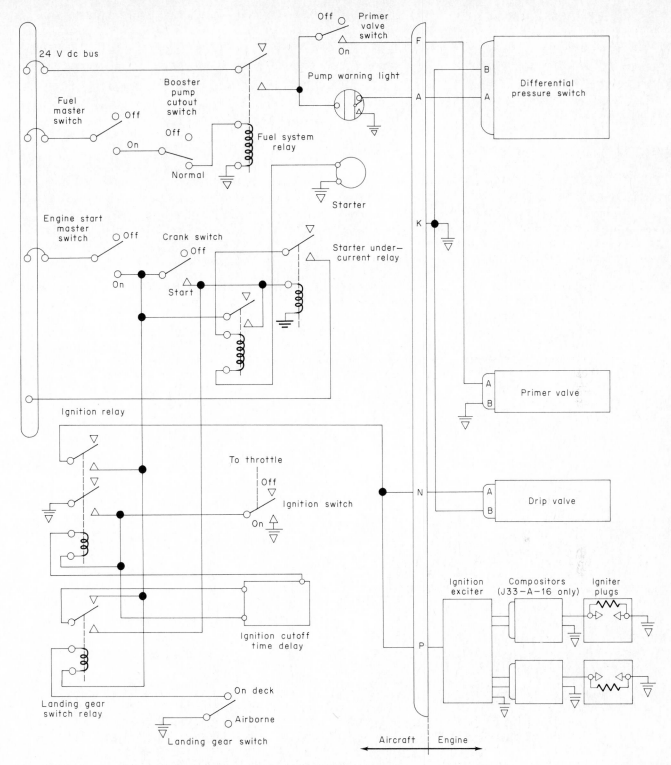

FIG. 17-3 An early electric motor starting and electrical system.

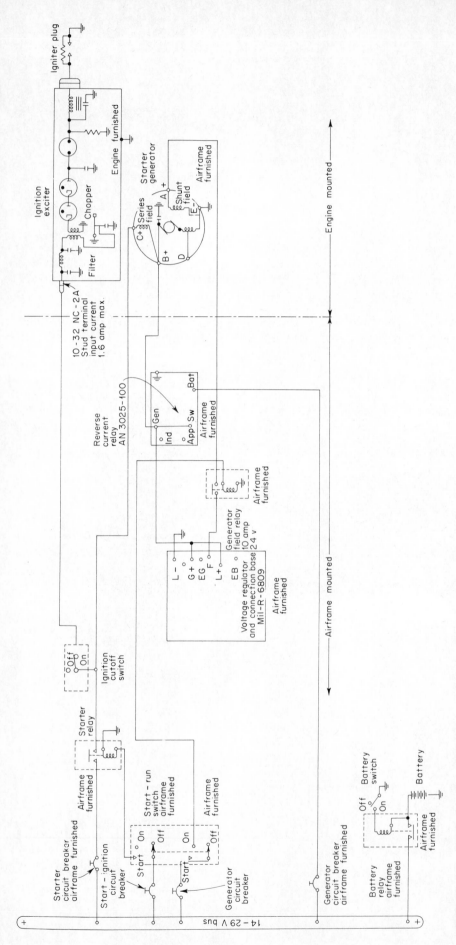

FIG. 17-4 A simple starter-generator system used on the Detroit Diesel Allison T63.

FIG. 17-5 A typical starter-generator.

field connection; and $E-$ or armature negative). The starting electrical load is very high. To limit the tremendous battery drain, some airplane electrical systems are arranged so that their two 28-V batteries can be placed in series for starting and in parallel operation for normal generator functions. This has the effect of providing the same amount of power (volts × amperes) to the starter with a reduced current flow.

AIR TURBINE STARTER
(FIGS. 17-6 and 17-7)

Models of this starter are installed in the Boeing 707, 720, 747, KC135, and B52, Douglas DC-8, DC-9, and DC-10, Convair 880 and 990, Lockheed Electra, General Dynamics F-111, and others. Its primary advantage is its light weight (about 20 to 25 lb) [9 to 11 kg] to torque ratio when compared with the electric motor starter and starter-generator. The principal disadvantage is that it requires a supply of high-volume airflow of approximately 40 lb/min [18 kg/min] at a pressure of about 50 psi [345 kPa]. Sources include compressed air from an auxiliary gas turbine engine carried on board the aircraft or maintained as a part of the airport facilities, compressed air bled from the other running en-

(a)

(b)

(c)

FIG. 17-6 Three Garrett AiResearch air turbine starters.
(a) This pneumatic starter is installed on the Boeing 727. Note the keyhole-type QAD flange.
(b) High (500 psig) [3448 kPa] or low (35 psig) [241 kPa] pressure may be applied to this ATS100-129 pneumatic starter used on many JT3D engines.
(c) The ATS-50 pneumatic starter for small gas turbine engines.

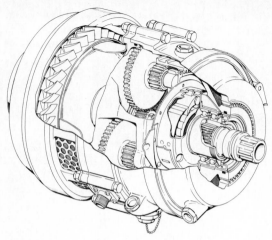

FIG. 17-7 The Hamilton Standard Division, United Technologies Corp., Model PS700-1 starter designed for use on the Pratt & Whitney Aircraft JT9D installed in the Boeing 747.

gine(s), or compressed air from an air storage system, as shown in Fig. 17-8. Very often one engine of a multiengine military airplane will be equipped with a cartridge, fuel-air combustion, or gas turbine starter, having self-start capabilities. Air bled from the running engine can then be supplied to the air turbine starters installed on the other engines, as shown in Fig. 17-8(e).

This starter and other types may be supplied with a quick-attach-detach (QAD) coupling, V band, or keyhold-type pad that attaches to a mounting flange which, in turn, is designed for direct attachment to a standard engine accessory drive.

The air turbine starter converts energy from compressed air to shaft power. To start the system, an air valve is opened by the "start" switch, after which the operation of the valve and starter is automatic. The same switch is used as a "stop" switch in emergencies. As air enters the starter inlet, the radial- or axial-flow turbine wheel assembly rotates. The reduction gears contained within the starter convert the high speed and low torque of the turbine wheel to a relatively low-speed and high-output torque. The reduction gears are lubricated with MIL-L-7808 oil, the same type used in the engine, by a splash-type oil system. Engagement of the starter is usually accomplished through a jaw or pawl and ratchet clutch. When a predetermined speed is reached, an internal governor assembly, through a switch, deactivates an electric circuit which closes the valve in the air supply line. The starter rotating assembly automatically disengages from the permanently engaged splined output shaft when the starter drive speed exceeds the starter output shaft speed.

For self-contained starting, independent of ground support equipment, some systems use air supplied from storage bottles installed in the airplane. A line combustor may heat the air to 700°F [371°C]. To minimize air consumption, water is injected during the starting cycle. (Refer to Fig. 17-8.)

An interesting variation of this form of starter, produced by the Garrett AiResearch Manufacturing Com-

pany, is the constant speed drive starter. The unit, installed in the British Aircraft Corporation (B.A.C.) One-Eleven transport, combines in one unit an air turbine starter plus constant-speed shaft power to drive the alternating current (ac) generator.

CARTRIDGE OR SOLID-PROPELLANT STARTER *(FIG. 17-9)*

Originally, starters of this type were constructed to operate solely by means of high-pressure, high-temperature gas generated by the burning of a solid propellant charge. Changes in the cartridge-type starter have added the additional capability of starting with compressed air from an external source.

A charge, about the size of a two pound coffee can, is inserted in the breech and ignited electrically. The relatively slow-burning propellant produces gases at approximately 2000°F [927°C] and 1200 psi [8274 kPa] to turn the starter for about 15 s.

In recent years the pneumatic-cartridge starter has achieved considerable use in the U.S. Air Force, primarily because of its inherent characteristics of a lightweight, self-contained system with the extremely high torque value of over 600 ft·lb [814 N·m] plus the option of quick starts from the high-pressure, high-temperature cartridge gases, or from low pressure supplied from a running engine, conventional ground support equipment, or airborne starting units (Fig. 17-10).

A more detailed examination of one make of pneumatic-cartridge jet engine starter used in the McDonnell-Douglas F4C follows. (Refer to Fig. 17-11.)

For a cartridge start, a standard Air Force type MXU-4 cartridge (Fig. 17-11) is first placed in the breech cap (2). Next, the breech cap is closed down on the breech chamber by means of the breech handle (3) and rotated a part-turn to engage the lugs between the two breech sections. This rotation allows the lower section of the breech handle to drop into a socket and completes the cartridge ignition circuit. (Up to this point, it would have been impossible to fire the cartridge.) As shown in Fig. 17-11, the cartridge is then ignited by applying voltage to the connector (4) at the base of the handle. This energizes the insulated ignition contact (5) at the top of the breech cap, which touches a point on the cartridge itself. The circuit is completed to ground by the ground clip (6) (a part of the cartridge) which contacts the inner wall of the breech cap. Upon ignition, the cartridge begins to generate gas. The gas is forced out of the breech to the hot gas nozzles (7), which are directed toward the buckets on the turbine rotor (8), and rotation is produced. Gas emerging from the opposite side of the wheel enters an exhaust ring (9), in the exhaust duct, is collected, and passes out of the starter via the overboard exhaust connector (10). However, before it reaches the nozzle, the hot gas passes an outlet leading to the relief valve (12). This valve ports hot gas directly to the turbine, bypassing the hot gas nozzle, as the pressure rises above the preset maximum. Therefore, the pressure of the gas within the hot gas circuit is maintained at the optimum level.

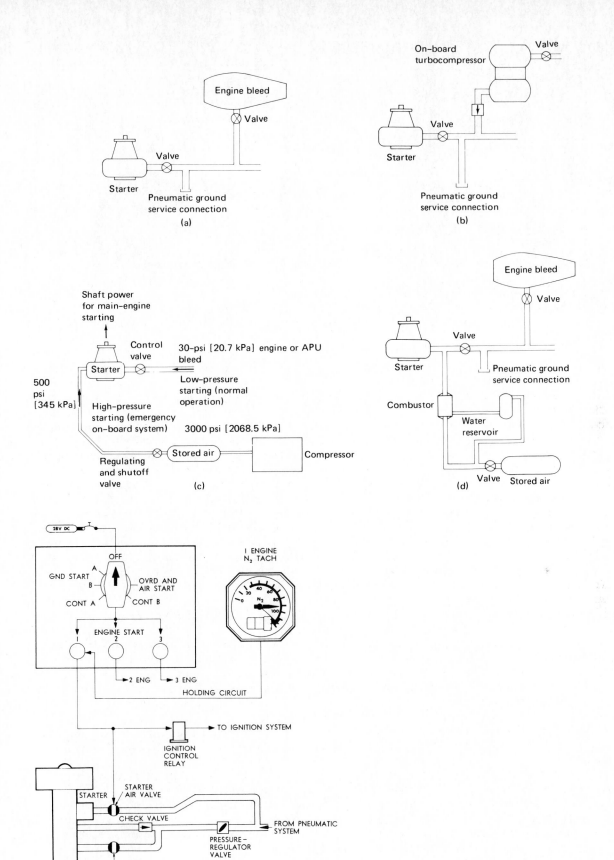

FIG. 17-8 (a), (b), (c), and **(d)** Simplified schematics showing several air sources.
(e) The General Electric CF6 engine starting system uses an air turbine starter.
Note that air from this engine can be used to start the other engines.

FIG. 17-9 A Garrett AiResearch Air/Cartridge starter.

FIG. 17-10 Cutaway view of the Sundstrand cartridge/pneumatic starter.

For a pneumatic start, compressed air from a ground cart is led by ducting on the aircraft to the compressed air inlet (13). It passes into the compressed air nozzle ring and is directed against the buckets of the turbine rotor by vanes placed around the ring. Rotation is thus produced in essentially the same manner as in the cartridge start. Compressed air leaving the turbine rotor collects in the same exhaust ring and is ported overboard via the overboard exhaust connector.

Whether starting is accomplished by cartridge or compressed air, some opposing force is required to keep turbine speed within safe limits. This opposing force is provided by the aerodynamic braking fan (14). The fan is connected directly to the turbine shaft. It is supplied with air from the aircraft nacelle and its output is carried off by an exhaust ring (16) concentric with, and located within, the turbine exhaust ring. Hot gas (or compressed air) exhaust and aerodynamic braking fan output are kept separate up to the overboard exhaust connector. At this point, they merge, the cool air from the fan cooling the hot exhaust gas.

The gearshaft (17) is part of a two-stage reduction which reduces the maximum turbine speed of 67,500 rpm to an output of approximately 4000 rpm.

The large gear of the final output turns the output spline shaft (24) through an overrunning clutch (18).

The clutch is situated in the output area between the gear shaft, on which the final drive gear is located, and the output spline shaft. It is a pawl or sprag type, one-way overrunning clutch, and its purpose is to prevent the engine, once operating under its own power, from driving the starter, thereby possibly driving the turbine rotor at a speed above its safe limit. The nature of a pawl or sprag clutch is such that it can transmit torque in only one direction. That is, the driving member will operate through the clutch, delivering full torque to the driven member. But the driven member cannot become the driver—even though revolving in the same direction—and transmit torque back into the original driver. Any tendency for it to do so would disengage the clutch. When the engine has been started and the starter has finished its cycle and stopped, only the output spline shaft and the outer (driven) part of the clutch will be revolving. The balance of the starter will be at rest.

The starter is equipped with an output spline shaft having a shear section that permits the shaft to shear if torque to the engine during the starting cycle is excessive. When the shaft shears, torque to the engine is stopped, thus preventing damage to the aircraft engine gearbox. The output spline shaft will also shear during the overrunning cycle (engine started and operating) if the starter malfunctions in such a manner as to develop a frictional resistance to torque from the aircraft engine gearbox.

In the event the clutch and spline shaft fail to operate and the turbine is driven beyond burst rpm by the aircraft engine, the containment clamp shown in Fig. 17-10 provides additional strength to the starter turbine area preventing damage to the aircraft.

A vent (23) through the clutch and output shaft eliminates internal pressure buildup. Centrifugal force caused by output rotation prevents oil leakage through the vent.

The starter is lubricated by a splash system. Oil slingers (25) attached to the clutch output race pick up oil from the sump (26) and distribute it throughout the interior of the starter as the output spline revolves. A catching cup construction in the housing carries the oil into the overrunning clutch and other difficult-to-reach areas. Since the part to which the slingers are attached is constantly spinning, even after the starter has completed its cycle, starter lubrication continues as long as the aircraft engine is operating. The oil sump contains a magnetic plug (27) to collect contaminants.

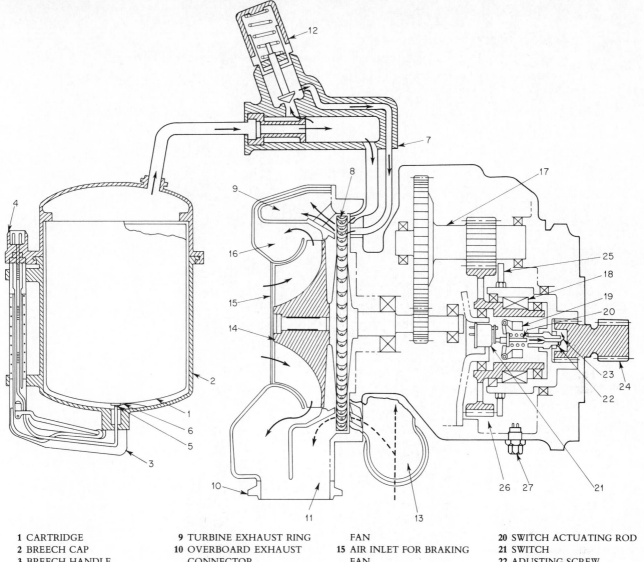

1 CARTRIDGE	9 TURBINE EXHAUST RING	FAN	20 SWITCH ACTUATING ROD
2 BREECH CAP	10 OVERBOARD EXHAUST	15 AIR INLET FOR BRAKING	21 SWITCH
3 BREECH HANDLE	CONNECTOR	FAN	22 ADUSTING SCREW
4 CONNECTOR	11 EXHAUST FROM TURBINE	16 FAN EXHAUST RING	23 GEARBOX VENT
5 IGNITION CONTACT	AND FAN	17 GEARSHAFT	24 SPLINE SHAFT
6 GROUND CLIP	12 RELIEF VALVE	18 OVERRUNNING SPRAG	25 OIL SLINGER
7 HOT-GAS NOZZLES	13 COMPRESSED AIR INLET	CLUTCH	26 OIL SUMP
8 TURBINE ROTOR	14 AERODYNAMIC BRAKING	19 FLYWEIGHT	27 MAGNETIC PLUG

FIG. 17-11 Sundstrand cartridge/pneumatic starter schematic.

FUEL-AIR COMBUSTION STARTER (FIG. 17-12)

The fuel-air combustion starter is essentially a small gas turbine engine, minus its compressor. It is completely self-contained, as is the cartridge starter system, but unlike the preceding system, requires no additional components to function. All fuel, air, and electric power needed for operation are carried on board the aircraft.

In addition to the turbine, the system consists of an air storage bottle, fuel storage bottle, and a combustion chamber, together with the necessary ignition and control components. During flight, an engine-driven compressor maintains 3000 lb [20,685 kPa] of air pressure in an airborne bottle. This permits engine starts without the necessity of recharging the air system from an external source. The usual high-pressure bottle will provide enough air for two restarts without recharging. Provision is also made to connect an external 600-psi [4137-kPa] air supply. In either case, the starter receives a reduced air pressure of 350 psi [2413 kPa].

In a typical system shown in Fig. 17-12, the starter is activated by a ground start switch in the cockpit. When the ground start switch is pressed, the starter air solenoid valve opens, admitting air from the storage bottle or from the external 600-psi connection into the

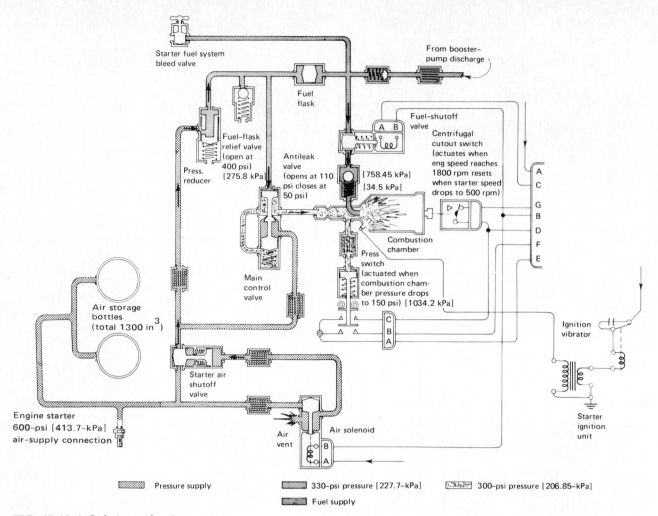

FIG. 17-12 A fuel-air combustion starter system.

combustion chamber. At the same time, the fuel valve opens to admit fuel from the accumulator, and the starter ignition system is momentarily energized, igniting the fuel-air mixture. This action causes a rapid expansion of air, which spins the starter turbine, which in turn accelerates the engine through the reduction gearing and clutch. When engine speed reaches about 21 percent rpm, starter fuel is exhausted, resulting in a dropoff of burner pressure. The pressure switch actuates, opening the air duct to the fuel and air valves. If starter speed exceeds 22.6 percent engine rpm before fuel is exhausted, the centrifugal switch will open, shutting off the fuel and air valves. Exhaust from the starter combustion chamber is directed through an exhaust duct at the lower side of the starter and into the engine air-guide section. A pressure reducer and an air-control valve in the starter reduce the supply air pressure to about 330 psi [2275 kPa] before it enters the fuel accumulator and the combustion chamber. The fuel accumulator contains enough fuel obtained from the airplane fuel system through a takeoff line at the engine fuel flow divider to operate the starter for about 4 s. The accumulator is pressurized with about 330 psi air pressure to ensure fuel flow to the combustion chamber. The starter has a safety clutch, which automatically disen-

gages the starter drive shaft from the engine drive spline to prevent the engine from driving the starter turbine to destructive overspeed. In case of failure of the starter clutch, the safety clutch must be manually reset which necessitates removing the starter from the engine. An air motoring switch allows the air solenoid valve to open while bypassing the fuel solenoid valve and starter ignition circuits for the purpose of motoring the engine to 4 to 6 percent.

THE GAS TURBINE STARTER (FIG. 17-13)

This is another of the completely self-sufficient starting systems. Relatively high power output is available for a comparatively low weight. The starter is actually a small free turbine engine complete with a gas generator section containing a centrifugal compressor, combustion chamber, and turbine to drive the compressor. It also contains its own fuel control, starter, lubrication pump and system, and ignition system. The gases flowing through the gas generator section drive the free turbine, which in turn, drives the main engine through a reduction gear and clutch mechanism to automatically

(a)

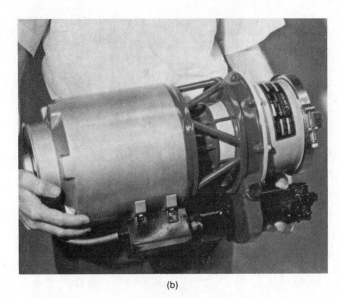

(b)

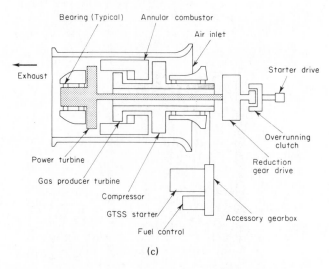

(c)

FIG. 17-13 The Garrett AiResearch and Solar gas turbine starters.

(a) Garrett AiResearch jet fuel or gas turbine starter JFS100.

(b) Solar gas turbine self-contained starter (GTSS).

(c) Schematic view of the Solar GTSS.

engage and disengage the starter's free turbine from the engine. The starter is itself started by using a small electric motor, compressed air, or hydraulic power from the aircraft system. Typical specifications are:

Type: Free turbine engine
Weight: 70 to 80 lb [31.8 to 36.3 kg]
Shaft speed: 0 to 8000 rpm
Performance: 20 to 30 s engine starting time
Fuel: Same as used in aircraft
Oil: Same as used in aircraft
Mounting: QAD

Inherent inefficiencies of transferring starting energy to the main engines through pneumatic, hydraulic, or electrical means are eliminated. The pilot has complete control of engine starting from the cockpit and the gradual application of starting torque extends the life of the main engine components. A further advantage of this system is that it can "cold crank" or "motor" the main engines for 10 min at a time to permit checking fuel, hydraulic, and electrical systems.

HYDRAULIC STARTERS

Hydraulic starting systems fall into two categories:

1. Energy limited
2. Power limited

The energy-limited system (Fig. 17-14) uses a highly pressurized accumulator and a large-positive-displacement motor. Examples of other starting systems that are also energy-limited are the electric motor, when supplied from a battery, and the cartridge starter. The energy-limited system is designed to complete the start in as short a time as possible in order to minimize the amount of stored energy required. The accumulator system is best suited to small engines up to 150 hp [112 kW].

A power-limited system (Fig. 17-15) uses an auxiliary power unit (sometimes a small gas turbine engine, which is itself started by an energy-limited system) to drive a pump which supplies the correct amount of flow and pressure to a variable-displacement hydraulic starter motor. The variable-displacement motor permits high torque to be applied without exceeding the power limits of the main engine at starter cutoff speed.

It is possible to adapt the hydraulic starter as a pump, but since the starter cutout speed is less than 50 percent of the normal engine operating speed, a two-ratio gearbox is necessary to provide proper speed for both pumping and starting. The hydraulic pump on the auxiliary power unit can also be used to supply power to the aircraft.

Figure 17-16 shows two typical hydraulic starter installations in current use.

The APU in the Sikorsky CH53A [Fig. 17-16(a)] is started hydraulically by means of stored energy in a 250 in³ [4.22 L] accumulator (4000 psi [27,580 kPa] maximum). The main engines receive their starting power

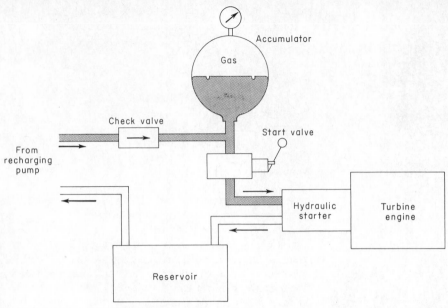

FIG. 17-14 The energy-limited or accumulator starting system.

from a pump mounted on the main accessory gearbox which is shaft driven by the Solar T-62T-12APU. [See Fig. 17-20(a).] The accessory pump also provides power for the winch and other utility functions. The T-64 main engine starters deliver 50 ft·lb [67.8 N·m] of torque at 3500 psi [24,132 kPa] and 21.5 gal/min [81.4 L/min]. Maximum starter speed is 7300 rpm, and cutout is accomplished by a mechanically actuated switch that senses motor displacement.

The starting system used in the Vertol CH47A [Fig. 17-16(b)] is somewhat similar to the one described above in that the Solar APU is started by means of stored energy in a 200-in³ [3.28-L], 3000-psi [20,685

kPa] hydraulic accumulator, and the main engines receive their starting power from a pump mounted on the main accessory gearbox. It is different in that the gearbox is driven by hydraulic power from the APU starter operating as a pump, driving a fixed-displacement motor mounted on the gearbox. During main engine operation an overrunning clutch isolates the accessory gearbox drive motor until the APU is again started for main engine starting or a system checkout. The accessory-gearbox-mounted dual-pressure pump supplies 9 gal/min [34.1 L/min] at 4000 psi [27,580 kPa] for starting the Avco Lycoming T-55 main engines on the ground, and 3000 psi [20,685 kPa] for air restarts. The variable

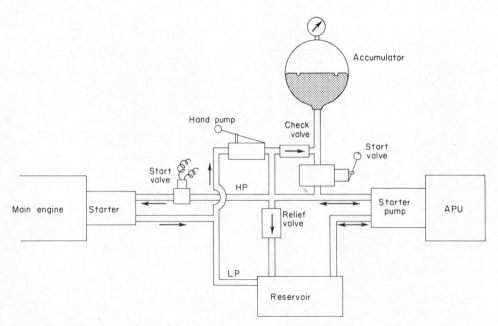

FIG. 17-15 The power-limited starting system.

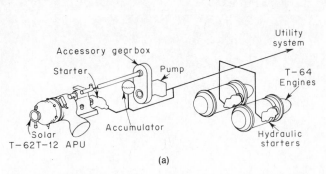

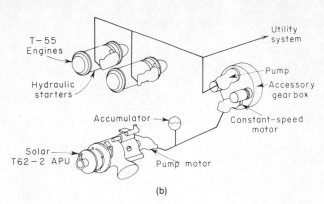

FIG. 17-16 (a) Vickers Corporation hydraulic starting system for the Sikorsky Division, United Technologies Corporation, CH53A helicopter.
(b) Vickers Corporation hydraulic system for the Boeing Vertol CH47A (Chinook) helicopter.

displacement maximum starter speed is 2800 rpm, and cutoff is accomplished by a tachometer signal from the engine.

LIQUID MONOPROPELLANT STARTER

In this system a charge of liquid monopropellant (a monopropellant fuel is one which requires no separate air supply to sustain combustion) is decomposed to produce the high-energy gas needed for turbine operation. Monopropellants which can be used include highly concentrated hydrogen peroxide, isopropyl nitrate, and hydrazine. All are difficult materials to handle, and principally because of this there has been no operational installation of such equipment in this country.

THE AIR-IMPINGEMENT STARTER *(FIG. 17-17)*

In many ways this starter system is the simplest of all starter types, consisting essentially of nothing more than a duct. An air supply from either a running engine or a ground power unit is directed through a check valve onto the turbine blades (most common) or the centrifugal compressor. Engines using this starting system are the Fairchild J44, on which the air is fed to the compressor, and some models of the General Electric J85 and J79. (See Chap. 2.) In the latter two engines, air is directed onto the rear or middle turbine wheel stages. (See Fig. 21-9.) Obviously, the advantage of this system is manifested in its extreme simplicity and light weight. It is best suited to smaller engines because of the high-volume air supply necessary for larger engines.

THE HAND-CRANK STARTER *(FIG. 17-18)*

This method of starting gas turbine engines is, of course, limited to very small units, on the order of 50 to 100 hp [37 to 75 kW]. As the name implies, starting is accomplished by turning a hand crank which, through a series of gears, turns the engine to the self-sustaining rpm. Hand-crank to engine-shaft speed ratios are in the order of 100:1.

GROUND OR AUXILIARY POWER UNITS *(FIG. 17-19)*

The ground or auxiliary power unit, while not an integral part of the primary aircraft engine, is nevertheless an important adjunct to it.

These units are small, lightweight, trouble-free gas turbine engines, completely automatic in their operation (Fig. 17-20). They are generally constructed with centrifugal compressors and axial- or radial-inflow turbines. Starting the GPU or APU is accomplished by means of a small electric or hydraulic motor, or by a hand crank. Typical of these units is the AiResearch gas turbine compressor GTC85, an electrically started, self-sufficient unit with a two-stage radial compressor and a turbine driven by the exhaust products of a single, tangentially located burner. Air is bled off from the compressor section and supplied to the main engine starter from this unit at a pressure ratio of approximately 3:1 and a temperature of 350°F [176.7°C]. The unit is approximately 38 in [965 mm] long, 18 in [457 mm] in diameter, and weighs 275 lb [125 kg]. Although these engines can operate on a wide variety of fuels, the units generally use the same fuel as the main engines.

Auxiliary power units have been used to drive ac and dc generators, hydraulic pumps and motors, other fluid pumps, and air compressors. The airborne units on conventional aircraft are usually located toward the rear, but, as shown in Fig. 17-21, can be installed in any location.

An unusual type of ground starting cart is used to start the Pratt & Whitney J58 engine installed in the Lockheed SR71. The cart is powered by two V-8 automobile engines driving a shaft which is mechanically coupled to the compressor section.

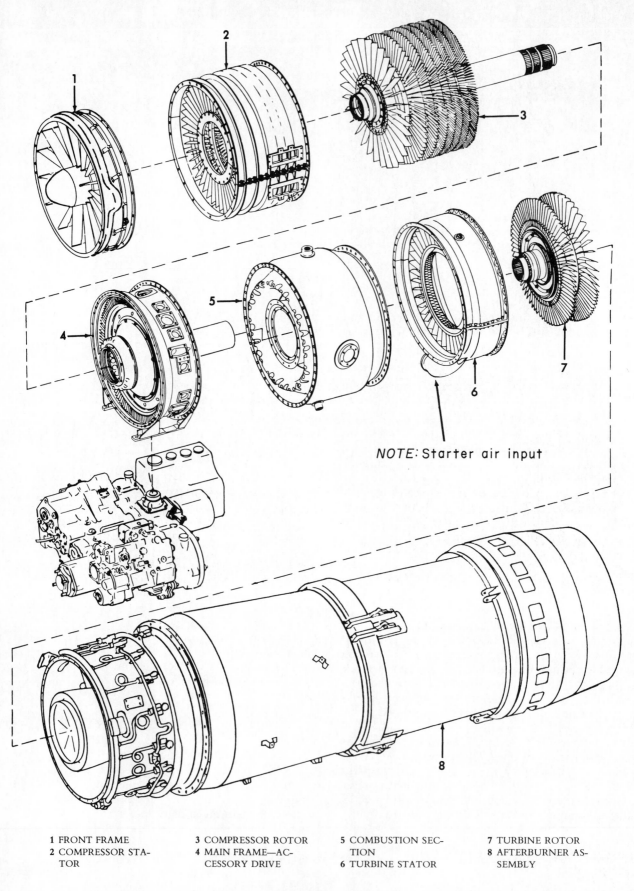

NOTE: Starter air input

1 FRONT FRAME
2 COMPRESSOR STA-
  TOR
3 COMPRESSOR ROTOR
4 MAIN FRAME—AC-
  CESSORY DRIVE
5 COMBUSTION SEC-
  TION
6 TURBINE STATOR
7 TURBINE ROTOR
8 AFTERBURNER AS-
  SEMBLY

FIG. 17-17 The General Electric J85 equipped with an air-impingement starting system.

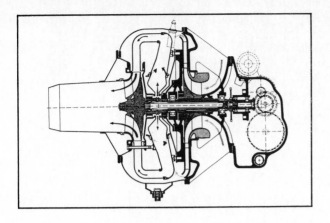 

(a)

(b)   (c)

FIG. 17-18 Some examples of engines equipped with hand-crank gas turbine starters.
(a) German engine.
(b) Japanese engine.
(c) British engine.

FIG. 17-19 These engines are installed in the widely used MA-1A U.S.A.F. starting cart.

(a) Garrett AiResearch GTC85-70-1.

(b) The MA-1A starting cart.

(c) The Teledyne CAE 141 external and cut-a-way view.

322 SYSTEMS AND ACCESSORIES

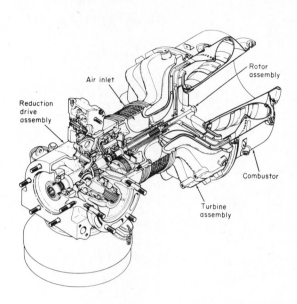

(a)

(b)

(c)

FIG. 17-20 Some auxiliary gas turbine units.

(a) The Solar T-62 Titan auxiliary gas turbine engine schematic and external view, used in several Sikorsky helicopters. Weight is 70 lb [31.8 kg.].

(b) Garrett AiResearch GTCP 85-98D used in the McDonnell Douglas DC-9.

(c) Garrett AiResearch GTCP 85-98 used in the Boeing 727.

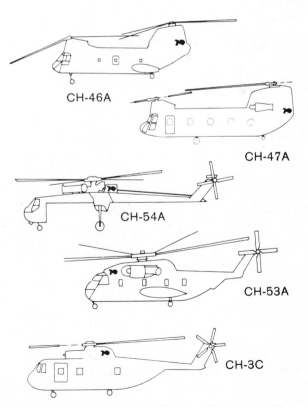

FIG. 17-21 Various locations for the Solar T-62 auxiliary gas turbine engine.

TABLE 17-1
Summary of the advantages and disadvantages of the starter types

Type	Advantages	Disadvantages
Electric motor starter	1 Self-contained starts possible for small engines 2 Engine may be motored for short periods without starter overheating	1 Limited to starting small engines 2 Relatively heavy for torque produced 3 Reduction gears necessary 4 Engaging mechanism necessary
Starter-generator	1 One less accessory drive necessary 2 No overrunning clutch, gearbox, or engaging mechanism necessary 3 Lighter than a single starter and generator 4 Self-contained starts possible for small engines	1 Limited to small engines 2 Relatively heavy for torque produced when operating as a starter
Air turbine starter	1 High torque-to-weight ratio (5 to 10 times higher than electric motor) 2 Engine may be motored at low or high speed 3 Can use air from a running engine	1 High-volume air supply required 2 Gearbox needed with self-contained oil supply 3 Electrical connections needed for speed control
Cartridge starter	1 Self-contained starts possible for large engines 2 Very high torque-to-weight ratio 3 Quick starts possible for military 4 Automatic starts possible	1 Cartridge needed for each start 2 Gearbox, clutch, and oil system necessary 3 No motoring possible for systems checkout
Fuel-air combustion starter	1 Completely self-contained 2 High torque-to-weight ratio 3 Automatic starts possible 4 Engine may be motored for short periods on internal air supply at low rpm	1 Relatively complex 2 Only two self-contained starts possible
Gas turbine starter	1 Completely self-contained starts possible 2 High torque-to-weight ratio 3 Long periods of engine motoring possible	1 One of the most complex of starter types in that it requires all of the systems of the main engine plus an overrunning clutch
Hydraulic starter	1 Compact in size 2 Can be self-contained for smaller engines 3 Can be adapted to function as a pump 4 Relatively uncomplicated	1 Requires external power for large engines or for continuous cranking (internal APU may be used)
Monopropellant starter	1 High starting-torque-to-weight ratio	1 Dangerous fuels 2 Complex system required
Air-impingement starter	1 Simplest of all types 2 Can be used to motor engine, but only with continuous air supply 3 Extremely light 4 Can use air from another running main engine	1 Requires a high-volume air supply (3 to 5 times the pneumatic energy requirements of the air turbine starter)
Hand-crank starter	1 Very reliable 2 Independent of external power systems, except muscle power 3 Lightweight	1 Limited to very small engines 2 Cranking handle must be stored

REVIEW AND STUDY QUESTIONS

1. Compare the starting requirements for the reciprocating and gas turbine engines.
2. What are some of the factors which influence the choice of the starting system?
3. List 10 types of starters. Very briefly describe each starter type.
4. Make a table listing the advantages and disadvantages of each starter type.
5. What form does the auxiliary, or ground power unit take? How can this engine be utilized?

PART FOUR
MAINTENANCE
AND
TESTING

CHAPTER EIGHTEEN
MAINTENANCE
AND
OVERHAUL
PROCEDURES

The length of time between overhauls (TBO) has increased from 10 h for the German Jumo 109-004B manufactured in 1945, to over 6000 h for the Pratt & Whitney Aircraft JT3D engine currently being produced. It should be kept in mind that between these major overhaul periods most engines are required to go through an intermediate "hot-section" inspection. This large improvement in TBO has been accomplished in the main through significant improvements in engine design, metallurgy, manufacturing, overhaul, inspection, and maintenance procedures.

Previous chapters in this book have dealt mainly with the design, metallurgical, and manufacturing aspects of the engine. The other three factors, overhaul, inspection, and maintenance procedures, will be discussed in this chapter.

OVERHAUL

The TBO varies considerably between engine types. It is generally established for civil aircraft by the equipment operator and the engine manufacturer, working in conjunction with the Federal Aviation Agency (FAA). With the exception of the FAA, the overhaul times for military aircraft are established in essentially the same manner. Taken into account are such factors as the type of operation and utilization, the servicing facilities and experience of maintenance personnel, and the total experience gained with the particular engine. As a specific model engine builds up operating time, and is sent to the overhaul agency, the parts are inspected for wear and/or signs of impending failure. If the critical parts seem to be wearing well, an extension of TBO may be approved. One of the most important factors in determining time between overhauls is the use to which the engine is put. Frequent starts and stops, or power changes (cycle changes), necessary on short-haul aircraft, result in rapid temperature changes which, in turn, will affect the TBO. Most manufacturers have adopted a system of permanently marking critical parts of the engine such as turbine disks and blades, which are subject to deterioration through cycle changes or time limits. A part must be removed from service when either the number of cycles or the time reaches the maximum limit.

Modern gas turbine engines are expensive, with some

versions costing over several million dollars. It is essential that the operators in the overhaul shops keep complete and accurate records to guarantee that a component be removed or modified when required, and, on the other hand, that parts are not discarded prematurely. In order to do this, most engine parts must be identifiable. The marking methods take several forms, determined by the desired permanency, the type of material being marked, and the location of the part. Temporary marking methods include:

1. Several brands of marking pencils. It is extremely important to *not* use any material that would leave a deposit of lead, copper, zinc, or similar material on any hot section part, as this might cause premature failure due to carburization or intergranular attack. (This includes grease and lead pencils.)
2. Chalk.
3. Several brands of ink.
4. Soapstone.

Permanent marks may be accomplished by:

1. Electrolytic etch applied through a stencil or with a special electrolytic pen. This is not the same as electric arc scribing which has been found unsuitable for the gas turbine engine. Electrolytic etch should not be used on anodized surfaces.
2. Metal stamping using a hammer, press, or roll. This method is limited to parts having less than a specific hardness.
3. Vibration peening produces characters by a vibrating radius-tipped tool.
4. Engraving with a rotating cutter or grinder.
5. Drag impression using a freely rotating radius-tipped conical tool.
6. Blasting with an abrasive substance through a stencil.
7. Branding used on nonmetallic parts such as plastic, bakelite, etc.

In all cases the manufacturer's recommendations must be followed.

The actual overhaul of the engine can be divided into:

1. Disassembly
2. Cleaning
3. Inspection

4. Repair
5. Reassembly
6. Testing
7. Storage

Disassembly

Disassembly can be accomplished on a vertical or horizontal disassembly stand (Fig. 18-1). Some engines can be disassembled by using either method, while others lend themselves to a particular procedure. After the engine is broken down into its major components, many of the subassemblies are then mounted on individual stands (Fig. 18-2) for further work. A large number of specialized tools are necessary to ensure dismantling without damage to the closely machined, highly stressed parts. A set of these tools often may cost as much as the engine.

Every manufacturer issues a complete and detailed overhaul manual which gives a step-by-step disassembly procedure and also shows where and how to use the special tools. Appropriate warnings and cautions where necessary to minimize possible injury to the worker and damage to the engine are included. Special instructions are given for the many parts such as the bearings and carbon seals which require special handling. Other parts must be reassembled in their original position, so they must be tagged and marked accordingly. Seals, other than the carbon rubbing types, are not reused. Metal-type seals will have been crushed, and many rubber-type seals are made to expand in contact with fuel or oil. Once this type of rubber seal has been removed, it will not fit back into its original groove.

Reproduced in Fig. 18-3 are two pages removed from the overhaul manual for an early General Electric J85-

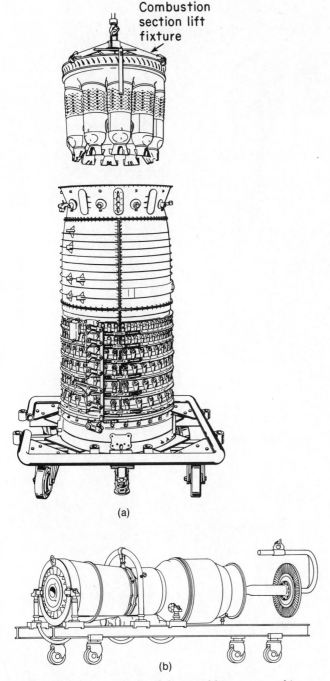

(a)

(b)

FIG. 18-1 Two methods of disassembling a gas turbine engine.
(a) Vertical disassembly of a General Electric J79.
(b) Horizontal disassembly of a Pratt & Whitney Aircraft JT3.

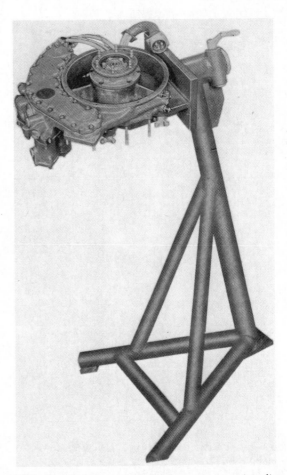

FIG. 18-2 Teledyne CAE J69 accessories section disassembly stand.

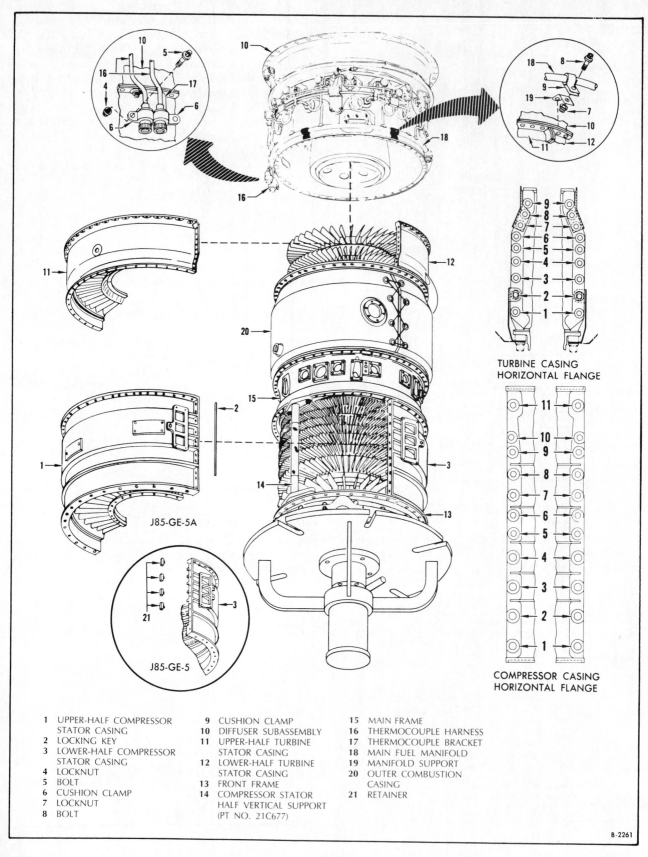

1	UPPER-HALF COMPRESSOR STATOR CASING	9	CUSHION CLAMP
2	LOCKING KEY	10	DIFFUSER SUBASSEMBLY
3	LOWER-HALF COMPRESSOR STATOR CASING	11	UPPER-HALF TURBINE STATOR CASING
4	LOCKNUT	12	LOWER-HALF TURBINE STATOR CASING
5	BOLT	13	FRONT FRAME
6	CUSHION CLAMP	14	COMPRESSOR STATOR HALF VERTICAL SUPPORT (PT NO. 21C677)
7	LOCKNUT		
8	BOLT		

15	MAIN FRAME
16	THERMOCOUPLE HARNESS
17	THERMOCOUPLE BRACKET
18	MAIN FUEL MANIFOLD
19	MANIFOLD SUPPORT
20	OUTER COMBUSTION CASING
21	RETAINER

B-2261

FIG. 18–3 Typical disassembly instructions taken from a General Electric J85 over-haul manual.

c. Seventeen bolts (16) and 16 locknuts (17) secure the forward flange of the upper-half compressor casing to the front frame aft flange.

Note

Five of the bolts were removed when the inlet-guide vane actuator ring was removed. Two locknuts were removed with the spark generator forward bracket. Two locknuts were removed with the anti-icing valve. One bolt hole under the right-hand actuator is not used. Remove the remaining bolts, locknuts, the junction box forward bracket (20) and the T₅ amplifier forward bracket (21).

d. Install the stator casing lifting device and remove the upper-half compressor casing (28). Remove the 4 retainers (22) for the compressor vane segments. Position the compressor stator-half vertical support at the 12 o'clock position and secure the support to the flange of both the mainframe and the front frame with 2 bolts and locknuts.

4–131. REMOVAL OF UPPER-HALF COMPRESSOR STATOR CASING, J85-GE-5. (See figure 4–48A.)

Note

For special tools, see figure 3–1, group 10.

a. Remove 6 locknuts and 6 washers from the body-bound bolts located in bolt holes 1, 5, and 11 in the horizontal flanges of the compressor stator casing.

b. Remove the 6 body-bound bolts and 4 washers by using a plastic drift pin to drive the bolts out of the holes.

```
CAUTION
```

Do not turn the bolt in the hole during removal; this would enlarge the hole and impair the alignment function.

c. Remove the remaining 16 bolts and 16 locknuts from the horizontal flanges of the compressor stator casing.

d. Remove the 2 supports and washers for the synchronizing cable conduit.

e. Seventeen bolts and 16 locknuts secure the forward flange of the upper-half compressor casing (1) to the front frame (13). Nine of the bolts were removed with the inlet guide vane actuator ring. Four locknuts were removed with the T₅ amplifier forward brackets. Three locknuts were removed with the anti-icing valve and stand-off bracket. One locknut was removed with the T₅ amplifier lead clamp. One bolt hole under the right-hand actuator is not used.

f. Remove the remaining locknuts and bolts, the junction box forward bracket, and the 2 offset brackets from the upper-half compressor casing forward flange.

g. Remove 26 locknuts and 26 bolts from the aft flange of the upper-half compressor stator casing.

h. Remove the junction box aft bracket, the afterburner control temperature (T₅) amplifier aft bracket, the 4 forward mounts and the 4 brackets.

i. Install the stator casing lifting device and remove the upper-half compressor casing. On J85-GE-5A engines, remove the 2 locking keys (2) for the compressor vane segments. On J85-GE-5 engines that have not been retrofitted to the J85-GE-5A configuration, remove the 4 retainers (21) for the compressor vane segments. Position the compressor stator half vertical support (14) at the 12 o'clock position and secure the support to the flange of both the mainframe (15) and front frame (13) with 2 bolts and nuts.

4–132. REMOVAL OF LOWER-HALF COMPRESSOR STATOR CASING, YJ85-GE-5. (See figure 4–49.)

Note

For special tools, see figure 3–1, group 10.

a. Twenty-six bolts (1) and 18 locknuts (2) secure the aft flange of the lower-half compressor casing (3) to the mainframe. Four bolts and locknuts, used for securing the trunnion brackets during shipping, were removed when the engine was removed from the shipping container. One bolt, locknut, and clamp were removed with the feedback cable. The 8 bolts that secure the forward end of the 2 gearbox mounting brackets were removed with the gearbox. Remove the remaining bolts, locknuts, 2 offset brackets (4, 5), and the fuel drain valve bracket (6).

b. Seventeen bolts (7) and 16 locknuts (8) secure the forward flange of the lower-half compressor casing to the front frame. Three of the bolts were removed when the inlet-guide vane actuator ring was removed. One locknut and 2 cushion clamps were removed with the actuator fuel lines. One locknut was removed with the left-hand actuator. One bolt under the left-hand actuator is not used. Remove the remaining bolts, locknuts, fuel filter support (10) and the check valve bracket (11).

c. Install the stator casing lifting device and remove the lower-half compressor casing. Position the compressor stator half vertical support at the 6 o'clock position and secure the support to the flange of both the mainframe and the front frame with 2 bolts and nuts.

4–133. REMOVAL OF LOWER-HALF COMPRESSOR STATOR CASING, J85-GE-5. (See figure 4–48A.)

Note

For special tools, see figure 3–1, group 10.

a. Seventeen bolts and 16 locknuts secure the forward flange of the lower-half compressor casing (3) to the front frame (13). Three of the bolts were removed with the inlet-guide vane actuator ring. One locknut was removed with the left-hand actuator. One locknut was removed with the afterburner high pressure filter clamp. One bolt hole under the left-hand actuator is not used. Remove the remaining bolts, locknuts and fuel hose bracket from the lower compressor casing forward flange.

GE-5 engine. The pages are typical of those found in a majority of these publications.

Cleaning

Engine cleaning is designed to accomplish several things.

1. Permit a thorough examination of components for the presence of service flaws and for changes in dimension through abrasion and wear.
2. Remove deposits which adversely affect the efficient functioning of the engine parts.
3. Prepare surfaces for applications of repair and salvage processes, such as plating, welding, and painting.
4. Remove various organic and inorganic coatings which require replacement either for inspection of the underlying surfaces, or to remove deteriorated coatings unsuitable for another engine run.

The selection of the cleaning materials and the processes used for each part are determined by the nature of the soil, type of metal, type of coating, and the degree of cleanliness necessary for a thorough inspection and the subsequent repair process. Not all parts need to be stripped down to the base metal, nor do all stains need to be removed from the plated parts. Furthermore, some cleaning solutions and/or procedures will strip or attack plated parts or cause undesirable reactions with the base metal. For example, titanium should not be cleaned with trichlorethylene or other chlorine-based compounds in order to avoid the possibility of stress corrosion associated with the entrapment of chlorine-containing materials in tight-fitting areas.

Cleaning solutions range from commonly used organic solvents such as petroleum washes and sprays for degreasing and general cleaning, vapor degreasing solutions such as trichlorethylene, and carbon solvents for hard carbon deposits, to less familiar cleaning materials. Steam cleaning can be used on parts not requiring mechanical or chemical cleaning and where paint and surface finishes need not be removed. Cleaning by tumbling is also an approved method for use on parts that are to be magnetically inspected. The tumbling process tends to obscure defects if the dye penetrant method of inspection is used. Most of these materials and methods are generally suitable for use on the cold section of the engine.

Hot-section cleaning requires processes involving a series of controlled acid or alkali baths and water rinses in various combinations. Grit blasting, wet or dry, is another method commonly used on both hot and cold sections of the engine.

Some parts such as ball and roller bearings require special handling. The bearing can become magnetized in service and may have to be demagnetized in order to properly clean it of magnetic particles. The bearing must not be allowed to spin during cleaning, and split bearings must be kept as units.

Since many of the solutions will attack the skin as readily as the part being cleaned, protective clothing and devices such as goggles, gloves, aprons, hand creams, etc., must be used while working with these products.

In all cases it is imperative to follow the manufacturer's recommendations in relation to materials and procedures used when cleaning gas turbine parts.

Inspection

When the engine is being manufactured, and during the overhaul process, it is, of course, necessary to check the quality of the various parts. The inspection section of the overhaul manual includes specific and detailed information, much of it gained through operating experience, outlining whether or not and the extent to which a part can be repaired and a table of minimum and maximum dimensional limits with which each part must comply. Special critical areas are called to the attention of overhaul personnel. Time and/or cycle limits are compared to the life limits of such parts as compressor and turbine blades and disks, and accurate records are kept of all work performed.

The inspection process can be divided into two broad groups.

NONDIMENSIONAL INSPECTION These methods include the use of fluorescent and nonfluorescent particles for those parts that can be magnetized, and fluorescent and nonfluorescent dye penetrants for those that cannot. Gages utilizing ultrasonic vibrations (Fig. 18-4) can be used to detect hidden flaws, and for critical parts x-rays are employed (Fig. 18-5). Included in this method is simple visual inspection for general condition (Fig. 18-6).

DIMENSIONAL INSPECTION This process includes the use of mechanical measuring tools such as

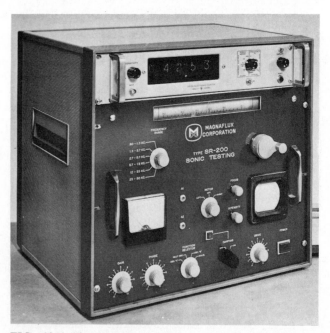

FIG. 18-4 The Magnatest SR-200 sonic testing machine. (*Magnaflux Corporation.*)

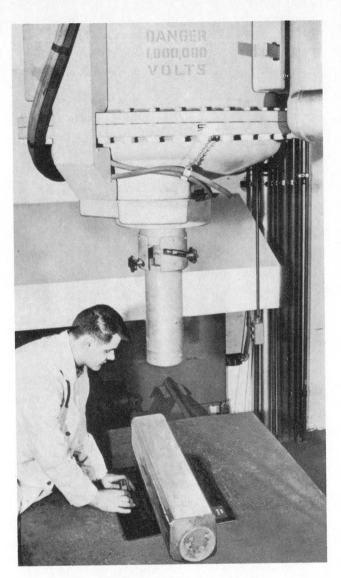

FIG. 18-5 A one-million-volt X-ray machine.

micrometers, dial indicators, and other specialized gages and plugs, and tools using light, sound, or air pressure as the measuring medium.

Magnetic-particle inspection (Fig. 18-7) is a nondestructive method of testing magnetizable ferrous materials for surface or subsurface cracks. When a part is magnetized by passing a current through it, or by placing it in a magnetic field, the two walls of any crack in the piece being tested will become weak secondary poles. If the part is immersed in a solution containing finely divided iron oxide magnetic particles, or if this solution is made to flow over the part, the suspended particles will tend to collect along the walls of the crack. The crack will be indicated by a dark red line or, if the suspension also contains fluorescent particles, as a bright line when viewed under an ultraviolet lamp. Each part to be tested must be magnetized using a current of correct magnitude applied in a specific direction (a process called circular or longitudinal magnetization) so that the magnetic lines of flux are likely to pass across a suspected crack at right angles (Fig. 18-8). Particular attention is paid to bosses, flanges, lugs, shoulders, splines, teeth, or other intricate areas. Some skill is necessary to properly interpret indications. After testing, the part must be completely demagnetized by passing it slowly through a coil in which alternating current is flowing. Complete demagnetization is important and may be tested by using a field-strength indicator or a good-quality compass. Specific procedures are generally outlined in the overhaul manual.

Dye penetrant inspection is a nondestructive method of testing nonmagnetizable materials such as aluminum and magnesium for *surface* cracks or imperfections only (Fig. 18-9). Although this process was developed for nonferrous materials, it can also be used to good advantage on products made of iron. Since this is a surface treatment, it is essential to have the surface absolutely clean and free of paint. A penetrating dye, which may or may not fluoresce, is painted or sprayed on the part.

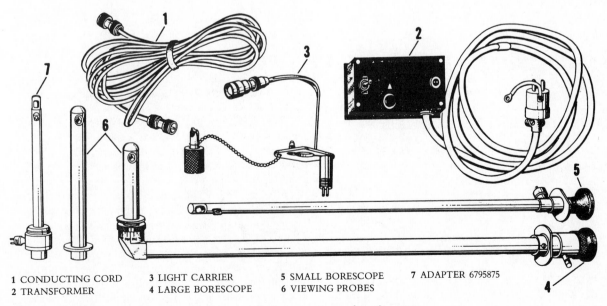

1 CONDUCTING CORD   3 LIGHT CARRIER   5 SMALL BORESCOPE   7 ADAPTER 6795875
2 TRANSFORMER   4 LARGE BORESCOPE   6 VIEWING PROBES

FIG. 18-6 Borescope components used to visually examine internal engine parts.

FIG. 18-7 A modern magnetic-particle testing machine (Magnaflux testing). (*Magnaflux Corporation.*)

FIG. 18-9 A crack found by using a dye penetrant may save time, money, and lives. (*Magnaflux Corporation.*)

The surface is then washed and a wet or dry developer is applied which will draw the penetrant from the defect. Cracks will appear as dark lines, or bright lines from the fluorescent material when viewed under the ultraviolet light. Again as with magnetic inspection, some degree of skill is necessary to interpret and evaluate the indications.

Visual inspection plays a most important part in the overhaul of the gas turbine engine. Some conditions likely to be found, and their causes, are listed below. (Fig. 18-10.)

1. *Abrasion*—A roughened area. Varying degrees of abrasion can be described as light or heavy, depending upon the extent of reconditioning required to restore surface. Usual cause is presence of fine foreign material between moving surfaces.

2. *Bend (bow)*—General distortion in structure as distinguished from a local change in conformation. Usual causes are uneven application of heat, excessive heat or pressure, or forces defined under stresses.

3. *Blistering*—Raised areas indicating separation of surface from base. Usually found on plated or painted surfaces. Associated with flaking or peeling. Usual cause is imperfect bond with base, usually aggravated by presence of moisture, gas, heat, or pressure.

4. *Break*—Complete separation by force into two or more pieces. Usual causes are fatigue, or sudden overload.

5. *Brinelling*—Indentations sometimes found on surface of ball or roller bearing parts. Usual causes are improper assembly or disassembly technique of the roller or ball bearings, by the application of force on the free race.

   NOTE: Bearings which do not have full, constant rotation and are subject to sudden loading have brinelling tendencies.

6. *Bulge*—An outward bending or swelling. Usual cause is excessive pressure or weakening due to excessive heat.

7. *Burning*—Injury to parts by excessive heat. Evidenced by characteristic discoloration or in severe cases, by a loss or flow of material. Usual causes are excessive heat due to lack of lubrication, improper clearance, or abnormal flame pattern.

8. *Burnishing*—Mechanical smoothing of a metal surface by rubbing, not accompanied by removal of material, but sometimes by discoloration around outer edges of area. Operational burnishing is not detrimental if it covers approximately the area carrying the load, and provided there is no evidence of pileup or burning. Usual cause is normal operation of the parts.

9. *Burr*—A sharp projection or rough edge. Usual causes are excessive wear, peening, or machining operation.

Defects shown by longitudinal field. Part magnetized in coil.

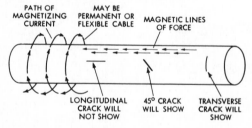

FIG. 18-8 Cracks will show best when in line with current flow and at right angles to the magnetic field. (*Magnaflux Corp.*)

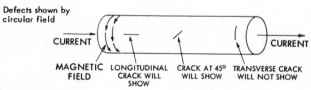

10. *Chafing*—A rubbing action between parts having limited relative motion. To be interpreted as an action which produces a surface condition rather than as a description of the injury.

11. *Chipping*—Breaking out of small pieces of metal. Do not confuse with flaking. Usual cause is a concentration of stress due to nicks, scratches, inclusions, peening, or careless handling of parts.

12. *Corrosion*—Breakdown of surface by chemical action. Usual cause is presence of corrosive agents.

13. *Cracks*—A partial fracture. Usual cause is excessive stress due to sudden overloading, extension of a nick or scratch, or overheating.

14. *Dent*—Small, smoothly rounded hollow in the surface. Usual causes are concentrated overloading resulting from peening, or presence of chips between loaded surfaces, or the striking of a part.

15. *Electrolytic action*—Breakdown of surfaces by electrical action between parts made of dissimilar metals. Usual cause is galvanic action between dissimilar metals.

16. *Erosion*—Carrying away of material by flow of hot gases, grit, or chemicals. See guttering. Usual causes are flow of corroding liquids, hot gases, or grit-laden oil.

17. *Fatigue failure*—Progressive yielding of one or more local areas of weakness such as tool marks, sharp indentations, minute cracks, or inclusions, under repeated stress. As working stress on the piece is repeated, cracks develop, at ends of which there are high concentrations of stress. Cracks spread, usually from the surface, or near the surface, of the area. After a time, there is so little sound metal left, the normal stress is higher than the strength of the remaining material, and it snaps. Failure is not due to crystallization of metal, as many mechanics believe. Appearance of a typical fatigue failure is easily explained. As failure proceeds, severed surfaces rub and batter each other, crushing grains of material and producing a dull or smooth appearance; the remaining unfractured portion preserves normal grain structure up to the moment of failure. The progressive nature of the failure is usually indicated by several more or less concentric lines, the center, or *focus*, of which discloses original point or line of failure. Usual causes are tool marks, sharp corners, nicks, cracks, inclusions, galling, corrosion, or insufficient tightening of studs or bolts to obtain proper stretch.

18. *Flaking*—Breaking away of pieces of a plated or painted surface. Usual causes are incomplete bonding, excessive loading, or blistering.

19. *Flowing*—Spreading of a plated or painted surface. Usually accompanied by flaking. Usual causes are incomplete bonding, excessive loading, or blistering.

20. *Fracture*—See break and chafing.

21. *Fretting corrosion*—Discoloration may occur on surfaces which are pressed or bolted together under high pressure. On steel parts the color is reddish brown and is sometimes called "cocoa" or "blood." On aluminum or magnesium, the oxide is black. Usual cause is rubbing off of fine particles of metal by slight movement between parts and subsequent oxidizing of these particles.

22. *Galling*—A transfer of metal from one surface to another. Usual cause is severe chafing or fretting action caused during engine operation by a slight relative movement of two surfaces under high contact pressure.

NOTE: Do not confuse with pickup, scoring, gouging, or scuffing.

23. *Glazing*—Development of a hard, glossy surface on bearing surfaces. An often beneficial condition. Usual cause is a combination of pressure, oil, and heat.

24. *Gouging*—Displacement of material from a surface by a cutting or tearing action. Usual cause is presence of a comparatively large foreign body between moving parts.

25. *Grooving*—Smooth, rounded furrows, such as tear marks whose sharp edges have been polished off. Usual causes are concentrated wear, abnormal relative motion of parts, or parts out of alignment.

26. *Guttering*—Deep, concentrated erosion. Usual cause is enlargement of a crack or defect by burning due to flame or hot gases.

27. *Inclusion*—Foreign material in metal. Surface inclusions are indicated by dark spots or lines. Usual cause is a discontinuity in the material.

NOTE: Both surface inclusions and those near the surface may be detected during magnetic inspection by grouping of magnetic particles. Examination of a fatigue fracture may reveal an inclusion at the focal point.

28. *Nick*—A sharp indentation caused by striking a part against another metal object. Usual causes are carelessness in handling of parts or tools prior to or during assembly, or sand or fine foreign particles in the engine during operation.

29. *Peening*—Deformation of surface. Usual cause is impact of a foreign object such as occurs in repeated blows of a hammer on a part.

30. *Pickup*—Rolling up of metal, or transfer of metal from one surface to another. Usual causes are rubbing of two surfaces without sufficient lubrication, presence of grit between surfaces under pressure during assembly, unbroken edges of press-fitted parts, or incipient seizure of rotating parts during operation.

31. *Pileup*—Displacement of particles of a surface from one point to another. Distinguished from pickup by the presence of depressions at the point from which the material has been displaced.

32. *Pitting*—Small, irregularly shaped cavities in a surface from which material has been removed by corrosion or chipping. Corrosive pitting is usually accompanied by a deposit formed by a corrosive agent on base material. Usual causes of corrosive pitting are breakdown of the surface by oxidation or some other chemical, or by electrolytic action. Usual causes of mechanical pitting are chipping of loaded surfaces because of overloading or improper clearance, or presence of foreign particles.

33. *Scoring*—Deep scratches made during engine operation by sharp edges or foreign particles; elongated gouges. Usual cause is presence of chips between loaded surfaces having relative motion.

34. *Scratches*—Narrow, shallow marks caused by move-

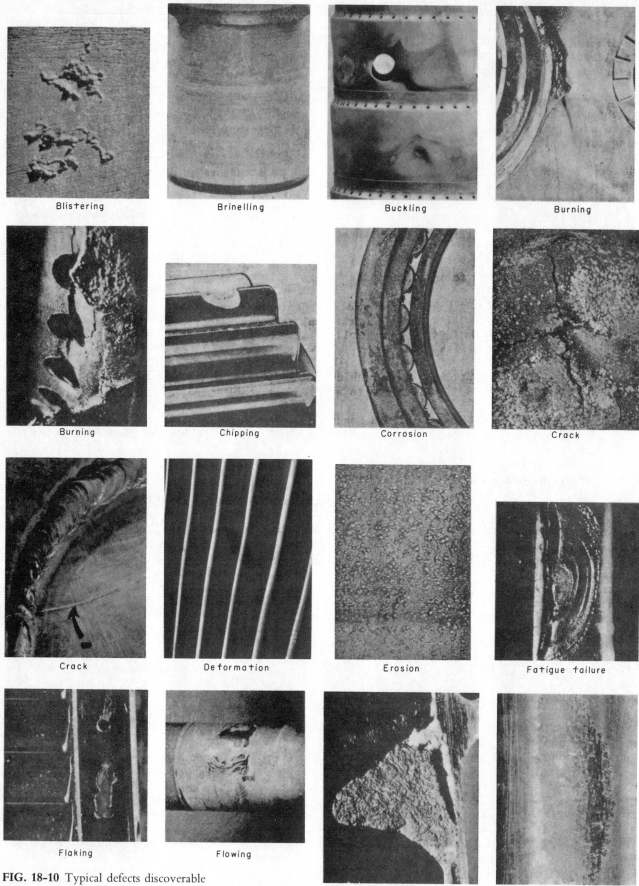

Blistering   Brinelling   Buckling   Burning

Burning   Chipping   Corrosion   Crack

Crack   Deformation   Erosion   Fatigue failure

Flaking   Flowing   Fracture   Fretting corrosion

FIG. 18-10 Typical defects discoverable through visual inspection.

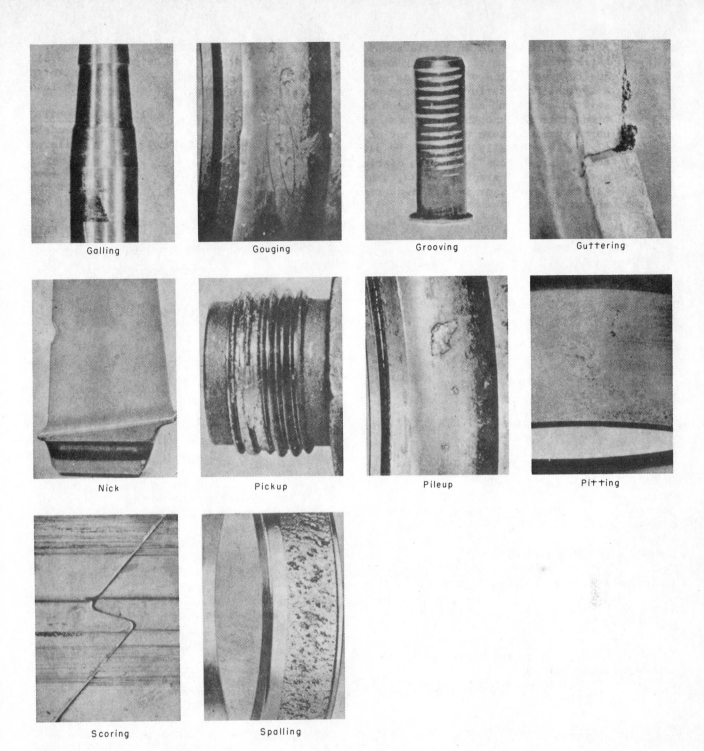

Galling   Gouging   Grooving   Guttering

Nick   Pickup   Pileup   Pitting

Scoring   Spalling

FIG. 18-10 (cont.)

ment of a sharp object or particles across a surface. Usual causes are carelessness in handling of parts or tools prior to or during assembly, or sand or fine foreign particles in the engine during operation.

35. *Spalling*—Sharply roughened area characteristic of progressive chipping or peeling of surface material. Do not confuse with flaking. Usual causes are surface cracks, inclusions, or any similar surface injury causing a progressive breaking away of surface under load.

36. *Stresses*—When used in describing the cause of failure of machine parts, stresses are generally divided into three groups—compression, tension, and shear. These forces are described as follows:
Compression—Action of two directly opposed forces which tend to squeeze a part together.
Tension—Action of two directly opposed forces which tend to pull apart.
Shear—Action between two opposed parallel forces.

Inspection is a vital part of engine overhaul. Without a quality inspection, the other overhaul procedures are essentially meaningless.

Repair

All serviceable engine parts must be repaired using methods approved by the manufacturer. Repair techniques vary widely. Welding is used extensively and is discussed in Chap. 10. Repairs on combustion chambers and many other parts of the engine are often made this way. After welding, it may be necessary to heat-treat the part in order to remove the stress induced through welding and to restore the original properties of the metal.

Other parts can be restored to their original dimensions by plating. This is also discussed in Chap. 10. Replating by electrochemical means or hard facing by plasma-sprayed coatings or by detonation flame coatings is used to build up hubs and disks, and to repair and protect parts of the engine that chafe on each other. For example, combustion-chamber outlet ducts on many engines are permitted limited movement to compensate for engine growth as the engine temperature changes. Repair methods involve operations of all kinds, including grinding, blending, and other abrasive processes, lathe work, boring, straightening, painting, etc. (Fig. 18-11). If the engine contains rivets, these are repaired or replaced as required. The bushings to be found in the accessories section and other parts of the engine are replaced if necessary. If any threaded holes are stripped, they are repaired at this time by drilling and tapping and installing a threaded bushing, an oversize stud, or a helicoil insert.

Once again, the overhaul manual gives a detailed and approved repair procedure which must be followed to assure a reasonable service life or TBO.

Reassembly

Reassembly is accomplished on the same horizontal or vertical stand that was used for disassembly. At this time all gaskets, packings, and rubber parts are replaced from fresh stock. Specific clearance and limits, for example, radial and axial position of the rotating turbine and compressor assemblies, blade-tip movement, or wear patterns on gear teeth, etc., are checked as reassembly continues.

Extreme care must be taken to prevent dirt, hardware, lockwire, or other foreign materials from entering the engine. If anything is dropped, all work must be stopped until the foreign object can be located and removed. All parts must be safetied in some manner. Standard safetying devices include lock or safety wire, plain and specialized lock washers, cotter pins used with castellated nuts, and fiber, nylon, or metal locknuts (Fig. 18-12). Some locknuts have a deformed section at the top to achieve the locking action. Lockwire, lockwashers, and cotter pins are never reused, but locknuts can be reused if they cannot be turned all the way on by hand. Standard and special torque values can be found in the overhaul manual and must be adhered to. (See the Appendix for a typical table of torque values.)

Bearings require special handling. Gloves must be worn to prevent contamination from skin oil and acid. Rubber or lint-free gloves are recommended. Carbon rubbing seals are also quite fragile and must be treated accordingly.

Synthetic lubricants can have an adverse effect on skin, and protective gloves or hand creams should be used if needed.

Accurate balancing of the rotating assemblies is most important because of the high rotational speeds involved. The two methods for determining out of balance are static and dynamic balancing. Static balancing, as the name implies, is done while the part is stationary. But it is quite possible to have a part in static balance and still experience considerable out of balance while rotating. Dynamic balancing is achieved while the assembly is rotating. Often the individual stages of the compressor and turbine are balanced separately and then rebalanced as assembled components. Other rotating parts such as shafts, couplings, etc., are also balanced.

Balancing is achieved by (see Chap. 5):

1. The addition or movement of weights riveted, pressed, or pinned, in place near the rim of the compressor or turbine rotor.
2. Balancing bolts.
3. Careful grinding or drilling in specified areas.
4. Shifting blades in the compressor. Each blade will have a coded letter or number indicating that a blade falls within a specified range of weights.

Many manufacturers will mark turbine blades with a *moment-weight* number to permit balancing during assembly. Turbine blades may also be weighed at the time of reassembly and assembled in the disk in the manner outlined.

The performance of the engine is strongly influenced by the area formed by the turbine nozzle vanes. On those engines in which the vanes are removable, the entire nozzle assembly must be built up out of vanes that will cause the turbine nozzle flow area to be within

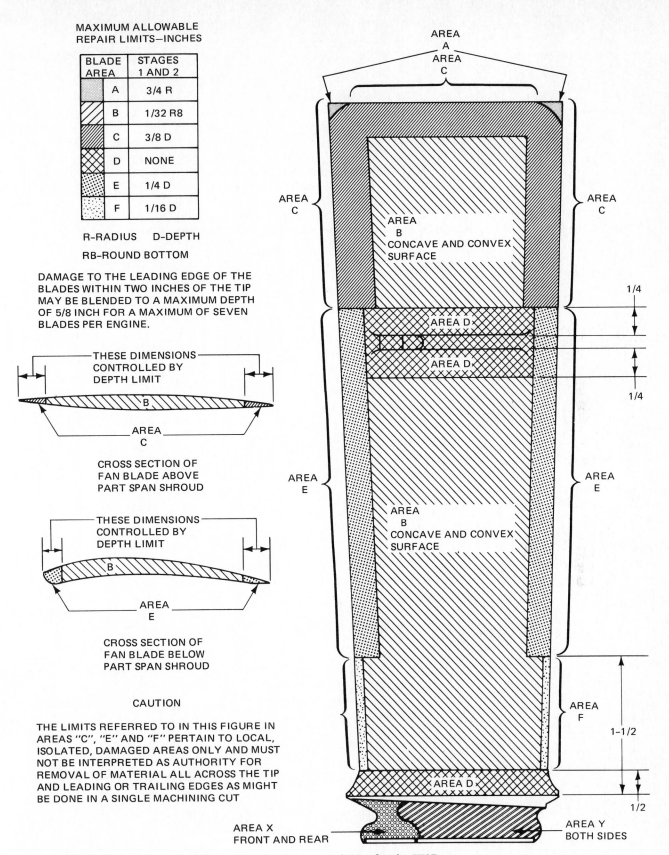

MAXIMUM ALLOWABLE
REPAIR LIMITS—INCHES

BLADE AREA		STAGES 1 AND 2
	A	3/4 R
	B	1/32 R8
	C	3/8 D
	D	NONE
	E	1/4 D
	F	1/16 D

R—RADIUS   D—DEPTH

RB—ROUND BOTTOM

DAMAGE TO THE LEADING EDGE OF THE BLADES WITHIN TWO INCHES OF THE TIP MAY BE BLENDED TO A MAXIMUM DEPTH OF 5/8 INCH FOR A MAXIMUM OF SEVEN BLADES PER ENGINE.

THESE DIMENSIONS CONTROLLED BY DEPTH LIMIT

B

AREA
C

CROSS SECTION OF
FAN BLADE ABOVE
PART SPAN SHROUD

THESE DIMENSIONS CONTROLLED BY DEPTH LIMIT

B

AREA
E

CROSS SECTION OF
FAN BLADE BELOW
PART SPAN SHROUD

CAUTION

THE LIMITS REFERRED TO IN THIS FIGURE IN AREAS "C", "E" AND "F" PERTAIN TO LOCAL, ISOLATED, DAMAGED AREAS ONLY AND MUST NOT BE INTERPRETED AS AUTHORITY FOR REMOVAL OF MATERIAL ALL ACROSS THE TIP AND LEADING OR TRAILING EDGES AS MIGHT BE DONE IN A SINGLE MACHINING CUT

AREA X
FRONT AND REAR

AREA
A
AREA
C

AREA
C

AREA
C

AREA
B
CONCAVE AND CONVEX
SURFACE

AREA D

AREA D

AREA
B
CONCAVE AND CONVEX
SURFACE

AREA
E

AREA
E

AREA
F

AREA D

AREA Y
BOTH SIDES

1/4

1/4

1-1/2

1/2

FIG. 18-11 Typical repair-section illustration showing repair limits for the JT3D fan blades.

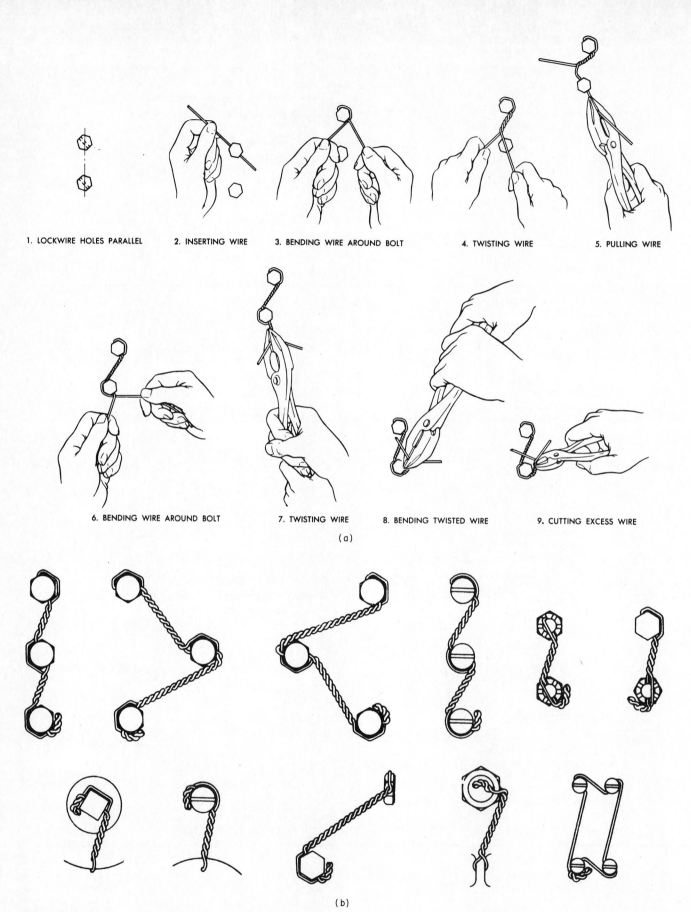

1. LOCKWIRE HOLES PARALLEL   2. INSERTING WIRE   3. BENDING WIRE AROUND BOLT   4. TWISTING WIRE   5. PULLING WIRE

6. BENDING WIRE AROUND BOLT   7. TWISTING WIRE   8. BENDING TWISTED WIRE   9. CUTTING EXCESS WIRE

(a)

(b)

FIG. 18-12 (a) Steps in applying safety wire.
(b) Typical safety wire patterns.

stated limits. On one model engine having fixed nozzle vanes, the nozzle area can be adjusted by bending the vanes with a special tool. The area is then checked with a gage.

Instructions for all phases of reassembly are set forth in the overhaul manual.

Testing

To ensure that the engine's performance meets that guaranteed to the customer, all engines are test-run to an established schedule before being shipped, stored, or used in an aircraft. Fuel and oil consumption is checked, pressure and temperature measurements are taken at several points, and thrust or horsepower is accurately measured. A more detailed discussion of engine testing will be found in Chap. 19.

Storage

The degree of preservation is determined by the anticipated length of time the engine is expected to be inactive. In order to protect the engine for long-term storage (usually three months or more), the following items are usually accomplished (Fig. 18-13):

1. All external openings in the engine are sealed with plugs and cover plates.

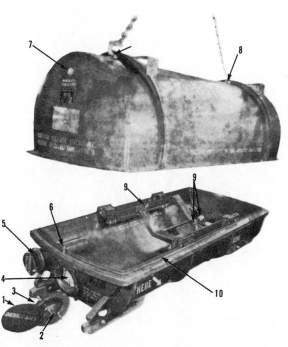

1	SERVICE RECEPTA-CLE COVER	6	CONTAINER LOCAT-ING PIN
2	AIR-RELEASE VALVE	7	HUMIDITY INDICATOR
3	SERVICE RECEPTA-CLE	8	CONTAINER LIFTING POINTS
4	DESICCANT	9	PINS, WASHERS, AND COTTER PINS
5	RECORD RECEPTA-CLE	10	CONTAINER GASKET

FIG. 18-13 A typical gas turbine shipping can.

2. A dehydrating agent, usually bags of silica gel (MIL-D-3464), is placed in the engine inlet and exhaust duct.
3. The oil system is drained, and may or may not be flushed with a preservative oil (MIL-C-8188).
4. The fuel system is drained and flushed with the preservative oil by motoring the engine (MIL-L-6081).
5. Some manufacturers recommend spraying oil into the compressor and turbine ends while motoring the engine, whereas others specifically caution against this practice on the grounds that dirt particles collecting on the blades will alter the airfoil shape and adversely affect compressor efficiency.
6. The engine is placed in a metal shipping container, and several bags of the desiccant are placed in a wire basket located within the shipping can. The container is then sealed and pressurized with approximately 5 psi [34 kPa] of dry air or nitrogen to exclude moisture. An observation port in the can allows an internal humidity indicator to be seen. A safe humidity level is indicated by a blue color. The unsafe color is pink.
7. If a shipping can is not used, the engine is generally wrapped in a protective vapor-proof storage bag made of foil, cloth, and plastic, and carefully sealed against moisture.

MAINTENANCE TECHNIQUES

Although engines are normally removed from the aircraft for overhaul in the shop, some new aircraft such as the Lockheed L1011 Tristar are equipped with handling rails which allow major engine modules to be changed while the engine is still installed in the aircraft (Fig. 18-14).

Maintenance practices for gas turbine engines differ little from those used on reciprocating engines. It is important to see that the engine compartment be kept as clean as possible because the high-velocity airflow through the engine will tend to draw any foreign objects into the compressor. All small parts, such as loose lockwire, nuts, bolts, etc., should be removed immediately.

The exterior of the engine should be inspected to see that all parts of the engine are secure and that there are no broken safety wires. Tubing should be checked for security, nicks, chafing, dents, and leaks. Controls should be checked for proper operation to ensure that they do not bind and that there is cushion in the flight deck controls. *Cushion* is obtained when the control lever on the unit hits its stop before the control lever in the cabin hits its stop.

Inspection of the gas turbine engine is made somewhat easier than that of a reciprocating engine because of the gas turbine's inherent cleanliness and the ready accessibility of many parts to visual inspection. The first several stages of most compressors can be inspected for cracks with the aid of a strong light. Also readily open to visual inspection for heat damage is the exhaust duct and the last stage or two of the turbine. The thermocouples and pressure probes are also available for inspection in this area, and in the case of the thermocouples, can be checked electrically and operationally with the proper equipment. As shown in Fig. 18-15, many

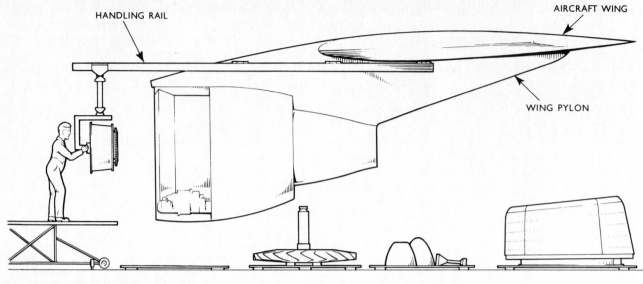

HANDLING RAIL

AIRCRAFT WING

WING PYLON

Wing Engine Modular Disassembly

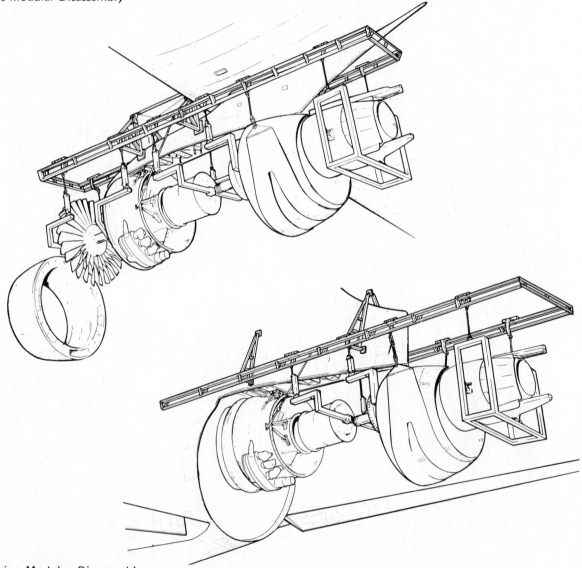

Center Engine Modular Disassembly

FIG. 18-14 Modular engine design of this Rolls-Royce RB·211 allows a high degree of on-wing repairability.

BORESCOPE INSPECTION PORTS

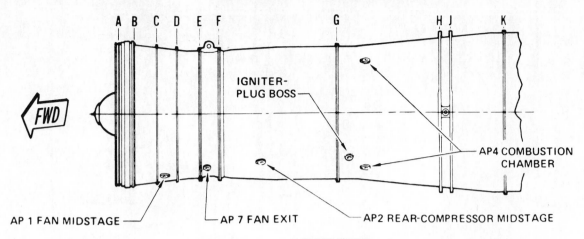

LEFT SIDE

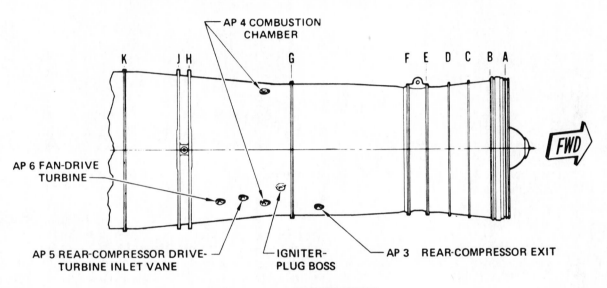

RIGHT SIDE

Engine Port	Port Location Location Between Flanges	Clock Position as Viewed From Rear	Nomenclature	Borescope View
AP 1	C and D	6:30	Fan midstage inspection port.	Second stage compressor blades trailing edge. Third stage compressor blades leading edge.
AP 7	E and F	7:00	Fan exit inspection port.	Third stage compressor blades trailing edge.
AP 2	F and G	6:45	Rear compressor midstage inspection port.	Sixth stage compressor blades trailing edge. Seventh stage compressor blades leading edge.
AP 3	F and G	5:00	Rear compressor exit inspection port.	Twelfth stage compressor blades trailing edge. Thirteenth stage compressor blades leading edge.
AP 4 (four ports)	G and H	1:00 7:00 5:00 11:00	Combustion chamber inspection port.	Combustion chamber first stage turbine stator vanes leading edge. Fuel nozzles.
Igniter plugs (two ports)	G and H	4:30 7:30		
AP 5	G and H	5:00	Rear compressor drive turbine inlet vane inspection port.	First stage urbine rotor blades leading edge.
AP 6	G and H	5:00	Fan drive turbine inspection port.	Second stage turbine rotor blades trailing edge. Third stage turbine rotor blades leading edge.

FIG. 18-15 Borescope inspection ports for the Pratt & Whitney Aircraft JT8D engine showing the location of the ports and the internal parts that can be viewed.

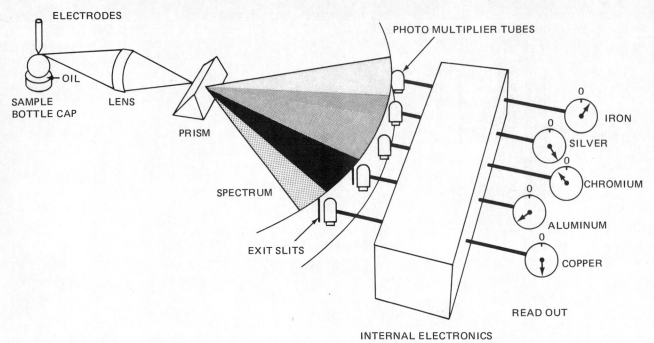

FIG. 18-16 Spectrometric oil analysis.

engines are equipped with special openings or ports for the insertion of a borescope (Fig. 18-6), making internal inspection of the engine even easier.

The oil system should be checked daily for proper oil level. It is recommended that the different brands of oil not be mixed when adding or changing oil. Oil filters are cleaned and fuel filters are replaced periodically. Since the engine fuel pumping elements are lubricated by the fuel itself, water or ice in the fuel may damage the fuel-system components and cause erratic engine operation. Therefore, the aircraft's fuel sumps should be checked daily for water. On those engines equipped with magnetic chip detectors, there should be no continuity through the detector. If continuity is found, the chip detector must be removed, and the source of metal contamination determined. The spectrometric oil analysis program (SOAP) is proving to be a reliable aid to preventing in-flight power-plant failures. In this process, periodic samples of used oil are taken and sent to an oil-analysis laboratory where the oil is burned by an electric arc. The light emitted is passed through a slit which is precisely positioned to the wavelength for the particular wear metal being monitored. Trends are observed and abnormal concentrations of metal are sought (Fig. 18-16).

An additional daily check should be made during coastdown after engine shutdown. Coastdown should be accomplished with no rubbing or abnormal noises. Some manufacturers will give a specific coastdown time.

ENGINE PERFORMANCE MONITORING [FIG. 18-17 (a)–(g)]

In recent years, a method of monitoring the gas turbine engine's day-to-day condition has been adopted by many operators. In this system the EPR (engine pressure ratio), rpm, F/F (fuel flow), EGT (exhaust gas temperature) and throttle position are used to determine the aerodynamic performance of the engine, while vibration amplitude and oil consumption (which may include periodic spectrometric oil analysis) is used to evaluate mechanical performance.

Although specific procedures will vary from operator to operator, in general, cockpit instrument readings are taken once a day or on every flight during cruise conditions. The recorded data is then processed in a variety of ways and compared with "normal" data established by the manufacturer or the operator as representing the normal performance of the engine. Trends in the operating parameters are then observable.

Engine performance monitoring is proving to be a very effective method of providing early warning information of ongoing or impending failures, thus reducing unscheduled delays and more serious engine failures.

Examples of several actual engine malfunctions that were detected using performance-monitoring techniques are shown in Fig. 18-17.

Trend Analysis Summary

Certain kinds of engine failures will result in specific changes in the monitored parameters. Pratt and Whitney Aircraft has summarized these failures for some of their engines as follows.

1. FAILURES RESULTING IN AIR LEAKAGE FROM THE COMPRESSOR CASE [FIG. 18-17(a)] A number of compressor section failures may be broadly classified as failures which result in the leakage of high-pressure air from the compressor section of

the engine. This leakage may be due to the failure of a bleed air duct external to the engine, a stuck overboard bleed valve, or failure of the engine casing itself.

In the cruise operating range, where engine monitoring is done, the turbine expansion ratio of the engine is fixed. Engine pressure ratio is therefore directly related to the pressure ratio across the compressors. The leakage of air from the compressor causes the compression ratio, and consequently the EPR, to drop if the throttle position is not changed.

To regain EPR, the throttle is pushed further forward increasing the fuel flow. This in turn increases the turbine inlet temperature, power generated by the turbine section, and rotor speeds. The increased power to drive the compressors will regain the compression ratio in spite of the air leak. Therefore, air leaks in the compressor section generally result in trend plots where all of the monitored performance parameters increase.

Air leakage may occur between the high and low compressors, at some intermediate stage, or from the diffuser case. This air may be discharged into the nacelle, overboard, or in the case of the JT8D, into the fan duct. The magnitude of change in the engine parameters is dependent on all the forementioned factors plus the size of the leak.

2. COMPRESSOR CONTAMINATION

[FIG. 18-17(b)] Engine performance will decrease due to compressor contamination. Contamination of the compressor may occur due to operation near salt water, the use of impure water for water injection, an oil leak in the forward part of the engine which may cause fine dust to adhere to the blades, or contamination from ingestion during reversing. Often the effects of compressor contamination can be eliminated by water washing or carboblasting the engine. (See p. 365–366.)

Contamination of the compressor blades changes their aerodynamic shape, roughens their surfaces, and reduces the airflow area. This reduces compressor efficiency and airflow capacity. When the compressors lose efficiency, more power and higher rotor speeds are required to achieve the desired compressor pressure ratio and hence EPR. This additional power is obtained by pushing the throttle further forward, increasing fuel flow, and increasing turbine inlet temperature.

The increase in speed of the low compressor relative to the increase in speed of the high compressor will be influenced by the type of contamination and whether the contamination is in the low-pressure compressor, high-pressure compressor, or both. Water contamination resulting from inlet water injection, for instance, normally settles out in the high-pressure compressor. Little or none of these deposits form on the low-pressure compressor. This is due to the combination of temperature and pressure existing in the high compressor, plus the time element involved in vaporizing the water.

3. MECHANICAL FAILURES

As is the case with compressor contamination, the loss of parts along the gas path has the effect of reducing the compressor efficiency. However, whereas contamination usually affects the whole compressor, mechanical failures normally involve only a few blades or vanes. The efficiency loss due to failures of this type is generally rather small and, for this reason, the failures are more difficult to detect in a monitoring program. More severe failures are usually first detected at high power settings when EGT limits are exceeded and/or compressor stalling occurs.

Broken blades and vanes usually result in N_1 and N_2 increasing to overcome the efficiency loss, and W_f and EGT increasing to provide additional energy. In some cases, however, one rotor system may actually drop in speed if metal contact is taking place.

In addition, the effect on engine performance due to a mechanical failure in the compressor may be further complicated by the fact that parts leaving the compressor may cause foreign object damage farther downstream. This in turn may cause a change in the performance of the burner section and/or the turbine section. The effects of compound failures within the engine are difficult to analyze and the number of ways in which performance may be affected are many.

4. COMBUSTION SECTION [FIG. 18-17(c)]

Generally speaking, of all the engine sections, the burner section is the least sensitive to failure detection using in-flight monitoring techniques. The majority of the combustion section problems experienced in the operation of an engine include failures such as blocked fuel nozzles, fuel line leaks, and failure of a burner can itself. Usually these problems must be detected by maintenance monitoring methods. It is only when the problem becomes severe enough to affect another section of the engine that in-flight monitoring becomes useful. The section most often affected by burner failures is the turbine section.

Pieces of the combustion chamber which become loose and eventually break away will most often cause some sort of damage further down the gas path. In some cases, the flame pattern becomes distorted sufficiently to burn or bow the inlet guide vanes immediately aft of this section. It is for this reason that monitoring plots of failures which include damage to the burner section most often resemble high-pressure turbine failures.

5. TURBINE FAILURES—GENERAL [FIG. 18-17(d) AND (e)]

Engine performance monitoring has proved to be very useful in detecting trouble areas in the high-pressure turbine. For example, loss of a first-stage turbine blade will most likely cause a marked shift in several of the performance parameters, whereas a similar single blade loss in either of the compressors could possibly go undetected. This is because there are relatively few blades in the system to develop the work required to drive the high-pressure compressor and a slight loss in turbine efficiency will be quite noticeable in engine operating performance. This sensitivity also applies to the components such as inlet guide vanes and turbine seals which also influence the turbine efficiency.

It is not possible to determine exact parameter shifts for failures in the high-pressure turbine. The degree of failure, the compounding effects of secondary damage to the low-pressure turbine, and the compressor and fuel-control design characteristics of the particular engine type, combine to determine the degree of perfor-

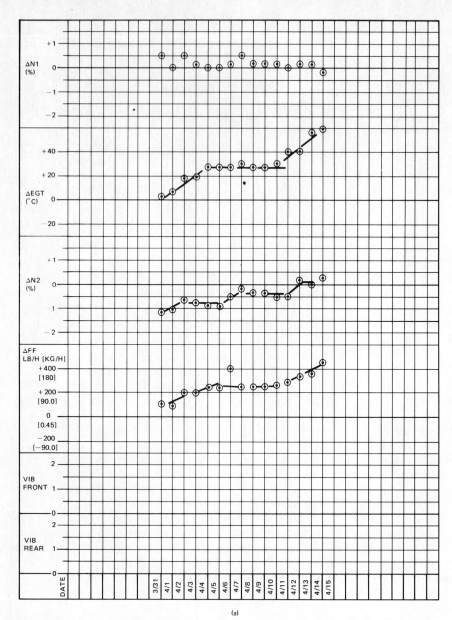

(a)

FIG. 18-17 (a) *Problem:* Cracked 13th stage bleed duct

Engine: JT8D

Malfunction: This illustrates the monitor plot associated with a 13th stage bleed (anti-ice manifold) failure. The assembly broke around the weld bead which joins the mount flange to the manifold.

Analysis: This failure progressed gradually as can be seen from the monitor plot, and finally resulted in the engine becoming EGT-limited on takeoff, with the throttle on the affected engine one knob ahead of the other two. Examination of the plot shows little or no effect on N_1, and an increase in EGT, N_2, and fuel flow. A slight engine mismatch occurs when air is bled from the compressor, due to both the change in turbine efficiency as the turbine inlet temperature changes, and the change in the ratio of airflow through the compressor to that through the turbine. In this case, the mismatch is such that the high compressor does most of the additional work required to regain EPR.

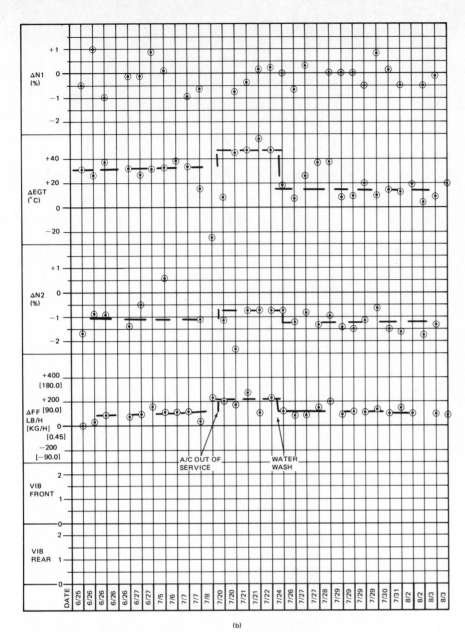

FIG. 18-17 (b) *Problem:* Compressor contamination and water wash

Engine: JT8D

Malfunction: This is an example of a monitor plot associated with compressor contamination. This engine was installed in an aircraft which was parked on a ramp, near salt water, for two weeks because of an airline strike. Apparently impurities from the air contaminated the compressor during this time.

Analysis: Compressor contamination was suspected when, after the engine had been out of service for two weeks, fuel flow and EGT increased significantly. A slight increase in N_2 was noted with little or no change in N_1. A water wash restored engine performance.

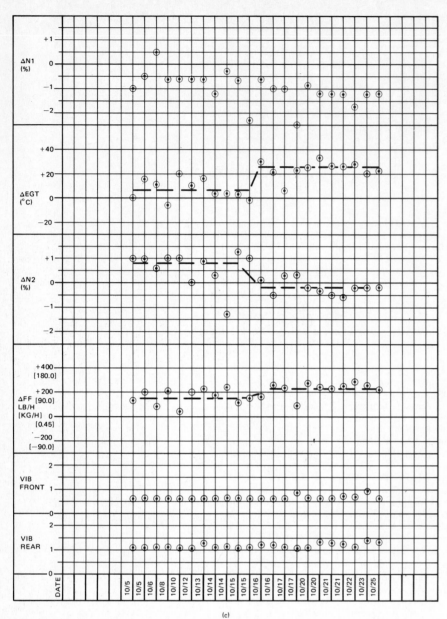

(c)

FIG. 18-17 (c) *Problem:* Burned and Deformed Combustion Chamber Outlet Duct

Engine: JT8D

Malfunction: This engine was removed entirely on the basis of the operator's in-flight data monitoring program trend implications. Despite the marked deterioration of engine performance, there had been no flight crew write-ups to indicate recognition of it. Investigation revealed that extensive deformation occurred in the engine inner combustion chamber outlet as well as considerable burning and bowing of first-stage nozzle guide vanes.

Analysis: The trend indications shown in this example are quite similar to others where a loss in turbine efficiency is involved. The failure first occurred in the combustion chamber and, for this reason, the example is included in this section of the report. However, it is felt that subsequent damage to the nozzle guide vanes was the dominating influence in the trend indications.

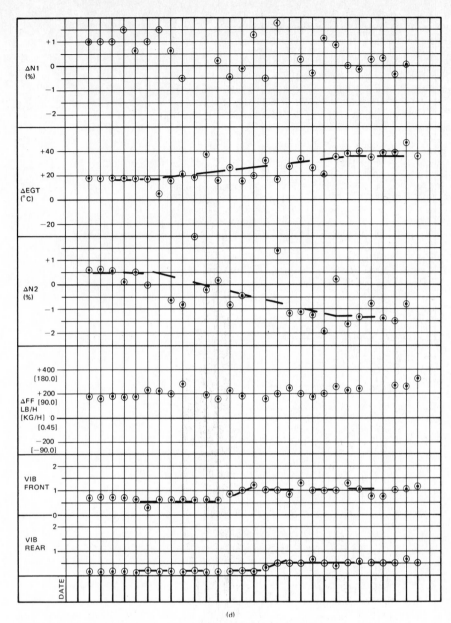

(d)

FIG. 18-17 (d) *Problem:* First-stage nozzle guide vanes failure

Engine: JT8D

Malfunction: The engine shown in this example was removed for inspection because of the performance deterioration noted in the monitoring plot. The tear-down revealed that first-stage nozzle guide vanes were excessively burned and bowed. No other damage was reported.

Analysis: The depression in N_2 and the increase in EGT are characteristic of most high-pressure turbine failures. The rate of deterioration in this case is basis for the assumption that the problem involves the gradual decay of some component in the high-pressure turbine assembly. As the deformation of the guide vane progressed in this example, the turbine inlet area continuously increased. This action generally has the effect of reducing the energy extracted per pound of air flow through the first turbine because of the smaller pressure drop across the turbine. The increased level of vibration remains unexplained by the information available on this example. It is assumed that the upset flow pattern around the deformed guide vanes changed the vibration signature of the installed engine. Note also the scatter of data as the failure started to develop. This condition often warns of an impending failure before any definite trends develop.

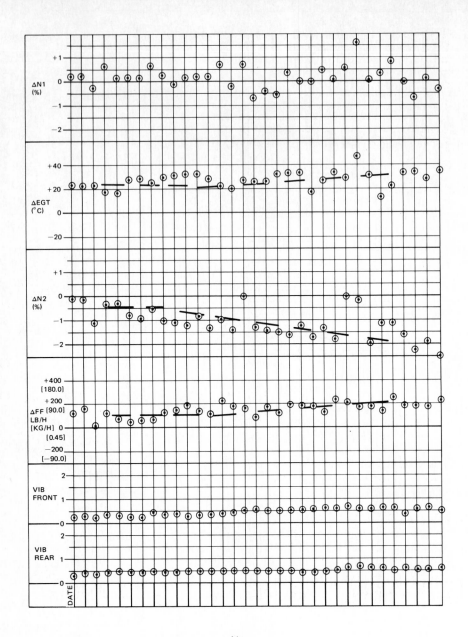

(e)

FIG. 18-17 (e) *Problem:* Turbine case separation

Engine: JT8D

Malfunction: The turbine case became partially separated from the nozzle guide vane case when 20 of the attaching bolts broke. Hot gases leaking outside the nozzle case ruptured the inner fan duct allowing the gases to escape into the fan exhaust stream.

Analysis: This example shows the familiar pattern of decreased N_2 and increased EGT and W_f associated with a high-pressure turbine efficiency loss. It is assumed that the nozzle guide vane shifted somewhat when the bolts sheared and the changed flow direction on the turbine caused the reduced efficiency. The monitoring trends indicate that the burn-through into the fan duct had minimal effect. Had the effect been greater, the trend would have begun to resemble that of a high-pressure bleed loss.

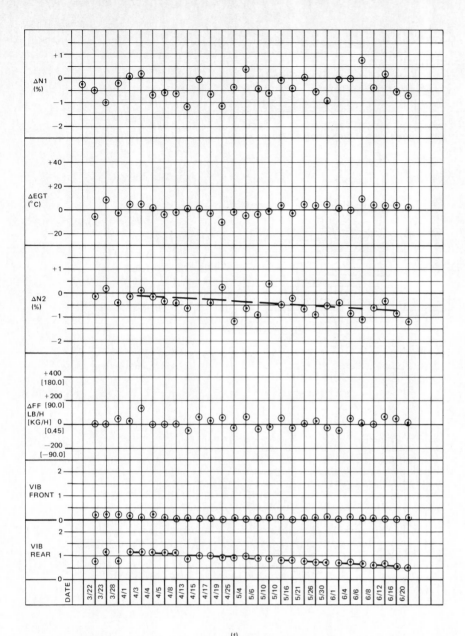

(f)

FIG. 18-17 (f) *Problem:* No. 3 bearing failure

Engine: JT8D

Malfunction: This engine was pulled when a routine oil filter check revealed metal on the screen.

Analysis: The AVM change which resulted from this failure was a decrease in vibration indicated on the turbine pickup. This is another good example of the change in vibration indication being toward a lower magnitude.

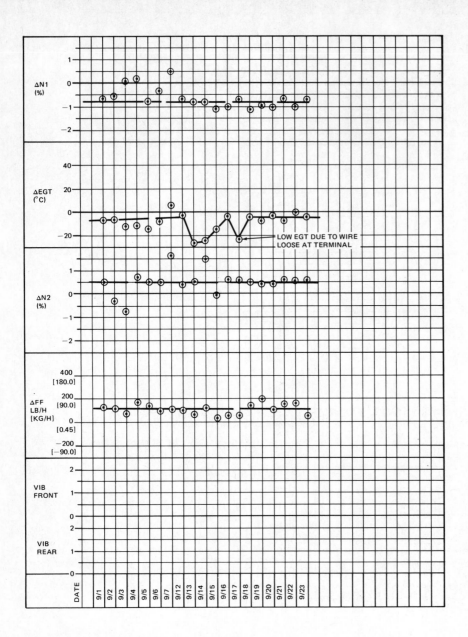

FIG. 18-17 (g) *Problem:* EGT system, wire loose at terminal

Engine: JT8D-1

Malfunction: The EGT system of this aircraft was examined when the EGT readings became low and erratic. A wire was found loose at a terminal.

Analysis: A change in magnitude of a reading or erratic readings without substantiation from the other parameters often indicates instrumentation problems.

mance change in each case. However, a general pattern usually appears in cases involving only the high turbine. A loss of turbine efficiency (broken blade or seal erosion) or an effective increase in turbine inlet area (bowed nozzle guide vanes) cause the turbine to absorb less than the designed amount of work resulting in a drop in N_2. For a given EPR, more energy is required and therefore, fuel flow and EGT increase. The change in N_1 is usually insignificant.

The conditions applying to the low-pressure turbine are much the same as the high-pressure turbine except that the work-per-blade ratio is quite a bit smaller. For this reason, the low turbine is less responsive to in-flight monitoring. Furthermore, damage to the low turbine is in most cases caused by debris from upstream failures.

The general pattern to be expected in failures affecting only the low-pressure turbine are the reverse of those described above. That is, N_1 normally shows a marked decrease. EGT and fuel flow again increase. N_2 will increase in this case; however, the increase may be slight

6. VIBRATION MONITORING [FIG. 18-17(f)] Engine malfunctions which exhibit themselves by a change in vibration level generally fall into two categories. The type of failure which produces an immediate unbalance, such as a broken turbine blade, will be evidenced by a sudden change in vibration level. The amount of change will depend upon the amount of unbalance. Turbine blade failures have occurred which increased the vibration level as little as one mil whereas others have resulted in full scale indicator readings. The other general type of malfunction is indicated by a progressive change in vibration level. This type of indication is usually more prevalent in bearing malfunctions where an initial unbalance can progress to an eventual failure of the bearing.

7. INSTRUMENTATION ERRORS [FIG. 18-17(g)] This section has been included to show how engine performance monitoring also can be used to detect instrumentation errors. The first, and most obvious, type of malfunction is the case where an individual engine instrument begins to give erroneous information. Then the performance plot for that instrument would show a deviation from the previously established base line. This malfunction is immediately suspected when a monitoring plot shows a trend in only one parameter, the reasoning here being that a malfunction affecting the gas path of an engine will cause trends in at least two of the performance parameters.

A second type of instrument malfunction is a bit more subtle. This is the type which affects more than one engine parameter and, in some cases, can resemble the indications of an actual engine failure. The instruments in this category are machmeter, EPR, and total air temperature.

SUMMARY

As has been indicated on the last several pages, cost and safety considerations have hastened the develop-

ment and adoption of a multitude of innovative maintenance and inspection procedures. These techniques have led to safer, lower-cost, and more efficient operation, with the expectation of reasonably long service life.

REVIEW AND STUDY QUESTIONS

1. Who determines the time between overhauls? What factors are taken into consideration in determining the TBO?
2. What is meant by a hot-section inspection?
3. Why is it necessary to mark gas turbine parts? List some temporary and permanent marking methods.
4. List the seven steps in the overhaul of any gas turbine engine.
5. What is contained in the overhaul manual that makes it so important to the overhaul process?
6. Briefly outline the disassembly process.
7. What disposition is made of all seals except the carbon rubbing types?
8. Why must engine parts be cleaned?
9. List some cleaning materials and processes for cleaning the cold section of the engine; the hot section. What protective measures should be taken while using some kinds of cleaning solutions?
10. What special precautions must be observed when cleaning and handling gas turbine bearings?
11. List some nondimensional inspection methods; dimensional inspection methods.
12. Describe the magnetic-particle inspection process; the dye-penetrant inspection process.
13. Make a list of some of the major conditions to be found by visual inspection. List the causes of these conditions.
14. Briefly discuss several methods of repair to gas turbine parts.
15. What is the purpose of safetying? List some safetying devices.
16. What precautions must be observed when working with synthetic lubricants?
17. What is meant by static and dynamic balance? How are rotating parts of the gas turbine balanced?
18. What marking systems are used to denote turbine blade weights?
19. What influence does the nozzle area have on engine performance? How is the nozzle area adjusted?
20. Briefly describe the tests performed on a gas turbine engine.
21. Describe the steps required to place an engine in storage.
22. What is the relationship between engine maintenance and engine cleanliness?
23. In what ways is the inspection of a gas turbine engine easier than that of a reciprocating engine?
24. Briefly describe some of the jobs to do when maintaining the gas turbine engine.
25. What is spectrometric oil analysis? How is this process used in the maintenance of the gas turbine engine?
26. Explain what is meant by performance monitoring, and discuss its implication to improved reliability of the engine.

CHAPTER NINETEEN
ENGINE TESTING AND OPERATION

All manufacturers run their engines in test cells before sending them to the users. After the test (green run), the manufacturer will usually disassemble one engine in several to ensure quality control. If an engine fails during a test run, that engine and a specific number of prior engines are disassembled to check for faults.

THE TEST CELL (*FIG. 19-1*)

Testing is done in a test cell or house, fully equipped to measure all of the desired operating parameters. Some of the larger installations cost several million dollars. The building is usually of concrete construction and contains both the control and engine room, although in some installations only the control or instrumentation room is enclosed. Most cells have silencers installed in the inlet stack for noise suppression and a water spray rig in the exhaust section for cooling. Many modern test cells incorporate computers to automatically record all instrument readings and correct them (see pages 354 to 356) to standard day conditions.

Testing of large modern engines has been a very real problem in that the amount of air required by the engine or its components was not readily available with existing equipment. New facilities have had to be built to simulate conditions encountered at very high Mach numbers and very high altitudes, and, in many cases, this has been as difficult as the development of the engine itself.

Test-cell instrumentation usually includes temperature gages to measure

1. Fuel and oil inlet temperature
2. Starter air temperature
3. Scavenge oil temperature
4. Compressor inlet temperature
5. Exhaust gas or turbine inlet temperature
6. Wet and dry bulb temperature
7. Ambient air temperature

and pressure gages and/or manometers to measure

1. Fuel inlet pressure
2. Lubrication-system pressure
3. Main and afterburner fuel-pump pressure
4. Nozzle pump inlet and rod end pressure (J79 engine)
5. Starter air pressure
6. Barometric or ambient air pressure
7. Sump or breather pressure
8. Turbine pressure or engine pressure ratio (EPR)
9. Water pressure
10. Turbine cooling air pressure

Additional instruments and controls include

1. Power lever and various control switches.
2. Vibration pickup and gage (usually taken at compressor and turbine planes).
3. Clock and stopwatch.

FIG. 19-1 A typical test cell and control room.

4. Tachometer-generator and readout device in actual rpm. It should be noted that in an aircraft installation "percent" rpm is used rather than actual rpm because there is a large difference in the actual rpm of the many different types and sizes of gas turbine engines. By using percent rpm, it makes it possible to have approximately the same percent rpm reading for the same power setting on a great variety of engines. In the United States, percent tachometers are designed to read 100 percent when the tachometer's drive shaft is turning at 4200 rpm. To find the actual rpm of any engine, simply divide the engine's tachometer drive gear ratio into 4200. For example, the gas generator (N_G) tachometer drive gear ratio on the Pratt and Whitney PT6 turboprop engine is 0.112, and $4200/0.112 = 37,500$ N_G rpm when the tachometer reads 100 percent. If the tachometer reads 90 percent, then $0.9 \times 37,500 = 33,750$ N_G rpm. (See Fig. 19-2 for another type of tachometer system.)

5. Fuel-flow transmitter and meter.
6. Thrust-measuring electronic or hydraulic load cell and readout, or torque readout.

When the engine is installed in the cell, a bellmouth inlet (Fig. 19-3) and screen are attached. The bellmouth inlet is a funnel-shaped tube having rounded shoulders which offers so little air resistance that the duct loss can be considered zero. The screen itself does offer some resistance and must be taken into account when extremely accurate data must be collected. Twenty-four-volt electric power is provided to operate the ignition system and any solenoid valves on the engine. One hundred fifteen volts, four hundred hertz may also be provided for some ignition systems and valves.

Test schedules vary with different model engines and manufacturers, but usually include instrument observations during starting and acceleration, and at the several thrust settings of idle, maximum cruise, maximum

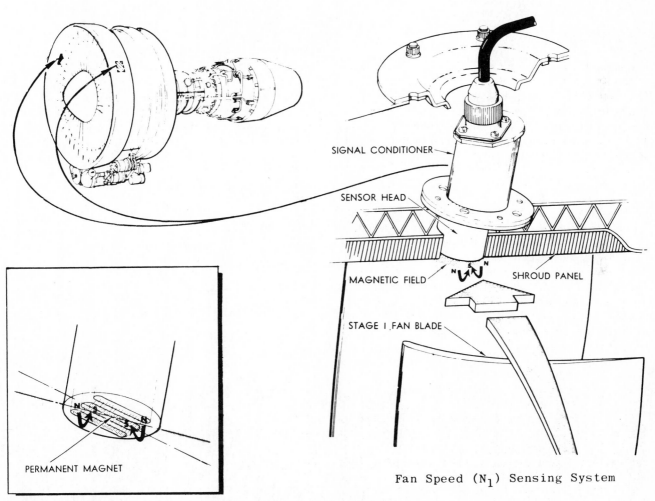

SIGNAL CONDITIONER

SENSOR HEAD

MAGNETIC FIELD

SHROUD PANEL

STAGE I FAN BLADE

PERMANENT MAGNET

Fan Speed (N_1) Sensing System

FIG. 19-2 This fan rpm indicating system is, in effect, a fan blade counting device, as opposed to the more common method of using a tachometer generator to measure rpm. The sensor heads mounted flush in the fan shroud panel contain permanent magnets. The passage of each fan blade disrupts the magnetic field set up by the sensor magnets, causing an electrical signal pulse. The frequency of the pulses is equal to the number of blades times the rpm, thus giving a signal frequency proportional to fan speed. The signal is amplified, conditioned and transmitted to the cockpit indicator to provide an N_1 readout in percent rpm.

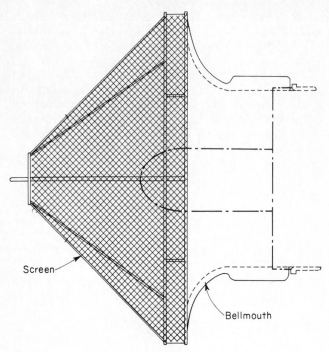

FIG. 19-3 A bellmouth inlet and screen.

climb, and maximum continuous takeoff. Acceleration time may also be recorded.

Most manufacturers will have an *engine log sheet* on which is recorded the following data in addition to the instrument readings (Fig. 19-4):

1. Date of run
2. Engine model and serial number
3. Serial number of components
4. Grade and specific gravity of fuel
5. Grade or specification of oil
6. Test-cell depression (pressure drop due to test-cell inlet restrictions)
7. Total time of test-cell runs
8. Reasons for unscheduled shutdowns
9. Repairs made to engine during test
10. Reasons for engine rejection (if applicable)
11. Oil consumption
12. Jet nozzle area
13. Overhaul agency (if applicable)
14. Test operator's and inspector's signatures

Correct engine performance is indicated by comparing corrected values (see Performance Testing that follows) with charts and graphs computed and drawn by the manufacturer guaranteeing minimum performance and values for the engine.

PERFORMANCE TESTING

As indicated earlier in this book, the performance of any engine is considerably influenced by changes in ambient temperature and pressure because of the way these parameters affect the weight of the air entering the engine. In order to compare the performance of similar engines on different days, under different atmospheric conditions, it is necessary to "correct" a given engine's performance to the standard day conditions of 29.92 inHg [101.3 kPa] and 59°F (519°R) [15°C (288°K)].

For example, the following conditions are known about a running engine:

1. rpm = 9465
2. EGT = 510°C (950°F or 1410°R). See note that follows.
3. W_f = 4000 lb/h [1814.4 kg/h]
4. W_a = 200 lb/s [90.7 kg/s]. (Although airflow is listed here, it is difficult, if not impossible, to measure the weight of airflow directly. Airflow can be determined indirectly through pressure measurements in the engine.)
5. F_n = 10,000 lb [4536 kg]
6. TSFC = 0.400

Barometric pressure	= 30.3 inHg [102.6 kPa]
Standard day pressure	= 29.92 inHg [101.3 kPa]
Ambient temperature	= 82°F [27.8°C]
Standard day temperature	= 59°F + 460° (519°R) [15°C + 273°C (288°K)]

NOTE: To convert degrees Fahrenheit to degrees Rankine, add 460 to the Fahrenheit reading.

Since these are all "observed" measurements, they must be "corrected" in order that valid comparisons can be made between engines. To change the observed operating parameters to the corrected values, i.e., the rpm, EGT, F/F, airflow (A/F), F_n, and TSFC that the engine would have if it were running under standard day conditions, it is necessary to apply a pressure correction factor, delta (δ), and a temperature correction factor, theta (θ).

$$\delta = \frac{\text{observed pressure (inHg)}}{\text{standard day pressure (inHg)}}$$

$$\theta = \frac{\text{observed temp. (°R)}}{\text{standard day temp. (°R)}}$$

For the atmospheric conditions stated above, delta and theta will be

$$\delta = \frac{30.3}{29.92} = 1.013$$

$$\theta = \frac{82 + 460}{59 + 460} = \frac{542}{519} = 1.045$$

$$\sqrt{\theta} = 1.022$$

NOTE: See pg. 97 for the reason the square root of theta is needed.

See the Appendix for tables listing the values of delta and theta for various pressures and temperatures.

To correct the observed data gathered for the above engine, the following formulas are used.

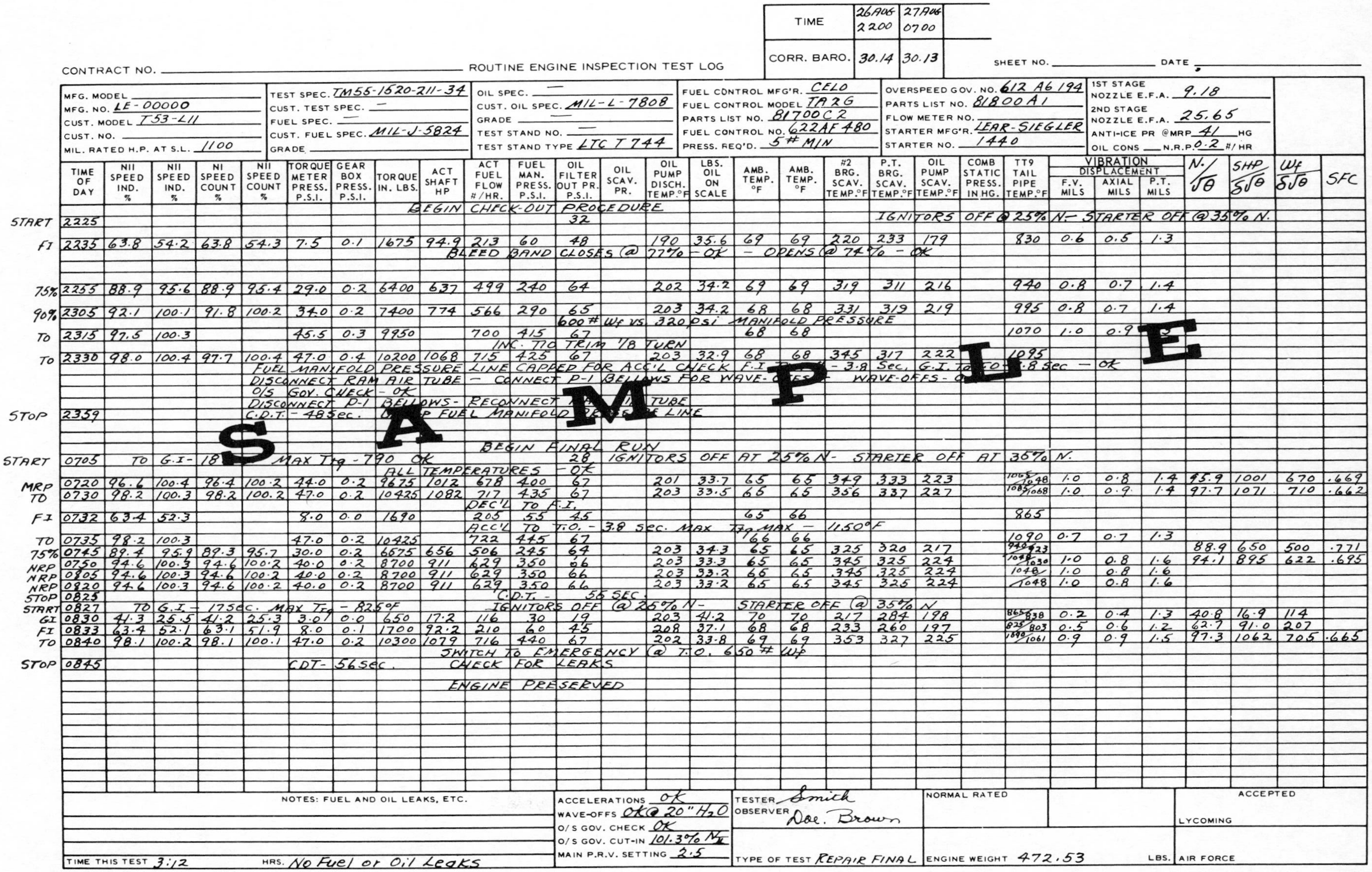

TIME: 26 AUG 2200 | 27 AUG 0700

CORR. BARO.: 30.14 | 30.13

CONTRACT NO. ______ ROUTINE ENGINE INSPECTION TEST LOG SHEET NO. ______ DATE ______

MFG. MODEL	TEST SPEC. TM55-1520-211-34	OIL SPEC. —	FUEL CONTROL MFG'R. CELO	OVERSPEED GOV. NO. 612 A6 194	1ST STAGE NOZZLE E.F.A. 9.18
MFG. NO. LE-00000	CUST. TEST SPEC. —	CUST. OIL SPEC. MIL-L-7808	FUEL CONTROL MODEL TA2G	PARTS LIST NO. 81800A1	2ND STAGE NOZZLE E.F.A. 25.65
CUST. MODEL T53-L11	FUEL SPEC. —	GRADE —	PARTS LIST NO. 81700C2	FLOW METER NO.	ANTI-ICE PR @MRP 41 HG
CUST. NO.	CUST. FUEL SPEC. MIL-J-5824	TEST STAND NO. —	FUEL CONTROL NO. 622AF480	STARTER MFG'R. LEAR-SIEGLER	OIL CONS ___ N.R.P. 0.2 #/HR
MIL. RATED H.P. AT S.L. 1100	GRADE	TEST STAND TYPE LTC T744	PRESS. REQ'D. 5# MIN	STARTER NO. 1440	

	TIME OF DAY	NII SPEED IND. %	NII SPEED IND. %	NI SPEED COUNT %	NII SPEED COUNT %	TORQUE METER PRESS. P.S.I.	GEAR BOX PRESS. P.S.I.	TORQUE IN. LBS.	ACT SHAFT HP	ACT FUEL FLOW #/HR.	FUEL MAN. PRESS. P.S.I.	OIL FILTER OUT PR. P.S.I.	OIL SCAV. PR.	OIL PUMP DISCH. TEMP.°F	LBS. OIL ON SCALE	AMB. TEMP. °F	AMB. TEMP. °F	#2 BRG. SCAV. TEMP.°F	P.T. BRG. SCAV. TEMP.°F	OIL PUMP SCAV. TEMP.°F	COMB STATIC PRESS. IN HG.	TT9 TAIL PIPE TEMP.°F	VIBRATION DISPLACEMENT F.V. MILS	AXIAL MILS	P.T. MILS	$N/\sqrt{\theta}$	$SHP/\delta\sqrt{\theta}$	$W_f/\delta\sqrt{\theta}$	SFC
START	2225	BEGIN CHECK-OUT PROCEDURE 32 — IGNITORS OFF @ 25% N- STARTER OFF @ 35% N.																											
FI	2235	63.8	54.2	63.8	54.3	7.5	0.1	1675	94.9	213	60	48		190	35.6	69	69	220	233	179		830	0.6	0.5	1.3				
		BLEED BAND CLOSES @ 77% - OK - OPENS @ 74% - OK																											
75%	2255	88.9	95.6	88.9	95.4	29.0	0.2	6400	637	499	240	64		202	34.2	69	69	319	311	216		940	0.8	0.7	1.4				
90%	2305	92.1	100.1	91.8	100.2	34.0	0.2	7400	774	566	290	65		203	34.2	68	68	331	319	219		995	0.8	0.7	1.4				
		600 # Wf vs. 320 psi MANIFOLD PRESSURE																											
TO	2315	97.5	100.3			45.5	0.3	9950		700	415	67				68	68					1070	1.0	0.9					
		INC. T/O TRIM 1/8 TURN																											
TO	2330	98.0	100.4	97.7	100.4	47.0	0.4	10200	1068	715	425	67		203	32.9	68	68	345	317	222		1095							
		FUEL MANIFOLD PRESSURE LINE CAPPED FOR ACC'L CHECK F.I. … - 3.8 Sec. G.I. TO … 8 Sec - OK																											
		DISCONNECT RAM AIR TUBE - CONNECT P-1 BELLOWS FOR WAVE-OFFS - WAVE-OFFS - OK																											
		O/S GOV. CHECK - OK																											
		DISCONNECT P-1 BELLOWS - RECONNECT [illegible] TUBE																											
STOP	2359	C.D.T. - 48 Sec. … FUEL MANIFOLD … LINE																											
		BEGIN FINAL RUN																											
START	0705	TO G.I. - 18 … MAX Trq - 790 OK — 28 IGNITORS OFF AT 25% N- STARTER OFF AT 35% N.																											
		ALL TEMPERATURES - OK																											
MRP	0720	96.6	100.4	96.4	100.2	44.0	0.2	9675	1012	678	400	67		201	33.7	65	65	349	333	223		1065/1048	1.0	0.8	1.4	95.9	1001	670	.669
TO	0730	98.2	100.3	98.2	100.2	47.0	0.2	10425	1082	717	435	67		203	33.5	65	65	356	337	227		1085/1068	1.0	0.9	1.4	97.7	1071	710	.662
		DEC'L TO F.I.																											
FI	0732	63.4	52.3			8.0	0.0	1690		205	55	45				65	66					865							
		ACC'L TO T.O. - 3.8 Sec. MAX T? MAX - 1150°F																											
TO	0735	98.2	100.3			47.0	0.2	10425		722	445	67				66	66					1090	0.7	0.7	1.3				
75%	0745	89.4	95.9	89.3	95.7	30.0	0.2	6575	656	506	245	64		203	34.3	65	65	325	320	217		940/923				88.9	650	500	.771
NRP	0750	94.6	100.3	94.6	100.2	40.0	0.2	8700	911	629	350	66		203	33.3	65	65	345	325	224		1048/1030	1.0	0.8	1.6	94.1	895	622	.695
NRP	0805	94.6	100.3	94.6	100.2	40.0	0.2	8700	911	629	350	66		203	33.2	66	65	345	325	224		1048	1.0	0.8	1.6				
NRP	0820	94.6	100.3	94.6	100.2	40.0	0.2	8700	911	629	350	66		203	33.2	65	65	345	325	224		1048	1.0	0.8	1.6				
STOP	0825	C.D.T. - 56 SEC.																											
START	0827	TO G.I. - 17 Sec. MAX Trq - 825°F — IGNITORS OFF @ 25% N- STARTER OFF @ 35% N																											
GI	0830	41.3	25.5	41.2	25.3	3.0	0.0	650	17.2	116	30	19		203	41.2	70	70	217	284	198		865/838	0.2	0.4	1.3	40.8	16.9	114	
FI	0833	63.4	52.1	63.1	51.9	8.0	0.1	1700	92.2	210	60	45		208	37.1	68	68	233	260	197		825/803	0.5	0.6	1.2	62.7	91.0	207	
TO	0840	98.1	100.2	98.1	100.1	47.0	0.2	10300	1079	716	440	67		202	33.8	69	69	353	327	225		1080/1061	0.9	0.9	1.5	97.3	1062	705	.665
		SWITCH TO EMERGENCY @ T.O. 650 # Wf																											
STOP	0845	CDT - 56 Sec. — CHECK FOR LEAKS																											
		ENGINE PRESERVED																											

NOTES: FUEL AND OIL LEAKS, ETC.

TIME THIS TEST 3:12 HRS. No Fuel or Oil Leaks

ACCELERATIONS OK
WAVE-OFFS OK @ 20" H₂O
O/S GOV. CHECK OK
O/S GOV. CUT-IN 101.3% N_II
MAIN P.R.V. SETTING 2.5

TESTER Smith
OBSERVER Doe. Brown
TYPE OF TEST REPAIR FINAL

NORMAL RATED
ENGINE WEIGHT 472.53 LBS.

ACCEPTED
LYCOMING
AIR FORCE

FIG. 19-4 Avco Lycoming T53 engine test log.

1. Corrected rpm = $\dfrac{\text{observed rpm}}{\sqrt{\text{temperature correction factor}}}$

or

$$\text{rpm}_{\text{corr}} = \dfrac{\text{rpm}_{\text{obs}}}{\sqrt{\theta}}$$

2. corrected EGT = $\dfrac{\text{observed EGT (°R)}}{\text{temperature correction factor}}$

or

$$\text{EGT}_{\text{corr}} = \dfrac{\text{EGT}_{\text{obs}}}{\theta}$$

3. Corrected fuel flow =

$$\dfrac{\text{observed fuel flow}}{\text{pressure correction factor} \times \sqrt{\text{temperature correction factor}}}$$

or

$$W_{\text{f,corr}} = \dfrac{W_{\text{f,obs}}}{\delta \sqrt{\theta}}$$

4. Corrected airflow

$$= \dfrac{\text{observed airflow} \times \sqrt{\text{temperature correction factor}}}{\text{pressure correction factor}}$$

or

$$W_{\text{a,corr}} = \dfrac{W_{\text{a,obs}} \sqrt{\theta}}{\delta}$$

5. Corrected thrust = $\dfrac{\text{observed thrust}}{\text{pressure correction factor}}$

or

$$F_{\text{n,corr}} = \dfrac{F_{\text{n,obs}}}{\delta}$$

6. Corrected TSFC

$$= \dfrac{\text{observed fuel flow}}{\text{observed thrust} \times \sqrt{\text{temperature correction factor}}}$$

or

$$\text{TSFC}_{\text{corr}} = \dfrac{W_{\text{f,obs}}}{F_{\text{n,obs}} \sqrt{\theta}}$$
$$= \dfrac{\text{TSFC}_{\text{obs}}}{\sqrt{\theta}}$$
$$= \dfrac{W_{\text{f,corr}}}{F_{\text{n,corr}}}$$

Additional corrections for humidity and variable-specific-heat fuels are also made on some engines.

Using the observed operating parameters given above, we find the corrected values to be

1. $\text{rpm}_{\text{corr}} = \dfrac{\text{rpm}_{\text{obs}}}{1\sqrt{\theta}} = \dfrac{9465}{1.022} = 9261$ rpm

2. $\text{EGT}_{\text{corr}} = \dfrac{\text{EGT}_{\text{obs}}}{\theta} = \dfrac{1410}{1.045} = 1349°\text{R} = 889°\text{F}$
   $= 476°\text{C}$

3. $W_{\text{f,corr}} = \dfrac{W_{\text{f,obs}}}{\delta \sqrt{\theta}} = \dfrac{4000}{1.013 \times 1.022} = 3864$ lb/h
   [1752.7 kg/h]

4. $W_{\text{a,corr}} = \dfrac{W_{\text{a,obs}} \sqrt{\theta}}{\delta} = \dfrac{200 \times 1.022}{1.013} = 202$ lb/s
   [91.6 kg/s]

5. $F_{\text{n,corr}} = \dfrac{F_{\text{n,obs}}}{\delta} = \dfrac{10,000}{1.013} = 9872$ lb [4477.9 kg]

6. $\text{TSFC}_{\text{corr}} = \dfrac{W_{\text{f,obs}}}{F_{\text{n,obs}} \sqrt{\theta}} = \dfrac{4000}{10,000 \times 1.022} = 0.391$

If one knows the corrected values (given in the manufacturer's performance specifications), engine operating parameters for *any* pressure and temperature can be computed as follows:

1. $\text{rpm}_{\text{obs}} = \text{rpm}_{\text{corr}} \sqrt{\theta}$
2. $\text{EGT}_{\text{obs}} = \text{EGT}_{\text{corr}} \theta$
3. $W_{\text{f,obs}} = W_{\text{f,corr}} \delta \sqrt{\theta}$
4. $W_{\text{a,obs}} = \dfrac{W_{\text{a,corr}} \delta}{\sqrt{\theta}}$
5. $F_{\text{n,obs}} = F_{\text{n,corr}} \delta$
6. $\text{TSFC}_{\text{obs}} = \text{TSFC}_{\text{corr}} F_{\text{n,obs}} \sqrt{\theta}$

GROUND OPERATING PROCEDURES (FIG. 19-5)

Although operation of the gas turbine engine has been greatly simplified as a result of automated-starting-sequence starting systems and more sophisticated fuel controls, it is still possible to seriously damage an engine or surrounding equipment through mismanagement of the engine's controls and improper positioning of the aircraft. Before starting any gas turbine engine, the operator must be familiar with the manufacturer's starting, operating, and stopping procedures and engine instrumentation, controls, and limitations (Fig. 19-6). Turbine temperatures are especially important in this respect and are more fully discussed on pages 361–363. Specific danger zones exist at the front and rear of the engine (Fig. 19-7). From a safety standpoint, gas turbine operaton is somewhat more hazardous than reciprocating engine operation. The whirling propeller at the front of the reciprocating engine is a clear and familiar hazard compared with the high-velocity airstream present at the front of an operating gas turbine engine. Several deaths and injuries have resulted from personnel being drawn into the inlet duct of an operating engine. Loose articles of clothing and other materials such as glasses, wipe-rags, caps, etc., have been snatched from people near an operating engine, causing severe damage to the engine.

AFFIRMATIVE;
CONDITION SATISFACTORY;
OK; TRIM GOOD, ETC.

Hold up thumb and fore-finger, touching at the tips, to form the letter "O."

NEGATIVE; CONDITION
UNSATISFACTORY;
NO GOOD, ETC.

With the fingers curled and thumb extended, point thumb downward to-ward the ground.

ADJUST UP (higher)

With the fingers extended and palm facing up, move hand up (and down), ver-tically, as if coaxing up-ward.

ADJUST DOWN (lower)

With the fingers extended and palm facing down, move hand down (and up), vertically, as if coaxing downward.

SLIGHT ADJUSTMENT

Hold up thumb and fore-finger, slightly apart (either simultaneously with the other hand when calling for an up or down adjustment, or with the same hand, immediately following the adjustment signal).

SHORTEN ADJUSTMENT
(as when adjusting
a linkage)

Hold up thumb and fore-finger somewhat apart (other fingers curled), then bring thumb and forefinger together in a slow, closing motion.

LENGTHEN ADJUSTMENT
(as when adjusting
a linkage)

Hold up thumb and fore-finger pressed together (other fingers curled), then separate thumb and forefinger in a slow, opening motion.

NUMERICAL READING (of an instrument
or to report a numerical value of any type)

Hold up appropriate number of fingers of either one or both hands, as necessary, in numerical sequence (i.e., 5, then 7 = 57).

NUMERICAL READING (of an instrument)

Same as "Numerical Reading," under General Signals

INCREASE TRIM SETTING
(or other adjustment)

Same as "Adjust Up (higher)," under General Signals

DECREASE TRIM SETTING
(or other adjustment)

Same as "Adjust Down (lower)," under General Signals

SLIGHT ADJUSTMENT

Same as "Slight Adjust-ment," under General Signals

TRIM GOOD

Same as "Affirmative," under General Signals

READ EXHAUST GAS
TEMPERATURE (EGT)

Make a motion with the hand, as if wiping perspi-ration from the brow.

CONNECT EXTERNAL POWER
SOURCE

Insert extended forefinger of right hand into cupped fist of left hand.

DISCONNECT EXTERNAL
POWER SOURCE

Withdraw extended forefinger of right hand from cupped fist of left hand.

START ENGINE

Circular motion with right hand and arm extended over the head. When nec-essary for multiengine air-craft, use a numerical fin-ger signal (or point) with the left hand to desig-nate which engine should be started. *

* NOTE: To use as an "all clear to start" signal, pilot or engine operator initiates the signal from the aircraft cockpit. Ground crewman repeats the signal to indicate "all clear to start engine."

FIG. 19–5 Hand signals used for safer operation of turbine aircraft. (Continued on pages 358 and 359.)

READ TURBINE DISCHARGE PRESSURE (P_{t7}) OR ENGINE PRESSURE RATIO (EPR) INDICATOR
Hold the palms of both hands together, with the fingers extended and pointing up, in front of the body. With the right hand, push the left hand back.

READ N₁ OR N₂ TACHOMETER
Point to nose with forefinger of right hand. When required, hold up one (1) or two (2) fingers of left hand to designate whether N_1 or N_2 is desired.

READ FUEL FLOW INDICATOR AT IDLE THRUST
(A) Point to eye with forefinger.
(B) With the fingers cupped and held close to the lips, make a drinking motion by tilting the head back.

READ FUEL FLOW INDICATOR AT MILITARY (OR TAKE-OFF) THRUST
(A) Starting at shoulder level, make circular motion upward with one hand, followed by full arm stretch.
(B) With the fingers cupped and held close to the lips, make a drinking motion by tilting the head back.

READ FUEL FLOW INDICATOR IN MANUAL OR EMERGENCY AT IDLE (fighter aircraft)
(A) Clasp the left wrist with the right hand.
(B) Point to eye with forefinger.
(C) With the fingers cupped and held close to the lips, make a drinking motion by tilting the head back.

READ FUEL FLOW INDICATOR IN MANUAL OR EMERGENCY AT MILITARY (fighter aircraft)
(A) With the right hand clasped over the left wrist, move clenched left fist forward, as if moving a throttle.
(B) With the fingers cupped and held close to the lips, make a drinking motion by tilting the head back.

READ OIL PRESSURE INDICATOR
Place hand on top of head, palm down.

CHECK OIL LEVEL
With the fingers of both hands extended, palms facing each other, place one palm a few inches above the other.

CHECK AFTERBURNER NOZZLE POSITION (fighter aircraft)

(A) REQUEST: Hold the palms of both hands together, with the fingers extended. Using the heels of the palms as a hinge, open and close the hands several times, in a flapping motion.

REPLY:
(B) AFTERBURNER NOZZLE OPEN: With both arms extended horizontally in front of the body, move the arms apart.
(C) AFTERBURNER NOZZLE CLOSED: With both arms extended horizontally to the sides of the body, bring the arms together in front of the body.

CHECK OVERBOARD AIRBLEED VALVE POSITION
(A) REQUEST: Hold out clenched fist, then open and close the fingers several times.

REPLY:
(B) VALVE CLOSED (OR CLOSING): With the fingers extended, slowly close the hand to form a "fist."
(C) VALVE OPEN (OR OPENING): With the fingers forming a "fist," slowly open the hand.

CHECK ANTI-ICING SYSTEM
REQUEST: Cover nose with cupped hand, as if to protect it from cold.
REPLY: Affirmative or negative signal, as applicable (refer to General Signals).

CHECK FOR FUEL OR OIL LEAK
Cup hand over an eye, telescope-fashion, moving the head as if looking for something.

PULL CIRCUIT BREAKER
With fingers of one hand cupped upward, draw hand down sharply, as if removing a plug.

REPLACE CIRCUIT BREAKER
With fingers of one hand cupped upward, push hand up sharply, as if replacing a plug.

PERSONNEL IN DANGER (for any reason). REDUCE THRUST & SHUT DOWN ENGINE(S).

(A) Draw right forefinger across throat. When necessary for multiengine aircraft, use a numerical finger signal (or point) with the left hand to designate which engine should be shut down.
(B) As soon as the signal is observed, cross both arms high above the face. The sequence of signals may be reversed, if more expedient.

FIRE IN TAILPIPE. TURN ENGINE OVER WITH STARTER.

(A) With the fingers of both hands curled and both thumbs extended up, make a gesture pointing upward.
(B) As soon as the signal is observed, use circular motion with right hand and arm extended over the head (as for an engine start). When necessary for multiengine aircraft, use a numerial finger signal (or point) with the left hand to designate the affected engine.

FIRE IN ACCESSORY SECTION. SHUT DOWN ENGINE AND EVACUATE THE AIRCRAFT.

(A) Draw right forefinger across throat. When necessary for multiengine aircraft, use a numerical finger signal (or point) with the left hand to designate which engine should be shut down.
(B) As soon as the signal is observed, extend both thumbs upward, then out. Repeat, if necessary.

IDLE RPM

Point to eye with forefinger. When tachometer reading is needed, engine operator holds up appropriate number of fingers in sequence to indicate numerical value.

ADVANCE THROTTLE TO MILITARY (OR TAKE OFF) THRUST

Starting at shoulder level, make circular motion upward with one hand, followed by full arm stretch.

JAM ACCELERATION
Short jab with clenched fist, as in boxing.

SLOW ACCELERATION
Same as for jam acceleration, except in slow motion.

REVERSE THRUST (commercial aircraft)

Swing out-stretched arm as far as possible behind the body and rotate in a circular motion.

ACTUATE THE EXHAUST SILENCER (commercial aircraft)

Place both hands over the ears, as if to protect them from noise.

RETRACT THE EXHAUST SILENCER (commercial aircraft)

With both hands placed over the ears, withdraw them outward, sharply.

OPERATE IN MANUAL OR EMERGENCY (fighter aircraft)

With the right hand clasped over the left wrist, move clenched left fist forward, as if moving a throttle.

RETURN TO NORMAL (fighter aircraft)

With the clenched left fist held forward, clasp the left wrist with the right hand, and draw left fist back toward the body.

START AFTERBURNER (fighter aircraft)

(A) Short, quick jab from the body with the left fist, as in boxing, followed by a pronounced movement of the fist to the left (outboard) (B).

STOP AFTERBURNER (fighter aircraft)

(A) Holding the left hand closed and arm horizontal, in front but slightly away from the side of the body, move the arm sharply to the right (inboard), then draw it quickly back, toward the body (B).

ACTUATE POP-OPEN NOZZLE (some fighter aircraft)

With the palms of both hands together and the fingers extended, open hands at the base, using finger tips as a hinge.

SHUT DOWN ENGINE

Draw right forefinger across throat. When necessary for multiengine aircraft, use a numerical finger signal (or point) with the left hand to designate which engine should be shut down.

(a)

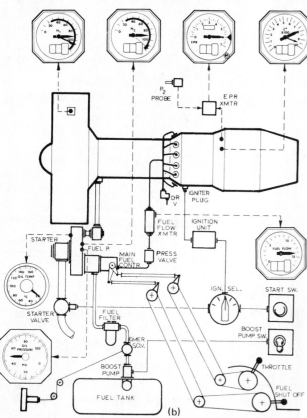

(b)

FIG. 19-6 Engine instrumentation for two high-bypass-ratio turbofan engines.

(a) The Lockheed L1011 incorporates a typical array of engine instruments and controls consisting of, from top to bottom,
  1. IEPR (integrated engine pressure ratio).
  2. N_1 rpm (high-speed spool speed).
  3. TGT (turbine gas temperature).
  4. N_3 rpm (fan speed).
  5. F/F (fuel flow).

(b) Basic engine instrumentation and controls for the General Electric CF6 series engine.

The high-velocity and high-temperature exhaust gases at the rear present a somewhat more obvious danger. They exit with a speed of over 1000 mph [1609 km/h], and can do serious damage to both personnel and equipment. To cite an example from one airline safety report, "A cart was blown over a ten foot fence, cleared the fence by another ten feet, and landed seventy-five feet beyond. The cart weighed seven hundred fifty pounds." The jet blast can also pick up and blow loose dirt, sizable rocks, and other debris a distance of several hundred feet and with considerable force.

The heat of the exhaust stream is serious only if the engine is operating at low rpm. The velocity of the jet blast at higher power settings is so great that a person entering the gas stream would be hurled a considerable distance without being burned. Of course the hot exhaust gases can have an irritating effect on lungs and eyes.

It is very important to correctly locate and position the airplane during ground operation. The aircraft should be run on concrete surfaces, since fuel and oil spillage will severely damage asphalt-type surfaces. The hot exhaust gas will also deteriorate the asphalt, especially if the engines are equipped with thrust reversers. An engine may occasionally *torch*. This is a condition where excess fuel accumulates during a starting attempt. It is not harmful to the engine if EGT limits are not exceeded, but it does point up the need to keep the area to the rear of the engine clear of flammable material. The airplane should face into the wind to reduce the dangers of starting overtemperatures and to obtain faster, smoother accelerations. Facing the airplane into the wind is a necessity if any adjustments are to be made.

Damage caused by foreign objects has been one of the principal reasons for premature engine removal. The axial-flow compressor is particularly sensitive to this type of injury. Foreign object damage (FOD) has cost the U.S. Air Force over fifty million dollars per year. Because FOD has been such a problem, military and civil airport operators have had to retrain their personnel to establish new working habits. It is no longer satisfactory to allow the end of a cotter pin or safety wire to remain where it falls. Special cleaners (Fig. 19-8) have been developed to continually vacuum ramps and runways. Some of these machines can pick up 1-lb [0.45-kg] chunks of steel while traveling at 25 mph [40 km/h].

Other dangers associated with the operation of the gas turbine are those resulting from noise (see Chap. 8), and the ignition system (see Chap. 16). No work should be performed on the ignition system while the engine is in operation. If the system must be worked on soon after it has been in operation, touch the end of the lead wire to the shell of the igniter to dissipate any residual energy. Caution must be exercised to avoid the chance of injury or death. Generally the system should be turned off for approximately five minutes, depending on the type, before the ignitor or leads are disconnected or removed. Many ignition systems incorporate radioactive discharge tubes and or beryllium oxide insulators in the ignitors, which should be disposed of in a special

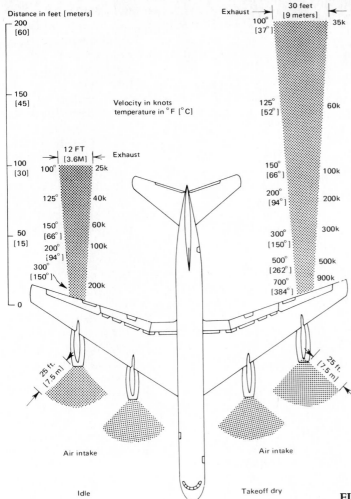

Distance in feet [meters]

Exhaust → | 30 feet [9 meters] | ← 35k

100° [37°]

200 [60]

Velocity in knots
temperature in °F [°C]

125° [52°]   60k

150 [45]

12 FT [3.6M] →| |← Exhaust

150° [66°]   100k

100° 25k
[30] 100°

200° [94°]   200k

125° 40k

300° [150°]   300k

150° [66°] 60k
50 [15] 200° [94°] 100k

500° [262°]   500k

300° [150°] 200k

700° [384°]   900k

0

25 ft. [7.5 m]

25 ft. [7.5 m]

Air intake

Air intake

Idle

Takeoff dry

FIG. 19-7 Typical engine inlet and exhaust hazard areas.

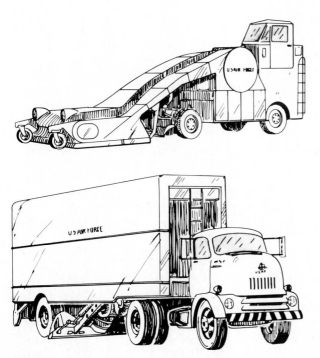

FIG. 19-8 Two types of runway and ramp vacuum cleaners.

manner. No work or inspection should be done in the area of the tailpipe for about ½ h or longer to reduce the chances of injury from hot metal parts or from the flashing of residual fuel in the combustion chambers or tailpipe. Spilled fuels and oils present a fire hazard, and direct contact may cause skin drying or other irritations on some people. Fuel and oil should be removed from the skin with soap and water as soon as possible.

Many engines are equipped with compressor bleed valves (Chap. 5). When checking bleed-valve operation or doing other work on or near the compressor bleed while the engine is running, care should be taken to stand clear during the period the bleeds are open. When the bleed valves open, high-pressure air at high velocity is dumped overboard.

STARTING A GAS TURBINE ENGINE

Starting procedures will, of course, vary, depending upon engine type and installation. Listed below is the starting procedure for the Pratt & Whitney Aircraft JT3D commercial turbofan engine equipped with a pneumatic starter (Fig. 19-9).

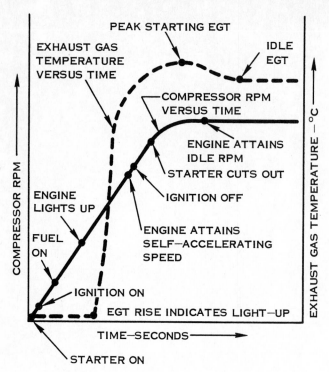

FIG. 19-9 Typical starting sequence for a gas turbine engine. (*Pratt & Whitney Aircraft.*)

1. *Power lever*—IDLE.
2. *Fuel shutoff lever*—CLOSED.

   CAUTION: Do not open the fuel valve before turning on the starter and ignition switches.

3. *Engine master switch*—ON.
4. *Fuel system shutoff switch*—OPEN.
5. *Fuel boost pump switch*—ON.
6. *Fuel inlet pressure indicator* (if the aircraft is so equipped)—5 psi [34.5 kPa] minimum (to ensure that fuel is being delivered to the engine fuel pump inlet).
7. *Engine starter switch*—ON. Check for oil pressure rise.
8. *Ignition switch*—ON.

   CAUTION: The ignition switch should not be turned on until the compressor begins to rotate.

When the N_2 tachometer indicates at least 10 percent rpm:

9. *Fuel shutoff lever*—OPEN. An engine light up will be noted by a rise in EGT. The engine should light up within 20 s or less, after it is pressurized by turning on the fuel.

   CAUTION: Insufficient air pressure to either a pneumatic starter or a combustion starter that is being used as a pneumatic starter may not supply enough starter torque to start the engine properly, resulting in a hot, hung, or torching start. Such a condition is highly undesirable and may normally be avoided. When air bleed from another engine is used to operate the starter, cau-

tion is necessary to ensure that the operating engine is turning over fast enough to provide an adequate supply of pressurized air to the engine being started. An engine should never be permitted to take longer than 2 min to accelerate to IDLE rpm. In the event of torching, higher-than-usual exhaust gas starting temperature, too long an acceleration time, or other abnormalities, discontinue the starting attempt and investigate (Fig. 19-9).

10. *Monitor the engine instruments* (See page 364)—until the engine stabilizes at IDLE, to ascertain that the start is satisfactory and that none of the limits stipulated in the engine check chart are exceeded.

When the engine attains IDLE rpm (if some means of automatic cutoff is not provided):

11. *Engine starter switch*—OFF
12. *Ignition switch*—OFF

   CAUTION: If the fuel is shut off inadvertently by closing the fuel-shutoff lever, do not reopen the fuel valve again in an attempt to regain the "light." Whenever the engine fails to light, shut off the fuel and ignition and continue turning the compressor over with the starter for 10 to 15 s to clear out trapped fuel or vapor. If this is not done, allow a 30 s fuel-draining period before attempting another start. Observe any starter cooling period or ignition cycle limitation that may be required. The starter may be reengaged at any time after the compressor has decelerated to 40 percent rpm or less.

Unsatisfactory starts can be categorized within the following three areas.

1. *Hot start*—The EGT goes above the manufacturer's specified limits as a result of a rich fuel/air ratio. Improper ratios can result from a malfunctioning fuel control, or ice, or other restrictions at the front of the compressor. Most manufacturers will list degrees of overtemperature in terms of time and temperature, rather than stating one specific overtemperature point. Figure 19-10 shows the starting overtemperature and time limits for the Pratt & Whitney JT12 (J60) engine installed in the Lockheed Jet Star and North American Sabreliner. Hot starts can very often be anticipated through experience by observing a greater-than-normal fuel flow or a faster-than-normal EGT rise. The operator should be prepared to abort the start, although some manufacturers recommend that the engine be run for a 5-min cool-down period unless it is obvious that the engine will be damaged by continued operation. A hot start may also be caused by a false or hung start. Experience has shown that there is a definite relationship between excessive exhaust gas temperatures and premature engine removals. The engine control system is designed so that exhaust gas temperature will normally be maintained within a safe margin. However, no system can be designed to compensate for operational malpractices. It is foolish to treat overtempera-

OVERTEMPERATURE PROCEDURES

A Shutdown. Check engine and determine cause of overtemperature.

B Perform visual inspection of all hot-section parts.
(1) Inspect exhaust duct for foreign particles.
(2) Inspect rear of the turbine for apparent damage.
(3) Inspect combustion section, turbine vanes, and front of turbine section for excessive distortion or damage.

C Perform teardown inspection of all hot-section parts.
(1) Inspect all turbine vanes for bowing, bend, and twist.
(2) Fluorescent penetrant inspect all turbine blades.
(3) Inspect all turbine blades for stretch.
(4) Inspect turbine disks for growth and hardness. (First- and second-stage disk hardness must be at least 66 on Rockwell A scale.)

D Perform complete overhaul inspection of all hot-section parts.
(1) Scrap all turbine blades.
(2) Fluorescent penetrant inspect all turbine vanes.
(3) Inspect all turbine vanes for bowing, bend, twist.
(4) Inspect turbine disks for growth and hardness. (First- and second-stage disk hardness must be at least 66 on Rockwell A scale.)

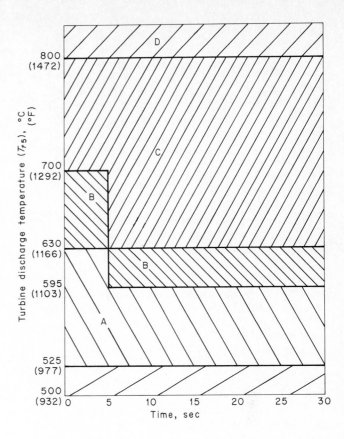

FIG. 19-10 Engine starting overtemperature limits for the Pratt & Whitney Aircraft JT12A-8 engine.

ture lightly. Just because the turbine does not fly apart or the engine "melt" away, is no reason to assume that the engine cannot be, or has not been, damaged. Several momentarily high overtemperatures may have as profound an effect on the engine as a single prolonged one of a lesser degree. Excessive internal temperatures aggravate such conditions as creep or deformation of sheet metal parts and shorten the life of the engine in general (Fig. 19-11).

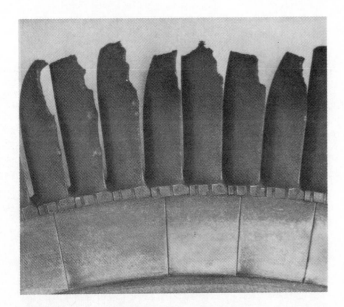

FIG. 19-11 The result of overtemperature.

2. *False or hung start*—After "light up" has occurred, the rpm does not increase to that of IDLE, but remains at some lower value. The EGT may continue to rise and again the operator should be prepared to abort before temperature limits are exceeded. A hung start could be the result of the starter receiving insufficient power or dropping out too soon.

3. *No start*—The engine does not light up within the specified time limit, indicated by no increase in rpm or EGT. Insufficient electric power, no fuel to the engine, problems in the ignition system, or a malfunctioning fuel control, all might lead to starting difficulties.

ENGINE OPERATION AND CHECKS

Just as starting procedures will vary with engine types, so will controls and instrumentation vary with airplane types.

Almost all gas turbine engine equipped airplanes will have the following levers, switches, and instruments to control and indicate engine operation:

1. Power lever
2. Fuel shutoff valve or switch
3. EPR gage
4. Percent rpm gage
5. EGT or TIT gage
6. Fuel-flow gage
7. Oil pressure and temperature gage
8. Torquemeter gage (turboprop or turboshaft only)
9. Starter switch

ENGINE TESTING AND OPERATION   363

10. Engine master switch
11. Fuel boost pump switch, pressure gage, and light
12. Ignition switch (may be activated as a function of power-lever position)

Depending upon the engine type, system, and aircraft, additional controls and instrumentation might be installed.

13. Water pump and injection switches
14. Anti-icing lights and switches
15. Nozzle area position indicator
16. Oil quantity gage
17. Feathering switches (turboprop only)
18. Emergency shutdown lever(s)
19. Free-air-temperature gage
20. Vibration indicator

Checking the gas turbine engine for proper operation consists primarily of reading engine instruments and then comparing the observed values with those given by the manufacturer for specific engine operating conditions, atmospheric pressure, and temperature.

Sudden throttle movements are to be avoided if possible in order to prevent cracking at the leading and trailing edges of the turbine blades. These parts of the blades are much thinner in cross section than the midspan areas and as a result heat and cool, and therefore expand and contract, faster than the thicker areas. Quick throttle movements cause exhaust gas temperatures to change very sharply and result in rapid expansions and contractions in the leading and trailing edges of the blades and slower changes in the thicker material. A 5-s acceleration would be better than a throttle burst and a slow deceleration is better than a throttle chop.

Good operating techniques are as follows:

1. Don't demand maximum power unless absolutely necessary. (See note.)
2. Start as fast as possible (high battery) to eliminate high temperatures for long periods of time.
3. Warm up and stabilize engine temperatures for a few minutes at IDLE to prevent rubbing and other damage.
4. Move the power lever slowly and smoothly.
5. Operate at less than "limits" and save a lot of money.
6. Cool the engine with a run faster than IDLE for a minute.
7. Calibrate the instruments for accurate readings.

NOTE: Some aircraft operators are limiting the maximum power to below the manufacturer's maximum rated power, in order to achieve extended reliabilty and engine life.

Early-model gas turbines usually used rpm as the sole engine operating parameter to establish thrust, while many present-day engines use EPR as the primary thrust indicator. On a hot day, compressor rpm for a given thrust will be higher than on a cold day. Furthermore, a dirty or damaged compressor will reduce thrust for a given rpm. EPR is used because it varies directly with thrust. It is the ratio of the total pressure at the front of the compressor to the total pressure at the rear of the turbine. The exhaust gas temperature is never used for setting thrust, although it must be monitored to see that temperature limits are not exceeded. Using EPR as the thrust indicator means that on a hot day it is quite possible for the engine rpm to exceed 100 percent, and on a cold day, desired thrust ratings may be reached at something less than 100 percent. Generally, thrust is set by adjusting the throttle to obtain a predetermined EPR reading on the aircraft instrument. The EPR value for given thrust settings will vary with ambient pressure and temperature.

On the newer, high bypass ratio fan jets such as the General Electric TF39 and CF6 engines used on the Lockheed C5A and Douglas DC10 respectively, the *fan* speed (see Fig. 19-2) is being used as the primary method of setting power because the large fan closely approximates the fixed pitch propeller and because of the fact that a large percentage of the total thrust is generated by the fan. While on the Rolls-Royce RB·211 engine used on the Lockheed L-1011 TriStar, the parameter used to indicate and manage thrust is the Integrated Engine Pressure Ratio (IEPR). This parameter is the integrated average of the fan and gas generator exhaust pressures (weighted by their respective nozzle areas) divided by the inlet total pressure. Rolls-Royce feels that because IEPR is based upon both the fan and gas generator exhaust pressure ratios, it is fundamentally related to engine gross thrust, and that IEPR provides the most accurate indication of engine thrust when considering total thrust scatter, engine ambient temperature sensitivity, altitude, and velocity, and the effect of engine component deterioration as compared to fan-speed (N_F), Turbine Gas Temperature (TGT), fuel flow (W_F), and gas generator pressure ratio (EPR). [See Fig. 19-6(a).]

ENGINE RATINGS *(FIG. 19-12)*

Turbojet and turbofan engines are rated by the pound of thrust they are designed to produce for takeoff, maximum continuous, maximum climb, and maximum cruise. The ratings for these operating conditions are published in the Engine Model Specification for each model engine. Takeoff and maximum continuous ratings, being the only two engine ratings subject to FAA approval, are also defined in the FAA Type Certificate Data Sheet. Engines installed in commercial aircraft are usually "part-throttle" engines; that is, takeoff-rated thrust is obtained at throttle settings below full-throttle position.

"Part-throttle" engines are also referred to as being flat rated, due to the shape of the takeoff thrust curves used for such engines. What is actually meant by the term "flat rating" is perhaps best described by comparing takeoff thrust settings on the military "full-throttle" engines with the "part-throttle" commercial engines.

The "full-throttle" engine is adjusted under sea-level standard conditions to produce full rated thrust with the throttle in full forward position. Ambient temperature changes occurring with the throttle in full forward position will cause thrust level changes. Temperatures ris-

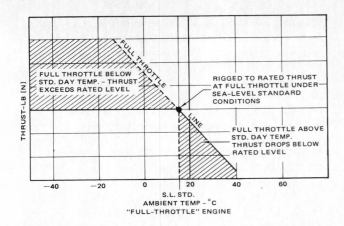

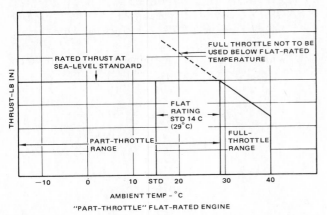

FIG. 19-12 Thrust rating versus ambient temperature for "full-throttle" and "part-throttle" engines. (*General Electric.*)

ing above the SL Std. 15°C will result in proportional thrust decrease, while at temperatures below standard, thrust will increase, exceeding the rated level as shown in Fig. 19-12.

For maximum reliability, better hot day performance, and economy of operation, commercial turbojet and turbofan engines are operated at the more conservative "part-throttle" thrust levels, thus in effect making them "flat rated." A flat-rated engine is adjusted under sea-level standard conditions to produce full rated thrust with the throttle at less than full forward position.

When ambient temperature rises above the SL Std. 15°C, rated thrust can still be maintained up to a given temperature increase by advancing the throttle. The amount of throttle advance available to keep the thrust level "flat rated" is determined by engine operating temperature limits.

As an example, the takeoff thrust of the General Electric CF6-6 high bypass turbofan engine is flat rated to sea-level standard day (15°C) plus 16°C = 31°C, at which point thrust becomes EGT-limited. Any further increase in ambient temperature will cause proportional decrease in thrust.

At ambient temperatures below SL Std., the thrust is held to the same maximum value as for a hot day. In this manner a flat-rated engine can produce a constant rated thrust over a wide range of ambient temperatures without overworking the engine.

Trimming

Operational engines must occasionally be adjusted to compensate, within limits, for thrust deterioration caused by compressor blade deposits of dirt or scale or other gas path deterioration. This process is called *trimming*. The word comes from the old practice of adjusting the engine's temperature and thrust by cutting or trimming the exhaust nozzle to size. Although the nozzle size on some engines can be varied by the insertion or removal of metal tabs called *mice,* the trimming process generally involves a fuel-control adjustment to bring the engine to a specific temperature, fuel flow, thrust, and engine pressure ratio. Manufacturer's instructions must be followed when performing trimming operations on any specific engine.

When the rated thrust cannot be restored without exceeding other engine limitations, the engine must either be field cleaned or removed and sent to overhaul (Fig. 19-13).

Field cleaning is accomplished by introducing a lignocellulose material into the air inlet duct while the engine is operating. The cleaning material known as Carboblast—Jet Engine Type is made by crushing apricot pits or walnut hulls. Specific steps to follow in cleaning any particular engine are to be found in the maintenance

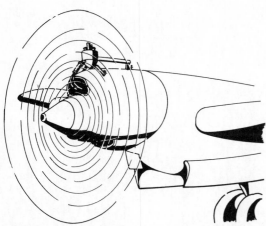

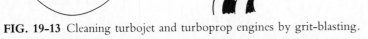

FIG. 19-13 Cleaning turbojet and turboprop engines by grit-blasting.

instructions for that engine. These steps generally include blocking some lines and ports and removing any equipment in the inlet duct that might be damaged by the cleaning material. The engine is then run at different speeds for stipulated periods of time while the Carboblast compound is fed into the inlet duct. After cleaning, the installation must be returned to its original configuration and the engine must be retrimmed.

On some engines, cleaning is accomplished by using a washing solution consisting of an emulsion of demineralized water, kerosene and other cleaning liquids such as Turco 4217. This type of field cleaning is done either as a desalination wash to remove salt deposits when operating in salt-laden air or as a performance recovery wash to remove dirt and other deposits which build up over a period of time depending on environment.

REVIEW AND STUDY QUESTIONS

1. Describe the construction on a typical gas turbine test cell.

2. List the instrumentation and controls necessary for the proper testing of the gas turbine engine.

3. What is the purpose of a bellmouth inlet?

4. What is the purpose of the engine log sheet? List some of the major items included on this log sheet.

5. What is the definition of the correction factors delta and theta? Why are these correction factors used?

6. Make a list of the precautions to observe before starting any gas turbine engine. Discuss the specific problems of the exhaust jet and of foreign objects.

7. Give the starting procedures and precautions to observe when starting the JT3D engine.

8. Describe a hot start; hung start; no start.

9. List the instruments and controls to be found in most gas turbine equipped airplanes.

10. What engine operating parameter is used as a primary indication of thrust? Why isn't rpm indication a good method of setting thrust?

11. What is engine trimming? How is it accomplished on most gas turbine engines?

12. What is meant by field cleaning? How is field cleaning accomplished?

PART FIVE
REPRESENTATIVE ENGINES

CHAPTER TWENTY
PRATT
& WHITNEY
AIRCRAFT
JT3D

The limited nature of this book does not allow a comprehensive treatment of all of the current gas turbine engines in use today. As a result, several engines which have gained wide use in either civil or military aircraft, have been selected for a somewhat more detailed examination of their construction. The Pratt & Whitney Aircraft JT3D engine is shown in Fig. 20-1.

SPECIFICATIONS

Number of compressor stages:	15
Number of turbine stages:	4
Number of combustors:	8
Maximum power at sea level:	17,000 lb [75,690 N]
Specific fuel consumption at maximum power:	0.52 lb/lbt/h [53 g/N/h]
Compression ratio at maximum rpm:	13:1
Maximum diameter:	53 in [135 cm]
Maximum length:	134 in [340 cm]
Maximum dry weight:	4150 lb [1884 kg]

The JT3D is an axial-flow turbofan engine having a 15-stage split compressor, an 8-can can-annular combustion chamber, and a 4-stage split turbine. Several models are available, differing only in minor details. In this engine the first two front-fan compressor stages are considerably larger than the remaining compressor stages. The fan provides two separate airstreams. (See Fig. 2-42.) The primary, or inner, airstream travels through the engine to generate pressures in the exhaust nozzle and thereby provide the propulsive force. The secondary, or outer, airstream is mechanically compressed by the fan and is ducted to the outside of the engine a short distance from the fan. This secondary airstream adds to the propulsive force and increases the propulsive efficiency, a process which is described in more detail in Chap. 3. Although the fan has the effect of a geared propeller, it is attached to and driven at the same speed as the front compressor.

OPERATION (FIG. 20-2)

The air enters the engine through the compressor inlet case assembly. The airframe inlet duct is attached to the front of the inlet case. This inlet case assembly is provided with vane-type multipurpose struts which transmit no. 1 bearing loads to the outer case structure, conduct anti-icing air and lubricating oil to the inner diameter of the engine, and direct air to the front-compressor section.

The compressor section is of the split type and consists of two rotor assemblies. Each rotor assembly is driven by an independent turbine, and each rotor is free to rotate at its best speed. Since it is necessary to rotate only one of these units during the starting operation, the selection of the small (rear) compressor permits the use of a smaller starter.

Primary air from the inlet case enters the front compressor, which consists of eight rotor stages and seven vane stages. The gas path of this compressor has a constant inside diameter and a decreasing outside diameter. This compressor is the larger of the two, and provides the initial compression of air. The rotating parts are connected by a drive shaft which passes through the inside of the rear-compressor rotor and drive shaft to the second-, third-, and fourth-stage turbines.

Between the front and rear compressors (and around the rear-compressor rotor and case) is the intermediate case. It has an automatic arrangement for bleeding front-compressor air. This is to improve the acceleration characteristics of the engine. (See Chap. 5.) The inlet vanes (ninth stage) transmit no. 2 and no. 3 bearing loads to the outer case, conduct supply and return oil to bearings, and direct the compressed air from the front compressor to the rear compressor.

The rear compressor has seven rotor stages (10 through 16) and six vane stages (10 through 15). It has a constant outside diameter and an increasing inside diameter, and is driven by the first-stage turbine through an independent shaft concentric with the front-compressor drive shaft. Just aft of the rear compressor is the diffusor case.

Compressor exit guide vanes (16th stage) at the front of the diffuser case straighten the air which is then expanded for entry into the combustion chambers. Struts in the case transmit the no. 4 and no. 5 bearing loads to the outer portion of the case. Also, these struts conduct bearing oil pressure and suction lines and provide high-pressure air for such engine functions as anti-icing. They also provide a source of clean air for aircraft pressure needs. On engines so equipped, the water injection manifold is mounted on the front flange. The water is dispersed through 16 curled tubes mounted on the periphery of the diffuser case, through mating holes in the

FIG. 20-1 The Pratt & Whitney Aircraft JT3D turbofan engine.

case, and injected into the airstream. The outside contours of the compressor and diffuser sections give the engine its "wasp waist" and at the same time provide a convenient location for the accessory section.

The fuel manifold, which consists of eight circular clusters of six fuel nozzles, is located in the diffuser case annulus and injects the fuel into the airstream in governed proportions. At this point, the air is channeled into eight portions for burning with the fuel in the eight combustion chambers.

The combustion section consists of eight separate cans arranged annularly. These chambers are connected by crossover tubes. The compressed air, its velocity increased, and with fuel injected, is lighted by spark igniters installed in two bottom combustion chambers. After lightup, the flame is perpetuated by the construction of the combustion chamber and the heat generated by previous combustion. The exhaust gases pass through the combustion-chamber outlet duct into the turbine nozzle case.

The turbine nozzle case houses the front stages of the four-stage turbine and is aft of the combustion section. The first stage drives the rear compressor, and the second, third, and fourth stages drive the front compressor. The turbine nozzle case also contains the four turbine exhaust nozzles. The nozzles are made up of a series of stationary vanes which direct the exhaust gases

through the turbine blades and into the turbine exhaust case.

To the rear of the turbine nozzle case is the turbine exhaust case which houses the fourth-stage turbine disk and blades and through which the exhaust gases are ejected from the engine. The turbine exhaust case also supports the no. 6 bearing and the no. 6 bearing oil sump.

FRONT ACCESSORY SECTION (FIG. 20-3)

The front accessory section consists of one assembly and related parts, the front accessory drive support. The case of this assembly is made of magnesium. Sixteen nuts secure the front accessory support to the no. 1 bearing support. The front accessory support has two mount pads on its front face. Only one is used. This is the mounting and drive for the compressor bleed-valve control. The other opening has provisions for mounting and driving a tachometer.

The no. 1 bearing scavenge pump is mounted on the lower rear face of the support. Air and oil for the bleed control are brought out of the no. 1 bearing support and tubed to the control. The front accessory drive gear, which is externally splined, is inserted, with an O-ring seal around it, into the front-compressor front hub engaging the internal spline in the hub. It is retained by the same nut that holds the no. 1 bearing inner race in position. The front accessory drive gear meshes with and drives the tachometer drive gear, the no. 1 bearing scavenge pump drive gear, and the compressor bleed-control drive gear.

NUMBER 1 BEARING SUPPORTS AND INLET CASE ASSEMBLY

These assemblies are discussed together since the supports are mounted in the inner diameter of the inlet case assembly, the vanes of which carry the structural load

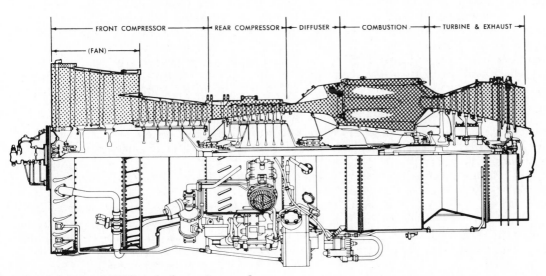

FIG. 20-2 Typical JT3D turbofan engine gas flow.

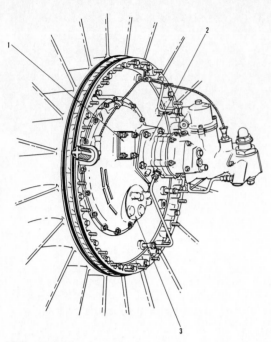

FIG. 20-3 Front accessories section (some models) showing (1) tachometer drive pad, (2) compressor bleed control mounting pad, and (3) oil scavenge pump area.

of the no. 1 bearing (front-compressor front) from the inner shroud to the outer shroud.

The inlet case assembly consists of 23 (or 28) hollow vanes incorporating foam-rubber stiffeners in the center bays, that are welded between hollow, double-walled titanium inner and outer shrouds. Each vane extends from the inner wall of the inner shroud to the outer wall of each shroud. There are holes in the side walls of the vanes that are between the shroud walls, and this forms a passage for the flow of anti-icing air. The left and right anti-icing tubes feed heated air, when desired, into the case and vane assembly, then forward into the inlet case.

Some of the vanes have tubes inside. These are for oil, breather, and bleed-valve control air lines. The no. 1 bearing oil pressure and oil return tubes are located in the seven o'clock and five o'clock position struts, respectively. The purpose of these vanes is to cause the air to enter the compressor rotor blades at the best angle for best compressor operation and to transport the structural load as discussed above.

The inner shroud, being of double-wall construction, provides for the passage of anti-icing air that has flowed inward through the guide vanes. The oil and air tubes that pass through the vanes have fittings mounted on the inner wall of the inner shroud.

On the front edge of the inner shroud, holes are drilled and tapped to hold the bolts that secure the no. 1 bearing front support and the compressor inlet airseal assembly. The no. 1 bearing housing holds the no. 1 bearing outer race and rollers in its inside diameter. The front accessory front support is bolted to its forward face. Anti-icing air outlet holes are also on the front edge of the shrouds. The no. 1 bearing rear support,

the inner section of which houses the no. 1 bearing oil seal, is bolted at its outer diameter to the compressor inlet vane inner-shroud rear flange.

The outer shroud, being also of double-wall construction, provides for the passage of anti-icing air. The outer wall of the outer shroud forms the support for the outer fittings of the tubes that pass through some of the vanes. A pressure probe fitting passes through both walls of the shroud to protrude into the airstream between the vanes. They are bolted to the outer walls.

The front rim of the outer shroud is drilled and tapped to receive the bolts which will hold the airframe air inlet duct. The rear rim of the outer shroud is drilled and tapped to receive the bolts that hold the shroud to the front-compressor case.

THE FRONT COMPRESSOR *(FIG. 20-4)*

The front compressor, which is housed in the front-compressor front case (and vane assembly) and rear case, consists of a rotor composed of eight rows of blades, with stators made up of seven rows of vane and shroud assemblies located between successive stages of blades. There is no second-stage stator, numbered as such, and there is no third-stage row of blades, designated as such. The first two rows of blades are considerably larger than the rest and are referred to as *fan* blades.

The airstream is separated into primary and secondary streams. The separation is achieved with the use of a spacer between the second- and fourth-stage blades. The primary airstream is directed internally by the third-stage stator while the secondary airstream is exhausted through the exit struts of the fan discharge case. The numbering of the blade stages from front to rear is 1, 2, and 4 through 9, and the numbering of the stator stages is 1 and 3 through 8. The ninth-stage stator vanes are incorporated in the intermediate case and will be discussed later.

The third-stage vane and shroud assembly is of single-piece construction, whereas the fourth through the eighth stages are of the split type. The spacers of the fourth through eighth stage are integral on most models. The inner shrouds form a seal ring for the two airseals on the outer diameter of each rotor disk spacer.

The compressor is driven by a shaft from the second-, third-, and fourth-stage turbines. Its rotational speed is roughly two-thirds the speed of the rear or high-speed compressor. It provides the initial compression to the air that passes through the engine and transmits this air to the rear or high-speed compressor. The stator vanes and rotor blades diminish in size and increase in quantity from the front to the rear of the compressor. This matches the decreasing volume of the air as the pressure of the air rises.

The front accessories are driven by a gear attached to the front hub of the front-compressor rotor. The rear hub of the front compressor contains the front-compressor drive turbine shaft coupling.

The rear hub of the front-compressor rotor is supported by a double ball bearing (no. 2). The two halves

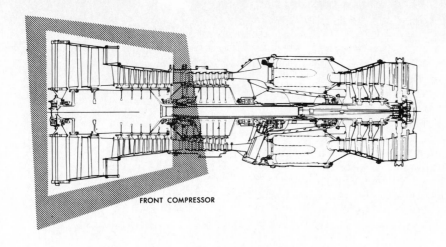

FRONT COMPRESSOR

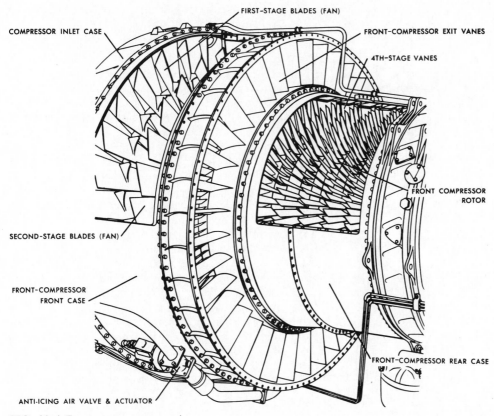

COMPRESSOR INLET CASE

FIRST-STAGE BLADES (FAN)

FRONT-COMPRESSOR EXIT VANES

4TH-STAGE VANES

FRONT COMPRESSOR ROTOR

SECOND-STAGE BLADES (FAN)

FRONT-COMPRESSOR FRONT CASE

FRONT-COMPRESSOR REAR CASE

ANTI-ICING AIR VALVE & ACTUATOR

FIG. 20-4 Front-compressor section.

of the no. 2 bearing (front compressor rear) are separated by an oil baffle in which drilled holes direct oil to the forward and aft sections of the double bearing. The no. 3 bearing (rear compressor front) inner race and rollers are mounted on the end of the rear hub. An oil seal on the rear end seals the bearing compartment from the engine airstream.

On the front hub, the rotor is supported by a roller bearing (no. 1), the liner of which is in the no. 1 bearing front support. An oil seal on the front end seals the bearing compartment from the engine airstream. The bearing support is secured to the inlet case assembly.

FRONT-COMPRESSOR CASES

There are three front-compressor cases, namely: the front-compressor case and vane assembly, the fan discharge case assembly, and the front-compressor rear case assembly. These will be discussed individually below. The front-compressor cases carry the structural load of the front of the engine. These cases decrease in diameter from front to rear to match the decreasing diameters of the stator rings. The front-compressor case and vane assembly and the fan discharge case are part of the front-compressor rotor and case assembly.

FRONT-COMPRESSOR CASE AND VANE ASSEMBLY

The front-compressor case and vane assembly attaches to the rear flange of the compressor inlet case and the front flange of the fan discharge case. In the approximate center of the case, riveted to internal flanges, are 55 (or 38) titanium first-stage stator vanes. At the titanium inner shroud of the vanes, a steel first-stage air-sealing ring is riveted. From the forward internal flange to the inlet case, a front airflow duct is inserted. Anti-icing air comes into the case through the reinforced openings and the outer wall, circulates, and flows forward into the inlet case. From the rearward internal flange to the fan discharge case forward flange, a rear airflow duct is inserted.

FAN DISCHARGE CASE ASSEMBLY

Attached to the rear flange of the front-compressor case and vane assembly by bolts and pinned to the rear airflow duct, previously discussed, is the fan discharge case. This case is constructed of stainless steel and consists of an outer shroud and an inner case with 38 steel ducts between them. The struts are secured in the inner case by a riveted locking plate at the rearward end and wired to the fairing retaining screws at the forward end.

FRONT-COMPRESSOR ROTOR AND CASE ASSEMBLY *(FIG. 20-5)*

The front-compressor rotor and case assembly consists of the front-compressor rotor, as described below, assembled to include the front-compressor case and vane assembly, the airflow ducts, the fan discharge case, and the third-stage vane and shroud assembly.

FRONT-COMPRESSOR ROTOR

The front-compressor rotor consists of 2 hubs, 7 disks, 7 spacer assemblies, 8 rows of blades, 2 sets (16 each) of tierods and associated hardware. Each row of blades is inserted into undercut slots in its disk. They are held in place by locks inserted under the blades and bent to secure. The blades do not have a tight fit, but rather are seated by centrifugal force during engine operation. The front hub forms the disk for the first row of blades, but the rear hub is a separate unit that is held fast to the rear face of the seventh-stage disk, or, on some models, to the forward face of the eighth-stage disk, by the tiebolts. The spacers between the disks are internally reinforced with tubes. The tiebolts run through these tubes and through the disks. These spacers have two knife-edges on their outside diameter, which run against the seal platforms on the stator vane inner shroud which was previously described. The exception to the above is the large spacer between the second- and fourth-stage compressor blades. This serves as a means of joining the fan section to the remainder

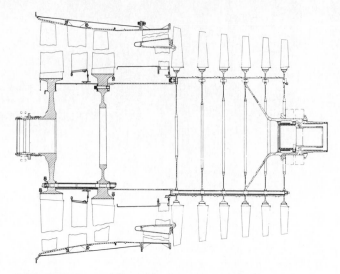

FIG. 20-5 Front-compressor case, vane assembly, and rotor.

of the compressor, and in so doing provides the needed space in that area. Each disk has 24 holes in its flange: 16 are for tierods and 8 are for balancing weights when needed. The front hub has two lips on its forward face and the second-stage disk has a similar lip on its rearward face to which balance weights can be added as required. The entire assembly is held together by the two sets of tiebolts. The smaller set holds the first two stages of blades together, while the longer set holds the third through eighth stages together. Both sections are joined together by the spacer, as previously mentioned. The rotor blades decrease in size from front to rear. The first two stages of blades (fan) are considerably larger than the rest. The angle of each row of blades is set to give the best efficiency at operating speed. The rear hub flange has large holes to allow some ninth-stage air into the compressor rotor. This air serves a dual purpose of providing bearing seal pressurizing and cooling. The front hub and the disks of the front-compressor rotor are titanium or steel. Most rear hubs are steel. The blades of the first two stages of all models are titanium. The remaining stages are of steel or titanium, depending on the engine model.

COMPRESSOR INTERMEDIATE SECTION *(FIG. 20-6)*

The forward mounting points are on the compressor intermediate case which is attached to the rear flange of the front-compressor case. A locating pin is used at top center between the two flanges. The intermediate case surrounds the rear compressor, but is not considered as part of it. It serves to separate the low-pressure front compressor from the high-pressure rear compressor and serves a structural function of joining the external cases. The intermediate case is of steel construction and has a double wall on the forward end. Guide vanes are welded from the outer wall, through the inner wall, and extended into a shroud ring. The double-mounting lugs are machined from rings. The case is welded to the

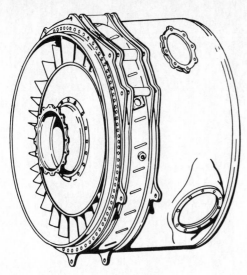

FIG. 20-6 Compressor intermediate case.

rings. The front flange of the case is drilled and tapped for bolts that attach it to the front-compressor case, and the rear flange is drilled.

An oil breather pad is located between the mounting flanges at approximately the two o'clock position. An oil-tube fitting is at the six o'clock position between the mounting flanges. Toward the rear of the case a hole is provided for air bleeding and the air-bleed valve mounting. The 30 vanes that are on the inside front of the case serve as air inlet guide vanes to the rear compressor. They are hollow steel vanes with openings in their side walls at the outer end. These openings, together with the double wall of the case, form the breather passage from the upper breather connections on the case to the lower opening. The lower opening connects directly to the main accessory case. Breather and oil tubes for the bearings located in the center of the case are through the hollow guide vanes. On many engines the inlet vanes are welded to the no. 2 bearing front support whose outer configuration is shaped to form an inner-shroud ring for these vanes. At the front (inner) and rear (outer) ends of this support, bolt circles are provided to receive bolts to hold the diagonal and rear bearing supports. On one model the no. 2 bearing front support is held in place with bolts. A hollow tubular seal is used at the rear (outer) bolt circle. On some model engines, the center support is part of the no. 2 bearing housing. On others, the center support is separate.

On some engines, in order to reduce absorption of heat by engine oil, heat shielding and thermal-blanket insulation is provided for the no. 2 bearing compartment. The bolts that secure the compressor intermediate front-bearing housing to the rear face of the front-bearing inner support also secure the compressor intermediate front-bearing front heat shield to the front face of the support.

The compressor intermediate rear-bearing heat shield consists of a front and rear heat shield, both of which are secured together and in turn are secured as a unit to the rear face at the inner bolt circle and outer bolt circle

of the compressor intermediate bearing rear support.

The thermal-blanket insulation is contained within the heat shield.

The rear support holds the no. 3 bearing seal assembly and the rear of the no. 2 bearing housing and is bolted to the rear of the center support and the vane inner shroud. A stepped-edge seal ring is bolted to the support at the outer diameter of the rear support plate. This ring, together with its mating knife-edge seal minimizes air leakage out of the front end of the rear compressor. The area inside the ring and behind the rear support plate is exposed to high-pressure air extracted from the 12th stage of compression. The air pressure outside the seal ring is from the ninth stage.

FRONT-COMPRESSOR REAR CASE ASSEMBLY

Attached to the rear innermost flange of the fan discharge case and to the forward flange of the intermediate case is the front-compressor rear case assembly. The front-compressor rear case assembly is of welded construction. Welded to it internally are the supports for the vane and shroud assemblies. A center external flange serves as a reinforcement and is also used to support the brackets that carry external items.

FRONT-COMPRESSOR VANE AND SHROUD ASSEMBLIES

The first-stage stator vanes were discussed under the front-compressor case and vane assembly. There is no second-stage stator. The third- and fourth-stage rows or stators are made of aluminum except the fourth stage of one model engine, which is steel. The fifth-through eighth-stage stator vanes are made of stainless steel and are welded into steel shroud rings. The stators are pinned to inner shrouds to which are riveted inner air-seal platforms. The third-stage stator vanes and shroud are built into a single circular assembly which is held stationary by a flange inserted between the fan discharge case and front-compressor rear flanges. The fourth-through eighth-stage stator vanes and shrouds are split and are prevented from rotating by being pinned to each other and to the third-stage vane and shroud. When assembled, the rings are held in the engine by shoulders on the inside of the front-compressor rear case.

The angle at which the vanes are mounted in the shrouds is set to feed air into the following row of rotor blades to give the best compressor efficiency at operating speed. Stator vanes decrease in size from front to rear to match the decreasing volume of air and the decreasing size of the rotor blades.

REAR COMPRESSOR *(FIG. 20-7)*

The rear, or high, compressor is driven by a hollow shaft from the first-stage turbine. Its function is to compress further the air delivered by the front compressor

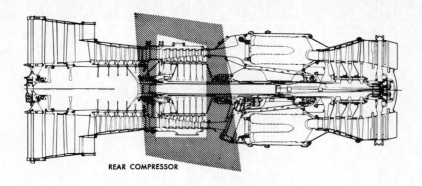

REAR COMPRESSOR

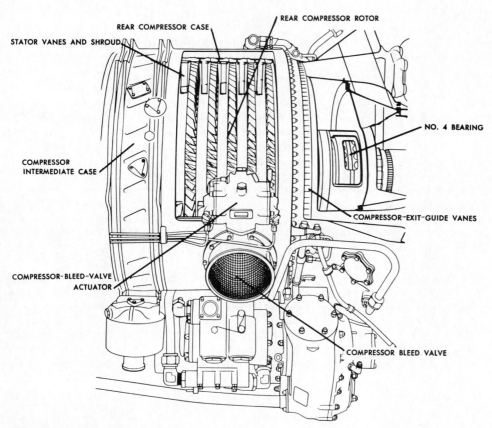

STATOR VANES AND SHROUD

REAR COMPRESSOR CASE

REAR COMPRESSOR ROTOR

NO. 4 BEARING

COMPRESSOR INTERMEDIATE CASE

COMPRESSOR-EXIT-GUIDE VANES

COMPRESSOR-BLEED-VALVE ACTUATOR

COMPRESSOR BLEED VALVE

FIG. 20-7 Rear-compressor section.

and then feed this air into the diffuser case and burners.

The rear compressor consists of a stator having 6 rows of vanes (stages 10 through 15) and a rotor having 7 rows of blades, stages 10 through 16. The exit guide vanes, stage 16, are mounted in the diffuser section. In function, these are part of the compressor, but because of their structural location, we will discuss them under the diffuser.

REAR-COMPRESSOR VANE AND SHROUD ASSEMBLIES

There are six vane and shroud assemblies in the rear compressor. The vanes reduce in height from front to rear of the compressor. The outer-side diameter of the air passage formed by these assemblies is constant; the decreasing size of the vanes is accomplished by increasing the diameter of the inner shroud rings. The vanes are steel and are brazed to the inner shroud and pierced through the outer shrouds. On some engines the 10th- through 14th-stage vane and shroud assemblies and the spacers separating the assemblies are integral. The 15th stage does not provide spacing. However, this is furnished by the 16th stage located in the diffuser case. Between each shroud, dowel pins are used to lock the series of shrouds together. The rear shroud is pinned to the outer shroud to prevent rotation. An edge of each spacer rests against a case shoulder to center the assembly.

AIRSEALS

The inside shroud of each row of vanes has a steel ring, with a small step on its inside face, riveted to it. Two knife-edge seals on each rotor spacer ride free of the steps forming an airseal between the compressor stages.

REAR-COMPRESSOR ROTOR *(FIG. 20-8)*

The rear-compressor rotor has 7 rows of blades on disks; 2 hubs, 6 spacer assemblies, and 16 tierods with nuts and washers. A knife-edge seal ring is riveted to a lip on the forward face of the 10th-stage disk. This seal ring rides free of the platformed seal which was referred to in the compressor intermediate case discussion. Each row of blades is inserted into undercut slots in its disk. They are held in place by locks inserted under the blades and bent to secure. The blades do not have a tight fit, but rather are seated by centrifugal force during engine operation.

Every disk has 32 holes in its flange: 16 are for tiebolts, and 16 are for balancing weights when needed. The entire assembly is held together by tiebolt heads on one end and a nut and washer on the other end.

The spacers between the disks are internally reinforced with tubes. The tiebolts run through the tubes and the spacer internal flanges as well as the disk flanges.

Neither the front nor rear hub is integral with a disk. On some models, the front hub is attached to the front face of the 12th-stage disk and the rear hub to the rear face of the 16th-stage disk. On others, the front hub is attached to the rear of the 11th-stage disk and the rear hub to the rear face of the 16th-stage disk. The hubs are secured to the rotor assembly by the steel tiebolts mentioned above. A steel tube runs from one hub to the other inside the rotor.

The tube is a force fit into the inside diameter of both hubs and permits breathing within its inside and keeps

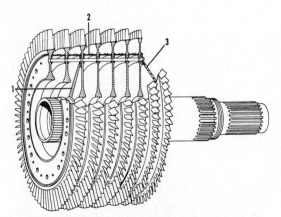

FIG. 20-8 Rear-compressor rotor showing (1) front hub, (2) spacer assembly, and (3) rear hub.

12th-stage pressure air from the no. 3 and no. 4½ bearings.

The front seal was mentioned previously. If required, a balance weight may be riveted to it.

The third spacer from the front has holes drilled in it. Through these holes 12th-stage air is bled into the rotor center where it is bled through holes in the front hub. The air is directed forward to the space just behind the no. 3 bearing rear support plate and pressurizes the no. 3 bearing oil seal. Since there are no holes in the rear hub disk, the pressure of the air against this disk counteracts part of the rotor forward thrust.

The rear compressor case carries no structural bearing loads and is thus made of relatively thin sheet metal. Its purpose is to hold the stator parts of the compressor and act as an air separator. The air pressure increases from front to rear and finally becomes 16th-stage pressure. Between this case and the intermediate case which envelops it, the pressure is ninth-stage air. The case has a flange at its rear end to which the screws that attach the case to the diffuser section are secured. Also, bolt holes are provided to hold the immediate intermediate case to the diffuser section.

DIFFUSER CASE *(FIGS. 20-9 and 20-10)*

The velocity of the air as it leaves the rear compressor is very high. This motion is both rearward and tangential around the engine. Exit guide vanes at the forward end of the diffuser case convert the tangential whirl into pressure energy. After the vanes, the high-pressure air will have a large rearward velocity. The gradual increasing area of the airflow passages provided by the diffuser case configuration decreases the airflow velocity to a suitable burning speed and increases the pressure.

The diffuser case shown in Fig. 20-10 is of welded corrosion- and heat-resistant steel construction with drilled flanges at both ends for bolting to the rear compressor and intermediate cases in the front and the combustion-chamber case in the rear.

The pad at the top center of the case is used for a breather connection. This leads to the bearing area through a strut. Airbleed outlet pads are on the outside of the case. An opening with mounting bosses at the bottom of the case provides for the angled accessory-drive system and the accessory-drive adapter which transmits high rotor motivation to the accessory-drive gearbox assembly. The rearmost pad at the bottom center of the diffuser case is for mounting the fuel pressurizing and dump valve, which is attached outside the case. The fuel manifolds attach to the inside of the case at this pad. The fuel manifold inlet distributor which connects to the fuel pressurizing and dump valve passes through a hole in the case at this location.

Two forward-facing and one rearward-facing bolt rings are mounted on the struts from the outer case. Sheet metal between these bolt rings and on the inside and outer case add to the strength and form the gradually increasing cross section of the diffuser. A slightly tapered diffuser inner inlet duct is bolted between an

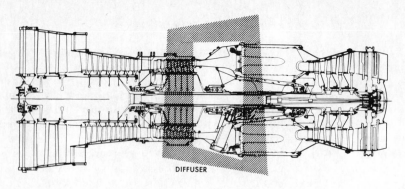

DIFFUSER

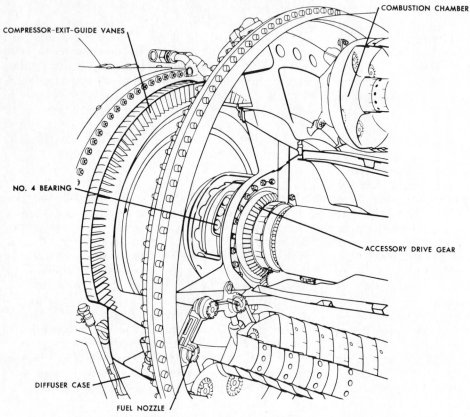

COMPRESSOR-EXIT-GUIDE VANES

COMBUSTION CHAMBER

NO. 4 BEARING

ACCESSORY DRIVE GEAR

DIFFUSER CASE

FUEL NOZZLE

FIG. 20-9 Diffuser section.

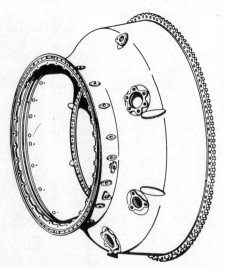

FIG. 20-10 Diffuser case.

inside flange on the inlet guide vane inner shroud and the outer, forward-facing, bolt circle inside the diffuser case. This piece forms a part of the diffuser ducting. The rivet circle that holds the diffuser inner inlet duct to the inlet vane ring extends forward to ride over two knife-edges that are on the rear disk of the rear compressor. This forms the airseal between the rear compressor and the bearing housing area at the center of the diffuser case. The outer shroud flange is clamped between the diffuser case and the rear compressor case when these units are assembled.

The optional water injection nozzles used on some engines are mounted around the periphery of the diffuser case. The nozzles permit the entry of water into the highly compressed airstream, thereby increasing its density and lowering its temperature prior to injection of the fuel from the fuel nozzle.

WATER-INJECTION MANIFOLD
(optional, on some models)

The water-injection manifold is the split type which is mounted on the diffuser case front flange. The water is dispersed through the 16 curled tubes mounted on the periphery of the diffuser case. The elbows of the tubes are aligned to the 16 holes in the diffuser case. Water is metered from the water-injection control through a supply tube to the tee fitting on the diffuser case front flange. Water then flows through the manifold into the airstream in the diffuser case.

NUMBER 4 BEARING HOUSING

The no. 4 bearing is in the diffuser section. This supports the rear-compressor rear hub. The bearing housing and its seal are fastened to the forward-facing inner bolt ring with studs and nuts. A synthetic rubber gasket is used just inside of the stud circle.

The no. 4 bearing housing has a heat-shield assembly bolted to it. The compressor rear-bearing heat-shield assembly consists of several individual heat shields welded together. The complete unit is bolted to the compressor rear-bearing housing and extends over the bearing housing, the breather tube, and the oil-pressure and scavenge tubes and along the towershaft.

The thermal-blanket insulation is contained within the heat shield.

NUMBER 4 AND NO. 5 BEARING OIL SUCTION PUMP

The no. 4 and no. 5 bearing oil suction pump and the accessory drive gears are in the lower rear of the diffuser.

BRACKETS

A quantity of brackets are attached to the external bolt circles for mounting external tubing and units.

FUEL MANIFOLD ATTACHMENT

As viewed from the rear, 16 brackets are welded to the inner structure of the diffuser. These brackets, and brackets that are welded to the inside rear of the diffuser outer case, support the fuel manifold.

FUEL MANIFOLD *(FIG. 20-11)*

The fuel manifold is split at the inner flange and between the nos. 1 and 8 clusters. The manifold assembly consists of a primary and secondary fuel manifold and eight clusters of fuel nozzles. There are six fuel nozzles in each cluster. The fuel manifold incorporates mounting lugs in the inner and outer diameter which are secured to the rear face of the diffuser case. Eight combustion-chamber positioning brackets are mounted on the outer-diameter mounting lug locations.

FUEL NOZZLES *(SEE FIG. 12-26)*

There are 24 nozzles mounted in each manifold, making a total of 48 per engine. There are two fuel outlets in each nozzle, a small center hole and a ring around the center hole. The ring sprays secondary fuel. This is done so that the small fuel flow that is used in idling will be broken up into a fine spray by being forced through a small outlet. The relatively larger outlet formed by the ring will generate a fine spray on the large flow of secondary fuel. Two screens are mounted in the rear of each nozzle, one for primary fuel and the other for secondary fuel. The primary screen is a small cylindrical type, whereas the secondary is a flat round type. Transfer of fuel from the tubes in the nozzle clusters is done by the nozzle body which contains internal passages that connect the nozzle center passage to the outer fuel tube and the nozzle outer passage to the inner fuel tube. Fuel nozzle seals, held under each nozzle, prevent leakage

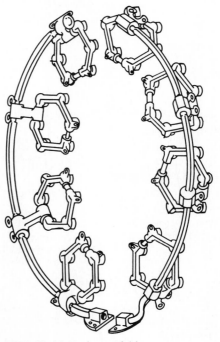

FIG. 20-11 Fuel manifold.

between passages and leakage to exterior. Each nozzle is held in place in the cluster by a threaded nozzle cap and a tab washer. There are holes around the nozzle cap wall to admit air for cooling and to aid in fuel vaporization.

Fuel is supplied to the primary and secondary orifices through separate manifold paths. The nozzle may be operated with spray from the primary orifice or with spray from both primary and secondary orifices as is the case at higher fuel flow. When both orifices are delivering fuel, their output is blended into a single spray.

The dual orifice nozzles are designed to discharge a predetermined amount of fuel when specified pressure heads are maintained across the primary and secondary stages. Extremely accurate calibration of each fuel nozzle minimizes flow variations between nozzles. Since the nozzle acts both as an atomizing and as a metering device, each nozzle must deliver a uniform spray pattern free from streaks and with an angle of spray which is held within close tolerances.

COMBUSTION SECTION (FIG. 20-12)

The combustion-chamber outer case is secured at the front flange to the rear flange of the diffuser case and at the rear flange to the turbine nozzle case. Eight combustion chambers, often called burner cans, the combustion-chamber inner case, the no. 5 bearing support, and the no. 5 bearing housing are located inside the combustion-chamber outer cases.

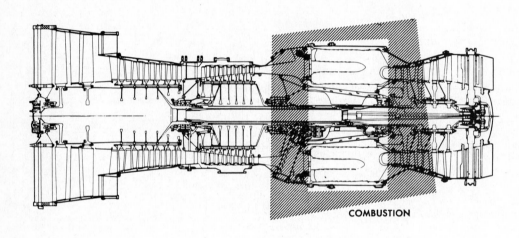

COMBUSTION

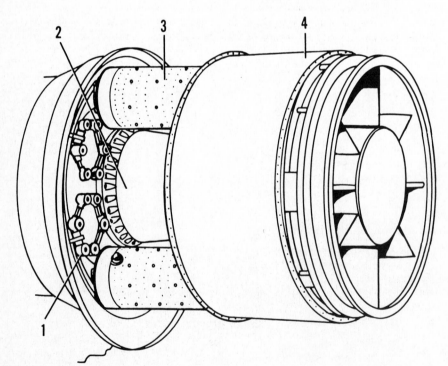

1 FUEL NOZZLES
2 INNER COMBUSTION CASE
3 COMBUSTION CHAMBER
4 OUTER COMBUSTION
   CASE

FIG. 20-12 Combustion section.

COMBUSTION-CHAMBER OUTER CASES

There are two combustion-chamber outer cases, the front and the rear. They are constructed of corrosion- and heat-resistant steel. The front flange of the combustion-chamber outer front case is bolted to the rear flange of the diffuser case. The front flange of the combustion-chamber outer rear case is bolted to the rear flange of the outer front case and to the front flange of the turbine nozzle case. The outer rear-case rear flange is internal. It lies behind the forward external flange of the turbine nozzle case and bolts to it. Thus, the outer cases (if bolted together) or the rear outer case can be slid rearward over the turbine nozzle case to give access to the combustion chambers. The outer front case has interlocking lips on its flanges to provide sealing when joined.

COMBUSTION CHAMBERS *(FIG. 20-13)*

The combustion chambers are radially housed inside the combustion-chamber outer case. The forward face of each combustion chamber presents six apertures which align the six nozzles of the corresponding fuel nozzle cluster.

There are three basic types of combustion chambers in service. Combustion chambers type I and type II are interchangeable. However, type I and type II chambers are *not* functionally interchangeable with type III chambers. Within type I and type II, chambers no. 1, 3, and 7 are interchangeable and numbers 2, 6, and 8 are interchangeable. In type III, nos. 2 and 3 are interchangeable, and nos. 6, 7, and 8 are interchangeable. Numbering is clockwise, as viewed from the rear, and starts with the one o'clock position chamber. Numbers 4 and 5 have spark-igniter sleeves and guides, and chamber no. 4 has an air-pressure transfer tube in it. Interconnecting flame tubes join all the chambers around the circle. The tubes on the odd-numbered type I and II chambers are female and have an air shield on the upstream side that overlaps the mating male flame tube. In type III chambers, no. 1 has female interconnecting flame tubes; no. 5 has male tubes; and nos. 2, 3, 4, 6, 7, and 8 have one male and one female flame tube. When removing types I and II, the even-numbered

chambers must be removed first. The removal sequence for type III is nos. 5, 6, 4, 3, 7, 2, 8, and 1.

The combusion chambers are a welded assembly. At the head of each chamber are six swirl vanes through which six fuel nozzles fit. The holes in the walls of the chamber meter the entering air. Near the head of the chamber most of the air is used as combustion air. Farther down the chamber the entering air is used to cool the hot gases to safe temperature for the chambers. The rows of smaller holes admit air that cools the can itself.

Each chamber is held in place by a positioning bracket and a clamp. The clamp holds the open end of the chambers to an outlet duct assembly which is fastened to the turbine case. Each positioning bracket is secured by two outer fuel manifold retaining bolts. The bracket hooks onto a retaining lug on the combustion chamber. The fuel nozzle inside the air-swirl guide holds the chamber in lateral movement.

THE FIRESEAL *(FIG. 20-14)*

The fireseal is held by the bolts that join the front and rear combustion-chamber outer cases. The outer rim of the fireseal fits tightly against the engine pod. Some engines use the narrow-band-type fireseal, while the others use the offset-lip type.

COMBUSTION-CHAMBER INNER CASE AND NO. 5 BEARING SUPPORT *(FIG. 20-15)*

The combustion-chamber inner case has reinforcements welded to its inside surface. Its prime function is to form the inner surface of the combustion section, but together with the no. 5 bearing support, it provides an inner passage for the 16th-stage air which will help cool the radiant heat from the combustion chambers.

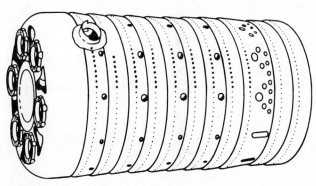

FIG. 20-13 Combustion chamber.

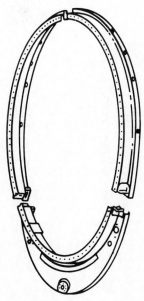

FIG. 20-14 Fireseal, narrow-band type.

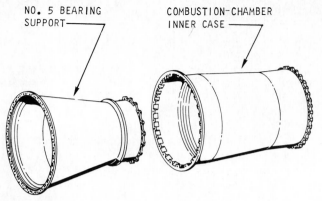

FIG. 20-15 Combustion-chamber inner case and no. 5 bearing support.

Holes in its forward flange, the diffuser, and the no. 5 bearing support permit this air to flow. A flange on the rear end attaches to the turbine nozzle inner case. This unit differs somewhat on different model engines.

TURBINE SECTION *(FIG. 20-16)*

The turbine section consists of a four-stage gas turbine. The first-stage rotor, also called the high-speed turbine, drives the rear compressor. The second-, third-, and fourth-stage rotors, also called the low-speed turbines, are mounted on a single shaft which drives the front compressor.

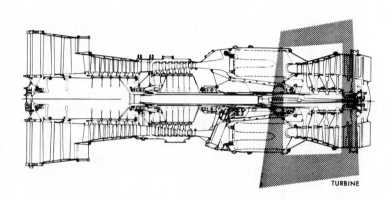

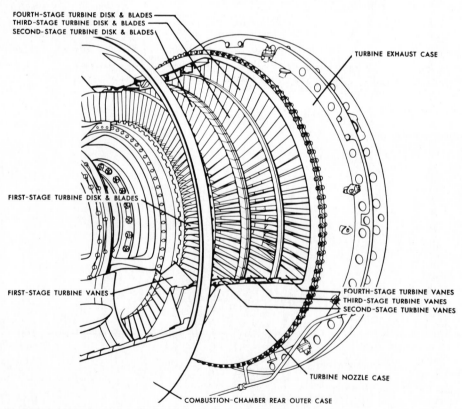

FIG. 20-16 Turbine section.

TURBINE NOZZLE CASE *(FIG. 20-17)*

The turbine nozzle case is constructed of corrosion- and heat-resistant steel. On its front flange, gang bolt assemblies are attached, and this flange bolts to the combustion-chamber outer case. The rear flange bolts to the turbine exhaust case using a spacer ring. Inside the turbine case, slots and ridges are machined and stator vanes and seals are inserted in these locations.

TURBINE-NOZZLE INNER CASE

On some engines, the front end of the turbine inner case is attached to the rear flange of the combustion-chamber inner case. On others, the front end is attached to the rear flange of the no. 5 bearing support. At the outer rear end, two shoulders hold the inner end of the first-stage turbine vanes. The front (largest) shoulder has pins for securing the vanes. On the inside diameter of the inner case, there are two knife-edged airseals. The innermost seal is riveted to a support welded to the approximate middle of the inner case. It rides free of an inner platform on the first-stage turbine disk. The outer airseal is riveted to the rear of the inner case. It rides free of an outer platform on the first-stage turbine disk.

COMBUSTION-CHAMBER OUTLET DUCT *(FIG. 20-18)*

This duct guides the hot gases to the first-stage nozzle. It is retained to the no. 5 bearing support. The duct is a welded assembly that is held in place by the combustion-chamber outer case-to-turbine nozzle case bolt circle. The forward flanges of the duct ports carry the rear edge of the clamps that hold the combustion chambers in place.

TURBINE VANES, SHROUDS, AND SEALS

First-stage vanes are cast from corrosion- and heat-resistant alloy with chromalize protective coating on all internal and external cast surfaces. Second-stage vanes are cast nickel base; third-stage vanes are cast cobalt base; and fourth-stage vanes are cast nickel base. First-stage vanes are held by pins between the recesses in the outer turbine case and are snapped onto their pins in the turbine nozzle inner case assembly to seat on the shoulders of the case.

A seal and a spacer are placed in the outer case just to the rear of the first-stage vanes. One knife-edge on the first-stage turbine blade shroud runs between the seal's two knife-edges.

Second-stage turbine vanes fit into the outer case just behind the first seal spacer. Extensions on the root platform of each vane hold them in place on their forward edges and a lock ring is used at the outer rear edge. The inner end of these vanes slip into slots in the shroud ring. The inside edge of the shroud ring has a turbine

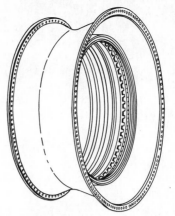

FIG. 20-17 Turbine nozzle outer case.

inner seal riveted to it. The seal ring carries four knife-edges on its inside flange. One of these knife-edges runs free against the shoulder on the rear-compressor drive turbine disk and the remaining three (rearmost) against the flange face of the front-compressor drive turbine hub. This forms a seal between the front and second-stage rotors to force the working gases to pass through the second-stage vanes. A lock ring, a stepped seal, and a seal spacer are in the outer case following the second-stage vanes. The two knife-edges of the second-stage blade run free against the stepped seal. Mounting of the third-stage turbine vanes is the same as the second stage. The inner seal ring, however, instead of having knife-edges has a stepped platform. The knife-edges are on the spacer between the second- and third-stage turbine disks. There are six of them set in pairs. Mounting of the fourth-stage turbine vanes is identical to the third stage. The outer seal is stepped and extends into the turbine exhaust case. There is no locking ring, but the fourth-stage vanes are held in the rear by the stepped seal, which is held by its flange that fits between the turbine nozzle case and the turbine exhaust case flanges.

FIG. 20-18 Combustion-chamber outlet duct.

REAR-COMPRESSOR DRIVE TURBINE ROTOR *(FIG. 20-19)*

The rear-compressor drive turbine rotor consists of a disk, blades, and a shaft. The no. 5 bearing inner race and seal assembly are located forward of the disk and secured on the shaft by a nut, a tab washer, and a lock ring during assembly. The turbine shaft splines into the rear-compressor rotor and is secured to the rotor hub by the rear-compressor drive turbine shaft coupling. The outer race of the no. 4½ bearing is located inside the turbine shaft.

TURBINE BLADES (FIRST STAGE) *(FIG. 20-20)*

The first-stage blades are made of forged nickel alloy. The blades have "fir tree" serrations as their roots. The blades can only be inserted by a forward and rear movement and are held in place by a rivet at the bottom of the "fir tree." The working part of the blade is a curved surface that changes the direction of the swiftly moving gases that pass over it. The reaction of these gases to change in direction of travel produces the force that turns the turbine rotor.

At the outward end of each blade is a wide flat section called a shroud. This shroud prevents gas leakage at the blade ends. Each blade has a stepped arrangement by which they lock to each other, and this forms a complete shroud ring when assembled. This ring gives the rotor greater strength as well as making up a gas seal. Because of the particular construction, it is impossible to remove one turbine blade only, and at least half of the blades at 90° on either side must be removed. A staggered movement of all these blades must be made before the center blade fir tree will clear the disk. There are 130 blades in the first-stage turbine rotor.

TURBINE DISK

The disk is made of corrosion- and heat-resistant nickel alloy. A drilled flange on the front face of the disk is bolted to the rear end of the rear-compressor drive turbine shaft. Balance holes in the disk flange are

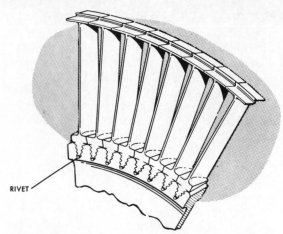

FIG. 20-20 First-stage turbine blades.

also provided. Fir tree serrations are cut in the rim matching the serrations in the turbine blades.

TURBINE SHAFT

The shaft is hollow with internal splines at its front end which mate with external splines of the rear-compressor rear hub. Turbine position in the turbine nozzle case is determined by a spacer ring that is between the end of the compressor hub shaft and an internal shoulder in the turbine shaft.

REAR-COMPRESSOR DRIVE TURBINE SHAFT COUPLING *(FIG. 20-21)*

A coupling holds the turbine shaft and the compressor rear hub together. Outside threads on the forward end of the coupling mate with threads on the inside of the compressor hub shaft, and a shoulder on the outside of the coupling butts against an internal shoulder in the turbine drive shaft. An internal spline at the rear of the coupling is used for wrenching. The coupling is locked in with a jam-nut having a left-hand thread, or by the no. 4½ bearing outer race and liner.

FRONT-COMPRESSOR DRIVE TURBINE ROTOR *(FIG. 20-22)*

Three disks with turbine blades, a rear hub, a spacer, and a drive shaft make up the rotor. The second-stage disk is bolted to the drive shaft and the rear hub. The third- and fourth-stage disks are held together and to the second-stage disks by 10 tierods.

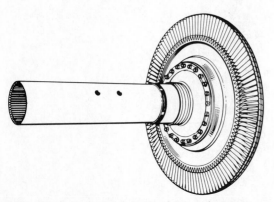

FIG. 20-19. Rear-compressor drive turbine rotor.

FIG. 20-21 Rear-compressor drive turbine shaft coupling.

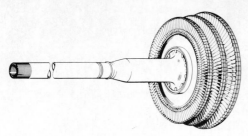

FIG. 20-22 Front-compressor drive turbine rotor.

TURBINE BLADES

The second- and third-stage turbine blades are cast from stainless steel. Fourth-stage blades are forged from nickel alloy material. They are assembled in the manner previously described in the section Rear-Compressor Drive Turbine Rotor and Turbine Blades. The second-stage blades are larger in height than those of the first stage, the third-stage blades are even larger, and the fourth-stage blades are the largest. The reason for this is to match the blade size to the increase in volume of gas as it expands while passing through the turbine. There are 114 blades in the second-stage disk and blade assembly, 108 blades in the third stage, and 80 blades in the fourth stage.

TURBINE DISKS

The second- and third-stage turbine disks are made from corrosion- and heat-resistant nickel alloy. The fourth-stage disk is made from corrosion- and heat-resistant steel. They are separated from each other by spacers and are attached to the front-compressor drive turbine shaft and the turbine rear hub by tierods as mentioned above. The second-stage disk has an integral spacer.

FRONT-COMPRESSOR DRIVE TURBINE SHAFT

The front-compressor drive turbine shaft is made from low-chrome steel alloy. It is a long, thin shaft, as it must pass through the first-stage turbine rotor shaft and must reach the front of the second-stage turbine disk to the rear hub of the front compressor. Its front end is supported inside the front-compressor hub, its middle by a bearing inside the rear-compressor turbine rotor shaft, and its rear end by the turbine assembly which, in turn, is supported by a bearing at the turbine rear hub.

The shaft is hollow and provides a passage for ninth-stage cooling air which passes through holes in the rear hub to cool the rear bearing seals, oil pump, and the rear face of the fourth-stage turbine disk. Oil is also piped through the shaft to its middle bearing. The forward end of the shaft is externally splined and internally threaded. The shaft external splines mate with the in-ternal splines on the front-compressor rear hub, and the shaft internal threads mate with the external threads on the front-compressor drive turbine coupling.

A spacer is used between the compressor rear hub shaft and the front edge of the drive shaft to position the shaft correctly.

FRONT-COMPRESSOR COUPLING (FIG. 20-23)

The coupling's external shoulder butts against an internal shoulder in the front-compressor hub. The internal splines in the coupling are for the wrench. The coupling front end has external splines with three short slots. A ring having external and internal splines and three inward projecting pins fits over the forward end of the coupling, the pins projecting through the slots. A compression spring is located just forward of the spline-pinned ring and is held in place by an internal shoulder of a collar. The collar is held in the coupling by splines and a snap ring. The spring keeps a rearward pressure on the ring which will lock the ring's external splines to the internal splines in the compressor hub. These splines are just forward of the shoulder, and as long as the ring is locked, the coupling cannot be unscrewed.

In order to unscrew the coupling from the drive shaft, the ring must first be moved forward against the spring pressure until its external splines disengage the small internal spline that is inside of the compressor hub ahead of the internal shoulder. The pins that project through the slots in the coupling are used to move the ring forward. A special wrench is used that can pull the pins forward and then engage wrenching splines to turn the coupling.

EXHAUST SECTION (FIG. 20-24)

The exhaust section of the engine includes all items following the turbine except the tailpipe and other rearward attachments which are furnished by the airframe manufacturer.

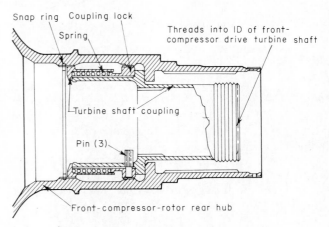

FIG. 20-23 Front-compressor drive turbine coupling.

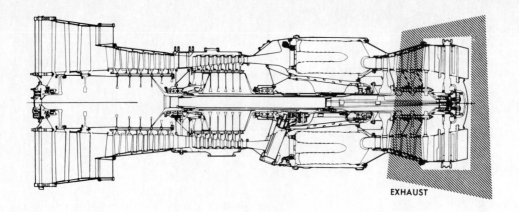

EXHAUST

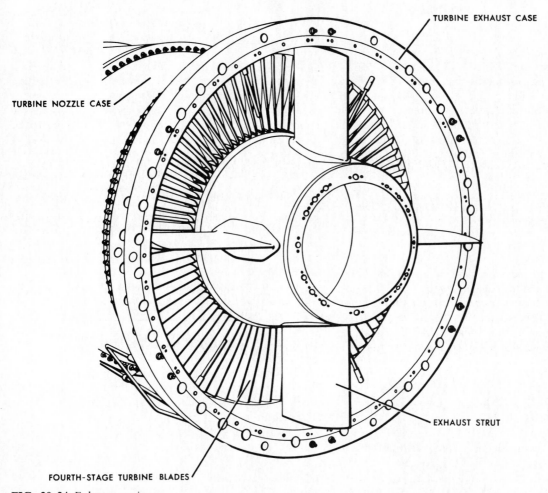

TURBINE EXHAUST CASE

TURBINE NOZZLE CASE

EXHAUST STRUT

FOURTH-STAGE TURBINE BLADES

FIG. 20-24 Exhaust section.

The outer turbine exhaust case is constructed of nickel alloy. It bolts to the rear of the turbine case at its front flange, and its rear flange bolts to the tailpipe. Two heavy external flanges near the center of the case are used for engine mounting.

The outer case supports all the parts used in the inner case. A few parts are mounted directly on the outer case. The fourth-stage turbine outer seal is located in the front of the outer case, the flange of which is positioned between the rear flange of the turbine nozzle case and the forward flange on the exhaust case.

A turbine exhaust fairing assembly, consisting of an inner duct and four turbine exhaust struts, is positioned in the exhaust case. Four no. 6 bearing support rods pass through the struts and through the holes in the outer case and are secured in strut supports between the engine mounting flanges mentioned above at the 12, 3, 6, and 9 o'clock positions. These rods are secured and positioned by a dual nut-locking arrangement in the strut supports. At the inner end, the rods are bolted to lugs on the no. 6 bearing housing. When assembled, the struts envelop the rods and the oil tubes (to be discussed) for streamlining. At the three o'clock position and to the rear of the support rod, the no. 6 bearing pressure oil tube passes from outside the exhaust case to the no. 6 bearing support. At the six o'clock position, the no. 6 bearing scavenge-oil tube is installed in the same manner.

A flange attached to the no. 6 bearing support holds the sump adapter to which the no. 6 bearing sump is attached. The entire sump area is enveloped by a heat shield. The fairing assembly area, internal to the struts, is also covered by a heat shield. An average pressure rake manifold is mounted around the case forward of the engine mounting flange. Just forward of the rear flange, the temperature thermocouples are installed. Attached to each are leads from a thermocouple harness which circles the case at this point.

NUMBER 6 BEARING SUPPORT AND SHIELDS

The no. 6 bearing seals buildup and no. 6 bearing inner race and rollers are mounted on the rear hub. The outer race is held in a housing which also serves as the seals housing. This housing is attached to and is held by the no. 6 bearing support. Four support rods from the outer case thread into the rear of the support. Attached to the rear of the support is the sump adapter to which the sump is bolted. The rear heat shield is bolted to the rear heat-shield flange which is retained in the support by the adapter. Bolted to the front of the no.

6 bearing support is the no. 6 bearing airseal support. A knife-edge is riveted to its forward extremity to ride on a shoulder which is riveted to the rear hub, thus forming a gas seal.

NUMBER 6 BEARING OIL SUMP

The sump contains the no. 6 bearing oil-scavenge pump which is driven by a pinion gear attached to the rear hub. Inserted through the pinion is the oil nozzle which provides lubrication for the no. 4½ bearing inside the turbine shaft. The sump bolts to the rear of the adapter and has a cover plate bolted to its rear side. O-ring seals are used between the sump adapter and the support and between the sump and the rear cover. The rear heat shield is bolted to the front heat shield at external flanges. The two heat shields cover all components inside the inner duct. All the metal parts are steel.

ACCESSORY- AND COMPONENT-DRIVES GEARBOX ASSEMBLY (FIGS. 20-25 AND 20-26)

The accessory- and component-drives gearbox consists of a front housing, a rear housing, and the internal gears and shaft gears. The gearbox assembly is mounted beneath the engine and secured to the diffuser case front flange. Power is supplied to the gearbox from the rear-compressor rotor shaft through an integral elbow on

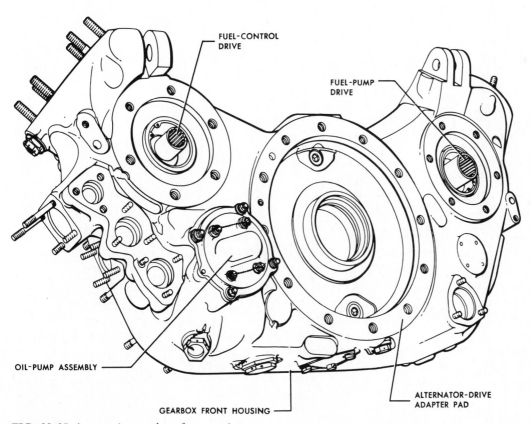

FUEL-CONTROL DRIVE

FUEL-PUMP DRIVE

OIL-PUMP ASSEMBLY

ALTERNATOR-DRIVE ADAPTER PAD

GEARBOX FRONT HOUSING

FIG. 20-25 Accessories gearbox front section.

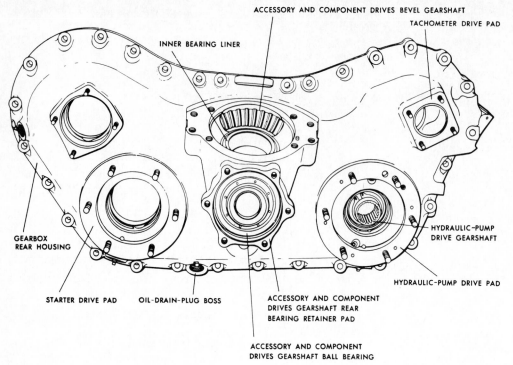

INNER BEARING LINER

ACCESSORY AND COMPONENT DRIVES BEVEL GEARSHAFT

TACHOMETER DRIVE PAD

GEARBOX REAR HOUSING

STARTER DRIVE PAD

OIL-DRAIN-PLUG BOSS

ACCESSORY AND COMPONENT DRIVES GEARSHAFT REAR BEARING RETAINER PAD

ACCESSORY AND COMPONENT DRIVES GEARSHAFT BALL BEARING

HYDRAULIC-PUMP DRIVE GEARSHAFT

HYDRAULIC-PUMP DRIVE PAD

FIG. 20-26 Accessories gearbox rear section.

the rear housing. The oil-pump assembly is located at the lower right front of the gearbox and contains both pressure and scavenge sections.

The engine fuel control mounts on a six-studded circular pad on the upper right front of the gearbox housing, and the fuel pump mounts on a similar pad on the upper left front. Gearing is provided to drive the fuel control, the fuel pump, rear-compressor tachometer, and a fluid pump. A six-studded accessory drive pad is positioned on an upward angle near the top on each side of the gearbox, and another is situated on the right rear face of the gearbox for the fluid pump. An optional accessory drive may be incorporated in either side upward drive.

Provisions are made for mounting either a constant-speed drive (for installation on Boeing aircraft) or a generator with an adapter (for installation on McDonnell Douglas aircraft) on the large alternator center front pad. At the left rear of the gearbox a six-studded pad is provided for a starter. The starter rotates the high-compressor rotor through the gearing and shafts.

ACCESSORY AND COMPONENT DRIVES GEARBOX FRONT SECTION

The oil-pump assembly, fuel control, fuel pump, starter drive shaft gears, and tachometer drive gear are located in the front housing. The duplex oil pump is mounted at the lower right front of the front housing. The pump is composed of two sections. The rear section is the pressure section, while the front section scav-

enges the gearbox assembly. The pressure pump section receives oil from the tank, pumps it through the filter and to all main bearings of the engine. The scavenge pump section clears the gearbox of scavenge oil and sends it to the oil tank. The right upward fluid drive (if incorporated) is driven by a gear bolted to the fuel control gearshaft and the left upward fluid drive (if incorporated) is driven by a gear bolted to the fuel pump drive gearshaft. An oil impeller rotary breather, a centrifugal device used to separate oil from air, is also attached to the fuel pump drive gear.

ACCESSORY- AND COMPONENT- DRIVES GEARBOX REAR SECTION

The rear housing contains the main accessory drive and idler gears, the tachometer drive, and the rear bearings for the horizontal gear trains.

Between the main accessory drive gear and the fuel-control drive gear is the idler gear through which power is transmitted to the right side of the gearbox. Splined to the alternator drive shaft gear is the main accessory drive gear which mates with the angled accessory drive gear in the elbow section of the housing. The angled accessory drive gear transmits power from the high-speed compressor to the gearbox. The alternator drive shaft gear transmits power to the left side of the gearbox assembly through the starter drive shaft gear and the fuel-pump drive gear. The tachometer drive is run by a tachometer drive gear internally splined to the fuel-control drive gear.

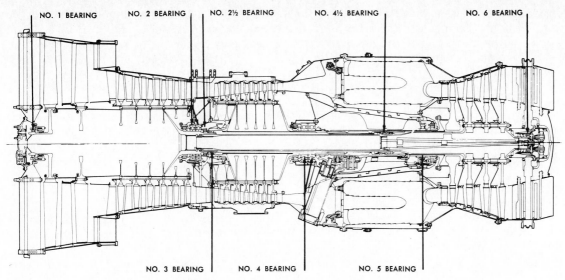

FIG. 20-27 Location of the main bearings.

BEARINGS *(FIG. 20-27)*

The bearings in the engine are of either the ball or roller type. Eight bearings are used on the principle rotating masses of the engine although in two instances, which we will discuss below, two bearings act as one. Twenty-five bearings are also used in fuel, oil, and accessory systems.

NUMBER 1 BEARING

The no. 1 bearing is of the roller type and is located at the front hub of the front compressor. Its outer race is held in a housing by means of a spanner nut and nut-locking rivet. The housing is supported by the no. 1 bearing front support and no. 1 bearing seal and support assembly at the center of the inlet guide vane circle. Its inner race has a pinch fit on the front-compressor front hub and is further held in place by a spacer, a seal plate, a spanner nut, and a nut-locking rivet. The inner race is larger than its rollers to allow for engine dimension changes with temperature. An oil-sealed assembly is used on the rear side of the bearing.

NUMBER 2 BEARING

The no. 2 bearing is of the ball type. It is a duplex bearing with two bearings acting as one. They are a matched pair. Their outer races are carried in a steel housing. This housing is supported by three steel pieces that make up the bearing support at the rear-compressor inlet vanes. These races are separated by an oil baffle and are held in place by an external shoulder in the housing and a spanner nut and tab washer. The inner races of both are split to aid installation. They are a pinch fit on the rear-compressor rear hub and are held in place by a hub shoulder, seal plate, oil baffle, and retaining nut. An oil-seal assembly is used on the front

side of the bearing. Bearings no. 1 and no. 2 are the front-compressor rotor supports.

In order to reduce absorption of heat by engine oil, heat shielding and thermal-blanket insulation are provided for the no. 2 bearing compartment on some engines. Bolts that secure the compressor intermediate front-bearing housing to the rear face of the front-bearing inner support also secure the compressor intermediate front-bearing front heat shield to the front face of the support.

The compressor intermediate rear-bearing heat shield consists of a front and rear heat shield, both of which are secured together and in turn are secured as a unit to the rear face of the compressor intermediate bearing rear support.

The thermal-blanket insulation is contained within the heat shield.

NUMBER 2½ BEARING

The no. 2½ bearing shown in Fig. 20-27 is a small, single-row ball bearing and serves as part of the support for the no. 2½ bearing housing. This housing is part of the seal system for the aft end of nos. 2, 2½, and 3 bearings. The outer race is held in the no. 2½ bearing housing by a snap ring. Its inner race is on the front-compressor rear hub and is held in place between the no. 2 bearing inner race retaining nut and a spacer. This spacer in turn is held in place by the no. 3 bearing inner race.

NUMBER 3 BEARING

The no. 3 bearing also shown in Fig. 20-27 is of the roller type and forms the front hub support for the rear compressor. Its outer race is a force fit into the inside diameter of the rear-compressor front hub. A shoulder and a snap ring hold the race in place. The inner race and rollers are shrunk onto the outside diameter of the front-compressor rear hub. A retaining nut, secured by a rivet, jams this inner race against the spacer just in the rear of the no. 2½ bearing inner race.

NUMBER 4 BEARING

The no. 4 bearing is of the ball type. It is a duplex bearing with two bearings acting as one. It forms the support for the rear hub of the rear compressor. The outer races are mounted in a housing which is supported inside of the engine diffuser case. They are held in place by a shoulder and a spacer under the diffuser-case bolt ring. There is an oil baffle between the outer races. The inner races are the split type, a pinch fit to the outside diameter of the rear-compressor rear hub. These races are positioned by a shoulder on the shaft, a seal plate, an oil baffle between the races, a spacer ring, a bevel gear, and a retaining nut with a lock and snap ring. There is an oil-seal housing on the forward side of this bearing.

To reduce swirl and heat rejection in one model engine, a compressor rear-bearing oil-scavenge baffle assembly is provided. The baffle assembly is mounted on the longer studs of the compressor rear bearing support.

To reduce absorption of heat by the engine oil, heat shielding and thermal-blanket insulation are provided for the no. 4 bearing compartment. The compressor rear-bearing heat shield assembly consists of several individual heat shields welded together. The complete unit is bolted to the compressor rear-bearing housing and extends over the bearing housing, the breather tube, the oil pressure and scavenge tubes, and along the towershaft.

The thermal-blanket insulation is contained within the heat shield.

NUMBER 4½ BEARING

The no. 4½ bearing is a small roller bearing which is located between the front- and rear-compressor turbine drive shafts to minimize the whip of the long (front) shaft. Its outer race is in the inside diameter of the rear-compressor drive turbine shaft. The race is held between a spacer on the shoulder on the inside diameter of the shaft and the turbine shafts bearing seals liner. The liner is torqued to specification. The bearing inner race and cage are shrunk onto the front-compressor drive turbine shaft. This race is held on the shaft by a shoulder on the shaft, the bearing seal assembly, a spacer and retaining nut, a tab washer, and a snap-ring assembly.

NUMBER 5 BEARING

The no. 5 bearing is a roller-type bearing, located just forward of the rear-compressor turbine. It forms the support for the rear-compressor drive turbine shaft. The bearing outer race housing and its seal housing are car-

ried by the turbine front-bearing support and the nozzle case and outlet duct assembly. The outer race carries the bearing cage and rollers. It is held in its housing by a retaining nut which is torqued and riveted. The bearing inner race is shrunk onto the outside diameter of the rear-compressor turbine shaft. This race is slightly wide to allow for engine temperature growth. The race is held in place by a shaft shoulder, airseal spacer, seal plate, retaining nut, and a retaining-nut tablock–snapping-ring assembly.

NUMBER 6 BEARING

The no. 6 bearing is a roller-type bearing which forms the support for the compressor turbine rear hub. The outer race is held in a housing which is bolted to the assembly. The support assembly is held by support rods which carry the load to the structure of the turbine exhaust case. The inner race, which holds the roller and cage, has a force fit on the front-compressor drive turbine rear hub. The inner race is located and held fast by the hub shoulder, two seal spacers, a seal plate, and the rear oil-scavenge-pump pinion flange. This flange is in turn bolted to a carrier sleeve which is threaded into the hub shaft.

OIL SYSTEM

The oil system for this engine is discussed in Chap. 15.

FUEL SYSTEM

A fuel system similar to the one used in the JT3D engine is described in Chap. 13.

IGNITION SYSTEM

The dc ignition system used on some engines is a high-energy, intermittent-duty, untimed electrical system designed to operate on an electrical input ranging between 14 and 29 V. The dc ignition system is not used continuously. Once the flame is ignited in the combustion chambers, there is no further need for the dc ignition system until the next time the engine is started. This could be at the beginning of the next flight, or in the air following some trouble that would extinguish the flame. Fuel spray into the engine combustion chambers is continuous. As long as the spark ignites the fuel, it makes no difference as to the exact instant of ignition. Thus, timing is not required. This system produces two sparks per second and is turned on just before the fuel is admitted to the engine. As soon as ignition occurs and the flame is established, the dc ignition system is turned off.

The ignition system receives its input electric energy from the aircraft source through a lead connected to the receptacle mounted on the right-hand-center flange of the front-compressor case. From here, it is conducted via a lead in the electrical harness assembly to a 20-J ignition exciter mounted on the upper-right-hand side of the engine between the front-compressor case front and middle flanges. The exciter stores energy converted from the dc input, and releases it as a fast high-voltage discharge. High-tension leads conduct the energy to two spark igniters installed into the no. 4 and no. 5 combustion chambers through holes in the lower rear of the diffuser case.The exciter provides the energy for the right and left spark igniters. This is done by means of a spark-dividing network.

Some models use an ac ignition exciter in addition to the dc unit. This is a 4-J unit which provides continuous ignition during flight to minimize the possibility of flame-out during engine operation under adverse ambient conditions. After the dc unit previously discussed has effected a satisfactory start, it is turned off. With the alternating current provided by the engine system, the ac ignition unit provides continuous ignition to the right spark igniter. The ac ignition unit has a nominal input of 115 V, 400 Hz ac, but will operate under the range of 90 to 124 V. It is a high-energy unit using stored energy. A radio suppressor built into the unit prevents radio frequencies generated within the unit from entering the supply circuit. (Refer to Chap. 16 for a more detailed examination of ignition systems.)

REVIEW AND STUDY QUESTIONS

1. Name several different airplanes using this engine.
2. List the engine's major specifications.
3. Where is the fan located? How is it driven?
4. Give a brief description of the engine operation and airflow.
5. Why are three turbines necessary to drive one of the compressors, while only one is needed to drive the other?
6. How is the chance for compressor stall reduced on this engine?
7. What type of combustion chamber is used? How does it differ from others of its type?
8. Describe the fuel manifold and fuel nozzle arrangement.
9. Very briefly describe the construction features of the following engine parts: front case, front- and rear-compressor rotors and cases, diffuser, combustion chamber and cases, turbine nozzle and case, turbine and case, exhaust duct, and accessories section.
10. How many main bearings are used? Name them and give their location.
11. Briefly describe the following engine systems: anti-ice, oil, fuel, and ignition.

CHAPTER TWENTY-ONE
GENERAL ELECTRIC J79

The General Electric J79 (Fig. 21-1) is a very highly produced engine used in such airplanes as the McDonnell-Douglas F-4 series, Lockheed F-104, and the North American RA-5C (Vigilante). Since so many models of this engine are in use, and since they all have a similar basic construction, the description that follows is for the J79 jet engine in general.

SPECIFICATIONS

Number of compressor stages:	17
Number of turbine stages:	3
Number of combustors:	10
Maximum power at sea level:	15,000 to 18,000 lb [66,786 to 80,143 N]
Specific fuel consumption at maximum power:	2.0 [203.9 g/N/h]
Compression ratio at maximum rpm:	12:1 to 13.5:1
Maximum diameter:	39 in [99 cm]
Maximum length:	202 to 208 in [513 to 528 cm]
Maximum dry weight:	3600 lb [1634 kg]

The J79 is an axial-flow turbojet engine with variable afterburner thrust. (See Fig. 2-22.) It incorporates a 17-stage compressor, of which the angles of the inlet guide vanes and the first 6 stages of stator vanes are variable; a combustion system, which consists of 10 individual combustion liners situated between an inner and outer combustion casing; a 3-stage turbine rotor, which is coupled directly to the compressor rotor; and an afterburner system, which provides afterburner thrust variation through fuel-flow scheduling and actuation of the variable-area, converging-diverging-type exhaust nozzle. The rotors are supported by three main bearings.

OPERATION

During operation, air enters the front of the engine and is directed into the compressor at the proper angle by the variable inlet guide vanes and variable stator vanes. The air is compressed and forced into the combustion section. A fuel nozzle extending into each combustion liner atomizes the fuel for combustion. The fuel-air mixture is initially ignited by a spark plug in one combustion liner, and is rapidly propagated to the remaining liners by cross ignition ducts which join adjacent liners. Once initiated, combustion is self-sustaining so ignition is turned off.

The gases which result from the combustion are directed into the turbine, which drives the compressor rotor. From the turbine, the exhaust gases flow into the afterburner where additional fuel may be injected and ignited to augment the thrust of the main engine. The exhaust gases then pass through, and are accelerated by, the exhaust nozzle.

The engine systems control engine thrust by regulating engine speed, stator angle, fuel flow, exhaust nozzle area, and exhaust gas temperature. Interconnecting signals integrate the various controls so that the systems function as a single unit in response to the throttle.

COMPRESSOR FRONT FRAME (FIG. 21-2)

The compressor front frame forms the air inlet passage for the engine and supports the front of the compressor rotor. The frame is made of stainless steel and consists of an outer shell, an inner hub, and eight evenly spaced, hollow struts. The frame assembly includes the no. 1 bearing area and the 20 inlet guide vanes and their actuating mechanism. The inner hub encloses the inlet gearbox assembly, the bearing seals, and an anti-icing air distribution manifold.

The outer shell of the frame contains three mounting pads to permit variations in aircraft mounting configuration, a gearbox mounting pad at the six o'clock position (Fig. 21-3) to support the transfer gearbox, and 20 evenly spaced holes which retain the spherical bearings assembled to the outer trunnions of the inlet guide vanes.

The inlet gearbox assembly, which is attached to the front face of the hub, becomes an integral part of the no. 1 bearing sump. A split-inlet guide vane inner support is attached to the rear face of the hub and retains the spherical bearings assembled to the inner trunnions of the inlet guide vanes. A manifold cover spans the space between the inner support and the no. 1 bearing oil seal to enclose the anti-icing air manifold.

The front frame struts provide passageways for sump cooling air, anti-icing air, supply-and-scavenge-oil tubes, and the radial drive shaft to the transfer gearbox (Fig. 21-4). The no. 1 strut encloses a tube which ducts ninth-stage air to the oil seal. The no. 4 strut encloses tubes which duct oil supply to—and scavenge oil

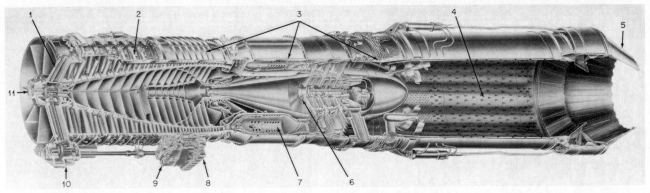

1 ANTI-ICED INLET CASE AND STRUTS
2 VARIABLE STATOR STAGES
3 SPLIT COMPRESSOR COMBUSTOR, AND TURBINE CASINGS
4 AFTERBURNER
5 VARIABLE-AREA CONVERGING-DIVERGING EXHAUST NOZZLE
6 THREE-STAGE TURBINE
7 COMBUSTION CANS
8 REAR GEARBOX
9 MAIN AND AFTERBURNER FUEL CONTROL
10 TRANSFER GEARBOX
11 FRONT GEARBOX FOR CARTRIDGE OR PNEUMATIC STARTER

(a)

(b)

FIG. 21-1 External and internal views of the General Electric J79-15 turbojet engine.
(a) Cutaway view.
(b) Left and right side views.

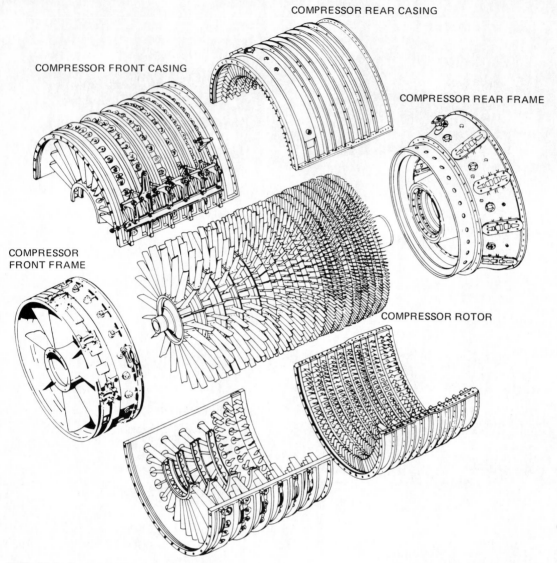

FIG. 21-2 Compressor assembly.

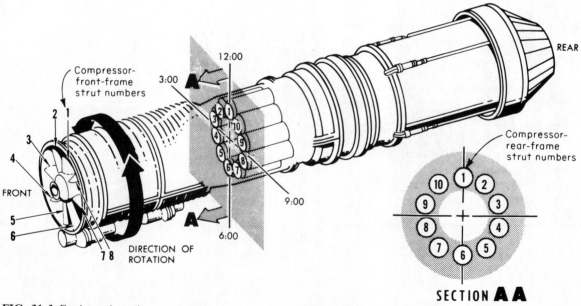

FIG. 21-3 Engine orientation.

392 REPRESENTATIVE ENGINES

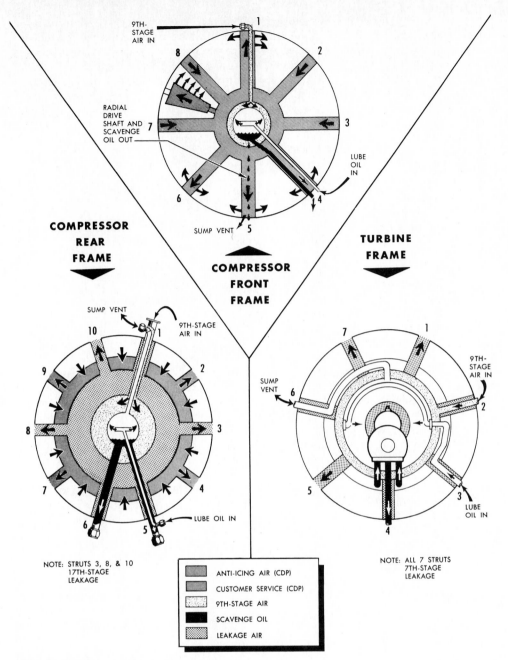

FIG. 21-4 Strut usage.

Legend:
- ANTI-ICING AIR (CDP)
- CUSTOMER SERVICE (CDP)
- 9TH-STAGE AIR
- SCAVENGE OIL
- LEAKAGE AIR

NOTE: STRUTS 3, 8, & 10 17TH-STAGE LEAKAGE

NOTE: ALL 7 STRUTS 7TH-STAGE LEAKAGE

from—the no. 1 bearing sump. Strut no. 5 encloses the radial drive shaft and provides a free-flow path for scavenge oil from the no. 1 bearing sump. Struts 2, 3, 7, and 8 duct anti-icing air into the hub; struts 1, 4, 5, and 6 contain a chamber near the leading edge that conducts the anti-icing air outward. The no. 1 strut contains a central passageway which is an extension of the no. 1 bearing sump.

NOTE: All struts are numbered in a clockwise direction beginning with the number 1 at the top or immediately to the right of the top, as the observer looks at the rear face of the frame.

COMPRESSOR CASING ASSEMBLIES

The compressor casing assemblies consist of two cylindrical, stainless steel casings, split along a horizontal line for removal. Their flanges interlock such that the rear casing half must be removed before the front; however, both casing halves can be removed as one piece if the front casing must be removed. The front casing assembly contains the six stages of the variable stator vanes and their actuating linkage and the seventh-stage stator vanes, which have a fixed angle. The outer ribs of the casings provide mounting adaptors for engine accessories.

The shanks of the variable stator vanes protrude through holes in the front casing; plastic bushings on the inside and outside of the casing provide an airseal and bearing surface for the vanes. The first four stages of stator vanes are shrouded at the inner end to reduce vibration and air leakage.

The levers, which are attached to the vane shanks, are pinned to half-rings. The two half-rings of each stage are connected at the horizontal lines to form a complete circle. Each circle is linked to two bellcranks. All of the bellcranks on each side, including the inlet guide vane bellcranks, are interconnected by a master rod so that the stator bellcrank and master rod assemblies actuate all stages of vanes simultaneously. The bellcranks are mounted on supports which are bolted to casing ribs, one at the 4 o'clock and the other at the 10 o'clock position on the casing.

The second-stage bellcrank on each support is attached to a vane actuator, which moves the stator vanes utilizing high-pressure fuel scheduled by the main fuel control. A permanent vane position indicator is attached to the fourth-stage actuating linkage at the seven o'clock position. The indicator is used to check the vane position and to rig the stator linkage.

The front casing has a manifold that conducts anti-icing air to the outer trunnions of the first-stage vanes. The air flows inward through the hollow trunnions and out through slots in the inner end of the vanes. Channel ribs near the rear flange of the front stator casing restrain the bases of the seventh-stage stator vanes. Each vane consists of an airfoil section welded to a hollow, T-shaped base.

The rear-compressor casing assembly includes the last 10 stages of stator vanes and one stage of exit guide vanes. Each stator vane consists of an airfoil section welded to a T-shaped base, similar to the 7th-stage vanes. An exit guide vane is mounted on the same base as each 17th-stage vane. The bases of the vanes slip into, and are restrained by, ribs in the casing.

The upper half of the rear casing has an air collection manifold. Ninth-stage air flows through holes in the casing and into the manifold, from which it is piped externally to the three bearing areas for sump cooling. The lower casing half contains mounting lugs for the rear gearbox.

COMPRESSOR ROTOR *(FIG. 21-5)*

The compressor rotor consists of a front stub shaft; 17 disks, spacers, and sets of blades; a 7th-stage air baffle and ducts; 4 torque cones; and a rear stub shaft.

The front stub shaft is bolted to the front of the first-stage disk and provides a surface for the no. 1 bearing and seal inner races. The hub of the shaft is internally splined to provide power to the accessory drive section.

The rotor blades are secured to the disks by single-tang dovetail connections. They are held in the dovetail slots by a blade retainer at the front of the rotor, the spacers throughout the rotor, and the 17th-stage airseal at the rear. The spacers transmit the torque forward from the 11th stage, and rearward to the 16th and 17th stages. The spacers between the 11th- and 15th-stage disks do not transmit torque, but continue the smooth contour of the rotor. The first four spacers form a mating surface for the shrouds on the stator vanes to form an airseal.

Holes through the seventh-stage spacer permit air to enter the rotor. An air baffle, bolted between the 7th- and 8th-stage disks, causes the air to continue rotating

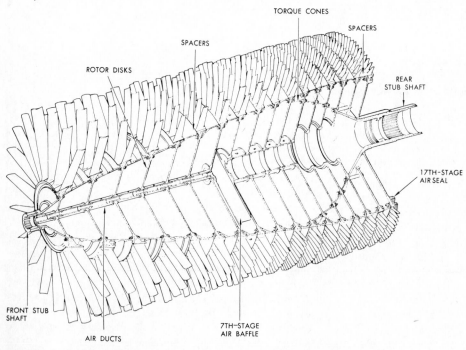

FIG. 21-5 Compressor rotor.

at the speed of the rotor. Air ducts allow the air to flow through the rotor to equalize the pressures on both sides of the disks and to duct it rearward through the stub shaft to cool the turbine rotor.

The torque cones transmit the torque from the small diameter at the 15th-stage disk outward to the large diameter of the 11th-stage disk. They also provide structural support for the larger disks and separate the different air pressures within the rotor. The rear stub shaft is bolted to the rear face of the 15th-stage disk. The inner races for air and oil seals and for the no. 2 bearing are assembled to the stub shaft. The shaft is internally splined and threaded to receive the turbine shaft and turbine bolt.

The 17th-stage airseal is bolted to the rear face of the 17th-stage disk to prevent compressor discharge air from leaking into the area behind the rotor. The seal consists of a double, grooved-type race (labyrinth) which mates with the seal on the front flange of the compressor rear frame.

COMPRESSOR REAR FRAME (FIG. 21-6)

The compressor rear frame absorbs the thrust loading of the rotors and the radial force of the compressor-turbine coupling. It also forms an annular diffuser for the compressor discharge air. The frame consists of an outer shell connected by 10 equally spaced, hollow struts to an inner shell. The outer shell provides 10 mounting pads for the fuel nozzles and 10 bosses

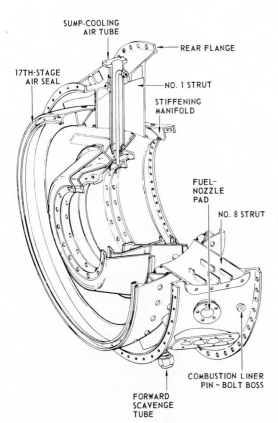

FIG. 21-6 Compressor rear frame.

through which the pin bolts that secure the combustion liners pass.

The inner shell contains a double seal on the front flange that mates with the 17th-stage airseal of the rotor and two manifolds on the inner surface that strengthen the shell. The manifolds are used to extract compressor discharge air for aircraft use. The sump is discussed in the Bearings and Seals section.

The struts support the bearing sump and provide various service passages (see Fig. 21-4). Numbers 2, 4, 7, and 8 serve as passageways for extraction air for aircraft purposes. Numbers 3, 8, and 10 vent 17th-stage seal leakage air overboard. Number 1 contains a pair of concentric tubes, one of which conducts sump cooling air from the sump, the other conducts 9th-stage air into the sump cooling cavity. The no. 5 strut also contains a pair of concentric tubes, one of which conducts oil supply to the sump, the other conducts scavenge oil from the rear portion of the sump. The no. 6 strut contains a tube which conducts scavenge oil from the front portion of the sump.

COMBUSTION OUTER CASING (FIG. 21-7)

The combustion outer casing is split on the horizontal line to permit easy removal for inspection and removal of the liners. The upper half contains a port at the 12 o'clock position for extraction of anti-icing air. The lower half contains two spark-plug bosses—the one at the no. 4 liner is used—and a combustion system drain. The drain allows excess fuel to drain from the combustion system. A port near the rear flange allows air to flow from the combustion casing to the pilot burner. A locking strip, which fits along the inside of the horizontal flange, strengthens the flange and prevents air leakage.

COMBUSTION LINERS (FIG. 21-8)

Each combustion liner consists of three parts riveted together; they are an inner liner, an outer liner, and a rear liner. The outer liner forms a snout which scoops compressor discharge air into the liner. Vanes in the snout produce a uniform distribution around the dome of the inner liner. A slot in the snout permits the fuel nozzle to extend into the inner-liner dome. The no. 4 liner has an igniter hole through the inner and outer liner. The adjacent liners are joined near the front end by cross-ignition tubes and the flanges of adjacent liners are held by V-band clamps to form a sturdy assembly. The liners are restrained by pin bolts in the compressor rear frame.

The rear liners are oval shaped at the rear and are oblique to facilitate their removal. They fit into the inlet ports of the annular transition duct and are supported by it. The liners have thimble holes through which air is introduced to complete the combustion, and louvers that provide a flow of cooling air along the inner surface of the liner.

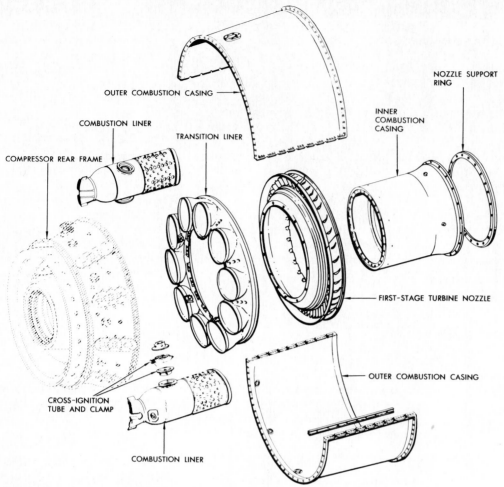

FIG. 21-7 Combustion section

COMBUSTION INNER CASING
(See FIG. 21-7)

The inner casing is an internally stiffened cylinder that bolts between the compressor rear frame at the front and the first-stage turbine nozzle at the rear. It absorbs the torque developed on the turbine nozzle, and it combines the combustion airflow to an annular passage around the liners. Holes near the front of the casing permit air to flow into the chamber around the turbine shaft to cool the shaft.

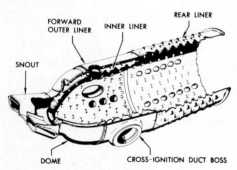

FIG. 21-8 Combustion liner.

ANNULAR TRANSITION DUCT
(See FIG. 21-7)

The transition duct provides a ring of 10 oval inlet ports and an annular exit, which is equal in area to the total of the 10 inlets. The duct is supported by the first-stage turbine nozzle, and is held in place by five pin bolts near the rear flange of the combustion inner casing.

The inlet ports of the transition duct support the rear end of the combustion liners. Small louvers between the inlet ports admit cooling air and allow flexibility to the duct to minimize cracking.

FIRST-STAGE TURBINE NOZZLE
(FIG. 21-9)

The first-stage turbine nozzle is assembled as a part of the combustion section and is bolted to the rear flange of the inner combustion casing. The rear flange of the outer band contacts the turbine stator rib and restrains the first-stage shroud. The outer band of the turbine nozzle is segmented to permit expansion.

The inner band of the nozzle is a one-piece structure

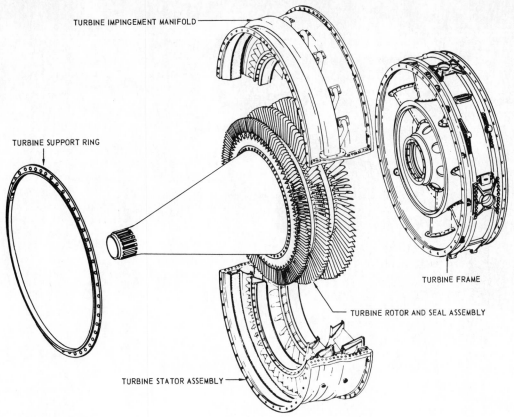

TURBINE IMPINGEMENT MANIFOLD

TURBINE SUPPORT RING

TURBINE FRAME

TURBINE ROTOR AND SEAL ASSEMBLY

TURBINE STATOR ASSEMBLY

FIG. 21-9 Turbine section.

from which the partitions are cantilevered. The inner band contains a corrugated baffle which mates with the transition duct and allows cooling air to flow across the band. The partitions are hollow airfoil sections with internal baffles so that cooling air, passing through the partitions, maintains an even skin temperature.

TURBINE STATOR ASSEMBLY
(See FIG. 21-9)

The turbine stator is split on a horizontal line for easy removal. It includes the three turbine shrouds and the second- and third-stage nozzles assembled into the turbine casing. On some models, a turbine impingement manifold encircles the upper half of the second-stage shroud to direct high-pressure air onto the turbine blades for starting. A check valve in the supply line prevents the turbine air from exiting into the manifold.

Flanges on the ribs within the casing hold the nozzles and shrouds. Each turbine nozzle is restrained by three pin bolts through each casing half. The partitions of the nozzle are supported by the outer rim. The rim protects the casing from the combustion gases. The inner band of the second- and third-stage turbine nozzles restrain the turbine seals that encircle the torque rings of the rotor. Some engines incorporate a turbine blade guard, consisting of a segmented soft metal half-ring that slips between the third-stage shroud and the upper turbine casing to prevent a fractured turbine blade from penetrating the casing and damaging the aircraft.

The nozzle partitions are hollow, with internal baffling. The second-stage partitions have cooling air routed through them. The third-stage partitions are not air-cooled.

The turbine shrouds also slip into the ribs of the casing and interlock with the flanges of the turbine nozzles. The shrouds are half-rings or 60° segments with a bonded honeycomb surface which reduces air leakage around the tips of the turbine blades. Lockstops, welded in the casing, prevent the shrouds from rotating.

A turbine support ring mounts between the turbine casing and the combustion outer casing to maintain the axial distance for the stator assembly in respect to the rotor.

TURBINE ROTOR (FIG. 21-10)

The turbine rotor produces the rotary power to drive the compressor. It consists of the turbine shaft; three turbine wheels, sets of blades, clips, and locking strips; two torque rings; two baffle assemblies; and a turbine airseal. Honeycomb seals encircling the torque ring are a functional part of the stator assembly.

The turbine shaft is a hollow, conical shaft. The shaft has an external spline that engages the compressor rotor rear stub shaft and is internally threaded to receive the turbine bolt. The rear flange of the shaft is bolted to the front of the first-stage turbine wheel.

The turbine wheels are bolted to the torque rings which separate them. The wheels consist of thin disks

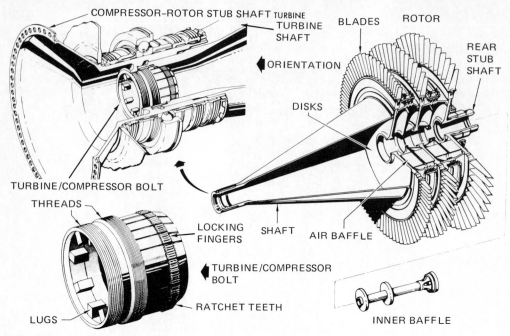

FIG. 21-10 Turbine rotor and turbine/compressor bolt.

with large open centers and widened rims. The rims have dovetail slots to retain the blades. The blades are assembled in pairs, two matching blades in each dovetail. The blade shanks keep the dovetail couplings out of the gas stream. Clips between the shanks prevent airflow between the blade pairs, and also damp their vibration. When bent, clips in the bottom of the dovetail retain the blades in the slots.

The torque rings transmit torque from the rear wheels forward. The rings have a machined, three-ring seal on the outer surface and circumferential cooling rings on the inside. The seal mates with the turbine honeycomb seals to minimize air leakage across the surface of the torque rings. The cooling rings expose a greater amount of the torque rings to the cooling air in the rotor.

Turbine rotor baffle assemblies consist of a ring support, bolted to each side of the second-stage turbine wheel, and a disk, with four baffles attached to each side, attached to each ring support. The disks cause the cooling air to flow outward to the inner surface of the torque rings while the baffles prevent a change in angular velocity as the air flows outward and successively inward within the rotor. The disks have large open centers to permit the locknut wrench, which loosens the turbine bolt, to be inserted through the rotor.

A turbine inner baffle fills the center of the turbine rotor baffles. The inner baffle (refer to Fig. 21-10) consists of two disks attached to a tube. The rear of the inner baffle contains a spoked disk which is locked to the third-stage wheel by a pin. The center of the spoked disk has an internal spline that drives the no. 3 bearing scavenge pump.

The turbine airseal is bolted to the rear face of the third-stage turbine wheel. The seal confines the sump cooling air to the cavity around the bearing sump. The hub of the third-stage wheel forms a stub shaft for the turbine rotor.

The turbine bolt is cylindrical and contains coarse threads, which engage threads within the compressor-rotor rear stub shaft; fine threads, which engage threads within the turbine shaft; serrated locking fingers, which engage serrations within the turbine shaft to lock the bolt; and lugs on the inside that permit the locknut wrench to turn the bolt. (Refer to Fig. 21-10.) Since both sets of threads are engaged at the same time, turning the bolt gradually pulls the shafts together into a solid coupling.

TURBINE FRAME

The turbine frame shown in Fig. 21-9 forms an exhaust diffuser, supports the rear of the turbine rotor, and provides the main engine-to-aircraft mounting structure. The frame consists of an outer cone, an inner cone, and a sump housing connected by seven equally spaced, hollow struts. The struts are housed within turbine frame vanes to shield them from exhaust temperature.

The outer cone of the frame contains 12 bosses which support the thermocouples that measure the turbine discharge temperature. Four sets of engine mounting supports are bolted to the ribs to provide support connections for various mounting configurations. Three spherical bearing housings are bolted to the ribs for thrust transmission. Strut ends are cast to add rigidity to the union of the strut and the cone.

The turbine frame inner cone encloses the sump housing. The inner exhaust cone bolts to the rear flange of the frame inner cone. The no. 3 bearing front airseal and spill baffle bolt to the front of the sump housing;

the no. 3 bearing rear airseal and the turbine cooling air baffle bolt to the rear.

All seven struts of the turbine rear frame (see Fig. 21-4) duct turbine cooling air to the engine compartment and, in addition, the no. 2 strut encloses a tube that conducts ninth-stage air to the sump cooling cavity. A tube in strut no. 6 vents the sump to the oil tank. A tube in strut no. 3 conducts lubricating oil to the sump. Strut no. 4 has a tube that conducts scavenge oil from the scavenge pump.

FORWARD EXHAUST DUCT ASSEMBLY *(FIG. 21-11)*

The forward exhaust duct is bolted to the rear flange of the turbine frame. It supports a smooth, inner liner which incorporates a retaining clip at the rear edge to interlock with the front of the no. 2 liner when the tailpipe is installed. The duct provides 21 mounting pads for the multijet fuel nozzles, and an opening for the pilot burner.

AFTERBURNER MANIFOLDS AND MULTIJET FUEL NOZZLES *(See FIG. 21-11)*

The four afterburner manifolds encircle the forward exhaust duct. Each manifold has 21 outlet ports, one for each fuel nozzle. The multijet fuel nozzles consist of four tubes, one for each of the four manifolds, that are fused into a probe. Holes in the sides of the tubes spray the fuel at right angles to the exhaust gas flow. The inner ends of the probes are retained by bosses imbedded in the rear inner cone.

The rear inner cone is ceramic coated to protect it from the high-temperature exhaust gases. It bolts to the rear flange of the turbine frame inner cone. The rear inner cone supports the flameholder through seven mounting brackets on its outer surface.

The flameholder consists of three concentric, V-gutter rings connected by seven equally-spaced radial links. The rings are staggered to ensure efficient burning without causing unnecessary airflow blockage.

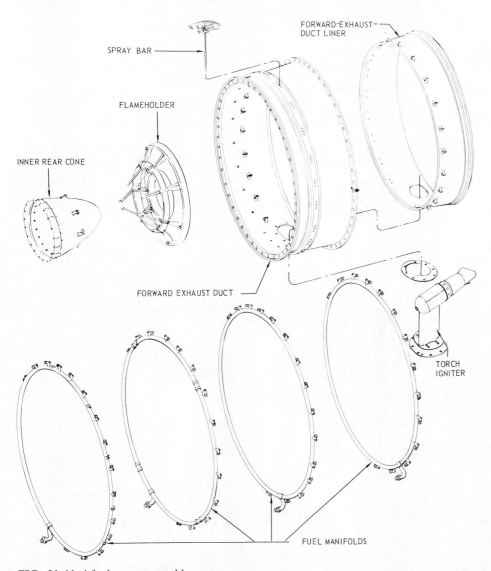

FIG. 21-11 Afterburner assembly.

PILOT BURNER

The pilot burner, or torch igniter, ignites the afterburner fuel in the exhaust section. The pilot burner attaches to the forward exhaust duct at the six o'clock position and extends into the inner and middle flameholder rings. A more complete description of the pilot burner is to be found on pages 409 and 410 and Fig. 21-23.

TAILPIPE ASSEMBLY *(FIG. 21-12)*

The tailpipe assembly consists of the rear exhaust duct, the numbers 2, 3, and 4 liners and the exhaust nozzle. The rear exhaust duct has four support brackets at the front for the exhaust nozzle actuators, and four brackets at the rear to support the exhaust-nozzle outer shroud. The liners are retained in the duct by tracks which engage clips on the liners.

The liners are ceramic coated to withstand the high afterburner temperatures. The no. 2 and no. 3 liners are corrugated and have cooling louvers to route cooling air along the inner surface of the liners. The liners are retained by clips that interlock with mating tracks in the duct.

The exhaust nozzle assembly consists of 24 nozzle flaps and seals interconnected by flap actuators and bellcranks to the support ring. Attached to the support ring are 24 shroud flaps and seals. The support ring telescopes into the outer shroud. Through this arrangement, movement of the support ring toward the rear of the engine causes a simultaneous increase in the opening area of the primary and secondary exhaust nozzles.

The nozzle flaps bolt to the rear flange of the rear exhaust duct and, with the seals, form the primary exhaust nozzle. The shroud flaps and seals form the secondary nozzle.

NUMBER 1 BEARING AREA

The no. 1 bearing, which is housed in the compressor front frame, is a roller bearing and restrains radial loads only; thus it permits the rotor to expand axially without transmitting stress to the surrounding structures. The

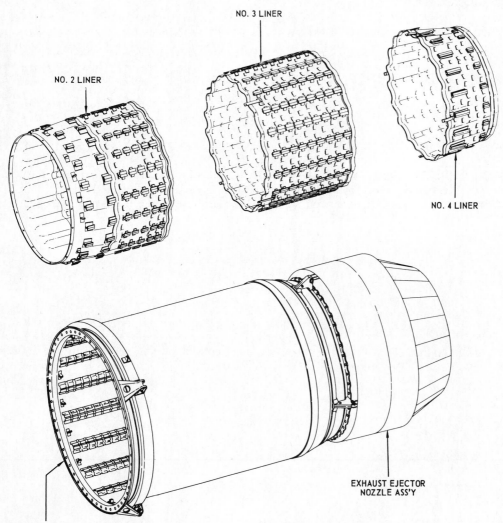

NO. 3 LINER

NO. 2 LINER

NO. 4 LINER

EXHAUST EJECTOR NOZZLE ASS'Y

REAR EXHAUST DUCT

FIG. 21-12 Exhaust or tailpipe assembly.

front gearbox is included in the sump area, so no engine oil and airseals appear in front of the no. 1 bearing. Behind the bearing is a dual, carbon-rubbing seal. Ninth-stage air is contained in the area between the seal rows to pressure load the seal segments against the race, to minimize oil loss from the sump. The forward seal-race contact surface is cooled by a spray of lubricating oil.

The oil that collects in front of the bearing flows through the open bottom strut of the front frame and into the transfer gearbox, while that behind the bearing is scavenged by a line leading to a separate element of the scavenge pump.

NUMBER 2 BEARING AREA *(FIG. 21-13)*

The no. 2 bearing, which is housed in the compressor rear frame, is a ball bearing and restrains both the radial and the thrust loading of the rotors. It is contained in a sump enclosed by carbon-rubbing oil seals, which consist of a single row of carbon segments mounted in a housing, and held circumferentially by a coil spring (see Fig. 21-13). The seal segments have a dam on the edge nearest the bearing that forms the seal, while positioning pads contact the race and maintain the proper sealing position of the segments. The seal-race contact surface is cooled by a spray of lubricating oil. The oil that is supplied to the sump is scavenged from the front and the back of the sump through separate tubes to provide positive scavenging.

The sump cooling air pressure is confined to a cavity surrounding the sump by an airseal on each side of the sump. The airseal consists of several rims (rotating part) that contact a soft metal on the seal (stationary part). The pressure drop across each rim reduces the amount

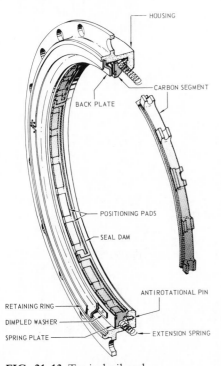

FIG. 21-13 Typical oil seal.

of air leakage from the cavity into the compressor rear frame. The air pressure thus contained minimizes oil loss from the sump.

NUMBER 3 BEARING AREA

The no. 3 bearing, which is a roller bearing, is housed in the turbine frame and restrains radial loads only. The sump area is formed by carbon-rubbing seals and is scavenged through two separate tubes. The air pressure around the sump is confined by the rear turbine airseal and the rear no. 3 bearing airseal. All of the seals are similar to those of the no. 2 bearing, except the carbon seals contain a second row of segments for backup.

FRONT GEARBOX *(FIG. 21-14)*

The front gearbox is housed within the hub of the front frame and is connected directly to the front stub shaft of the compressor rotor through a spline. It contains a spline on the front which drives the aircraft constant-speed drive, and it is connected to the transfer gearbox by a radial drive shaft. The gearbox housing contains an anti-icing pad at the top to supply anti-icing air to the aircraft nose dome.

TRANSFER GEARBOX *(See FIG. 21-14)*

The transfer gearbox is mounted at the bottom of the compressor front frame, and receives power from the front gearbox through a radial drive shaft housed in the no. 5 strut of the compressor front frame. It converts a radial drive to several horizontal drives and supplies the power to drive aircraft hydraulic pumps, a tachometer-generator, engine fuel pumps, a control alternator, and an oil scavenge pump. A combination cartridge/pneumatic starter is mounted on the rear face. A horizontal shaft transmits power to the rear gearbox.

DRIVE SHAFT SUPPORT BEARING *(See FIG. 21-14)*

The horizontal drive shaft support bearing is spline-connected to the front and rear horizontal drive shafts to prevent deflection of the shafts. The bearing housing is bolted to the bottom of the compressor front casing.

REAR GEARBOX *(See FIG. 21-14)*

The rear gearbox is attached to the bottom of the compressor rear casing and is hinge-mounted to compensate for the difference in the rates of expansion of the casing and the rear gearbox. It supplies power to drive the engine oil pumps, a nozzle hydraulic pump, an oil scavenge pump, and the main fuel control. It receives power from the transfer gearbox through the horizontal drive shaft.

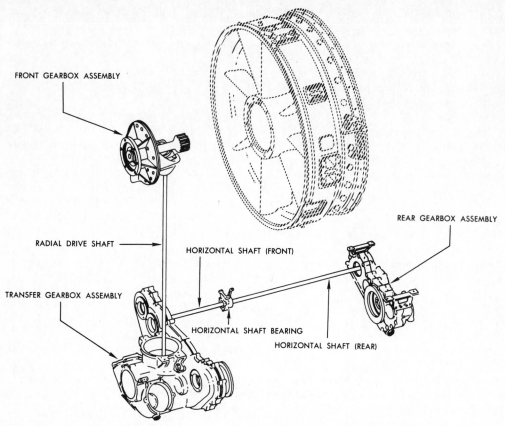

FRONT GEARBOX ASSEMBLY

RADIAL DRIVE SHAFT

HORIZONTAL SHAFT (FRONT)

REAR GEARBOX ASSEMBLY

TRANSFER GEARBOX ASSEMBLY

HORIZONTAL SHAFT BEARING

HORIZONTAL SHAFT (REAR)

FIG. 21-14 Accessory drive system.

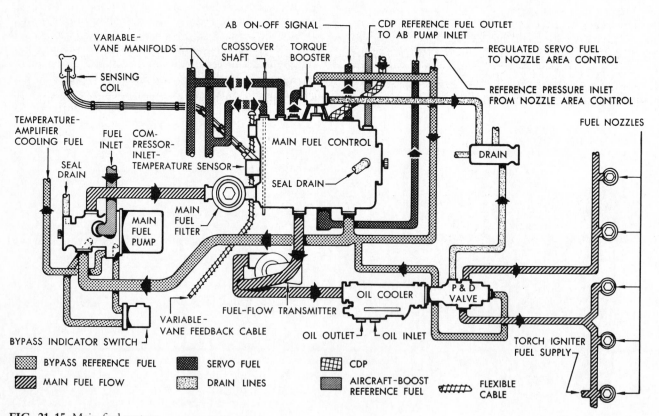

FIG. 21-15 Main fuel system.

MAIN FUEL SYSTEM *(FIG. 21-15)*

The main fuel system regulates the flow of fuel that is sprayed into the combustion section of the engine. In addition to regulating fuel flow, the system produces signals that (1) schedule the position of the compressor variable vanes, which govern the amount of airflow through the compressor, and (2) prevent afterburner system operation until engine speed and throttle position are proper. The system also supplies servo fuel at regulated pressure to the nozzle area control, and provides fuel for the afterburner-system torch igniter.

The path of fuel through the main fuel system is as follows: The fuel flows from the airplane fuel supply system, through the engine inlet connector, and into the main fuel pump. The pump filters the fuel and delivers it at high pressure to the main fuel filter. The fuel flows from the filter to the main fuel control, which regulates the amount of fuel that will be used to operate the engine, and bypasses the excess fuel back to the main fuel pump. The metered fuel flows from the control through the fuel-flow transmitter, the main oil cooler, the pressurizing and drain valve, and the fuel nozzles. The nozzles spray the fuel into the combustion section of the engine.

MAIN FUEL PUMP *(FIG. 21-16)*

The main fuel pump is bolted to the right-hand pad on the rear side of the transfer gearbox. The pump filters the low-pressure fuel and delivers it at high pressure for use in the main fuel system.

The pump consists of an impeller-type boost pumping element, a low-pressure filter with a bypass valve that allows fuel to bypass the filter if filter inlet pressure exceeds filter discharge pressure by 33 psi [227.5 kPa], a gear-type main pumping element, and a pressure relief valve that limits pump discharge pressure to 1125 psi

[7756.9 kPa] above the main pumping element inlet pressure.

Fuel flows through the pump inlet into the boost pumping element; the impeller increases the pressure and delivers the fuel to the outside of the low-pressure filter; it then flows inward through the filter and into the main pumping element which discharges the fuel at high pressure. A small amount of fuel at main pumping element inlet pressure is ported from the pump to cool the temperature amplifier. Excess fuel that is bypassed by the main fuel control is returned to the main pumping element inlet for recirculation through the fuel system.

BYPASS INDICATOR SWITCH

The bypass-indicator-switch mounting bracket is bolted near the four o'clock position on the forward end of the front-compressor casing. A set of electrical contacts in the switch closes when the low-pressure filter in the main fuel pump is becoming clogged.

The differential pressure switch senses the filter fuel inlet and outlet pressure. If the inlet pressure exceeds outlet by 25 psi [172.4 kPa] the contacts close; this completes a 28-V circuit that actuates an indicator in the cockpit, so the filter can be inspected and cleaned before the following flight.

MAIN FUEL FILTER

The main fuel filter is bolted to the forward side of the main fuel control, which is clamped to the mounting pad on the forward side of the rear gearbox. The filter removes contamination from the high-pressure fuel in the main fuel system.

The filter contains a main element that filters all the fuel which enters the filter inlet, and a smaller servo

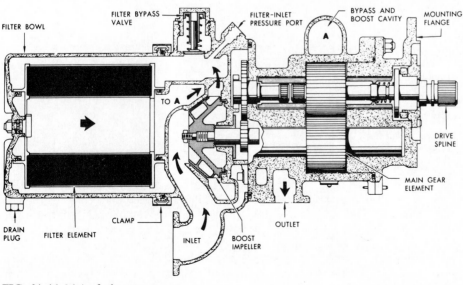

FIG. 21-16 Main fuel pump.

element that refilters the fuel that is to be used as a servo fluid in various engine controls. Each element has a bypass valve that allows fuel to bypass its filter if the pressure differential across the element exceeds 25 psi [172.4 kPa]. The filter discharges the servo fuel and the main engine fuel through separate ports into the main fuel control.

MAIN FUEL CONTROL

The main fuel control is clamped to the mounting pad on the forward side of the rear gearbox. (See Chap. 12.) The control uses five signal inputs—throttle position, engine speed, compressor inlet temperature CIT, compressor discharge air pressure (CDP), and the position of the compressor variable vanes—to control the engine main fuel scheduling and the amount of airflow through the compressor section.

The functions of the control are:

1. Governs engine speed.
2. Cuts off fuel when the throttle is in the OFF position.
3. Limits the least amount of fuel that can flow to operate the engine when the throttle is at or above the IDLE position. This allows the thrust output of the engine to be reduced at altitude without combustion flameout, and also provides a suitable amount of fuel during an engine automatic start.
4. Schedules fuel flow during acceleration and deceleration, to prevent compressor stall, combustion flameout, and excessive speed fluctuation.
5. Increases engine idle speed, when the CIT is high, to maintain airflow through the compressor.
6. Decreases engine scheduled top speed, when the CIT is low, to limit pressure in the engine.
7. Reduces fuel flow when CDP is high, to prevent an excessive amount of airflow through the compressor.
8. Controls the position of the compressor variable vanes to maintain compressor efficiency under various operating conditions.
9. Prevents afterburner-system operation until main engine conditions are suitable.

Fuel that enters the servo inlet port of the control is used to actuate the variable vanes and the torque booster, and is also used as a servo fluid for positioning valves and pistons inside the main fuel control, the compressor-inlet-temperature sensor, and the nozzle area control.

Fuel that enters the main fuel inlet port is divided by the metering valve into bypass or excess fuel and metered fuel. The bypass fuel is returned to the main fuel pump; the metered fuel flows to the fuel nozzles to be sprayed into the combustion section.

COMPRESSOR-INLET-TEMPERATURE SENSOR

The compressor-inlet-temperature sensor transmits a signal representing inlet air temperature to the main fuel control. The sensor consists of a sensing coil that mounts in the compressor front frame and extends into the compressor inlet near the eight o'clock position, a sensor unit that is attached to the forward side of the main fuel control, and two metal-covered flexible tubes.

The sensing coil, two bellows in the sensor unit, and both tubes are filled with a temperature-sensitive fluid. Variations in temperature cause the volume of the fluid to change and expand or contract the bellows. The position of the bellows, an indication of inlet temperature, is transmitted through linkage to the main fuel control.

The flexible tube that is not connected to the sensing coil is used to counterbalance any effect that temperature inside the engine compartment has on the other tube.

A detailed examination of the Woodward 1307 fuel control is made in Chap. 12.

PRESSURIZING AND DRAIN VALVE [refer to FIG. 12-36]

The pressurizing and drain valve is bolted to the rear side of the main oil cooler, near the four o'clock position on the compressor rear casing.

During an engine start, the valve prevents fuel from flowing to the fuel nozzles until fuel pressure is high enough to operate the flow-scheduling mechanisms in the main fuel system. At engine shutdown, the valve allows fuel in the fuel nozzle manifold to drain, to prevent post-shutdown fires in the combustion section, but keeps the rest of the main fuel system primed.

Metered fuel from the main fuel control flows from the main oil cooler into the forward side of the valve; fuel at control bypass pressure is ported to the rear side of the valve. When metered fuel pressure is 90 psi [620.6 kPa] higher than bypass fuel pressure, the valve allows metered fuel to flow to the fuel nozzles and cuts off flow from the drain. If the pressure differential becomes less than 90 psi, the valve shuts off metered fuel flow and allows fuel in the fuel nozzle manifold to drain.

FUEL NOZZLES [refer to FIG. 12-27(a)]

The 10 fuel nozzles, which are bolted to flanges around the compressor rear frame, spray fuel into the forward end of the combustion liners. Each nozzle contains a filter, a flow divider, and two separate passages from which the fuel is discharged in a single cone-shaped spray.

When inlet fuel pressure to the nozzle is less than 90 psi [620.6 kPa], the fuel flows through the filter, into the smaller of the passages, and is discharged into the combustion liner.

When inlet pressure reaches 90 psi, the flow divider valve opens and allows filtered fuel to flow into the larger passage too. The fuel discharged from both passages converges at the nozzle tip and is sprayed into the liner.

A shroud around the end of the nozzle deflects a small amount of air onto the tip of the nozzle to cool it and to retard the accumulation of carbon on the face of the nozzle.

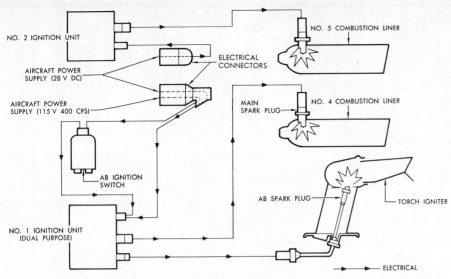

FIG. 21-17 Main and afterburner ignition systems.

MAIN IGNITION SYSTEM (*FIG. 21-17*)

The main ignition system consists of two circuits that are independent of each other except that a single cockpit switch energizes them both. The circuits, which operate only while the engine is being started, furnish the ignition for the fuel-air mixture in the combustion liners. Each circuit has its own ignition unit, shielded cables, and spark plug.

The two boxes containing the main ignition units are mounted next to each other near the four o'clock position on the compressor rear casing. The rear box energizes the spark plug in the no. 5 combustion liner; the forward box energizes the plug in the no. 4 liner. (The forward box also contains the circuit for transforming alternating current to energize the spark plug in the torch igniter of the afterburner system. This circuit is unrelated to the main ignition unit circuit.)

When the main engine ignition switch is closed, 28-V direct current flows to both of the main ignition units. (See Fig. 16-10.) Each unit includes a step-up transformer controlled by a vibrator to produce the desired voltage; a half-wave rectifier to control current flow; a storage capacitor to accumulate sufficient energy to produce the spark potential; and a gap which ionizes at approximately 3000 V and allows the high-potential current to flow across it.

The main spark plugs (Fig. 21-18), which ignite the fuel-air mixture in the combustion liners during engine starts, screw into bosses in the outer combustion casing. The inner end of one plug extends into the no. 4 combustion liner, the other into the no. 5 liner.

The spark-plug electrodes are shunted by a ceramic semiconductor that ionizes the spark gap when the output of the ignition unit is impressed across it. This creates a low-resistance path across which the built-up en-

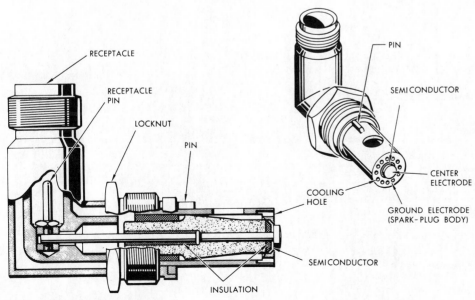

FIG. 21-18 Main spark plug.

ergy in the ignition unit storage capacitor is discharged. This discharge is rapid and, consequently, the spark intensity is high.

AFTERBURNER FUEL SYSTEM
(FIG. 21-19)

The afterburner (AB) fuel system increases the thrust output of the engine by spraying and igniting fuel in the exhaust tailpipe. Heating and expanding the exhaust gases increases their velocity.

Fuel from the aircraft supply system is furnished to the AB system through the engine inlet fuel connector. The valve in the inlet of the AB fuel pump prevents the fuel from entering the pump except during AB-system operation. When the engine throttle lever is positioned in the AB range (above the 76° position) and engine speed exceeds 90.3 percent, the main fuel control transmits a fuel signal that opens the inlet valve and allows fuel to enter the AB fuel pump.

The engine-driven AB fuel pump, which rotates continuously during engine operation but pumps fuel only while the inlet valve is open, discharges the high-pressure fuel to the AB fuel filter. The fuel flows from the filter into the AB fuel control.

The control regulates the amount of fuel for AB operation and divides its output into two separate flows, core and annulus. The core fuel from the control flows through the AB oil cooler to the pressurizing valve. The annulus fuel flows from the control directly to the pressurizing valve.

The fuel pressurizing valve consists of four valves which divide the core fuel flow into primary core and secondary core and the annulus fuel flow into primary annulus and secondary annulus. This division ensures that adequate pressures are maintained to prevent vaporization within the spraybar tubes. Each pressurizing

valve ports fuel to a fuel manifold, which delivers fuel to the 21 spraybars.

The AB fuel-air mixture is ignited by a torch igniter, which extends into the forward exhaust duct. The flame of the torch igniter is provided by combining fuel which is piped from the main fuel system; air, which is piped from the outer combustion casing; and ignition, provided by the AB ignition system. Continuous ignition is provided during AB operation to ensure satisfactory burning.

Torch igniter fuel is scheduled or interrupted by the ON-OFF valve in response to AB pump discharge pressure increase or termination. The fuel is metered within the ON-OFF valve and is routed through a check valve prior to entering the torch igniter.

AFTERBURNER FUEL PUMP
(FIG. 21-20)

The AB fuel pump is located at the seven o'clock position on the rear face of the transfer gearbox. The gear-driven impeller-type pump supplies fuel, under pressure, to the AB fuel system.

A valve, located in the inlet of the pump, regulates the passage of fuel from the fuel supply system into the pump. The valve, in turn, is controlled by an actuator, which is positioned according to pressure differential across the piston. When the differential exceeds 80 psi [551.6 kPa], the valve opens to allow fuel to enter the pump.

When AB operation is terminated, the main fuel control vents the actuator high-pressure (ON-OFF) signal line to reference pressure. Since both sides of the piston then are subject to an equal pressure, the actuator spring moves the piston and closes the valve. The inlet valve is fully closed when the fuel pressure differential on the piston decreases to less than 20 psi [137.9 kPa].

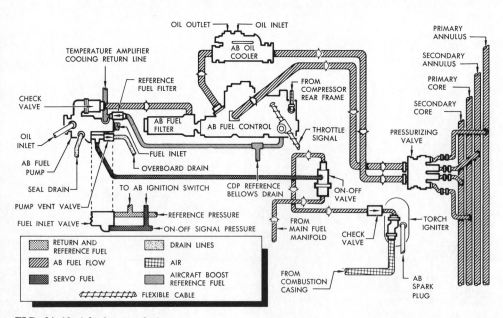

FIG. 21-19 Afterburner fuel system.

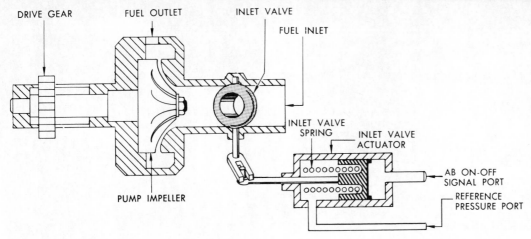

FIG. 21-20 Afterburner-fuel-pump schematic.

PUMP VENT VALVE
(See FIG. 21-19)

The pump vent valve is assembled to the signal-hose tee installed in the vent port of the AB fuel pump.

The pump vent valve drains the AB fuel pump cavity and vents the torch igniter ON-OFF valve signal line to ambient pressure during non-afterburning operation. It closes to prevent fuel from escaping during AB-system operation.

AIRCRAFT REFERENCE FUEL FILTER
(See FIG. 21-19)

The aircraft reference fuel filter is located at the tee connector in the reference fuel return port of the AB fuel pump. The filter prevents contamination from entering the main and AB fuel systems during reverse fuel flow that occurs during engine start.

During normal engine operation, reference pressure fuel is returned to the AB fuel pump from the main fuel control and the AB fuel control. Reference pressure fuel returning to the AB fuel pump is bypassed around the filter element. When the engine is being started, a reverse flow of fuel occurs in the reference fuel return line. During this reverse flow, the fuel from the aircraft fuel supply system that flows to the control system is filtered.

AFTERBURNER FUEL FILTER
(See FIG. 21-19)

The AB fuel filter is located at the seven o'clock position on the compressor rear casing. This filter removes contamination from the afterburner fuel.

The filter is an in-line filter. It incorporates a pressure-relief valve that allows fuel to bypass if the filter element becomes clogged.

AFTERBURNER FUEL CONTROL
(FIG. 21-21)

The AB fuel control is located at the eight o'clock position on the compressor front casing and regulates fuel flow for the AB as a function of throttle angle and CDP.

The AB fuel control regulates AB fuel flow between the minimum necessary for combustion and the maximum allowable, and divides the flow into core and annulus supplies. The fuel control varies fuel flow by regulating the area of a metering orifice while maintaining a constant pressure differential across the orifice. Core fuel from the control flows through the AB oil cooler and to the pressurizing valve.

When the pressure differential between the total flow and the core flow reaches 40 psi [275.8 kPa], the control schedules the additional fuel to the annulus manifold. The fuel/air ratio within the core area is maintained by scheduling the size of the core throttling valve orifice proportionate to CDP. The annulus fuel flow is piped directly to the AB fuel pressurizing valve.

AFTERBURNER FUEL-PRESSURIZING VALVE (See FIG. 21-19)

The AB fuel-pressurizing valve keeps the system components filled with fuel and divides the flow between primary and secondary flows. It is located at the six o'clock position on the turbine casing.

The valve consists of four pressurizing pistons. Fuel is directed to two of them by the core fuel line and to the other two by the annulus fuel line. Fuel is admitted to the primary core manifold when core fuel pressure exceeds 121 psi [834.3 kPa], and to the secondary core manifold when it exceeds 265 psi [1827 kPa]. Fuel is admitted to the primary annulus manifold when annulus fuel pressure exceeds 101 psi [696.4 kPa], and to the secondary annulus manifold when it exceeds 245 psi [1689.3 kPa].

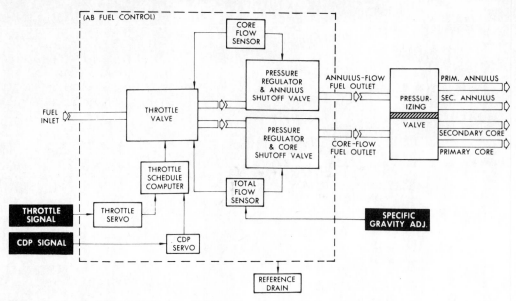

FIG. 21-21 Afterburner-fuel-control block diagram.

When CDP drops below 76 psi [524 kPa], the fuel pressures are so low that only the primary core and primary annulus manifolds discharge fuel.

FUEL MANIFOLDS AND SPRAYBARS (FIG. 21-22)

The fuel manifolds and spraybars, located just aft of the turbine frame, direct sprays of fuel into the exhaust gas stream.

Fuel is distributed by 4 fuel manifolds to 21 spraybars. Each nozzle incorporates four tubes, one for each manifold. Holes in the sides of the tubes spray fuel into the

AB. Core tubes spray fuel near the center of the tailpipe; the annulus tubes spray fuel near the outside of the tail-pipe.

TORCH-IGNITER ON-OFF VALVE (See FIG. 21-19)

The torch-igniter ON-OFF valve controls torch-igniter fuel flow to permit torch-igniter combustion only during AB operation.

The torch-igniter ON-OFF valve is located at the bottom of the compressor rear frame, in the torch-igniter fuel supply line. It is a two-position, hydraulically operated fluid metering valve. It contains a relieving-type inlet filter and a flow-metering cartridge.

The torch-igniter ON-OFF valve is actuated by discharge fuel pressure from the AB fuel pump. When the engine throttle is advanced to afterburner range, a pressure signal from the main fuel control actuates the AB ignition switch and opens the inlet valve, admitting fuel to the AB fuel pump. The pump discharge pressure closes the pump vent valve and applies the pressure signal to the torch-igniter ON-OFF valve operating piston. Opposing this piston movement is engine main fuel manifold pressure plus a compression spring. The signal pressure overcomes these opposing forces and moves the piston to open the valve. As the piston travels to its stop, fuel displaced by the piston is discharged from the valve and fills the downstream piping to the torch igniter. With the piston in its extreme travel position, fuel flow to the torch igniter is metered by the cartridge orifice pack to support combustion in the torch igniter.

When the throttle is retarded below the AB range, fuel flow to the pump is interrupted and the pump discharge is vented overboard by the pump vent valve. This action reduces the signal pressure to the torch-igniter ON-OFF valve and allows the valve to close

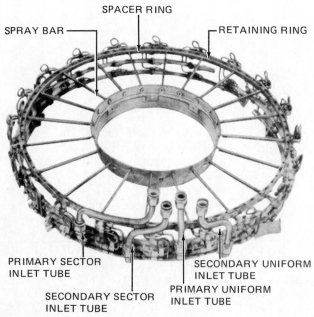

FIG. 21-22 Fuel manifold and spraybars.

from the combined pressures of the compression spring and inlet fuel against the piston. With the piston in the closed position, the valve is again primed for refilling the downstream piping the next time AB operation is called for.

TORCH-IGNITER CHECK VALVE
(See FIG. 21-19)

The torch-igniter check valve permits fuel to flow to the torch igniter but prevents post-shutdown fires in the torch igniter.

The check valve is set to open at a pressure differential of 6 to 8 psi [41.4 to 55.2 kPa]. It will permit fuel flow to the torch igniter at the rate of 200 lb/h [90.7 kg/h] with a pressure drop of 10 psi [68.95 kPa] maximum.

TORCH IGNITER *(FIG. 21-23)*

The torch igniter is located at the bottom of the tailpipe directly behind the spraybars. During AB operation, the torch igniter provides an intense flame that ensures positive combustion of AB fuel.

Fuel metered by the torch-igniter ON-OFF valve enters the fuel nozzle, which is located in a small combustion chamber. Some of the air, which is piped to the torch igniter, flows through a tube, to the fuel nozzle where it forces the fuel into the torch-igniter liner in a fine spray. This forced spray reduces the formation of carbon deposits on the fuel nozzle face.

The remainder of the air that is piped to the torch igniter enters the strut and flows to the inner liner, where it is mixed with the atomized fuel. The resultant mixture is ignited by a spark plug in the combustion chamber, and the burning mixture emerges from the open end of the liner as an intense flame. The rear of the torch-igniter liner spreads the flame and reduces the velocity of the surrounding fuel-air mixture. A tab on the rear of the torch-igniter liner conducts the flame to the middle flameholder ring.

Torch-igniter operation is continuous during afterburning. Because both fuel and air are scheduled to the torch igniter on the basis of CDP, the intensity of the torch-igniter flame is unaffected by altitude and airspeed.

AFTERBURNER IGNITION SYSTEM
(See FIG. 21-17)

The AB ignition system furnishes the ignition for the fuel-air mixture in the torch igniter. The resultant combustion ignites the fuel sprayed into the tailpipe by the AB spraybars. The system, which operates continuously during AB operation, consists of an ignition switch, an ignition unit, shielded cables, and a spark plug.

AFTERBURNER IGNITION SWITCH

The AB ignition switch is located near the rear end of the compressor front casing, just below the left horizontal bolt flange. During non-afterburner engine op-

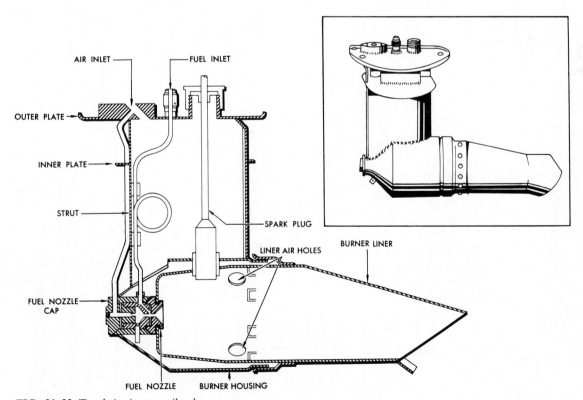

FIG. 21-23 Torch igniter or pilot burner.

eration, main-fuel-system bypass pressure holds the switch contacts open. When the throttle is moved to the AB range and engine speed is within limits, the main fuel control transmits a high-pressure fuel signal to open the fuel inlet valve of the AB fuel pump and to initiate AB ignition. When the high-pressure signal exceeds reference pressure by 70 psi [482.6 kPa] the contact points in the switch close, completing the circuit to the AB ignition unit. When the pressure differential drops to less than 40 psi [275.8 kPa], the contact points in the switch open and break the circuit.

AFTERBURNER IGNITION UNIT
(refer to FIG. 16-11)

The box containing the AB ignition unit is mounted near the four o'clock position on the compressor rear casing. (The box also contains the main ignition unit circuit for transforming direct current, to energize the spark plug in no. 4 combustion liner. This circuit is unrelated to the AB ignition unit circuit.)

When the AB ignition switch is closed, 115-V alternating current flows to the AB ignition unit. The unit includes a step-up transformer to produce the desired voltage; a full-wave rectifier to control current flow; a

storage capacitor to accumulate sufficient energy to produce the spark potential; and a gap which ionizes at approximately 3000 V and allows the high-potential current to flow across it.

AFTERBURNER SPARK PLUG

The AB spark plug is located in the torch igniter, which is mounted at the six o'clock position of the tailpipe. The spark-plug electrodes are shunted by a ceramic semiconductor that ionizes the spark gap when the output of the ignition unit is impressed across it. The ionization creates a low-resistance path across which the built-up energy in the ignition unit storage capacitor is discharged. This discharge is rapid and, consequently, the spark intensity is high.

VARIABLE-NOZZLE SYSTEM
(FIG. 21-24)

The variable-nozzle system controls the engine thrust output and protects the engine parts from overtemperature damage by regulating the position of the exhaust nozzle flaps.

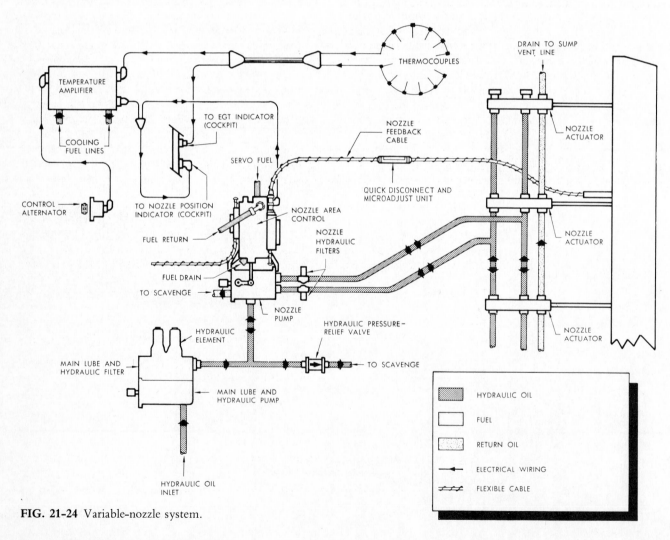

FIG. 21-24 Variable-nozzle system.

Movement of the throttle during engine operation is transmitted through a cable to the nozzle area control. The control uses this mechanical signal of throttle position to schedule the position of the nozzle. However, when the EGT limit is exceeded, or during a rapid acceleration, the temperature amplifier transmits an electrical signal to the nozzle area control that overrides the throttle-controlled schedule. A cable attached to the exhaust nozzle support ring continuously provides a feedback signal, which indicates the position of the nozzle flaps, to the nozzle area control.

When the signals to the nozzle area control indicate that a change in the position of the nozzle flaps is required, the control output rod moves the lever arm on the nozzle pump. When the lever arm moves (clockwise movement opens the nozzle), the pump sends high-pressure oil to either the head end (open) or rod end (close) of the nozzle actuators. As the actuator pistons move the nozzle, the feedback signal is nullifying the movement signal; when the nozzle flaps reach the desired position, the output of the pump maintains the actuator at this setting.

The components in the variable-nozzle system are: control alternator, thermocouples, temperature amplifier, nozzle area control, nozzle pump, nozzle actuators, lube and hydraulic oil pump and filters, pressure-relief valve, and nozzle hydraulic oil filters.

CONTROL ALTERNATOR
(See FIG. 21-24)

The control alternator is mounted on the rear side of the transfer gearbox near the seven o'clock position. The engine-driven alternator furnishes the power supply and speed signal for the temperature amplifier. The voltage and frequency of the alternator output are proportional to engine speed.

THERMOCOUPLES *(FIG. 21-25)*

Two semicircular harnesses, each containing six dual thermocouples, are mounted around the turbine frame. The thermocouples, which extend into the exhaust gas stream just to the rear of the third-stage turbine blades, produce thermoelectric currents that are proportional to temperature. The output of 12 of the thermocouple loops (6 in each harness) is used as a temperature signal in the temperature amplifier; the output of the other 12 energizes the exhaust gas temperature indicator in the cockpit.

TEMPERATURE AMPLIFIER
(See FIG. 21-24)

The temperature amplifier is mounted on the forward end of the compressor front casing near the eight o'clock position. A small amount of fuel from the main fuel system is used to cool the components inside the amplifier.

Electrical signals provided by the thermocouples and the control alternator furnish indications of the engine EGT, the rate of change of EGT, engine speed, and the rate of change of engine speed. The amplifier uses these four indications of engine operating conditions to produce an electrical output signal that is used in the nozzle area control to determine the proper position for the exhaust nozzle flaps.

A reference temperature adjustment on the temperature amplifier is set to limit EGT, as sensed by the thermocouples, to 625°C at 100 percent engine speed. (To improve acceleration during flight at altitude, the limiting temperature decreases when engine speed is less than 100 percent. For example, at 80 percent engine speed, the temperature at which limiting begins, has dropped to approximately 535°C.) The temperature signal transmitted by the thermocouples is compared with the reference temperature signal. When the thermocouple signal is less than the reference signal, the amplifier transmits a signal to the nozzle area control, but the signal has no effect on nozzle flap position. However, if the thermocouple signal exceeds the reference signal, the amplifier output overrides the throttle-position schedule, and the amplifier output signal causes the nozzle area control to open the nozzle flaps enough to lower the EGT to the limiting temperature.

The amplifier output signal also (1) compensates for lag in the variable-nozzle system during a rapid accel-

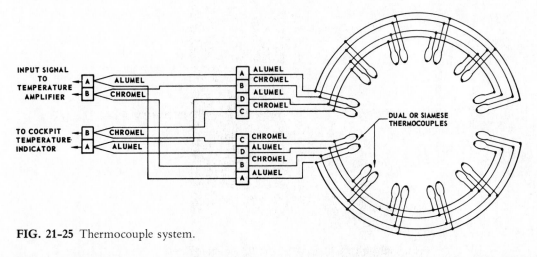

FIG. 21-25 Thermocouple system.

eration, (2) reduces rollback of engine speed when afterburner operation is initiated, and (3) reduces the amount of overspeed when afterburner operation is terminated.

NOZZLE AREA CONTROL
(See FIG. 21-24)

The nozzle area control, which is mounted near the eight o'clock position on the rear end of the compressor rear casing, controls the position of the exhaust nozzle flaps by regulating the output of the nozzle pump.

The control receives mechanical signals that indicate the position of the throttle and of the exhaust nozzle flaps, and an electrical signal that is transmitted by the temperature amplifier.

Movement of the engine throttle rotates a cam in the control; the contour of the cam represents the mechanically scheduled nozzle flap position. Any position of the throttle corresponds to a scheduled setting for the nozzle flaps. However, the electrical signal can override this mechanical schedule and cause the nozzle flaps to move to a more open position when necessary.

Throttle movement or the electrical signal from the temperature amplifier, that necessitates a change in nozzle flap position is transmitted through gearing and servo mechanisms to the output rod of the nozzle area control. As the nozzle flaps move, a feedback signal, which indicates the position of the flaps, is canceling the movement signal. When the flaps reach the required position, the output rod is scheduling the nozzle pump to maintain an output that holds the nozzle actuators at this position.

Regulated pressure fuel from the main fuel control is used as the fluid in the nozzle area control servo mechanisms. A potentiometer in the control transmits a signal indicating position of the nozzle flaps to an indicator in the aircraft cockpit.

NOZZLE PUMP *(See FIG. 21-24)*

The nozzle pump, which is bolted to the rear side of the rear gearbox near the seven o'clock position, is a variable displacement, variable pressure, reversible flow, piston-type pump. It supplies engine oil to the nozzle actuators, to maintain the nozzle flaps at a position or to reposition them as necessary.

The output of the pump is controlled by a lever arm that is attached to the output rod of the nozzle area control. The direction of lever arm moves (clockwise—open) determines whether the output of the pump will go to the head end (open) or rod end (close) of the nozzle actuators. The amount the lever arm moves varies the pressure and volume of the oil flowing to the actuators. The lever arm is spring-loaded so it will return to a midtravel position when there is no signal from the nozzle area control; in the midtravel position the pump output is zero.

NOZZLE ACTUATORS *(FIG. 21-26)*

The four nozzle actuators are attached to brackets on the forward end of the exhaust tailpipe assembly. The actuators, which are located near the 2, 5, 8, and 11 o'clock positions on the tailpipe, move the nozzle flaps as scheduled by the nozzle area control.

Oil from the nozzle pump flows to either the head or rod end of each of the actuators. The fluid pressure moves the actuator piston to open or close the exhaust nozzle flaps.

Flexible shafts inside the head-end pressure tubes, which connect adjacent actuators, ensure simultaneous and parallel movement of all the actuator pistons: movement of the piston rotates a worm gear in the forward end of each of the actuators; as the worm gear turns, it rotates the flexible shafts which transmit the movement to the adjacent actuators.

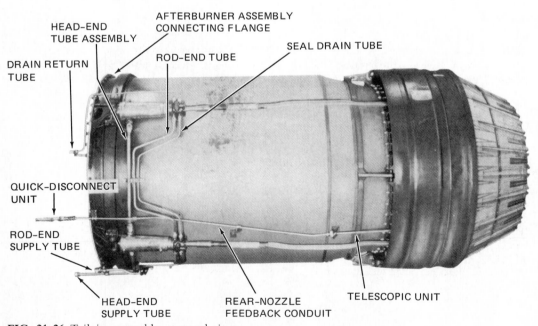

FIG. 21-26 Tailpipe assembly, external view.

An orifice through the piston of each actuator allows a small amount of oil to flow from one side of the piston to the other, to reduce the temperature of the oil in the actuator.

HYDRAULIC RELIEF VALVE
(See FIG. 21-24)

The discharge end of the hydraulic relief valve is attached to the outlet fitting of the rear-gearbox scavenge pump. The pump is bolted to the rear side of the rear gearbox near the six o'clock position.

The poppet-type check valve prevents pressure of the inlet oil to the nozzle pump from becoming too high. If the pump inlet pressure exceeds scavenge-oil pressure by 95 psi [655 kPa], the valve opens and dumps excess oil into the scavenge system. The valve also prevents scavenge oil from entering the nozzle system.

NOZZLE HYDRAULIC OIL FILTER
(See FIG. 21-24)

The nozzle hydraulic oil filters prevent contamination from passing with the hydraulic oil from the exhaust nozzle actuators to the nozzle pump, or from the pump to the actuators. The two identical filters are bolted to a bracket on the rear side of the nozzle pump—one filter is attached to the pump head-end port, the other to the pump rod-end port.

Each filter allows oil to flow through its filter element in either direction. A bypass valve in each filter allows unfiltered oil to bypass the filter element if the element becomes clogged.

VARIABLE-VANE SYSTEM *(FIG. 21-27)*

The variable-vane system schedules the position of the inlet guide vanes and the other six stages of variable vanes to maintain compressor efficiency during various operating conditions.

The variables that are used in scheduling the vane position are the CIT and engine speed. When the air-flow requirements for operating the engine are low (during low engine speed and/or high CIT), the vanes are scheduled to a position that limits the amount of airflow through the compressor. As engine speed increases and/or CIT decreases, the engine requires a larger volume of air and the vanes are opened to minimize airflow restriction. Between approximately 65 and 95 percent corrected engine speed (measure of airflow) the vanes are positioned in accordance with the variable-vane schedule. Below 65 percent corrected engine speed the vanes are fully closed; above 95 percent the vanes are fully opened.

The variable vanes are positioned by two actuators, which use high-pressure fuel from the main fuel control to move the vane linkages.

A variable-vane schedule cam inside the main fuel control is positioned by forces that indicate engine speed and CIT. When a change in vane setting is required, the resultant of these forces on the cam opens a valve that allows high-pressure fuel to flow to the actuators. As the actuators vary the position of the vanes, a signal indicating vane position is transmitted through a flexible cable back to the main fuel control. When the vanes reach the desired position, the feedback signal has cancelled out the movement signal and the valve cuts off flow to the actuators. Some engines incorporate a vane closure valve to prevent engine stall when the aircraft guns are fired. During this period the CIT increases because of the ingestion of hot gas from the guns. The vane closure valve closes the vanes 4°, thus bringing the compressor away from stall.

VARIABLE-VANE ACTUATORS

The two variable-vane actuators are attached to brackets near the 4 and 10 o'clock positions on the compressor front casing (see Fig. 21-2). The actuators move the compressor variable vanes as scheduled by the main fuel control.

High-pressure fuel from the main fuel control flows to the head or rod end of each of the actuators. The fluid pressure moves the actuator piston to open or close the variable vanes. An orifice through the actuator piston allows a small amount of fuel to flow from one side of the piston to the other, to cool the actuator.

LUBRICATION SYSTEM *(FIG. 21-28)*

The lubrication system ensures adequate lubrication for the bearings and gears of the engine. The system can be divided into three subsystems:

1. The lube supply subsystem delivers filtered oil under pressure to the three main bearing areas and the three gearboxes. The oil is sprayed onto the bearings, gears, and oil seals to cool and lubricate them. Lubrication oil flows from the oil tank to the lube and hydraulic pump. The lube pump discharges the oil into the lube filter element of the lube and hydraulic filter. The oil flows from the filter to the bearing areas and gearboxes,

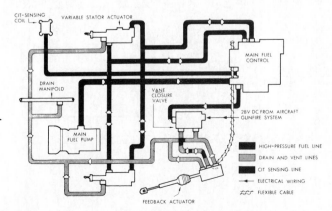

FIG. 21-27 Variable-vane system.

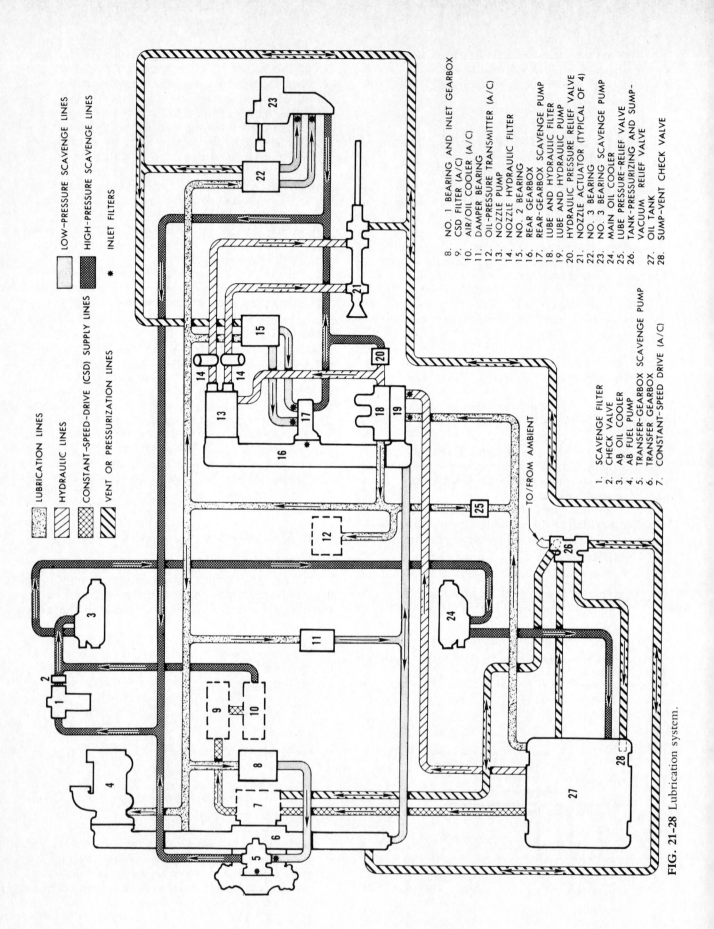

LUBRICATION LINES

HYDRAULIC LINES

CONSTANT-SPEED-DRIVE (CSD) SUPPLY LINES

VENT OR PRESSURIZATION LINES

LOW-PRESSURE SCAVENGE LINES

HIGH-PRESSURE SCAVENGE LINES

INLET FILTERS

*

1. SCAVENGE FILTER
2. CHECK VALVE
3. AB OIL COOLER
4. AB FUEL PUMP
5. TRANSFER-GEARBOX SCAVENGE PUMP
6. TRANSFER GEARBOX
7. CONSTANT-SPEED DRIVE (A/C)

8. NO. 1 BEARING AND INLET GEARBOX
9. CSD FILTER (A/C)
10. AIR/OIL COOLER (A/C)
11. DAMPER BEARING
12. OIL-PRESSURE TRANSMITTER (A/C)
13. NOZZLE PUMP
14. NO. 2 HYDRAULIC FILTER
15. NO. 2 BEARING
16. REAR GEARBOX
17. REAR-GEARBOX SCAVENGE PUMP
18. LUBE AND HYDRAULIC FILTER
19. LUBE AND HYDRAULIC PUMP
20. HYDRAULIC PRESSURE RELIEF VALVE
21. NOZZLE ACTUATOR (TYPICAL OF 4)
22. NO. 3 BEARING
23. NO. 3 BEARING SCAVENGE PUMP
24. MAIN OIL COOLER
25. LUBE PRESSURE-RELIEF VALVE
26. TANK-PRESSURIZING AND SUMP-
    VACUUM RELIEF VALVE
27. OIL TANK
28. SUMP-VENT CHECK VALVE

TO/FROM AMBIENT

FIG. 21-28 Lubrication system.

where it is sprayed through nozzles onto the gears, bearings, and oil seals.

2. The scavenge subsystem recovers the oil from the lubrication areas and from the variable-nozzle system. The scavenged oil is filtered, cooled, and returned to the oil tank for reuse. Each bearing sump and gearbox area has two scavenge-oil outlets—one in the forward end and one in the rear—to ensure removal of the oil whether the aircraft is diving, climbing, or flying level. The oil scavenged from these areas flows to three pumps: one pump is located on the rear side of the transfer gearbox, another on the rear side of the rear gearbox, the third to the rear of the no. 3 bearing on the turbine frame. The output of the three scavenge pumps, plus oil returned from the exhaust nozzle actuation system, is delivered to the scavenge-oil filter. The oil flows from the filter through an aircraft furnished air-oil cooler, the AB fuel-oil cooler, and the main fuel-oil cooler, and back to the oil tank.

3. The pressurization subsystem (Fig. 21-29) regulates air pressure inside the oil tank, bearing sumps, and gearboxes, to ensure proper operation of the oil seals and to keep pressure in these areas at a safe level. Because the total capacity of the three scavenge pumps exceeds the output capacity of the lube pump, the scavenge pumps remove air, in addition to the oil, from the lubrication areas. This scavenge pump overcapacity causes a relatively low air pressure in the sumps and gearboxes, and high-pressure air leaks across seals into the areas. The amount of seal leakage varies between engines. If seal leakage air exceeds the amount required to maintain proper sump and gearbox pressure, excess air flows from the sumps, through the sump vent check valve, oil tank, tank pressurizing valve, and sump pressurizing valve to ambient pressure. However, if seal leakage is not sufficient to maintain the internal operating pressure, the sump and tank pressurizing valve allows enough air to enter the system to maintain the proper pressure.

OIL TANK

The oil tank, which is mounted between the 12 and 3 o'clock positions on the compressor front casing, stores the oil used in the lubrication system, the variable-exhaust-nozzle system, and the aircraft-furnished constant-speed drive.

Oil for the lubrication system flows through a port in the bottom of the tank into the lube pumping element of the lube and hydraulic pump. During normal flight, oil is supplied to the system as long as any remains in the tank. However, during inverted flight no oil is supplied to the lubrication system.

Oil for the nozzle system is contained in a compartme..t in the lower part of the tank. A flexible tube picks up the oil, which flows out of the tank to the hydraulic pumping element of the lube and hydraulic pump. Dur-

ing normal flight, oil flows freely from the tank into the compartment; however, if the quantity of oil in the tank drops to less than 0.9 gal [3.4 L], no more oil flows into the compartment and the supply to the nozzle system stops. During inverted flight a check valve prevents oil in the compartment from flowing back into the rest of the tank, and enough oil remains to continue to operate the nozzle system for about 30 s.

A flexible tube, located near the center of the tank, picks up oil for the constant-speed drive. During normal flight if the amount of oil in the tank drops to less than 2.5 gal [9.46 L], the level of oil is below the inlet of the tube and oil supply for the drive is terminated. The flexible tube picks up oil for the constant-speed drive regardless of flight attitude.

The scavenge oil, which is being returned to the tank, contains a large amount of air that is removed in a deaerator chamber inside the tank. The air in the tank is regulated to reduce oil foaming and to provide a positive inlet pressure for the supply pumps.

A check valve in the sump vent port fitting prevents air in the tank, which normally has a higher pressure than sump pressure, from flowing into the sumps. However, if sump pressure becomes higher than tank pressure, the check valve allows airflow into the tank.

LUBE AND HYDRAULIC PUMP

The lube and hydraulic pump, which is bolted to the rear side of the rear gearbox near the five o'clock position, is a positive-displacement, rotary-vane-type pump. It contains two pump elements, one for the lubrication system and the other for the variable-nozzle system.

Lubrication system oil flows from the oil tank into the element nearest the drive-shaft end of the pump. The pump discharges the oil into the lube-oil filter of the lube and hydraulic filter assembly. The capacity of the lube pump element is 11.8 gal/min [44.7 L/min].

Nozzle-system oil flows from the oil tank into the rear element. The pump, which has a capacity of 4.1 gal/min [15.5 L/min], discharges the oil into the hydraulic oil filter of the lube and hydraulic filter assembly. The pump element contains a relief valve that prevents its discharge pressure from exceeding inlet pressure by more than 110 psi [758.4 kPa].

LUBE AND HYDRAULIC FILTER

The lube and hydraulic filter is an assembly of two filters—one filters lubrication system oil, the other filters variable-exhaust-nozzle system oil. The two filtering elements are removable and cleanable, but cannot be interchanged.

The filter for each system incorporates a bypass valve, a check valve, and a shutoff valve. Normally, oil flows from the lube and hydraulic pump through the filter elements and out the discharge ports. However, if either filter element becomes clogged, its bypass valve will open and allow unfiltered oil to flow to its discharge port. The check valves close when the engine is stopped to prevent oil leakage from the tank into the systems during shutdown. The shutoff valves close when the filter elements are removed for inspection or cleaning to prevent loss of oil from the systems.

A fitting on the lubrication-system side of the filter provides a tap for connecting a line to the aircraft-furnished lube pressure transmitter.

LUBE PRESSURE-RELIEF VALVE

The lube pressure-relief valve, which is attached to the lube oil inlet fitting of the lube and hydraulic pump, protects the aircraft-furnished oil-pressure transmitter from high oil pressures.

When the engine is started during cold weather, lube oil pressure is high due to the increased viscosity of the oil. If the differential between lube pump inlet and discharge pressures exceeds 95 psi [655 kPa], the relief valve opens to prevent the pressure in the transmitter line from damaging the transmitter. However, the relief valve has a very little effect on actual system pressure, because a restriction in the fitting at the lube filter allows very little oil to flow into the transmitter line. When the oil warms and becomes more fluid, the relief valve closes and the transmitter measures actual lube-system oil pressure.

TRANSFER-GEARBOX SCAVENGE PUMP

The transfer-gearbox scavenge pump, which is mounted on the rear side of the transfer gearbox at the six o'clock position, contains two positive-displacement, gerotor-type pumps. The number 1 pump element, which is nearest the drive shaft end, has a capacity of 11 gal/min [38.3 L/min]. It scavenges oil from inside the transfer gearbox. The other pump element has a capacity of 2.7 gal/min [10.2 L/min], and scavenges oil that flows into the oil outlet at the rear of the no. 1 bearing sump. Each pump has a separate inlet, but both discharge oil through the same outlet.

REAR-GEARBOX SCAVENGE PUMP

The rear-gearbox scavenge pump is mounted on the rear side of the rear gearbox at the six o'clock position.

It contains three positive-displacement, vane-type pumps. The no. 1 pump element, which is nearest the drive-shaft end, has a capacity of 11 gal/min [38.3 L/min]. It scavenges oil from inside the rear gearbox. The no. 2 pump element has a capacity of 5.2 gal/min [19.7 L/min] and scavenges oil that flows into the rear oil outlet of the no. 2 bearing sump. The no. 3 pump element has a capacity of 3.9 gal/min [14.8 L/min] and scavenges oil that flows into the forward oil outlet of the no. 2 bearing sump. Each pump has a separate inlet, but all three discharge oil through the same outlet.

NUMBER 3 BEARING SCAVENGE PUMP

The no. 3 bearing scavenge pump, which is mounted on the rear of the turbine frame hub, contains two positive-displacement, vane-type pumps, each with a capacity of 3 gal/min [11.4 L/min]. The forward pump element scavenges oil that flows into the forward oil outlet of the no. 3 bearing sump. The rear element scavenges oil that flows into the rear oil outlet of the sump. The two elements discharge the oil into the same outlet.

SCAVENGE-OIL FILTER

The scavenge-oil filter, which is located near the nine o'clock position on the compressor front casing, removes contamination from the scavenge oil before it flows to the aircraft-furnished air-oil cooler.

Oil from the three scavenge pumps and the nozzle actuators flows into the filter inlet, through the filtering element, and out the discharge port. If the filtering element becomes clogged, a relief valve opens and allows unfiltered oil to flow out the discharge port. A check valve, which is assembled into the discharge port of the filter, prevents oil from flowing from the oil tank into the scavenge system while the engine is shut down.

AFTERBURNER FUEL-OIL COOLER

The AB fuel-oil cooler is mounted on the compressor rear casing near the nine o'clock position. Fuel that is discharged from the "core" outlet port on the AB fuel control is used in the cooler to reduce the temperature of the scavenge oil.

The fuel enters the fuel inlet port, flows through thin-walled tubes and out the fuel outlet port. Scavenge oil enters the oil inlet port, flows around the fuel-filled tubes and out the oil outlet port.

A bypass valve allows oil to flow from the oil inlet directly to the oil outlet, without going through the fuel tube portion of the cooler, if the temperature of the inlet oil is less than 110°F [43.3°C] or if the pressure of the

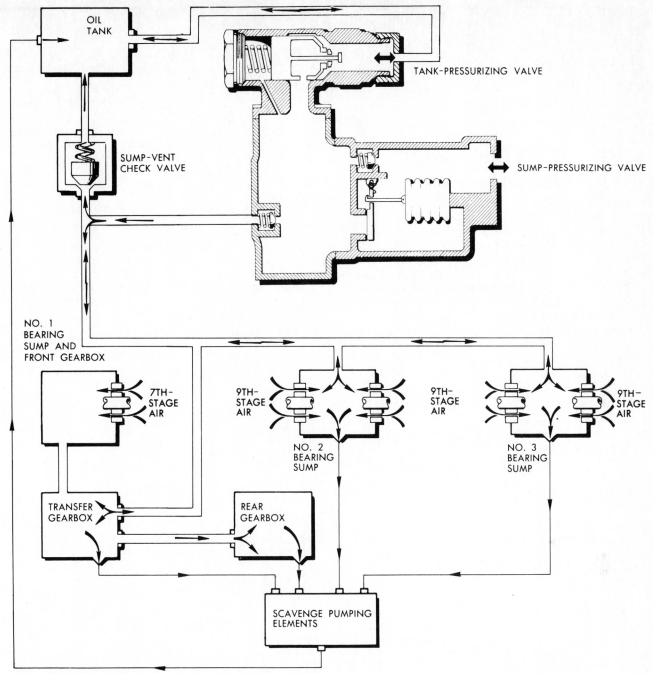

FIG. 21-29 Sump-pressurizing system.

inlet oil exceeds outlet oil pressure by more than 40 psi [275.8 kPa].

MAIN FUEL-OIL COOLER

The main fuel-oil cooler is mounted on the compressor rear casing near the four o'clock position. Metered fuel from the main fuel system is used in the cooler to cool the scavenge oil. The main fuel-oil cooler is identical to the AB fuel-oil cooler.

TANK PRESSURIZING AND SUMP VACUUM RELIEF VALVE *(FIG. 21-29)*

The tank pressurizing and sump vacuum relief valve is located on the compressor rear casing near the two o'clock position. It contains two relief valves that regulate air pressure inside the oil tank, gearboxes, and bearing sumps. One valve vents air from the oil tank whenever tank air pressure exceeds ambient pressure by more than 4.5 psi [31 kPa]; the other valve allows ambient air to bleed into the sumps and gearboxes if ambient pressure exceeds sump pressure by more than 2.5 psi [17.2 kPa].

ANTI-ICING SYSTEM *(FIG. 21-30)*

The anti-icing system directs compressor discharge air into the struts of the compressor front frame, the inlet guide vanes, and the first-stage compressor vanes. The temperature of this air prevents the formation of ice that might prove damaging to the engine.

Compressor discharge air passes through a port in the outer combustion casing into the anti-icing air tube, which directs it to the anti-icing valve. When the anti-icing control switch is closed, the anti-icing valve opens and the anti-icing air flows through struts no. 2, 3, 7, and 8 into a manifold within the hub of the front frame.

The air passes from the manifold through a port in the front-gearbox casing, struts no. 1, 4, 5, and 6, and into the 20 inlet guide vanes. The air is discharged into the primary airstream through holes near the outer end of the four struts and through openings in the trailing edge of the inlet guide vanes.

At the compressor front casing, the anti-icing air flows through five tubes into a manifold located over the first-stage vanes. The air passes from the manifold through the variable vanes and returns to the primary airstream through two slots at the inner end of each vane.

ANTI-ICING VALVE

The anti-icing valve, which is mounted at the top of the compressor rear casing, regulates the flow and pressure of anti-icing air.

When the engine anti-icing switch in the airplane cockpit is positioned at DE-ICE, a solenoid in the valve is energized; this allows the main poppet valve to be unseated and anti-icing air to flow through the valve.

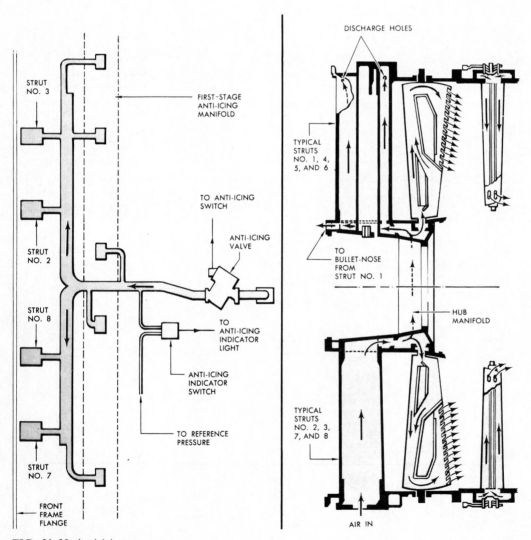

FIG. 21-30 Anti-icing system.

The valve regulates the pressure of the anti-icing air to below 28 psi [193 kPa], and a pressure-relief valve prevents the pressure in the valve from exceeding approximately 35 psi [241.3 kPa].

ANTI-ICING INDICATOR SWITCH

The anti-icing indicator switch, which is mounted near the 11 o'clock position on the compressor front casing, furnishes a signal to indicate when the anti-icing valve is open and allowing anti-icing air to flow to the forward parts of the engine. The pressure of the air inside the anti-icing air tube forward of the anti-icing valve actuates a set of contacts inside the switch. When the anti-icing valve is closed, no anti-icing air flows past the valve and the contacts are open. When the valve is opened, the pressure of the anti-icing air causes the set of contacts to close; this completes a 28-V circuit that activates the indicator in the airplane cockpit.

CONTROL LINKAGE SYSTEM (FIG. 21-31)

The control linkage system transmits input and feedback signals to the engine controls. The system is composed of four linkages, throttle, variable-vane feedback, variable-nozzle feedback, and nozzle area control to nozzle pump.

THROTTLE LINKAGE

The aircraft throttle lever is connected through a linkage to the torque booster, which is attached to the throttle input shaft of the main fuel control. Also attached to the input shaft is a sheave inside the cable box housing. A flexible shaft that passes around the sheave also engages sheaves in the cable boxes on the nozzle area

control and afterburner fuel control. As the throttle lever is moved, the movement is transmitted through the cable and sheaves to the three controls to synchronize the various systems' schedules.

TORQUE BOOSTER

The torque booster, which is mounted on the main fuel control, assists the pilot in positioning the throttle linkage by amplifying the force applied to the booster input shaft.

High-pressure fuel from the main fuel control is the actuating fluid in the torque booster. Rotation of the booster input shaft moves a pilot valve that directs the high-pressure fluid to one end of a piston and relieves pressure on the other end. The fuel pressure moves the piston, which rotates the booster output shaft. When the output shaft reaches the desired setting, the pilot valve is positioned so it cuts off flow to the piston, and movement ceases.

VARIABLE-VANE FEEDBACK LINKAGE

The variable-vane feedback linkage conveys a signal of variable-vane position back to the main fuel control, which schedules movement of the vanes. The signal is transmitted through a flexible cable; one end of the cable is connected to a variable-vane bellcrank; the other end engages a sheave that is attached to a spring-loaded shaft on the main fuel control.

When the control schedules a change of vane position, a valve is opened that allows high-pressure fuel to flow to the vane actuators. As the actuator pistons move the vanes, the flexible cable feedback signal is neutralizing the movement by closing the valve. When the vanes reach the scheduled position, the valve is closed and fuel flow to the vane actuator stops.

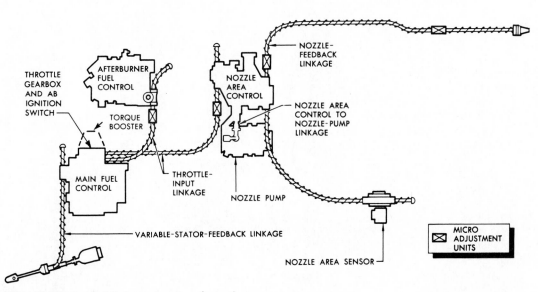

FIG. 21-31 Control-linkage-system schematic.

VARIABLE-NOZZLE FEEDBACK LINKAGE

The variable-nozzle feedback linkage transmits a signal of the position of the variable-nozzle flaps back to the nozzle area control, which schedules movement of the flaps. The signal is conveyed through a two-piece flexible cable: one end of the cable is fastened to the exhaust nozzle support ring; the other end engages a sheave that is attached to the feedback shaft on the nozzle area control. The sheave is spring-loaded to maintain tension on the flexible cable.

When the nozzle area control schedules a change of nozzle flap position, the output rod of the control extends (open nozzle) or retracts (close nozzle) and rotates the lever arm on the nozzle pump. The pump sends high-pressure oil to the head-end (open) or the rod-end (close) of the nozzle actuators. As the actuator pistons move the nozzle, the flexible cable feedback signal is canceling the movement signal in the control. When the nozzle flaps reach the scheduled position, the pump lever arm has returned to a setting that maintains the actuators at this position.

NOZZLE AREA CONTROL TO NOZZLE PUMP LINKAGE

The nozzle area control to nozzle pump linkage consists merely of the ball-socket-joint coupling that is screwed onto the output rod of the nozzle area control, and the bolt, washer, and nut that connect the coupling to the lever arm on the nozzle pump.

ENGINE AIRFLOW DESCRIPTION

During engine operation, air is drawn or rammed into the inlet, where it is directed onto the first-stage rotor blades by the inlet guide vanes. The successive stages of the compressor increase the pressure of the air, while forcing it to the rear. The angles of the first six stages of vanes and the inlet guide vanes are variable to maintain the efficiency of the compressor over a wide range of operating conditions.

The pressure rise of each stage of compression depends upon the speed of rotation of the rotor, which is reflected by engine speed; the density of the air, which is reflected by CIT; and the angle at which the air strikes the blades and vanes. The position of the vanes during any engine-speed/CIT condition is established by the main fuel control.

As the air travels to the rear, some of it bleeds inward through holes in the seventh-stage rotor spacer, and some outward through holes in the ninth-stage vane bases in the upper casing half. The seventh-stage air extraction is used as pressure-equalization air in the compressor rotor and turbine-cooling air. The ninth-stage air extraction becomes sump cooling air for the three bearing sumps.

As the air leaves the rear of the compressor, it is straightened by the outlet guide vanes to prevent swirling in the combustion section. The compressor rear frame is a diffuser which decreases the air velocity and increases its pressure. The rear frame contains air extraction manifolds to supply air to the aircraft.

The air within the combustion section supports the combustion of fuel and cools engine assemblies. A snout on each of the combustion liners directs air into the outer liners. The vanes in the snout distribute the air uniformly around the domes of the inner liners. Some of the air passes through the cowls on the fuel nozzles to hold the flame away from the nozzle tips, some through the louvers in the dome of the liners to atomize the fuel, and some through louvers in the liner to separate the flame from contact with the surface of the liner. Thimble holes in the inner liner direct the air inward to center the flame.

The remainder of the air continues to flow to the rear, surrounding the combustion liners. A small amount flows inward through holes in the inner casing, to cool the turbine shaft. It flows across a baffle on the inner rim of the first-stage turbine nozzle and onto the front face of the first-stage turbine blade shanks.

Air enters through louvers in the combustion rear liners to keep the flame from contacting the inner surface, and through thimble holes to ensure complete combustion. Combustion is completed at a point early enough to prevent the flame from being directed onto the first-stage turbine nozzle. The air surrounding the combustion liners cools the outer surface of the transition liner and some of it passes through louvers between the ports.

Air flows through the baffles on the rear flange of the combustion liners and along the inner surface of the transition-liner ports. It also enters a baffle on the rear of the transition liner and flows over the inner and outer bands of the first-stage turbine nozzle.

Some air enters the outer end of the first-stage nozzle vanes, passes inward through the vane, and is directed onto the front of the first-stage turbine blade shanks. Some air continues to flow rearward through holes in the inner rib of the turbine casing and inward through second-stage turbine nozzle vanes. It is directed onto the rear of the first-stage turbine blade shanks.

The first-stage nozzle vanes increase the velocity of the gas stream from the combustion section and direct it onto the first-stage turbine blades. The second-stage vanes reduce the swirling, again increase the velocity of the gas stream, and direct it onto the second-stage blades. The third-stage vanes reduce the swirling, accelerate the gas stream, and direct it onto the third-stage blades. The energy extracted, as a reaction to the high-velocity gases striking the blades, produces the rotary motion that drives the compressor.

Turbine shrouds and turbine seals prevent excessive leakage of gases around the tips of the blades and over the torque rings of the rotor. Strut covers, surrounding the struts of the turbine frame, reduce the swirling of the gases entering the exhaust section. The turbine frame diffuses the gas stream as it enters the tailpipe.

Thermocouples, mounted in the turbine frame, produce a signal that is proportional to the temperature of the turbine discharge (exhaust gas) temperature. The signal is used for cockpit indication and for control of the exhaust-nozzle area.

The air in the exhaust section is diffused between the inner rear cone and the forward exhaust duct liner. The air is divided into cooling air, which flows between the liners and the duct, and exhaust air, which flows through the liners. The cross section of the airstream changes from annular to circular within the duct.

Spraybars, extending into the exhaust-gas stream, add fuel which is ignited to augment the thrust of the basic engine when afterburner operation is selected. The flameholder produces a turbulence that enhances burning of the fuel. The torch igniter, receiving its air supply from the outer combustion casing, maintains the flame.

The cooling air between the liners and the ducts passes through louvers in the liner to shield its inner surface from direct contact with the flame. It also flows along the inner surface of the primary nozzle flaps and seals.

The exhaust nozzle causes the velocity of the airstream to increase by restricting its flow. The velocity of the exhaust gases past the throat (smallest area) is limited to the speed of sound within the gases. Since the speed of sound increases in proportion to an increase in temperature, the afterburner produces thrust by increasing both the temperature and the velocity of the gases.

The converging portion of the nozzle, formed by the primary flaps, accelerates the gases to a sonic velocity. The diverging portion of the nozzle, formed by secondary air directed by the shroud flaps, controls the rate of expansion, and thus in effect accelerates the gases beyond the throat. The nozzle area is determined by the throttle position, until the temperature of the exhaust gases reaches the reference temperature schedule of the engine; then the throttle control of the nozzle is overridden by a temperature-limiting system to maintain the exhaust-gas temperature according to the reference temperature schedule. The high velocity of the exhaust gases, passing from the throat of the nozzle to the exit, acts as an aspirator to cause air to flow along the outside of the engine. This is called secondary air.

The secondary air cools the engine and accessories, and forms the diverging (aerodynamic) exhaust nozzle. The air enters the engine compartment, around the engine inlet, and passes along the outside of the engine. It removes any fumes or leakage air from the engine compartment. The air is drawn into the outer shroud, between the primary and the secondary (shroud) nozzle flaps, and forms a nozzle around the expanding exhaust gases. The air mixes with the primary air and increases the total amount of airflow. Secondary air controls the rate of expansion of the exhaust gases beyond the primary nozzle throat and reduces the gas temperature. The shroud flap seals, and a seal between the fuselage and the outer shroud, reduce air leakage from around the outside of the aircraft, which would reduce the secondary airflow.

AIR EXTRACTION (refer to FIG. 21-4)

Air is extracted from the compressor for many uses. It is piped or ducted to supply air for turbine cooling, anti-icing, and for airplane uses. Some air passes inward through holes in the seventh-stage rotor spacer. It is ducted throughout the rotor to equalize the air pressures on the disks. It passes through the rear stub shaft and into the turbine shaft to cool the shaft and the front of the first-stage turbine wheel. The air flows through the center of the first-stage wheel, outward on the rear face of the wheel, across the inside of the first-stage torque ring, and inward on the front face of the second-stage wheel. The air flows through the center of the second-stage wheel, outward on the rear face of the wheel, across the inside of the second-stage torque ring, and inward on the front face of the third-stage wheel. The air flows through the center of the wheel, into the turbine cooling air baffle, and outward, through all seven struts of the turbine frame, into the engine compartment.

Cooling air for the three bearing areas flows outward from the compressor into the ninth-stage air manifold. It is externally piped to a tube in the no. 1 strut of the compressor front frame, to a tube in the no. 1 strut of the compressor rear frame, and to one in the no. 2 strut of the turbine frame. The air minimizes oil leakage from the bearing sumps, surrounds the no. 2 bearing sump to prevent the heat of the compressor discharge air from being transmitted to the sump, and surrounds the no. 3 bearing sump to prevent the heat of the turbine discharge air from being transmitted to the sump. The air that leaks across the oil seals enters the bearing sumps.

The sump cooling air is confined in the compressor rear frame cavity by the no. 2 bearing front and rear airseals. The seal-leakage air enters the 17th-stage seal leakage air cavity. The cooling air is confined in the turbine frame cavity by the rear turbine airseal at the front and the no. 3 bearing airseal at the rear. The seal leakage enters the turbine discharge air from the front seal and the turbine cooling air at the rear seal.

Air, for use in the airplane, is extracted from the compressor discharge. The air enters the stiffening manifolds in the compressor rear frame inner shell. It is ducted through the nos. 2, 4, 7, and 9 struts of the rear frame.

ANTI-ICING AIR (refer to FIGS. 21-4 and 21-30)

The anti-icing air prevents or removes ice formation in the engine inlet. The air is extracted through a port at the top of the outer combustion casing. It is piped to the anti-icing valve, which regulates the flow and pressure. It is piped from the valve to the first-stage compressor vanes anti-icing manifold, and to pads on the front frame, where it is ducted into the hub by four struts. The air is confined by a manifold cover in the hub of the frame and is distributed through the hollow shanks of the inlet guide vanes and the remaining four struts of the front frame. A port at the top of the gear-box casing conducts anti-icing air to the aircraft nose dome.

REVIEW AND STUDY QUESTIONS

1. Name several airplanes using this engine.
2. List the engine's major specifications.
3. Give a brief description of the engine operation and airflow.
4. Very briefly describe the construction features of the following engine parts: front frame, compressor rotor and cases, diffuser, combustion liner and cases, turbine stator and rotor, afterburner, and variable-area exhaust nozzle.
5. How is the chance for compressor stall reduced on this engine?
6. How many main bearings does the engine have?
7. How is excessive oil consumption in the bearing compartment controlled?
8. Briefly describe the following systems: anti-icing, oil, fuel, and ignition.
9. List the functions of the fuel control.
10. Describe the operation of the afterburner.
11. What is the purpose of the vane closure valve? Why is it needed?
12. Describe how and why the fuel flow, throttle position, afterburner operation, and nozzle position are all integrated.
13. Where is air extracted on this engine? For what purpose is the air used?

CHAPTER TWENTY-TWO
AVCO
LYCOMING
T-53

The Lycoming T-53 (Fig. 22-1), installed in the Grumman OV-1 and several models of the Bell UH-1 and 204 helicopter, has been selected for review because of its several distinctive features. A later, more powerful model, the T-55, has many of the same construction features and is used in the Vertol CH47A, B, and C helicopter and the YAT-28E.

SPECIFICATIONS

Number of compressor stages:	5 axial, 1 centrifugal
Number of turbine stages:	2 or 4
Number of combustors:	1
Maximum power at sea level:	1100 to 1400 SHP
Specific fuel consumption at maximum power:	0.58 to 0.68 lb/SHP/h [263 to 309 g/SHP/h]
Compression ratio at maximum rpm:	6:1
Maximum diameter:	23 in [58 cm]
Maximum length:	47 in [119 cm]
Maximum dry weight:	550 lb [250 kg]

The Lycoming T-53 gas turbine engine is a free-turbine power plant. Basically, all models of these engines are of the same configuration, but differ in some parts or assemblies. A major difference on later versions of the T-53 is the use of two gas-producer turbines and two free-power turbines instead of one of each type. The description and information given applies to all models except where noted.

The engine described here is a shaft turbine design with a single-stage, free-type power turbine and a single-stage gas-producer turbine that drives a combination axial-centrifugal compressor. The combustor is an external, reverse flow annular, vaporizing type. Five major sections of the engine are air inlet, compressor, diffuser, combustor, and exhaust.

One model difference is the incorporation of a transient airbleed in conjunction with a faster acceleration schedule, modified airbleed actuator system, spring-mounted no. 1 bearing, and bypass fuel filter.

OPERATION (FIG. 22-2)

Air enters the inlet housing assembly and flows into the compressor section, where it is compressed by the five-stage axial-compressor rotor assembly and the cen-

trifugal-compressor impeller. The compressed air flows through the diffuser housing and into the combustion chamber, where it mixes with fuel from the vaporizers. Air is then discharged through the turbines and exhaust diffuser.

Combustion gases drive two separate and independent turbine stages. The first-stage turbine drives the compressor rotor (N_1), which compresses and forces air rearward into the combustion chamber where it mixes with fuel. The second-stage turbine supplies the driving force for the power output gearshaft through reduction gearing.

The engine is started by energizing the starter, the starting-fuel solenoid valve, and the ignition system. Starting fuel, flowing into the combustion chamber through two starting-fuel nozzles, is ignited by the two igniter plugs adjacent to the nozzles at the four and eight o'clock positions in the combustion chamber. At 8 and 13 percent N_1 speed, the fuel-regulator foot valve opens and main fuel flows into the combustion chamber through 11 fuel vaporizers and is ignited by the burning starting fuel.

Rotor speed (N_1) increases as the additional fuel mixes with compressed air and burns. When rotor speed reaches ground idle (40 to 44 percent), the igniter plugs are deenergized and the solenoid valve is deactivated, shutting off the starting fuel. During this period, the starter should be deenergized.

Combustion gases pass through the first-stage turbine nozzle assembly, impinge upon the blades of the first-stage turbine rotor, and pass through the second-stage turbine nozzle and cylinder assembly and onto the blades of the second-stage turbine rotor assembly. Approximately two-thirds of the gas energy passing through the first-stage turbine rotor is used to drive the compressor rotor assembly; the rest is used by the second-stage turbine rotor to drive the power shaft.

The second-stage power turbine rotor is splined to the power shaft and secured to it by the power-shaft bolt. At the inlet end of the engine, the power shaft is splined into the sun shaftgear, which drives the output reduction gears and power-output gearshaft.

DIRECTIONAL REFERENCE (FIG. 22-3)

The following directional references are used in this section.

1. *Front*—End of engine from which output power is extracted.

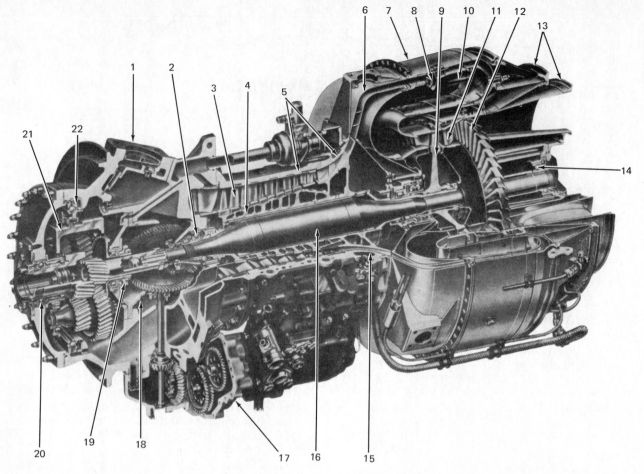

1 INLET HOUSING ASSEMBLY
2 NO. 1 MAIN BEARING
3 COMPRESSOR VANE
4 COMPRESSOR ROTOR AS-
   SEMBLY
5 INTERSTAGE BLEED COM-
   PRESSOR AND IMPELLER
   HOUSING
6 DIFFUSER HOUSING

7 COMBUSTION-CHAMBER
   ASSEMBLY
8 SCOOPS
9 FIRST-STAGE TURBINE
10 FUEL VAPORIZER
11 SECOND-STAGE TURBINE
   NOZZLE AND CYLINDER
   ASSEMBLY
12 SECOND-STAGE TURBINE

13 EXHAUST DIFFUSER
14 NO. 4 MAIN BEARING
15 MANIFOLD
16 POWER SHAFT
17 ACCESSORY DRIVE GEAR-
   BOX
18 ACCESSORY DRIVE CAR-
   RIER ASSEMBLY
19 OVERSPEED GOVERNOR

   AND TACHOMETER DRIVE
   SUPPORT AND GEAR AS-
   SEMBLY
20 OUTPUT SHAFT BEARING
21 OUTPUT REDUCTION
   CARRIER AND GEAR AS-
   SEMBLY
22 TORQUEMETER REAR
   PLATE AND CYLINDER

FIG. 22-1 The Avco Lycoming T-53.

2. *Rear*—End of engine from which exhaust gases are expelled.

3. *Right and left*—Determined by observing the engine from the exhaust end.

4. *Bottom*—Determined by location of accessory drive gearbox.

5. *Top*—Directly opposite, or 180° from, the accessory drive gearbox. (The hot-air solenoid valve is at the top of the engine.)

6. *Direction of rotation*—Determined as viewed from the rear of the engine. Direction of rotation of the compressor rotor and first-stage turbine is counterclockwise. The second-stage turbine and the power-output gearshaft rotate clockwise.

7. *O'clock*—Position expressed as viewed from rear of the engine.

The major assemblies of the engine (Fig. 22-4) include the:

Inlet housing
Overspeed governor and tachometer drive assembly
Accessory gearbox assembly
Output reduction carrier and gear assembly
Compressor and impeller housing assemblies
Compressor rotor assembly
Diffuser housing
Combustor turbine assembly
Piping and accessories

The following descriptions of the major engine assemblies are coordinated with Fig. 22-4.

INLET-HOUSING ASSEMBLY

The inlet-housing assembly (5) is divided into two principal areas. The outer housing, supported by air struts, forms the outer wall of the air inlet area and

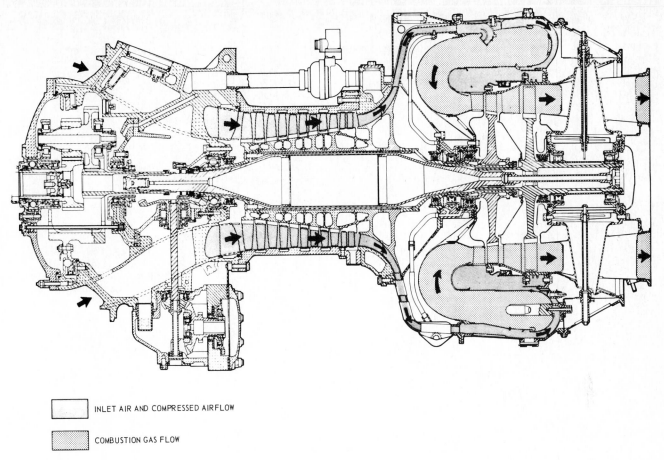

INLET AIR AND COMPRESSED AIRFLOW

COMBUSTION GAS FLOW

FIG. 22-2 Engine airflow.

houses the deicing manifold. The inner housing forms the inner wall of the air inlet area. The inlet-housing assembly encloses the output reduction carrier and gear assembly, the output shaft bearing, the no. 1 main bearing, the overspeed governor and tachometer drive support and gear assembly, the accessory drive carrier assembly, and the torquemeter rear plate and cylinder. The inlet housing provides mounting for the accessory drive gearbox. The overspeed governor and tachometer drive assembly is mounted on the outer-housing left side.

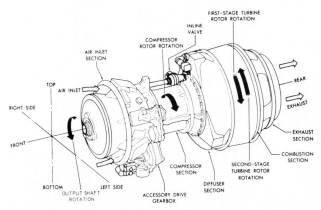

FIG. 22-3 Directional references.

OVERSPEED GOVERNOR AND TACHOMETER DRIVE ASSEMBLY

The overspeed governor and tachometer drive assembly (4) generates the indication and control signals for the power turbine. The output of the tachometer-generator, mounted on the overspeed governor and tachometer drive assembly, is transmitted to an airframe tachometer indicator, which reflects power turbine speed. The overspeed governor, in conjunction with the fuel control, maintains the power turbine speed within permissible limits. The assembly is mounted at the nine o'clock position on the exterior of the inlet housing and is driven through shafts and gearing from the power turbine power-output shaft. The assembly also provides a mount for and drives the torquemeter booster pump.

ACCESSORY DRIVE GEARBOX ASSEMBLY

The accessory drive gearbox assembly (6) is mounted at the six o'clock position on the exterior of the inlet housing. It is driven by a shaftgear mated to a driving gear on the compressor rotor. The gearbox provides the drive for the oil pump, fuel control, compressor rotor tachometer-generator, and starter-generator (not part of the engine). A magnetic-chip-detector drain plug is installed in the bottom of the gearbox.

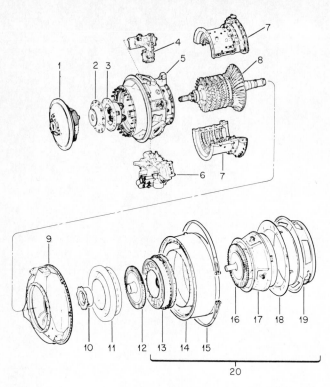

FIG. 22-4 Engine major assemblies.

OUTPUT REDUCTION CARRIER AND GEAR ASSEMBLY

The output reduction carrier and gear assembly (1), mounted at the front of the inlet housing, consists of the support housing, carrier assembly, three planet-reduction-gear assemblies, a torquemeter, and power-output gearshaft. The sun shaftgear, which is splined to the power shaft, drives the three helical planet-reduction gears mounted in the carrier and gear assembly. The reduction gears, in turn, drive the power-output gearshaft. The rear plate of the torquemeter is attached to the carrier.

COMPRESSOR AND IMPELLER HOUSING ASSEMBLIES

The compressor and impeller housings (7) each consist of two halves. The housings enclose the five-stage axial compressor and the single-stage centrifugal compressor impeller of the compressor rotor assembly. The axial-compressor stator vane assemblies are bolted to the compressor housing halves. An airbleed actuator is mounted on the right side of the impeller housing. An airbleed-connecting manifold and adapter assembly ducts bleed air from the diffuser housing to the impeller housing. Customer air is available from this adapter assembly, which also directs engine anti-icing air through a port in the impeller housing.

COMPRESSOR ROTOR ASSEMBLY

The compressor rotor assembly (8) consists of five compressor rotor disk and blade assemblies, five compressor rotor spacers, and one centrifugal compressor impeller assembly, all retained upon the compressor rotor sleeve. The first-stage turbine rotor is mounted on the rear-compressor shaft. The compressor rotor assembly encloses, but is not connected to, the power shaft.

DIFFUSER HOUSING

The diffuser housing (9) conducts air from the compressor to the combustion chamber. It supports the no. 2 main bearing (compressor rotor rear bearing) and the first-stage turbine nozzle and flange assembly. Air is bled from the aft face of the diffuser vanes, through a connecting manifold, to supply anti-icing and customer bleed air.

COMBUSTOR TURBINE ASSEMBLY

The combustor turbine assembly (20) consists of the combustion-chamber assembly, power turbine nozzle and cylinder assembly, second-stage turbine support assembly, fire shield, and support cone. The combustion-chamber assembly consists of the combustion-chamber housing, combustion-chamber liner, and fuel vaporizers. The second-stage turbine support is made up of the second-stage turbine rotor assembly and the exhaust diffuser. The second-stage turbine rotor assembly consists of the second-stage turbine disk and blades, the nos. 3 and 4 bearings, and the no. 3 bearing seal. The welded exhaust diffuser contains hollow struts through which cooling air is supplied to the nos. 3 and 4 bearings housings and the rear face of the second-stage turbine disk.

PIPING AND ACCESSORIES

Piping and accessories include the following:

1. Main and starting fuel manifolds and associated hoses
2. Fuel control, temperature-sensing element, and inlet-air-pressure sensing hose
3. Main wiring harness
4. Starting-fuel solenoid valve
5. Ignition unit, ignition lead and coil assembly, and igniter plugs
6. Hot-air solenoid valve and airbleed components
7. Interstage airbleed piping
8. Lubrication manifold
9. Pressure and scavenge-oil hoses
10. Torquemeter booster pump
11. Oil pump
12. Oil filter
13. Clamps, brackets, screws, and other attaching parts
14. Bypass fuel filter (on some models)
15. Combustion-chamber drain valve
16. Exhaust thermocouple harness
17. Starting-fuel nozzles
18. Exhaust diffuser cover

The engine systems are as follows:

1. Lubrication
2. Internal cooling
3. Pressurizing and anti-icing
4. Fuel, fuel control
5. Electrical
6. Interstage airbleed

LUBRICATION SYSTEM *(FIG. 22-5)*

The engine lubrication system consists of the main oil-pressure supply system and the scavenge-oil system. The principal components of the lubrication system are the main oil pump, oil filter, torquemeter booster pump, overspeed governor and tachometer-drive scavenge pump, and associated external lines and internal passages. The maximum oil inlet temperatures for various models run from 200°F [93°C] to 210°F [99°C]. The engine lubrication system will operate satisfactorily with an engine oil inlet temperature between −65 and 210°F [−54 and 99°C]. Recommended oils meet MIL-L-7808 or MIL-L-23699 specifications.

MAIN OIL-PRESSURE SUPPLY SYSTEM

An aircraft-mounted oil tank supplies engine-lubricating oil. Oil enters the oil pump, mounted on the accessory drive gearbox, and discharges through internal passages to the oil filter. Filtered oil is directed into two main flow paths.

One oil flow passes through internal passages in the inlet housing to supply lubricating oil to the front section of the engine, including the reduction gearing, torquemeter, accessory drive gearing, and the no. 1 main bearing. The second oil-flow path through external lines to the rear section of the engine lubricates the nos. 2, 3, and 4 main bearings.

In the inlet-housing section, oil is directed through the accessory drive carrier flanges into an annular passage located in the rear support flange of the carrier. Oil from this passage is directed for forced-feed spray lubrication to the carrier gears and power-shaft support bearing. A transfer tube from the accessory drive carrier assembly lubricates the no. 1 main bearing and accessory drive pinion gear. Oil under constant pressure lubricates the support bearing.

A third transfer passage from the manifold is directed through an inlet-housing strut to the torquemeter pump. A pressure-regulating valve limits the rotary-pump output pressure by circulating the excess pressurized oil back to the inlet housing. The pressure oil from the torquemeter pump is directed back through an inlet-housing strut to the torquemeter valve.

An offset passage in the overspeed governor and tachometer drive assembly mounting flange supplies oil to the strainer and metering cartridge in the overspeed governor, which directs metered oil to the overspeed governor and tachometer drive gear train. An additional transfer passage from the output gearshaft rear-support bearing directs oil to the power takeoff mounting flange by means of internal passages in the inlet housing.

Oil flow to the rear section of the engine is supplied from an oil pressure port at the five o'clock position on the inlet housing through an external flexible oil line to an external, rigid manifold mounted on the forward face of the diffuser housing. Oil is directed from the right side of this manifold through a strainer mounted on the diffuser housing to the no. 2 main bearing. From the left side of the manifold, oil is directed through a flexible hose, strainer, and the oil-supply nozzle through the upper strut in the exhaust diffuser to lubricate the nos. 3 and 4 main bearings.

SCAVENGE-OIL SYSTEM

All internal scavenge oil from the inlet-housing section drains through hollow support struts to the bottom strut in the inlet housing, through a screen and transfer tube, and into the accessory drive gearbox. Scavenge oil from the output reduction carrier and gear assembly flows by gravity into the hollow inlet-housing struts. Scavenge oil from the no. 1 main bearing is pumped to the inlet-housing struts by a paddle pump mounted on

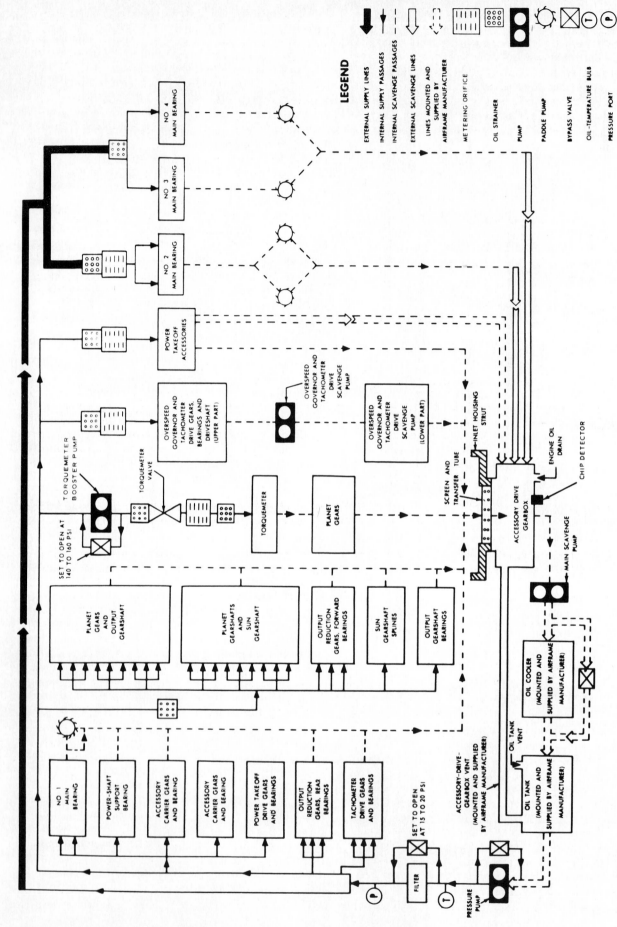

FIG. 22-5 Lubrication system schematic.

the rear of the bearing. Scavenge oil from the no. 2 main bearing, aided by two impellers mounted on each side of the bearing, flows through a scavenge-oil tube in the diffuser housing and is directed to the accessory drive gearbox by an external scavenge-oil line. Scavenge oil from nos. 3 and 4 bearings, aided by two impellers located in the bearing housing, returns to the accessory drive gearbox through an external scavenge-oil line and an oil tube that extends through the bottom of the exhaust diffuser. The scavenge oil flows from the accessory drive gearbox through the aircraft oil cooler and back to the oil storage tank.

TORQUEMETER

The torquemeter is a hydromechanical torque-measuring device located in the reduction gearing section of the inlet housing. The torquemeter uses lubricating oil, but is not part of the lubrication system. It consists of a stationary plate, a movable plate attached to the planet gear carrier, and 18 steel balls positioned in conical pockets located in both plates. Rotation of the planetary gears causes the carrier-mounted plate to rotate slightly. The torquemeter balls are displaced from their individual pockets, forcing the rear torquemeter plate to move rearward. The rearward motion of the plate unseats a spring-loaded poppet valve that allows high-pressure oil to enter the torquemeter cylinder chamber, equalizing the force exerted by the displaced carrier. Torquemeter oil pressure from the cylinder and gearbox air pressure are directed to the aircraft torque-pressure transmitter, which indicates differential torque oil pressure in pounds per square inch. The differential torque oil pressure is proportional to the torque delivered to the output gearshaft.

OIL PUMP

Some model engines use a two-element gear-type lube and scavenge-oil pump, driven by a single, splined drive shaft. One element supplies main lubricating oil pressure; the other element returns scavenge oil to the aircraft-mounted oil tank. Others use a power-driven rotary lube and scavenge pump (vane type). The vane-type pump can also be used as an alternate for the gear-type pump. A common splined shaft drives both elements. A pressure-relief valve in the oil pump is adjusted to deliver a minimum of 60 to 80 psi [413.7 to 551.6 kPa] oil pressure (measured at the oil filter discharge port) at sea level, normal rated power. This setting is rated for a maximum inlet oil temperature of 200°F [93°C] or 210°F [99°C]. At pressures below relief-valve setting, oil pressure is directly proportional to compressor rotor speed. Oil pressure also varies with altitude.

OIL FILTER

The wafer-disk-type oil filter is bolted to the accessory drive gearbox. The filter contains a bypass valve, set to open at a 15 to 20 psi [103.4 to 137.9 kPa] differential pressure, that allows the oil flow to bypass the filter elements and supply oil to the engine if the filter is clogged.

TORQUEMETER BOOSTER PUMP

The torquemeter booster pump, containing the pressure and the scavenge elements, is mounted on and driven by the overspeed governor and tachometer drive assembly. Each element is an individual pumping unit and draws oil from a separate source. The pressure element receives engine-lubricating oil at 60 to 80 psi [413.7 to 551.6 kPa] and delivers it, through a filter, to the torquemeter valve at a pressure of 140 to 160 psi [965.3 to 1103 kPa]; excess oil flows back to the inlet side of the pump. A relief valve in the overspeed governor and tachometer drive assembly sets the outlet pressure. The scavenge element receives oil from the overspeed governor and tachometer drive gear housing and delivers it to the oil return passages in the inlet-housing assembly.

CHIP DETECTOR

A chip detector is installed in the lower right side of the accessory drive gearbox. The chip detector will provide an indication of the presence of metal particles in the engine lubrication system when a continuity check is performed.

INTERNAL COOLING AND PRESSURIZATION (FIG. 22-6)

The internal cooling system provides cooling air to internal engine components and pressurizes the nos. 1, 2, and 3 main bearing and intershaft oil seals. Compressed bleed air from the centrifugal compressor section flows through internal passages to the front of the no. 2 bearing housing, then under the air deflector to the rear of the bearing housing to pressurize the rear seal on the no. 2 main bearing. Some of the air flows out through openings to cool the forward face of the first-stage turbine rotor. It then passes into the gas stream.

Some of the compressed air, bled from the centrifugal compressor, flows through holes in the compressor rotor rear shaft into the space between the compressor rotor shaft and the power shaft. The air bleeds rearward between the shafts, then up across the rear face of the first-stage turbine rotor. A part of the air passes through the second-stage turbine sealing ring, upward across the forward face of the second-stage turbine rotor, and into the exhaust stream.

The remainder of the compressed air (bled from the centrifugal compressor), passes through holes in the compressor rotor rear shaft into the space between the compressor rotor sleeve and the power shaft. Some of this air flows forward into the center of the seal behind

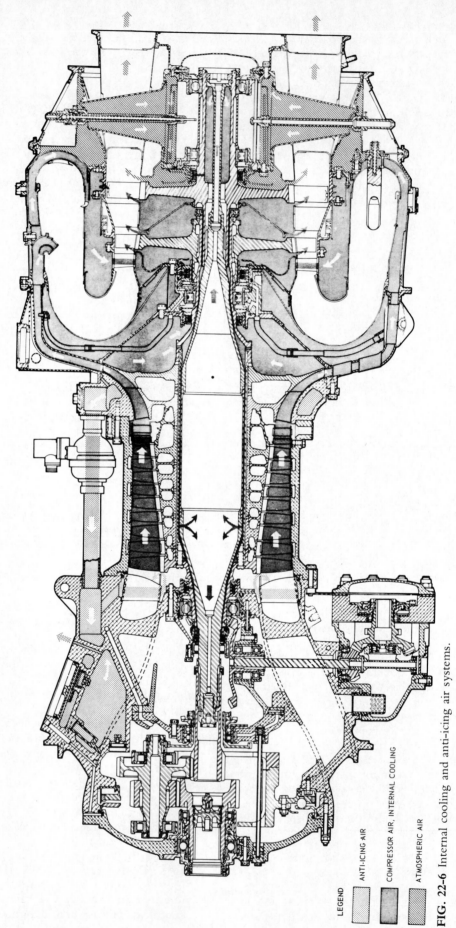

LEGEND

ANTI-ICING AIR

COMPRESSOR AIR, INTERNAL COOLING

ATMOSPHERIC AIR

FIG. 22-6 Internal cooling and anti-icing air systems.

the no. 1 bearing and to the intershaft seal located forward of the no. 1 bearing. The remainder of this air flows into the power shaft. This air flows rearward through a hole drilled in the power shaft throughbolt and then into the hollow interior of the second-stage turbine rear shaft. The air bleeds through holes in the second-stage turbine shaft to pressurize the seal in front of the no. 3 bearing.

The rear face of the second-stage turbine rotor and the housing for nos. 3 and 4 bearings are cooled by ambient air entering between the exhaust diffuser support cone and inner cone. The air moves forward through holes in the exhaust diffuser support cone, through exhaust diffuser struts, into the area around the nos. 3 and 4 bearing housing. The air moves forward around the nos. 3 and 4 bearing housing baffle, around the power-turbine cooling air deflector, past the second-stage turbine wheel rear face, and into the exhaust stream.

ANTI-ICING SYSTEM *(See FIG. 22-6)*

The anti-icing system supplies hot air under pressure to prevent icing of the inlet-housing areas when the engine is operating under icing conditions. Pressurized hot air from the air diffuser flows through holes in the aft face of the diffuser vanes and collects in the air diffuser internal bleed air manifold, where it passes to an external manifold located at the one o'clock position on the diffuser housing. A connecting manifold, consisting of an external elbow and tubing, is attached to the external bleed air manifold and to an adapter located on top of the impeller housing. The connecting manifold passes air through the impeller housing to the hot-air solenoid valve.

The hot-air solenoid valve is mounted on top of the compressor and impeller housing assembly. The solenoid-operated valve controls the flow of anti-icing hot air from the diffuser housing to the inlet housing to prevent the formation of ice. During engine operation, the hot-air solenoid valve is normally energized in the CLOSED position by manually actuating a switch in the cockpit. In the event of electrical power failure, the fail-safe, spring-loaded valve returns to the OPEN position to provide continuous anti-icing air.

After leaving the hot-air solenoid valve, anti-icing air flows forward through a regulator tube into a hollow annulus (port) on top of the inlet housing. This hot air is then circulated through five of the six hollow inlet-housing support struts to prevent ice formation in the inlet housing area. Hot air also flows into an annulus in the rear of the inlet housing where it passes through the hollow inlet guide vanes to prevent icing. After passing through the inlet guide vanes, the air flows into the compressor area. In the event of electrical power failure, anti-icing becomes continuous. Hot scavenge oil, draining through the lower strut into the accessory drive gearbox, prevents ice formation in the bottom of the inlet-housing area. The anti-icing system is designed to accommodate air at static pressure and to reduce the possibility of the entrapment of solid or liquid particles.

FUEL SYSTEM

The fuel system consists of the

1. Fuel control
2. Starting-fuel solenoid valve
3. Main- and starting-fuel manifolds and hoses
4. Bypass fuel filter (on some models)
5. Starting-fuel nozzles
6. Fuel vaporizers
7. Pressure-operated drain valve

STARTING-FUEL SYSTEM *(FIG. 22-7)*

During engine start, the starting-fuel system delivers starting fuel to the combustion chamber. Energizing the primer fuel switch opens the starting-fuel solenoid valve, allowing scheduled fuel from the fuel regulator to flow through the starting-fuel manifold, two starting-fuel nozzles, and into the combustion chamber where it is ignited by two igniter plugs. At ground idle speed, the ignition system is deenergized, causing the solenoid valve to close and stop the flow of starting fuel. The starting-fuel nozzles are self-purging and automatically remove excess fuel.

MAIN-FUEL SYSTEM *(See FIG. 22-7)*

The main-fuel system delivers metered fuel from the fuel regulator to the main-fuel manifold where it is discharged through 11 fuel vaporizers into the combustion chamber. Main fuel is ignited by the burning starting fuel.

FUEL-CONTROL ASSEMBLY

The fuel-control assembly consists of a fuel regulator, which incorporates an emergency (manual) control system, and an overspeed governor. The fuel regulator is a hydromechanical device containing a dual-element fuel pump, compressor rotor speed governor, an acceleration and deceleration control, an airbleed signal mechanism, and a fuel shutoff valve. Functionally, the fuel regulator is divided into a flow-control section and a computer section. The flow-control section consists of the components that meter engine fuel flow. The computer section comprises elements that schedule the positioning of the metering valve of the flow-control section as a function of the input signals to the regulator. The overspeed governor provides the fuel control with an N_2 speed signal. Fuel flow is then metered to maintain the desired N_2 speed plus or minus 50 rpm. The overspeed governor also serves as an override mechanism to reduce fuel flow in case of a power-turbine overspeed.

STARTING-FUEL SOLENOID VALVE

The starting-fuel solenoid valve is mounted on a bracket that is secured to the compressor housing at the

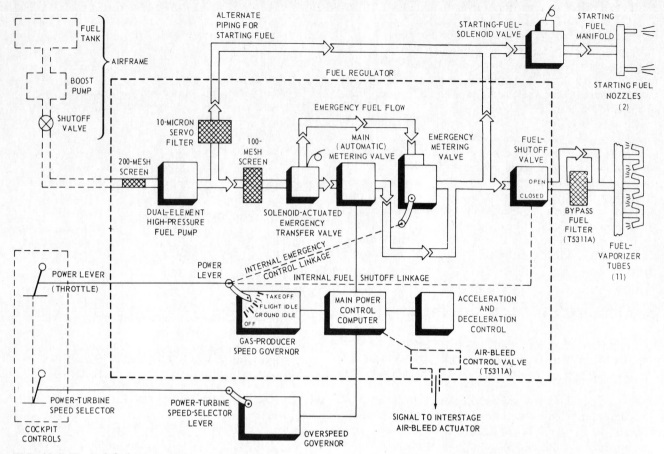

FIG. 22-7 Typical fuel system schematic.

10 o'clock position. When energized, the valve allows starting fuel from the fuel control to flow to the starting-fuel nozzles. Under normal conditions, the starting-fuel solenoid valve is energized until ground idle speed is reached.

STARTING- AND MAIN-FUEL MANIFOLDS

The starting- and main-fuel manifolds are bracketed together and mounted at the rear of the combustion-chamber housing. The starting-fuel manifold receives fuel from the fuel regulator, through the starting-fuel solenoid valve, and delivers it to the starting-fuel nozzles. The main-fuel manifold receives fuel from the fuel regulator, through the main-fuel hose and bypass filter (some models only), and delivers it to the 11 fuel vaporizers.

BYPASS FUEL FILTER
(some models only)

A bypass fuel filter, located on the lower left side of the combustion-chamber housing, filters fuel from the

fuel regulator by means of a corrosion-resistant, 200-mesh, stainless-steel element. As fuel passes through the filter element, contaminants are deposited on the inner wall of the filter element. In the event that clogging of the element occurs, fuel will bypass around the element into a hollow annulus within the filter housing and supply the main fuel manifold.

STARTING-FUEL NOZZLES

Two starting-fuel nozzles, located at the four and eight o'clock positions in the rear of the combustion-chamber housing, deliver atomized fuel to the combustion chamber during starting. A ball check valve permits air from the combustion chamber to purge the nozzle when starting fuel stops flowing.

FUEL VAPORIZERS

Eleven equally spaced fuel vaporizers deliver main fuel to the combustion chamber. The fuel vaporizers receive fuel from the main-fuel manifold, combine it with compressed air, and deliver vaporized fuel and air to the combustion chamber.

COMBUSTION-CHAMBER DRAIN VALVE

The combustion-chamber drain valve is located at the six o'clock position on the combustion-chamber housing. The drain valve is spring-loaded open to allow drainage of residual fluids after engine shutdown. Internal pressure during engine operation keeps the valve closed.

FUEL-CONTROL SYSTEM
(See FIG. 22-7)

The fuel-control system consists of a primary control for the gas-producer section and an overspeed governor for the power-turbine section. An integral dual fuel pump and an emergency (manual) control system are incorporated in the primary control unit. The fuel control incorporates acceleration and deceleration controls and a droop-type governor for steady-state speed control. The main metering valve of the fuel regulator is the controlling unit by which the main fuel flow is metered to the engine. Its position is determined by the action of the gas-producer speed governor, the power-turbine overspeed governor, or the acceleration-deceleration control, depending upon engine requirements. In regulating the main metering valve, the governor or control that demands the least fuel flow overrides all others, except the deceleration control, to ensure a minimum fuel-flow rate.

The functions of the gas-producer speed control are to govern ground and flight idle operations, to limit the maximum power of the engine, and to maintain steady-state conditions through all power regimes. The gas-producer speed governor is driven, through gears, at a speed proportional to the gas-producer rotor speed. It regulates gas-producer rotor speed to the value selected by the power lever. Acceleration fuel-flow limits are scheduled over the entire operating range by scheduling maximum fuel flow as a function of gas-producer rotor speed and compressor inlet pressure and temperature. The absolute maximum fuel flow for acceleration or steady-state operation is determined by the maximum fuel-flow stop setting. The deceleration fuel-flow limits are scheduled as a function of gas-producer rotor speed and compressor inlet pressure. The absolute minimum fuel flow for deceleration or steady-state operation is determined by the minimum fuel-flow stop setting.

The power-turbine rotor is protected against overspeed operation by the power-turbine overspeed control (overspeed governor). The power-turbine governor is driven, through gears, at a speed proportional to power-turbine rotor speed. Limits for the power turbine governor are set by adjustable stops.

FUEL FLOW

Fuel enters the dual fuel pump after passing through the inlet screen. It is then pumped through the check valves and the outlet screen to the transfer valve. With the transfer valve in the normal position for automatic operation, fuel flows to the main metering valve at a pressure controlled by the main pressure-regulating valve. The position of the main metering valve and hence the flow of fuel is automatically controlled by the computer section of the fuel control. The metered fuel flows through the open shutoff valve and the fuel discharge port to the engine main-fuel manifold and the fuel vaporizer tubes in the combustion chamber. When the transfer valve is in the emergency position, fuel flows through and is metered by the emergency (manual) metering valve. Fuel pressure is controlled by the emergency pressure regulating valve, and fuel is delivered through the open shutoff valve to the fuel discharge port and to the engine. The area of the valve opening and the resulting flow of fuel are determined by the position of the power lever controlled from the cockpit.

OPERATION OF THE EMERGENCY (MANUAL) FUEL SYSTEM
(See FIG. 22-7)

If the automatic fuel-control system fails, a changeover to the emergency (manual) fuel system should be made in accordance with the airframe instructions. When the emergency system is in operation, the main metering valve is bypassed and fuel is metered to the engine by the manual-system metering valve, which is positioned from the pilot's compartment by the power lever. Acceleration and deceleration control is not provided in the emergency system; therefore the power lever should not be moved rapidly when the emergency fuel system is in operation. Engine overspeed or possible flame-out could result. The emergency fuel system does not affect the operation of the starting-fuel system if engine restart is required.

FUEL-CONTROL POWER LEVER

The power lever on the fuel control modulates the engine from OFF to TAKEOFF power. The total travel of the lever is 100°. There is a 3° dwell at GROUND IDLE position and a 4° dwell at FLIGHT IDLE position.

OFF:	0°
GROUND IDLE:	23° to 26°
FLIGHT IDLE:	38° to 42°
NORMAL:	83°
TAKEOFF:	100°

FUEL

These engines are designed to operate on Grade JP-4 or JP-5 fuel, military specification MIL-J-5624, or fuel types A or B, commercial specification ASTM D1655.

ELECTRICAL SYSTEM AND MAIN WIRING HARNESS (FIG. 22-8)

The main wiring harness contains connections for the ignition unit, hot-air solenoid valve, starting-fuel solenoid valve, inlet-oil temperature bulb, fuel-control transfer solenoid valve, and power-turbine and gas-producer tachometer-generators. Quick-disconnect plugs are incorporated on the harness.

IGNITION SYSTEM (See FIG. 22-8)

The high-energy, medium-voltage ignition system consists of an ignition unit, an ignition lead and coil assembly, and igniter plugs. The system requires 14-V dc minimum input at 3.0 A.

The ignition unit is attached to a bracket located at 10 o'clock on the impeller-housing rear flange. The ignition unit converts low voltage through a vibrator transformer to a high voltage that passes through the ignition lead and coil assembly. The high voltage that is produced ionizes a gap in each igniter plug to produce a spark.

The ignition lead and coil assembly transmits high voltage from the ignition unit to the igniter plugs in the combustion chamber. The spark splitter coil, located below the ignition unit, distributes electric current equally to each igniter plug.

Two igniter plugs are installed in receptacles in the aft end of the combustion chamber at the four and eight o'clock positions. The igniter plugs produce high-voltage sparks to ignite the fuel-air mixture in the combustion chamber.

INLET-OIL TEMPERATURE-SENSING BULB (See FIG. 22-8)

An inlet-oil temperature-sensing bulb is installed in the main oil pump. This bulb is connected through the wiring harness to a cockpit indicator.

EXHAUST THERMOCOUPLE HARNESS (See FIG. 22-8)

An exhaust thermocouple harness, consisting of an electrical connector, shielded manifold, and three chromel-alumel thermocouples, is provided with the engine. The thermocouples, inserted through the exhaust diffuser into the path of exhaust gas at the 2, 4, and 10 o'clock positions, transmit exhaust-gas temperatures to a cockpit indicator.

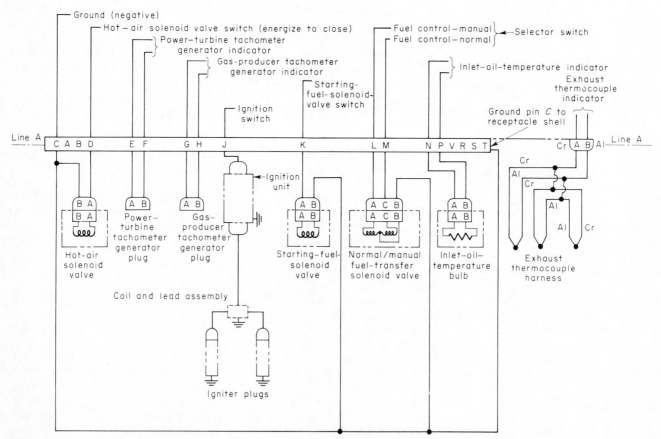

FIG. 22-8 Electrical system schematic. Note: typical airframe wiring diagram is shown above line *A*. The engine wiring diagram is shown below line *A*.

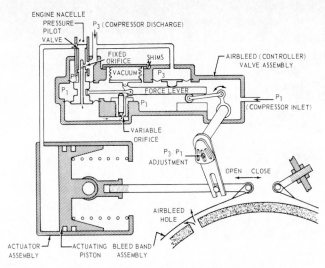

FIG. 22-9 Schematic of one form of airbleed actuator.

INTERSTAGE AIRBLEED SYSTEM (FIG. 22-9)

An interstage airbleed system is provided to facilitate acceleration of the compressor rotor assembly. Principal components of the system are an airbleed actuator, an airbleed valve (on some engines), and a bleed band assembly. The actuator controls operation of the compressor bleed air by tightening or loosening the bleed bands that encircle a ring of bleed air holes in the compressor housings at the exit guide vane location. The airbleed system on one model of this engine incorporates a transient control feature. The bleed bands open during all engine accelerations and open at speeds below 70 percent N_1 speed at standard day sea level conditions as directed by the sensors in the fuel control. On some

engines, the bleed bands are closed at speeds above 75 to 80 percent N_1.

INTERSTAGE AIRBLEED SYSTEM COMPONENTS *(See FIG. 22-9)*

The airbleed actuator is mounted on the right side of the compressor-housing assembly. Air pressure for operation of the airbleed actuator is obtained from a bleed port on the right side of the air-diffuser housing.

The airbleed valve (some models) is mounted on the airbleed actuator and senses the ratio of compressor discharge pressure to the compressor inlet air pressure. When the ratio reaches a preset point, the airbleed valve provides compressor discharge air pressure for operation of the airbleed actuator.

The bleed band assembly consists of two band halves bolted together. It is positioned around the rear portion of the axial-compressor housings and secured by clips bolted to the compressor housings. The looped ends of the bleed band assembly are attached to the airbleed actuator.

REVIEW AND STUDY QUESTIONS

1. What airplanes use this engine?
2. List the engine's major specifications.
3. Give a brief description of the engine and its operation.
4. Very briefly describe the construction features of the following parts: inlet-housing assembly, accessories section, compressor rotor and housing, diffuser, combustor, and turbine assembly.
5. How many main bearings are used?
6. How is the chance for compressor stall reduced in this engine?
7. Briefly describe the following systems: oil, fuel, ignition, torquemeter, and anti-icing.

CHAPTER TWENTY-THREE
DETROIT DIESEL ALLISON 501-D13

The Allison Division of the General Motors Corporation has produced the 501-D13 (Fig. 23-1) in various configurations for several years. It is used in the Lockheed Electra and Convair Conversion, while a more powerful version, the T56 is installed in the Lockheed C130 and P-3B.

SPECIFICATIONS

Number of compressor stages:	14
Number of turbine stages:	4
Number of combustors:	6
Maximum power at sea level:	3750 ESHP
Specifc fuel consumption at maximum power:	0.53 lb/ESHP/h [241 g/ESHP/h]
Compressor ratio at maximum rpm:	9.2:1
Maximum diameter:	41 in [104 cm]
Maximum length:	146 in [371 cm]
Maximum dry weight:	1645 lb [747 kg]

The Allison 501-D13 engine is rated at 3750 eshp at standard day conditions with an rpm of 13,820 (100 percent), and a turbine inlet temperature of 1780°F [971°C]. The engine consists of a power section, a reduction gear, and a torquemeter assembly. The power section and reduction gear are connected and aligned by the torquemeter assembly, and added rigidity is provided by two tie struts. The propeller is mounted on a single-rotation number 60A propeller shaft.

INTRODUCTION

The power section is composed of a compressor assembly, accessory drive housing assembly, combustion assembly, and turbine assembly. The compressor assembly and the accessory drive housing assembly are referred to as the *cold section* of the engine, while the combustion assembly and turbine assembly are called the *hot section*. The power section includes the oil, electrical, airbleed, anti-icing, fuel, and control systems.

The compressor assembly is an axial-flow type, incorporating a 14-stage compressor rotor and vane assembly encased in a 4-piece compressor casing (quadrant). An air inlet housing, secured to the forward end of the compressor casing, receives air from the aircraft duct and directs this air to the compressor rotor. Provisions are made for the anti-icing of the air inlet-housing struts and the inlet anti-icing vanes which guide air into the rotor. A compressor diffuser, secured to the rear end of the compressor casing, guides the air from the rotor into the combustion assembly.

The accessory drive housing assembly gear train receives its drive from the compressor extension shaft. Through a gear train, the power section rpm is reduced to that required to drive certain accessories. The speed-sensitive valve, speed-sensitive control, and oil pump are mounted on the front cover of the accessory drive housing assembly. The fuel control and fuel pump are mounted on the rear side of the accessory drive housing.

The combustion assembly, attached to the compressor diffuser, incorporates six cylindrical-shaped combustion liners positioned between inner and outer combustion casings. The combustion liners mix the fuel and air, support combustion, and guide the exhausting gases into the turbine assembly.

The turbine assembly, attached to the inner and outer combustion casings, consists of a four-stage turbine rotor and vane assembly encased in the turbine inlet casing, turbine vane casing, and turbine rear-bearing support. The rotor absorbs the necessary energy from the expanding exhaust gases to drive the compressor rotor, accessories, reduction gear assembly, and propeller. The turbine inlet casing supports the turbine front bearing, thermocouples, and houses the first-stage turbine vane assemblies. The turbine vane casing houses the second-, third-, and fourth-stage vane assemblies. The turbine rear bearing is supported and retained by the rear-bearing support which guides the exhausting gases into the aircraft tailpipe.

The torquemeter assembly, attached between the air inlet housing and the reduction-gear assembly, consists of a housing and shaft assembly. The housing serves as the structural support which aligns the power section with the reduction-gear assembly. The two tie struts provide the necessary rigidity to maintain this alignment. The shaft assembly transmits torque from the power section to the reduction-gear assembly. A pickup assembly, attached in the forward end of the housing, detects the torque transmitted through the shaft assembly. The pickup-assembly signals are directed to a flight-deck torquemeter indicator which registers the torque in shaft horsepower delivered into the reduction-gear assembly.

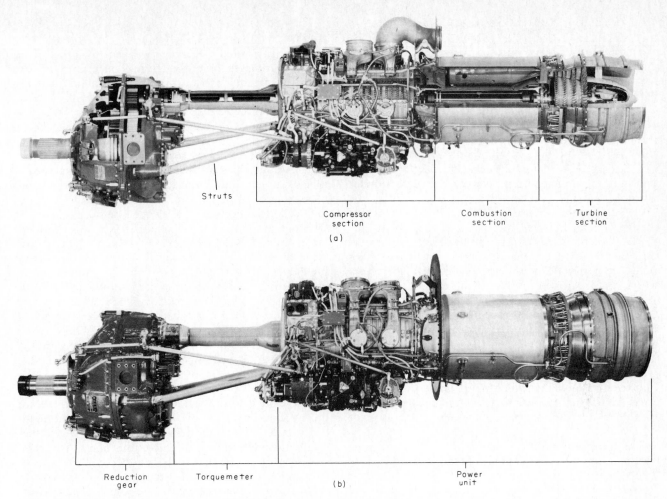

FIG. 23-1 (a) Sectioned and **(b)** external view of the Detroit Diesel Allison 501–D13 (T56) turboprop engine.

The reduction gear incorporates a single propeller drive shaft, a negative torque signal system, a thrust-sensitive signal system (auto feather), a propeller brake, a two-stage reduction-gear train, an accessories-drive gear train, and an independent dry sump oil system.

The overall reduction-gear ratio of the two stages of reduction is 13.54:1, and thus the propeller rotates at an efficient rpm. Aircraft accessories are mounted on the rear side of the reduction-gear assembly, as well as an engine-furnished reduction-gear oil pump and filter assembly.

DIRECTIONAL REFERENCES AND DEFINITIONS

1. *Front*—The propeller end.
2. *Rear*—The exhaust end.
3. *Right and left*—Determined by standing at the rear of the engine and facing forward.
4. *Bottom*—Determined by the power-section accessories drive housing, which is located at the forward end of the power section.
5. *Top*—Determined by the breather located at the forward end of the power section.

6. *Rotation*—The direction of rotation is determined when standing at the rear of the engine, and facing forward. The power-section rotor section turns in a counterclockwise direction, and the propeller rotation is clockwise.
7. *Accessories rotation*—Determined by facing the mounting pad of each accessory.
8. *Combustion-liner numbering*—The combustion liners are numbered from 1 through 6 in a clockwise direction when viewing the engine from the rear. The no. 1 liner is located at the top vertical center line.
9. *Compressor- and turbine-stages numbering*—Numbered beginning from the forward end of the power section and progressively moving rearward. The compressor has 1 through 14 stages, and the turbine 1 through 4 stages.
10. *Main-rotor-bearings numbering*—Numbered beginning at the forward end of the power section, moving rearward, with no. 1 being at the forward end of the compressor rotor, no. 2 at the rear of the compressor rotor, no. 3 at the forward side of the turbine rotor, and no. 4 at the rear of the turbine rotor.
11. *Igniter-plug location*—There are two igniter plugs used, and these are located in combustion liners nos. 2 and 5.

COMPRESSOR ASSEMBLY

This assembly consists of a compressor air inlet-housing assembly, compressor housing assembly, compressor rotor assembly, diffuser, and diffuser scavenge-oil pump assembly.

The compressor air inlet housing (Fig. 23-2) is a magnesium-alloy casting, and is designed to direct and distribute air into the compressor rotor. It also provides the mounting location for the front-compressor bearing, the engine breather, the accessories drive housing assembly, the anti-icing air valves, the torquemeter housing, and the inlet anti-icing vane assembly. The inlet anti-icing vane assembly is mounted on the aft side of the air inlet housing, and is utilized to impart the proper direction and velocity to the airflow as it enters into the first stage of the compressor rotor. These vanes may "ice up" under ideal icing conditions. Therefore, provisions are made to direct heat to each of the vanes. Air, which has been heated due to compression, may be extracted, if so desired, from the outlet of the compressor (diffuser) and directed through two tubes to the anti-icing valves mounted on the compressor air inlet housing. The inlet anti-icing vanes are hollow, and mate with inner and outer annuli, into which the hot air is directed. From the annuli, the air flows through the vanes, and exits into the first stage of the compressor through slots which are provided in the trailing edge of the vanes.

The compressor rotor is an axial-flow type and consists of 14 stages. It is supported at the forward end by a roller bearing, and at the rear by a ball bearing (Fig. 23-3).

The entire compressor housing assembly (Fig. 23-4) is fabricated from steel and consists of four quarters permanently bolted together in halves. The split lines of the housing are located 45° from a vertical center line. The compressor vane assemblies are installed in channels in the compressor housing, and are securely located and held in position by bolts. The inner ring of the vane assemblies supports the interstage airseals which form a labyrinth seal, thus preventing air from one stage bleeding back to the previous one. Between each of the vane assemblies, the housing is coated with a special type of sprayed aluminum to provide a minimum compressor rotor blade-tip clearance, thus increasing compressor efficiency. The outlet vane assembly consists of an inner and an outer ring supporting two complete circles of vanes. These are used to straighten airflow prior to entering the combustion section. The compressor housing is ported around the circumference. Four bleed air valves are mounted on the outside of the compressor case at the fifth stage, and four at the tenth stage. Those valves at the fifth stage are manifolded together, as are those at the tenth stage. These valves are used to unload the compressor during the start and acceleration, or when operating at low-speed taxi idle. (Refer to Chap. 5.)

The compressor diffuser, of welded-steel construction, is bolted to the flange at the aft end of the compressor housing assembly. It is the midstructural member of the engine (Fig. 23-5). One of three engine-to-aircraft mountings is located at this point. Six airfoil struts form passages which conduct compressed air from the outlet of the 14th stage of the compressor to the forward end of the combustion liners. These struts

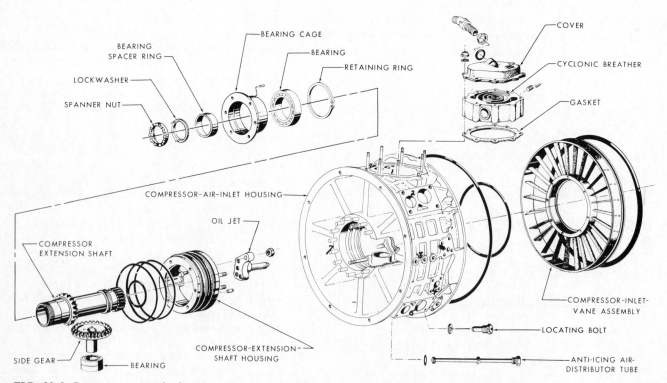

FIG. 23-2 Compressor air inlet housing.

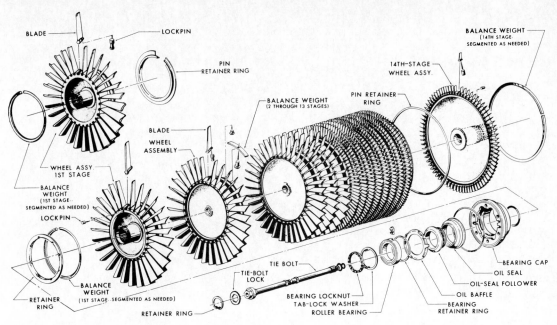

FIG. 23-3 Compressor rotor.

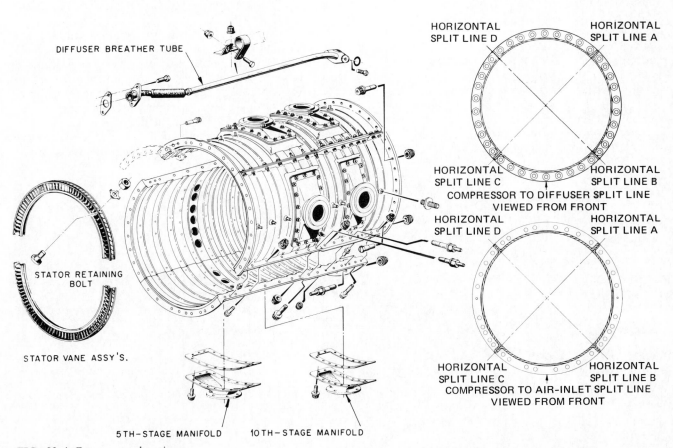

FIG. 23-4 Compressor housing.

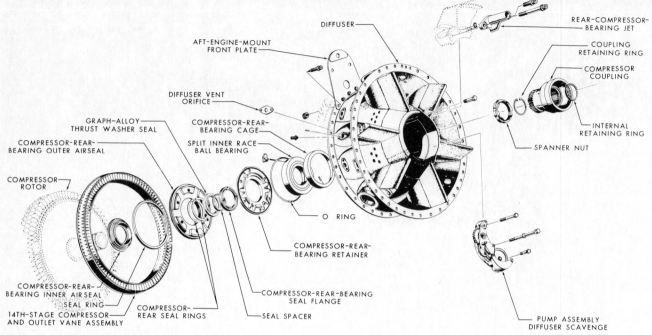

FIG. 23-5 Compressor diffuser.

also support the inner cone which provides the mounting for the rear-compressor bearing (ball), the seals, the rear-compressor bearing oil nozzle, the diffuser scavenge-oil pump, and the forward end of the inner combustion chamber. Air is extracted from ports on the diffuser for anti-icing and operation of the 5th- and 10th-stage bleed air valves and the 14th-stage bleed air valve. During the starting cycle, air is bled from the diffuser through the 14th-stage bleed air valve to promote better starts. Bleed air is also extracted from this point by the airframe manufacturer for aircraft anti-icing, and for cross-feeding from one engine to another for engine starter operation. The six fuel nozzles are mounted on and extend into the diffuser, and a fire shield is provided at the rear splitline.

COMBUSTION ASSEMBLY (FIG. 23-6)

This assembly consists of outer and inner combustion chambers that form an annular chamber in which six combustion liners are located. Fuel is sprayed continuously during operation into the forward end of each combustion liner. During the starting cycle, two igniter plugs, located in combustion liners no. 2 and 5, ignite the fuel-air mixture. All six liners are interconnected near their forward ends by crossover tubes. Thus, during the starting cycle after ignition takes place in nos. 2 and 5 combustion liners, the flame propagates to the remaining liners. Liners, which do not utilize an igniter plug, have at the same location a liner-support assembly

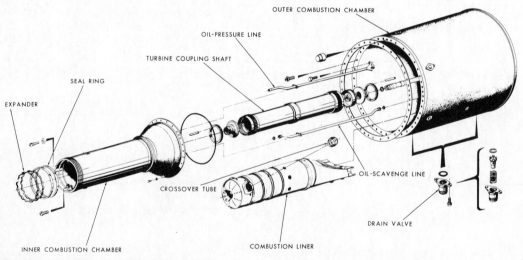

FIG. 23-6 Combustion section.

that positions the combustion liner and retains it axially. The outer combustion chamber provides the supporting structure between the diffuser and the turbine section. Mounted on the bottom of the outer combustion chamber are two combustion-chamber drain valves, which drain fuel after a false start or at engine shutdown.

TURBINE UNIT ASSEMBLY
(FIG. 23-7)

This assembly includes six major items, namely: turbine inlet casing, turbine rear-bearing support, turbine rotor, turbine scavenge-oil pumps, turbine vane casing, and turbine vane assemblies.

The turbine inlet casing is attached at its forward end to the outer and inner combustion chambers. It is designed to locate and house the forward turbine bearing (roller), the seal assembly, front-turbine-bearing oil jet, and the turbine front scavenge-oil pump. The casing is divided into six equal passages by six airfoil struts. Each of these passages provides the means of locating and supporting the aft end of a combustion liner. Located around the outer casing are 18 holes, with one thermocouple assembly positioned in each. Thus, three thermocouple assemblies are available at the outlet of each combustion liner. These 18 thermocouple assemblies are dual, and thus two complete and individual circuits are available. One is used to provide a temperature indication (referred to as turbine inlet temperature) to the flight deck, the other is used to provide a signal to the electronic fuel-trimming system (part of the fuel system). The circuits measure the average temperature of all 18 thermocouples, and thus a very accurate indication of the gas temperature entering the turbine section is at all times available. This is important, as the power being produced under any given set of conditions is dependent upon turbine inlet temperature.

The turbine rotor assembly (Fig. 23-8) consists of four turbine wheels which are splined on a turbine shaft. The entire assembly is supported by a roller bearing at the forward end and a roller bearing at the aft end. A turbine coupling shaft assembly connects the turbine rotor to the compressor rotor, and thus power, extracted by the four stages of the turbine, is transmitted to the compressor rotor, driven accessories, reduction-gear assembly, and propeller. All four stages of blades are attached to the wheel rims in broached serrations of five-toothed, "fir tree" design. The first-stage turbine wheel has the smallest blade area, with each succeeding stage becoming larger.

The turbine vane casing encases the turbine rotor assembly, and retains the four stages of turbine vane assemblies. It is the structural member for supporting the turbine rear bearing support. The vanes are airfoil design, and serve two basic functions. They increase the gas velocity prior to each turbine wheel stage, and also direct the flow of gases so that they will impinge upon the turbine blades at the most efficient angle.

The turbine rear bearing support (Fig. 23-9) attaches to the aft end of the turbine rear vane casing. It houses the turbine rear bearing (roller), the turbine rear scavenge pump and support, and the inner exhaust cone and insulation. It also forms the exhaust (jet) nozzle for the engine.

ACCESSORIES-DRIVE HOUSING ASSEMBLY *(FIG. 23-10)*

This is a magnesium-alloy casting mounted on the bottom of the compressor air inlet housing. It includes the necessary gear trains for driving all power-section-driven accessories at their proper rpm in relation to engine rpm. Power for driving the gear trains is taken from the compressor extension shaft by a vertical shaft

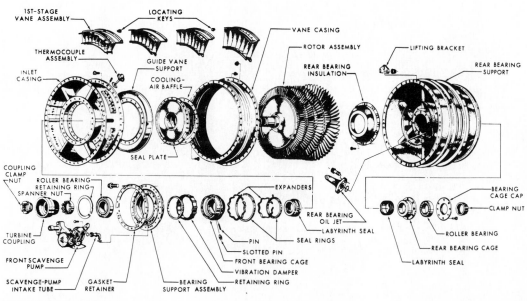

FIG. 23-7 Turbine unit.

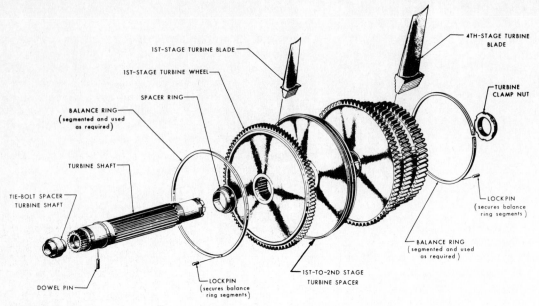

FIG. 23-8 Turbine rotor.

gear. The following accessories are driven from this housing:

1. Speed-sensitive control
2. Speed-sensitive valve
3. Oil pump
4. Fuel control
5. Fuel pump

A number of nondriven accessories and components are furnished with the engine. These may be broadly classified into fuel system, airbleed, ignition, oil, and torquemeter systems (Fig. 23-11).

Fuel System

1. High-pressure fuel filter
2. Low-pressure fuel filter
3. Primer valve
4. Fuel manifold pressure switch
5. Coordinator
6. Relay box (aircraft-mounted)
7. Temperature datum control (aircraft-mounted)
8. Temperature datum valve
9. Fuel nozzles (6)
10. Drip valve
11. Drain valves (2)

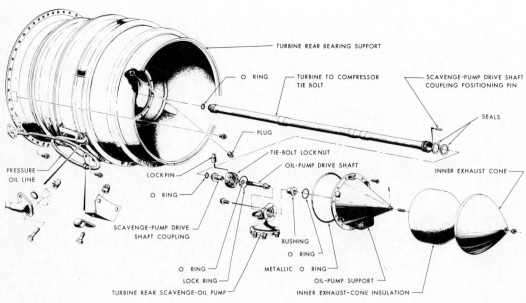

FIG. 23-9 Turbine rear bearing support.

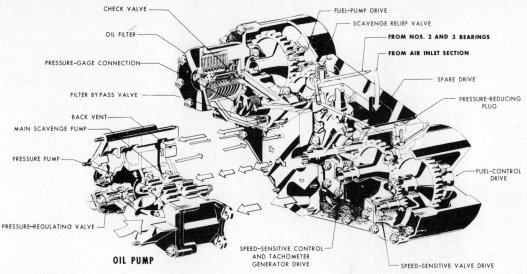

FIG. 23-10 Accessories case and oil pump.

Airbleed System

1. Anti-icing solenoid valve
2. Anti-icing air valves (2)
3. Fifth- and tenth-stage compressor bleed air valves (8)
4. Fourteenth-stage starting bleed air valve
5. Fourteenth-stage bleed air control valve

Ignition System

1. Ignition exciter
2. Ignition relay
3. Igniter plugs (two)

Oil System

1. Oil filter (power section)
2. Oil filter (reduction gear)

Torquemeter System

Torquemeter indicator

REDUCTION-GEAR ASSEMBLY (FIG. 23-12)

The prime function of the reduction-gear assembly is that of providing the means of reducing power-section rpm to the range of efficient propeller rpm. It also provides pads on the rear case for mounting and driving the following aircraft-furnished accessories:

1. Starter
2. Cabin supercharger (engine nos. 2 and 3)
3. Alternator (115 V, 400 Hz)

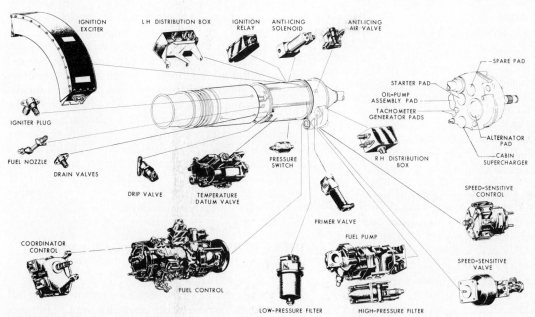

FIG. 23-11 Accessories location.

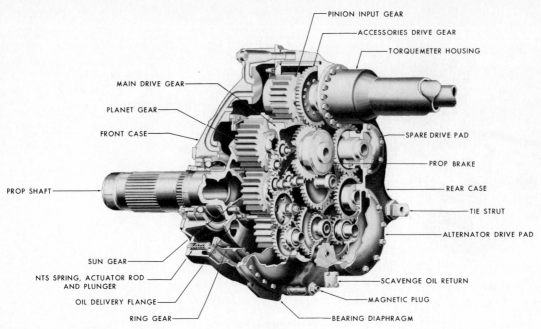

FIG. 23-12 Reduction gearbox.

4. Tachometer-generator
5. Propeller alternator (engine-propeller rpm signal)
6. Hydraulic pump or dc generator (if required)

NOTE: In addition to the aforementioned accessories, there is an engine-furnished oil pump, which is mounted on the rear case.

The reduction gear has an independent lubrication system, which includes a pressure pump and two scavenge pumps. Oil supply is furnished from an aircraft-furnished tank which also supplies the power section.

The reduction gear assembly is remotely located from the power section and is attached by a torquemeter housing and two tie struts. The remote location offers a number of decided advantages, which include the following:

1. Better air inlet ducting, which increases engine efficiency and performance
2. The opportunity of readily mounting the gearbox offset up or down for high- or low-wing aircraft
3. The advantage of additional space for mounting driven accessories without affecting frontal area
4. Containing the engine in the minimum frontal area
5. The ability to utilize an electronic torquemeter

The reduction-gear housing is a magnesium-alloy casting. It has an overall reduction-gear ratio of 13.54:1, and this is accomplished through a two-stage stepdown. The primary stepdown is accomplished by a spur gear train having a ratio of 3.125:1, and the secondary stepdown is by a planetary gear train with a ratio of 4.333:1. The propeller shaft, size SAE 60A, rotates in a clockwise direction when viewed from the rear. In addition to the reduction gears and accessories drives, the reduction-gear assembly includes the following major units:

1. *Propeller brake*—Used to stop windmilling of feathered propeller, and to reduce time for propeller to come to rest after ground shutdown.
2. *Negative torque signal (NTS) system*—Designed to prevent excessive propeller drag.
3. *Thrust-sensitive signal (TSS)*—A device which will provide automatic feathering when armed during takeoff.
4. *Safety coupling*—A safety device backing up the NTS system.

PROPELLER BRAKE *(FIG. 23-13)*

The propeller brake is designed to prevent the propeller from windmilling when it is feathered in flight, and to decrease the time for the propeller to come to a complete stop after ground shutdown. It is a friction-type brake, consisting of a stationary inner member and a rotating outer member which, when locked, acts upon the primary-stage reduction gearing. During normal engine operation, reduction-gear oil pressure holds the brake in the released position. This is accomplished by oil pressure (hydraulic force) which holds the outer member away from the inner member. When the propeller is feathered or at engine shutdown, as reduction-gear oil pressure drops off, the effective hydraulic force decreases and a spring force moves the outer member into contact with the inner member.

NEGATIVE TORQUE SIGNAL SYSTEM (NTS) *(FIG. 23-14)*

The NTS system is designed to prevent the aircraft from encountering excessive propeller drag. This system is part of the reduction gear, and is completely mechanical in design and automatic in operation. A

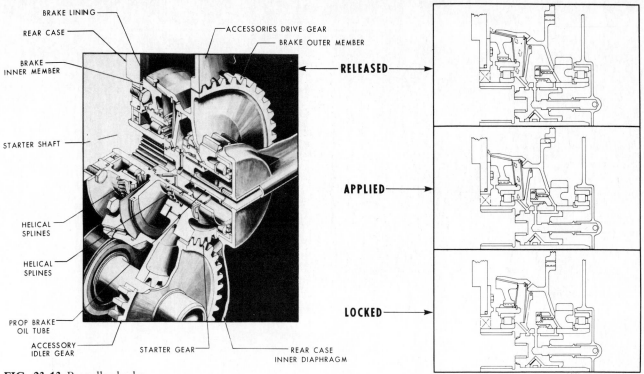

FIG. 23-13 Propeller brake.

negative torque value in the range of 250 to 370 hp [186 to 276 kW], transmitted from propeller into the reduction gear, causes the planetary ring gear to move forward, overcoming a calibrated spring force. As the ring gear moves forward, it actuates two rods which move forward through openings in the reduction-gear front case. Only one rod is used to actuate the propeller NTS linkage. When actuated, the propeller increases blade angle (toward feather) until the abnormal propeller drag and resultant excessive negative torque are relieved. The propeller will never go to the feather position, when actuated by the NTS system, but will modulate through

a small blade-angle range such that it will not absorb more than approximately 250 to 370 hp [186 to 276 kW]. As the negative torque is relieved, the propeller returns to normal governing.

THRUST-SENSITIVE SIGNAL (TSS) (FIG. 23-15)

The TSS provides for initiating automatic feathering at takeoff. The system must be armed prior to takeoff if it is to function, and a blocking relay is provided to

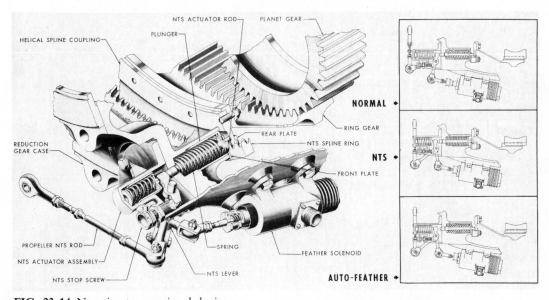

FIG. 23-14 Negative torque signal device.

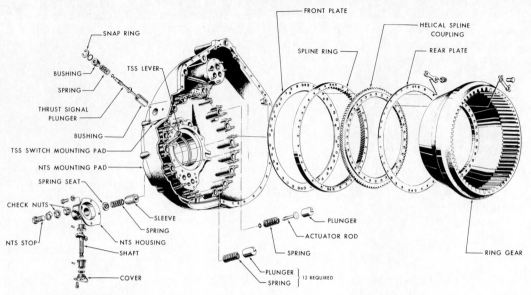

FIG. 23-15 Negative torque and thrust-sensitive assembly.

prevent auto feathering of more than one propeller. The system is armed by the *auto feather arming switch* and a throttle-actuated switch. The setting of the throttle switch is such that, if operation is normal, the propeller will be developing considerably in excess of positive 500 lb of thrust [2224 N]. This prevents auto feather except when a power failure occurs.

The system is designed to operate (if armed) when the propeller is delivering less than 500 lb of positive thrust. The propeller shaft tends to move in a forward axial direction as the propeller produces thrust. Axial travel is limited by mechanical stops. Forward movement of the shaft compresses two springs. As power decreases to 500 lb of thrust, the springs' force moves the shaft axially in a rearward direction. This movement

is multiplied through mechanical linkage, and transmitted mechanically to a pad on the left side of the reduction-gear front case. An electrical switch mounted on the case, when actuated, energizes the feathering circuit.

SAFETY COUPLING *(FIG. 23-16)*

The safety coupling could readily be classified as a backup device for the NTS system. It has a negative torque setting of approximately 1500 hp [1119 kW]. In the event that the NTS system or propeller would not function properly, the safety coupling would uncouple the reduction gear from the power section. By so doing,

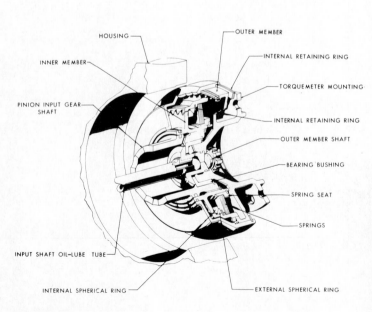

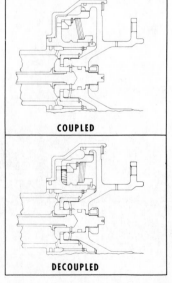

FIG. 23-16 Safety coupling.

the drag effect would be greatly reduced. The safety coupling is located and attached to the forward end of the torquemeter shaft, which transmits power-section horsepower into the reduction-gear assembly. During normal operation, the safety coupling connects the torquemeter shaft (power being produced by power section) to the reduction-gear assembly by helical splines. Aiding in the normal windup of these splines, as induced by power input and direction of rotation, is a set of four springs. When negative torque occurs, in excess of the preset value, an unwinding force overcomes the spring force and the helical splines move apart. This action disengages the power section from the reduction gear. The safety coupling is designed to reengage when power-section and reduction-gear rpm are approximately the same.

TORQUEMETER ASSEMBLY AND TIE STRUTS *(FIG. 23-17)*

The torquemeter housing provides alignment, and two tie struts provide the necessary rigidity between the power section and the reduction-gear assembly. The tie struts are adjustable through two eccentric pins which are located at the reduction-gear end. These pins are splined to enable a positive locking method after proper alignment is established by the torquemeter housing. The torquemeter provides the means of accurately measuring shaft-horsepower input into the reduction-gear assembly. It has an indicated accuracy of ±35 hp [±26.1 kW] from zero to maximum allowable power, which represents ±1 percent actual horsepower at standard day static takeoff power. The torquemeter consists of the following major parts.

1. Torquemeter inner shaft (torque shaft)
2. Torquemeter outer shaft (reference shaft)
3. Torquemeter pickup assembly (magnetic pickup)
4. Torquemeter housing
5. Phase detector
6. Indicator

The principal operation of the torquemeter is that of measuring electronically the angular deflection (twist), which occurs in the torque shaft, relative to the zero deflection of the reference shaft. The actual degree of angular deflection is measured by the pickup assembly, and transmitted to the phase detector. The phase detector converts the pickup signal into an electrical signal, and directs it to the indicator located on the instrument panel.

LUBRICATION SYSTEMS *(refer to FIG. 15-11)*

The 501-D13 power section and reduction-gear assembly have separate and independent lubrication systems which utilize a common airframe-furnished oil-supply system.

The engine manufacturer supplies the airframe manufacturer with the amount of oil flow required by the reduction gear and the power section, and also the heat rejection from the reduction gear and power section. The airframe manufacturer, with this information, designs an aircraft oil-supply system which will provide the required volume flow and the necessary oil cooling. In addition, the airframe manufacturer must provide the following cockpit indications for each engine.

1. Power-section oil pressure
2. Reduction-gear oil pressure
3. Oil inlet temperature (inlet to engine lubrication system)
4. Oil quantity

POWER-SECTION LUBRICATION SYSTEM *(FIG. 23-18)*

The power section contains an independent lubrication system, with the exception of airframe-furnished parts which are common to power section and reduction gear. The power-section system includes the fol-

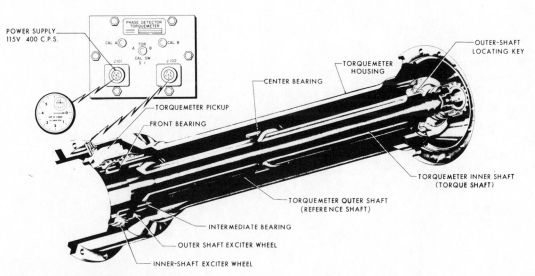

FIG. 23-17 Torquemeter assembly.

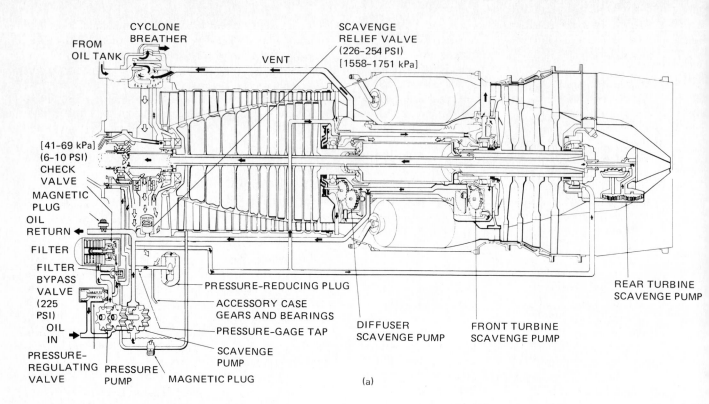

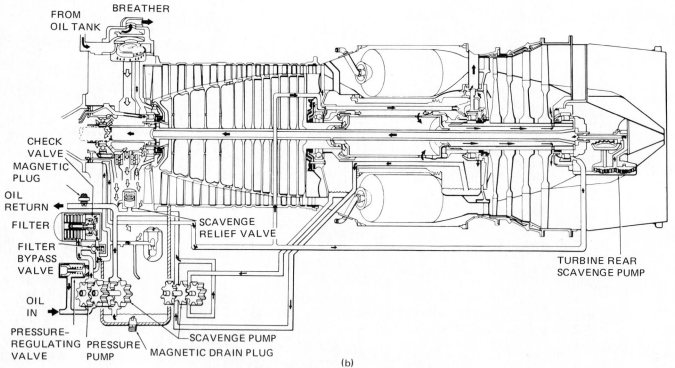

FIG. 23-18 Two types of power unit oil systems.
(a) Power unit oil system with internal scavenge pumps.
(b) Power unit oil system with external scavenge pumps.

lowing, with each of their respective locations as indicated.

1. *Main oil pump*—This includes the pressure pump, a scavenge pump, and the pressure-regulating valve which is located on the forward side of the accessories-drive housing assembly.
2. *Oil filter*—Located on the forward side of the accessories drive housing assembly.
3. *Check valve*—Located in the oil-filter assembly.
4. *Bypass valve (filter)*—Located in the accessories drive housing assembly.
5. *Three scavenge pumps*—Located in the diffuser, turbine inlet casing, and the turbine rear-bearing support.

   NOTE: Late-model engines locate the scavenge pumps externally [Fig. 23-18(b)].

6. *Scavenge relief valve*—Located in the accessories drive housing assembly.
7. *Breather*—Located on top of the air inlet housing.

Oil is supplied from the aircraft tank to the inlet of the pressure pump. Before the oil is delivered to any parts requiring lubrication, it flows through the oil filter. System pressure (filter outlet pressure) is regulated to 50 to 75 psi [344.8 to 517.1 kPa] by the pressure-regulating valve. A bypass valve is incorporated in the system in the event that the filter becomes contaminated, thereby obstructing oil flow. A check valve prevents oil from seeping into the power section whenever the engine is not running.

The scavenge pump, which is incorporated in the main oil pump, and the three independent scavenge pumps are so located that they will scavenge oil from the power section in any normal attitude of flight. The scavenge pump, located in the main oil pump, scav-enges oil from the accessories drive housing. The other three scavenge oil from the diffuser, and from the front and rear sides of the turbine. The outputs of the diffuser and the front-turbine scavenge pumps join that of the main scavenge pump. The output of the rear-turbine scavenge pump is delivered to the interior of the turbine-to-compressor tie bolt and the compressor rotor tie bolt. This oil is directed to the splines of the turbine coupling shaft assembly, and to the splines of the compressor extension shaft. Thus, the output of the rear-turbine scavenge pump must be rescavenged by the other three scavenge pumps. A scavenge relief valve is located so that it will prevent excessive pressure buildup in the power-section scavenge system. The combined flows of scavenged oil from the power section and reduction-gear scavenge systems must be cooled, and returned to the supply tank. A magnetic plug is located on the bottom of the accessories drive housing, and another at the scavenge-oil outlet on the forward side of the accessories drive housing.

REDUCTION-GEAR LUBRICATING SYSTEM *(FIG. 23-19)*

The reduction-gear lubricating system includes the following, with each of their respective locations as indicated:

1. *Pressure pump*—Located on the left rear side of the reduction gear.
2. *Filter*—Located in pump body assembly.
3. *Filter bypass valve*—Located in pump body assembly.
4. *Check valve*—Located in pump body assembly.
5. *Two scavenge pumps*—One located in the bottom of the rear case; the other in the front case below the prop shaft.

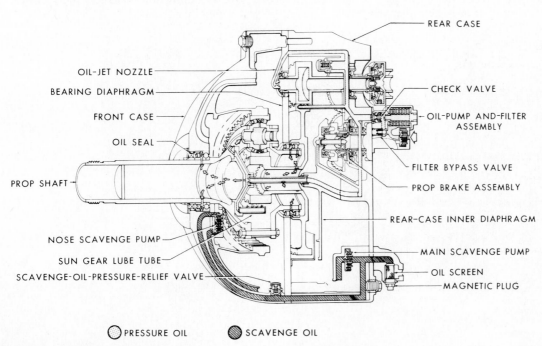

FIG. 23-19 Reduction-gear oil system.

6. *Two pressure-relief valves*—One for the pressure system; the other for the scavenge system. The scavenge relief valve is located in the common outlet of the scavenge pumps; the other in the rear-case housing near the oil filter outlet.

Oil flows from the pressure pump through a filter, and to all parts within the reduction gear which require lubrication. In addition, oil pressure is used as hydraulic pressure in the propeller brake assembly. A filter bypass valve guarantees continued oil flow in the event that the filter becomes contaminated. A check valve prevents oil flow into the reduction gear after engine shutdown. A relief valve, which is set at 180 psi [1241 kPa] to begin opening and to be fully open at 250 psi [1724 kPa], prevents excessive system pressure. This valve is not to be construed as being a regulating valve, as its only function is that of limiting pressure.

The location of the scavenge pumps provide for scavenging in any normal attitude of flight. The output of the two scavenge pumps returns the oil by a common outlet to the aircraft system. A relief valve, which is set at the same values as the one in the reduction-gear pressure system, limits the maximum scavenge pressure. A magnetic plug is located on the bottom rear of the reduction-gear assembly and provides a means of draining it.

ANTI-ICING SYSTEM *(FIG. 23-20)*

The system includes an anti-icing solenoid valve located on the top of the compressor housing, two anti-icing-valve assemblies located one on either side of the compressor air inlet housing, and the necessary lines and passages from the compressor diffuser to the anti-icing valves. The system is entirely manual in operation, being selected by the crew from the flight deck by a switch called the *engine air scoop and inlet vanes anti-icing*

switch. When selected, compressor discharge air, which has been heated due to compression, will flow to the two anti-icing valves. From this point, the air flows to the inlet anti-icing vane assembly, the compressor air inlet-housing struts, the fuel-control temperature probe deicer (located in the air inlet housing below the left horizontal strut), and the upper half of the torquemeter housing shroud. The fuel-control total pressure probe, located in the left horizontal strut of the air inlet housing, is anti-iced by heat conduction. Whenever engine anti-icing is selected, each engine will be independent of the other, since each will have a switch. During anti-icing, approximately 1 percent of air will be bled which will result in a horsepower decrease of approximately 3 percent.

FIFTH- AND TENTH-STAGE BLEED AIR SYSTEM (ACCELERATION BLEED SYSTEM) *(FIG. 23-21)*

This is an entirely automatic system which bleeds air from the fifth and tenth stages during engine start and acceleration, and at low-speed taxi. It is used to unload the compressor from 0 to 13,000 rpm in order to prevent engine stall and surge. The system includes four pneumatically operated bleed air valves located at the fifth stage and four located at the tenth stage, a speed-sensitive valve mounted on the forward side of the accessories housing assembly, and the necessary manifolding and plumbing. The bleed air valves at the fifth stage are manifolded together with the outlet being provided through the nacelle, forward of the engine baffle assembly (firewall at diffuser). The tenth-stage bleed air valves empty into another manifold, which is ducted to the aft side of the engine baffle assembly. The speed-sensitive valve is a flyweight type, which responds to engine rpm. When running at less than 13,000 rpm, the valve is so positioned that all bleed air valve piston heads

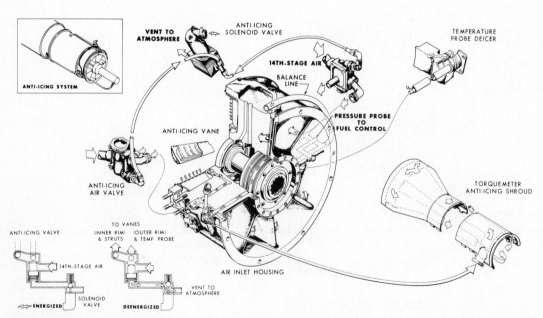

FIG. 23-20 Anti-icing system.

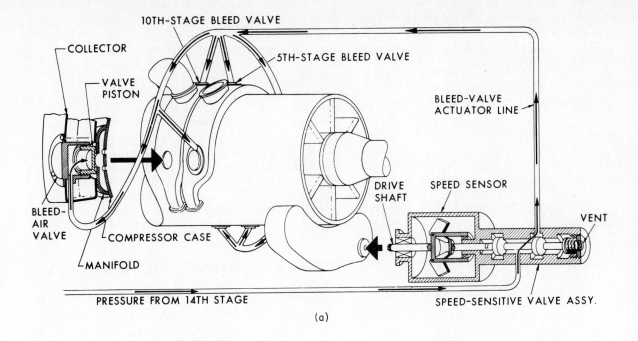

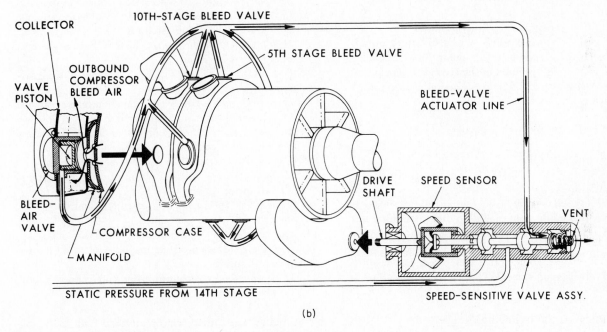

FIG. 23-21 Fifth- and tenth-stage bleed valve operation.
(a) Bleed valve closed.
(b) Bleed valve open.

are vented to the atmosphere. This allows the compressor fifth- and tenth-stage pressures to move the pistons to their open position, bleeding air overboard. When running at 13,000 rpm or better, the speed-sensitive valve directs 14th-stage air to the bleed air valve piston heads. Since 14th-stage pressure is always greater than 5th- or 10th-stage pressures, the bleed air valve pistons move to the closed position, thus preventing airbleed from the 5th and 10th stages. During low-speed taxi operation, the 5th and 10th stage bleed air valves will be in the open position, thus bleeding air.

FOURTEENTH-STAGE BLEED AIR SYSTEM (STARTING BLEED SYSTEM)

To facilitate the ignition of fuel and air during the starting cycle and to aid in initial acceleration after "light-off," a 14th-stage bleed is utilized. The system includes the 14th-stage bleed air valve and bleed air control valve which are mounted on the compressor diffuser. The 14th-stage bleed air valve is spring-loaded in the open position, and closed by compressor discharge pressure directed from the 14th stage by the bleed air control valve. At approximately 5000 engine rpm, air pressure from the 14th stage moves the bleed control valve to a position which allows 14th-stage air pressure to close the 14th-stage bleed air valve.

SPEED-SENSITIVE CONTROL

The speed-sensitive control is mounted on the forward side of the accessories drive housing assembly. The control is a flyweight type which incorporates three microswitches. At certain settings, electrical circuits are "made" or "broken" which make the entire engine starting procedure an automatic one. At 2200 rpm, the following take place:

1. The fuel control cutoff valve actuator opens the cutoff valve at the outlet of the fuel control.

   NOTE: Fuel and ignition switch must be on in the cockpit.

2. *Ignition system*—ON, provided it is armed by fuel and ignition switch.
3. *Drip valve*—Energized to closed position.
4. *Fuel-pump paralleling valve*—CLOSED, fuel pumps placed in parallel; engine fuel pump light should go ON, indicating operation of the secondary pump.
5. *Primer valve*—Opens; this takes place only when primer switch is held to the ON position.

At 9000 rpm, the following occurs:

1. *Ignition system*—OFF.
2. *Drip valve*—DEENERGIZED, remains closed due to fuel pressure.
3. *Paralleling valve*—OPEN, fuel pumps placed in series; engine fuel pump light should go OFF, indicating operation of the primary pump.

   NOTE: If fuel primer valve were used, it will automatically close at 50 psi [344.8 kPa] fuel manifold pressure.

At 13,000 rpm, the following occurs:

1. The electronic fuel-trimming system is changed from temperature-limiting with a maximum temperature of 871°C to temperature-limiting with a maximum temperature of 977°C.
2. Resets maximum possible "take" of fuel by the temperature datum valve to 20 percent, rather than previous 50 percent.

IGNITION SYSTEM

Ignition is required only during the starting cycle, since the combustion process is continuous. Once ignition takes place, the flame in the combustion liners acts as the ignition agent for the fuel-air mixture.

This ignition system is classified as a condenser-discharge, high-energy type. The system includes an exciter and an ignition relay which are mounted on the top of the compressor housing, the lead assemblies, and two igniter plugs. It operates on 14 to 30 V dc input. Actually there are two independent systems, as the exciter is a dual unit with individual leads going to the two igniter plugs. During the starting cycle as rpm reaches 2200, the speed-sensitive control automatically completes an electrical circuit to the ignition relay. This closes the circuit to the exciter, thus providing electric energy to the igniter plugs. When engine rpm reaches 9000, these circuits are deenergized through the action of the speed-sensitive control. Operation of the ignition system requires that the fuel and ignition switch in the cockpit be in the ON position. (Refer to Fig. 16-4).

FUEL SYSTEM *(refer to FIG. 13-4)*

The fuel system includes the following, and their locations are as indicated (Fig. 23-22).

1. *Fuel pump*—Right rear side of accessories housing
2. *Low-pressure fuel filter*—Attached to right forward side of compressor-housing assembly
3. *High-pressure fuel filter*—Attached to bottom of fuel pump
4. *Fuel control*—Left rear side of accessories housing assembly
5. *Fuel-primer valve*—Bracket-mounted on center rear side of accessories drive housing assembly
6. *Manifold pressure switch*—Left side of compressor housing above fuel control
7. *Coordinator*—Attached to rear side of fuel control
8. *Temperature datum valve*—Bottom of compressor housing, aft of fuel pump
9. *Temperature datum control*—Airframe mounted
10. *Relay box*—Airframe mounted
11. *Fuel nozzles (six)*—Mounted to compressor diffuser
12. *Drip valve*—Bottom of fuel manifold, aft end of compressor housing
13. *Drain valves (two)*—Bottom, front, and rear outer combustion chamber

The fuel system must deliver metered fuel to the six fuel nozzles, as required to meet all possible conditions of engine operation either on the ground or in flight. This imposes a number of requirements on the fuel system. Some of these requirements are as follows:

1. The capability of starting under all ambient conditions.
2. Requirements for rapid changes in power.
3. A means of limiting the maximum allowable turbine inlet temperature.
4. A system which will enable the operator to select a desired power setting (turbine inlet temperature) and have it automatically maintained regardless of altitude, free-air temperature, forward speed, and fuel Btu content.
5. A system which incorporates an rpm-limiting device in the event of propeller governor malfunction.
6. A system which must control fuel flow during the rpm

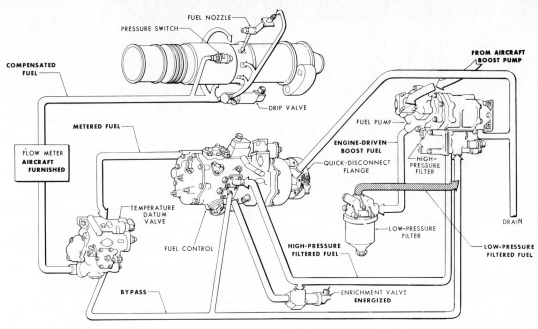

FIG. 23-22 Fuel-system schematic.

range in which the engine compressor is susceptible to stall or surge.

7. A system which coordinates propeller blade angle during ground operation (taxi range—start, taxi, and reverse—0 to 34° coordinator quadrant) with fuel flow.

8. A system which is capable of operating, if necessary, on the hydromechanical fuel control. However, if this is necessary, closer monitoring of throttle and engine instruments will be required.

Rather than attempt a flow description, each accessory or component will be covered in a normal sequence of flow. Since a few of the units are not associated with flow, these will be brought up in a logical sequence.

FUEL PUMP AND LOW-PRESSURE FUEL FILTER *(refer to FIG. 12-23)*

Fuel is supplied to the engine fuel pump from the aircraft system. It enters into a boost element and is then directed to the low-pressure filter.

The low-pressure fuel filter is a paper-cartridge-type filter that incorporates two bypass valves (relief safety valves) which open in the event of unusual fuel contamination. The paper cartridge is of the type that must be replaced at certain inspection periods.

The fuel-pump assembly includes, in addition to the boost element, two spur-gear-type high-pressure pumps (Fig. 23-23). These are commonly referred to as the

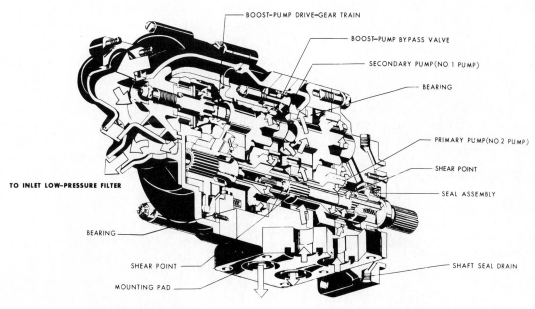

FIG. 23-23 Fuel pump.

primary and the secondary elements. During normal operation, these pumps are in series. However, during engine starting (2200 to 9000 rpm), the pumps are placed in parallel by the action of a paralleling valve in the high-pressure filter. The paralleling of the pumps is used to increase fuel flow during low rpm. Failure of either primary or secondary pump will not affect normal operation, as either pump has sufficient capacity of fuel flow for takeoff power.

HIGH-PRESSURE FUEL FILTER (FIG. 23-24)

The high-pressure fuel-filter assembly consists of two check valves, a paralleling valve, a fuel filter, a pressure switch, and a bypass valve. The fuel-filter assembly accomplishes the following:

1. Filters the output of primary and secondary pumps
2. Connects the two pumps in parallel during the starting cycle (2200 to 9000 rpm)
3. Connects the two pumps in series during normal operation, with the primary pump supplying high-pressure fuel flow to the power section
4. Will automatically enable the secondary pump to "take over" upon failure of the primary pump
5. Provides a means of checking primary- and secondary-pump operation during the starting procedure
6. Provides a means of indicating primary-pump failure (by the engine fuel pump light)

FUEL CONTROL (refer to FIG. 12-10)

The fuel control is a hydromechanical metering device which is designed to meet the following requirements:

1. Changes fuel flow with factors affecting air density as sensed at the engine inlet.

2. Meters fuel flow during starting (in conjunction with the temperature datum valve).
3. Meters fuel flow during engine acceleration to aid in preventing compressor stall or surge, and is scheduled to prevent excessive turbine inlet temperature.
4. Controls power available in reverse.
5. Meters fuel to assist in controlling rpm during low- and high-speed taxi.
6. Provides an overspeed governor for ground operation, and for flight operation in the event of propeller governing malfunction.
7. Provides for manual selection of power (turbine inlet temperature) by movement of the throttle.
8. Permits any selection of power in the flight range and turbine inlet temperature (34 to 90° coordinator control) to be automatically maintained regardless of altitude, free-air temperature, and forward speed.
9. Meters 120 percent of engine fuel requirements, based upon compressor inlet air temperature and pressure, rpm, and throttle setting.
10. Allows cutoff of fuel flow manually or electrically.

PRIMER VALVE AND MANIFOLD PRESSURE SWITCH (FIG. 23-25)

The fuel primer system includes a primer valve which is solenoid-actuated, a manifold pressure switch, and necessary aircraft wiring and primer switch. This system may be used during the starting cycle. It is placed in operation by the spring-loaded primer switch on the flight deck. If the primer system is used, it will provide an increased initial fuel flow. This is accomplished by permitting fuel to flow through the primer valve, bypassing the metering section of the fuel control and entering just prior to the fuel-control cutoff valve. This fuel flows through the cutoff valve and is directed through the temperature datum valve, then to the manifold and fuel nozzles. The pressure switch, which senses manifold fuel pressure, breaks the electrical circuit to the primer-valve solenoid when the fuel pressure

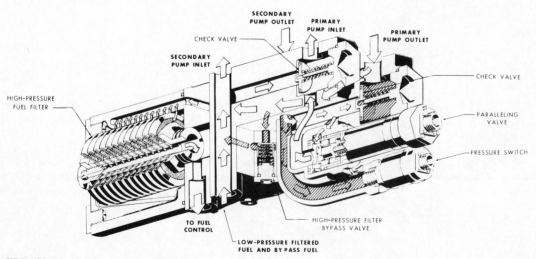

FIG. 23-24 High-pressure fuel filter.

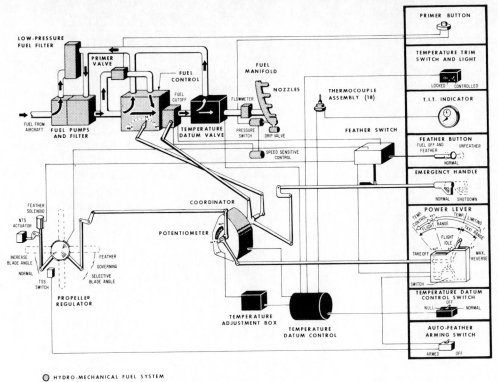

FIG. 23-25 Control-system schematic.

reaches 50 psi [344.8 kPa]. An electrical interlock in the control system prevents energizing of the primer system after the engine is once started.

COORDINATOR (*See FIG. 23-25*)

This coordinates the fuel control, propeller, and the electronic fuel-trimming system. The operation of the coordinator is controlled by mechanical linkage from the flight deck, normally through the throttle, and, for special conditions, by the emergency shutdown handle.

The coordinator includes a discriminating device, two microswitches, a temperature datum control scheduling potentiometer, a coordinator lock, and the necessary gears and electrical wiring. Throttle movement controls the main shaft of the coordinator, which in turn, controls the following:

1. The amount of fuel flowing from the fuel control under any given set of conditions (by mechanical linkage to the fuel control).
2. The propeller blade angle during all ground operation (taxi range: 0 to 34° on coordinator quadrant). This is accomplished by mechanical linkage to the propeller.
3. Controls potentiometer output signal from 65 to 90° (coordinator quadrant). This schedules turbine inlet temperature.
4. Controls two microswitches which are set, one at 65° and the other at 66°. The 65° switch transfers the electronic fuel-trimming system from temperature limiting to temperature control. The one set at 66° arms the

temperature-trim switch which permits locking of the temperature datum valve brake. This enables the operator to "lock in" a fuel correction with any power setting above 66°.

5. Propeller beta follow-up mechanism.

The discriminating device permits the use of the same mechanical linkage between the coordinator and the propeller for throttle or emergency shutdown handle operation. The discriminator allows the throttle to position the propeller linkage at all times other than emergency shutdown. When the emergency shutdown handle is pulled, the propeller linkage will always be actuated to feather by means of the discriminator regardless of throttle setting.

ELECTRONIC FUEL-TRIMMING SYSTEM

The electronic fuel-trimming system contributes the following:

1. Positive overtemperature protection during starting and acceleration.
2. Allows engine to operate closer to the maximum turbine inlet temperature because of accurate monitoring of fuel scheduling.
3. Permits selection of any desired turbine inlet temperature in the control range 760 to 971°C to be automatically maintained without any throttle change.
4. Permits use of kerosene, Allison specification EMS-

64A, or JP-4 fuel without requirement for rerigging or recalibration of the fuel control.

5. Permits use of power unit bleed air for anti-icing purposes without the necessity of changing power settings to avoid the possibility of overtemperature.
6. Will trim fuel flow to compensate for erroneous compressor inlet air temperature or pressure sensing by the fuel control which could be caused by aircraft installation.
7. Provides a more uniform throttle setting for all engines.
8. Provisions for "locking in" a fuel correction prior to landing for a more balanced power from all engines.

The system trims the 120 percent fuel flow from the fuel control as required for any condition of engine operation. There are two ranges of operation, namely *temperature limiting* and *temperature control*.

Temperature limiting serves to prevent the possibility of exceeding critical turbine inlet temperature limits during starting or acceleration. Whenever operating with the throttle in the 0 to 65° position (coordinator quadrant), the engine is operating in limiting. Temperature limiting also occurs when operating with a locked-in fuel correction above 65° (coordinator quadrant).

Two different limits are required: a lower one of 871°C needed below 13,000 rpm when the fifth- and tenth-stage compressor bleed air valves are open; a higher temperature limit of 977°C permits locking-in of a fuel correction at any throttle setting above 65° (coordinator quadrant) without having an overtemperature signal. Thus, the following operational conditions exist:

1. Temperature limit of 871°C when starting and accelerating up to 13,000 rpm
2. Temperature limit of 871°C when operating at low-speed taxi
3. Temperature limit of 977°C during high-speed taxi
4. Temperature limit of 977°C up to 65° (coordinator quadrant)
5. Temperature limit of 977°C when rpm is greater than 13,000 and temperature trim switch is in locked (locked-in fuel correction)

Temperature control permits the use of the throttle to schedule a desired turbine inlet temperature when operating above 65° (coordinator quadrant). Temperature control requires rpm in excess of 13,000 with temperature trim switch in CONTROLLED (*no* locked-in fuel correction).

COMPONENTS REQUIRED FOR OPERATION OF THE ELECTRONIC FUEL-TRIMMING SYSTEM (*FIG. 23-26*)

A number of engine and airframe components are required for operation of the electronic fuel-trimming system, and include the following:

1. Temperature datum valve
2. Temperature datum control
3. Relay box
4. Thermocouples (18 in parallel)
5. Coordinator (potentiometer and microswitches)
6. Speed-sensitive control (13,000-rpm switch)
7. Throttle (flight deck)
8. Temperature trim light (flight deck)
9. Temperature trim switch (flight deck)
10. Engine-temperature datum control switch (flight deck)

The temperature datum valve receives 120 percent of engine fuel flow requirement. Obviously some fuel must be bypassed by the temperature datum valve. When 20 percent is bypassed, the term *null* is used to describe this condition. When more than 20 percent is bypassed, a *take* condition exists, and if less than 20 percent is bypassed, a *put* condition exists.

In order for the electronic fuel-trimming system to function, the temperature datum control must receive a temperature signal and a reference signal. The temperature signal always comes from 18 thermocouples at the turbine inlet. The reference signal can be from one of three sources: start 871°C from a potentiometer in the temperature datum control; normal 977°C from a potentiometer in the temperature datum control; or from a variable potentiometer in the coordinator (approximately 760 to 971°C turbine inlet temperature). The variable potentiometer in the coordinator is positioned by movement of the throttle on the flight deck in the range of 65 to 90° (coordinator quadrant). When the temperature signal (18 thermocouples) is less than the reference signal (start or normal), the temperature datum control sends no signal to the temperature datum valve. Hence, the valve remains in the *null* position since there is no overtemperature condition (bypass 20 percent).

If the temperature signal (18 thermocouples) is greater than the reference signal (start or normal), the temperature datum control sends a signal to the temperature datum valve to *take* fuel (bypasses more than 20 percent).

During temperature control, if the temperature signal exceeds the reference signal from the coordinator potentiometer, the temperature datum control sends a signal to the temperature datum valve to *take* fuel (bypasses more than 20 percent).

When operating in the control range and the temperature signal is less than the reference signal from the coordinator potentiometer, the temperature datum control sends a *put* (bypasses less than 20 percent) signal to the temperature datum valve.

Thus, in temperature limiting, the signal to the temperature datum valve can only be one which will cause the valve to move to a *take* position. However, in temperature control, the signal may result in placing the valve in either a *take* or *put* position.

The following are conditions of engine operation which will result in *null*, *put*, or *take*:

1. When no correction is necessary, a null condition exists (bypass 20 percent).
2. When turbine inlet temperature is less than desired in

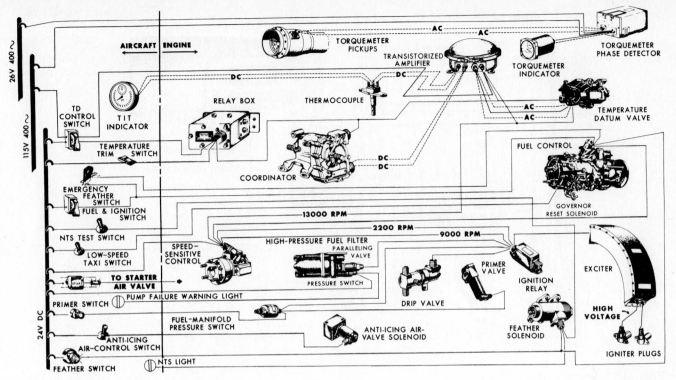

FIG. 23-26 Electrical-system schematic.

the temperature-control range, a maximum of 15 percent put is possible (bypass 5 percent).

3. During starting, and acceleration up to 13,000 rpm, the maximum take may be as high as 50 percent (bypass 70 percent). However, when engine rpm reaches 13,000, the speed-sensitive control deenergizes a circuit which resets the maximum possible take to 20 percent (bypass 40 percent).

NOTE: During starting and engine acceleration, it is necessary to provide a means of taking a greater percentage of fuel than when the engine is operating at, or near, its maximum rpm.

TEMPERATURE DATUM VALVE
(FIG. 23-27)

The temperature datum valve is an electrically operated fuel-trimming device. It is located in the fuel system so that all fuel, flowing from the fuel control to the nozzles, must pass through it. The valve operates on ac power fed from the temperature datum control. Fuel, in excess of that required by the engine, is bypassed and returned with excess fuel from the fuel control to the inlet of the primary and secondary fuel pumps. As previously mentioned, it is necessary to have available, two distinct ranges of fuel *take*. To accomplish this, a solenoid-operated control valve is an integral part of the temperature datum valve. The solenoid is energized by 28 V dc through the speed-sensitive control when the engine is operating in the 0- to 13,000-rpm range. At

13,000 rpm, the solenoid is deenergized, which resets the *take* stop to 20 percent of the nominal fuel-flow requirements. A brake is also included in the temperature datum valve. When the temperature trim switch is placed in the locked position and the throttle is in the range of 65 to 90° (coordinator quadrant—approximately 760 to 971°C), the valve will then supply fuel-flow correction in the 0 to 65° (coordinator quadrant) range. In the event that an overtemperature condition occurs while the switch is in locked, the control system automatically unlocks the brake and bypasses more fuel to prevent an overtemperature condition. (Refer to Fig. 12-15.)

TEMPERATURE DATUM CONTROL
(See FIG. 23-25)

The temperature datum control is an electronic control, using 115-V, 400-Hz alternating current. The control may be thought of as a comparator in that it compares actual with desired or limited turbine inlet temperature signals. If an out-of-balance condition exists, the control signals the temperature datum valve to *put* (increase) or *take* (decrease) fuel flow as required to bring the temperature back to that which is scheduled. Operation of the temperature datum control requires having the engine-temperature datum control switch (located on the flight deck) in the NORMAL position.

The temperature datum control contains four potentiometers, the necessary electrical wiring, and four external adjustments. The external adjustments are re-

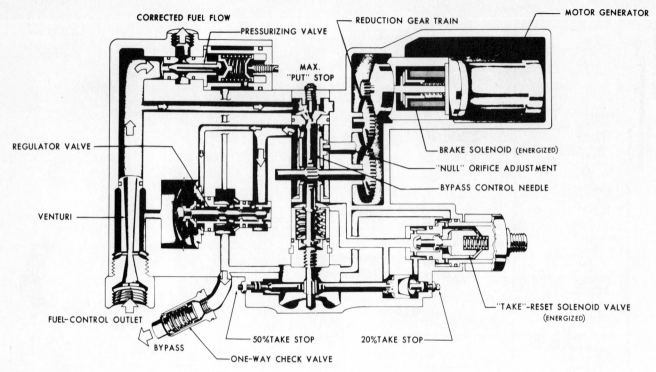

FIG. 23-27 Temperature-datum-valve schematic.

ferred to as the *bias, slope, start limiting*, and *normal limiting*. The bias and slope adjusts the temperature schedule for engine operation between 65 and 90° (coordinator quadrant) which is known as the control range. The other two adjust the start-limiting temperature and the normal-limiting temperature.

RELAY BOX

The relay box contains the relays which are required by the engine for proper sequencing of all control components.

THERMOCOUPLES *(FIG. 23-28)*

There is a total of 18 dual thermocouples, forming two individual circuits. One circuit provides turbine inlet temperature to the flight-deck instrument, and the other provides an actual temperature indication to the temperature datum control. Due to the fact that each circuit is a parallel one made of 18 thermocouples, the temperature indication is an average of all 18.

COORDINATOR POTENTIOMETER AND MICROSWITCHES

The coordinator potentiometer is a variable one and provides the desired turbine inlet temperature signal to the temperature datum control in the range of 65 to 90° (coordinator quadrant—approximately 760° to 971°C.

The variable signal is a result of throttle movement on the flight deck positioning this potentiometer.

The microswitch, set at 65°, transfers the electronic fuel-trimming system from temperature limiting to temperature control. The one at 66° arms the temperature trim switch which allows locking of the temperature datum valve brake.

SPEED-SENSITIVE CONTROL (13,000-rpm SWITCH)

At 13,000 rpm, the speed-sensitive control deenergizes the solenoid-operated control valve in the temperature datum valve, thus switching from a maximum *take* of 50 percent to one of 20 percent.

THROTTLE *(See FIG. 23-25)*

The throttle provides the means of complete power control during all normal conditions of operation. Movement of the throttle actuates mechanical linkage to the coordinator which has a total quadrant travel of 0 to 90°. Ground operation, which includes start, ground idle, taxi, and reverse, is in the range of 0 to 34°. Flight operation is between 34 and 90° at all times. Temperature limiting schedule is 0 to 65° and temperature control schedule is 65 to 90°.

NOTE: All of the aforementioned ranges are in coordinator quadrant degrees. Total flight station throttle travel is somewhat less.

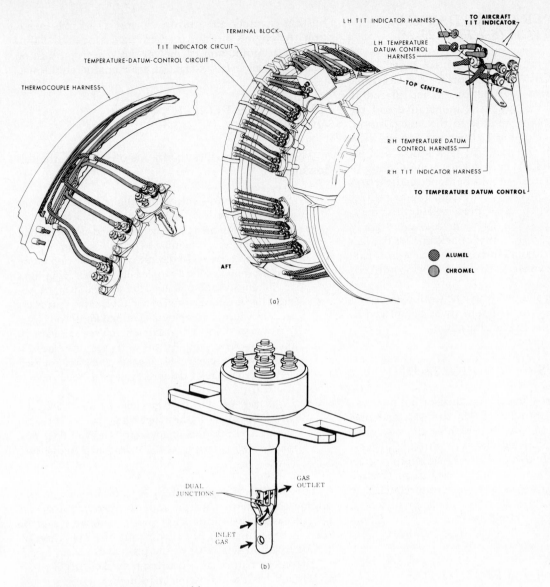

FIG. 23-28 Thermocouple assembly.
(a) Thermocouple harness.
(b) Typical thermocouple.

TEMPERATURE TRIM SWITCH
(See FIG. 23-25)

The temperature trim switch, when placed in the locked position, causes the datum valve brake to move to locked. However, this occurs only with the throttle in a position greater than 66° (coordinator quadrant). When the switch is moved to the controlled position, it releases the brake.

TEMPERATURE TRIM LIGHT
(See FIG. 23-25)

The temperature trim light will be on from 0 to 65° (coordinator quadrant), indicating operation in the tem-

perature-limiting range. From 65 to 90° (coordinator quadrant), it will be off, indicating operation in the temperature control range. It will also be off from 90 to 0° (coordinator quadrant) when the temperature-datum-valve brake is in the locked position. However, if an overtemperature condition occurs with the temperature-datum-valve brake locked at any power setting, the light will come on. When the temperature trim light is off, it indicates that the electronic fuel-trimming system is making a fuel flow correction (put or take); and when the light is on, it indicates that no correction (null) is being made or there is an overtemperature condition.

NOTE: If the engine temperature datum control switch is in emergency NULL or OFF, the light indication should be disregarded.

ENGINE-TEMPERATURE DATUM CONTROL SWITCH *(See FIG. 23-25)*

The engine-temperature datum control switch must be placed in the NORMAL position for the electronic fuel-trimming system to function. When placed in emergency NULL, the system is inoperative and the temperature datum valve returns to the null position, bypassing 20 percent of the 120 percent furnished by the fuel control. The metering of fuel is now accomplished solely by the fuel control. Closer monitoring of turbine inlet temperature should be done, and the operator should remember that overtemperature protection is lost. The switch has a third position, namely, OFF. When in this position, the same conditions apply as in the emergency NULL position, except for the fact that the temperature datum valve is locked into whatever position it was prior to the switch being moved.

NOTE: The switch should always be turned on before engine starting. The switch should always be placed in the OFF position after engine shutdown.

FUEL NOZZLES *(refer to FIG. 12-32)*

The fuel nozzles are duplex-type, dual-orifice nozzles. The six nozzles are mounted in the diffuser, and one extends into the forward end of each of the six combustion liners. The fuel nozzle must provide a controlled pattern of fuel flow, and also a maximum degree of atomization. At the tip of the nozzle, an air shroud surrounds the dual orifices. The air shroud contains a number of air holes through which air is circulated at high velocity, thus preventing the formation of carbon around the orifices.

DRIP VALVE *(refer to FIGS. 23-11 and 23-22)*

The drip valve is located at the lowest point in the fuel manifold. It is designed to drain the manifold at engine shutdown, thus preventing fuel from draining into the combustion liners after the fuel-control cutoff valve is closed. The drip valve is a solenoid-operated valve, which is closed by completion of the electrical circuit by the 2200-rpm switch in the speed-sensitive control. At 9000 rpm, the electrical circuit is broken and fuel manifold pressure, acting upon the valve, continues to hold it in the closed position. At engine shutdown, when manifold pressure drops to a value of 8 to 10 psi [55.2 to 68.95 kPa], the valve opens due to spring force.

DRAIN VALVES *(refer to FIG. 23-11)*

There are two drain valves which are located at the forward and aft ends on the bottom of the outer combustion chamber. These valves are set at 2 to 4 psi [13.8 to 27.6 kPa] air pressure, and are held closed by combustion-chamber air pressure during all engine operation. At engine shutdown, these valves open and thus prevent accumulation of fuel in the outer combustion chamber after a false start, or after engine shutdown.

THROTTLE POSITION (FIG. 23-25)

The throttle provides the means of selecting the following:

1. *Start blade angle*—This is the minimum torque blade angle and is desired for best starting characteristics.
2. *Ground idle*—This is a blade angle and power setting which requires a minimum of wheel braking.
3. *Maximum reverse*—The throttle is in the full aft position which produces a blade angle and power setting for a maximum aircraft braking after touchdown. Movement of the throttle forward and toward the start blade angle produces lesser amounts of braking force.
4. *Taxi range*—From the maximum reverse position to the flight-idle detent. At this time, the propeller is a multiposition, selective blade-angle propeller. For each and every change of the throttle, a new blade angle is selected.
5. *Flight range*—From the flight-idle detent to the full-forward position. In this range, the propeller is a nonselective, automatic rpm governor. The throttle in this range serves primarily as the means of changing fuel flow.

The throttle is connected by aircraft linkage to the engine coordinator control, and any movement of the throttle will move this linkage. The coordinator control then has the function of coordinating the operation of the fuel system with that of the propeller.

The coordinator has a quadrant marked off from 0 to 90° and has the following markings.

1. Maximum reverse: 0°
2. Ground idle: 9°
3. Flight idle: 34°
4. Takeoff: 90°

Coordinator-quadrant degrees are usually called throttle degrees since the throttle actually sets the coordinator at any setting that it may have. The cockpit throttle quadrant has less than 90° travel and, therefore, measuring the degrees that the throttle moves in the cockpit will not correspond to the throttle degrees which one may read about in the publications of the engine manufacturer.

The coordinator-quadrant degree indications for the various ranges are as follows:

1. High-speed taxi range: 0°-34°
2. Low-speed taxi range: 9°-30°
3. Flight range: 34°-90°
4. Temperature-limiting range: 0°-65°
5. Temperature-control range: 65°-90°

NOTE: Power-unit rpm in the low-speed taxi range is 10,000 + (30/−10), and power-unit rpm in the high-speed taxi range is in excess of 13,000. To obtain low-speed taxi, the coordinator pointer must be between 9 and 30° and the cockpit low-speed taxi switch must be in the low position.

REVIEW AND STUDY QUESTIONS

1. List several airplanes that use this engine.
2. Give a brief description of this engine and its operation.
3. Very briefly describe the construction features of the following parts: compressor assembly, combustion assembly, turbine unit assembly, and accessories-drive assembly.
4. Discuss the reduction-gear assembly including the description of the propeller brake, negative torque system, thrust-sensitive signal, and safety coupling.
5. What is the function of the torquemeter? Describe its operation.
6. How many main bearings does this engine have?
7. Briefly describe the following systems: lubrication (reduction gear and main), anti-icing, bleed air, ignition, and fuel, including the electronic trim system.
8. Describe the coordinator control and its operation.
9. List the throttle positions and tell what mode of engine operation each position selects.

CHAPTER TWENTY-FOUR
TELEDYNE
CAE
J69-T-25

The construction of this engine (Fig. 24-1) is somewhat different from the larger turbojet engines in use at this writing, due principally to the novel design of the combustion chamber and fuel distribution system. Two J69's are installed in the Air Force's primary trainer, the Cessna T-37. Other models of this engine are installed in several target drones and special-purpose aircraft.

SPECIFICATIONS

Number of compressor stages:	1
Number of turbine stages:	1
Number of combustors:	1
Maximum power at sea level:	1025 lbt [4559 N]
Specific fuel consumption; at maximum power:	1.14 lb/lbt/h [116.2 g/N/h]
Compressor ratio at maximum rpm:	4:1
Maximum diameter:	24.9 in [632 cm]
Maximum length:	50 in [127 cm]
Maximum dry weight:	364 lb [165 kg]

The J69 engine consists of the following sections, accessories, and parts (Fig. 24-2):

1. Accessory case
2. Compressor housing
3. Turbine housing

External accessories include (Fig. 24-3):

1. Starter-generator (airframe supplied)
2. Starting-fuel system
3. Ignition system
4. Fuel pump
5. Fuel control
6. Oil pump
7. Oil filter

In addition to these parts, there are fuel, air, and oil lines. The engine is mounted with one top mount at the rear and two front mounts, approximately on the shaft axis.

OPERATION (FIG. 24-4)

NOTE: All numbers refer to Fig. 24-4.

In starting, the starter-generator drives through gears to spin the turbine shaft assembly (1). Air drawn into the inducer and compressor rotors (11, 12) is flung outward radially into the radial and axial diffusers (14, 16) which convert the velocity of the air into pressure. This pressure developed at the inlet to the turbine housing (21) forces the air into the combustion chamber (20). Main fuel sprays from the fuel distributor (2) into the combustion chamber to mix with this air. When ignition of the air-fuel mixture takes place, the hot gases flow out through the combustion chamber and are then directed by fixed vanes of the turbine inlet nozzle (24) to impinge upon the blades of the turbine rotor (25).

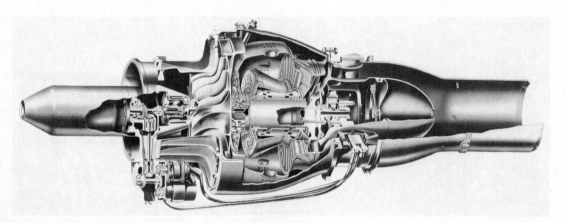

FIG. 24-1 Cutaway view of the Teledyne CAE J69-T-25.

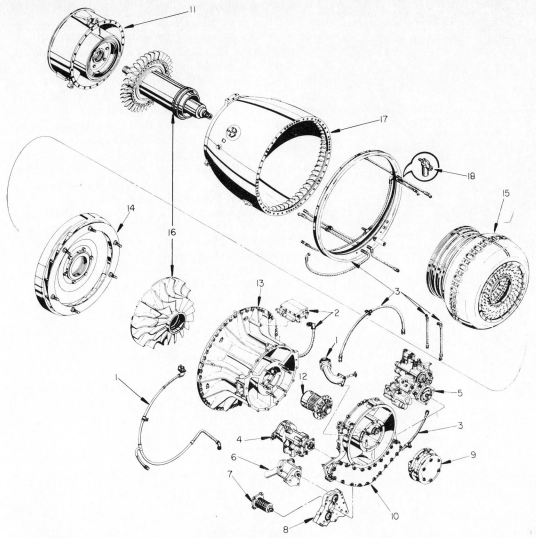

FIG. 24-2 Major engine components.

1 ELECTRICAL CABLE
 GROUP
2 IGNITION GROUP
3 FIREWALL AND HOSES
 GROUP
4 FUEL PUMP GROUP
5 FUEL CONTROL GROUP
6 OIL-PUMP GROUP

7 OIL-FILTER ASSEMBLY
8 OIL-PUMP DRIVE GROUP
9 STARTER-GENERATOR
 DRIVE GROUP
10 ACCESSORY CASE GROUP
11 EXHAUST DIFFUSER AND
 REAR-BEARING GROUP
12 ACCESSORY DRIVE GEAR-

SHAFT AND FRONT BEAR-
 ING CAGE GROUP
13 COMPRESSOR HOUSING
 GROUP
14 RADIAL DIFFUSER AND
 COMPRESSOR COVER
 GROUP
15 COMBUSTOR SHELL AND

INLET-NOZZLE GROUP
16 TURBINE AND COMPRES-
 SOR SHAFT ASSEMBLY
17 TURBINE HOUSING
 GROUP
18 COMPRESSED AIR-FILTER
 ASSEMBLY

The resultant torque speeds up the turbine shaft assembly to draw in and compress additional air. This new air enters the combustion chamber to mix with fuel. The mixture burns in the presence of the previously established flame so that the cycle is continuous. After the engine is working under continuous-flame operation (3500 rpm), the starting-fuel solenoid valve and the ignition system must be deenergized. The starter is cut out at about 5000 rpm. The hot products of combustion pass through the exhaust diffuser (29) and through the aircraft tailpipe to produce thrust.

An ignition system and a separate fuel system are provided for starting. Fuel is fed from the fuel control through a solenoid valve to the starting-fuel nozzles (17) installed in the lower portion of the turbine housing.

Adjacent to each of the two nozzles is an igniter plug (18) which is energized by an ignition coil. These components are located at the outlet from the axial diffuser assembly, outside of the combustion chamber. Starting fuel is injected from the nozzles into the incoming air, and this mixture is ignited by the spark at the plugs. The flame follows air path B (described later) into the combustion chamber. This starting combustion occurs at 1500 to 2000 rpm. As the shaft speeds up, main fuel issues from the rotating fuel distributor to build up flame in the combustion chamber. After flame is established in the combustion chamber, a control switch must cut off the ignition and the starting fuel. The ignition-control switch and starter-control switch are not supplied with the engine.

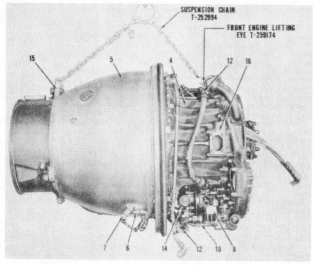

1	STARTER-GENERATOR	9	FUEL CONTROL
	MOUNT LOCATION	10	OIL PUMP
2	AIR INTAKE	11	OIL FILTER
3	ACCESSORY CASE	12	FUEL LINES
4	COMPRESSOR HOUSING	13	AIR LINES
5	TURBINE HOUSING	14	OIL LINES
6	STARTING FUEL SYSTEM	15	REAR MOUNT
7	IGNITION SYSTEM	16	FRONT MOUNT
8	FUEL PUMP		

FIG. 24-3 External parts of the Teledyne CAE J69-T-25

CONSTRUCTION (See FIG. 24-4)

The turbine shaft assembly (1) is carried in a single ball bearing (3) in the compressor housing (15) at the intake end and in a single roller bearing (30) at the exhaust end. The rear bearing is supported on three tangentially arranged link-type supports (26) designed to ensure centering under temperature-expansion conditions by a small rotation of the rear housing. The supports are encased in three streamlined struts (28) of the exhaust diffuser. These streamlined struts are hollow in order to carry cooling air to the rear bearing area. The cooling air passes around the outside of the rear-bearing

housing and also through passages in the turbine shaft assembly to provide cooling of the hollow rear section of this shaft assembly. This airflow is induced by tubular passages (31), located at the back of the turbine rotor, which act as an ejector-type centrifugal fan. One of the streamlined struts contains oil passages (27) carrying oil to and from the rear bearing. It also has an atmospheric vent for balancing pressures at the rear bearing. The front portion of the turbine shaft assembly is hollow. Through this the main fuel supply for the engine is carried through a fuel seal (4) to the fuel distributor (2) at the combustion chamber (20), where the centrifugal effect of the rotating distributor ejects the fuel into the combustion chamber.

The turbine housing (21) is the structural frame of the engine. The annular combustion-chamber structure (22, 23) is installed inside the turbine housing over the center portion of the turbine shaft assembly. The products of combustion are directed from the combustion chamber into the turbine inlet nozzle (24) (stationary vanes), through the turbine rotor (25), and then pass on to the exhaust diffuser. The front of the turbine housing is enclosed by the compressor housing (15), installed over the compressor rotor (12) and the inducer rotor (11). The compressor housing carries the front bearing (3) and the gear train which drives the various accessories at the intake end. The starter-generator and its drive (7) are mounted at the forward center of the accessory case (6) which is mounted to the front of the compressor housing.

Mounted on the accessory case are the fuel pump, oil pump, oil filter, and fuel control. The tachometer drive is taken from this area. Provisions are also available for mounting a hydraulic pump. (The pump is not supplied with the engine.) A gear train (5) serves to drive the various accessories mounted on the accessory case.

The turbine-shaft-assembly parts are machined. The blades of the compressor and inducer rotors are integral with the hub metal, while the turbine rotor blades are removable and replaceable. The turbine housing, turbine inlet nozzle, radial diffuser, and exhaust diffuser are of welded built-up construction, whereas the compressor housing, accessory case, its cover, oil-pump drive adaptor, starter-generator adaptor, axial diffuser, and rear-bearing housing, its supports, the various labyrinth seals, the front bearing cage, and the oil pump are machined items. The combustor shells and structure are of sheet metal construction in nature, but use high-temperature-resistant alloy metal.

AIRFLOW (See FIG. 24-4)

Air entering the inlet (9) flows past the three struts (10) in the compressor housing air passage to the inducer rotor (11). The inducer rotor whirls the air in the direction of rotation of the compressor rotor (12) just before the air enters the compressor rotor. In the compressor rotor, the air is flung outward radially at high velocity to pass through the radial diffuser assembly (14). The airflow is then turned 90° and directed through the axial diffuser assembly (16) which is parallel

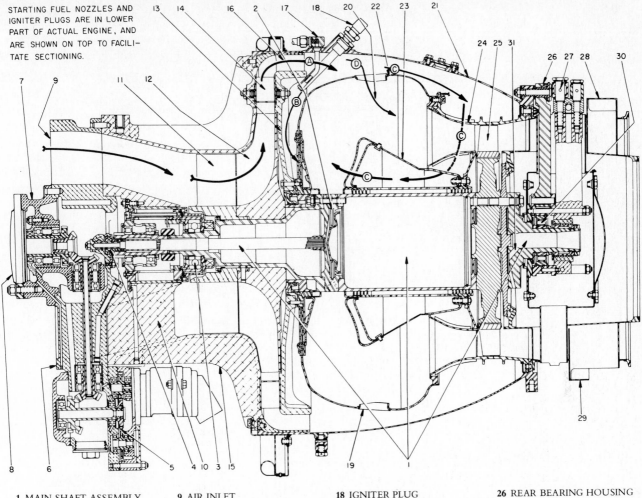

STARTING FUEL NOZZLES AND
IGNITER PLUGS ARE IN LOWER
PART OF ACTUAL ENGINE, AND
ARE SHOWN ON TOP TO FACILI-
TATE SECTIONING.

1 MAIN SHAFT ASSEMBLY	9 AIR INLET	18 IGNITER PLUG	26 REAR BEARING HOUSING
2 FUEL DISTRIBUTOR	10 COMPRESSOR HOUSING	19 AIR INLET TUBE	SUPPORT
3 FRONT BALL BEARING	STRUT	20 COMBUSTION CHAMBER	27 OIL PASSAGE
4 FUEL SEAL	11 INDUCER ROTOR	21 TURBINE HOUSING	28 STREAMLINE STRUT
5 ACCESSORY GEAR TRAIN	12 COMPRESSOR ROTOR	22 OUTER COMBUSTOR	29 EXHAUST DIFFUSER
6 ACCESSORY CASE	13 COMPRESSOR COVER	SHELL	30 REAR ROLLER BEARING
7 STARTER-GENERATOR	14 RADIAL DIFFUSER	23 INNER COMBUSTOR	31 TUBULAR AIR PASSAGE
DRIVE	15 COMPRESSOR HOUSING	SHELL	
8 STARTER-GENERATOR	16 AXIAL DIFFUSER	24 TURBINE INLET NOZZLE	
REPLACEMENT COVER	17 STARTING FUEL NOZZLE	25 TURBINE ROTOR	

FIG. 24-4 Sectional view. Note: The starting fuel nozzle (17) and the igniter plug
(18) are in lower part of engine, but are shown on top to facilitate sectioning.

to the turbine shaft assembly. After the reduction in velocity in the diffusers and the resultant increase in pressure, the air now enters the interior of the turbine housing at A.

Part B of the air coming out of the axial diffuser passes down behind the compressor cover (13) to enter the combustion chamber through the louvers in the outer combustor shell (22) just forward of the fuel distributor (2). These louvers are spaced and designed to provide thorough mixing of air and fuel. Part C of the incoming air passes around the outer combustor shell (22) of the combustion chamber, following its shape approximately, and is thereby led to the hollow vanes of the turbine inlet nozzle (24). This air cools the vanes while passing into the chamber formed by the inner combustor shell (23) and the turbine shaft assembly.

The front end of the inner combustor shell is perforated so the air C (heated in the chamber) flows into the combustion chamber just aft of the fuel distributor. The air following paths B and C is known as *primary* air and it is this air which forms the air-fuel mixture in the combustion chamber. As this mixture heats in the presence of the established flame, it ignites. The combustion chamber is approximately L-shaped in cross section. As the products of combustion turn the corner of the L, air inlet tubes (19) feed incoming air D directly into the flow. This dilution holds down peak temperatures. The air following path D is known as *secondary air*, since it does not directly enter the combustion process. The total volume now passes through the turbine inlet nozzle (cooled by primary air), to the turbine rotor (25) from which it passes on out the exhaust diffuser (29).

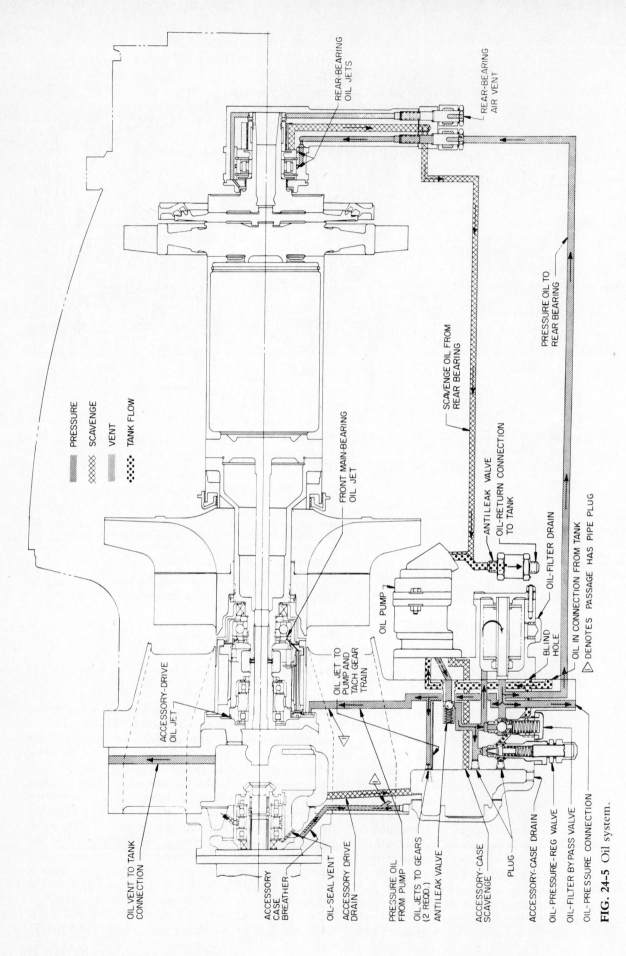

REAR-BEARING OIL JETS

REAR-BEARING AIR VENT

SCAVENGE OIL FROM REAR BEARING

PRESSURE OIL TO REAR BEARING

ANTILEAK VALVE

OIL-RETURN CONNECTION TO TANK

OIL-FILTER DRAIN

BLIND HOLE

OIL IN CONNECTION FROM TANK

△ **DENOTES PASSAGE HAS PIPE PLUG**

PRESSURE
SCAVENGE
VENT
TANK FLOW

FRONT MAIN-BEARING OIL JET

OIL JET TO PUMP AND TACH GEAR TRAIN

OIL PUMP

ACCESSORY-DRIVE OIL JET

OIL VENT TO TANK CONNECTION

ACCESSORY CASE BREATHER

OIL-SEAL VENT

ACCESSORY DRIVE DRAIN

PRESSURE OIL FROM PUMP

OIL JETS TO GEARS (2 REQD.)

ANTILEAK VALVE

ACCESSORY-CASE SCAVENGE

ACCESSORY-CASE DRAIN

OIL-PRESSURE-REG VALVE

PLUG

OIL-FILTER BYPASS VALVE

OIL-PRESSURE CONNECTION

FIG. 24-5 Oil system.

OIL SYSTEM (FIG. 24-5)

Oil from the engine oil tank (not supplied with the engine) is led to the main oil pump where the pressure section develops main oil pressure. The output of the pressure pump is led through an antileak valve to the main oil filter. This filter system incorporates a pressure-regulating arrangement as well as bypass provisions to pass oil beyond the filter element if it should become clogged. From the oil-filter output, oil for the rear bearing is carried by an external hose. Another external hose picks up return oil from the rear bearing to carry this oil to the rear-bearing scavenge section of the oil pump. The rear-bearing housing incorporates a vent passage as well as passages to feed the oil to and from the rear bearing. At the front of the engine, oil is led from the main oil-filter output through passages to the front bearings and front-end gears. Oil is also fed to the accessory gear train that fans out across the lower part of the compressor housing. Oil from the front bearings and upper gears drains down to the accessory case from which one scavenge section of the oil pump pulls return-oil. All scavenge sections of the oil pump lead return-oil back to the engine oil tank through an antileak valve. This valve prevents tank oil from draining back into the engine after engine shutdown. The front-end section is vented by a passage to the top of the upper gear housing.

FUEL FLOW (FIG. 24-6)

The aircraft fuel system includes the fuel tanks, booster pump(s), a fuel strainer, a shutoff valve, and fuel flowmeters in the flow of fuel up to the engine. The engine fuel system starts with the fuel pump. This pump, driven off the accessory gear train, has a centrifugal booster stage which is intended to provide boost pressure if the boost provisions in the aircraft system should fail. It also reduces vapor effects by raising total boost pressure. The centrifugal booster stage feeds two gear-pump pressure sections operating in parallel. Either section will provide full pressure and flow for the engine, and each section is independent of the other. From the main fuel pump, fuel is carried by a hose to the fuel filter built into the fuel-control unit. This filter incorporates two separate filtering elements with provisions to bypass the fuel if the elements should become clogged. A differential pressure indicator is installed in one of the filtering elements to sense the increase in back pressure as the elements become dirty. An easily read, calibrated button protrudes from the indicator to show the amount of back pressure. When the button is full out (40 to 45 psi [275.8 to 310.3 kPa]), the filters must be cleaned and the button reset. A manually operated flushing valve permits closing off the rest of the fuel system when reverse flushing of the filter is accomplished. This valve is at the filter outlet.

FUEL CONTROL

Within the fuel control there are two separate fuel paths.

1. From the flushing valve outlet, starting fuel is led through a starting-fuel filter to a pressure regulator, then to the starting-fuel solenoid valve. From this valve, starting fuel passes through adjustable bleed valves to the external piping that leads the fuel to the starting-fuel nozzles.

2. The main fuel path feeds to the acceleration control, to the governor valve, then to the cutoff valve from which flow is to the pressurizing valve and thence into the engine fuel tube. The acceleration control is designed to influence fuel input during acceleration and also to compensate for change of altitude or other ambient air conditions. The governor valve influences flow to hold the speed called for by the throttle-lever setting. The governor valve is servo-operated and responds to pressure signals developed in the speed-sensing element. The latter also sends pressure signals to the bypass valve. The function of the bypass valve is to maintain a design pressure differential across the metering elements (which are the acceleration control and the governor valve). This pressure differential is maintained by bypassing fuel back to the fuel-pump inlet. Since the design pressure differential must change with speed, the bypass valve is made responsive to a signal from the speed-sensing element. The pressurizing valve is designed to open only above a minimum pressure and so prevents "dribble" of fuel or drainage of the control unit when the engine comes to a stop.

The fuel control also contains check valves, "trim" provisions, and passages for return of fuel bleed-off or seepage.

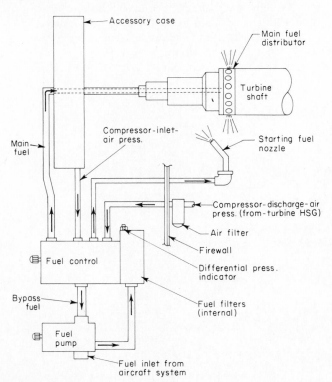

FIG. 24-6 Fuel system schematic.

ENGINE CONTROL

The fuel control is the key element affecting engine control. Provided the proper volume and pressure of fuel are fed into the fuel-control unit, it regulates and meters engine fuel input to cover all operating conditions automatically.

The following main conditions are controlled:

1. *Starting*—The separate starting-fuel path sets up fuel flow to the starting-fuel nozzles. The fuel-control starting-fuel solenoid valve opens this path and closes it in response to signals from a control element not supplied with the engine.
2. *Acceleration to idle*—The acceleration control sets up fuel flow to the main fuel distributor such as to speed the engine up from starting speed to idle without surge or overtemperature. As the engine reaches the speed set by the throttle-lever position, the governor valve will come into action to hold the speed as set.
3. *Above idle*—The acceleration control is designed to control fuel input for all changing conditions for all operations from idle up to full speed without allowing surge or overtemperature. It compensates for acceleration, for change of altitude, and for other changes of ambient air characteristics. The governor valve, in all cases, acts to hold engine speed to the value set by the throttle-lever position.

ELECTRICAL SYSTEM (*FIG. 24-7*)

The electrical system of the engine includes the starter-generator and its lead assembly; the ignition coil, igniter plugs, and their leads; the starting-fuel solenoid valve incorporated in the fuel control; the tachometer; and the accessory electrical lead assembly. The remaining elements that constitute the complete electrical system are provided by the aircraft manufacturer.

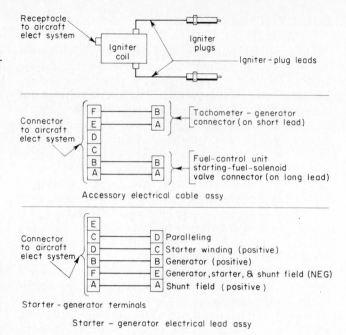

FIG. 24-7 Electrical system schematic.

REVIEW AND STUDY QUESTIONS

1. What airplane uses this engine?
2. List the engine's major specifications.
3. Give a brief description of the engine and its operation.
4. How many main bearings does this engine have?
5. Very briefly describe the construction features of the following parts: accessory case, compressor housing and rotor, diffuser, combustor liner, turbine housing, turbine and shaft, and exhaust diffuser.
6. What provision is made to reduce compressor stall?
7. Briefly describe the following systems: lubrication, ignition, fuel, and electrical.

CHAPTER TWENTY-FIVE
GENERAL ELECTRIC CF6

The General Electric CF6-6 and CF6-50, Fig. 25-1, are second-generation engines incorporating advanced technology in which the main design features are:

A high thrust-to-weight ratio
A significantly reduced specific fuel consumption
Simplicity of construction and improved materials
Advanced manufacturing techniques resulting in better maintainability and reliability
A lower initial cost
A lower operating cost
A built-in growth potential
A low noise level and smokeless operation

The basic design of the General Electric CF6 is based on the TF39 engine shown in Fig. 2-26. Comparative cross-section views of the TF39/CF6-6, and the CF6-6 and CF6-50 are shown in Fig. 25-2. The core for the two engines is nearly identical. The CF6-6 five-stage low-pressure (LP) turbine is similar in mechanical design to the six-stage TF39 turbine. The differences (five stages vs. six stages) are required to properly match the smaller CF6-6 fan.

The CF6-50 engine represents the growth version of the CF6. A 25 percent increase in takeoff thrust with the same frame size is achieved by increasing the overall pressure ratio of the fan component. A three-stage low pressure compressor is added and the pressure ratio for the fan, plus low-pressure compressor, is increased to 2.35. The overall compressor ratio (fan, low-pressure compressor, and high-pressure compressor) is increased from 24:1 to 30:1. Turbine inlet temperatures, to be consistent with the CF6-6, are maintained by reducing the number of high-pressure (HP) compressor stages from 16 to 14. In addition, this "high-flowing" of the

core provides more energy to drive the LP turbine than is necessary. Therefore, in order to maintain a proper work balance, the LP turbine stages are reduced from five to four.

SPECIFICATIONS (CF6-6)

Number of fan stages: 2
Number of compressor stages: 16
Number of turbine stages: 2 plus 5
Number of combustors: 1
Maximum power at sea level: 41,000 lbt
                                         [182,368 N]

Specific fuel consumption at   0.35 lb/lbt/h
  maximum power: [35.68 g/N/h]

Compressor ratio at maximum rpm: 24.7:1
Maximum diameter: 94 in [239 cm]
Maximum length: 188 in [478 cm]
Maximum dry weight: 7765 lb [3525 kg]

SPECIFICATIONS (CF6-50C)

Number of fan stages: 4
Number of compressor stages: 14
Number of turbine stages: 2 plus 4
Number of combustors: 1
Maximum power at sea level: 51,000 lbt
                                         [226,846 N]

Specific fuel consumption at   0.39 lb/lbt/h
  maximum power: [39.76 g/N/h]

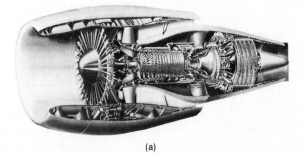

(a)

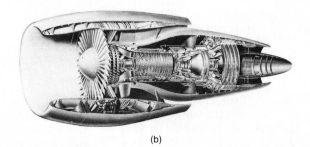

(b)

FIG. 25-1 (a) Cutaway view of the G.E. CF6-6 engine.
(b) Cutaway view of the G.E. CF6-50 engine.

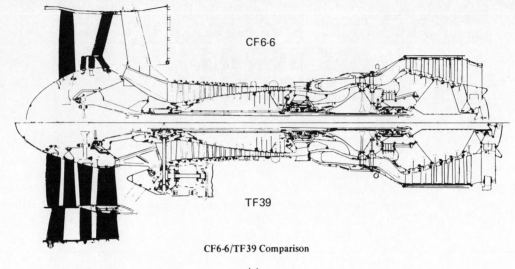

CF6-6

TF39

CF6-6/TF39 Comparison

(a)

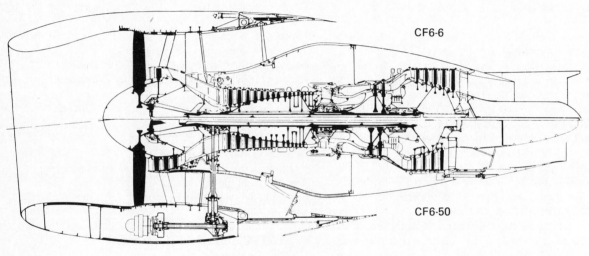

CF6-6

CF6-50

CF6-6 / CF6-50 Comparison

(b)

FIG. 25-2 (a and **b)** Comparison of the G.E. TF39, CF6-6, and CF6-50 engines.

Compressor ratio at maximum rpm:	29.5:1
Maximum diameter:	94 in [239 cm]
Maximum length:	183 in [465 cm]
Maximum dry weight:	8355 bb [3793 kg]

GENERAL DESCRIPTION

CF6-6 and CF6-50 series engines as used on the McDonnell Douglas DC-10 (-10 and -30) and the French Airbus A300B, are dual-rotor, high-bypass-ratio turbofan engines incorporating a variable-stator, high-pressure-ratio compressor, an annular combustor, an air-cooled core engine turbine, and a coaxial front fan with a low-pressure compressor driven by a low-pres-

sure turbine. In addition to a fan reverser, the core engine incorporates a turbine reverser to produce reverse thrust during landing roll (Fig. 25-3).

By incorporating minor cycle and structural changes and utilizing improved turbine cooling, the CF6-6 and CF6-50 engines have both been offered with two growth steps. The CF6-6D is the basic 40,000 lb [177,920 N] engine with planned growth to 43,000 lb [191,264 N]. Likewise, the CF6-50A is 49,000 lb [217,952 N] with growth to over 55,000 lb [244,640 N].

Engine Sections

Basically, the engine consists of a fan section, compressor section, combustion section, turbine section,

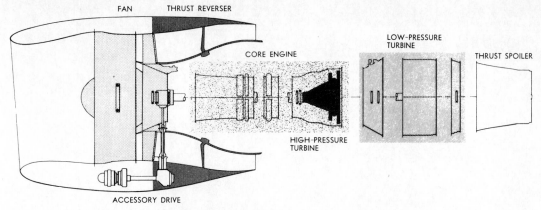

FIG. 25-3 The G.E. CF6 engine can be disassembled into engine maintenance units (EMUs). This modular design permits both sectionalized repair and overhaul either on the aircraft or at line maintenance stations.

and accessory-drive sections. These basic sections are shown in Fig. 25-4. The following present a general description of the engine by sections.

Fan—Thirty-eight wide-chord titanium blades, individually replaceable on the wing, provide tolerance to erosion and FOD. Controls and accessories are mounted on the fan case for improved maintenance and a cooler environment. The fan assembly for the CF6 engine is shown in Fig. 25-5.

High-pressure compressor—The single-rotor, high-pressure compressor has variable stator vanes for high efficiency and rapid accelerations, and it operates with a large stall margin

to avoid compressor stalls. The horizontal flange compressor casing design permits access to the compressor blades and vanes by removing a compressor casing half without complete engine disassembly. Borescope ports are provided for every stage of the compressor and every turbine stage as well.

Combustor—Long turbine life is made possible because the film-cooled annular combustor provides a uniform temperature distribution. Thirty fuel nozzles and air-mixing devices distribute the fuel evenly to give favorable temperature profiles with minimum carbon particle generation and no visible smoke. The combustor can be removed without

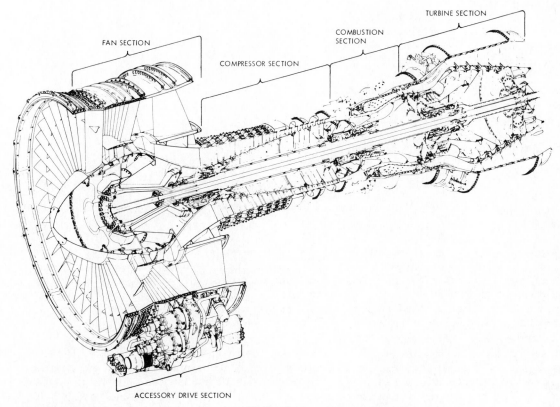

FIG. 25-4 CF6-6D engine cutaway showing the five basic sections.

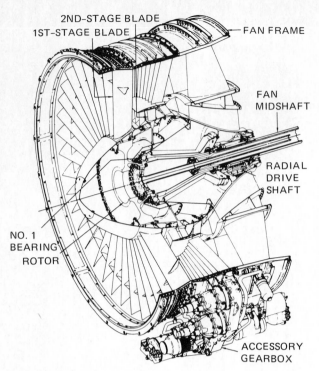

FIG. 25-5 The fan assembly for the CF6-6 engine.

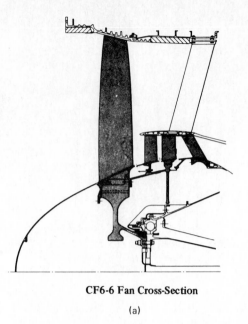

CF6-6 Fan Cross-Section

(a)

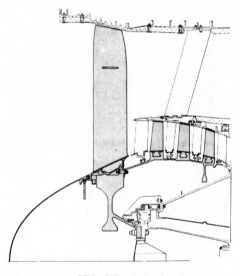

CF6-50 Fan Cross-Section

(b)

FIG. 25-6 One of the essential differences between the CF6-6 and CF6-50 engines.

disturbing the fuel nozzles, thus saving maintenance time. The combustor can also be inspected through six borescope ports.

High-pressure turbine—Turbine cooling technology is the key to the high performance levels and ruggedness of the engine. Development of the air-cooled turbine blade has produced reliable long-life blading with combinations of convection, impingement, and film cooling. The René 80 blade material is resistant to sulfidation and corrosion.

Low-pressure turbine—Turbine blades on the low-pressure turbine are shrouded for improved efficiency and ruggedness. The turbine rotor is supported between bearings mounted in the turbine mid and rear frames to control running stability and also to improve maintenance.

Accessories—The accessories, located on the engine fan casing, are grouped together and are optimized for easy maintenance.

FAN

The CF6-6 series fan assembly [Fig. 25-6(a)] is composed of a single-stage fan and a ¼ or "booster" stage which handles about 16 percent of the main fan flow aft of the fan first stage.

The CF6-50 series engines [Fig. 25-6(b)] utilizes a three-stage low-pressure compressor aft of the same fan in a space provided in the original design. The LP compressor supercharges the high pressure core so that it pumps 55 percent more air than in the basic design. Variable bypass valves are provided aft of the LP compressor to discharge air into the fan stream to establish proper flow matching between the low- and high-pressure spool during transient operation. The engine au-

tomatic control and bleed system provides good stall margin and engine handling characteristics.

There are no inlet guide vanes in the inlet of the full diameter fan stage. Canted outlet guide vanes are used to reduce swirl velocity of fan air downstream of the rotor to keep noise levels at a minimum.

Two bearings support the fan rotor assembly for tip clearance control and ease of maintenance. The forward bearing is a thrust bearing to provide greater safety in the event of shaft malfunction allowing the low-pressure turbine (LPT) rotor to move aft to engage with the stator thus avoiding hazardous overspeeding. The rear bearing is a roller bearing.

The CF6 fan consists of 38 titanium fan blades which have 21 or 22 drilled holes at the tip to reduce weight and to assure that critical system resonances occur outside the engine operating speed range. The rotor and stator have also been designed to minimize noise.

Fan Material

The fan rotor is made of forged titanium blades, titanium disks, aluminum platforms and spacers, and a forged and machined marage or B5F5 steel stub shaft. Conventional dovetail-type blade attachments modified to permit individual blade removal are used. These parts are shown in Figs. 25-7 and 25-8.

The rotor shaft, disk, and spacers are attached by close-fitting dowel bolts whose alignment and long-time rotor stability capabilities have been successfully proven in the TF39 engine fan and the J79 compressor.

FOD Resistance

The CF6-6D/-50A inlet systems are designed to provide optimum protection against foreign objects entering the main engine inlet duct. The ability to separate foreign objects is the result of:

1. The slope of the fan hub flow path in relationship to the main compressor inlet which makes it difficult for foreign objects to enter the main compressor inlet duct.
2. The rotation of the fan, which imparts a force on the foreign objects and centrifuges the objects toward the outside of the fan flow paths. This centrifugal force, due to the angular velocity imparted by the fan blades, forces solid particles to travel outward across the streamline to a point where very few are captured by the high-pressure compressor inlet even if the particles are introduced at the root of the fan inlet blades.

Water, ice, birds, etc. have been ingested by the CF6 with no hazardous or unreliable consequences to the engine.

CF6 COMPRESSOR

The CF6-6D HP compressor is a 16-stage, 16.8:1-pressure-ratio component at the design point (35,000 ft [10,668 m], Mach = 0.85). The pressure ratio is 15:1 at S.L. takeoff and 84°F [32.5°C]. The aerodynamic efficiency is greater than 86 percent and it has ample stall margin (over 20 percent) with a corresponding broad tolerance to inlet distortion during takeoff, landing, and reverser operation. The takeoff rating corrected airflow is 131 lb/s [59.5 kg/s] (185 lb/s [84 kg/s] physical airflow.)

The CF6-50A HP compressor is aerodynamically and structurally similar to the CF6-6. The compressor contains fourteen stages and provides (at sea level, static, 84°F) a corrected airflow at takeoff power of 133 lb/s [60 kg/s] (266 lb/s [120.8 kg/s] physical airflow) at 12:1 pressure ratio (13.0 at the design point).

All of the CF6 HP compressors are designed for an

operational life of 30,000 hours. The HP compressor incorporates the following features:

1. Split casings for easy removal for inspection, repair, or blade replacement.
2. All stages of stator vanes and rotor blades can be replaced by removal of a compressor casing half.
3. Corrosion-resistant alloys—titanium, A286, and Inco 718 are used throughout the compressor.
4. The rotor has a minimum of bolted joints, and these are joined with closefitting rabbets for excellent rotor stability and low vibration.

The HP compressor uses variable stators on the inlet guide vanes and first six stator rows to maintain a generous stall margin over the entire operating range. This gives the engine good transient characteristics and makes it tolerant to inlet distortion.

The first stage hub radius ratio of 0.48 was used to permit a high flow per frontal area, resulting in a smaller-diameter, lightweight compressor. At the same time, the hub radius ratio is sufficiently high to prevent hub stalling tendencies at the low corrected speeds where the front stages have to operate at higher aerodynamic loadings.

Since efficiency is strongly a function of tip clearance/blade height ratio, the low radius ratio is beneficial in obtaining high efficiency because the blades are longer than for a high-radius-ratio compressor having the same flow.

The aerodynamic blade and vane loadings are low in the front stages. The loadings gradually increase through the first stages to a moderate level and then remain about constant through the remaining stages. The loading level of the last stage is slightly higher than the rest of the compressor, but well within the limits of demonstrated good operation.

The moderate aerodynamic loadings of the compressor are conducive to good stall margin and high efficiency and enhance the ability of the HP compressor to tolerate inlet distortions.

The HP compressor rotor (Fig. 25-9) is a combined spool and disk structure utilizing axial dovetails (stages 1 and 2) and circumferential dovetails for the remaining stages. Compressor rotor blade installation is shown in Figs. 25-10 and 25-11.

Compressor Material

Blading materials for the CF6-6 series are 6-4 titanium for stages 1 through 12, 6-2-4-2 titanium for stages 13 and 14, and A286 for stages 15 and 16. On the CF6-50 series, blading materials are 6-4 titanium for stages 1 through 7, 6-2-4-2 titanium for stages 8 and 9, and A286 steel for stages 10 through 14. The rotor is designed so that individual blades can be replaced without rotor disassembly.

The 14-stage spool/disk structure on the CF6-50 series and the 16-stage spool/disk on the CF6-6 series consists of only seven major elements connected with three rabbeted flange joints. Forward disks are made of titanium and rear disks are made of steel.

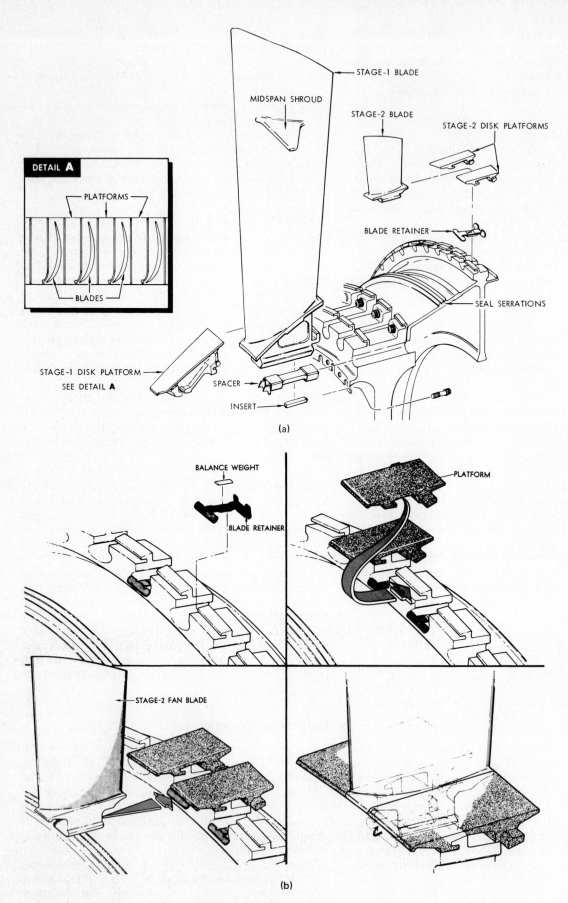

FIG. 25-7 **(a)** Stage 1 fan blade removal and installation.
(b) Stage 2 fan blade removal and installation.

474 REPRESENTATIVE ENGINES

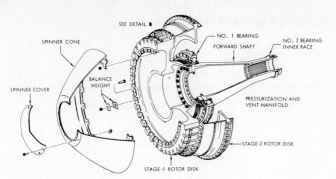

FIG. 25-8 Fan rotor for the CF6-6 engine.

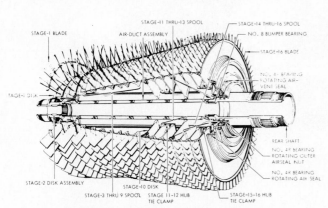

FIG. 25-9 The CF6-6 compressor rotor assembly.

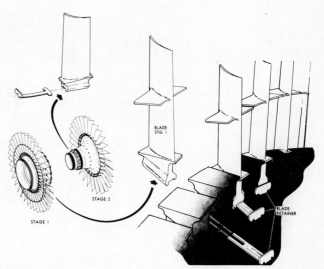

FIG. 25-10 Compressor rotor blade installation for stages 1 and 2 (axial dovetails).

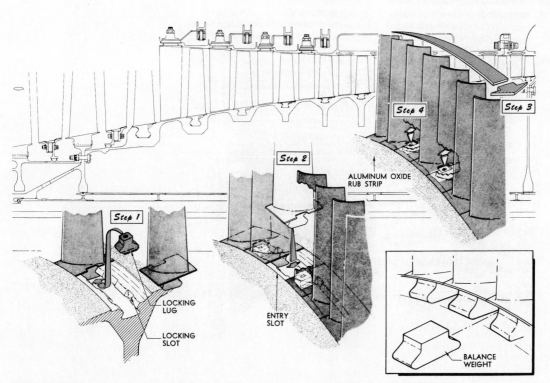

FIG. 25-11 Compressor rotor blade installation for stage 3 and onward (circumferential dovetails).

Compressor Stator

The CF6-50 series HP compressor stator comprises one stage of inlet guide vanes (IGVs) and 14 stages of stator vanes. The IGVs and stages 1 through 6 are variable. The CF6-6 series engine is basically the same construction but with 16 stages of stator vanes. Stator vanes are also made from titanium and steel.

Casing Manifold Systems

On the CF6-50 series engines, stage 8 customer air (Fig. 25-12) is bled from the inside diameter of the air passage through hollow stage 8 stator vanes, through the vane bases, and then through round holes in the casing skins into a pair of manifolds. Engine air is extracted at stage 7 for turbine midframe cooling. The air

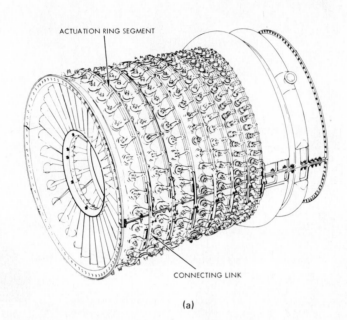

(a)

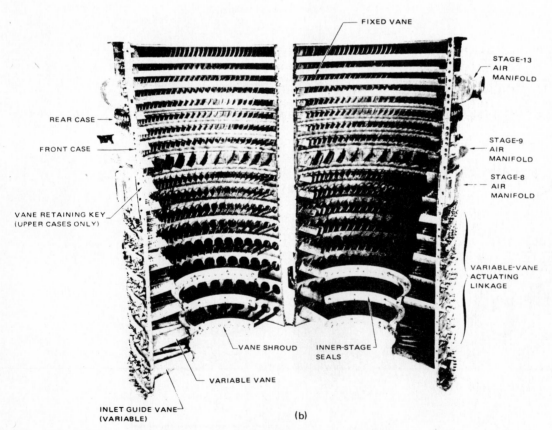

(b)

FIG. 25-12 **(a)** Assembled view of the front stator assembly.
(b) Compressor front and rear casing components split along the parting surface.

passes through semicircular slots in adjacent stage 7 vane bases, and through round holes in the casing skins to the manifold.

Engine air is also extracted at the stage 10 stator for the second-stage high-pressure turbine cooling and nose cowl anti-icing. Similarly, this air passes through semicircular slots in the adjacent stage 10 vane base, and through round holes in the casing skins to a pair of manifolds. These manifolds are welded to casing circumferential ribs at the stage 10 stators.

On the CF6-6 series engines the system function is the same with the exception that stage 9 is used instead of stage 7, and stage 13 is used instead of stage 10 to extract engine air for cooling and nose cowl anti-icing. Stage 8 continues to extract customer bleed air.

Three separate bleed manifolds are designed integrally with the casing to supply engine cooling air as well as airframe cabin-conditioning air. Fig. 25-13 shows compressor stage 8 bleed points.

Compressor Casing Design

The variable IGVs and stages 1 through 6 are mounted in the conical portion of the forward casing. The variable-vane bearing seats are formed by radial holes and counterbores through circumferential sup-

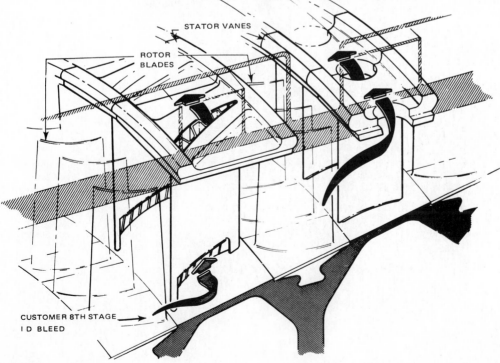

FIG. 25-13 Compressor stator air extraction.

(a)

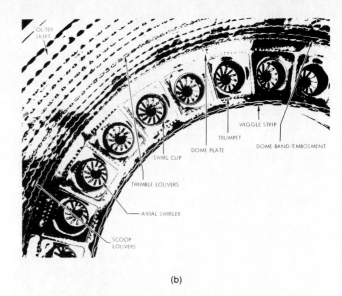

OUTER
SKIRT

SWIRL CUP

THIMBLE LOUVERS

AXIAL SWIRLER

SCOOP
LOUVERS

DOME PLATE

TRUMPET

WIGGLE STRIP

DOME BAND /EMBOSMENT

(b)

FIG. 25-14 (a) The CF6-6 combustor liner.
(b) Construction details of the CF6-6 combustor liner.

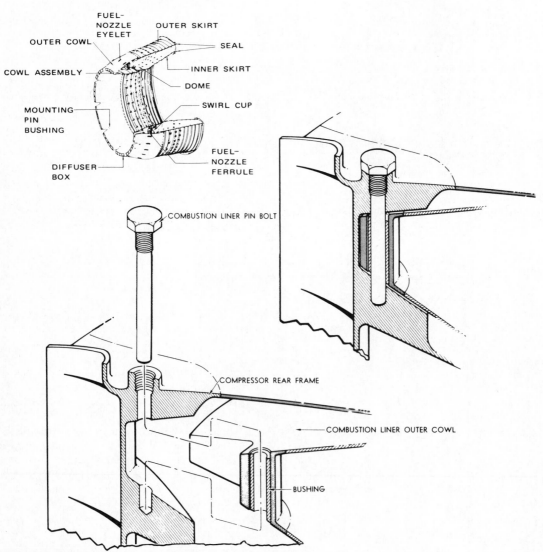

FUEL-
NOZZLE
EYELET

OUTER COWL

COWL ASSEMBLY

MOUNTING
PIN
BUSHING

DIFFUSER
BOX

OUTER SKIRT

SEAL

INNER SKIRT

DOME

SWIRL CUP

FUEL-
NOZZLE
FERRULE

COMBUSTION LINER PIN BOLT

COMPRESSOR REAR FRAME

COMBUSTION LINER OUTER COWL

BUSHING

FIG. 25-15 Combustion liner installation.

porting ribs. These are similar to those used on the variable stators of other General Electric engines. The fixed stages 7 through 11 are mounted in the cylindrical portion of the forward casing. The fixed base vanes are forged in one piece and are mounted in circumferential tracks.

On the aft stator casing of the CF6-6, fixed vane stages 12 through 15, and on the CF6-50, fixed vane stages 12 through 13 and outlet guide vanes (OGVs), are also mounted in circumferential tracks in the casing. The vane track slots are tapered like the slots in the front casing. The aft casing material is an Inco 718 forging, chosen for its good machining characteristics, excellent strength in this temperature range, good weldability, and high resistance to corrosion.

All fixed-base-vane track slots are designed so that the vanes cannot be improperly assembled. Also, vanes of similar size and appearance from stage to stage have noninterchangeable bases to prevent incorrect assembly. Both the forward and aft stator casings can be removed from the engine.

COMBUSTOR

The CF6 annular combustor provides efficient, smokeless operation over the entire spectrum of engine operating conditions. The combustor design permits uniform turbine-inlet-temperature distribution and minimum pressure losses, reliable relight characteristics, and long life.

The combustor (Fig. 25-14) consists of four sections which are riveted together into one unit and spot welded to prevent rivet loss: the cowl (diffuser) assembly, the dome, the inner skirt, and the outer skirt.

The unit fits around the compressor rear frame struts where it is mounted at the cowl assembly by 10 equally spaced radial mounting pins.

Cowl Assembly

The cowl assembly is designed to provide the diffuser action required to establish uniform flow profiles to the combustion liner in spite of irregular flow profiles which might exist in the compressor discharge air. Forty box sections welded to the cowl walls form the aerodynamic diffuser elements as well as a truss structure to provide the strength and stability of the cowl ring section. The combustor mounting pins (Fig. 25-15) are completely enclosed in the compressor rear frame struts and do not impose any drag losses in the diffuser passage. Mounting the combustor at the cowl assembly provides control of diffuser dimensions and eliminates changes in the diffuser flow pattern due to axial thermal growth.

Combustor Dome

The dome contains 30 vortex-inducing axial swirler cups, one for each fuel nozzle. The swirl cups are designed to provide the airflow patterns required for flame stabilization and proper fuel-air mixing. The dome de-

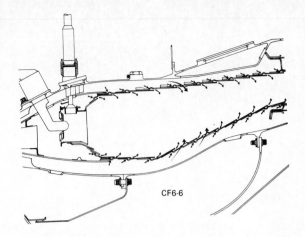

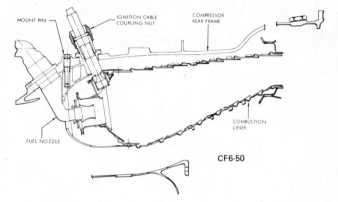

FIG. 25-16 Combustion liner for the CF6-6 and -50 engines.

sign and swirl cup geometry, coupled with the fuel nozzle design, are the leading factors which contribute to the CF6 smokeless combustion achievement. The axial swirlers serve to lean out the fuel-air mixture in the primary zone of the combustor which helps eliminate the formation of the high-carbon visible smoke which normally results from overrich burning in this zone. The dome is continuously film cooled. The cooling flow path is shown in Fig. 25-16.

Combustor Skirts

The inner and outer skirts, shown in Fig. 25-17, comprise the combustion liners of the unit. Each skirt consists of a series of circumferentially stacked rings which are joined by resistance-weld and brazed joints. The liners are continuously film-cooled by primary combustion air which enters each ring through closely spaced circumferential holes. The primary zone hole pattern is designed to admit the balance of the primary combustion air and to augment the recirculation for flame stabilization. Three axial planes of dilution holes on the outer skirt and five planes on the inner skirt are employed to promote additional mixing and to lower the gas temperature at the turbine inlet. Combustion liner/turbine nozzle air seals provided on the trailing edge of the skirt allow for thermal growth and accommodate manufacturing tolerances. The seals are coated with wear resistant material.

In addition to the smokeless combustion achieve-

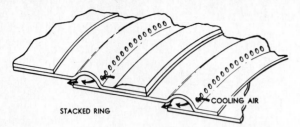

FIG. 25-17 Combustion liner cooling configuration.

ment, the CF6 combustor system represents accomplishments in aerodynamic and mechanical design.

The principal aerodynamic achievements lie in the combustor-inlet-diffuser design. Large-radius turns and smooth wall contour are utilized to minimize diffuser total pressure losses and to provide uniform and consistent exit flow patterns. The diffuser design concept avoids flow separation during normal operating conditions, and during conditions when the nominal flow patterns are disrupted by large compressor bleed extractions or when the compressor-discharge-velocity profile changes. Variations in circumferential flow patterns due to compressor rear frame strut wakes are minimized by design of the strut profiles, cowl leading edges, and combustor passage contours.

The mechanical design stresses combustor durability while still satisfying the high performance requirements of the combustion system. The emphasis on durability is most prominent with the selection of the annular combustor for the CF6. The annular combustor is inherently easier to cool than is the cannular design since there is less metal surface to cool.

A preferential cooling scheme is used to provide continuous cooling-air distribution to the combustor liner. This scheme minimizes circumferential temperature gradients by incorporating larger circumferential cooling-air holes in those areas where component tests have revealed higher temperatures.

The cooling slot design also imparts resistance to vibration and thermal fatigue. The combustor material has excellent oxidation resistance and high ductility combined with high temperature strength.

Dimple-type supports have been utilized on cooling slots to eliminate slot distortion. Stiffening bands have been added to the outer skirt for stability. .

HIGH-PRESSURE TURBINE

The high-pressure turbine is a two-stage, high-inlet-temperature, air-cooled, high-efficiency turbine. The high-pressure turbine extracts energy to drive the high-pressure compressor. Fig. 25-18 shows the high-pressure turbine.

Design features include:

1. Complete structure cooled with continuous flow of compressor-discharge air.
2. Internal convective cooling for both stages of blades. Additional film and impingement cooling are provided for first-stage blades.

FIG. 25-18 The high-pressure turbine rotor.

3. First-stage nozzle vanes employ convective and film cooling. CF6-6 second-stage vanes are convective cooled; CF6-50 second-stage vanes are impingement cooled.
4. CF6-6 blades are paired with long shanks to provide thermal isolation of dovetails, airflow paths, and low disk-rim temperatures.
5. CF6-50 blades are single blades with cast internal cooling passages in the airfoil shank and dovetail.
6. CF6-50 has a lightweight damping system.
7. Smooth abrasion-tolerant shrouds used for high turbine efficiency.
8. Blade material selected for high resistance to hot corrosion, good ductility, and fatigue strength.

First-stage High-pressure Nozzle Assembly

The first-stage high-pressure nozzle (Figs. 25-19 and 25-20) is air cooled by convection, impingement, and film cooling. The purpose of the nozzle is to direct high-pressure gases from the combustor onto the first-stage high-pressure turbine blades at the proper angle and velocity.

The major components of the nozzle assembly, as shown in Fig. 25-19, are the nozzle support (1), pressure balance seal support (8), vanes (3), and inner (4) and outer seals (2). The nozzle is bolted at its inner diameter to the first-stage support and receives axial support at its outer diameter from the second-stage nozzle support.

The first-stage nozzle support is a sheet-metal and machined-ring weldment. In addition to supporting the first-stage nozzle, it forms the inner flow path wall from the compressor rear frame to the nozzle.

First-stage vanes (Fig. 25-21) are coated to improve erosion and oxidation resistance. The vanes are cast individually and welded into pairs to decrease the number of gas leakage paths and to reduce the time required for field replacement.

These welds are partial penetration welds (50 percent) to allow easy separation of the two vanes for repair and replacement of individual halves. The vanes are cooled

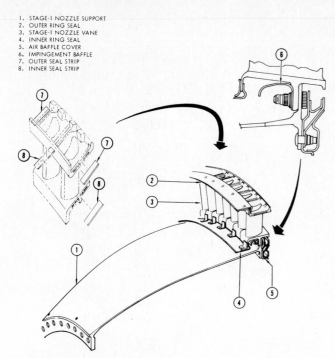

1. STAGE-1 NOZZLE SUPPORT
2. OUTER RING SEAL
3. STAGE-1 NOZZLE VANE
4. INNER RING SEAL
5. AIR BAFFLE COVER
6. IMPINGEMENT BAFFLE
7. OUTER SEAL STRIP
8. INNER SEAL STRIP

FIG. 25-19 The stage 1 high-pressure turbine nozzle.

by compressor discharge air which flows through a series of leading edge holes and gill holes located close to the leading edge on each side. Air flowing from these holes forms a thin film of cool air over the length of the vane.

Internally, vanes are divided into two cavities. Air flowing into the aft cavity of the CF6-6 is discharged through trailing edge slots. Aft cavity air exits the CF6-50 vanes through trailing edge slots and a row of pressure side holes.

Second-stage High-pressure-turbine Nozzle Assembly

The purpose of the nozzle is to direct high-pressure gases exiting from stage 1 blades onto stage 2 turbine blades at the proper angle and velocity. The CF6-6 vanes are convectively cooled with 13th-stage compressor air. The CF6-50 stage 2 vanes are impingement cooled.

The major components of the second-stage high-pressure nozzle assembly (Fig. 25-22) are the second-stage nozzle vane segments (7), nozzle support (1), first (2) and second (6) stage turbine shrouds and the interstage seal (8).

The nozzle support is a conical ring, and is bolted rigidly between the flanges of the compressor rear frame and the turbine midframe. The support mounts the nozzle vane segments, cooling air feeder tubes, and the first- and second-stage turbine shrouds.

The CF6-6 nozzle vane leading edges (Fig. 25-23) are cooled by internal impingement air (13th stage) which enters through the cooling-air tubes. This air is then used for convective cooling of the midchord region. The CF6-50 vane leading edge, pressure and suction side walls are impingement cooled. Both vanes discharge air through the trailing edge and into the stage

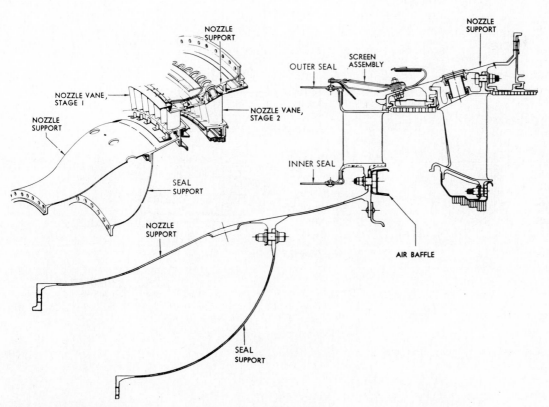

FIG. 25-20 The high-pressure turbine stator assembly stackup.

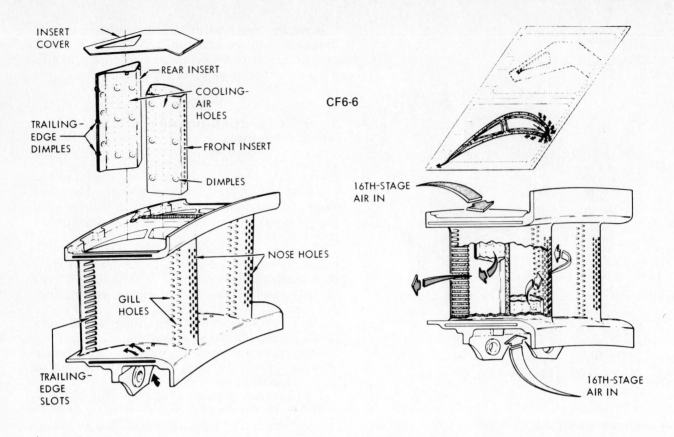

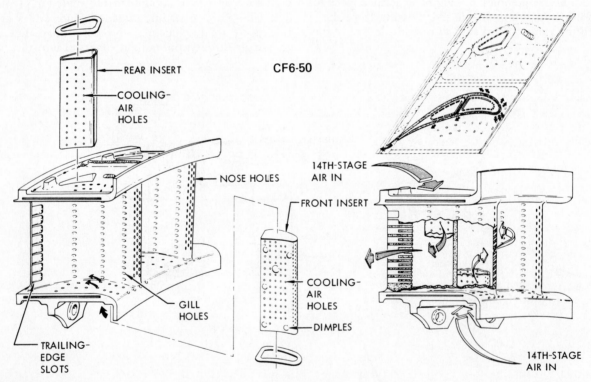

FIG. 25-21 The stage 1 high-pressure turbine nozzle vane for the CF6-6, and -50 engines.

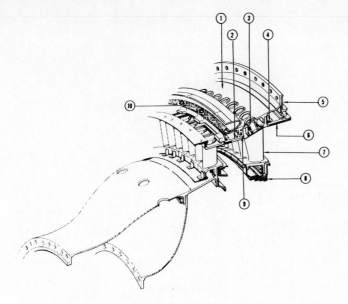

1 STAGE-2 NOZZLE SUPPORT
2 STAGE-1 SHROUD (24 SEG-
   MENTS)
3 COOLING-AIR-FEED TUBES
   (66)
4 STAGE-2 SHROUD FOR-
   WARD SUPPORT (11 SEG-
   MENTS)
5 STAGE-2 SHROUD REAR
   SUPPORT

6 STAGE-2 SHROUD (11 SEG-
   MENTS)
7 STAGE-2 NOZZLE VANE (33
   PAIRS)
8 INTERSTAGE AIR SEAL (6
   SEGMENTS)
9 STAGE 1 SHROUD REAR
   SUPPORT (24 SEGMENTS)
10. STAGE 2 NOZZLE SUP-
   PORT AIR FILTER

FIG. 25-22 The stage 2 high-pressure-turbine nozzle assembly. Stage 1 is also shown for clarity.

1 aft wheel space for interstage seal leakage and rotor cooling.

The nozzle segments are cast and then coated. The vanes (two per segment) direct the gas stream onto the second-stage turbine blades. The inner ends of the segments form a mounting circle for the innerstage seal attachment.

The innerstage seal is composed of six segments of approximately 60° which bolt to the vane segments. The function of the seal is to minimize the leakage of gases between the inside diameter of the second stage nozzle and the turbine rotor. The sealing diameter has four consecutive steps for maximum effectiveness of each sealing tooth. The seal backing material and the open-faced honeycomb sealing surface are made of Hastelloy X.

The turbine shrouds form a portion of the outer aerodynamic flow path through the turbine. The shrouds are located axially in line with the turbine blades and form a pressure seal to prevent high-pressure-gas leakage or bypass at the blade tip end. The sealing (rubbing) surfaces are a nickel aluminide compound. Retention of the compound is accomplished with integrally cast pegs in the first stage and brazed-on honeycomb in the second stage. The first stage consists of 24 segments; the second stage consists of 11 segments.

High-pressure-turbine Rotor Assembly

The high-pressure-turbine rotor (Fig. 25-24) consists of a conical forward turbine shaft, two turbine disks, two stages of turbine blades, a conical turbine rotor spacer, a catenary-shaped thermal shield, aft stub shaft, and precision fit rotor bolts. The rotor is cooled by a continuous flow of compressor discharge air drawn from holes in the nozzle support. This flow cools the inside of the rotor and both disks before passing between the paired dovetails and out to the blades.

The conical forward turbine shaft transmits energy to the compressor. Torque is transmitted through the female spline at the forward end of the shaft. Two seals attached to the shaft at the forward end maintain compressor discharge pressure in the rotor/combustion chamber plenum to furnish part of the corrective force necessary to minimize the unbalanced thrust load on the high-pressure-rotor thrust bearing. The inner rabbet diameter on the rear flange provides positive radial location for the stage 1 retainer and a face seal for the rotor internal cooling air. The outer rabbet diameter on the flange provides radial location for the stage 1 disk and stability for the rotor assembly.

The turbine rotor spacer is a cone which serves as the structural support member between the turbine disks and transmits the torque from stage 2. (See Figs. 25-24 and 25-25.)

The CF6-6 blades (108 in stage 1; 116 in stage 2) (Fig. 25-26) are brazed together in pairs with side-rail doublers added for structural integrity. CF6-50 blades (80 in stage 1; 74 in stage 2) are single-shank blades with integral case shank cooling. Channel-shaped squealer tip caps are inserted into the blade tips and held by crimping the blade tip and brazing. In both engines, both stages of blades are cooled by compressor discharge air. Stage 1 blade cooling is a combination of internal convective, impingement, and external film cooling. The convective cooling of the midchord region is accomplished in serpentine passages. The leading edge circuit provides internal convective cooling by impingement of air against the inside surface and by flow through the leading edge and gill holes. Convective cooling of the trailing edge is provided by air flowing through the trailing-edge exit holes. Stage 2 blades (Fig. 25-27) are entirely cooled by convection. All of the cooling air is discharged at the blade tip.

The disk rim incorporates local bosses around the rim bolt holes on both sides of each stage to provide resistance to low-cycle fatigue. The bottom tang and bottom of the slot are cooled by compressor discharge air. The rabbeted construction provides the required rotor alignment and eliminates the need for close-tolerance bolt/hole design.

The catenary-shaped thermal shield contains turbine-rotor cooling air and provides the rotating portion of the interstage seal. The seal is rabbeted to the rotor structure to reduce bending stress at the flange neck.

The rear shaft, which bolts to the second-stage disk, supports the aft end of the turbine rotor. The shaft incorporates integral air seals.

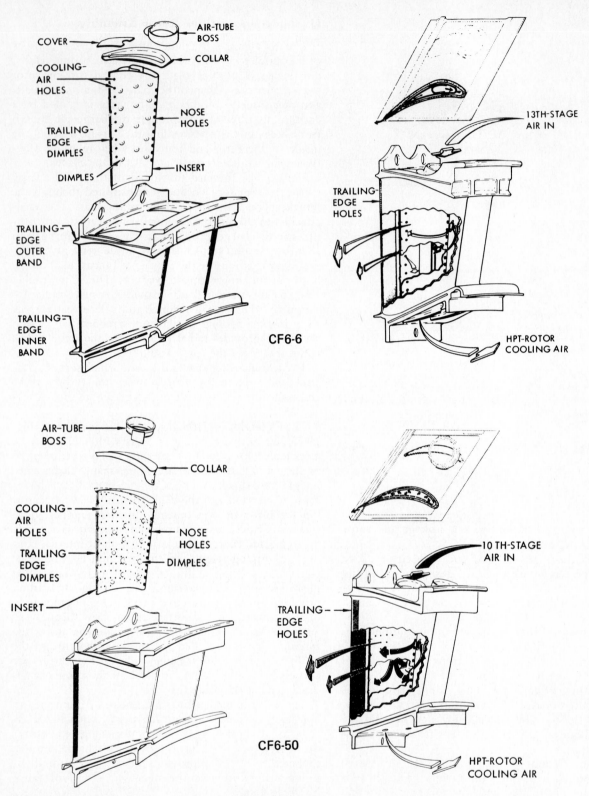

COVER

AIR-TUBE BOSS

COLLAR

COOLING-AIR HOLES

NOSE HOLES

TRAILING-EDGE DIMPLES

DIMPLES

INSERT

TRAILING-EDGE OUTER BAND

TRAILING-EDGE INNER BAND

CF6-6

13TH-STAGE AIR IN

TRAILING-EDGE HOLES

HPT-ROTOR COOLING AIR

AIR-TUBE BOSS

COLLAR

COOLING-AIR HOLES

NOSE HOLES

TRAILING EDGE DIMPLES

DIMPLES

INSERT

CF6-50

10 TH-STAGE AIR IN

TRAILING-EDGE HOLES

HPT-ROTOR COOLING AIR

FIG. 25-23 The high-pressure turbine stage 2 nozzle vanes for the CF6-6, and -50 engines.

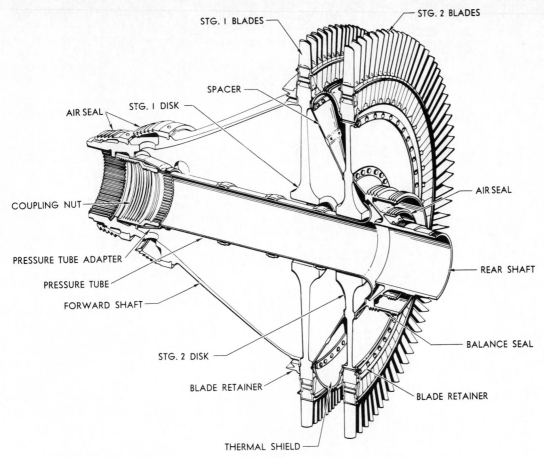

FIG. 25-24 The high-pressure turbine rotor.

The blade retainers serve two primary functions: they prevent the blades from translating axially under gas and maneuver loads, and they seal the forward face of the first-stage rim dovetail and the aft face of the second-stage rim dovetail from the leakage of cooling air. An additional function is to cover the rotor bolt ends at the rotor rim thus preventing a substantial drag loss. These retainers are a single piece and are held on by the same bolts that attach the forward shaft and thermal shield to the turbine disks.

The pressure tube serves to separate the high-pressure-rotor internal cooling-air supply from the region of the fan midshaft which is concentric to the rotor. It is threaded into the front shaft and bolted to the rear shaft.

Turbine Materials

René 80, used on stages 1 and 2 blades and stage 2 vanes, is a G.E.-developed nickel-base alloy which has improved high-temperature strength from both a stress rupture and cyclic life standpoint. This alloy improves the metal temperature capability of the stage 1 bucket by 70°F [38.8°C] metal temperature, and will, with the highly effective cooling system, improve the gas temperature capability by 140°F [77.8°C] over the previously used René 77 material.

The primary rotor structure is Inco 718. This nickel-base alloy provides strength and ductility to metal temperatures in excess of 1200°F [648.9°C]. The catenary, or heat shield, between the stage 1 and 2 turbine disks is René 41. This alloy was chosen because of its higher temperature capabilities since there is a possibility that it could be exposed to higher temperatures than the disk. The basic stator structure is also Inco 718. The stage 1 nozzle vane is cast of X-40, a cobalt-base alloy.

Experience with aircraft engines has shown that life is enhanced when the airfoil surfaces have protective oxidation and corrosion coatings. The blades and vanes are coated with Codep (General Electric trademark). Codep is one of a series of G.E.-developed coatings used specifically for blades and vanes. This is a coating that can be used and applied by the airlines on General Electric parts.

All flange surfaces and dovetail attachments in the high-pressure turbine are shot-peened to provide fretting resistance and improved cyclic life capability. The turbine shroud has a Bradalloy rubbing surface. Bradalloy has a microballoon structure of nickel and aluminum. This provides a smooth, low-aerodynamic-loss surface to the gas stream which can sustain blade rubs without loss of capability.

Turbine Cooling

Convective, film, and impingement cooling and combinations of these three cooling methods are used in this engine.

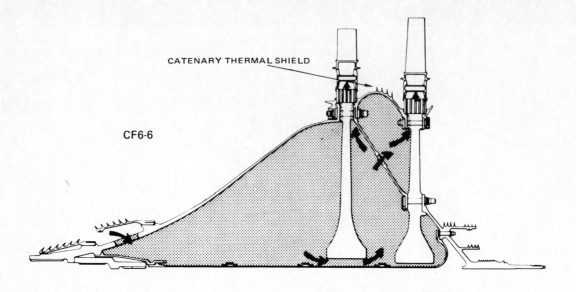

CATENARY THERMAL SHIELD

CF6-6

CATENARY THERMAL SHIELD

CF6-50

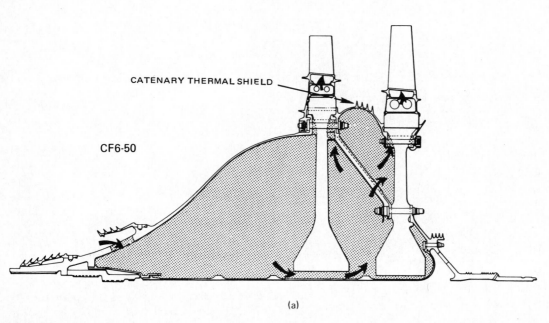

(a)

FIG. 25-25 (a) Cross section of the high-pressure turbine for the CF6-6, and -50 engines. (Continued on the following page.)

Turbine cooling has been limited in the past by the ability to manufacture blades and vanes with advanced cooling systems. Simple convective systems were first used in vanes. These were hollow sheet metal or castings with baffles inserted to increase the velocity so some small amount of cooling could be realized.

The advent of the shaped-tube electrolytic machining (STEM, General Electric trademark) process permitted the drilling of small holes in the walls of blades and vanes. The velocities of the coolant through the vane were then higher and the amount of convective cooling increased substantially.

Film insulation is a highly effective means of cooling.

The STEM process was first applied to film-cooled vanes in production engines to solve problems associated with peak local combustor temperatures in engines which did not have rotor cooling capability. Film passages were put into the leading- and trailing-edge regions. These designs have been in military and commercial service (CJ805, CJ610, CF700) for a number of years.

Electrostream (General Electric trademark) drilling made the adoption of film-cooling principles to rotor blades a practical step. A number of advanced engines take advantage of film cooling in the rotor. This has proven to be a reliable system which provides substan-

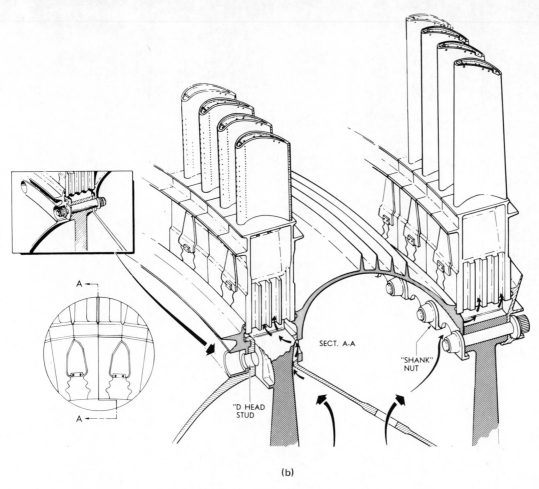

SECT. A-A

"SHANK" NUT

"D HEAD STUD"

(b)

FIG. 25-25 (cont.) **(b)** Cooling airflow detail for the high-pressure turbine used in the CF6-6 engine.

tial life improvement and the reduced cooling flows permitted by film cooling provides improved engine performance as well.

Dirt Ingestion

HP compressor discharge air and bleed air are used to cool the vanes and blades of the high-pressure turbine.

The HP compressor discharge air flowing around the outside and inside diameter of the combustor is used to cool the first-stage nozzle guide vane. Contaminants in the outside diameter air are filtered by a 10-mil [0.254-mm] screen wrapped around the case above the vane. Smaller contaminants are allowed to flow through this screen and are discharged through the trailing-edge cooling holes that are of sufficient size to easily pass these contaminants.

The HP compressor discharge air flowing around the inside diameter of the combustor will be measurably cleaner but, additional steps have been taken to provide that the cooling air to the rotating blades has contaminants of very low micron rating. The flow path takes

two 180° turns before entering the "static spiral separator" adjacent to the rotating HP turbine shaft. The "separator" is a series of small nozzles that swirl the cooling flow in the direction of rotation and, as a result, any contaminant particles are centrifuged to the outside of the stream where small holes are provided to duct them overboard before passing into the primary rotor structure. These turns and the separator reduce the contaminant levels significantly. The contaminants that are discharged through the separator, reenter the gas stream in front of the first-stage rotor. In its passage from the separator to the individual blades, the air passes through a number of dirt traps which utilize the centrifugal field to collect and disperse any particles before they reach the blades. The cooling air to the blades must now flow under the first-stage rotor. The centrifugal field of the rotor accelerates the cooling air to the same velocity as the first stage blades in order to gain maximum cooling effectiveness and then enters the root of the blades. A portion of this same cooling air is directed to the root of the second-stage blades. This tortuous flow path coupled with the dust bleed holes in the blade tips prevents plugging of the turbine cooling holes.

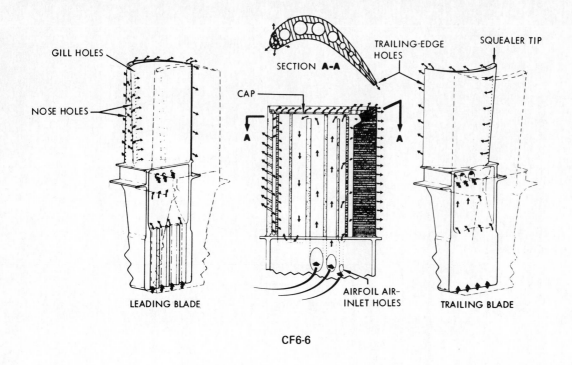

GILL HOLES

NOSE HOLES

SECTION **A-A**

TRAILING-EDGE HOLES

SQUEALER TIP

CAP

A

A

LEADING BLADE

AIRFOIL AIR-INLET HOLES

TRAILING BLADE

CF6-6

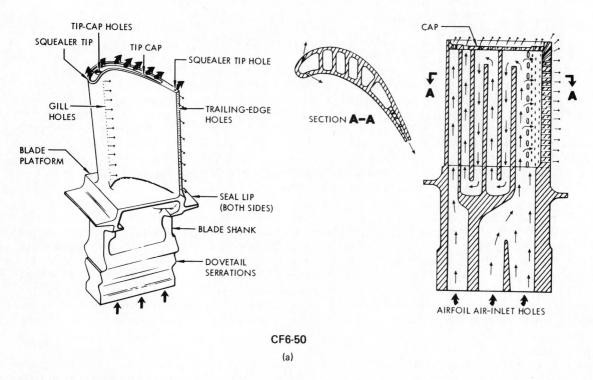

TIP-CAP HOLES

SQUEALER TIP

TIP CAP

SQUEALER TIP HOLE

GILL HOLES

TRAILING-EDGE HOLES

BLADE PLATFORM

SEAL LIP (BOTH SIDES)

BLADE SHANK

DOVETAIL SERRATIONS

SECTION **A-A**

CAP

A

A

AIRFOIL AIR-INLET HOLES

CF6-50

(a)

FIG. 25-26 (a) The stage 1 bucket high-pressure turbine for the CF6-6, and -50 engines. (Continued on the following page.)

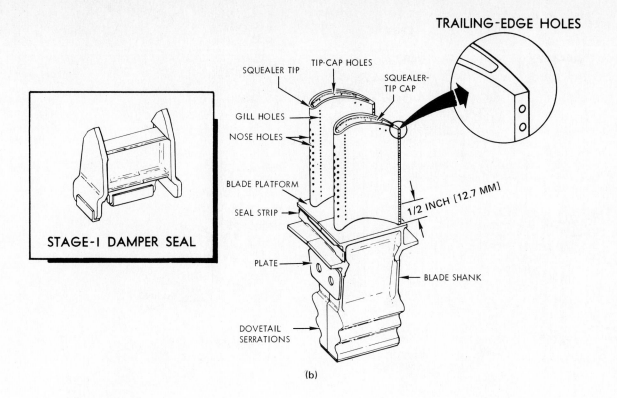

FIG. 25-26 (cont.) **(b)** Stage 1 high-pressure turbine blade pair and damper seal for the CF6-6 engine.

Thirteenth-stage HP compressor bleed is used to cool the CF6-6 second-stage stator vane (the CF6-50 uses 10th stage air) and is finally discharged into the gas stream.

LOW-PRESSURE TURBINE

The LPT (Fig. 25-28) for all CF6 series engines utilizes the same technology and design concepts, but the different aerothermodynamic conditions and work output requirements of the CF6-6 and CF6-50 turbines result in design differences.

Both CF6-6 and CF6-50 low-pressure turbines utilize a rotor supported between roller bearings mounted in the turbine midframe and the turbine rear frame. A horizontally split low-pressure-turbine casing containing stator vanes is bolted to these frames to complete the structural assembly. This provides a rigid, self-contained module which can be precisely and rapidly interchanged on the engine without requiring a subsequent engine test run. The LPT shaft engages the long fan drive shaft through a spline drive and is secured by a lock bolt. The forward flange of the turbine midframe is bolted to the aft flange of the compressor rear frame, after installation of the high-pressure turbine, to complete the engine assembly.

Common elements of CF6-6 and CF6-50 LPT modules include the turbine midframe (not including the liner), "C" sump and the no. 6 bearing, and the "D" sump and no. 7 bearing.

Both CF6-6 and CF6-50 low-pressure turbines have high-aspect-ratio shrouded blades which operate at low tip speeds and at moderate turbine-stage loading factors. Because turbine inlet temperature is relatively low (1500 to 1700°F) [815.6 to 926.7°C], turbine blade cooling is not required.

Bleed air is used to cool the first- and second-stage LPT disks to reduce thermal gradients. Ninth-stage air is used on the CF6-6 and seventh-stage air on the CF6-50.

The CF6-6 LPT has five stages and the CF6-50 LPT has four stages. Increased flow and higher wheel speed of the CF6-50 would result in very low stage loadings if a five-stage LPT were used on the -50 engine. The decision to use a four-stage LPT on the CF6-50 was the result of studies that considered aerodynamic efficiency, engine length, weight, cost, and commonality.

First-stage Low-pressure-turbine Nozzle Assembly

The first-stage low-pressure-turbine nozzle of the CF6-6 consists of 14 segments (Fig. 25-29) each containing 6 vanes. The CF6-50 LP turbine uses 12 of the 6 vane segments. The segments are supported at their inner and outer ends by the turbine midframe and the LP turbine casing respectively; this provides low vane-bending stress and freedom to expand or contract thermally without thermal loads and stresses.

The other stages of LPT vanes are cantilevered from the LPT stator casing.

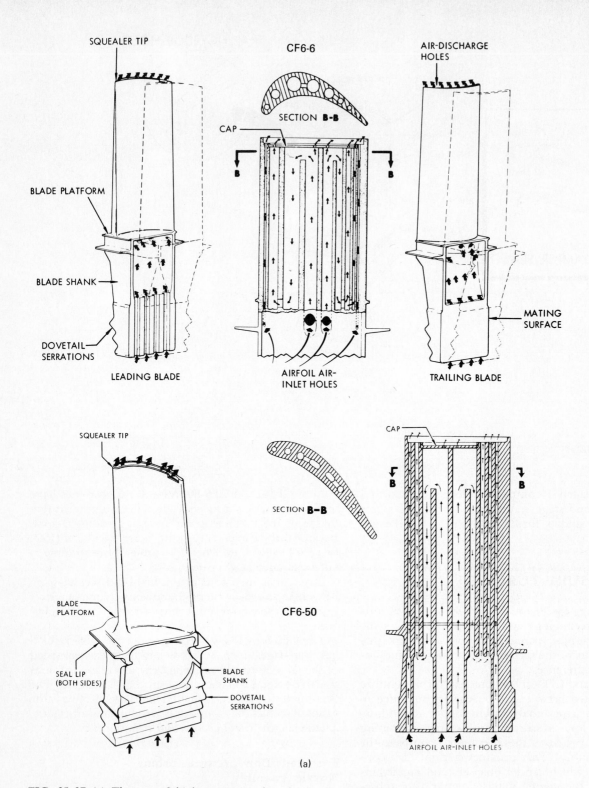

SQUEALER TIP

CF6-6

AIR-DISCHARGE HOLES

SECTION **B-B**

CAP

BLADE PLATFORM

BLADE SHANK

DOVETAIL SERRATIONS

LEADING BLADE

AIRFOIL AIR-INLET HOLES

MATING SURFACE

TRAILING BLADE

SQUEALER TIP

SECTION **B-B**

CAP

BLADE PLATFORM

SEAL LIP (BOTH SIDES)

BLADE SHANK

DOVETAIL SERRATIONS

CF6-50

AIRFOIL AIR-INLET HOLES

(a)

FIG. 25-27 (a) The stage 2 high-pressure turbine bucket for the CF6-6, and -50 engines. (Continued on the following page.)

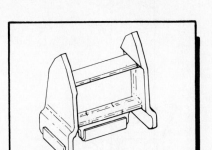

STAGE-2 DAMPER SEAL

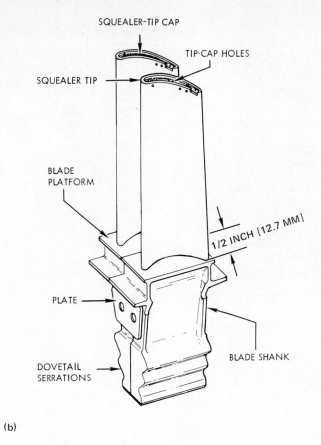

SQUEALER-TIP CAP

TIP-CAP HOLES

SQUEALER TIP

BLADE PLATFORM

1/2 INCH [12.7 MM]

PLATE

BLADE SHANK

DOVETAIL SERRATIONS

(b)

FIG. 25-27 (cont.) **(b)** Stage 2 high-pressure turbine blade pair and damper seal for the CF6-6 engine.

Low-pressure-turbine Stator

The LPT stator (Fig. 25-30) is designed for ease of maintenance, accessibility, and life in excess of 50,000 h.

The major parts of the stator are the nozzle stages (CF6-6, five stages; CF6-50, four stages), a split casing, shrouds, and interstage air seals. Each of the nozzle stages is composed of cast segments of multiple vanes per segment, any segment of which can be replaced with simple tools. The shrouds are in segments held in place by projections mating with slots formed by the casing and nozzles. The interstage seals are bolted to the ID flange of the nozzles. Both shrouds and seals have abradable honeycomb sealing surfaces to allow close clearance without risk of rotor damage caused by unusual rubs. The interstage seals partially restrain the inner vane ends to provide damping which results in low vane stresses.

Low-pressure-turbine Rotor Assembly

The LPT rotor (Fig. 25-31) drives the fan rotor through the fan midshaft. As a result of low rotor speed and interlocking integral blade-tip shrouds, rotor blade stresses are low.

The shroud interlocks are in contact at all rotor speeds due to an interference fit caused by pretwisting the blades. This provides adequate damping at low speeds and damping proportional to rotor speed at higher speeds. Shroud interlocks are hard-coated for wear resistance.

LPT rotor construction (Fig. 25-32) consists of separate Inco 718 disks having integral torque ring extensions, each of which is attached to the adjacent disk by close-fitting bolts. Bolt holes through the disk webs have been eliminated by locating them in the flanges where stresses are low. Front and rear shafts are attached to the disks between stages 2 and 3 and 4 and 5 (CF6-6) or 3 and 4 (CF6-50), respectively, to form a stiff rotor structure between bearings. All stages of blades contain individual interlocking tip shrouds to lower vibratory stresses. Blades are attached to the disks by means of multitang dovetails. Replaceable rotating seals, mounted between disk flanges, mate with stationary seals to provide interstage air seals.

By virtue of its low length-to-diameter ratio and the accurately fitted bolts used to fasten the disks together, the LPT rotor is highly stable. In the event of a midshaft failure, the LPT rotor would move aft until blade rows interfered with the stators, avoiding any catastrophic overspeed. The two bearing supports permit clearance control during maneuver loads. Together with low rotational speed, this makes the rotor relatively insensitive to normal unbalance.

The disks have high radius ratios which reduce the

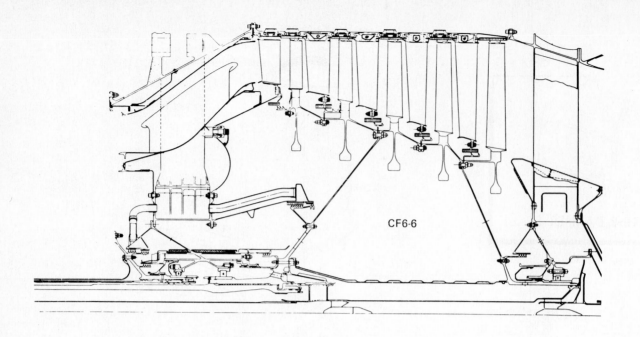

CF6-6

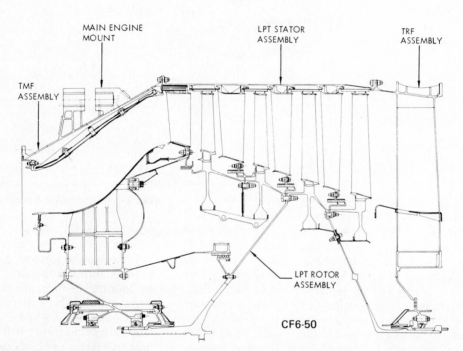

FIG. 25-28 The low-pressure turbine section for the CF6-6 and -50 engines.

radial temperature gradients and parasitic thermal stresses. This high radius ratio is made possible by eliminating bolt holes in the disks and putting them in flanges at the ends of spacers where the stresses are low. The result is more uniform disk stresses and the removal of the major cause of low cycle-fatigue limitations.

Turbine Case Cooling

LPT case cooling was developed during the TF39 program to improve turbine efficiency by controlling clearances. It is not required to meet life requirements of the case.

The CF6-6 LPT employs ninth-stage bleed air to cool the flanges which retain the stationary shrouds in the LPT casing. This reduces the clearance between these seals and the rotating blade tip shrouds and provides an improvement in LPT efficiency which results in an improved cruise SFC. In the CF6-6 this case cooling system is internal to the LPT. (Fig. 25-30)

In the CF6-50 an external impingement system is used to cool the LPT casing, employing fan discharge

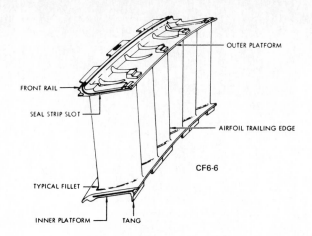

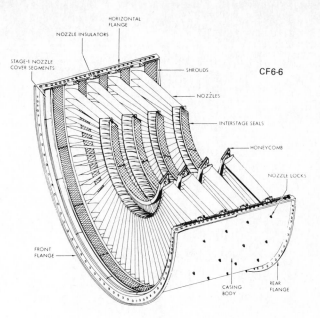

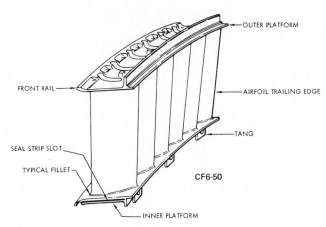

FIG. 25-29 Stage 1 LP nozzle vane segment for the CF6-6 and -50 engines.

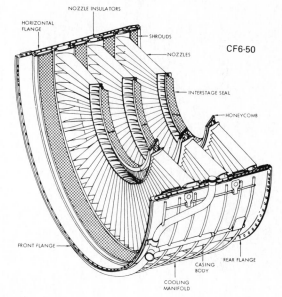

FIG. 25-30 Low-pressure turbine stator assembly for the CF6-6 and -50 engines.

air bled from the inner fan flow path. Utilizing fan air instead of ninth-stage compressor bleed results in a lower chargeable airflow and provides a further improvement in SFC.

LPT Module

The LPT module is designed for easy removal and reinstallation, either with the engine installed or after removal from the airplane. The only quick engine change (QEC) items requiring removal are the thrust spoiler and drives, EGT electrical harness, and condition-monitoring leads, if installed. Once these items are removed, the LPT module may be removed and replaced in 4 elapsed hours, requiring 28 worker-hours. The turbine midframe may be removed as part of the module, or the LPT may be separated behind the turbine midframe, leaving it installed on the engine. Shaft alignment is readily maintained during reinstallation, and rebalance of the complete low-pressure-rotor system is not required.

Condition of the LPT may be determined in place by using the borescope ports incorporated for that purpose. More thorough inspection is possible by removing one

of the split casing halves. Therefore, the LPT module should only rarely require unscheduled removal except in cases of confirmed upstream damage in the flow path.

SUPPORT STRUCTURES

Rotor stability and blade-tip clearance control is achieved through the use of four bearing support frames. The frames provide two bearing supports for each rotating mass. This results in rigid rotor support. The frames are:

1. *Fan frame*—includes the forward main engine mount, provides support for the fan rotor and stator, cowl,

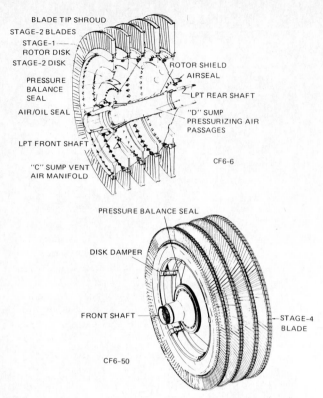

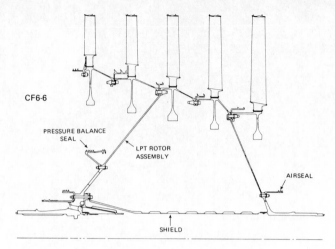

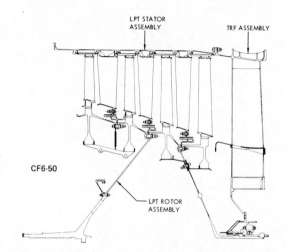

FIG. 25-31 Low-pressure turbine rotor for the CF6-6 and -50 engines.

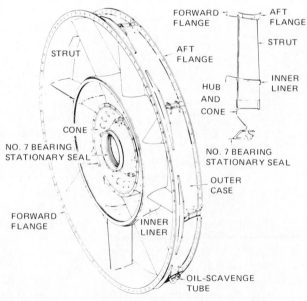

FIG. 25-32 The low-pressure turbine rotor for the CF6-6 and -50 engines. (Sectioned view.)

thrust reverser, front of engine cowl, and front of high-pressure rotor and stator. In the CF6-50, it also contains the bypass valves. The CF6-6 has no bypass valves.

2. *Compressor rear frame*—provides the housing for the engine combustor and supports the middle of the high-pressure shaft.

3. *Turbine midframe*—includes the rear main engine mount. Provides support for the rear of the high-pressure rotor and the front of the low-pressure-turbine rotor.

4. *Turbine rear frame*—provides support for the rear of the low pressure turbine rotor. Also provides the primary system exhaust nozzle and spoiler.

The four frames are similar in CF6-6 and -50 engines. Certain changes were necessary because of the higher thrust, flow, and pressure ratio of the CF6-50 series engines.

Fan Frame

The fan frame is a major support structure with 17-4 PH steel struts, inner and intermediate hub, and an aluminum outer case. The fan frame (Fig. 25-33) supports the entire fan section.

It supports the forward end of the compressor, fan rotor, fan stator, the forward engine mount, the radial drive shaft, transfer gearbox, horizontal drive shaft, and accessory gearbox.

FIG. 25-33 The fan section.

The fan frame has 12 struts equally spaced at the leading edge. The struts bolt to the aft outer casing with eight bolts per strut. The 6 and 12 o'clock position struts are the leading edges of the upper and lower pylons, and are shaped to fair into the pylon sidewalls. The radial shaft is enclosed within the lower pylon and is bolted to the six o'clock position strut. The fan frame houses the A sump, which includes the nos. 1, 2, and 3 bearings. The A sump pressurizing air enters through the leading edge of struts no. 4, 5, 9, and 10 in the fan stream.

The forward engine mount attaches to the aft flange of the fan frame and an auxiliary mount linkage connects to the fan casing.

The bypass valves on the CF6-50 engines consist of 12 bypass valve subassemblies which are mounted between 12 fan-frame radial struts. Each bypass valve subassembly consists of a door hinged in an opening in the frame.

The valves are made of aluminum castings for light weight and low cost. All moving surfaces have replaceable Vespel inserts which have low weight and a low coefficient of friction. These valve subassemblies are fastened to the frame by two bolts allowing easy installation and removal for work on the bench.

Compressor Rear Frame

The frame assembly (Fig. 25-34) is made up of the mainframe structural weldment, the inner combustion casing and support, the compressor discharge air seal and the B sump housing. Axial and radial loads are taken in the rigid inner ring structure and transmitted in shear into the outer casing.

Inco 718 is used for the frame because of high strength at temperature, corrosion resistance, and good repairability relative to the other high-temperature nickel alloys.

Bearing axial and radial loads and a portion of the HPT first-stage nozzle loads are taken in the inner ring or "hub" and transmitted through the 10 radial struts to the outer shell. The inner ring or hub of the frame is a casting which contains approximately half of the radial strut length. The cross-sectional shape of the hub is a box to provide structural rigidity.

The outer strut ends are castings which, when welded to the hub, complete the formation of the struts. Combining the hub and the outer strut casting in this manner forms a smooth strut-to-ring transition with the minimum concentration and no weld joints in the transition area. The 10 radial struts are airfoil-shaped to reduce aerodynamic losses and are sized to provide adequate internal area for sump service lines and bleed airflow. The hub and outer strut end assembly is then welded into the outer shell which is a sheet-metal and machined-ring weldment defining the outer flow annulus boundary as well as providing the structural load path between the high-pressure-compressor casing and turbine midframe.

To provide for the differential thermal growth between sump service tubing and the surrounding structure, the tubes are attached only at the sump, and slip joints are used where tubes pass through the outer strut ends.

Turbine Midframe

The turbine midframe (Fig. 25-35) consists of the outer casing reinforced with hat section stiffeners, the link mount castings, strut and castings, cast hub, eight semitangential bolted struts, the C sump housing and a one-piece flow-path liner. The frame casing and cast hub operate cool enough (less than 1100°F) [593.3°C] at all conditions to permit the use of Inco 718.

The eight partially tangential frame struts are secured to rings by bolts. The tangential struts are used to control thermal stress in the structure itself (thermal differentials between the struts and outer and inner rings produce rotation of the hub which imposes bending moments about the strut minor axis). Because the struts are relatively flexible around their minor axis, loads and stresses are low at the strut ends compared with radial strutted frames for the same thermal gradient. Because of the bolted feature, all major parts of the frame may be replaced or repaired without disturbing any structural welds. The inner structural ring or hub is an open U-shaped, one-piece casting with flanges provided to support the bearing cone, stationary seals, liner support cones, and eight gussetted pads between the sides of the U for attaching the struts. The outer ring consists of eight castings, butt-welded into the outer casing skin between which are fabricated sheet metal hat-shaped sections. These hat sections are butt-welded to the casting and seam-welded to the casing. The hats then essentially become continuous structural rings. The outer ends of the struts are bolted through these casing castings. The outer casing itself is a conical shell with a machined flange butt-welded on each end. The forward flange supports the HPT casing and the rear flange supports the LPT casing.

The bearing support cone and sump housing for the no. 5 and no. 6 bearings is bolted to the forward flange of the frame inner hub for ease of replacement, maintenance, and manufacture. Tubing (321 stainless steel) through the frame structural struts for sump service is secured to the sump by bolted flanges or B nuts to make the sump completely separable from the frame structure. The sump housing is double-walled construction so that the wall of the sump when wetted by oil is cooled by fan discharge air, keeping its temperature below 350°F [176.7°C].

The tubes, with wear sleeves at contact points, pass through slip joints at both the inner and outer ends of the struts to allow for differential thermal expansion between the frame and tubes. Tube configuration and clamping is established so that tubing resonant frequencies are kept out of the engine operating speed range to preclude high vibratory loading.

The Hastelloy X flow-path liner assembly is fabricated from an outer liner, inner liner, and eight airfoil-shaped strut fairings butt-welded to both liners.

The liner assembly is supported by a cone from the

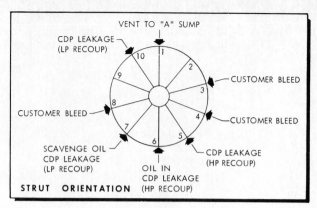

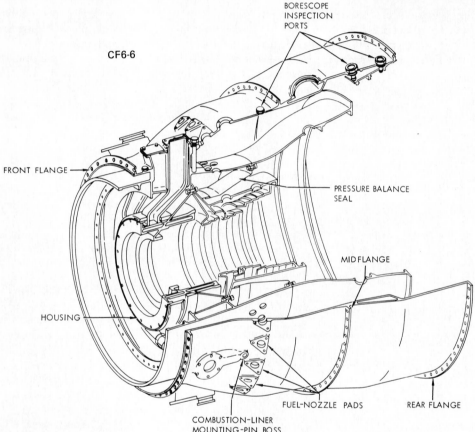

FIG. 25-34 Compressor rear frame for the CF6-6 and -50 engines.

forward side of the inner structural ring and guided and sealed at the aft inner end. Seals at the forward and aft outer portions of the liner assembly are provided to eliminate the possibility of hot gases circulating behind the liner.

Turbine Rear Frame

The frame assembly (Fig. 25-36) can be divided into the main frame structure, the inner flow-path liner, the D sump, and the sump service piping.

The assembly of the turbine rear frame is similar to the turbine midframe, but without bolted struts. Although there are significant maintainability advantages for a bolted structure, it is important to note that it is used only where there is a distinct configuration requirement because of the inherent weight penalty of the mechanical joints. This turbine rear frame is a welded Inco 718 structure with eight equally spaced, partially tangential struts supported on two axially spaced rings at both the hub and outer ring. With the tangential struts and 500°F [277.8°C] lower gas temperature than the turbine midframe, the rear frame can be designed without flow-path liner or strut fairings. The main advantage of this type of construction is the accessibility of structural welds for visual inspection without disassembly.

To ensure long tubing life there are no fixed attachments between the frame and the tubes. All frame tube joints have heavy wear sleeves. The vent tubes are dou-

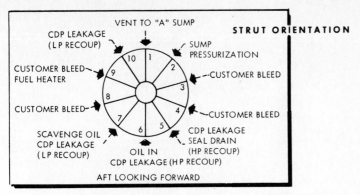

STRUT ORIENTATION

VENT TO "A" SUMP
CDP LEAKAGE (LP RECOUP)
SUMP PRESSURIZATION
CUSTOMER BLEED FUEL HEATER
CUSTOMER BLEED
CUSTOMER BLEED
CUSTOMER BLEED
SCAVENGE OIL CDP LEAKAGE (LP RECOUP)
CDP LEAKAGE SEAL DRAIN (HP RECOUP)
OIL IN
CDP LEAKAGE (HP RECOUP)

AFT LOOKING FORWARD

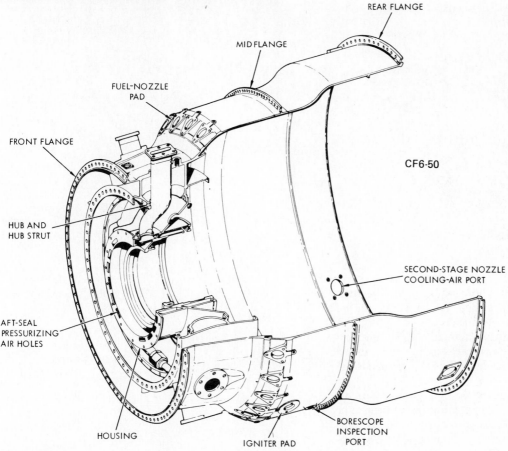

REAR FLANGE

MID FLANGE

FUEL-NOZZLE PAD

FRONT FLANGE

CF6-50

HUB AND HUB STRUT

SECOND-STAGE NOZZLE COOLING-AIR PORT

AFT-SEAL PRESSURIZING AIR HOLES

HOUSING

IGNITER PAD

BORESCOPE INSPECTION PORT

FIG. 25-34 (cont.)

ble-walled to insulate and prevent oil coking. All tube fittings have wrenching surfaces to prevent tube damage due to torquing. The tubes are also supported or clamped to drive all tube resonant frequencies well above the engine operating speed range.

Bearings and Seals (see Lubrication System, page 511)

The CF6-6/-50 engine bearings, bearing arrangement, seals, and coupling shaft are generally identical. The CF6-50 low-speed shaft has been strengthened to carry more torque.

The seals around the bearings are used to provide cavities which will ensure a cool blanket of air around each engine sump. Cool air bled from fan discharge is circulated through internal passages in the engine. This cool air is protected from the local ambient temperature and pressure around the bearing by incorporating a cavity which is vented to a low-pressure source of air. This source is chosen so that flow will always be away from the bearing in the outer sump pressurization seal. This design ensures a flow of cool air into the oil cavity to prevent leakage through the oil seal adjacent to the bearing, and prevents the hot ambient air from coming in contact with the sump walls. This design concept has been used throughout the engine.

The lube system on the CF6-6, discharges sump vent air and oil vapor out the end of the primary nozzle plug through a pipe that is an extension of the center vent

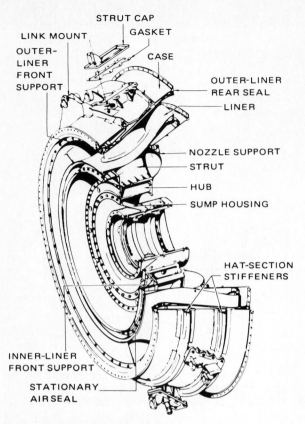

FIG. 25-35 The turbine mid frame.

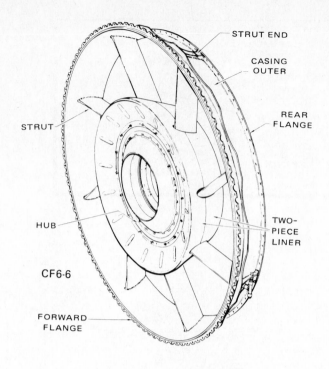

CF6-6

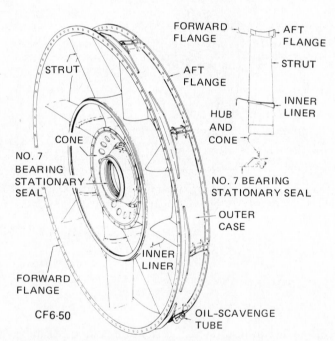

CF6-50

FIG. 25-36 The turbine rear frame for the CF6-6 and -50 engines.

system in the rotor shaft. The vent air enters the shaft at the center of a centrifugal field which expels the oil prior to discharging the air overboard.

Labyrinth seals, used throughout the lubrication system, have well-vented sumps to ensure positive inward flow and provide a positive dependable controlled-wear system.

Each labyrinth seal is composed of a stationary member (rub strip) and a rotating member (labyrinth). The rotating member contains sharp-edged teeth for improved performance. The tip clearance of the sharp-edged teeth is chosen so that controlled clearances are assured under normal operational conditions. The seal stator incorporates a rub material, chosen to satisfy the specific temperature requirements of the seal location. All seals that operate at relatively cool temperature (less than 600°F [315.6°C] use either an epoxy or a silver rub strip material. Seal stators of higher temperature, such as the compressor discharge pressure seal, employ open or filled cell honeycomb for ease of "rub in."

In each of the sumps, provisions have been made to prevent oil leakage from the seals at low pressure conditions, such as are encountered at idle, in windmilling, or at extremely high altitude conditions. A slinger has been provided at the entrance of each oil seal to prevent oil from "crawling" along the shaft and entering the seal. In addition, a screwthread "windback" is used, which consists of a helical thread with the spiral direction chosen to push the oil back into the sump. These two features are provided on each of the oil seals.

The engine has four basic sumps, which are designated A, B, C, and D. The nos. 1, 2, and 3 bearings and inlet gearbox are contained in the A sump. The nos. 4R and 4B bearings are contained in the B sump. The nos. 5 and 6 bearings are in the C sump, and the no. 7 bearings are in the D sump.

The roller bearings at all locations are supported by cone structures attached to the engine frame members. The cones provide the stiffness for the engine rotor mounting system, and also act to attenuate the thermal deflections and to avoid overstressing the housing connection points to the frame. They provide a structural support for main engine bearings because of their radial and axial stiffness combined with the capability of accommodating thermal deflections at each end.

Each of the bearings is lubricated with two oil jets so that if one jet becomes plugged, oil will continue to be supplied to the bearing. These jets are mounted in a one-piece machined lube-jet housing. Another feature is the incorporation of a small collection cavity just behind the oil jet, so that in the event any foreign material finds its way into the lube system, it is not likely to become lodged in a jet orifice. The bearings are arranged as shown in Fig. 25-37.

Roller bearings are used predominantly in order to accept the axial thermal differential motion between the rotor and stator members of the engine. Thrust is taken from each of the rotor systems by a single ball-thrust bearing. Single-row ball-thrust bearings rather than tandem (two-row) ball-thrust bearings, are used to obtain reliability and to avoid problems of load-sharing tandem ball bearings. A roller and a ball bearing are used in combination on the high-speed rotor. The ball bearing is mounted on cantilever fingers so that it can carry thrust loads, but will not be loaded radially because of the low radial spring rate of the support. This design ensures that the load on this bearing will be predictable and will be thrust load only. The roller bearing is located near the forward compressor discharge seal, and positions and seals accurately in a radial direction, as well as accepting all the radial loads from the compressor rotor shaft.

Rotation of the outer races is prevented by one of two methods. On all of the roller bearings and the low-pressure thrust bearing the outer race is flanged and bolted to the housing. Rotation of the 4B ball bearing is prevented by providing a heavy axial clamp across the outer race. This bearing does not carry appreciable radial load.

Coupling Shaft

The coupling shaft (Fig. 25-38) transmits power from the low-pressure turbine to the fan. The shaft is supported at the forward end on the nos. 1 and 2 rotor bearings and on the aft end by the low-pressure turbine shaft. The shaft configuration permits the removal of the fan disks, or removal of the low-pressure turbine, without disturbing the shaft.

The forward end of the shaft is splined directly to the fan stub shaft. This two-piece shaft arrangement is made possible by locating the thrust bearing in the no. 1 position. Although the absolute stress level of the spline shaft is similar to previous designs, the added yield strength of the maraged steel shaft improves strength margin and reliability.

Main Engine and Ground Support System Mounts

As shown in Fig. 25-39, there are two main flight mounts and a number of ground handling, hoist, and support mounts located around the engine. Fig. 25-40 shows a complete CF6-6 wing installation.

ACCESSORY DRIVE

All driven accessories for the engine and airframe are mounted on a single engine accessory gearbox (Fig. 25-41). The location of the accessory drives and the position of the engine and airframe accessories are optimized for maintenance and installation performance considerations. The gearbox is mounted on the engine fan casing at the bottom of the engine.

Power to drive the accessories is extracted from the HP compressor front stub shaft and transmitted through a large-diameter hollow shaft, which encircles the fan drive shaft, to the bevel gear set in the inlet gearbox. The radial shaft carries the power from the inlet gearbox to the outside of the fan through a radial strut and housing at the engine bottom center line to the transfer gearbox. The transfer gearbox drives the accessory gearbox through a horizontal shaft.

The plug-in concept (Fig. 25-42) is utilized on all accessory gearbox pads and idler gears. With this concept, an entire gear, bearing, seal, and pad assembly may be removed and replaced without disassembling the gearbox.

Alignment of bearing bores does not require line-bor-

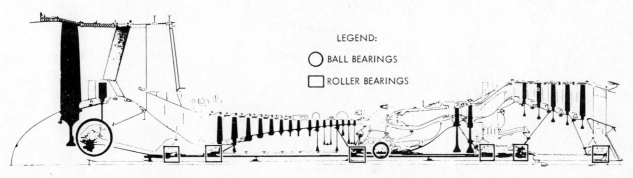

LEGEND:

○ BALL BEARINGS

▢ ROLLER BEARINGS

(8 INDIVIDUAL BEARINGS IN 7 LOCATIONS- NO DIFFERENTIAL BEARINGS)

FIG. 25-37 Bearing arrangements.

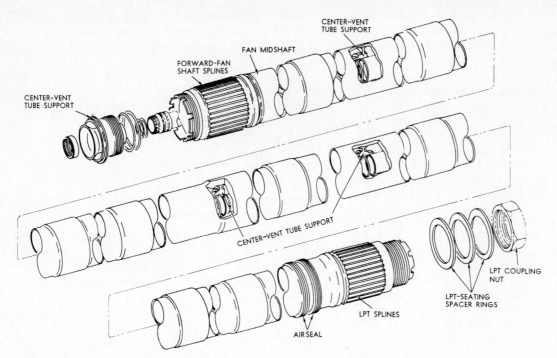

FIG. 25-38 The CF6-6 coupling shaft assembly.

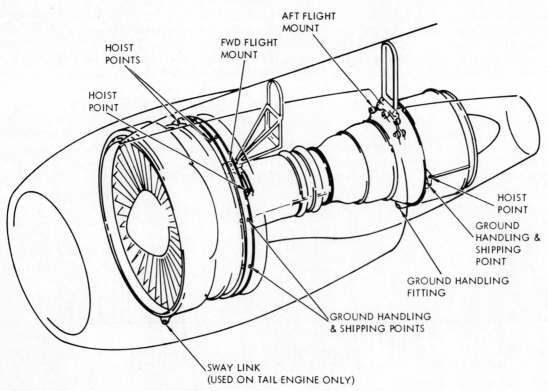

FIG. 25-39 Engine mounts and handling points.

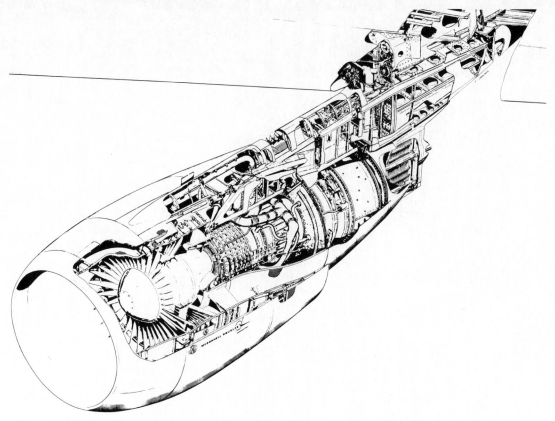

FIG. 25-40 A CF6-6 wing installation.

ing mating parts, which permits this simple replacement.

The power extraction and drive from the engine to the accessory gearbox consists of:

1. A right-angled bevel gearbox, located inside the engine front frame and mounted on a flange in the hub of the frame.
2. The transfer bevel gearbox, mounted on the engine fan casing at the bottom of the engine, outside of the fan flow path.
3. Drive shafts, which connect the various gearboxes through involute splines.
4. The accessory gearbox mounted forward of the transfer bevel gearbox on the outside of the fan casing.

Both spiral bevel gears and involute spur gears are used in the accessory-drive system. The bevel gears incorporate a 35° spiral angle to obtain a maximum contact ratio, consistent with the smooth operation for aircraft gearing. High contact ratio means that more teeth are sharing loads, and dynamic tooth loads are reduced. Spur gears are used throughout the accessory box. The involute tooth form is highly tolerant of center-distance variations and has a relatively lower cost to manufacture and mount. Gear meshes are designed with "hunting tooth" ratios. This feature prevents repetitive contact on each revolution between the same teeth in the gear mesh, and is used throughout the gearboxes to improve life.

Ball and roller antifriction bearings are used to support the shaft of the gears in the accessory-drive system. Roller bearings are used where possible because of their higher capacity and smaller size. Where it is important to hold axial position, ball bearings are used. Duplex bearings mounted face to face are used as thrust bearings on the bevel gears to provide maximum control of gear contact patterns and backlash. Bearing material is M50 or SAE 52100 vacuum-degassed steel.

All gearbox oil seals are carbon-face rubbing seals. The carbon element rubs on a hardened flat mating ring, which rotates with the shaft. For ease of removal, each seal is retained from the outside of the gearbox and can be replaced without teardown of the gearbox assembly.

Splined drive shafts are manufactured from AISI 4350 alloy steel. The splines are lubricated to decrease wear and increase life. A shear section in the radial shafts has been incorporated near the inboard end to prevent shaft whip in the event of a shaft failure caused by overload. Critical speed of the shaft has been set so that calculated values are not less than shaft speed at 120 percent engine speed.

The gearbox casings are AMS 4218 and AMS 4219 cast aluminum. Aluminum was chosen in place of the

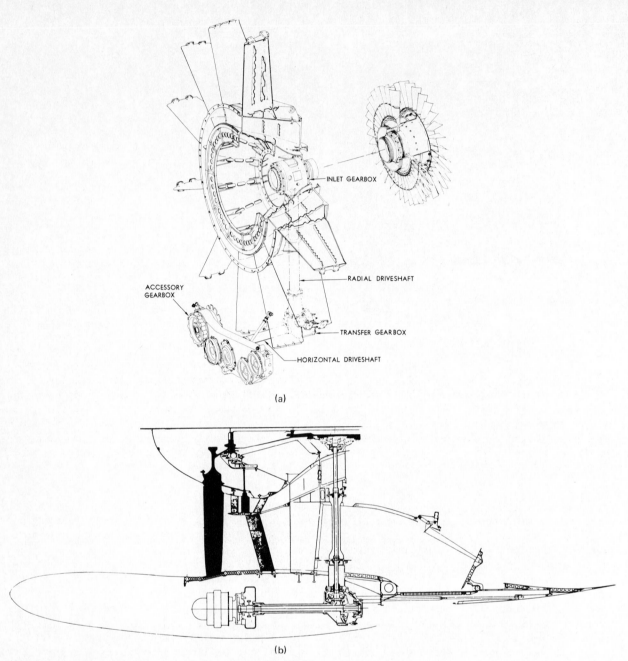

FIG. 25-41 (a) The inlet, transfer, and accessory gearboxes.
(b) Placement of the accessory drive.

lighter-weight magnesium to obtain maximum corrosion protection for the gear casings and improved stability and ease of case repairs.

A pressurized dry sump system is used for lubricating the accessory-drive gears and bearings. The system uses engine oil from the main lube pump to jet-lubricate critical gear meshes and bearings. The bevel gear meshes are lubricated on both the incoming and leaving sides where practical to ensure adequate lubricant and cooling flow for proper operation. Jet lubrication is used on the more heavily loaded bevel-gear bearings. On the more lightly loaded spur-gear bearings, mist and splash lubrication from oil directed at the gear meshes is used.

Accessory Arrangement

All driven accessories are mounted on the gearbox by pads which incorporate female drive splines to accept accessory quill shafts, and a pilot to locate the accessory with respect to the spline center line (Fig. 25-43). The use of quick attach/detach (QAD) connections for accessories to facilitate removal and replacement of components has been used successfully in both commercial and military applications. Aeronautical standard (AS) specification for standard QAD pads also provides an additional benefit in the female spline configuration which can be sealed using simple preformed (O-ring)

FIG. 25-42 The plug-in gearbox.

packings, thus permitting more reliable lubrication of the spline.

Provisions have been made for QAD pads for both engine- and aircraft-mounted accessories. The starter and fuel pump have interrupted flange type QADs, the lub/scavenge pump has a V-band-type QAD. The aircraft-supplied accessories can be made with an adaptor pad where the QAD is a part of the adaptor pad.

Maintenance Considerations

The CSD and alternator units are separately mounted on the gearbox, thus permitting individual removal of the components. This also reduces the amount of overhung moment.

The location of the accessory drives and the position of the engine and airframe accessories were optimized for engine maintenance and installation performance considerations. Gearbox and accessory removal from the engine is readily accomplished.

THRUST REVERSER SYSTEM

The engine fan/turbine reverser system for the DC-10 (Fig. 25-44) is designed to provide a minimum reverse thrust of 40 percent of maximum takeoff forward thrust. This reverse thrust is achieved by two systems: a core engine turbine reverser which provides 5 percent reverse of core engine thrust on CF6-6 engines and about 30 percent reverse thrust on CF6-50 engines, and a fan reverser which provides about 48.5 percent reverse thrust of the fan stream or secondary exhaust system.

In the forward-thrust mode, the fan and turbine reversers must function as exhaust nozzles and inherently must have high thrust coefficients. This requires smooth and properly contoured flow paths with minimum drag and leakage losses. The importance of this is illustrated by the fact that 1 percent penalty in fan-nozzle thrust

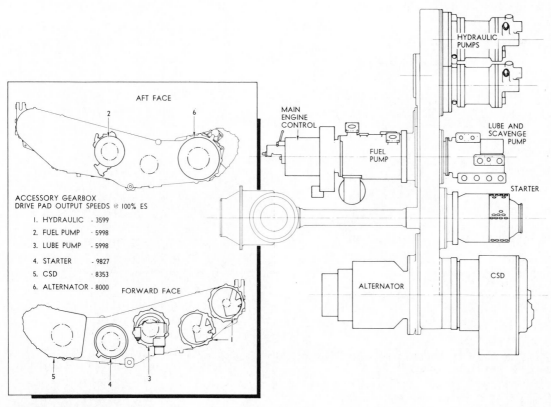

FIG. 25-43 Accessory drives.

ACCESSORY GEARBOX
DRIVE PAD OUTPUT SPEEDS @ 100% ES

1. HYDRAULIC - 3599
2. FUEL PUMP - 5998
3. LUBE PUMP - 5998
4. STARTER - 9827
5. CSD - 8353
6. ALTERNATOR - 8000

FIG. 25-44 The CF6-50 fan and turbine reverser shown in the deployed position.

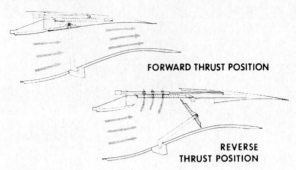

FORWARD THRUST POSITION

REVERSE THRUST POSITION

FIG. 25-45 Fan reverser positions.

coefficient results in a 2.2 percent increase in specific fuel consumption at cruise flight conditions.

The fan and turbine reversers are modular units which can be assembled and rigged off the engine and aircraft. In addition, the engine can be removed and replaced without removing the fan reverser from the aircraft. The fan reverser is split at the bottom and can be opened like the cowling for easy access to the core engine.

Fan Reverser

This fan-reverser design (Fig. 25-45) is being used on both the CF6-6 and -50 series engines. As shown, a series of airfoil-shaped turning vanes are mounted in cascades around the outer circumference of the fan nacelle. These vanes are surrounded by a cowl which provides a smooth flow path for both internal fan exhaust flow and the external airstream. The outer cowl extends aft and in conjunction with the cowl around the core engine forms an annular convergent plug nozzle. Aft translation of the outer cowl uncovers the turning vanes; and a series of doors (16 total), flush-mounted by hinges to the cowl and attached to links extending from the inner cowl, are automatically pivoted inward to block the flow through the fan exhaust nozzle. The fan exhaust is then directed radially outward and forward through the vanes, thus providing a reverse-thrust force. The cowl translation (Fig. 25-46) is accomplished by rotating ball screws which are driven by flexible shafts connected to a pneumatic motor mounted in the pylon.

The engine noise suppression requirements include the treatment of the fan reverser/nozzle flow path with a sizable surface area of sound suppression material. This requirement was a factor in defining the length of

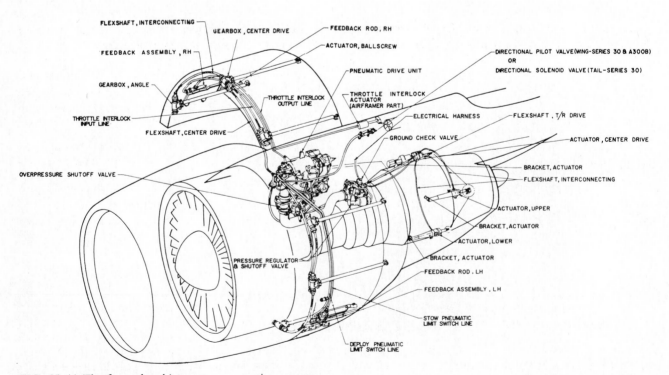

FIG. 25-46 The fan and turbine reverser actuation system.

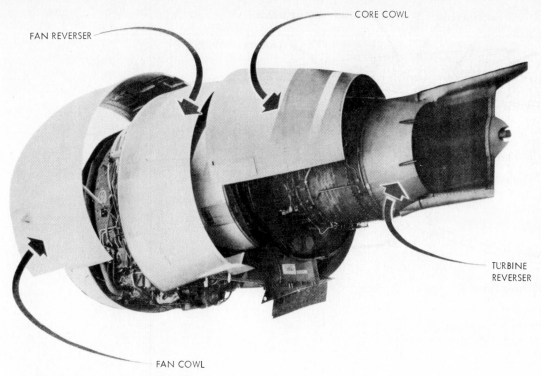

FAN REVERSER

CORE COWL

TURBINE REVERSER

FAN COWL

FIG. 25-47 The CF6-6 thrust reverser.

the fan reverser/exhaust nozzle system. Because of the nozzle length, servicing of the core engine requires access behind (or inside) major elements of the reverser. Thus, a split fan reverser employing a bifurcated duct was selected. This feature permits ready access to the core engine as shown in Fig. 25-47.

Quick-release handles operating latches at the bottom and at the top forward support structure are the only fan-reverser-related fasteners that must be manipulated to open a reverser duct half. As a further aid to maintainability the split fan reverser ducts are opened by a power actuation system and the fan reverser transcowl actuation system can be completely rigged off the engine.

The translating cowl which surrounds the static structure is also split into halves. During forward-thrust operation, the outer cowl forms the basic pressure vessel in conjunction with the duct side wall, the inner cowl, and forward outer static structure.

The fan reverser, with the exception of the areas having noise treatment, is composed primarily of aluminum sheet and honeycomb with castings and extrusions being used in transition and concentrated-load areas.

Turbine Exhaust Performance

Both the CF6-6 and -50 series engines employ the same basic fan-reverser design for the fan airflow (Fig. 25-48). Attenuation of the turbine exhaust of these two engine series differs, however, as a result of the higher core engine air-flows in the CF6-50 series engine. In order to meet the FAA airport noise requirements, it was necessary to provide sound treatment on the tur-

bine exhaust nozzle on the CF6-50 series engines. A converging-diverging (C-D) nozzle was also incorporated to give optimum engine performance at cruise and takeoff power settings.

Turbine Reverser

The CF6-6 turbine reverser [Fig. 25-49(a)] is a translating fairing, hinged-cascade-type design as illustrated.

Two cascade elements (turning vanes) are mounted on a fixed pivot aft of the core nozzle exit and are enclosed in a fairing which forms an airfoil-shaped plug nozzle. Aft translation of the fairing uncovers the cascades which open across the nozzle exit and divert the exhaust flow radially outward and slightly forward in the horizontal direction.

Deployment of the turbine reverser is accomplished by two ball screws which are powered through a cable by the reverser-actuation-system motor. The two screws translate the cascade fairing to the aft position. Cam tracks which are mounted on the fairing and engage the two cascade doors open the cascades as the fairing moves aft. Stowing is effected by the reverse process.

The CF6-50 series engines employ a C-D plug-type turbine exhaust nozzle. A C-D turbine exhaust nozzle provides overall optimum performance during cruise and takeoff. In order to comply with the FAA noise criteria, the inner flow path walls are sound treated.

Fig. 25-49(b) shows a cutaway of the CF6-50 C-D plug-type nozzle and turbine reverser.

The CF6-50 plug-type turbine-reverser nozzle incorporates a similar concept of reversing as the fan reverser

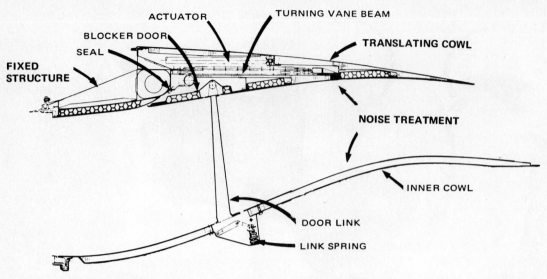

FIG. 25-48 The translating fan reverser cowl schematic.

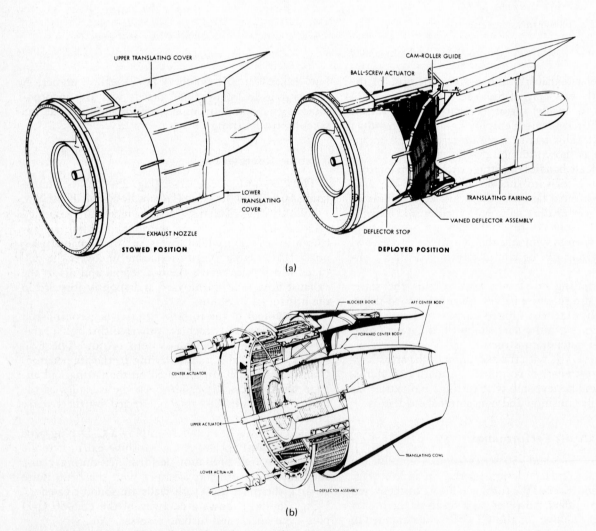

(a)

(b)

FIG. 25-49 (a) The CF6-6 forward- and reverse-thrust positions.
(b) CF6-50 reverse thrust position.

506 REPRESENTATIVE ENGINES

using blocker doors activated by a translating cowl tracked on a structure of fixed cascade segments bolted to a ring and mounted to the turbine rear frame.

The nozzle plug is fixed to the turbine frame struts and provides the inner fixed pivot for the blocker door links like the inner wall on the fan reversers.

Reverser Actuation and Control System

The reverser actuation system employs three ball-screw-type actuators to drive the translating cowl and blocker doors. These actuators are integral to torque-converting gearboxes which receive high-speed, low-torque power via a flexible shaft from the thrust-reverser system air motor located in the aircraft pylon to one of the actuator gearboxes. The gearboxes are interconnected with cables and transform the air motor output into the low speed, high torque required for turbine reverser actuation.

The reverser system (Fig. 25-50) utilizes a bidirectional pneumatic motor as the power source. Air for the motor is obtained from the aircraft system. A mechanical cam attached to the throttle control cables position a small poppet valve which serves as a pilot for a position selector and an air supply valve. Movement of the power lever to the reverse position opens the poppet valve which positions the above valves, thus supplying air to the motor as well as establishing the direction of rotation. The power lever is restricted from advancing to full power in the reverse position by an interlock. This interlock keeps the power lever at the IDLE position until the reverser is approximately 90 percent deployed, as indicated by the engagement of the translating cowl with an interlock valve. This engagement opens the valve, which in turn removes the interlock stop and permits advancement of the power lever.

In addition, at 90 percent the air motor is stopped and supplied with lower-pressure air which then is released within a fraction of a second to move at a slower speed to full deploy.

At 98 percent full deploy a snubber valve is actuated in the exit air line of the motor. The resulting back pressure provides a breaking force and permits the reverser to reach the end of the stroke with a relatively low impact force. The actuator screws contain a limit stop for cowl deployment, and seating against the static structure provides the limit stop in the stow position. The reverse of the above action takes place in stowing.

The pneumatic motor contains a brake which serves as a lock in the stowed and deployed position and comes on after some snubbing torque is applied onto the actuator stops. This brake is released when the air supply valve is opened. As noted in the reverser design discussion, failure of this brake will not result in deployment of the reverser.

The turbine reverser is also actuated by the same motor through a flexible cable which runs aft from the motor to the gearbox of one of the ball screws that positions the turbine reverser. Interconnecting flexible shafts connect to the other two ball-screw gearboxes.

FUEL SYSTEM

Fuel from the aircraft fuel system (Fig. 25-51) enters the engine at the engine main-fuel-pump inlet. Pressure fuel from this two-stage pump flows through the fuel-oil heat exchanger, through the fuel filter, and into the main engine control. Metered fuel from the fuel control flows through the pressurizing valve, through the customer-furnished flowmeter, through the fuel manifold, and into the 30 fuel nozzles. A fuel manifold overboard drain eliminator (eductor and valve assembly), is supplied as optional equipment. The eductor and valve assembly pumps the fuel that drains from the fuel manifold into the aircraft drain can back into the aircraft fuel system via the fuel pump eductor during engine shutdown. The engine fuel control also schedules pressure fuel to actuate the variable stator vane system and the CF6-50 variable-bypass doors. Bypass fuel from the control is returned to the fuel pump interstage.

The fuel system is characterized by the clustering of the primary system components (Fig. 25-52) around the fuel pump and by mounting this cluster on the accessory gearbox. This feature has several advantages.

1. Locates the fuel system components in a low-temperature environment of minimum fire hazard.
2. Eliminates the complex bracketry and plumbing required for individual components by providing internal, flange-mated parts for each component.
3. Enables easy maintenance because of easy access.

The combined effects of this feature serve to enhance the overall reliability of the fuel system.

Main Fuel Pump

The main fuel pump provides high-pressure fuel to the main engine control for combustion and for use as hydraulic fluid for the compressor variable-geometry systems. The pump is composed of a centrifugal element, a vapor eductor, and a positive displacement high-pressure gear element. A 30- × 30-in mesh screen and integral bypass valve are located in the pump interstage section. A high-pressure relief valve is provided

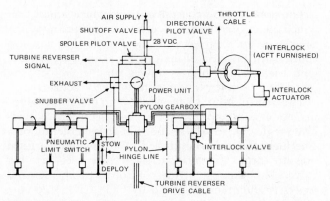

FIG. 25-50 The thrust reverser control and actuation system.

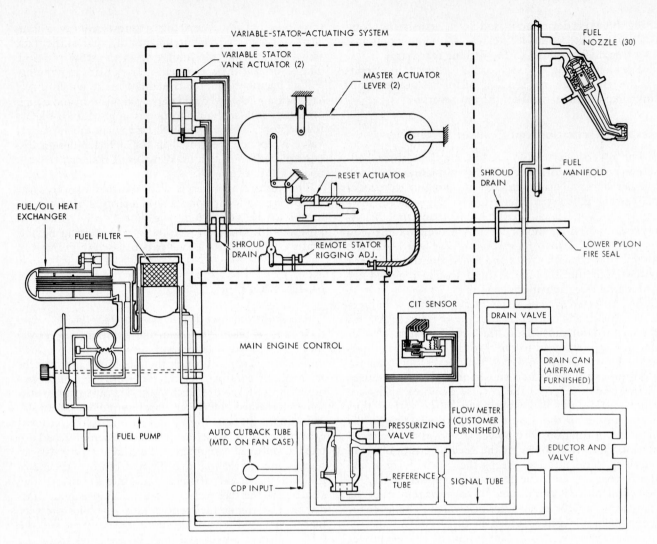

FIG. 25-51 Fuel system schematic.

to limit pump pressure rise to protect downstream fuel-system components.

The internal features of the CF6 fuel pump are shown in Fig. 25-53. The pump housing provides mounting pads and flange ports for the fuel filter, fuel-oil heat exchanger, and the main engine control. The pump also provides the drive input for the fuel control, thus eliminating the requirement for a separate accessory gearbox drive pad. Mounting the fuel system components directly on the pump casting minimizes the number of external fuel lines and connections, thereby reducing the possibility of fuel leakage and improving component accessibility.

Fuel-oil Heat Exchanger

The fuel-oil heat exchanger is a lightweight, high-pressure aluminum shell and tube unit which serves to cool the engine lubrication oil and to heat the fuel above 35°F [1.7°C]. The exchanger is constructed internally so that the fuel flows in two passes through tubes; and the surrounding oil flows in six passes. The unit is mounted on the left side of the fuel pump where the fuel enters

and exits through two adjacent ports on a common flange. The design and location of the heat exchanger eliminates the requirement for a fuel heater on the CF6. Operation of the heat exchanger is automatic and requires no pilot actuation or airframe interface. (See Fig. 25-58 in the following section.)

Main Engine Control

The main engine control is a hydromechanical unit which meters combustion fuel flow to maintain the desired engine speed selected by the throttle. The unit also controls the position of the variable stator vanes (VSV) (and the variable bypass valves on the CF6-50) by scheduling high-pressure fuel to the VSV actuators.

The control is an isochronous-speed governor which maintains constant core speed N_2 at constant power-lever angle in spite of variations in ambient conditions. The fuel schedule is controlled by maintaining a fixed pressure drop across a variable-orifice fuel-metering valve. The response of the control to power demand inputs is continuously biased by CIT, CDP, and core engine rotor speed N_2. Core speed is controlled and

FIG. 25-52 Fuel system components.

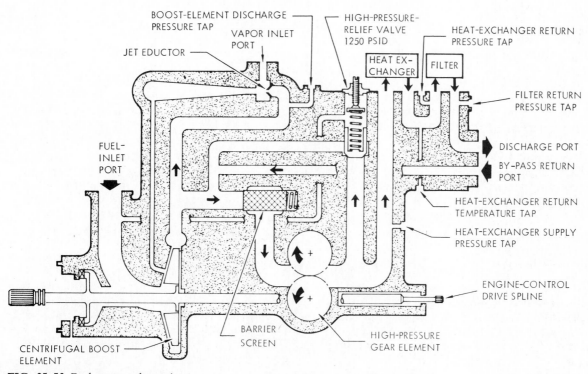

FIG. 25-53 Fuel pump schematic.

limited by the fuel control, while control of the low pressure rotor speed N_1 is accomplished indirectly by controlling N_2. A flight/ground idle solenoid is provided to obtain CIT-biased ground or flight idle schedule. Ground idle provides the low thrust requirement to minimize aircraft braking during taxi operations. Flight idle provides the higher initial power setting that enables rapid in-flight acceleration. The solenoid is energized manually by an airframe electrical signal to obtain ground idle. The solenoid is, therefore, fail-safe to flight idle. Internally the control has a self-washing filter which requires no maintenance between overhauls.

Feedback-cable Reset Actuator

The feedback-cable reset actuator transiently repositions the VSV feedback cable during engine acceleration. When takeoff power is set, the VSVs are reset CLOSED initially and then gradually returned to the scheduled position. This reset action provides partial compensation for the inherent tendency of exhaust gas temperature to overshoot until the turbine reaches stabilized operating temperature.

Fuel Filter *(FIG. 25-54)*

The fuel filter is a lightweight, high-pressure unit with aluminum head and bowl and disposable filter ele-

ment. The filter head houses a bypass relief valve and the bowl houses the disposable filter element. The bypass relief valve is designed to ensure fuel flow in the event that the element becomes clogged. Positioning the filter on the fuel pump renders the filter most accessible for replacement of the element.

Pressurizing and Drain Valve Unit *(FIG. 25-55)*

The pressurizing and drain valve serves three purposes: ensures that adequate fuel servo pressure is maintained in the fuel control, ensures that the metered fuel pressure is sufficiently high to actuate the variable stator vanes and variable bypass valves, and the drain valve serves to drain the fuel manifold upon engine shutdown to prevent fuel from leaking into the combustor and to prevent fuel coking in the nozzles. During engine operation, the drain valve is held closed by a metered-fuel-pressure signal. During the engine shutdown, the signal pressure drops and a spring force unseats the valve, causing fuel from the manifold to drain through the outlet port to an airframe-furnished drain tank. The valve is opened by metered fuel pressure which acts against a spring-loaded piston seated on the inlet port, and reference pressure from the main engine control. During shutdown, the pressurizing valve closes as the inlet pressure decays. The pressurizing valve is mounted on the discharge port of the fuel control.

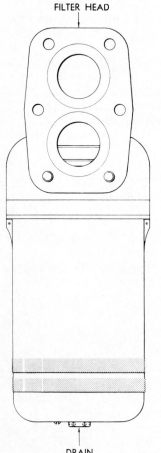

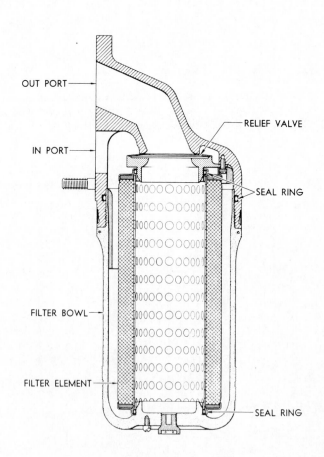

FIG. 25-54 Fuel filter.

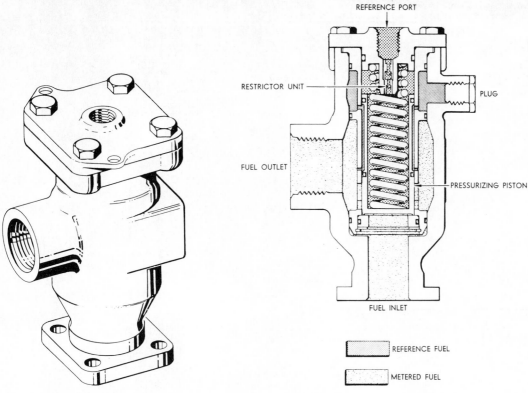

REFERENCE PORT

RESTRICTOR UNIT

PLUG

FUEL OUTLET

PRESSURIZING PISTON

FUEL INLET

REFERENCE FUEL

METERED FUEL

FIG. 25-55 Pressurizing valve.

Fuel Manifold *(see FIG. 12-29)*

The fuel manifold is a single-tube unit which distributes the metered fuel to the 30 fuel nozzles. The unit, including its 30 feeder tubes, is shrouded for protection against fire and high-pressure leaks. It is divided into right and left halves, each of which supplies 15 feeder tubes. The manifold is supplied by a single tube which runs into the core engine compartment from the fan accessory compartment through a sealed junction trap. The unit is mounted on the compressor rear frame by eight circumferential brackets which are designed to provide sufficient damping, rendering the plumbing less susceptible to vibrations. The fuel manifold shrouds are drained to a customer-furnished drain tank.

Fuel Nozzles *(see FIG. 12-29)*

The fuel nozzles are the dual-orifice type with integral flow divider. The CF6 fuel system has 30 fuel nozzles which are individually inserted through pads in the compressor rear frame and into the axial swirlers of the combustor dome. The dual-orifice nozzle system provides primary and secondary flows for proper fuel atomization during all phases of engine operation. Design of the fuel nozzle for compatibility with the combustor design contributes to the CF6 smokeless combustion operation.

The dual-orifice nozzle's primary portion is designed to provide the good atomization necessary for starting and idle conditions. The additive secondary portion is designed to provide the high flow capability and ul-

trafine atomization for the clean, efficient combustion requirements at higher power settings. The nozzle design incorporates tube inserts and a protective heat shield to guard against fuel coking.

The integral flow divider is a slide-type valve with sharp edges and close clearances. As compared with the single flow divider, the integral flow divider system provides substantial advantages in circumferential fuel distribution and acceleration rate. The head effect between top and bottom fuel nozzles is reduced to an insignificant fraction of the fuel-nozzle inlet pressure. Further, the single fuel manifold minimizes system volume and simplifies system plumbing.

LUBRICATION SYSTEM

The lubrication system is completely self-contained and is designed to operate independently of the airframe systems. The features of this system are the integral lube and scavenge pump, the center ventilation system, the all-labyrinth bearing seals, and the simplified plumbing arrangement.

The lubrication system (Fig. 25-56) is a dry sump system composed of four major subsystems: lube supply, lube scavenge, oil seal pressurization, and sump ventilation. Oil from the engine oil tank is distributed to the lubrication areas by the lube element of the integral lube and scavenge pump. The oil is removed from these areas by the scavenge element of the pump, and is filtered (Fig. 25-57) and cooled (Fig. 25-58) before it is returned to the tank. Labyrinth seal pressurization

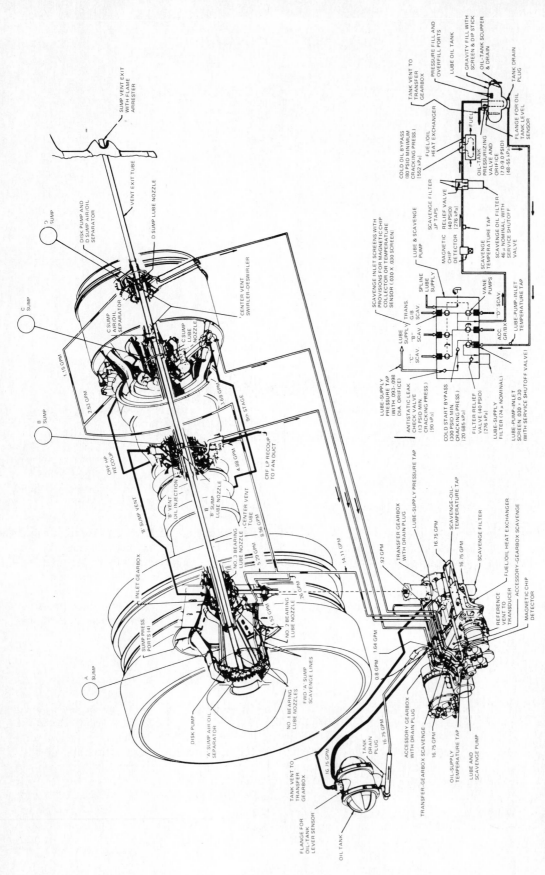

FIG. 25-56 The CF6-6 lube system.

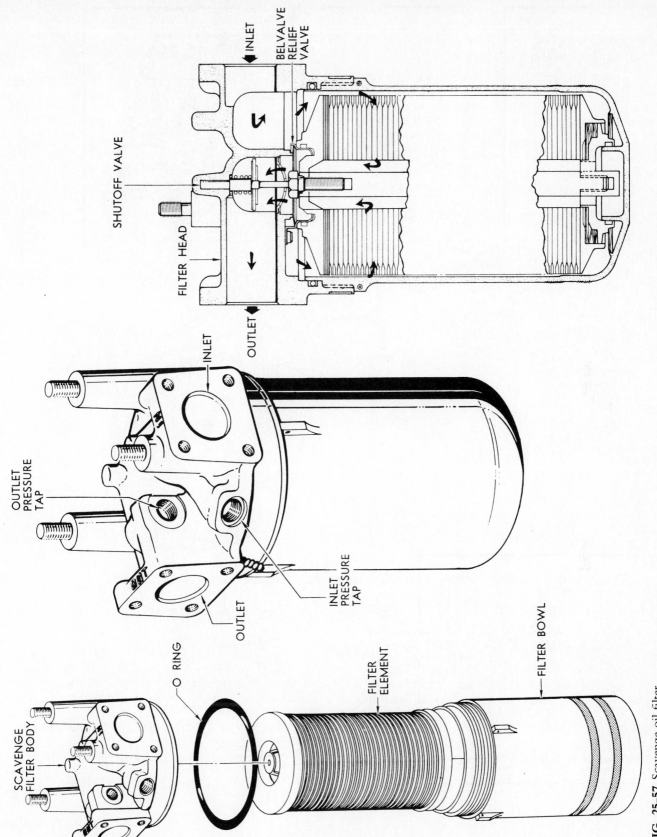

INLET

BELVALVE
RELIEF
VALVE

SHUTOFF VALVE

FILTER HEAD

OUTLET

INLET

OUTLET PRESSURE TAP

INLET PRESSURE TAP

OUTLET

O RING

FILTER ELEMENT

FILTER BOWL

SCAVENGE FILTER BODY

FIG. 25-57 Scavenge oil filter.

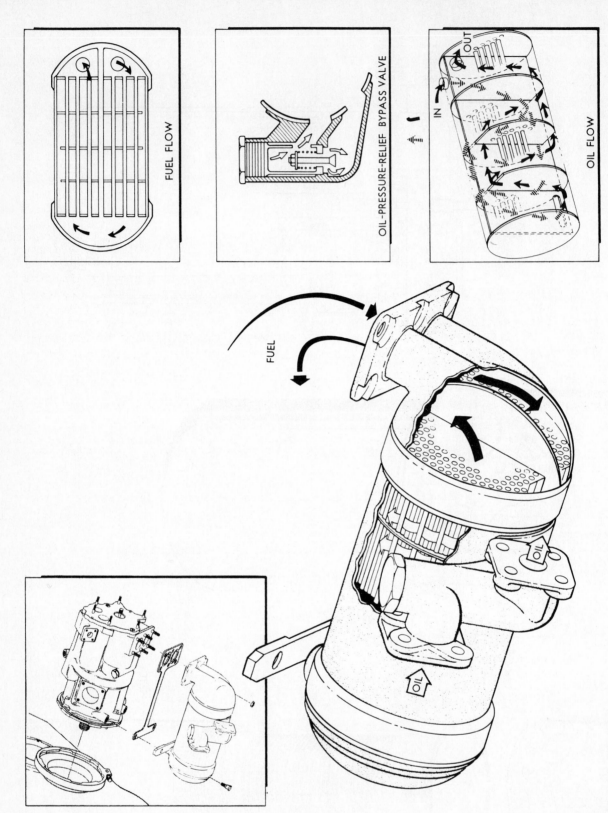

FUEL FLOW

OIL-PRESSURE-RELIEF BYPASS VALVE

OUT

IN

OIL FLOW

FUEL

OIL

OIL

FIG. 25-58 Fuel-oil heat exchanger.

and sump venting complete the operation of the lubrication system.

Lube Supply Subsystem

The lube supply subsystem consists of the oil tank, the lube supply element of the lube and scavenge pump, the lube discharge nozzles and jets, and related lube supply plumbing. Oil is gravity-fed from the oil tank and is directed to the supply element of the lube and scavenge pump through a 26-mesh screen in the inlet supply port. The pressure oil from the supply element is forced through an internal 74-μ filter before it is discharged from the pump. The oil leaving the pump is routed through the lube supply lines to the discharge nozzles and jets, located in the main shaft bearing areas, in the accessory and transfer gearboxes, and in the gearbox drive trains.

The features of the lube supply subsystems are:

1. The cylindrical oil tank is removable and can be disassembled for cleaning. The unit is silicone-coated as a fire-prevention measure; and it is isolation-mounted to alleviate susceptibility to vibrations.
2. The integral lube and scavenge pump (Fig. 25-59) is located in a cool, accessible environment for greater reliability and easy maintenance. The integral pump concept simplifies the lube system plumbing and eliminates the additional plumbing and bracketry required to support separate scavenge pumps. Check valves in the pump permit removal of the integral filter element without loss of oil. The lube pump also provides the drive input for the core speed sensor.
3. Positive lubrication of the main bearing areas is assured with the incorporation of two lube supply jets for each bearing. These jets are mounted in a common housing, thereby reducing problems with fatigue by raising the resonant frequency of the assembly.
4. Lube supply jets are also provided for the splines in the engine gearbox.
5. An antileak check valve is provided to prevent leakage of oil into the gearbox after shutdown.
6. The lube supply has no pressure regulator. The system has been designed to provide oil flow and pressure, proportional to engine speed, throughout the engine operating regime. With the absence of a regulating device, it is easier to detect pressure excursions associated with lube system malfunctions.

Lube Scavenge Subsystem

The lube scavenge subsystem consists of the five scavenge elements of the lube and scavenge pump, the scavenge oil filter, the fuel-oil heat exchanger, and the related plumbing. Oil from the B, C, and D sumps is suction-fed through separate tubes and is directed to the respective scavenge elements of the pump. Oil from the A sump is channeled down the vertical shaft to the transfer gearbox and is suctioned to its respective scavenge element. The oil from the accessory gearbox is suctioned to the remaining element.

The scavenge oil from each element is combined in the pump and is discharged through a common port. This common oil is then directed through the master chip detector, the 46-μ scavenge oil filter, the fuel-oil heat exchanger, and finally to the oil tank. During cold-start operations the oil is bypassed from the fuel-oil heat exchanger through a valve in the exchanger inlet. Upon return to the oil tank, the oil is fed through a vortex-type deaerator where entrained air is removed and vented to the transfer gearbox.

The same features applied to designing the lube supply subsystem are duplicated in the design of the scavenge subsystem.

1. Each scavenge element of the integral lube and scavenge pump is fitted with a 26-in mesh inlet screen for fault isolation. Provisions are also made for a magnetic chip detector in each element.
2. The scavenge oil is directed to a common line upstream of the scavenge filter; and this filter is equipped with a bypass valve to ensure oil flow should the element become clogged. Check valves permit removal of the filter without loss of oil. A master chip detector is also provided upstream of the filter for early detection of metal particles in the engine oil.

Oil Seal Pressurization Subsystem

The oil seal pressurization subsystem consists of the plumbing and air passages which route fan discharge air to the main shaft oil seals. The pressurized air is utilized to prevent the oil from leaking through the seals, and to cool the bearing sumps. (Fig. 25-60)

The fan discharge air is extracted at the leading edge of the compressor front frame and is diffused and distributed internally to each engine oil seal. The air in the sumps is removed directly by the sump vent subsystems, while air that is mixed with the oil is removed by the vortex deaerator in the oil tank.

The internal air passages are designed to provide positive flow through each seal. Balance piston seals and plenum cavity seals are used to minimize air leakage from the supply.

The sumps (Fig. 25-61) are encased in protective air jackets which prevent excessive heat fluxes from reaching the oil-wetted walls, thereby preventing coking and thermal deterioration of the oil. Insulation of the A, C, and D sumps is accomplished by directing the seal pressurization air through an insulating air cavity. The cavity around the A sump absorbs heat rejected by the sump oil, while the cavities around the C and D sumps absorb heat from the high-temperature surroundings.

Sump Vent Subsystem

The sump vent subsystem is composed of the plumbing necessary to remove the air from the engine sumps and to vent it overboard through the tubing in the LPT shaft. The A, C, and D sumps are vented internally to the center vent tube, while B sump is vented to A sump through a tube that runs externally along the compressor case. Employment of internal sump ventilation minimizes external engine plumbing and eliminates airplane

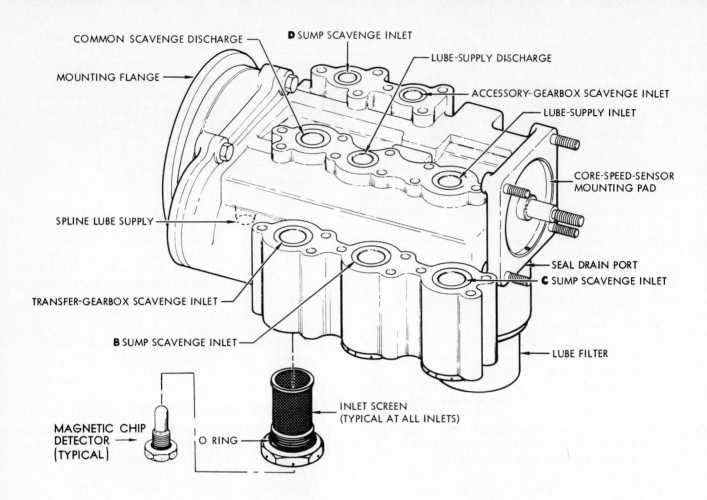

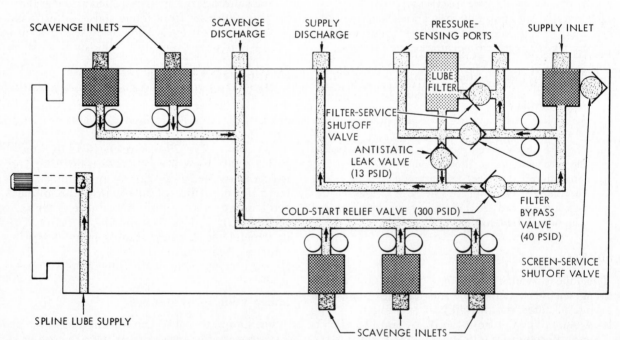

FIG. 25-59 The CF6-6 lube and scavenge pump external view and internal schematic.

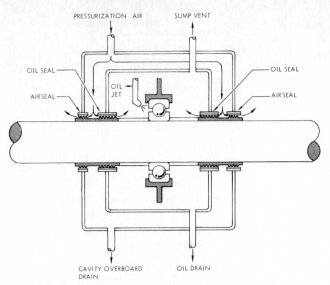

FIG. 25-60 Typical sump sealing arrangement.

interface requirements for venting. Windbacks are incorporated on the shaft in the bearing areas to prevent outward flow of oil during operation of low fan-discharge pressure. Slinger disks are incorporated in the A and D sumps to enhance scavenge oil flow during climb, dive, and roll attitudes. Coking of the center vent tube is prevented with the flow of cool fan discharge air through the shaft enroute to the C and D sumps. Coking of the B sump vent line is prevented by oil injection into the line.

ELECTRICAL SYSTEM

This section describes engine components (Fig. 25-62) that rely on electricity for their operation. Included as electrical equipment, however, are the following items, where (1) indicates systems not included in engine standard equipment, (2) indicates airframe-furnished equipment, and (3) indicates parameter available for indication, but not utilized by airframer:

1. Ignition system
2. Exhaust gas temperature
3. Fan speed sensor
4. Core speed sensor (1)
5. Fuel pump interstage pressure (2)
6. LPT inlet pressure $Pt_{5.4}$
7. EPR (2)
8. Oil supply pressure (2)
9. Scavenge oil temperature (2)
10. Usable oil quantity (2)
11. Scavenge filter pressure drop (2)
12. Fuel flow (2)
13. Fuel-oil heat fuel discharge temperature (2) (3)
14. Low fuel pump interstage pressure (2)
15. Fuel filter pressure drop (2)
16. Starter air valve (1)
17. Main engine control flight/ground idle reset signal (2)
18. Variable-stator-vane power trim motor and solenoid operated by power lever rig pin (1)

Ignition System

The purpose of the ignition system (Fig. 25-63) is to ignite the fuel-air mixture during the starting cycle and to provide continuous ignition during takeoff, landing, and adverse weather conditions. These functions are accomplished with two independent systems, each composed of a high-energy ignition exciter, shielded ignition lead, and an igniter plug.

High-energy Ignition Exciter

The two ignition exciters are located on the fan frame at approximately the four o'clock position. Each exciter has three connectors: input, output, and fault isolation.

Input:	115 V, 400 Hz, 100 VA
Output:	Pulsed output, designed to provide 2 J per spark at two sparks per second
Fault isolation:	Monitors quality of output.

The function of the exciter is to transform the 115-V, 400-Hz input current into a pulsed high-energy output. It is capable of storing 14.5 to 16.0 J with a 15 to 20 kilovolt (kV) output.

Shielded Ignition Lead

The ignition leads serve to deliver the 2-J energy to the igniter plug. The leads run from the exciter on the outside of the fan frame to the core engine via the no. 7 fan casing strut. Each lead is about 13 ft [3.9 m] in length. They are constructed of silicone insulated wire in sealed flexible conduit having a copper inner braid and nickel outer braid.

Igniter Plug

The igniter plugs are mounted in the compressor rear frame at the four and five o'clock positions and extend into two combustion liner swirl cups. The plugs are mounted on a threaded adapter which receives the lead end.

The plugs deliver 2 Joules per spark at 100,000 W peak power. Their construction is a surface-gap-type with a large center electrode (0.200 to 0.240 in) [0.06 to 0.072 cm]). Estimated continuous duty sparking life is 100 h.

Shown in Fig. 25-63 is a schematic of the dual ignition system with its cockpit control. Alternating between systems 1 and 2 is recommended to prolong the life of the igniter plugs, and thus reduce the maintenance requirement.

Exhaust Gas Temperature

The EGT indicating system consists of four thermocouple harnesses (Fig. 25-64) and probe segments and two thermocouple leads.

Three of the thermocouple harness segments (Fig. 25-65) have three dual immersion probes and a fourth segment has two dual immersion probes. The aft thermocouple lead has four electrical connectors for attach-

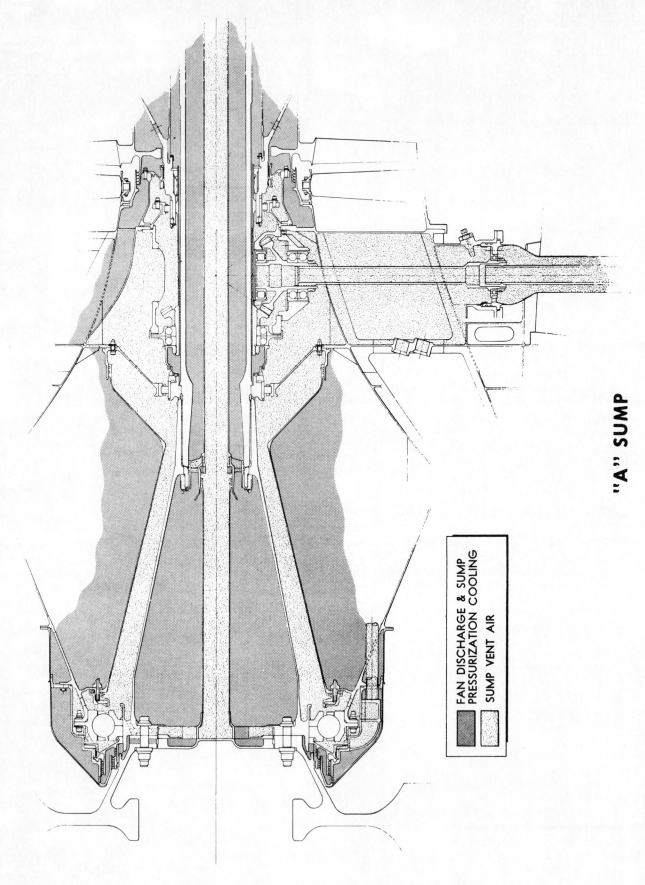

"A" SUMP

FAN DISCHARGE & SUMP
PRESSURIZATION COOLING

SUMP VENT AIR

FIG. 25-61 The A, B, C, and D sump areas.

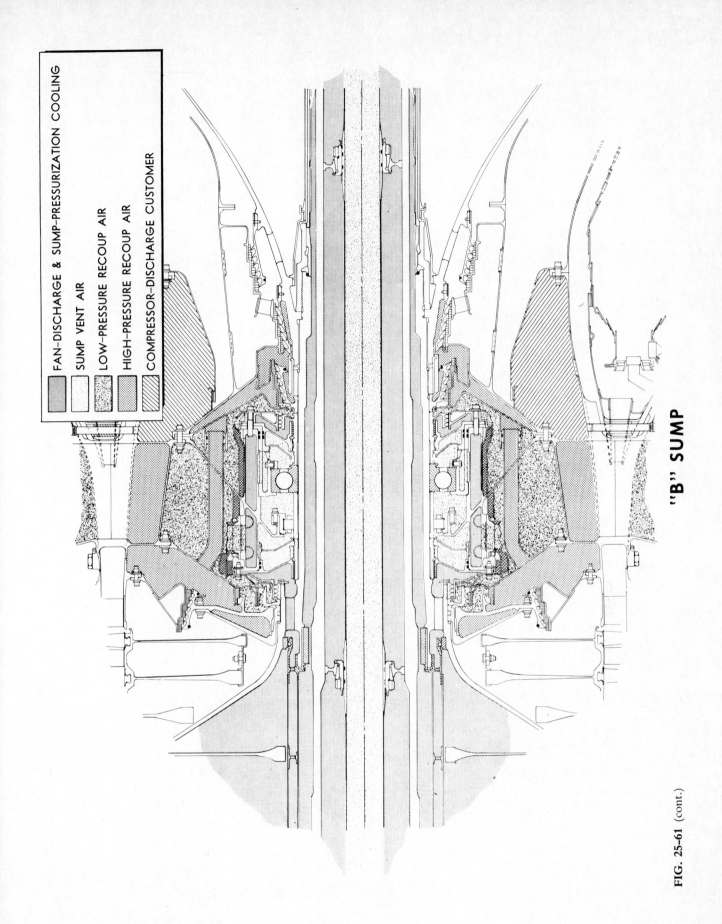

FAN-DISCHARGE & SUMP-PRESSURIZATION COOLING

SUMP VENT AIR

LOW-PRESSURE RECOUP AIR

HIGH-PRESSURE RECOUP AIR

COMPRESSOR-DISCHARGE CUSTOMER

"B" SUMP

FIG. 25-61 (cont.)

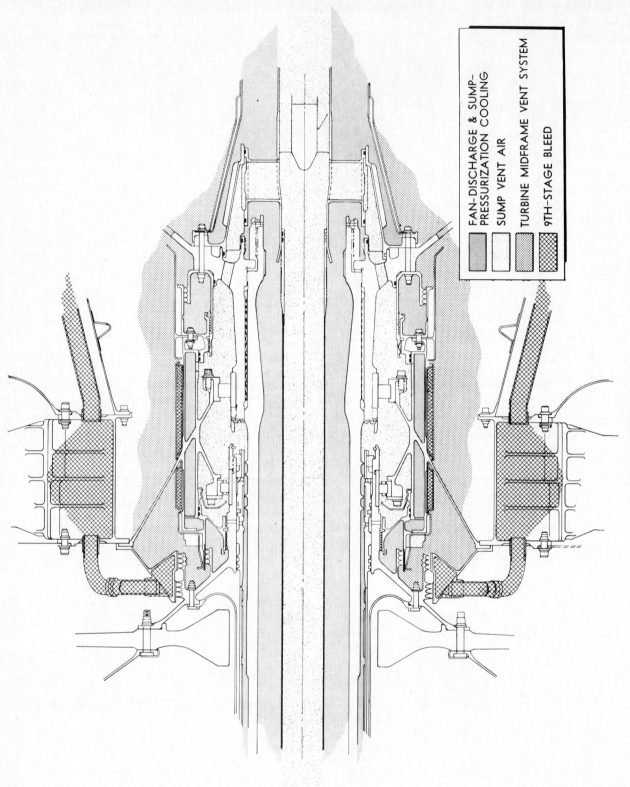

FAN-DISCHARGE & SUMP-
PRESSURIZATION COOLING

SUMP VENT AIR

TURBINE MIDFRAME VENT SYSTEM

9TH-STAGE BLEED

"C" SUMP

FIG. 25-61 (cont.)

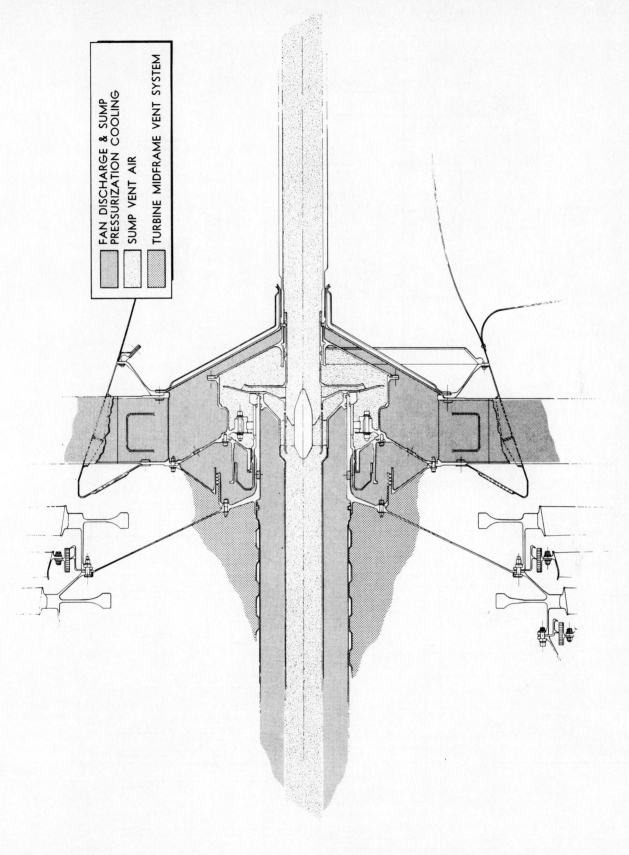

FAN DISCHARGE & SUMP
PRESSURIZATION COOLING

SUMP VENT AIR

TURBINE MIDFRAME VENT SYSTEM

"D" SUMP

FIG. 25-61 (cont.)

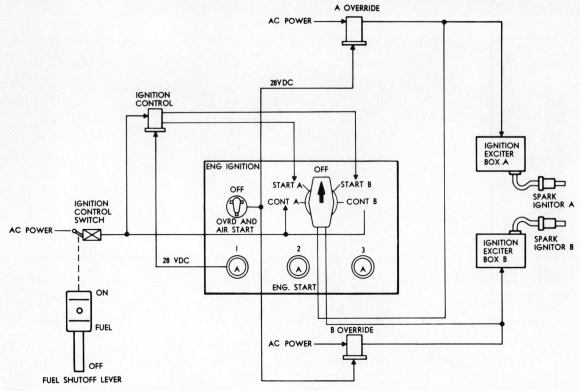

FIG. 25-62 The electrical system.

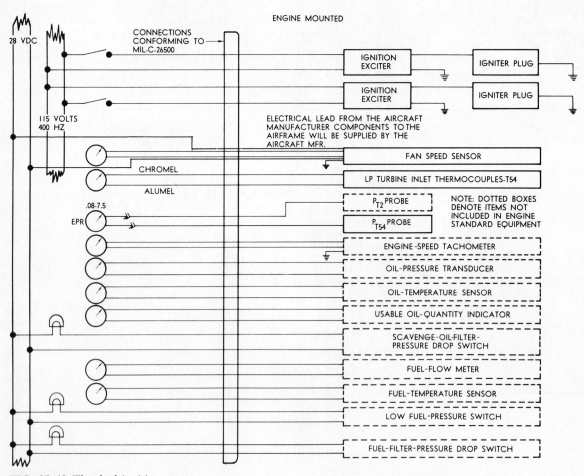

FIG. 25-63 The dual ignition system.

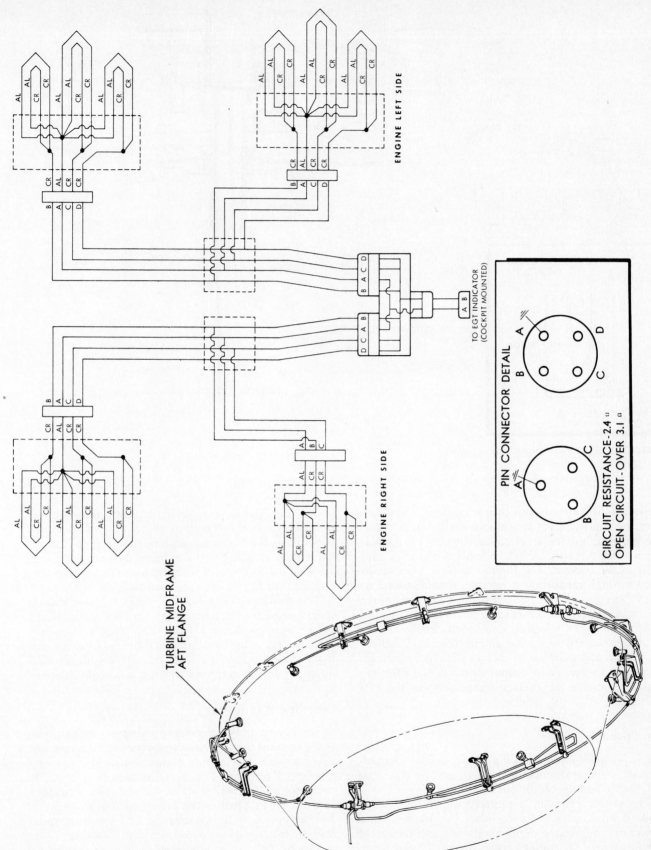

ENGINE LEFT SIDE

ENGINE RIGHT SIDE

TURBINE MID FRAME
AFT FLANGE

TO EGT INDICATOR
(COCKPIT MOUNTED)

PIN CONNECTOR DETAIL

CIRCUIT RESISTANCE-2.4 Ω
OPEN CIRCUIT-OVER 3.1 Ω

FIG. 25-64 Thermocouple harness.

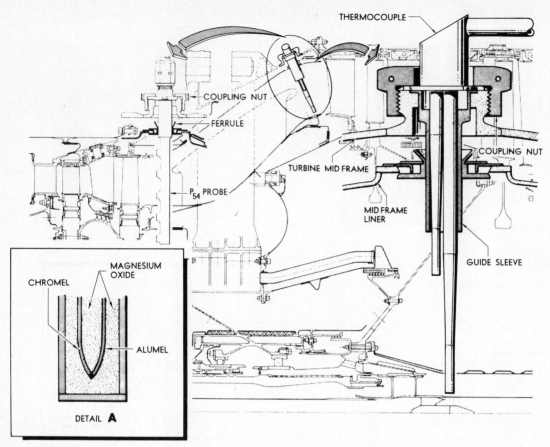

FIG. 25-65 EPR and thermocouple probes.

ment to the four thermocouple segments and one electrical connector for attachment to the forward lead. The forward lead has an electrical connector for attachment to the aircraft lead which leads to the EGT indicator in the cockpit. The thermocouple probes are located through the turbine midframe, spaced around the engine circumference. The circuitry of the thermocouple harness permits the reading of individual probes which facilitates the detection of open circuits and the locating of hot streaks in the engine. The EGT indicating system uses a common junction and is geometrically balanced. The resultant output signal of the four harnesses represents the average temperature of the LPT inlet gas.

Fan Speed Indicator

The N_1 speed sensors (see Fig. 19-2) are eddy-current-type self-contained units mounted at the 10 o'clock and 2 o'clock positions on the fan case.

The sensor provides a primary signal for cockpit readout of fan speed and a secondary 1/revolution signal to identify the passage of one modified fan blade on the fan rotor for "on engine" balancing of the fan rotor.

The N_1 speed sensor is mounted on the fan case and penetrates the rub strip material in the plane of blade rotation. The passage of each fan blade disrupts the flux field set up by the sensor causing an electrical signal

pulse. These pulses are equal in frequency to the number of blades times the rpm, thus giving a signal frequency proportional to fan speed, with no inaccuracy. This signal is then amplified and conditioned to provide a 0 to 10 V signal to the cockpit indicator. The input power requirement for the N_1 speed sensor is 28 V dc.

The secondary signal pulse results when a slug of special material (located in the tip of one fan blade) passes through the flux field. This pulse occurs only once per fan rotor revolution and differs from the primary pulse because the slug material has higher conductivity and/or permeability than fan blade material.

Core Speed Indicator

The N_2 or core-engine-speed indicating system consists of a bearingless-tachometer-type core speed sensor that generates an electrical signal to a cockpit core-engine-speed readout. This signal is generated by a 31-tooth rotor. It provides at 100 percent N_2 a frequency approximately equal to the fan speed sensor output at 100 percent N_1. This permits use of a common speed readout for N_1 and N_2 except for dial marking.

The N_2 or core-engine-speed indicating system incorporates a generator which emits a single-phase ac signal. The generating unit is mounted on the forward end of the lube and scavenge pump and is driven by an extension of the lube and scavenge pump drive shaft.

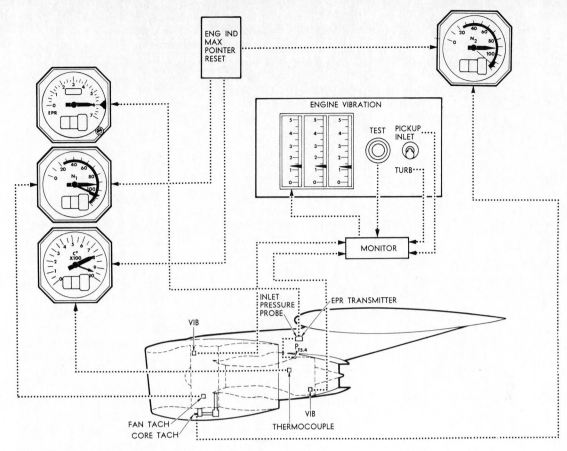

FIG. 25-66 Engine indicating functional diagram.

Low-pressure-turbine Inlet Pressure

The LPT inlet pressure probe ($Pt_{5.4}$) is a single-tube, closed-end probe with four equal-diameter orifices equally spaced along the working length of the probe. The four orifices permit averaging the LPT inlet pressure. The probe is mounted on the outside of the turbine midframe for easy accessibility and the position of the probe is fixed by locating lugs on the probe flange.

The pressure probe extends into the gas stream in the LPT inlet. The $Pt_{5.4}$ pressure is transmitted to the airframe-furnished pressure ratio transmitter to be compared with fan inlet total pressure (Pt_2). Fig. 25-66 shows the relationship of the various transmitters and probes.

REVIEW AND STUDY QUESTIONS

1. What aircraft use this engine?
2. Why does the LPT have five stages on the CF6-6 model, and four stages on the -50 model engine?
3. Discuss some of the new technology that has been used on this engine to allow higher TIT.
4. Name, and discuss some of the features incorporated into this engine to make inspection, maintenance, and overhaul easier.
5. Smokeless combustion is a feature of this engine. How is this accomplished?
6. The labyrinth seal is used extensively in this engine. Discuss the philosophy behind this method of sump sealing.
7. Briefly explain the construction features of the fan and turbine reversers.
8. In addition to the control of fuel, what other function does the main engine control (MEC) have?
9. List all the parameters used to indicate correct engine operation.
10. List the CF6-6 and -50 engine's major specifications.
11. Give a brief description of airflow through the fan and core portions of this engine.
12. List the number and location of the main bearings.
13. What are the advantages and disadvantages of the variable-geometry compressor used on this engine?

CHAPTER TWENTY-SIX
PRATT & WHITNEY
AIRCRAFT
JT8D

The last engine to be discussed in detail is the highly produced Pratt & Whitney JT8D turbofan engine (Fig. 26-1) used in the several versions of the Boeing 727, Boeing 737, and McDonnel Douglas DC-9 aircraft. Since this engine has been in production for a number of years, many developmental changes have taken place that have resulted in increased thrust, reliability, and service life. Improvements include cooled turbine parts, reduced-smoke combustion chambers and strengthened and redesigned parts throughout. (See Fig. 2-43)

SPECIFICATIONS

Number of fan stages:	2
Number of compressor stages:	11
Number of turbine stages:	4
Number of combustors:	9
Maximum power at sea level:	14,000 lb [62,272 N] to over 17,000 lb [75,616 N]
Specific fuel consumption at maximum power:	0.60 lb/lbt/h (61.2 g/N/h)
Compression ratio at maximum rpm:	16:1 to 17:1
Maximum diameter:	45 in (114.3 cm)
Maximum length:	124 in (315 cm)
Maximum dry weight:	3250 lb (1475.5 kg)

GENERAL DESCRIPTION *(FIG. 26-2)*

This engine operates similarly to all turbojet versions of a gas turbine engine in that it derives its propulsive force through the application of Sir Isaac Newton's third law, which states that for every action there is an equal and opposite reaction. The engine cases form the backbone of the engine when bolted together, and support all of the inner parts of the engine through struts and bearings. The fan discharge air is ducted outside the inner cases because the air has already been accelerated by the fan and has therefore served its purpose of providing additional thrust, the same kind of additional thrust that would be gained from air passing through the propeller of a turboprop or reciprocating engine (Fig. 26-3).

The JT8D engine is an axial-flow front turbofan engine having a 13-stage split compressor, a 9-can (can-annular) combustion chamber, and a split 4-stage reaction impulse turbine. The engine is equipped with a full-length annular fan discharge duct. The low-pressure system is made up of the front compressor rotor and the second-, third-, and fourth-stage turbine rotors and is mechanically independent of the high-pressure system which consists of the rear compressor rotor and the first-stage turbine rotor. The engine is mounted from two points. The front mount is located at the fan discharge intermediate case. The engine rear is located at the turbine exhaust section outer duct.

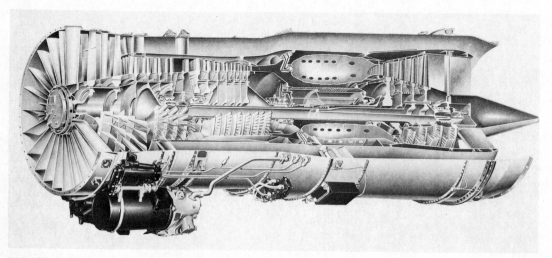

FIG. 26-1 Cutaway view of the Pratt & Whitney Aircraft JT8D Turbofan engine.

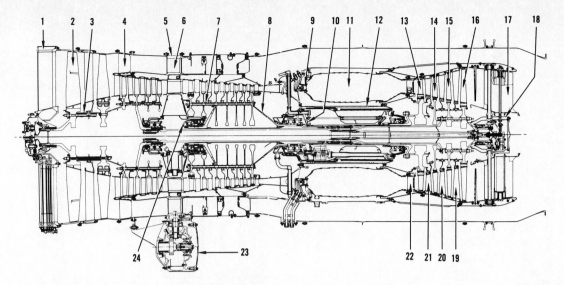

1 FAN INLET CASE
2 FIRST-STAGE FAN BLADES
3 FRONT COMPRESSOR ROTOR
4 FAN-DISCHARGE VANES
5 FAN-DISCHARGE INTERMEDIATE CASE
6 FAN-DISCHARGE-INTERMEDIATE-CASE STRUTS
7 REAR COMPRESSOR ROTOR

8 REAR-COMPRESSOR REAR HUB
9 FUEL NOZZLE
10 NO. 4 BEARING OIL NOZZLE
11 COMBUSTION CHAMBER
12 COMBUSTION-CHAMBER INNER CASE
13 FIRST-STAGE TURBINE BLADES

14 SECOND-STAGE TURBINE BLADES
15 THIRD-STAGE TURBINE BLADES
16 FOURTH-STAGE TURBINE BLADES
17 EXHAUST STRUT
18 NO. 6 BEARING HEAT SHIELD
19 FOURTH-STAGE TURBINE VANES

20 THIRD-STAGE TURBINE VANES
21 SECOND-STAGE TURBINE VANES
22 FIRST-STAGE TURBINE VANES
23 GEARBOX
24 GEARBOX DRIVE BEVEL GEAR

FIG. 26-2 Sectioned view showing major parts.

AIR INLET SECTION

Fan Inlet Case Assembly *(FIG. 26-4)*

The air enters the engine through the compressor inlet case. The inlet case and its vanes, with one thicker vane at the bottom carrying engine tubing, direct air to the face of the compressor.

The no. 1 bearing front support assembly is mounted in the center of the compressor inlet case. Behind the front support is the no. 1 bearing rear support. Mounted on the front of the inlet case, in the center, is the front accessory-drive support.

The fan inlet case contains 19 equally spaced vanes, 18 of which have 2 internal ribs running the length of

the vane. These ribs divide the hollow portion of the vanes into three passageways. The vane at the six o'clock position also contains the two internal ribs running the length of the vane; however, in this instance all three passageways are used to conduct tubing. The center passageway of this vane is filled with a rubber compound to dampen tube vibration.

The vanes are brazed between titanium inner and outer shroud cases. A welded ring, brazed in place, adds structural rigidity and forms the outer wall of the outer case. Studs in rear flange are engaged by locknuts to hold the case to the front compressor case.

Five tubes are routed through the three passageways between the outer and inner case in the vane at the six o'clock position: The no. 1 bearing oil pressure tube,

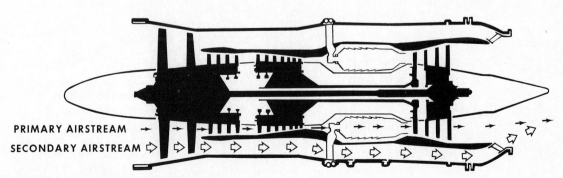

PRIMARY AIRSTREAM
SECONDARY AIRSTREAM

FIG. 26-3 Primary and secondary airflow.

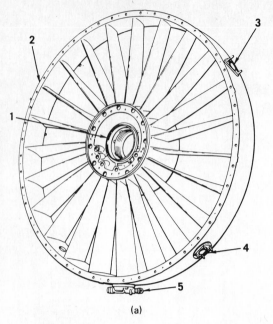

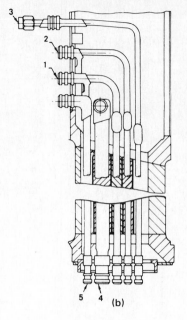

1	NO. 1 BEARING FRONT SUPPORT	4	BOSS—TEMPERATURE PROBE, FAN INLET CASE
2	FAN INLET CASE	5	CONNECTOR ASSEMBLY
3	BOSS—AIR, FAN INLET CASE LEFT		

1	OIL SCAVENGE TUBE
2	OIL-PRESSURE TUBE
3	P_{t2} TUBE
4	BREATHER TUBE
5	TACHOMETER WIRE TUBE

FIG. 26-4 (a) Fan inlet case and No. 1 bearing front support.
(b) Fan inlet master vane.

no. 1 bearing oil scavenge tube, no. 1 bearing breather tube, the tachometer conduit tube, and the compressor inlet air pressure tube which senses P_{t2} from the airframe nose cone.

A five-passage tube connector is secured to the bottom of the case, as is a water drain screen and plug assembly.

Two bosses (four studs each) are located on the outer case near the bottom, one on each side. The engine pressure probe is located in the boss on the right and the temperature probe boss is located on the left.

There are two anti-icing air bosses (three studs each), at the approximate 10 and 2 o'clock positions, brazed in the outer wall of the case. Anti-icing air passages are formed within the case to permit the air to flow between the outer wall and outer shroud, inward through the hollow vanes, and discharge forward through the front of the inner shroud case.

No. 1 Bearing Front and Rear Support
(see FIG. 26-39)

The no. 1 bearing front-support assembly is mounted in the center of the compressor inlet case behind the front accessory-drive support. It holds the no. 1 bearing outer race secured in a steel bushing in its inner diameter (ID) by a nut and a flared rivet. Three puller slots in the rear lip of the bushing facilitate removal of the bearing outer race. The support, of cast aluminum, also holds an aluminum multistepped seal ring behind the bushing.

The seal ring is secured in place by a flange near the front and a flared lip at the rear.

Holes in the front flange of the support accommodate the oil and anti-icing air tubes.

Behind the front support is the no. 1 bearing rear support, cast from aluminum alloy in the shape of an open dish. This unit snap-fits on the rear side of the front support and is bolted and lockwired to the rear flange of the inlet case inner shroud. It shares the supporting loads of the no. 1 bearing with the front support.

Three puller lugs on the ID of the rear flange provide a means of removing the support from the inlet case.

Front Accessory-drives Support

Mounted on the front of the inlet case, in the center, is the front accessory-drives support. This case magnesium support incorporates a four-stud N_1 tachometer pad on the upper front face. A pressure-oil passage in the support carries oil from the rear of the outer flange into the center, then rearward through the no. 1 bearing oil nozzle. A scavenge-oil passage carries oil from a pump boss, on the lower rear face of the support cavity, back toward the outside of the support then to another opening in the rear of the outer flange.

The front accessory-drives support has a machined flat on the lower mounting lugs to accommodate a bracket assembly to be used for the N_1 tachometer equipment.

The N_1 tachometer drive gearshaft and the scavenge pump gearshaft are driven by the front accessory-drives gearshaft located in the front hub of the front compressor rotor.

The no. 1 bearing oil scavenge pump mounts on the pump boss inside the front accessory-drives support.

COMPRESSOR SECTION (FIG. 26-5)

Fan Section (FIG. 26-6)

To the rear of the inlet case, enclosing the fan section, are the front and rear fan cases. The fan is not a separate unit but is formed by the outer diameter of the first two stages of the front compressor and is described in greater detail with it. At the rear of the compressor section the compressed air enters the diffuser section.

Front Compressor Section (FIG. 26-7)

The axial-flow front (low-pressure) compressor partially compresses the air that passes through the primary (inner) air stream of the engine, then delivers this air to the rear (high-pressure) compressor. In addition, the larger first- and second-stage blades also accelerate the secondary (outer) air stream, which then passes through the fan discharge vanes and rearward through the annular duct.

The front compressor rotor (N_1) is driven by the second-, third-, and fourth-stage turbines through the front compressor drive (long) turbine shaft at N_1 rotational speed. The front-compressor drive turbine shaft is splined at the front into the front compressor rotor rear hub. The shaft and hub are held axially by a coupling threaded to the shaft and fixed in position by a coupling lock inside the front-compressor-rotor rear hub.

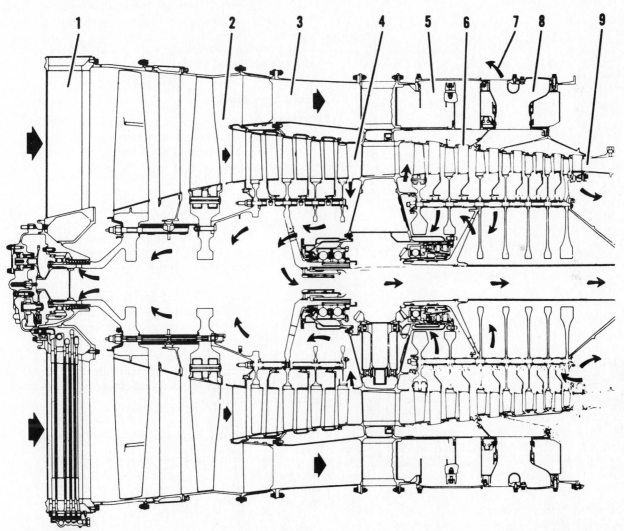

1 FAN INLET AIR (P_{t2}, T_{t2})
2 LOW-PRESSURE COMPRESSOR (N_1)
3 FAN-DISCHARGE AIR
4 LOW-PRESSURE COMPRESSOR DISCHARGE AIR (P_{t3}, T_{t3})
5 LOW-PRESSURE (N_1)BLEED
AIR
6 HIGH-COMPRESSOR (N_2)
7 ANTI-ICING AIR
8 EIGHTH-STAGE BLEED AIR
9 HIGH-PRESSURE COMPRESSOR-DISCHARGE AIR (P_{t4}, T_{t4})

FIG. 26-5 Gas flow diagram, compressor section.

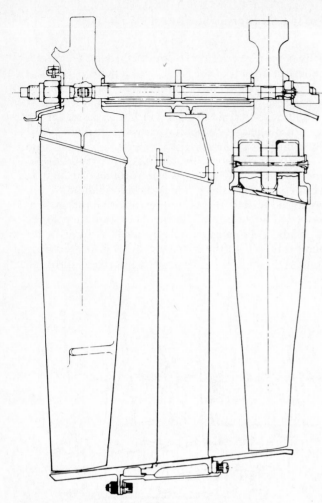

FIG. 26-6 N_1 compressor fan rotors.

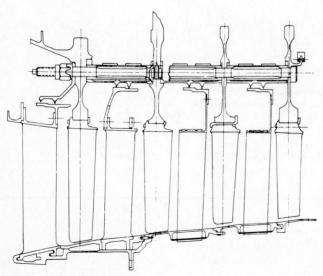

FIG. 26-7 N_1 compressor rotors.

Each compressor rotor is driven by a separate turbine, and each rotor is mechanically independent of the other.

Rear Compressor Section

The purpose of the rear compressor is to further compress the air delivered by the front compressor and to then feed this air into the diffuser case and combustion section.

The rear compressor rotor is driven by the first stage of the turbine through the rear-compressor drive turbine (short) shaft. The turbine shaft splines onto the rear compressor rear hub and is retained by the turbine shaft coupling.

Front and Rear Fan Cases

Behind the inlet case, and enclosing the fan section, are the front and rear fan cases, of decreasing diameter from front to rear. Both cases are constructed of steel and the rear case has a shoulder at the front to accommodate the first-stage vanes.

Antirotation positioning pins in the flanges ensure correct positioning of the cases in the engine.

Front Compressor Stators

There are six stages of stator vanes in the front compressor. Five of the stages are within the rotor and stator assembly. The sixth is in the front of the compressor (intermediate) case and is described with it. Stage 1 is titanium with an aluminum seal ring. Stages 2 and 3 are either aluminum with riveted vanes, or are steel with aluminum seal rings and steel strip stock vanes. (Most stage 2 and 3 steel strip stock vane stators incorporate silicone compound in the outer box shroud between the vanes.) Stages 4 and 5 are steel, with a box-type outer shroud. All of stages 1 through 5 are of continuous ring construction.

The stators are designed to resist the torque loads transmitted by the aerodynamic forces along the vanes and to absorb the bending moments imposed by the pressure differentials across the vanes. The vanes decrease in size from front to rear and the angle at which they are mounted is set to feed air into the following row of rotor blades to give best compressor efficiency at operating speed.

Front-compressor Rotor and Stator Assembly

The front-compressor rotor and stator assembly consist of the six-stage front-compressor rotor, the front-compressor front and rear cases, and the vanes and shrouds for stages 1 through 5. The sixth-stage vanes are in the compressor intermediate case and will be described with it.

The front fan case encloses the first stage blades and the front-compressor fan case encloses the first-stage vanes and the second-stage blades. The rear stages of the rotor are enclosed within the inner shroud of the fan-discharge-case vane assembly and the front-compressor section inner duct.

The numbering of the blade stages from front to rear is 1 through 6. The first- and second-stage blades are considerably larger than the rest and are also referred to as "fan blades." The vane stages, to the rear of their blade stages, are numbered in the same manner, 1 through 5 in the rotor and stator assembly and stage 6 in the compressor intermediate case.

Front-compressor Rotor

The front-compressor rotor has a front hub (which serves as the first-stage disk), a rear hub (which serves as the fourth-stage disk), four rotor disks, six stages of blades secured in the hubs and disks, five rotor disk spacers, and two sets of tierods. Some models have only four separate rotor disk spacers. The third-stage disks of these models incorporate an integral spacer. This rotor is driven by the front-compressor drive turbine rotor.

The front hub, the first-to-second-stage spacer, the second-stage disk, and the second-to-third-stage spacer are held together by 16 front tierods. The second-to-third-stage spacer, the third-stage disk, the rear hub, and the fifth- and sixth-stage disks are held together, with the spacers between, by 12 rear tie rods. The front and rear sections of the rotor are held together at the front and rear flanges of the second-to-third-stage spacer.

The rotor disk spacers each have two knife-edge airseals on their outer diameter (OD). These knife-edges rotate just inside matching seal rings on the ID of the vane and shroud assemblies. The knife-edge airseal of the second-to-third-stage spacer is incorporated in the rear flange of the spacer and matches the ring inside the second-stage vanes.

The first-stage-compressor blades (fan) are dovetailed into matching grooves in the front hub rim and are retained by a tab at the leading edge of the blade root that prevents rearward movement and a positioning ring that prevents forward movement. The positioning ring is retained by the rotor tierods. The second-stage blades (fan) are held in the disk by a pin-joint attachment with a flared rivet.

The third through the sixth stages of blades are fastened to the disks by dovetail root sections which fit into broached slots in the disk rim. Tablocks fit in the bottom of the blade root (and disk slots) and are bent inward at the tab to effect blade locking.

The front hub and the second-stage disk are titanium. The third-stage disk is steel; the rear hub and the fifth- and sixth-stage disks are titanium.

The first through sixth stage blades are titanium.

Front-compressor Rotor Rear Hub Coupling (FIG. 26-8)

The front-compressor drive turbine shaft splines into the front-compressor rotor rear hub. The shaft and the hub are held axially by a hollow front-compressor drive coupling, threaded to the shaft and fixed in position by a coupling lock inside the front-compressor-rotor rear hub.

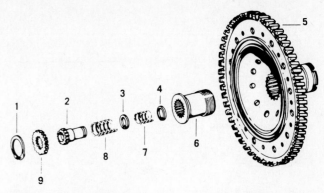

1 RING—LOCKRING RETAINING	SOR ROTOR, REAR
2 LOCK	6 COUPLING
3 RING RETAINING	7 SPRING—REAR
4 RING RETAINING	8 SPRING—FRONT
5 HUB—FRONT COMPRES-	9 LOCKRING

(a)

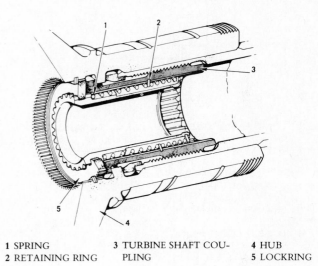

1 SPRING	3 TURBINE SHAFT COU-	4 HUB
2 RETAINING RING	PLING	5 LOCKRING

(b)

FIG. 26-8 (a) Front-compressor coupling—exploded view. **(b)** Front-compressor coupling—detailed sectioned view.

Compressor (Intermediate) Case Assembly (FIG. 26-9)

At the rear of the front compressor rotor is the welded steel compressor intermediate case. This case forms the outer wall of the basic inner engine from the fan discharge vanes to the diffuser-case front flange.

The sixth-stage steel vanes and the no. 3 bearing housing are welded inside the case. The no. 2 bearing housing is bolted to the front face of the case. A steel support and a support bushing below it are positioned at the bottom center of the case to accommodate the main accessory-drive gearshaft bearing housing. Positioned inside the bearing housing is the main accessory drive bevel gearshaft, in a roller bearing at the top and a ball bearing at the bottom.

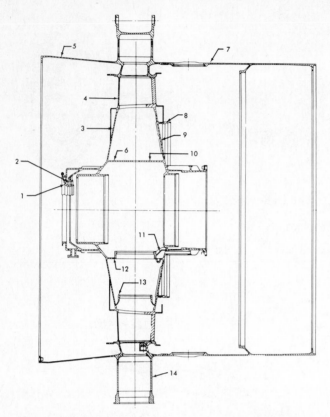

1 NO. 2 BEARING AIRSEAL RING	8 REAR-COMPRESSOR ROTOR FRONT AIRSEAL RING
2 NO. 2 BEARING OIL-SEAL RING	9 NO. 3 BEARING SUPPORT
3 NO. 2 BEARING SUPPORT	10 NO. 3 BEARING HOUSING
4 SIXTH-STAGE VANE	11 SEAL-BLEED MANIFOLD SEGMENT
5 FAN-DISCHARGE FRONT-COMPRESSOR INNER DUCT	12 ACCESSORY DRIVES SUPPORT
6 NO. 2 BEARING HOUSING	13 GEARSHAFT-BEARING HOUSING GUIDE
7 FAN-DISCHARGE REAR-COMPRESSOR INNER DUCT	14 COMPRESSOR-INTERMEDIATE FAN CASE

FIG. 26-9 Compressor intermediate case construction.

Fan-discharge Rear-compressor Section Inner Duct

Riveted to the rear of the compressor intermediate case is the fan-discharge rear-compressor section inner duct. This cylindrical steel duct forms the inner wall of the fan discharge air passage at this location and the flanges are fabricated on the inside surface to accommodate the airflow. The rear flange of this case is equipped with self-locking nuts flared securely in place.

Two double-hole low-pressure air bleed bosses are near the front, each approximately 45° above the horizontal center line.

At the rear are two single-hole eighth-stage air bleed bosses, each approximately 45° above the horizontal center line. On engines equipped with an eighth-stage bleed system, a boss is provided for an eighth-stage bleed valve at the six o'clock position on the inner duct.

Compressor Intermediate Fan Case

Secured to the rear flange of the front-compressor section outer duct, is the compressor intermediate fan case. The compressor intermediate fan case is an integral part of the outermost diameter of the compressor (intermediate) case.

The compressor intermediate fan case incorporates streamlined struts between the intermediate case and the outer diameter of the engine. The larger six o'clock position strut accommodates the accessory gearbox main drive shaft.

The front mounting points of the engine are on the outer flange of this case and the accessory gearbox is secured to it at the bottom.

Main Accessory-drive Bevel Gearshaft and Bearing

The main accessory-drive bevel gearshaft is driven by the gearbox drive bevel gear which is inside the no. 3 bearing, splined to the front hub of the rear compressor rotor, and rotates at N_2. The main accessory drive bevel gearshaft, in turn, rotates the gearbox drive shaft through the splined drive-shaft coupling.

Rear Compressor Section (FIG. 26-10)

The purpose of the rear compressor is to further compress the air delivered by the front compressor and to then feed this air into the diffuser case and combustion section.

The rear compressor rotor is driven by the first stage of the turbine through the rear compressor drive turbine (short) shaft. The turbine shaft splines onto the rear compressor rear hub and is retained by the turbine shaft coupling.

Rear-compressor Rotor and Stator Assembly

The rear compressor utilizes a rotor having seven stages of disks and blades, separated by disk spacers, and six stator vane stages.

The blade stages are numbered 7 through 13 from front to rear. The vane stages are numbered correspondingly behind their blade stages, 7 through 12 in the rotor and stator assembly, with the 13th stage and compressor exit vanes in the front end of the diffuser case.

Rear-compressor Stators

The 7- through 12-stage stator assemblies are each of single-piece continuous ring construction incorporating box outer shrouds and stainless steel vanes. The outer shroud of each stage extends forward around the blades. The inner shroud has an airseal ring which provides a mating surface for the knife-edge airseals of the rotor.

Three locking straps retain the 10th-, 11th-, and 12th-stage vane shrouds at their OD. The seventh, eighth, and ninth stages are retained securely in position by

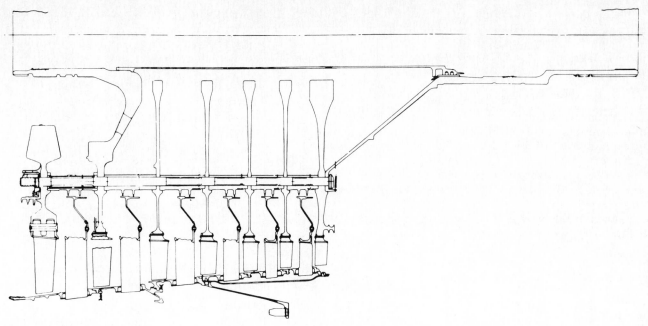

FIG. 26-10 N_2 compressor.

lockwired stator locks on the shroud lugs (or lockwired lugs) and an extended rear flange on the ninth stage outer shroud. A tube and baffle on the ninth-stage stator directs eighth stage air into the no. 4 bearing seal air system.

At the 13th-stage compressor exit, vanes are secured in the exit stator assembly mounted in the forward end of the diffuser case. These vanes are aerodynamically part of the rear compressor but because of their location are described with the diffuser case.

Rear-compressor Rotor

Twelve tierods fasten the disks, spacers, and front and rear hubs together axially. The front hub is positioned at the ninth-stage disk so that the seventh- and eighth-stage disks are cantilevered forward. Knife-edge airseals on the disk spacers are positioned just inside an airseal ring on the inner shroud of each corresponding vane stage. A triple knife-edge airseal is secured to the front of the 7th-stage disk and a four-edge airseal is integral with the rear of the 13th-stage disk.

Blade attachment to the disks is accomplished by a dovetailed lock at the blade root, with the exception of the seventh stage, which uses a pin-joint attachment.

The seventh-, eighth-, and ninth-stage blades of the rear-compressor rotor are titanium. The 10th- through 13th-stage blades are steel. The rear-compressor rotor disks are steel except for the 13th-stage disk, which is made of nickel alloy. The 13th-stage disk incorporates an airsealing configuration on the rear having four knife-edges. The knife-edges match the steel 13th-stage airsealing ring positioned inside the diffuser case.

An oil sealing sleeve extends from the rear of the front hub ID to a steel bushing in the bore of the rear hub. Metal seal rings are positioned in two grooves in the rear end of the sleeve inside the bushing.

The rear compressor is balanced dynamically as a unit. Counterweights may be riveted to the front of the seventh-stage disk, the front hub, and to the OD of the rear hub adjacent to the point of attachment.

The no. 4 bearing inner races and oil baffle are secured on the OD of the rear hub by an inner race retaining nut locked in place with a keywasher. The no. 4 and 5 bearing oil suction-pump drive gear is held on the rear of the hub by two lockrings.

The rear-compressor rotor is driven by the first stage of the turbine through the rear-compressor drive turbine (short) shaft. The turbine shaft splines onto the rear-compressor rear hub, and is retained by the turbine shaft coupling.

DIFFUSER SECTION

General

The function of the diffuser section is to straighten the air flow from the rear compressor and to diffuse the flow to the proper velocity for entry into the combustion chamber. The air passes through the last row of rear compressor blades at a fast rate of speed. This motion is both rearward and circular in pattern around the engine. Two rows of radial straightening exit guide vanes, made of steel and located at the entrance of the diffuser case, slow the circular whirl pattern and convert the whirl velocity energy to pressure energy. After passing through these straightening vanes the air still has a strong rearward velocity. This velocity is so high that it would be nearly impossible to maintain a flame in the air stream. A gradually increasing cross section of the air passage decreases the velocity of the airflow and at the same time converts the velocity energy to pressure energy.

Diffuser Case (FIG 26-11)

The main structural member of this section is the steel diffuser case. The forward part of this case houses the rearmost portion of the rear compressor.

The exit stator assembly is bolted to flanges in the front openings of the diffuser case. This unit contains an inner shroud, outer shroud and small vanes brazed in place.

Located in the divergent section of the case are nine hollow struts having small circular openings on either side which supply compressor discharge air to a manifold around the diffuser case. The manifold provides the discharge air for anti-icing and airframe use through two ports (upper left and upper right) on its outer perimeter.

Between the nine hollow struts, located radially near the rear of the case, are nine fuel-nozzle-support mounting pads. Behind the mounting pads are nine mounting lugs for the front of the individual combustion chamber.

The pressure sensing boss is located at approximately the two o'clock position on the right side outer surface of the diffuser case. Locknuts are incorporated on the rear face of the intermediate front flange and gangnuts are riveted to the inner rear flange to facilitate assembly and disassembly.

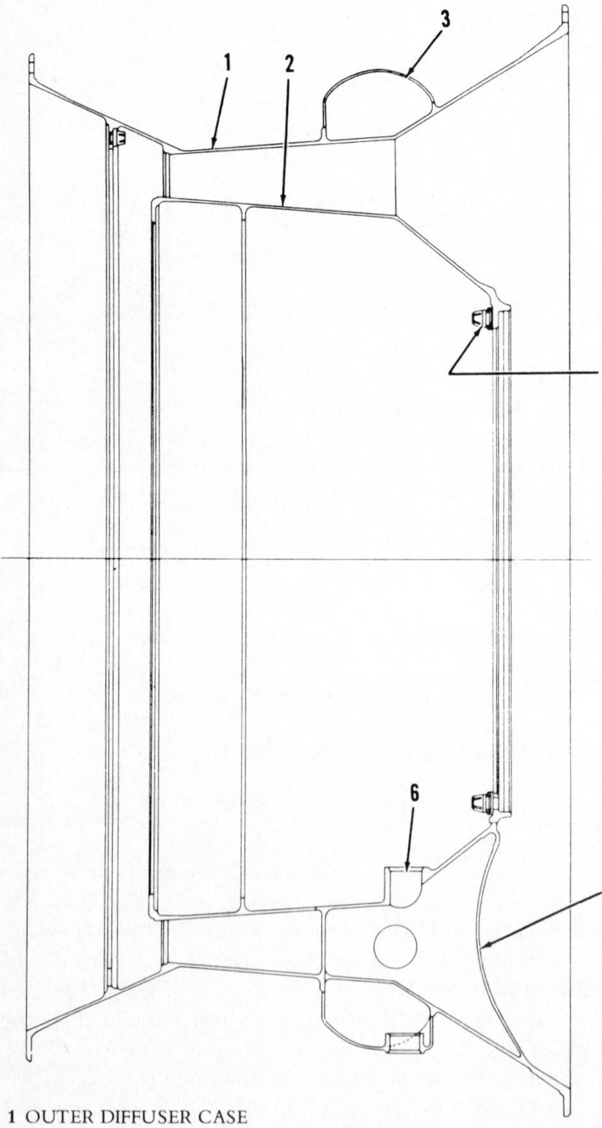

No. 4 Bearing Compartment and No. 4 Bearing Seal Air System (see FIG. 26-41)

The compartment houses the no. 4 bearings in its ID-positioned bearing support. Heat shields are bolted and lockwired in front of the no. 4 bearing compartment to minimize the temperature within. In addition, a tubing system brings eighth-stage discharge air to the annulus between the second and third labyrinth seal units, and bleeds air from the annulus between the first and second labyrinth units to the fan discharge path. The tubes, secured to the openings in the no. 4 bearing airseal ring assembly, hold down the bearing compartment temperature by bleeding hot air before it can reach the compartment.

No. 4 and 5 Bearing Oil Scavenge Pump

The no. 4 and 5 bearing oil-scavenge-pump assembly located inside the diffuser case has two stages driven by a gear mounted on the rear-compressor rear hub.

No. 4 Bearing Housing

Positioned in the center of the diffuser case, within the bearing compartment, is the no. 4 bearing housing. The no. 4 bearing outer races are held in the ID bore of the housing by a large retaining nut, riveted in place. The rear portion of the housing encloses no. 4 and no. 5 bearing oil scavenge pump.

COMBUSTION SECTION (FIG. 26-12)

General (FIG. 26-13)

In the combustion section, fuel is mixed with air at the proper ratio, and the resultant fuel-air mixture is burned, adding energy to the air passing through the engine. The fuel is routed through left and right semi-circular manifolds secured around the outside of the diffuser case at the rear. Nine individually supported nozzles inside the diffuser case deliver fuel into the combustion chambers.

1 OUTER DIFFUSER CASE
2 INNER DIFFUSER CASE
3 DIFFUSER-CASE AIR MANI-
  FOLD
4 NO. 4 BEARING SUPPORT

BOLT CIRCLE
5 DIFFUSER-CASE STRUT
6 NO. 4 BEARING AIR-BLEED
  TUBE OPENING

FIG. 26-11 Diffuser case construction.

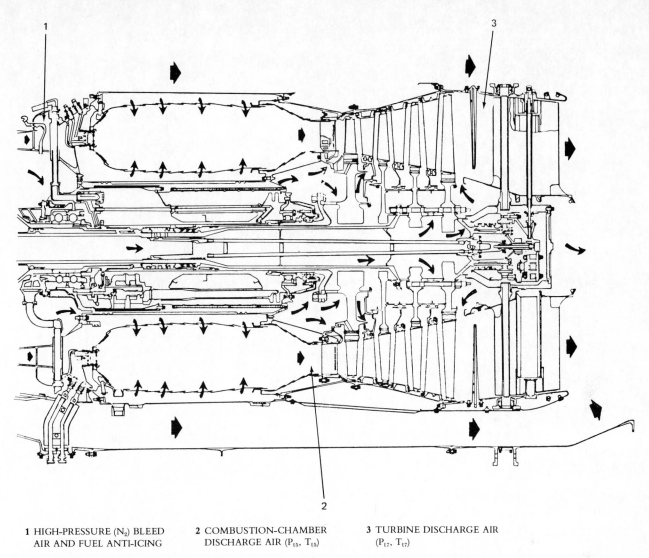

1 HIGH-PRESSURE (N₂) BLEED
AIR AND FUEL ANTI-ICING

2 COMBUSTION-CHAMBER
DISCHARGE AIR (P_{t5}, T_{t5})

3 TURBINE DISCHARGE AIR
(P_{t7}, T_{t7})

FIG. 26-12 Gas flow diagram, combustion section and turbine section.

Combustion Chamber Inner Case

The combustion chamber inner case is secured to the diffuser case inner rear flange and to the outer flange of the no. 5 bearing housing. It forms the inner wall of the combustion chamber and serves to position the no. 5 bearing through the bearing housing.

Combustion Chamber Rear Support and Outlet Ducts

Positioned inside the rear of the combustion chamber area is a welded combustion chamber rear support. This large circular plate has nine openings around a single larger central opening, and holds the rear of the individual combustion chambers in place. The rear outer flange is equipped with bolts held to the support by stops and rivets.

Fitted behind the support are the combustion chamber inner and outer outlet ducts which feature air deflector ducts which divide the cooling air in both the inner and outer duct into two streams. The hot gases pass through the nine support openings and are guided to the first-stage nozzle between the outlet ducts.

Turbine Shafts and No. 4½ Bearing Heat Shields (FIG. 26-14)

Located within the combustion chamber inner case and bolted to the rear of the no. 4 bearing support are the turbine shaft heat shields. The oil scavenge pump shield is cylindrical in shape and is held in place by the same bolts holding the no. 4½ bearing heat shield assembly inside it.

The no. 4½ bearing heat shield assembly is equipped with support tubes around its OD and has a reinforced wasp-waist shape. It is designed to accommodate axial movement of the bearing supporting structure. Self-locking nuts are riveted to the rear flange. A pin in the rear flange and an offset hole in the front flange ensure correct positioning in the engine.

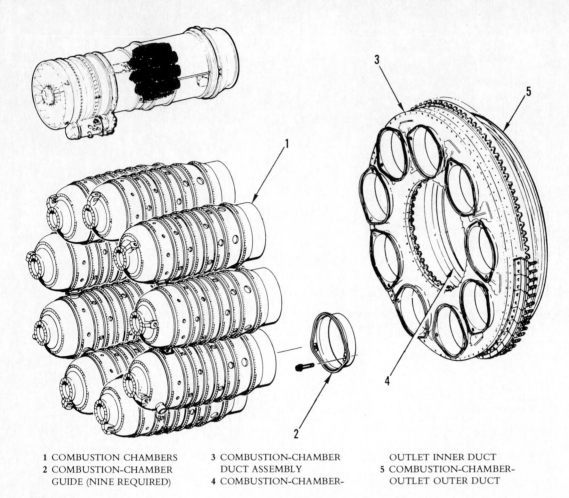

1 COMBUSTION CHAMBERS
2 COMBUSTION-CHAMBER
GUIDE (NINE REQUIRED)
3 COMBUSTION-CHAMBER
DUCT ASSEMBLY
4 COMBUSTION-CHAMBER-

OUTLET INNER DUCT
5 COMBUSTION-CHAMBER-
OUTLET OUTER DUCT

FIG. 26-13 Combustion chambers and combustion chamber duct assembly.

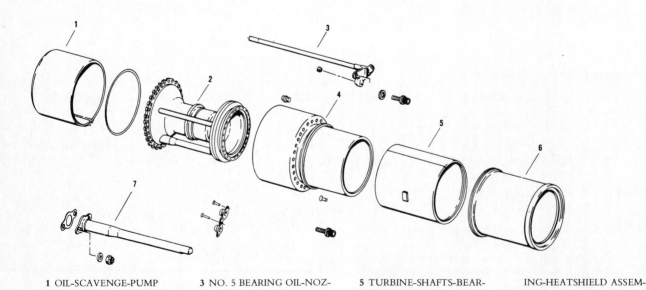

1 OIL-SCAVENGE-PUMP
HEATSHIELD ASSEMBLY
2 NO. 4½ BEARING HEAT-
SHIELD ASSEMBLY
3 NO. 5 BEARING OIL-NOZ-
ZLE ASSEMBLY
4 OIL-SCAVENGE-PUMP
SHIELD
5 TURBINE-SHAFTS-BEAR-
ING HEATSHIELD ASSEM-
BLY
6 TURBINE-SHAFTS-BEAR-

ING-HEATSHIELD ASSEM-
BLY
7 NO. 5 BEARING OIL SCAV-
ENGE TUBE

FIG. 26-14 Turbine shafts and no. 4½ bearing heat shields.

Combustion Chamber Outer Case

The combustion chamber outer case is secured to the rear flange of the diffuser case and the front flange of the turbine nozzle case and encloses the combustion chamber. It forms the inner wall of the annular duct at this location.

The fuel drain valves are located on the bottom center line of the case, one at the front and one at the rear. A fuel drain manifold carries any drain fuel to the outside of the outer duct. Both flanges of this case turn inward and the front flange is scalloped. Two bosses, each with a single threaded hole, are located at the four and eight o'clock positions near the front of the case. The case is constructed of corrosion- and heat-resistant steel, with nickel-cadmium and baked-on aluminum enamel at the flanges to ensure against corrosion.

Combustion Chambers (FIG. 26-15)

Nine one-piece combustion chambers (or cans) are located between the combustion chamber outer case and the combustion chamber inner case in a can-annular arrangement. Chamber no. 1 is at the 12 o'clock position, and the chambers are numbered clockwise around the engine as viewed from the rear.

Each complete bullet-shaped combustion chamber is of welded construction, having a series of round liners. The chambers are equipped with positioning brackets.

The rear of the chambers fits into the nine openings in the front of the combustion chamber rear support (Fig. 26-16). The chambers fit onto the nozzle of the fuel manifold at the front, where they are held by lock-wired bolts and pins through the positioning brackets. Interconnecting flame tubes between the chambers serve to spread the flame uniformly to all the chambers.

All the chambers have one male and one female flame tube. In addition, chambers 4 and 7 have a spark igniter opening. Nine two-bolt interconnector tubes connect the male and female flame tubes of the chambers. The chambers are equipped with cooling deflectors (or air scoops) at the flame tubes.

TURBINE SECTION

General

The turbine section contains two cases (turbine front case and turbine rear case), an inner case and seal, four stages of turbine vanes, and the turbine rotors with their drive shafts.

Turbine Front Case

The turbine front case is secured to the rear of the combustion chamber outer case and is of decreasing

FIG. 26-15 Smoke-reduction combustion chamber.

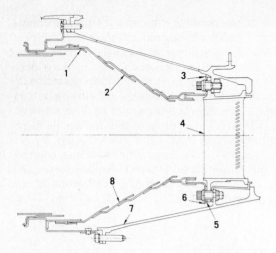

1 COMBUSTION-CHAMBER
  REAR SUPPORT
2 COMBUSTION-CHAMBER
  INNER OUTLET DUCT
3 COMBUSTION-CHAMBER-
  OUTLET-DUCT INNER
  REAR SUPPORT
4 FIRST-STAGE TURBINE VANE

5 SPACER PLATE
6 COMBUSTION-CHAMBER-
  OUTLET-DUCT OUTER
  REAR SUPPORT
7 TURBINE OUTER FRONT
  CASE
8 COMBUSTION-CHAMBER
  OUTER OUTLET DUCT

FIG. 26-16 Louvered combustion chamber outlet duct.

diameter from front to rear. It is constructed of corrosion- and heat-resistant steel. The first-stage turbine vanes are bolted into this case.

Turbine Rear Case

Secured to the rear of the turbine front case is the larger turbine rear case, constructed of steel. The diameter of this case increases from front to rear to accommodate the second-, third-, and fourth-stage turbine vanes. The rear flange of this case is bolted with the fourth-stage turbine outer seal ring, to the front flange of the turbine exhaust case. Holes located in the front inner flange allow cooling air to flow by the turbine rotor airseal and turbine outer rear case. The vanes fit in machined grooves (in the ID of the case) which have antitorque lug slots for the blade shroud seal rings.

A pin in the front flange (six o'clock position) and an offset hole in the rear flange position the case correctly in the engine. (Viewed from the rear, the clockwise hole in the lug having only two holes is offset.) Anchor nuts are secured to the inside of the front flange.

Turbine Nozzle Inner Case and Seal Assembly

Secured to the outer flange of the no. 5 bearing housing and extending rearward is the turbine nozzle inner case and seal assembly. Riveted to its rear inner flange is the turbine rotor inner first-stage airseal, which matches the integral shoulders on the front of the first-stage turbine disk.

A groove in the rear outer flange accommodates the inner rear shroud of the first-stage turbine vanes. Segmented multiple turbine vane shroud nuts are riveted to the forward outer flange.

First-stage Turbine Nozzle (FIG. 26-17)

The first-stage turbine vanes (and inner duct positioning supports) are held at the inner end to the segmented multiple nuts of the inner case and seal assembly by bolts and lockwire. Later models' first-stage vanes are air-cooled by an internal tube and a series of air exit holes in the airfoil trailing edge on the concave side.

Bolts at the outer end hold the vanes in the turbine outer front case. A ring of segmented supports, under the boltheads at the front outer shroud of the vanes, positions the combustion chamber outer outlet duct. The outer shroud of the vanes fits against the Z-shaped first-stage stator seat positioned inside the turbine case. The vanes are removable from the front.

To the rear of the first-stage vanes' outer shroud is the first-stage turbine rotor outer airseal. This airseal has multiple seals on its ID.

The first-stage turbine rotor outer airseal damper (a gapped steel ring), between the first-stage outer airseal and the turbine case, controls cooling airflow through this area.

Second-stage Turbine Nozzle

There are 95 second-stage vanes. The vanes are installed in a grooved shoulder in the ID of the turbine case, held in place by pins in the groove. In front of the vanes, at the outer end, is the gapped second-stage vane retaining ring, against the rear of the first-stage outer airseal. To the rear of the vanes' outer end is the second-stage turbine vane lock plate. Behind the lock plate is the second-stage outer airseal ring and turbine damper. This ring has two stepped platforms which match the two knife-edge seals on the outer shroud of the second-stage turbine blades.

At the inner end, the vanes seat in slots in the second-stage turbine vane inner shroud assembly. At its ID, this shroud assembly has three knife-edge seals, two of which match the ID of a flange on the second-stage disk and one of which matches the OD of an inner flange on the rear of the first-stage turbine rotor. Another airseal is mounted on the front portion of the shroud ID and matches the OD of an outer flange on the rear of the first-stage turbine rotor.

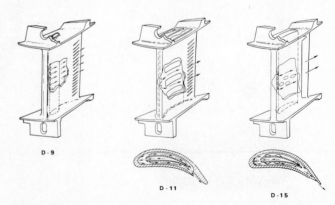

FIG. 26-17 Various configurations of air-cooled first stage turbine nozzle vanes used on several models of the JT8D engine.

Third-stage Turbine Nozzle

The 79 third-stage vanes are held in the turbine case, as the second-stage vanes are. There is a gapped retaining ring in front, a lock plate behind, and a stepped third-stage outer airseal ring to the rear of the lock plate.

At the inner end, the third-stage vanes seat in slots in the turbine rotor third-stage inner airseal ring assembly. At its ID, this assembly has a double-platform ring which matches the knife-edge seals on the OD of the turbine rotor second-to-third-stage inner airseal.

Fourth-stage Turbine Nozzle

There are 77 fourth-stage vanes positioned in the rear of the turbine case. These vanes mount on pins in the case and against a gapped retaining ring in front. They are held against a rear shoulder in the case by the fourth-stage turbine rotor outer airseal ring, which is secured to the rear flange of the turbine case by flathead screws. The ring has platforms to match the two seals on the outer end of the fourth-stage turbine blades.

At the inner end, the fourth-stage vanes seat in the turbine fourth-stage inner airseal assembly. This inner airseal ring assembly has a double-platform inner airseal ring which matches the knife-edge seals on the OD of the turbine rotor third-to-fourth-stage inner airseal.

Turbine Rotors

There are two separate drive turbine rotors, the rear-compressor drive (first stage) and the front-compressor drive (second, third, and fourth stages). Except for necessary bearing support they are not mechanically connected. They are aerodynamically coupled, since the gases that exhaust from the first-stage turbine rotor pass through the second, third, and fourth stages.

No. 5 Bearing Housing (see FIG. 26-42)

Secured to the rear flange of the turbine shaft's inner heat shield is the no. 5 bearing housing and the no. 5 bearing seal support assembly. The steel no. 5 bearing housing holds the no. 5 bearing outer race in its ID bore with a retaining nut and rivet. The housing is attached to the rear flange of the combustion chamber inner case, the front flange of the turbine nozzle inner case and seal assembly, and the inner rear flange of the turbine shaft's heat shield. The outer front flange of the housing also provides a point of attachment for the inner flange of the nine-hole combustion chamber support.

Rear-compressor Drive Turbine Rotor (FIG. 26-18)

In early engines, the rear-compressor drive turbine rotor consists of an integral rear-compressor drive turbine shaft and first-stage disk. In later engines, this shaft and disk are separate and are held together by 18 equally spaced tiebolts, tabwashers, and nuts. In all engines the first-stage blades are held in fir-tree slots by rivets and washers. Later engines, and some others which incorporate a back-up carbon seal at the no. 5 bearing, are

equipped with a spacer and an airseal between the seal face plate and the shoulder of the turbine shaft.

Counterweights may be secured to the rear flange of the shaft to obtain optimum rotor balance.

At assembly, the turbine shaft splines onto the rear compressor rear hub and is retained by the rear-compressor drive turbine shaft coupling.

Turbine position in the turbine case is determined by a ring-shaped spacer that is between the rear face of the compressor rear hub shaft and an internal shoulder in the turbine shaft.

Rear-compressor Drive Turbine Shaft Coupling

A steel rear-compressor drive turbine shaft coupling secures the rear-compressor drive turbine shaft to the rear compressor. The rear-compressor rear hub has an OD spline which mates with the rear-compressor drive turbine shaft front ID. The coupling has a left-hand thread at the front and a shoulder which holds the shaft to the rear-compressor rear hub. It has a wrench spline in its ID and multiple oil holes at the rear. The coupling is silver-plated.

Front-compressor Drive Turbine Rotor (FIG. 26-19)

The front-compressor drive turbine rotor includes the front-compressor drive turbine shaft, the second-, third-, and fourth-stage turbine disks and blades, and the spacers and airseals between the disks. Twelve tie rods secure the disks and spacers to each other and to the rear flange of the rotor shaft, and a trumpet-shaped turbine bearings pressure- and scavenge-oil tubes assembly is positioned inside the shaft.

The blades are secured in the disks with rivets: 88 blades in the second stage, 92 blades in the third stage, and 74 blades in the fourth stage. Provisions are made for rotor counterweights on the front face of the second-stage disk and the rear face of the fourth-stage disk.

The second-stage disk has 88 fir-tree serrated slots around the OD. The disk is made of steel and has a scalloped flange on the front which accommodates rotor counterweights. A flange on the rear has 12 tie-rod holes, one of which is offset and is identified by an adjacent dimple, and nine counterweight holes.

The third-stage disk, manufactured of steel, has 92 fir-tree serrated slots around the OD. The disk has 12 tierod holes, one of which is offset and is identified by an adjacent dimple. Twelve counterweight holes are located between the tierod holes. The disk has shoulders near the OD for the mating inner airseals and spacers.

Lugs on the second- and third-stage turbine rotor inner airseals mate with slots on the second- and third-stage disks and prevent rotation between the disks and seals.

At the fourth stage, the turbine rotor rear hub, made of steel, incorporates an integral disk having 74 fir-tree serrated slots around the OD. The hub has 12 tie-rod holes (one offset) and 12 counterweight holes through the thick portion of the disk web. A scalloped flange at the rear has holes for mounting rotor balance counter-

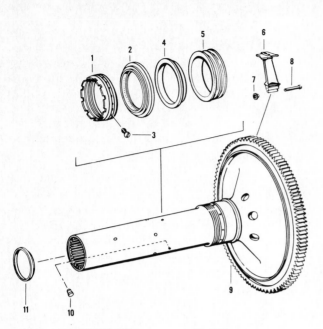

1 NO. 5 BEARING INNER
  RACE NUT
2 SEAL SEAT (PLATE)
3 NUT RETAINING SCREW (2)
4 SPACER
5 AIRSEAL
6 FIRST-STAGE TURBINE
  BLADE
7 WASHER
8 TURBINE BLADE RIVET
9 REAR-COMPRESSOR-
  DRIVE TURBINE SHAFT
10 POSITIONING PLUG
11 TURBINE SHAFT SPACER

(a)

1 NO. 5 BEARING INNER-
  RACE RETAINING NUT
2 SEAL SEAT
3 BEARING SPACER
4 LABYRINTH SEAL
5 FIRST-STAGE TURBINE
  DISK
6 WASHER
7 FIRST-STAGE TURBINE
  BLADE
8 RIVET
9 COUNTERWEIGHT
10 COUNTERWEIGHT
11 RIVET
12 TIEBOLT
13 KEYWASHER
14 TIE-ROD NUT
15 POSITIONING PLUG
16 TURBINE SHAFT SPACER
17 REAR-COMPRESSOR-
  DRIVE TURBINE SHAFT
18 RETAINING SCREW

(b)

FIG. 26-18 (a) Rear-compressor drive turbine rotor (integral hub and shaft).
(b) Rear-compressor drive turbine rotor assembly (separable hub and shaft).

weights. There are eight equally spaced threaded holes on the rear face of the hub used to secure the no. 6 bearing oil scavenge gearshaft in place.

The no. 6 bearing spacer mounts on the small diameter of the hub with the no. 6 bearing behind it. Both are held on the hub by the oil scavenge gearshaft, bolted and locked with keywashers.

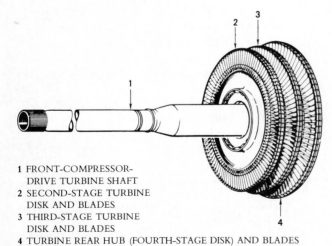

1 FRONT-COMPRESSOR-
  DRIVE TURBINE SHAFT
2 SECOND-STAGE TURBINE
  DISK AND BLADES
3 THIRD-STAGE TURBINE
  DISK AND BLADES
4 TURBINE REAR HUB (FOURTH-STAGE DISK) AND BLADES

FIG. 26-19 Front-compressor drive turbine rotor.

Front-compressor Drive Turbine Rotor and Stator Assembly (Unit Turbine) (*FIG. 26-20*)

Later engines and some others are equipped with a front-compressor drive turbine rotor and stator assembly (unit turbine). This turbine differs from a standard turbine in that the turbine case parting surfaces are arranged so that the low-pressure turbine may be assembled and installed in the engine as a unit. The assembly includes a rear turbine case, the second-, third-, and fourth-stage turbine vanes and inner stator shrouds, the front-compressor drive turbine shaft, second- and third-stage turbine disks and blades, fourth-stage hub and blades, shaft-to-third- and third-to-fourth-stage spacer, and airseals between the disks.

The front-compressor drive turbine shaft is made of steel. It is a long shaft which is positioned inside the rear-compressor drive turbine shaft and extends from the second-stage turbine disk to the front-compressor rotor rear hub.

Located on the OD of the shaft, about one-third of the way forward, is a machined diameter, on which are installed the no. 4½ bearing (turbine intershaft) inner race, seals, and seal spacers. The seals and the bearing are held on the shaft by a large nut with oil holes, secured with a tablock and snap ring.

At the front of the machined diameter are two groups of five oil holes, equally spaced, through the wall of the

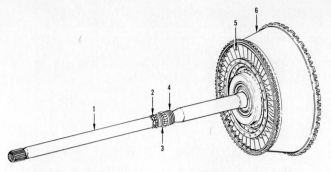

1 FRONT-COMPRESSOR-
  DRIVE TURBINE SHAFT
2 NO. 4½ BEARING RETAIN-
  ING NUT
3 NO. 4½ BEARING INNER
  RACE AND ROLLERS

4 NO. 4½ BEARING CARBON
  SEALS
5 SECOND-STAGE TURBINE
  VANES
6 TURBINE NOZZLE CASE

FIG. 26-20 Front-compressor drive turbine rotor and stator assembly (unit turbine).

shaft. At the rear of the machined diameter is a group of five smaller holes which pass through the shaft wall at an angle. Oil flows forward through a trumpet-shaped turbine bearings pressure- and scavenge-oil tubes assembly, then through these holes to the no. 4½ bearing area.

The flange at the rear of the shaft has 12 tierod holes, and three equally spaced holes for screws which secure the second-stage disk to the shaft. In addition, the flange has nine shallow holes to accommodate counterweights as required on the rear flange of the second-stage disk.

Turbine Bearings Pressure- and Scavenge-oil Tubes Assembly

This unit, shaped like a long, slender trumpet, is positioned inside the front-compressor drive turbine rotor. Two long scavenge tubes and a shorter pressure tube carry oil between the no. 4½ and no. 6 bearing areas as described for the lubrication system.

Front-compressor Drive Turbine Shaft Coupling Arrangement

The front-compressor drive turbine shaft splines into the front-compressor rotor rear hub. The shaft and hub are secured together by the arrangement described for the front-compressor rotor rear hub coupling in the compressor section.

Turbine exhaust case (FIG. 26-21)

At the rear of the basic inner section of the engine, bolted to the rear flange of the turbine case, is the welded steel turbine exhaust case. This case decreases in diameter from front to rear, has a double flange mount ring encircling it near the center, and has outer flanges at the front and rear. The front flange has a snap diameter and boltholes with one hole offset next to the six o'clock position. The rear flange is scalloped and has locknuts held securely in place by rivets.

Eight thermocouple inner bosses are welded on the OD of the case just behind the mount ring flanges and strut retaining pin bosses are located at the three, six, and nine o'clock positions. An oil pressure tube boss is

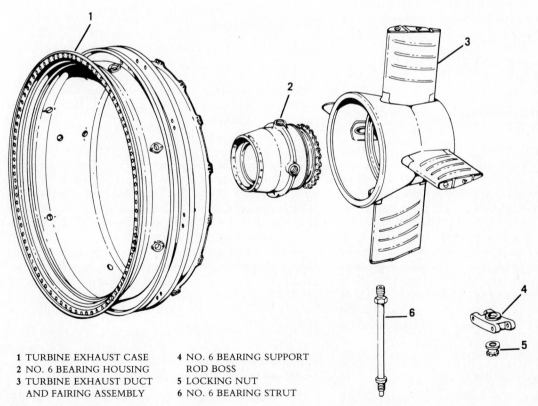

1 TURBINE EXHAUST CASE
2 NO. 6 BEARING HOUSING
3 TURBINE EXHAUST DUCT
  AND FAIRING ASSEMBLY

4 NO. 6 BEARING SUPPORT
  ROD BOSS
5 LOCKING NUT
6 NO. 6 BEARING STRUT

FIG. 26-21 Turbine exhause case and fairing assembly.

located behind the mount ring flanges at the 12 o'clock position. Six pressure probe bosses are located forward of the mount ring flanges. The case incorporates an internal grooved ring near the front flange which engages and locks the rear edge of the fourth-stage turbine rotor outer airseal ring.

The turbine average pressure sensing manifold is mounted around the turbine exhaust case, forward of the mounting flanges. It connects to the pressure probes bolted to the case and, at the outer end, to a fitting near the bottom of the fan discharge turbine exhaust outer duct.

Turbine Exhaust Strut Assembly

The steel turbine exhaust strut assembly, an inner exhaust duct with four turbine exhaust struts welded to its outer surface, is positioned inside the turbine exhaust outer case. The inner exhaust duct rear flange is equipped with anchored locknuts which facilitate assembly and disassembly.

Four no. 6 bearing support rods pass through the struts, and through holes in the turbine exhaust case, and are secured in strut supports between the exhaust case mounting flanges at the 12, 3, 6, and 9 o'clock positions. The rods are held securely by a dual nut locking arrangement in the strut supports. At the inner end the rods thread into the no. 6 bearing support and are held securely with keywashers.

At the 12 o'clock position, to the rear of the support rod, the no. 6 bearing pressure oil tube passes from outside the exhaust case through the strut to the no. 6 bearing support.

The no. 6 bearing oil scavenge pump is bolted to the rear of the no. 6 bearing support (see Fig. 26-45).

ACCESSORY- AND COMPONENT-DRIVES GEARBOX HOUSING SECTION

Accessory- and Component-drives Gearbox Assembly (FIG. 26-22)

The accessory- and component-drives gearbox assembly (accessory-drives gearbox) consists of the gearbox housing, the gearbox rear housing, and the internal gears and shaft gears. The gearbox assembly is mounted beneath the engine, secured to the fan discharge intermediate case flanges and, at the front, to another flange. Power is supplied to the gearbox from a bevel gear splined to the front of the rear compressor drive turbine rotor shaft. An oil pump assembly is located in the bottom of the gearbox, left of center, and contains both pressure and scavenge sections. The gearbox incorporates pinned carburized bearing liners except for the towershaft drive gear roller bearings liners which are retained by bolts.

Gearbox Protective Coating

Gearbox assemblies are painted in two ways. Early models are painted with gray lacquer; subsequent

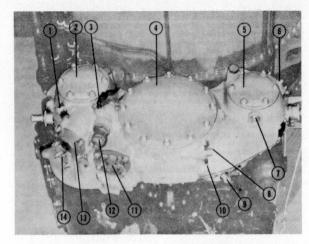

1 DRAIN	**7** DRAIN
2 STARTER DRIVE PAD	**8** DRAIN
3 OIL COOLER TO RELIEF-VALVE TUBE (OIL PRESSURE SIGNAL)	**9** DRAIN
4 CONSTANT-SPEED-DRIVE ALTERNATOR MOUNTING PAD	**10** GEARBOX MAIN OIL DRAIN
	11. MAIN OIL PUMP
5 HYDRAULIC-PUMP MOUNTING PAD	**12** OIL PRESSURE-REGULATING VALVE
6 N_2 TACHOMETER DRIVE PAD	**13** OIL STRAINER
	14 OVERBOARD-BREATHER MOUNTING PAD

FIG. 26-22 Accessory and component drives gearbox.

models are sprayed with the recommended aluminized epoxy paint, silver in color.

Accessory- and Component-drives Gearbox Rear Housing

On the rear face of the gearbox rear housing there is a starter drive pad on the left, a 10-in constant-speed drive (CSD) and alternator drive pad in the center, and a hydraulic pump drive pad on the right. On the right end of the gearbox housing is a standard four-stud pad for the N_2 tachometer drive.

Near the bottom of the housing, to the left of the oil pump, is an oil pressure relief valve. The main oil strainer is located to the left of the oil pressure relief valve. An oil strainer bypass valve is located in the center of the main oil strainer assembly. An integral boss is provided on the side of the main oil strainer boss to accommodate the airframe manufacturer's oil filter bypass pressure warning switch.

Mounted on the starter drive gearshaft are two rotary breather impellers, used to separate oil from air. A replaceable spline adapter is splined inside the starter drive gearshaft and held in place with a locknut and anchor bolt. The CSD incorporates a replaceable splined coupling inside the gearshaft.

Accessory- and Component-drives Gearbox Front (Cover) Housing

The engine fuel pump mounts on a six-stud circular pad on the right front face of the main gearbox cover and the fuel control mounts on the front of the fuel pump. The oil tank (optional) mounts on the left front face of the gearbox front housing.

Power Lever Cross Shafts and Linkage

A power lever outer cross-shaft assembly, which is hollow and has an inner shaft running through it, is mounted in bushings at the top of the gearbox housing cover. The fuel-control power linkage arm is secured on the outer shaft at the right and, further outboard, the fuel control shutoff arm is held securely on the inner shaft. A stop plate and locking plate are mounted on the housing. The airframe control arms or pulleys may be mounted on either end of the cross shafts.

FAN DISCHARGE SECTION (FIG. 26-23)

General

Behind the fan exit case, and enclosing the engine, is a series of fan discharge outer ducts and cases. These outer ducts and the outside surface of the inner ducts and inner engine cases form the annular duct air passage for fan discharge air to flow to the rear of the engine.

Fan Discharge Case Assembly
(and Fan Discharge Vanes)

Bolted to the rear flange of the fan rear case is the fan exit case, extending from flange D to E. This case incorporates matched upper and lower halves of aluminum, bolted together. Each half has 28 aluminum vanes brazed to an inner shroud and the outer case wall, with aluminum plugs at the outer end.

A rectangular hole at the mating surfaces, nine o'clock position, of the case matches a pin flared in place on the rear flange of the front-compressor case.

Fan Discharge Front-compressor Outer Duct

The aluminum fan discharge front-compressor outer duct is secured to the rear of the fan discharge front cases and forms the outer wall of the annular duct from flange E to F. Forming the inner wall is the front-compressor section fan discharge inner duct extending forward from, and integral with, the compressor intermediate case assembly. The front-compressor outer duct incorporates a pin in the front flange nine o'clock location and a hole in the rear flange approximate six o'clock location to ensure correct positioning.

Fan Discharge Rear Compressor
Section Outer Duct

Between flanges G and H, and secured to the rear outer flange of the compressor intermediate fan case, is the fan discharge rear-compressor section outer duct, cylindrical in shape and of aluminum construction. This duct mates at the rear to the front flange of the fan discharge diffuser section outer duct. Holes at the bottom of the front and rear flanges match the pins in the adjacent cases.

Both upper quadrants of this duct have a large three-hole bleed boss 45° above the horizontal center line. Each of these two bosses has two round holes toward the front for low-pressure bleed air and an elongated hole toward the rear for anti-icing and eighth-stage bleed air. The bosses incorporate replaceable helical coil inserts.

Diffuser Outer Fan Duct

From flange H to J is the fan discharge diffuser section outer duct, cast of aluminum alloy. It has two oval-shaped high-pressure bleed pad bosses, one between the 10 and 11 o'clock positions and another between the 1 and 2 o'clock positions. There are two round bosses at the five and seven o'clock positions with openings for the fuel manifolds. Three studs at the seven o'clock position accommodate the fuel pressurizing and dump valve and four studs just below the four o'clock point are used for mounting the pressure ratio bleed control. Three bosses at the eight o'clock location accommodate, from front to rear, elbows for the breather, oil pressure, and oil scavenge tubes to the no. 4 bearing area.

The duct is equipped with antirotation positioning pins at the six o'clock location on the front flange and nine o'clock location on the rear flange.

Diffuser Inner Fan Duct

Two semicircular aluminum duct segments, one left and one right, are positioned around the diffuser case. These diffuser inner fan duct segments form the inner wall of the annular duct at this location. Holes and cutouts in each segment accommodate the tubing running from the outer duct to the diffuser case.

The duct segments are joined together by screws and anchored locknuts. They are held at the front flange of the diffuser case by screws secured to two semicircular removable gang nut assemblies. These gang nuts are held, in turn, by the diffuser case front flange bolts. The duct segments are held at the rear by a metal strap arrangement featuring overlapping ends and attached by screws.

Combustion Chamber and Turbine Fan Duct
Assembly (Combustion Section Fan Duct)

Between flanges J and K is the combustion section fan duct assembly composed of upper and lower ducts, each in the shape of a half cylinder. Both halves are aluminum and are joined together along the sides of the engine by bolted flanges. The upper and lower ducts together form a single matched set. Holes at the nine o'clock position on the front and rear flanges match positioning pins in the mating cases.

The lower duct has a combustion chamber outer drain boss riveted in place at the bottom midway between the front and the rear flanges. Certain JT8D engines have an additional drain boss located on the bottom forward part of the intermediate flange. An

LETTERS ARE
FLANGE DESIGNATIONS

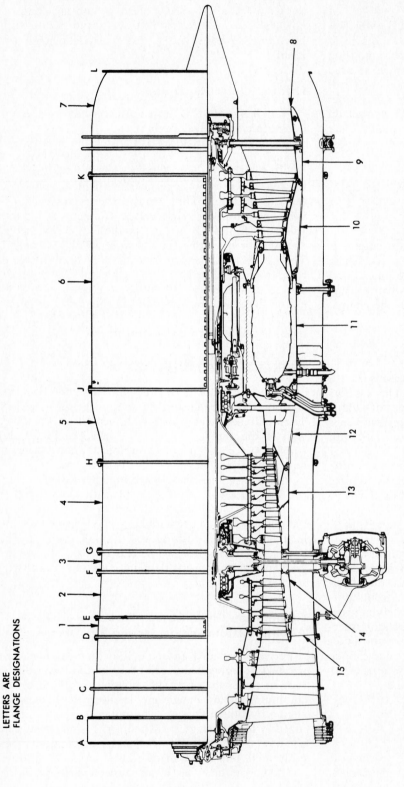

1. FAN-DISCHARGE FRONT
   CASE
2. FAN-DISCHARGE FRONT-
   COMPRESSOR OUTER
   DUCT
3. FAN-DISCHARGE INTER-
   MEDIATE CASE
4. FAN-DISCHARGE REAR-

COMPRESSOR SECTION
   OUTER DUCT
5. FAN-DISCHARGE DIF-
   FUSER-SECTION OUTER
   DUCT
6. FAN-DISCHARGE COM-
   BUSTION-SECTION
   OUTER DUCT

7. FAN-DISCHARGE TUR-
   BINE-EXHAUST SECTION
   OUTER DUCT
8. FAN-DISCHARGE TUR-
   BINE-EXHAUST INNER
   REAR DUCT
9. FAN-DISCHARGE TUR-
   BINE-EXHAUST INNER

FRONT DUCT
10. FAN-DISCHARGE TUR-
    BINE-SECTION INNER
    DUCT
11. COMBUSTION-CHAMBER
    OUTER CASE
12. FAN-DISCHARGE DIF-
    FUSER-SECTION INNER

DUCT
13. FAN-DISCHARGE REAR-
    COMPRESSOR SECTION
    INNER DUCT
14. COMPRESSOR INTERME-
    DIATE CASE
15. FAN-DISCHARGE VANES

FIG. 26-23 Fan discharge ducting.

intermediate flange, designated J1, supports the ignition exciter rear brackets. Two large fan airbleed openings equipped with studs are at the rear, just below the engine center line, one on each side. A smaller boss on the lower right-hand side near the front may be utilized to provide fan air for cooling an airframe ac generator.

Two igniter plug bosses are riveted to the wall of the lower duct at the four and eight o'clock positions near the front. A streamlined right igniter cable fairing is riveted inside at the four o'clock position. The left igniter cable fairing is not secured to the duct but is bolted to the rear of the no. 4 bearing tubes fairing. Both igniter cable fairings are fiberglass.

The upper duct, like the lower, has flanges on both the sides and the ends.

Fan Turbine Inner Duct

Two identical aluminum duct halves form the fan turbine inner duct, positioned around the turbine nozzle section of the basic inner engine. Each half is semicircular and has locknuts riveted along one side flange. Screws secure the halves together.

Two semicircular gang nut assemblies are secured to the rear face of the combustion chamber case rear flange. They have an L-shaped cross section and riveted anchor nuts to which the inner duct is held at the front by screws. At the rear, the inner duct fits closely over the fan discharge turbine exhaust inner duct front flange.

Fan Discharge Turbine Exhaust Section Outer Duct

Positioned outside the turbine exhaust outer case at the rear of the engine, from flange K to M, is the fan discharge turbine exhaust section outer duct assembly, constructed of steel. The engine rear mount ring, which has holes for ground-handling provisions, is an integral welded part of this duct.

One Pt_7 tube boss is located to the left of the six o'clock position behind the mount ring flanges. Eight streamlined struts in the ID secure this duct to the turbine exhaust outer case. An antirotation-type positioning pin is located in the front flange at the nine o'clock point.

Fan Discharge Turbine Exhaust Inner Ducts

Positioned around the front portion of the turbine exhaust case is the fan discharge turbine exhaust inner front duct. This aluminum duct, shaped like a circular band, has screw holes through the front flanges and cutouts in the rear flange. It is held by screws to a two-piece gang nut assembly on the front face of the rear flange of the turbine outer rear case.

Around the rear of the turbine exhaust case is the fan discharge turbine exhaust inner rear duct. The outer surface of this circular aluminum duct decreases in diameter from front to rear then flares forward to form a smoothly rounded trailing edge. The duct fits closely inside the front duct rear edge and is secured to the rear flange of the exhaust case by bolts. Cutouts in the duct accommodate the engine tubing.

Rear Compressor Air Bleed Manifolds and Tubes (FIG. 26-24)

Bolted into the fan discharge rear compressor section outer duct are two Y-type air bleed manifolds, one on each side 45° above the horizontal center line. Each manifold has a single opening at the inner end which

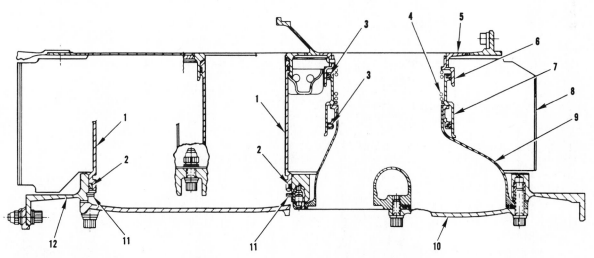

1 BLEED-AIR-TRANSFER TUBE, TWO REQUIRED AT EACH SIDE OF ENGINE.
2 SEAL RING, TWO TO EACH TRANSFER TUBE
3 SEAL RING, ONE EACH FOR BLEED MANIFOLDS AND LINERS
4 COIL SPRING
5 REAR-COMPRESSOR INNER FAN DUCT
6 EXPANSION-JOINT SLEEVE (RIVETED TO INNER FAN DUCT)
7 EXPANSION-JOINT LINER
8 FRONT FAIRING
9 BLEED MANIFOLD, ONE AT EACH SIDE OF ENGINE
10 BLEED-MANIFOLD COVER
11 TRANSFER TUBE RETAINING RING, ONE TO EACH TUBE
12 REAR-COMPRESSOR OUTER FAN DUCT (BETWEEN FLANGES G AND H)

FIG. 26-24 Rear-compressor air bleed manifolds and tubes.

receives eighth-stage air. Liners and sleeves plated on the inner sealing surfaces and metal seal rings accommodate the expansion and contraction of the engine. The manifold divides into two openings at the outer end, the forward opening for engine antiicing air and the larger rear opening for the eighth stage bleed air pad.

In front of each air manifold, two straight transfer tubes conduct low-pressure air from the sixth stage to the double-hole low-pressure bleed pads. When blanked off, an oval-shaped gasket and cover are used over each double-hole pad.

Diffuser Section Air Bleed Manifolds (FIG. 26-25)

Bolted into the diffuser section outer duct, one between the ten and eleven o'clock positions and another between the one and two o'clock positions, are two similar diffuser air supply manifolds. Each steel manifold has a single opening with a ring groove at the inner end to fit into the diffuser case, and divides into two openings at its outer boss. Liners and sleeves plated on

the inner sealing surfaces and metal seal rings accommodate the expansion and contraction of the engine. High-pressure bleed air is carried directly from the diffuser case to outside the diffuser section outer duct through these left and right manifolds.

Rear-compressor Section Fan Duct Fairings

In the fan discharge air path behind the fan discharge intermediate case, two fiber glass fairings are bolted in place, each positioned 45° above the horizontal center line. These fairings streamline the flow of air around the air bleed manifolds between the rear compressor section and diffuser section inner and outer ducts. Each fairing has a full front segment and two rear segments.

No. 4 Bearing Tubes Fairing (FIG. 26-26)

The no. 4 bearing tubes fairing, a streamlined fiber glass two-piece unit, is bolted in place at the eight o'clock position in the fan air stream between the diffuser case and diffuser section outer duct. It streamlines the flow of air around the no. 4 bearing pressure scavenge, and breather tubes located within it.

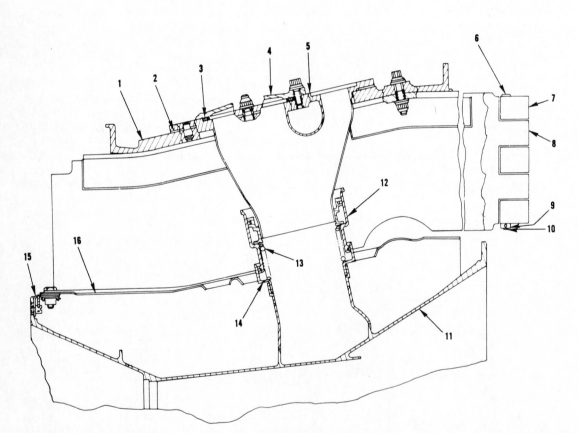

1 DIFFUSER OUTER FAN DUCT	6 FAIRING PIN	10 COTTERPIN	15 DIFFUSER-INNER-FAN-DUCT MOUNTING FLANGE
2 BOSS GASKET	7 RIGHT-REAR FAIRING SEGMENT	11 DIFFUSER CASE	
3 COVER GASKET	8 LEFT-REAR FAIRING SEGMENT	12 LINER (WITH SEAL RING AT INNER END)	16 DIFFUSER INNER FAN DUCT
4 COVER	9 FAIRING PIN WASHER	13 SPRING	
5 MANIFOLD OUTER BOSS		14 SLEEVE	

FIG. 26-25 Diffuser section air bleed manifold.

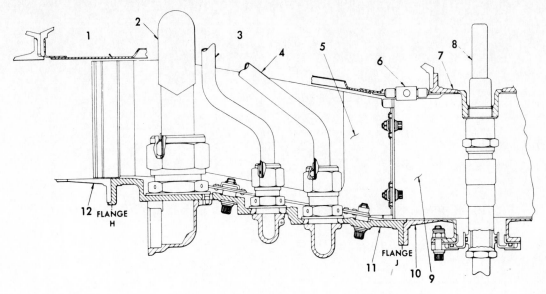

1 DIFFUSER INNER DUCT
2 NO. 4 BEARING BREATHER MANIFOLD
3 OIL-PRESSURE LINE
4 OIL-SCAVENGE LINE
5 NO. 4 BEARING TUBES FAIRING
6 DIFFUSER FAN DUCT STRAP
7 COMBUSTION-CHAMBER
OUTER CASE
8 IGNITER PLUG
9 LEFT IGNITER-PLUG-CABLE FAIRING
10. COMBUSTION-SECTION
FAN DUCT
11. DIFFUSER-SECTION OUTER DUCT
12 REAR-COMPRESSOR OUTER FAN DUCT

FIG. 26-26 Section at no. 4 bearing tubes fairing.

ENGINE FUEL AND CONTROL SYSTEM (FIG. 26-27)

General

The engine fuel distribution and control system of the JT8D engine consists basically of an engine-driven fuel pump and fuel control, an optional fuel anti-icing system, a fuel pressurizing and dump valve, and a split fuel manifold delivering fuel to nine individual fuel nozzles.

Engine Fuel Distribution System

1. *Fuel pump*—Fuel is supplied from the tanks through the necessary strainers and valves to the engine-driven fuel pump which is supplied with the engine. From here it is pumped to the fuel control where it is metered in the proper quantities. Excess fuel is returned to the pump. A fuel filter is integral with the pump.

2. *Fuel anti-icing system (optional)*—The optional fuel anti-icing system is located between the boost and main stages of the engine-driven fuel pump and consists primarily of an air-fuel heater, air shutoff valve, differential fluid pressure switch, and the necessary tubing. The differential fluid pressure switch provides a means of indicating icing conditions or a clogged filter. The fuel anti-icing system functions to prevent the restriction, or even stoppage, of fuel flow caused by the formation of ice within any of the components of the fuel system through which fuel may subsequently pass.

Refer to page 550 for additional descriptive information on the fuel and anti-icing system.

3. *Fuel pressurizing and dump valve*—From the fuel control, the fuel flows through the fuel-flow meter and the fuel coolant oil cooler to the pressurizing and dump valve. The pressurizing valve schedules flow to the secondary fuel nozzles as a function of pressure drop across the primary nozzles. The dump valve is a two-position valve which is hydraulically operated by primary fuel pressure during engine operation. At shutdown the dump valve opens and allows fuel in the manifold to drain.

4. *Fuel manifolds and fuel nozzle and support assemblies*—The divided fuel flow from the fuel pressurizing valve is delivered through the annular duct to the dual-tube fuel manifolds mounted on the diffuser case. The fuel then enters the nine fuel nozzle and support assemblies. The nozzle support flange of each assembly is bolted to the diffuser case, and the support positions the nozzle in the combustion chamber.

Engine Fuel Control

The fuel control is provided with two control levers; one to control the engine speed during all forward- and reverse-thrust operations, and the other to control engine starting and shutdown, by operating the fuel shutoff lever-operated pilot valve and thus the manifold drain valve in proper sequence. The fuel control accurately governs the steady-state selected speed, acceleration, and deceleration, and it indirectly governs the maximum turbine temperature of the engine during both forward- and reverse-thrust operation.

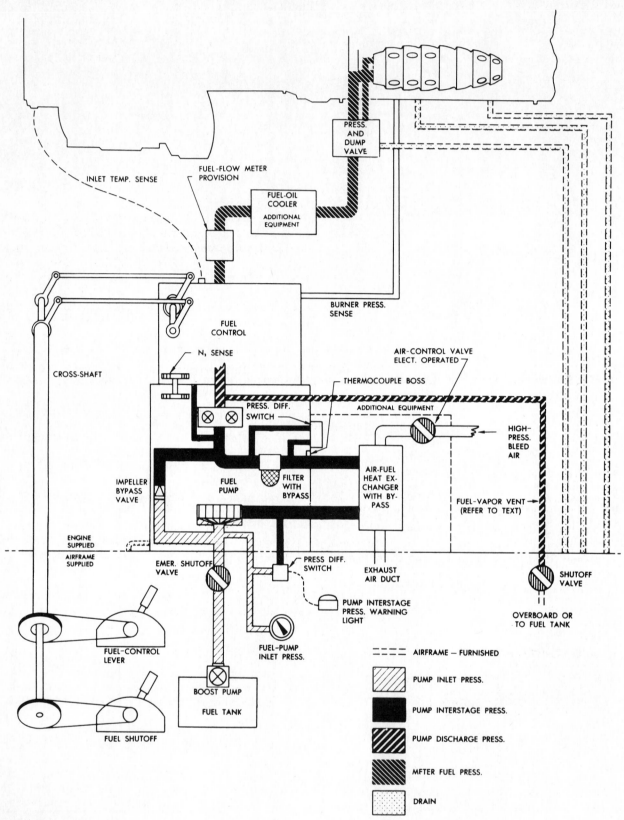

INLET TEMP. SENSE

FUEL-FLOW METER PROVISION

PRESS. AND DUMP VALVE

FUEL-OIL COOLER
ADDITIONAL EQUIPMENT

BURNER PRESS. SENSE

FUEL CONTROL

N₁ SENSE

CROSS-SHAFT

AIR-CONTROL VALVE ELECT. OPERATED

THERMOCOUPLE BOSS

ADDITIONAL EQUIPMENT

PRESS. DIFF. SWITCH

HIGH-PRESS. BLEED AIR

IMPELLER BYPASS VALVE

FUEL PUMP

FILTER WITH BYPASS

AIR-FUEL HEAT EX-CHANGER WITH BY-PASS

FUEL-VAPOR VENT (REFER TO TEXT)

ENGINE SUPPLIED
AIRFRAME SUPPLIED

EMER. SHUTOFF VALVE

PRESS DIFF. SWITCH

EXHAUST AIR DUCT

SHUTOFF VALVE

PUMP INTERSTAGE PRESS. WARNING LIGHT

OVERBOARD OR TO FUEL TANK

FUEL-CONTROL LEVER

FUEL-PUMP INLET PRESS.

AIRFRAME — FURNISHED

PUMP INLET PRESS.

PUMP INTERSTAGE PRESS.

PUMP DISCHARGE PRESS.

MFTER FUEL PRESS.

DRAIN

BOOST PUMP
FUEL TANK

FUEL SHUTOFF

FIG. 26-27 Engine fuel system schematic.

Engine Fuel Indicating Systems

Information on the airframe-mounted engine fuel indicating systems will be found in the airframe manufacturer's maintenance manual.

(a)

Fuel Pump *(FIG. 26-28)*

The main fuel gear pump consists of a single-element gear stage with a high-speed centrifugal boost stage. A cartridge-type relief valve is incorporated to limit the pressure rise across the gear stage. The unit provides a rigid mounting pad arrangement and a rotational splined drive for the fuel control. An integral fuel filter containing a replaceable micronic barrier filter element is located between the discharge of the centrifugal stage and the inlet of the gear stage. Should the pressure drop across the filter exceed a predetermined limit, a bypass valve directs flow into the gear stage. A mounting pad is provided on the filter housing to permit the use of a remote-reading differential pressure warning device. An accessible and removable cover forms the lower portion of the sump area of the filter housing. This cover contains a plug-type valve for draining both the sump and center tube of the filter element. In the event of a malfunction of the boost stage, a bypass valve opens into the inlet passage of the pump to direct flow into the gear stage. This is held normally closed by a light spring force and remains closed due to boost-stage pressure. Outlet and return ports are provided between the boost stage discharge and the filter inlet for installation of an external fuel heater. A drive shaft seal drain is located in the lower extremity of the mounting flange.

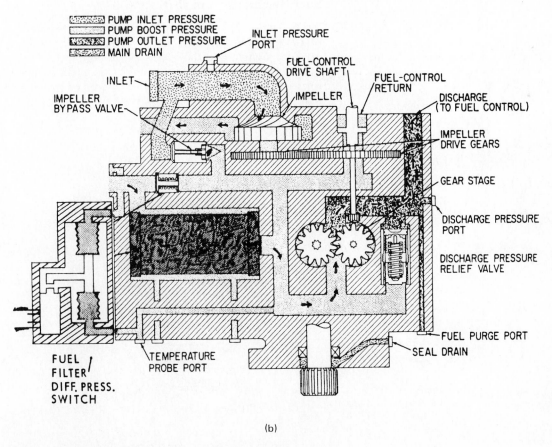

(b)

FIG. 26-28 **(a)** External view of the main fuel gear pump.
(b) Sectioned view of the fuel pump.

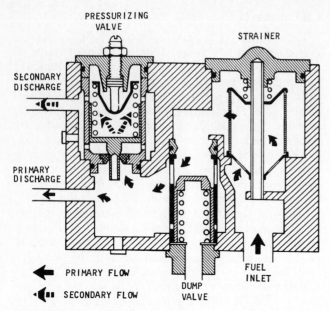

FIG. 26-29 Fuel pressurizing and dump valve schematic.

Pressurizing and Dump Valve *(FIG. 26-29)*

The fuel pressurizing and dump valve is located downstream of the fuel control and is connected to the primary and secondary fuel manifolds to which it discharges its fuel. The essential parts of the fuel pressurizing and dump valve include a 200-mesh fuel inlet screen, a dump (manifold drain) valve, and a pressurizing (flow-dividing) valve.

The fuel pressurizing and dump valve is located on the lower left side of the fan discharge diffuser section outer duct. Locknuts secure the valve to three studs in the duct.

Fuel entering the fuel pressurizing and dump valve is filtered by the 200-mesh fuel inlet screen. The screen unseats at approximately 11.7 psi [80.7 kPa] in the event of clogging, and permits bypass fuel flow. As the fuel pressure in the strainer chamber increases, spring pressure behind the dump valve is overcome and the valve is forced into the closed position. This movement of the dump valve exposes the port through which fuel then flows to the primary chamber of the pressurizing valve. The entire flow then discharges to the primary manifold. When primary chamber fuel pressure surrounding the pintle of the pressurizing valve becomes sufficient to overcome the force of the valve spring, the valve unseats, and fuel flows into the secondary chamber. Flow from the secondary chamber then discharges to the secondary manifold. The contour of the pintle is designed to divide the primary and secondary flows to give satisfactory nozzle spray characteristics at all fuel flow conditions.

Fuel Nozzle and Support *(FIG. 26-30)*

There are nine fuel nozzle and support assemblies. Fuel for each nozzle passes from the manifold through the support. The nozzle and support assemblies are positioned inside the front of the diffuser case with each nozzle facing rearward into its combustion chamber (Fig. 26-31).

Fuel Deicing System *(FIG. 26-32)*

The optional fuel deicing system consists primarily of an air-fuel heater, air shutoff valve, differential fluid pressure switch, and the necessary tubing.

In this system, hot compressor discharge air is piped forward from the diffuser section air bleed manifold rear openings (on the diffuser section outer duct)

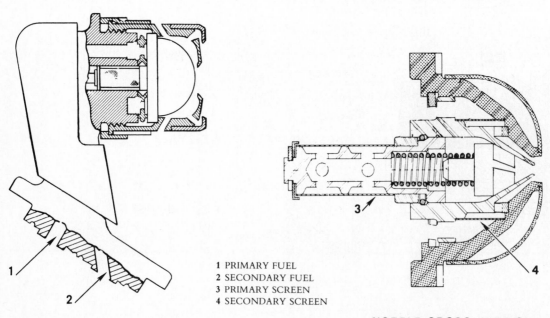

1 PRIMARY FUEL
2 SECONDARY FUEL
3 PRIMARY SCREEN
4 SECONDARY SCREEN

NOZZLE CROSS SECTION

FIG. 26-30 Fuel nozzle and support.

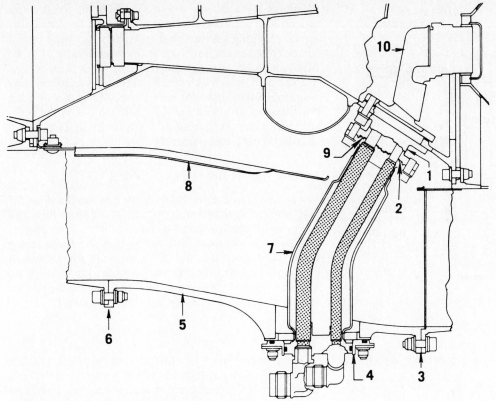

FIG. 26-31 Fuel inlet tube assemblies.

1 DETAIL OF FUEL MANI-
FOLD
2 METAL GASKET
3 FLANGE J
4 PACKING HOLDER
5 FAN-DISCHARGE DIF-
FUSER-SECTION OUTER
DUCT
6 FLANGE H
7 FUEL-MANIFOLD INLET
TUBE ASSEMBLY
8 FAN-DISCHARGE DIF-
FUSER SECTION INNER
DUCT
9 METAL GASKET (LARGE)
10 FUEL NOZZLE AND SUP-
PORT ASSEMBLY

through a deicing manifold on the top of the engine to an electrically operated air-shutoff valve and actuator on the right of the engine. After leaving the valve, the air is piped down to the fuel deicing heater, adjacent to the fuel pump, and then vented overboard.

The pressure-drop warning switch mounted on the fuel filter indicates when the filter is iced. When the cockpit fuel heater switch is activated, an electrically actuated air-shutoff valve located in the bleed air line at the inlet of the heater opens to permit high compressor discharge air to flow through the heater.

The fuel deicing heater and filter are installed in the fuel system between the boost and main stages of the engine-driven fuel pump. All of the engine fuel flow passes through the fuel deicing heater at all times. The fuel is heated, however, only when the air-shutoff valve

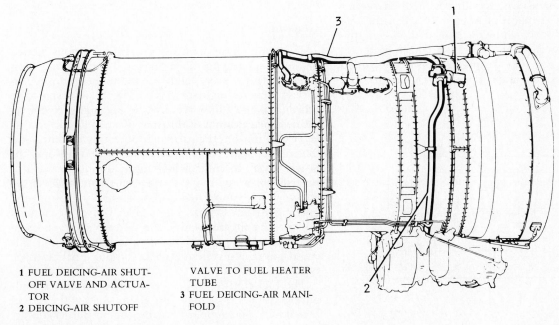

1 FUEL DEICING-AIR SHUT-
OFF VALVE AND ACTUA-
TOR
2 DEICING-AIR SHUTOFF

VALVE TO FUEL HEATER
TUBE
3 FUEL DEICING-AIR MANI-
FOLD

FIG. 26-32 Fuel deicing system.

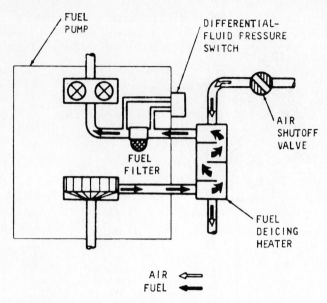

FIG. 26-33 Fuel deicing system schematic.

is opened, allowing high compressor discharge air to flow through the air side of the heater (Fig. 26-33).

Fuel Deicing Heater *(FIG. 26-34)*

The fuel deicing heater is located between the boost and main stages of the engine-driven fuel pump. The fuel deicing heater functions as an air-fuel heat exchanger to protect the engine fuel system from ice. It consists of a housing containing a core composed of over 150 soda-straw-like tubes through which compressor bleed air passes and around which the entire engine fuel flow is circulated, a series of baffles within the core which direct the flow of fuel around the tubes so that the fuel is uniformly heated, and a bypass valve which permits fuel flow in the event of clogging. The fuel deicing heater uses high compressor discharge air as a source of heat, and functions only when the air-shutoff valve is open, allowing high compressor discharge air to flow through the air side of the heater.

Operation of the fuel deicing heater is controlled manually. A differential pressure switch on the fuel

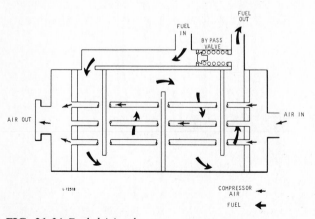

FIG. 26-34 Fuel deicing heater.

deicing filter activates a warning light in the cockpit when there is a pressure drop across the filter caused by ice or clogging. A fuel heater switch can then be actuated, to open the fuel heater air-shutoff valve. Engine bleed air passes through the tubes of the heater, warming the fuel which is baffled around these tubes. The resulting warm fuel will melt any ice formation within the filter and the warning light will go out as the pressure drop across the filter is decreased. The heater should be used intermittently.

Differential Fluid-pressure Switch

The differential fluid-pressure switch mounted on the fuel deicing filter assembly measures the filter inlet and filter outlet fuel-pressure difference. (See Fig. 26-33).

When icing conditions exist within the fuel deicing filter, ice collects on the surface of the filter element causing a pressure drop in the housing. Upon reaching a predetermined pressure differential, the fluid-pressure switch activates a warning light in the cockpit.

Fuel Tubes *(FIG. 26-35)*

The divided fuel flow (primary and secondary) from the fuel pressurizing and dump valve is delivered through two short external manifolds to two points on the exterior of the diffuser case outer fan duct at the five and seven o'clock positions. These tubes lead through the annular duct to the left and right fuel manifold tubes which are mounted on the exterior of the diffuser case. The left manifold tubes supply fuel to the nozzles for combustion chambers 6, 7, 8, 9, and 1. The right manifold tubes supply fuel to the nozzles for chamber nos. 2, 3, 4, and 5.

FUEL CONTROL—SENSORLESS TYPE *(FIG. 26-36)*

General

Two types of fuel controls are used on various models of the JT8D engine: those having a sensor-type pressure-regulating valve, and those equipped with a sensorless pressure-regulating valve. Both are designed to schedule the fuel flow required by the engine to deliver the designed amount of thrust as dictated by the power-lever position and the particular operating conditions of the engine. Two control levers are provided—one, the power lever, to control the engine during forward or reverse operation and the other, the shutoff lever, to effect engine shutdown and starting by closing and opening a fuel-shutoff valve. The control accurately governs the engine steady-state selected speed and regulates acceleration and deceleration fuel flows. The speed-governing system is of the proportional or droop type.

The fuel control consists of a metering and a computing system. The metering system selects the rate of fuel flow to be supplied to the engine burners in accordance with the amount of thrust demanded by the

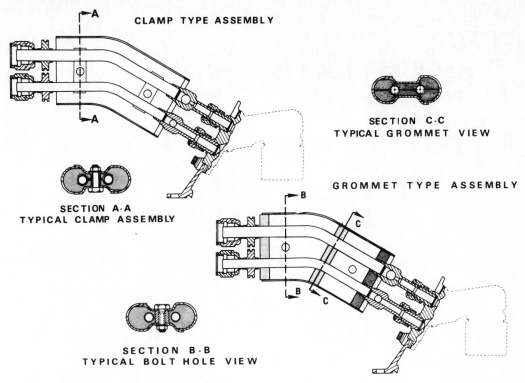

CLAMP TYPE ASSEMBLY

SECTION C-C
TYPICAL GROMMET VIEW

GROMMET TYPE ASSEMBLY

SECTION A-A
TYPICAL CLAMP ASSEMBLY

SECTION B-B
TYPICAL BOLT HOLE VIEW

FIG. 26-35 Fuel manifold tube assembly.

pilot, but subject to engine operating limitations as scheduled by the computing system as a result of its monitoring various engine operational parameters. The computing system senses and combines the various parameters to control the output of the metering section of the control during all regimes of engine operation.

Metering System

High-pressure fuel is supplied to the control from the engine drive fuel pump. This fuel is passed through the coarse filter which protects the metering system against large particles of fuel contaminants. Fuel going to the various servos is passed through a second filter of finer mesh which protects the computing system against solid contaminants. This fine filter is self-cleaning due to the presence of a flow deflector in the main flow stream which causes the flow velocity through the axis of the cylinder to be significantly greater than that of the flow through the mesh supplying the servo control valves. Both coarse and fine filters are protected by relief valves that open to allow fuel to bypass in the event the screens become clogged.

The main fuel-flow stream then flows to the metering valve (throttle valve) across which a constant pressure differential is maintained by the pressure-regulating system. The throttle valve is a window-type valve and is positioned by a half-area servo. This valve consists essentially of a fixed and a movable sleeve and an extension spring which is attached to the multiplying lever. The movable sleeve (piston) position is controlled by a rotating pilot valve which is displaced from its hydraulic null (steady-state) position by compressor dis-

charge pressure, engine speed, compressor inlet temperature, power-lever position, or any combination of these parameters. These actuating signals work in conjunction with each other to produce a net torque on the multiplying lever. A balancing torque is created by the throttle valve extension spring load varied with the valve position. As long as the resultant torque is zero, the throttle valve maintains a constant position. However, any change in the signal torque will displace the pilot valve and cause motion of the throttle valve piston until the unbalanced signal torque is balanced by the new throttle valve position and corresponding spring force. By virtue of the constant pressure drop maintained across the throttle valve, fuel flow is proportional to the position of the piston. The fuel flow rate for the full travel of the throttle valve is externally adjustable by rotation of the outer sleeve relative to the piston. An adjustable stop is provided which is preset for a specific maximum value of fuel flow. An adjustable stop is also provided to limit the motion of the piston in the decrease-fuel-flow direction to permit selection of the proper minimum fuel flow.

In order for the control to ensure a predictable flow through the selected valve-window opening, the pressure drop across the throttle valve is maintained at 40 psi [275.8 kPa] (nominal) by a bypass-type regulating valve. All high-pressure fuel in excess of that required to maintain the pressure differential is bypassed to pump interstage. The bypass valve utilizes the impulse bucket principle to control the flow forces resulting from this bypass flow. The lower end of the pressure-regulating valve is subjected to upstream throttle valve pressure. Balancing this on the upper side is a downstream throt-

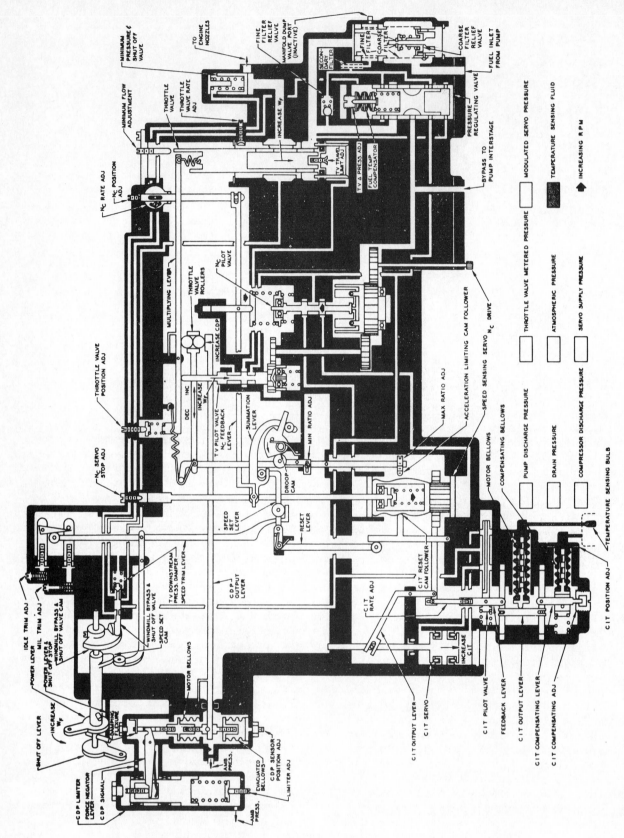

FIG. 26-36 The Hamilton Standard JFC60-2 fuel control.

tle valve pressure nominally 40 psi lower than the lower end and a spring force equivalent to 40 psi. The valve bucket configuration is such that changes in spring load are compensated by the flow forces, thus maintaining the 40-psi pressure drop across the throttle valve. Variations in fuel temperature are compensated for by the action of bimetallic disks working on the 40-psi (nominal) spring at the upper end of the valve. The pressure regulating valve is capable of bypassing 16,000 lb/h [7258 kg/h] with a pressure drop across the bypass ports below 135 psi [931 kPa]. This low pressure drop prevents the pump from operating against an excessively high head or causing large increases in fuel temperature.

Fuel leaving the throttle valve passes through the minimum pressure and shutoff valve on its way to the engine. This valve is essentially a plunger-type valve, spring-loaded to the closed position, and is designed to shut off the flow of metered fuel to the engine when the pilot moves the shutoff lever to the OFF position. When it is actuated for the shutoff function, high-pressure is directed to the spring side of the valve by the action of the windmill bypass and shutoff valve. This pressure closes the valve and allows the spring to keep it in the shutoff position. When the shutoff lever is moved to the ON position, the high pressure on the spring side of the valve is replaced by pump interstage pressure and when metered fuel pressure has increased sufficiently to overcome the spring and interstage fuel pressure force, the valve opens and fuel flow to the engine is initiated. Thereafter, the valve will provide a minimum operating pressure within the fuel control, ensuring that adequate pressure is always available for operation of the servos at low flow conditions.

The windmill bypass and shutoff valve, in addition to supplying the high-pressure signal for the shutoff function, also provides a windmill bypass feature. This valve is plumbed to a line leading to the spring side of the pressure-regulating valve and is positioned by a shutoff-lever-operated cam so that signals are generated at the desired shutoff-lever positions. Movement of the shutoff lever toward the shutoff position displaces the valve, thereby porting the pressure on the spring side of the pressure-regulating valve to pump interstage. The pressure regulating valve now operates as a relief valve to handle the full windmilling fuel flow.

Computing System

The computing system positions the throttle valve to control fuel flow during steady-state operation, acceleration, and deceleration by using the ratio of metered fuel flow to engine compressor discharge pressure (Wf/P_{s4}) as a control parameter.

The positioning of the throttle valve by means of the Wf/P_{s4} parameter is accomplished through a multiplying system whereby the Wf/P_{s4} signal for acceleration, deceleration, or steady-state speed control is multiplied by a signal proportional to compressor discharge pressure to provide the required fuel flow.

The compressor discharge pressure-sensor assembly consists essentially of a pair of matched bellows, the evacuated and the motor bellows, and a sensor lever.

The motor bellows is externally exposed to compressor discharge pressure and is referenced to an evacuated bellows of equal size to produce a resultant force which is proportional to absolute compressor discharge pressure. This force is transmitted through a sensor lever to a set of rollers whose position is proportional to the required Wf/P_{s4} ratio. These rollers ride between the sensor lever and a multiplying lever. The force is transmitted through the rollers to the multiplying lever. Any change in the roller position or the compressor discharge pressure signal results in an unbalanced torque which will displace the rotating throttle-valve pilot valve from its hydraulic null position, thereby repositioning the throttle valve. The movement of the throttle valve extends or relaxes the throttle valve feedback spring which will return the multiplying lever to its equilibrium position when the throttle valve reaches the required fuel flow position. Both the motor and evacuated bellows are located in a chamber vented through an orifice to ambient pressure so that in event of an evacuated bellows failure, the fuel flow error is only the difference between the flow required for the absolute pressure reading and that required for a gage pressure reading. In the event of a motor bellows failure, compressor discharge pressure will be sensed on the external surface of the evacuated bellows and the system will continue to function.

The compressor discharge pressure limiter group consists of a clevis housing, spring housing, linear ball bearings, rolling diaphragm and piston, pushrod, two adjusting setscrews, transfer tube, limiter lever and spring, clevis support flexure, and a bellows clevis rod. At values of P_{s4} below the limiting pressure, the compressor discharge pressure limiter spring forces the diaphragm retainer against the pushrod housing. When P_{s4} exceeds the limiting value, the spring force is overcome and the diaphragm retainer becomes unseated.

The pushrod, which is attached to the diaphragm retainer, then engages the compressor discharge pressure limiter lever which in turn engages the compressor discharge pressure bellows stem clevis, thereby reducing the bellows output force.

The speed-sensing governor consists essentially of a rotating pilot valve, flyweights, and flyweight head. The engine speed signal is transmitted from the engine-driven drive shaft through a gear train to the centrifugal-type flyweight governor. This governor controls movement of the speed servo (three-dimensional or 3D cam) by displacing the rotating pilot valve from its hydraulic null position. When the speed changes, the flyweight force varies and the pilot valve is displaced, causing motion of the speed servo. This motion of the speed servo repositions the pilot valve, through the action of a feedback lever working on a spring, until the speed-sensing governor returns to null position at the new speed servo position. The position of the speed servo is, therefore, indicative of actual engine speed. The 3D cam is contoured to protect against the condition of a broken speed-sensing drive shaft. In the event of such a failure, the speed servo bottoms in its bore while the pushrod on the speed servo bottoms on an adjustable stop set near the zero speed position to eliminate feedback to the speed governor pilot valve. The 3D cam then places the limiting linkage at the Wf/P_{s4} ratio corresponding to the value selected for this failure condition.

Acceleration control is provided by adjustment of the roller positioning linkage to effect a maximum Wf/P_{s4} ratio stop for a particular value of speed and compressor inlet temperature. The maximum Wf/P_{s4} ratio value at the stop is controlled by a 3D cam which is translated by a signal proportional to engine speed and rotated by a signal proportional to compressor inlet temperature. The 3D cam is so contoured as to define a schedule of Wf/P_{s4} versus compressor inlet temperature which is used as a limiting value for each speed throughout the acceleration transient. This combination will permit engine acceleration within the overtemperature and surge limits of the engine. When the acceleration limiting lever is in operation to control the maximum value of the Wf/P_s4 ratio, it overrides the speed setting linkage.

Deceleration control is provided by the constant radius portion of the droop cam and by adjustment of the roller positioning linkage to limit the travel of the rollers toward decreasing fuel flow, thereby effecting a minimum Wf/P_{s4} ratio. This provides a linear relationship between fuel flow and compressor discharge pressure which results in blowout-free deceleration.

Engine speed control is accomplished by comparing the actual speed, as indicated by the position of the speed servo, with the desired speed value required for the power selected by the pilot through a power lever positioning the speed set cam. The power lever actuates the speed set cam to select a governor droop line. The position of the droop line is biased by compressor inlet temperature. The deviation of desired speed from the actual speed (speed error) causes movement of the speed servo. This movement of the speed servo is transmitted through a lever and results in the repositioning of the droop cam. The rollers in the multiplication system are positioned through the action of the droop cam to be a function of the speed error. The repositioning of the rollers then provides the required steady-state Wf/P_{s4} ratio setting.

The temperature-sensor bellows assembly consists of the motor and compensating bellows unit, a feedback lever, a pilot valve, and output lever and pushrod, a compensating lever and pushrod, and the temperature-sensor housing. Compressor inlet temperature is sensed by a liquid-filled bulb mounted in the compressor inlet and connected to a liquid-filled bellows in the control. The liquid expands with increased temperature and the extra volume travels through a capillary tube to the liquid-filled motor bellows in the control. The bellows length changes and through levers displaces a four-way pilot valve which results in movement of the temperature servo piston. The servo piston is connected through a linkage to a rack which meshes with the spline on the 3D cam, and motion of the piston rotates the cam. The feedback lever is attached to the rack and as the rack moves to rotate the cam, it also repositions the pilot valve in order to return the valve to the steady-state position. The rotation of the 3D cam, acting through a linkage, resets the governor droop and ac-

celeration line. Any ambient air or fuel temperature variation which acts on the motor bellows and capillary tube also acts on a compensating bellows and dead-ended capillary tube causing the fixed pivot of the motor bellows lever system to move to a new position so that the net result of the variation is not sensed at the pilot valve.

MAIN SHAFT BEARINGS *(FIG. 26-37)*

General

The front compressor system (low pressure) with its related turbine rotor, is supported by four antifriction bearings—one (no. 1) in front of and one (no. 2) behind the front-compressor rotor, and one (no. 4½) in front of and one (no. 6) behind the front-compressor drive turbine rotor. The no. 2 bearing assembly consists of a matched pair of ball thrust bearings and locates the front-compressor system axially.

The rear compressor system (high pressure) and related turbine rotor, is mounted on three antifriction bearings—one (no. 3) in front of and one (no. 4) behind the rear-compressor rotor, and one (no. 5) in front of the rear compressor drive turbine rotor. The no. 4 bearing assembly consists of a matched pair of ball thrust bearings and locates the rear-compressor system axially.

The main accessory drive gearshaft rotates inside the upper roller and lower ball bearing inside the intermediate section of the compressor case. See Fig. 26-37 for the location of all the main shaft bearings and Fig. 26-39 through 26-45 for details of each bearing arrangement.

Structure

All the roller bearings employ a one-piece cage, a recessed race ring and a plain raceway ring. Either the inner or the outer ring of a bearing may be the recessed race ring, the determining factor being assembly and disassembly requirements. The rollers in the bearings are crowned in a conventional fashion.

Engines equipped with an oil-dampened no. 1 bearing configuration have a larger bearing outer race (ring), lugs on the front face for locking with a retaining plate, and grooves in the outer race OD to hold metal seal rings. The outer race is held by four bolts rather than bearing retaining nuts.

The main shaft thrust bearings (no. 2 and no. 4) are used in tandem pairs in order to obtain a better safety margin. The bearings are manufactured in matched pairs and assembled in the engine in a manner to obtain the best possible distribution of load for this type of arrangement. The bearing oil baffle, which is positioned between the bearing outer races, serves to ensure the proper distribution of oil to each of the two bearings, and also, in case of failure of one bearing, prevents a flow of chips into the second bearing. The inner races of each bearing are split to permit a maximum ball complement as well as a one-piece cage.

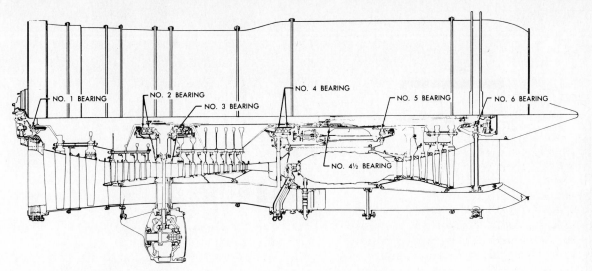

FIG. 26-37 Bearing location.

Main Shaft Oil and Airseals

The main shaft seals for the nos. 1, 2, 3, and 4 bearing locations are labyrinth type. Split-ring-type seals are used at the no. 4½ and no. 6 bearing locations. A face-type seal with two expanding metal seal rings is used at the no. 5 bearing location.

The following table lists the main shaft seals by companion bearing number and type.

Oil and airseals table

Number	Type	Location
1	Labyrinth seal	Rear of no. 1 bearing
2	Labyrinth seal	Forward of no. 2 bearing
3	Labyrinth seal	Rear of no. 3 bearing
4	Labyrinth seal	Forward of no. 4 bearing
4½	Split-ring seal	Rear of no. 4½ bearing
5	Face-type seal	Rear of no. 5 bearing
6	Split-ring seal	Forward of no. 6 bearing

PRESSURE OIL SYSTEM (FIG. 26-38)

General

The engine lubrication system is of a self-contained, high-pressure design consisting of a pressure system which supplies lubrication to the main engine bearings and to the accessory drives, and a scavenge system by which oil is withdrawn from the bearing compartments and from the accessories, and then returned to the oil tank. A breather system connecting the individual bearing compartments and the oil tank completes the lubrication system.

Oil is gravity-fed from the oil tank into the main oil pump within the gearbox. The pressure section of the main oil pump forces oil through the main oil strainer located immediately downstream of the pump dis-charge. The main oil strainer filter element is a stacked-disk type, a reusable filter element, or a disposable filter element. A bypass valve is incorporated in the center of the filter element. If the filter element becomes clogged, the bypass valve will move off its seat and the oil will bypass through the center of the filter.

Proper distribution of the total oil flow to the various locations is maintained by metering orifices and clearances. The main oil pump is regulated by a valve to maintain a specified pressure and flow. Pressure, relative to internal engine breather pressure (tank pressure), and flow are essentially constant with changes in altitude and engine speed.

Oil leaves the gearbox and flows to the fuel-coolant oil cooler (optional). If the cooler is blocked, an oil cooler bypass valve opens to permit the continuous flow of oil. Oil leaves the cooler (or passes through the valve) and flows into the oil pressure tubing to the main bearing compartments. The pressure sense line maintains a constant oil pressure at the bearing jets, regardless of the pressure drop of the oil at the fuel-oil cooler.

No. 1 Bearing Lubrication and Seal [FIG. 26-39(a)]

Oil for the no. 1 bearing enters the inlet case through a tube in the bottom vane.

For engines equipped with oil-dampened no. 1 bearing [Fig. 26-39(b)], a transfer tube from the front accessory support leads back into the bearing support to supply oil to a cavity around the bearing outer race.

The remainder of the oil moves up the tube and is then routed, through a small strainer in the front accessory-drives support, into the accessory-drives gearshaft. It moves to the outer wall of the gearshaft and through holes in that wall; then through holes in the front hub and inner race retaining nut to the front of the no. 1 bearing.

At the no. 1 bearing, a stainless steel seal with multiple knife-edges is mounted on the front hub of the front compressor rotor. This seal rotates inside an alu-

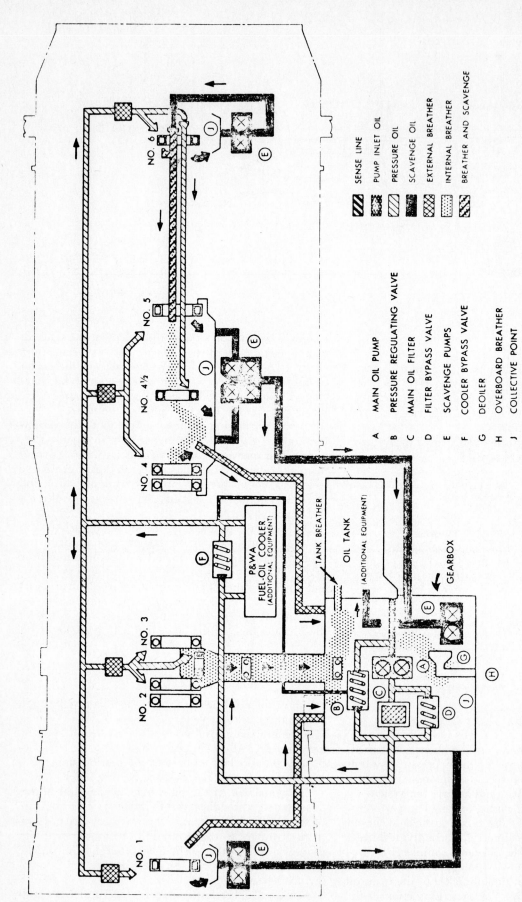

SENSE LINE
PUMP INLET OIL
PRESSURE OIL
SCAVENGE OIL
EXTERNAL BREATHER
INTERNAL BREATHER
BREATHER AND SCAVENGE

A MAIN OIL PUMP
B PRESSURE REGULATING VALVE
C MAIN OIL FILTER
D FILTER BYPASS VALVE
E SCAVENGE PUMPS
F COOLER BYPASS VALVE
G DEOILER
H OVERBOARD BREATHER
J COLLECTIVE POINT

NO. 6
NO. 5
NO. 4½
NO. 4
NO. 3
NO. 2
NO. 1

P&WA
FUEL-OIL COOLER
(ADDITIONAL EQUIPMENT)

TANK BREATHER

OIL TANK
(ADDITIONAL EQUIPMENT)

GEARBOX

FIG. 26-38 Oil system schematic.

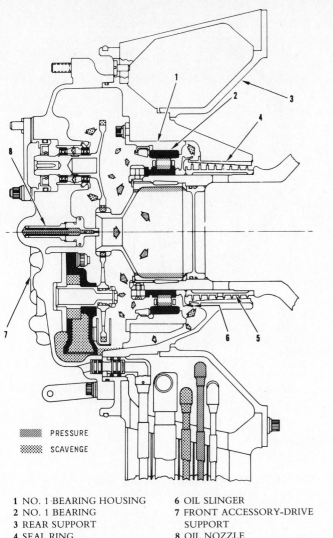

PRESSURE

SCAVENGE

1 NO. 1 BEARING HOUSING	6 OIL SLINGER
2 NO. 1 BEARING	7 FRONT ACCESSORY-DRIVE
3 REAR SUPPORT	SUPPORT
4 SEAL RING	8 OIL NOZZLE
5 BEARING SEAL	

(a)

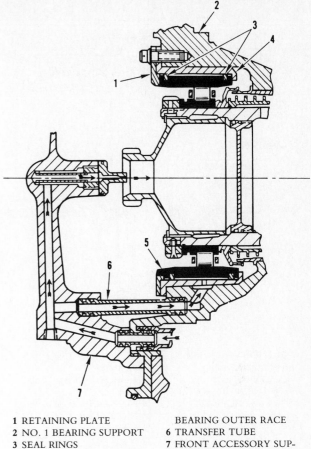

1 RETAINING PLATE	BEARING OUTER RACE
2 NO. 1 BEARING SUPPORT	6 TRANSFER TUBE
3 SEAL RINGS	7 FRONT ACCESSORY SUP-
4 NO. 1 BEARING HOUSING	PORT
5 OIL-DAMPENED NO. 1	

(b)

FIG. 26-39 (a) No. 1 bearing, seals and lubrication (non-oil-damped bearing).
(b) No. 1 bearing lubrication (oil-damped bearing).

minum multiplatform seal ring positioned inside the no. 1 bearing front support. In front of the knife-edge seal, a steel oil slinger is positioned behind the no. 1 bearing inner race on the front hub.

No. 2 and No. 3 Bearings Lubrication Seals and Air Tubes [*FIG. 26-40(a)*]

Oil enters the no. 2 and no. 3 bearing compartment through a small strainer and is sprayed onto the bearings through a three-legged oil nozzle assembly. A front leg (or nozzle) directs oil toward the no. 2 bearing, a second toward the no. 3 bearing, and a third toward the gearbox drive shaft upper bearing. Oil flows through holes in the rear hub to the ID of the no. 2 bearing. Flow through the gearbox drive bevel gear holes carries oil to the ID of the no. 3 bearing.

Mounted on the front compressor rotor rear hub (forward of the no. 2 bearings) from front to rear, are a multiple knife-edge airseal, a multiple knife-edge oil seal, and an oil baffle. The oil seal is cantilevered forward so that it is concentric, and outside, the smaller airseal.

Both the oil seal and airseal rotate inside stationary, stepped seal rings, riveted to the front of the no. 2 bearing seal ring support. At the bottom of the seal ring support, an air bleed boss accommodates the no. 2 bearing air bleed tube which vents into the no. 2 and 3 bearing seal air system.

The oil baffle and oil seal oppose the action of the oil to come through the labyrinth. The air pressure behind the labyrinth airseals further opposes the oil. Any oil which may seep by the seals will be carried off as oil vapor through the air bleed tube.

Welded between the no. 2 and no. 3 bearing housings at the 9, 12, and 3 o'clock positions are three air tubes. See Fig. 26-40(b). These three curved tubes run between the compressor intermediate inner case and the bearing housing: air pressure to the 12 o'clock position on the housing, air bleed from the 3 o'clock and the 9 o'clock

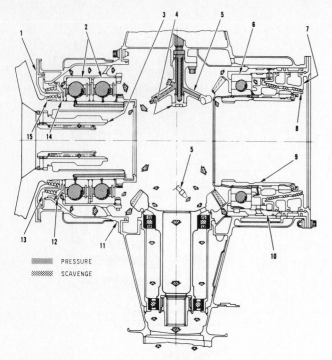

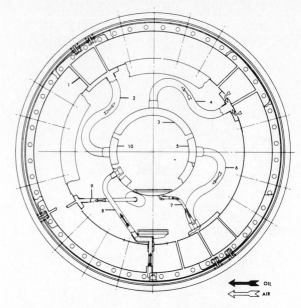

1 NO. 2 BEARING INNER
  AIRSEAL RING
2 NO. 2 BEARING
3 FRONT-COMPRESSOR RO-
  TOR REAR HUB COU-
  PLING
4 OIL-STRAINER ELEMENT
  ASSEMBLY
5 OIL-NOZZLE ASSEMBLY
6 NO. 3 BEARING
7 NO. 3 BEARING AIRSEAL-
  ING RING

8 NO. 3 BEARING SEAL AS-
  SEMBLY
9 GEARBOX-DRIVE BEVEL
  GEAR
10 NO. 3 BEARING HOUSING
11 NO. 2 BEARING HOUSING
12 OIL BAFFLE
13 NO. 2 BEARING OUTER
   AIRSEAL RING
14 NO. 2 BEARING AIR/OIL
   SEAL
15 NO. 2 BEARING AIRSEAL

(a)

1. SIXTH-STAGE VANE (24)
2 SEAL AIR TUBE
3 SEAL MANIFOLD SEG-
  MENT
4 SEAL AIR TUBE
5 SEAL-BLEED-MANIFOLD
  SEGMENT
6 SEAL AIR TUBE

7 MIXED OIL/AIR DIS-
  CHARGE TUBE
8 OIL-PRESSURE TUBE (REF-
  ERENCE)
9 NO. 2 BEARING REAR AIR-
  BLEED TUBE
10. SEAL-BLEED-MANIFOLD
   SEGMENT

(b)

FIG. 26-40 (a) No. 2 and 3 bearings, seals and lubrication. **(b)** No. 2 and 3 bearing seal air tubes.

positions on the housing. A smaller air bleed tube runs to the eight o'clock position of the inner case from a boss on the housing at the seven o'clock position.

Air enters the two o'clock position by two drilled holes in the inner wall of the vane support. The air passes inward through this tube then through the labyrinth seal compartments, cooling them. It passes outward through the other tubes, through the vanes at the 5, 8, and 11 o'clock positions and vents into the fan discharge air path.

A no. 3 bearing oil-air tube for any oil which may bypass the seal runs from the bottom of the bearing housing down to the five o'clock vane and into the air bleed tube at that location.

To the rear of the no. 3 bearing, a steel multiple-knife-edge oil and airseal is mounted on the gearbox drive bevel gear. This seal incorporates a steel spacer brazed inside and oil holes between the front groove and the front inner surface.

Bolted to the no. 3 bearing end of the intermediate case is the no. 3 bearing housing and airseal ring assembly. A packing and two seal rings on the OD of both the housing and seal ring assembly prevent leakage. On the ID, it incorporates sealing platforms in the ID and,

at the inner (large ID) end, a platform-type oil seal ring riveted in place. The knife-edge seal on the bevel gear rotates inside the platform-type ring assembly.

No. 4 *(FIG. 26-41)* and 5 *(FIG. 26-42)* Bearings Lubrication, Seals, and Air Tubes

Pressure oil for the no. 4 and 5 bearings locations flows into the engine through a tube at the eight o'clock location, on the left side of the fan discharge diffuser outer duct. It then flows upward around the diffuser case to the ten o'clock position and inward (through the inner passage of dual concentric pressure and breather tubing) to the no. 4 bearing support. Here it is directed rearward through an elbow and flows into the multi-passage no. 4 bearing oil nozzle assembly.

The no. 4 bearing oil nozzle assembly has an inlet passage, outlet holes at the bottom directing oil toward the no. 4 bearing, and an outlet passage toward the rear. An oil strainer is positioned inside the inlet passage. The outlet passage toward the rear accommodates the long oil tube of the no. 5 bearing oil nozzle assembly. Oil passes rearward through this tube and is then directed through the no. 5 bearing oil nozzle assembly. From

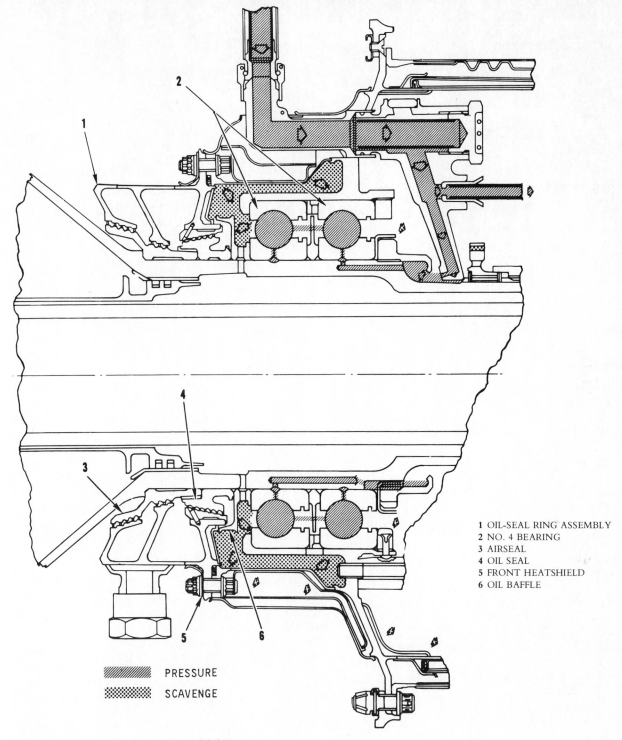

1 OIL-SEAL RING ASSEMBLY
2 NO. 4 BEARING
3 AIRSEAL
4 OIL SEAL
5 FRONT HEATSHIELD
6 OIL BAFFLE

▨▨▨ PRESSURE
▨▨▨ SCAVENGE

FIG. 26-41 No. 4 bearing, seals and lubrication.

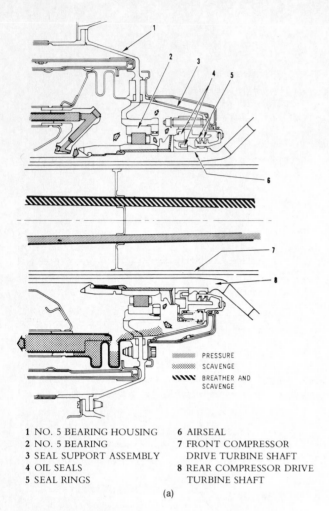

1 NO. 5 BEARING HOUSING   6 AIRSEAL
2 NO. 5 BEARING   7 FRONT COMPRESSOR
3 SEAL SUPPORT ASSEMBLY   DRIVE TURBINE SHAFT
4 OIL SEALS   8 REAR COMPRESSOR DRIVE
5 SEAL RINGS   TURBINE SHAFT

(a)

PRESSURE
SCAVENGE
BREATHER AND SCAVENGE

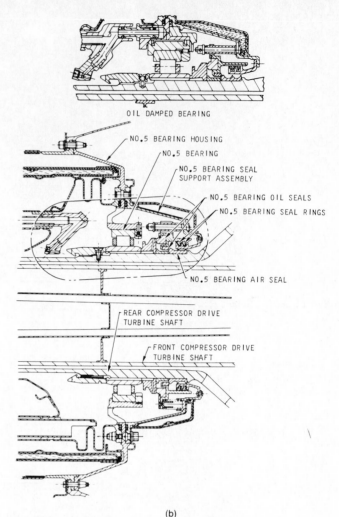

OIL DAMPED BEARING

NO.5 BEARING HOUSING
NO.5 BEARING
NO.5 BEARING SEAL SUPPORT ASSEMBLY
NO.5 BEARING OIL SEALS
NO.5 BEARING SEAL RINGS
NO.5 BEARING AIR SEAL
REAR COMPRESSOR DRIVE TURBINE SHAFT
FRONT COMPRESSOR DRIVE TURBINE SHAFT

(b)

FIG. 26-42 (a) No. 5 bearing, seals and lubrication.
(b) No 5 bearing with oil-damped insert.

the oil nozzle it passes under the bearing race and through the seal plate to the no. 5 bearing compartment. See Figure 26-43.

The no. 4 bearing airseal, mounted on the rear compressor rotor rear hub, is steel and has two groups of knife-edge seals on its OD. These knife-edges rotate

inside multistepped platform rings which are an integral part of the no. 4 bearing oil seal ring assembly, bolted to the front of the no. 4 bearing support.

The oil seal ring assembly incorporates three seal tube openings to which are connected the no. 4 bearing air tubes. The tubes bring eighth-stage discharge air to the annulus between the second and third labyrinth seal units and bleed air from the annulus between the first and second labyrinth units to the fan discharge path. The moving air holds down the bearing compartment temperature by bleeding hot air before it can reach the bearing compartment.

Aft of the airseal, also mounted on the rear hub, is a steel knife-edge oil seal which rotates inside the multistepped oil seal ring pinned in place in the rear of the no. 4 bearing oil seal ring assembly. Mounted on the rear hub behind the airseal is a steel ring-type oil baffle which opposes the action of the oil to enter the seal labyrinth.

The no. 5 bearing seal (face type) consists of a hard-faced (flame-coated) seat, mounted on the rear-compressor drive turbine rotor, which rides against one of two carbon seals mounted in a spring-loaded support. The forward face of an airseal on the rotor shaft rides

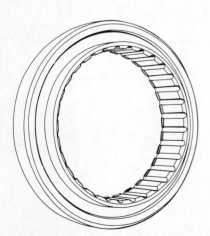

FIG. 26-43 Underrace oil grooves.

against the other carbon seal. The seal support incorporates two metal seal rings inside the rear of the seal assembly and a heat shield over its outer surface.

No. 4½ (FIG. 26-44) and No. 6 (FIG. 26-45) Bearings Lubrication and Seals

Oil flows to the no. 6 bearing area through a tube located in the upper turbine exhaust strut and down into the no. 6 bearing scavenge-pump housing. In the scavenge-pump housing oil passes through a small strainer then down into the outer passage of the no. 6 bearing oil nozzle assembly.

For engines having oil-damped no. 6 bearings, oil flows from the oil scavenge pump through a tube to the no. 6 bearing housing. The oil is then distributed to a cavity formed between the housing and the bearing outer race. Seal rings around the bearing outer race help contain oil in the cavity.

The oil flows forward in the oil nozzle outer passage and divides into two streams. One stream flows outward, through small holes on the OD of the nozzle outer-passage tube to lubricate the no. 6 bearing area. From the same nozzle outer-passage tube the other stream continues forward through holes on the nozzle outer front face and into the outer passage of the turbine bearings oil pressure and scavenge tubes assembly (oil trumpet) inside the front-compressor drive turbine rotor. The oil continues forward through the single (short) pressure tube in the oil trumpet to the no. 4½ bearing area.

The outward stream for the no. 6 bearing area, as previously mentioned, flows into the front-compressor drive turbine rotor rear hub. Through two sets of holes in the hub it flows to the no. 6 bearing seals and to the no. 6 bearing inner race.

The pressure oil in the oil trumpet flows forward and out through an oil baffle then through holes in the long turbine shaft to cool the no. 4½ bearing seal spacers and lubricate the no. 4½ bearing.

The no. 4½ seal consists of bonded-graphite ring seals mounted between spacers on the front-compressor drive turbine rotor shaft and rotating within the bore of the rear-compressor drive turbine rotor shaft. The split-ring seals are forced outward against the bore of the rear-compressor drive turbine shaft by centrifugal force and by gas pressure acting on the inner diameter. The gas pressure acting on the face of each seal ring forces the other face against the adjacent spacer.

The no. 6 bearing seal (split-ring type) consists of bonded-graphite ring seals mounted between spacers on the front-compressor drive turbine rotor rear hub and rotating within the bore of a stationary seal housing welded inside the no. 6 bearing support. The split-ring seals are forced outward toward the bore of the stationary housing by centrifugal force and by gas pressure acting on the inner diameter.

The gas pressure acting on one face of each seal ring forces the other face against the adjacent spacer. Relief holes through the sealing faces of the carbon rings prevent the pressure differential across a seal from giving

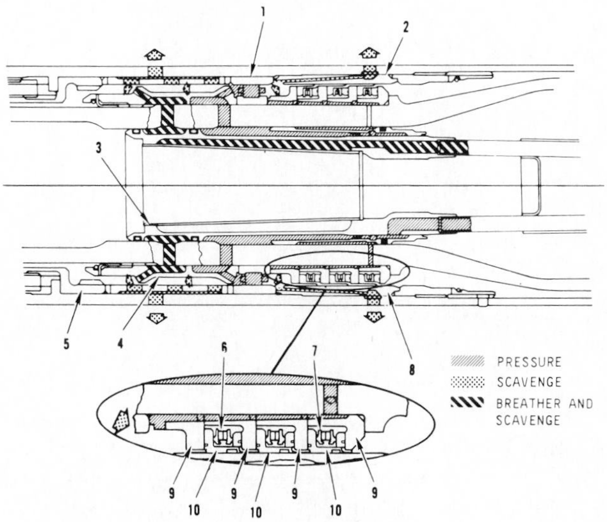

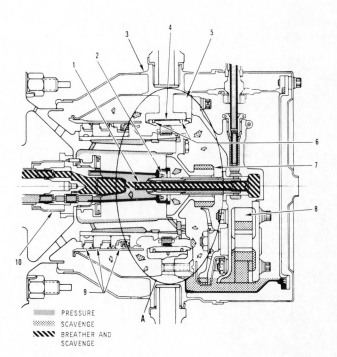

PRESSURE
SCAVENGE
BREATHER AND SCAVENGE

1 NO. 4½ BEARING	5 COUPLING
2 SEAL HOUSING	6 SPRING (3)
3 OIL-PRESSURE AND SCAVENGE TUBE ASSEMBLY	7 THRUSTRING (6)
4 INNER-RACE RETAINING NUT	8 SEAL HOUSING
	9 SPACER
	10 SEAL

FIG. 26-44 No. 4½ bearing, seals and lubrication.

PRESSURE
SCAVENGE
BREATHER AND SCAVENGE

1 OIL NOZZLE	SHAFT
2 SEAL	8 NO. 6 BEARING OIL-SCAVENGE PUMP
3 NO. 6 BEARING HOUSING ASSEMBLY	9 NO. 6 BEARING SEALS
4 NO. 6 BEARING	10 NO. 4½ AND 6 BEARING SHIELD AND TUBE ASSEMBLY
5 RETAINING PLATE	
6 SEAL RINGS	
7 SCAVENGE-PUMP GEAR-	

FIG. 26-45 No. 6 bearing seals and lubrication (oil-damped bearing).

an excessive closing force. There is also a steel-reinforced rubber oil seal mounted in the end of the oil pressure tube. This seal fits around the OD of the nozzle assembly and confines pressure oil within the oil pressure tube.

SCAVENGE OIL SYSTEM
(Refer to FIG. 26-38)

General

The scavenge oil system of the engine includes four gear-type pumps (five pump stages) which scavenge the main bearing compartments and deliver the scavenged oil to the engine oil tank.

No. 1 Bearing Compartment
(Refer to FIG. 26-39)

The single stage scavenge pump for the no. 1 bearing compartment is located in the cavity of the front accessory-drive housing. The pump is driven by the front accessory-drives gearshaft located in the front hub of the front-compressor rotor. The pump picks up the oil and sends it outward through a passage in the housing then down a tube located in the bottom vane of the inlet case.

No. 2 and 3 Bearing Compartment
(Refer to FIG. 26-40)

The second pump is located in the scavenge stage of the main oil pump assembly in the accessory-drive gearbox. Gearbox drive shaft bearings and no. 2 and no. 3 bearing scavenge oil, which drains down the outside of the accessory-drive shaft, is pumped from its collection point in the gearbox. One bevel gear drives both the scavenge and pressure stages in this pump.

No. 4, No. 4½, and No. 5 Bearings Area
(Refer to FIGS. 26-41, 26-42, and 26-44)

The third pump, with two stages driven by the same gear, is the no. 4 and no. 5 bearings oil scavenge-pump assembly, located inside the diffuser case. Together, the two stages of the pump scavenge the oil from the no. 4, no. 4½, and no. 5 bearing areas. In addition, scavenge oil from the no. 6 bearing area, after flowing forward through the two long scavenge tubes in the oil trumpet, flows into this compartment. A tube in the combustion chamber heat shield allows passage of the oil forward from the no. 5 bearing cavity.

The discharge from the pump is carried forward into the scavenge adapter at just below the nine o'clock position in the no. 4 bearing support, then outboard. It flows outboard, through the inner tube of dual concentric tubing (shared with a breather passage) to the outside of the diffuser case. From the outside of the diffuser case the oil flows downward to the eight o'clock position where it is then routed through the fairing to the outside of the diffuser outer duct.

No. 6 Bearing Compartment (Refer to FIG. 26-45)

The fourth scavenge pump is located in the no. 6 bearing scavenge pump housing where it is driven by a gearshaft bolted to the rear of the turbine rotor fourth-stage rear hub. It scavenges oil from the no. 6 bearing compartment and pumps it upward into the inner passage of the no. 6 bearing oil nozzle assembly.

The oil flows forward in the oil nozzle inner passage and is discharged through the center hole in the front of the nozzle. It passes forward into the inner passage of the oil trumpet and continues forward through the two long scavenge tubes as previously mentioned. At the front of the trumpet, the oil flows outward through holes in the front compressor drive turbine shaft and in the front of the no. 4½ bearing inner race retaining nut. The oil is then spun outward through holes in the rear-compressor drive turbine shaft and into the no. 4 bearing cavity.

Return oil passed forward by the two rearmost pumps, as well as that from the front oil suction pump, is directed into the gearbox cavity. From here the oil is pumped, by the scavenge stage of the gearbox pump, to the oil tank. Within the tank, the oil passes through a deaerator where the major part of the entrapped air is removed.

Breather System (Refer to FIG. 26-38)

To ensure proper oil flow and to maintain satisfactory scavenge pump performance during operation, the pressure in the bearing cavities is controlled by the breather system. The atmosphere of the no. 2 and 3 bearing cavity vents into the accessory gearbox. Breather tubes in the compressor inlet case and diffuser case discharge through external tubing into the accessory drive gearbox. Breather air from the no. 6 bearing compartment comes forward through the oil pressure and scavenge tubes assembly (oil trumpet), with the scavenge oil from that compartment, to the diffuser case cavity.

In the gearbox, vapor-laden atmosphere passes through rotary breather impellers, mounted on the starter drive gearshaft, where the oil is removed. The relatively oil-free air reaching the center of the gearshaft is conducted overboard.

AIR SYSTEMS (Refer to FIGS. 26-5 and 26-12)

Cooling Air System

Cooling air for the interior of the combustion chamber and turbine area is 13th-stage compressor air which passes between the multiedged airseal on the rear of the 13th-stage disk and the mating seal rings at the inner shroud of the exit vanes. This air enters the diffuser case inner cavity, formed by the rear compressor rear hub and the diffuser inner inlet duct, and goes rearward between the combustion chamber inner case and the turbine shafts heat shield. From this area it passes through holes in the no. 5 bearing housing.

A tubing system from the diffuser case inner cavity to the upper left boss on the diffuser outer fan duct provides a means to measure the turbine cooling air pressure.

A portion of the air passes between the first-stage turbine disk (front side) and airseal and through holes in the first-stage disk and blades. The rest of the air passes through holes located just aft of the disk front flange cooling the disk rear face. A portion of sixth-stage compressor air escapes via the interstage airseal, while the remainder of cooling air enters the front-compressor drive shaft and circulates around the turbine disks before passing outward through the rear hub. Cooling air passes around the no. 6 bearing support housing and the sump heat shield as well as joining the engine exhaust gas flow ahead of the exhaust struts.

At the no. 4 bearing compartment, the seal air tubes hold down the bearing compartment temperature by bleeding hot air before it can reach the compartment.

Labyrinth Seal Air System

The nos. 2, 3, and 4 bearing seals utilize an air system to facilitate their function. The air tubes, and their function, are described in Cooling Air System, and in Main Shaft Oil and Airseals in this section.

Internal Bleed Air System

In order to achieve a partial balance of thrust and maintain pressure on oil sealing areas, some compressor air is allowed to fill interior areas.

A small amount of sixth-stage air passes between the no. 2 bearing support and the rear of the sixth-stage disk. This air maintains a pressure upon the front of the no. 2 bearing seals and also passes forward through holes in the front compressor rear hub then forward into the front hub to maintain pressure at the no. 1 bearing seal.

Ninth-stage air is admitted through holes in the interstage spacer, aft of the interstage seal, to the interior cavity of the rear-compressor rotor. Because of holes in the rotor front hub and a three-edged seal on the front face of the seventh-stage disk, air pressure is maintained upon the no. 3 bearing seals.

The maintenance of pressure approximately equal to the ninth-stage discharge pressure on the area within the seal diameter at the forward end of the rear compressor achieves a partial balance of the thrust on the rear-compressor rotor. Aft of the rear compressor, the engine cooling air also functions to maintain a pressure on the seals.

IGNITION SYSTEM *(FIG. 26-46)*

20/4 J Exciter—General

The ignition exciter is a capacitor discharge system designed to provide ignition for the JT8D turbofan engine. This ignition exciter serves the dual purpose of providing intermittent-duty starting ignition and con-

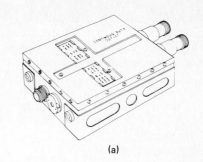

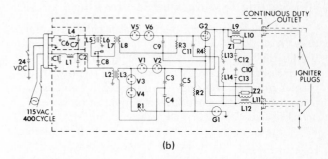

FIG. 26-46 (a) Ignition exciter (20/4 J).
(b) Ignition system schematic.

tinuous-duty ignition which is used as required after starting. Two different input voltages are required for the exciter. The intermittent-duty starting circuit requires an input of 28 V dc nominal, while the continuous duty requires an input of 115 V ac, 400 Hz. The intermittent-duty starting circuit discharges through both outlets, firing two igniter plugs. The continuous-duty circuit discharges only through the outlet marked CONTINUOUS DUTY OUTLET, firing one igniter plug. Spark gaps prevent current from flowing in one circuit when the other circuit is in operation. The ignition exciter is contained in one compact housing, with one input power connection and two output connections. An optional system, installed on some engines, consists of two independent 20-J, ac-powered, intermittent-duty capacitor-discharge circuits.

Specification Data

	Intermittent duty circuit	Continuous duty circuit
Input connector pins	B positive, A ground	C positive, D ground
Input voltage	14 to 29 V dc	90 to 124 V ac 350 to 440 Hz
Input current	5.0 A dc max.	2.5 A rms max.
Duty cycle	Intermittent	Continuous
Number of plugs fired	2	1
Stored energy	20 J	4 J
Spark rate	0.5 sparks/s min.	0.7 sparks/s min.
Ionizing voltage	22 to 26 kV	22 to 26 kV
Ambient temperature	−65°F to 275°F [−53.8°C to 135°C]	
Operating altitude	70,000 ft [21,336 m] max.	

WARNING: Ignition voltage is deadly. Do not touch igniter plugs if ignition is ON. Do not test ignition system when personnel are in contact with the igniter plugs or when inflammables are nearby.

The operation of the starting and continuous duty circuits is described separately in the following paragraphs, since it is not intended that the two systems operate simultaneously.

20-J Starting System (Dual Igniter Plugs)

When the ignition control switch is closed, electrons flow from the 24-V dc power supply through ground within the ignition exciter to the normally closed vibrator contacts, vibrator coils L5 and L6, power transformer primary winding L7, filter coil L4 and back to power supply. Electron flow through vibrator coils L5 and L6 cause the normally closed contacts to open. When the contacts open, electrons cease to flow through power transformer winding L7 and vibrator coil L5, and the vibrator contacts close. This repeated operation changes the 24 V dc input to a pulsating dc voltage which is applied across primary winding L7. Capacitor C8 prevents excessive arcing and burning of the vibrator contacts.

As a result of the pulsating dc voltage applied to the primary winding L7, an increased ac potential is produced across the secondary winding L8. When the top of L8 is negative, no electrons flow, since tubes V5 and V6 will not conduct. When the top of L8 is positive, electrons flow from L8 to C9, V6, V5, and back to L8. Each time tubes V5 and V6 conduct, an additional charge is built up on storage capacitor C9. C11 in parallel with C9 also builds up a charge. Resistor R3 has a high resistance and will not affect the charging of storage capacitor C9. The function of R3 is to discharge C9 when the ignition system is turned OFF.

The gap between electrodes 1 and 2 of G2 is set to break down between 2900 and 3100 V. The gap between electrodes 2 and 3 is set to break down at approximately 4500 V. When the charge across C9 and C11 builds up to 2900 to 3100 V, the gap between electrodes 1 and 2 of G2 ionizes and electrons flow from C11 to L11, electrodes 1 and 2 of G2 and back to C11. R4 limits electron flow from C9 across electrodes 1 and 2. The flow of electrons between electrodes 1 and 2 and G2 ionizes the gap between electrodes 2 and 3. This produces a low-impedance path between electrodes 2 and 3. Electrons now flow from C9 and C10, L10, electrodes 3 and 2, and back to C9.

The current in the primary windings L10 and L11 is a high-frequency oscillating current caused by C10, L10, and C11, L11 respectively. This induces a high-frequency high voltage across secondary winding L9 and L12.

This high-frequency high voltage ionizes the gaps of the igniter plugs. The current required to ionize the igniter plugs is very small. Therefore, the energy stored in capacitor C9 is virtually unchanged up to this point.

With the gap G2 and the igniter plugs both ionized, a low-impedance path exists from C9, through coil L12, lower igniter plug, ground, upper igniter plug, coil L9, electrodes 3 and 2 of G2, and back to C9. This results in a heavy flow of electrons. The positive voltage of the upper plate of C9 drops very rapidly to zero. However, the flow of electrons does not stop instantly. Instead, excessive electrons flow into the upper plate of storage capacitor C9, producing a negative voltage much smaller than the original value. Because the gap G2 and igniter plugs are still ionized, electrons flow from the upper plate of C9, to electrodes 2 and 3, coil L9, upper igniter plugs, ground, lower igniter plugs, coil L12 and back to C9. Several such oscillations occur until the voltage across C9 is no longer sufficient to restrike the arc between the electrodes of the igniter plugs.

Coils L13 and L14 are saturable inductors. If either of the igniter plugs becomes open-circuited, a larger current will flow through coil L13 or L14. The larger the current flow through the coil, the lower the resistance of the coils becomes. If the top igniter plug becomes open-circuited, electrons can flow from C9 to L12, lower igniter plug, ground, L13, electrodes 3 and 2 of G2, and back to C9. If the bottom igniter plug becomes open-circuited, electrons will flow from C9 to L14, ground, top igniter plug, L9 electrodes 3 and 2 of G2, and back to C9.

This feature allows the series discharge circuit (normally firing two igniter plugs) to fire one igniter plug even though the other plug is open-circuited.

Z1 and Z2 in parallel with primary windings L10 to L11 limit the circuit through the primary windings. This prevents the high-frequency voltage across L9 and L12 from going too high. Gap G1 isolates the continuous duty circuit from the intermittent duty circuit because of its high breakdown voltage.

4-J Continuous System (Single Igniter Plug)

When the control switch is closed, alternating current flows from the 115-V ac, 400-Hz power supply through primary winding L2, filter coil L1, and back to the power supply. The voltage applied to primary winding L2 induces an increased voltage of same frequency across secondary winding L3. When the top of L3 is positive, electrons flow from L3 to C3, V2, V1, and back to L3. When the top of L3 is negative, electrons flow from L3 to V3, V4, R1, C4, and back to L3. Capacitor C3 is charged on one half of the cycle and C4 is charged on the other half. Capacitor C5 in parallel with C3 and C4 is charged to a potential equal to that of C3 plus C4.

Gap G1 is set to break down between 3500 and 3600 V. When the voltage across C5 reaches this value, G1 will ionize and electrons will flow from C5 to G1, ground, C12, L10 and Z1, and back to C5. The current through primary winding L10 is a high-frequency oscillating current caused by C12 and L10 which induces a high-frequency high voltage across secondary winding L9.

This high-frequency voltage is sufficient to ionize the gap of the igniter plug. With G1 and the igniter plug ionized, a low-impedance path exists from C5 to G1, ground, igniter plug, L9, and back to C5. Because only

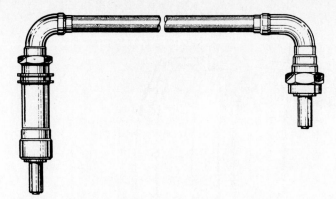

FIG. 26-47 Igniter plug lead assembly.

a small amount of the energy of C5 is used to ionize the igniter plug, most of the energy of C5 will flow through this low-impedance path. The positive potential at the upper plate of C5 drops very rapidly to zero. However, the flow of electrons does not drop instantly and excessive electrons flow into the upper plate of C5, producing a negative voltage which is much smaller than initial breakdown voltage. Because G1 and the igniter plug are still ionized, electrons flow from C5 to L9, igniter plug, ground, G1, and back to C5. Several such oscillations occur until C5 no longer has sufficient energy to restrike the arc at the igniter plug.

Resistor R1 limits the current flow through tubes V1, V2, V3, and V4 when the charge on top of C5 is negative. Resistor R2 discharges capacitors C3, C4, and C5 when the circuit is shut off. Gap G2 isolates the intermittent-duty circuit from the continuous-duty circuit because of the high breakdown voltage between electrodes 2 and 3.

High-tension Leads (FIG. 26-47)

The igniter plug lead assemblies are installed between the ignition exciter and igniter plugs. These lead assemblies carry the electrical energy from the ignition exciter to the igniter plugs. The left lead assembly is approximately 30 in long and the right lead assembly is ap-

proximately 51 in long. Figure 26-47 shows an illustration of a typical igniter plug lead assembly.

The chamfered washer/rubber bushing at both terminations must be replaced at the maintenance level during every lead installation.

Igniter Plugs (FIG. 26-48)

There are two igniter plugs which are mounted on the lower front of the combustion chamber outer case. One projects into the no. 4 combustion chamber and the other projects into the no. 7 combustion chamber.

The igniter plug provides the gap across which the electrical spark passes to ignite the fuel-air mixture. The igniter plug gap is ionized and becomes conductive by the surge of very high voltage from the high-frequency coils of the ignition exciter; then the storage capacitor discharges its accumulated energy across the ionized igniter plug gap. This results in a capacitive spark of very high energy, capable of vaporizing globules of fuel and overcoming carbon deposits.

ENGINE AIR SYSTEMS

General

The engine air systems are the anti-icing air system, compressor bleed air system, cooling air system, labyrinth seal air system, internal bleed system, and the fuel deicing air system. The cooling air system, labyrinth seal air system, and the internal bleed system are integral with the engine and are described in detail elsewhere in this chapter. The fuel deicing air system is also described elsewhere. The two remaining systems, i.e., the anti-icing and compressor bleed air systems, are discussed here.

Anti-icing Air System (FIG. 26-49)

There are three valve and actuator assemblies on the engine which can be used to permit passage of high-pressure air for air-inlet anti-icing and fuel deicing. The

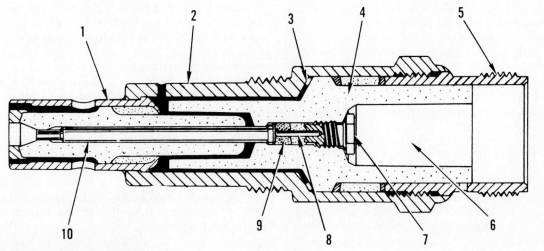

1 LOWER SHELL
2 UPPER SHELL
3 GASKET
4 INSULATOR
5 COUPLING THREAD
6 TERMINAL WELL
7 TERMINAL SCREW
8 SEALING WIRE
9 CEMENT
10 CENTER ELECTRODE

FIG. 26-48 Igniter plug.

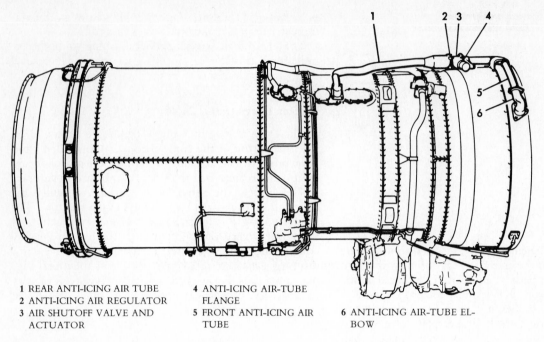

1 REAR ANTI-ICING AIR TUBE
2 ANTI-ICING AIR REGULATOR
3 AIR SHUTOFF VALVE AND
   ACTUATOR
4 ANTI-ICING AIR-TUBE
   FLANGE
5 FRONT ANTI-ICING AIR
   TUBE
6 ANTI-ICING AIR-TUBE EL-
   BOW

FIG. 26-49 Engine anti-icing air system.

butterfly valves are electrically actuated to be turned ON or OFF. One valve and actuator is located within the fuel deicing system and the other two are in the left and right air-inlet anti-icing systems.

In order to prevent undesirable icing of the engine air inlet surfaces, an anti-icing air system is incorporated in the engine. The principal components of the system are the two air shutoff valve and actuator assemblies, the two regulators, and appropriate tubing.

Hot air (eighth stage) is bled from each side of the rear compressor and is piped forward to the inlet section. From the outer annulus of the compressor inlet case, into which the heated air is piped, the air flows inward through the hollow compressor inlet vanes to an inner annulus.

During periods of anti-icing system operation, restrictive metering orifices in the left and right anti-icing air tubes control hot airflow into the engine inlet and minimize loss of engine thrust. The metering orifices are in the form of either separate metering plugs assembled at the anti-icing air regulator flanges, or restrictive openings incorporated in each of the rear anti-icing air tubes.

Compressor Bleed System *(FIG. 26-50)*

The compressor bleed air system is primarily designed to permit operational flexibility by allowing high compressor discharge air to bleed into the fan discharge duct. Spaced around the diffuser case at the four o'clock and seven o'clock locations, adjacent to the 13th-stage vanes, are two compressor bleed valves. On certain engines, an additional single eighth-stage bleed valve is located at the six o'clock position on the compressor

fan discharge inner duct; this 8th-stage bleed valve operates in unison with the 13th-stage bleed valves.

In the static position (engine not running) bleed valves may be either open or closed, depending upon gravity and/or drag caused by contact of the valves with the cylinder walls. During periods of engine operation, compressor discharge air pressure exerted on the valve faces acts to force valves into the open position. When P_{s3} pressure on one side of the diaphragm in the pressure ratio bleed control increases to the point where it overcomes combined P_{t2} and spring pressure forces, the poppet valves in the control reverse position, the muscle valve transfers, and P_{s4} actuating air is directed to the back side of bleed valves. This P_{s4} air acting on the larger area of the back side of the valves is sufficient to overcome compressor discharge air acting on the valve faces and the valves close. When the P_{s3}/P_{t2} differential pressures become high enough, the procedure is reversed. P_{s4} actuating pressure on the back side of the valves is reduced to ambient, and internal engine pressure forces the bleed valves open.

The pressure-ratio bleed control is located on the lower right side of the engine at the diffuser section.

Compressor discharge air (P_{s4}) is routed to the pressure-ratio bleed control mounted on the outside of the engine. This unit operates so as to schedule the bleed valve operation as a function of the pressure rise across the front compressor. Senses utilized are inlet total pressure (P_{t2}) and compressor discharge static pressure (P_{s3}).

P_{s3} is ported through a two-stage nozzle system and the resultant sense, upstream of the second nozzle, moves a diaphragm against P_{t2} and spring pressure. Any change from the schedule position of the dia-

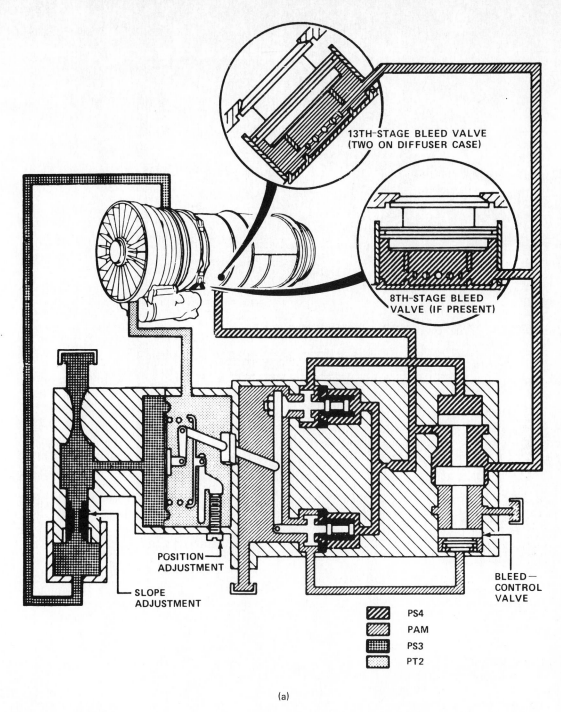

13TH-STAGE BLEED VALVE
(TWO ON DIFFUSER CASE)

8TH-STAGE BLEED
VALVE (IF PRESENT)

POSITION
ADJUSTMENT

SLOPE
ADJUSTMENT

BLEED—
CONTROL
VALVE

▨	PS4
▨	PAM
▦	PS3
▦	PT2

(a)

FIG. 26-50 (a) Bleed system schematic—closed. (Continued on the following page.)

phragm produces a corrective action by varying the low-pressure bleed valves. This is accomplished by a yoke which transmits the signal from the diaphragm to a transfer valve assembly consisting of poppet valves linked together. Movement of these valves directs high compressor discharge air (P_{s4}) to a "servo valve" which in turn moves to port actuating air to the manifold which supplied the individual bleed valves.

ENGINE INDICATING *(FIG. 26-51)*

General

The two engine indicating systems furnished with the JT8D engine are the turbine exhaust temperature-indicating system and the turbine exhaust total-pressure-

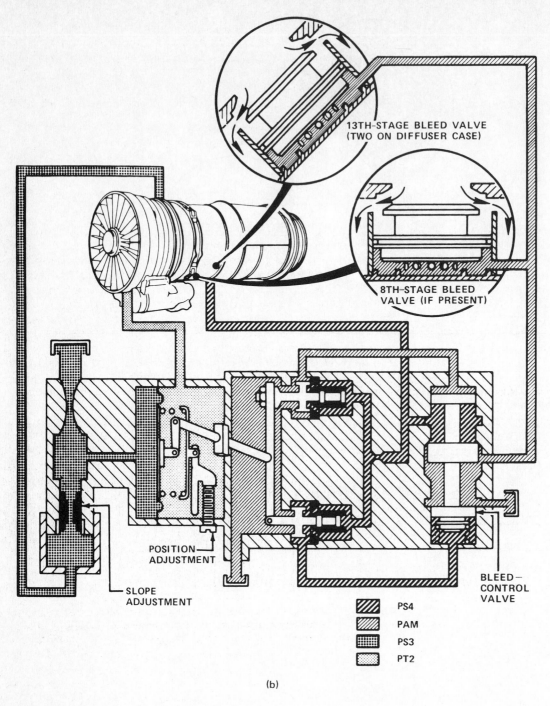

POSITION
ADJUSTMENT

SLOPE
ADJUSTMENT

13TH–STAGE BLEED VALVE
(TWO ON DIFFUSER CASE)

8TH–STAGE BLEED
VALVE (IF PRESENT)

BLEED—
CONTROL
VALVE

	PS4
	PAM
	PS3
	PT2

(b)

FIG. 26-50 (cont.) **(b)** Bleed system schematic—open.

indicating system. The temperature-indicating system consists of eight single-junction thermocouple probes connected by a branched cable to provide individual thermocouple temperature indications. A special-purpose lead connected to the branch cable provides an average EGT (T_{t7}) indication at the lug-type terminal block. The pressure-indicating system consists of six pressure-sensing probes connected by tubes to measure the average pressure in the turbine discharge area (P_{t7}).

Turbine Exhaust Pressure-indicating System

Six averaging total-pressure probes, manifolded together forward of the mount ring on the turbine exhaust outer case, measure the turbine outlet total pressure. In using these pressure probes, connection is made to the manifold at a single external point (bottom left of the turbine exhaust outer duct) to measure the average pressure at the turbine discharge (P_{t7}).

The probes are located at the approximate 2, 4, 7, 8, 10, and 12 o'clock positions.

Many engines are equipped with an engine pressure ratio (EPR) system. The EPR system is composed of six pressure probes, transmitter, indicator, and associated wiring (Fig. 26-52). The EPR transmitter converts the exhaust pressure (P_{t7}) and the inlet pressure (P_{t2}) into a ratio, and generates an electrical signal corresponding to pressure changes in the engine, while the EPR indicator provides a visual indication of the engine exhaust and inlet pressure ratio (P_{t7}/P_{t2}). There is one indicator for each engine. The indicators are located on the pilots' engine instrument panel.

The system operates on ac power in the following manner. The engine exhaust and inlet pressure are sensed by the pressure probes. These pressures act on the bellows assembly of the pressure ratio transmitter, causing bellows movement whenever pressures change. The generated electrical signals are transmitted to their respective pressure ratio indicators over a three-wire system. The indicator converts the electrical signals into the pointer shaft rotation or indicator pointer movement corresponding to the pressure change in the engine.

Turbine Exhaust Temperature-indicating System

The averaging temperature measurement provided by this system is an operating limit on the engine and is used to monitor the mechanical integrity of the turbines as well as to check engine condition during op-

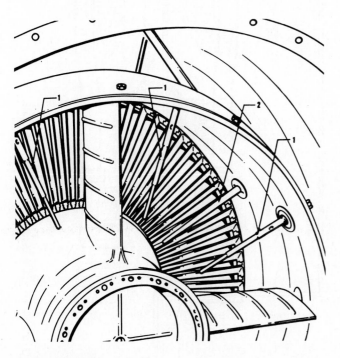

1 THERMOCOUPLE PROBE
(T_{t7})
2 PRESSURE PROBE (P_{t7})

FIG. 26-51 Pressure and temperature probes.

eration. Each of the eight single-junction thermocouple probes contains one thermocouple junction. A reading of each thermocouple may be obtained at the junction box on the branched thermocouple cable. An average reading is obtained at the end of a special-purpose lead, one end of which attaches to the junction box. The other end of the lead terminates at an engine-supplied, lug-type terminal block.

The eight thermocouple probes are mounted on the turbine inner rear case and extend into the turbine discharge passage.

The thermocouple assembly consists of a probe and a head (Fig. 26-53). Two thermocouple junctions are contained in the probe. These junctions are in parallel and electrically averaged in the head of the thermocouple assembly. They are located directly in the gas path in different positions within the probe. Two terminal studs are located at the head end of the thermocouple assembly. The larger diameter stud is alumel material and is connected to the alumel wires that make up one side of the thermocouple junctions. The smaller stud is chromel material and is connected to the chromel wires that make up the other side of the thermocouple junctions. The voltage across the chromel and alumel studs is an average of the voltage induced at each of the junctions and is proportional to the average temperature of the two junctions.

The probe portion of the thermocouple assembly is located in the exhaust gas path. The assembly is fastened to the turbine exhaust case with a spannernut over the threaded thermocouple head. Proper orientation of the probe is assured by a key slot in the probe head. The thermocouple assembly is electrically connected to the wiring harness via harness terminal lugs fastened to the thermocouple assembly terminal studs. The functioning of a thermocouple circuit is primarily dependent upon a difference of materials at the thermocouple junctions and a difference in temperature between the hot junction and cold junction. The hot junction is located in the area to be measured, and the cold junction is generally located in the cockpit meter. One thermocouple material is alumel, which if color-coded is green. Alumel attracts a magnet and has a negative polarity. The other thermocouple material is chromel, which if color-coded is white. Chromel does not attract a magnet and has a positive polarity. With this difference in material and the temperature difference between the hot junction and the cold junction, a voltage is induced in the circuit which is proportional to the temperature difference and deflects the cockpit meter.

The thermocouple branched cable assembly consists of an electrical conduit mounted on the outer circumference of the turbine exhaust case. This cable assembly contains eight sets of branch leads which attach to the connections on the eight single-junction thermocouples installed in the turbine exhaust case. This cable assembly is designed to provide individual readings of the eight thermocouples. The branched cable assembly is connected to a junction box which averages the individual thermocouple readings through the use of a bus bar. Individual thermocouple readings are obtained by re-

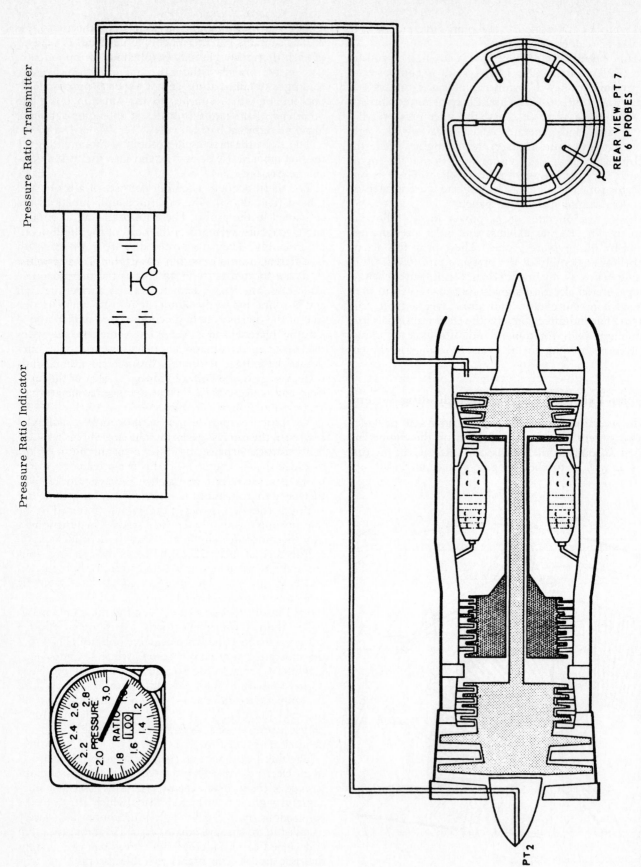

Pressure Ratio Transmitter

Pressure Ratio Indicator

REAR VIEW PT 7
6 PROBES

2.4 2.6 2.8 3.0
2.2 PRESSURE RATIO .2
2.0 1.00 .4
1.8 1.6 1.4

PT 2

FIG. 26-52 Engine pressure ratio system schematic.

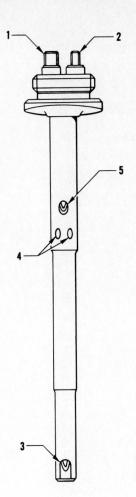

1. ALUMEL STUD (LARGER)
2. CHROMEL STUD (SMALLER)
3. OUTERMOST SAMPLING PORT
4. MIDDLE SAMPLE PORTS
5. INNERMOST SAMPLING PORT

moving this bus bar. A special-purpose lead transmits the average reading at the junction box to an engine-supplied, lug-type terminal block (Fig. 26-54).

REVIEW AND STUDY QUESTIONS

1. What aircraft use this engine?
2. Name some other engines that incorporate full-length fan ducts besides the JT8D.
3. How many stages are in the compressor and turbine of each spool?
4. Why do so many models of this engine have two ignition systems?
5. Name the number and location of the main bearings.
6. List four purposes for which air, taken from the compressor, may be used.
7. Name some of the materials used in the construction of the
   (a) Cases
   (b) Compressor section
   (c) Combustion chamber
   (d) Turbine section
8. What is the purpose of the no. 4½ bearing?
9. What is the function of the compressor bleed air system? What pressures does this system sense?
10. Name the main components incorporated in the engine fuel distribution system, and trace the flow of fuel. Include the function of each component.
11. Briefly describe the flow of oil through this engine in both the pressure and scavenge systems. Why and how is pressurized air used in the oil system?
12. Describe the temperature- and pressure-indicating systems.

FIG. 26-53 Thermocouple.

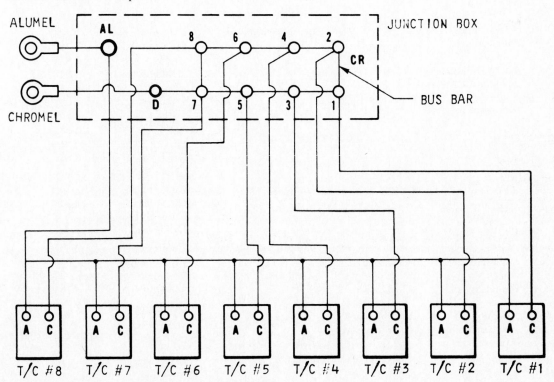

FIG. 26-54 Thermocouple lead and cable assembly.

To Convert From—	To—	Multiply By—	To Convert From—	To—	Multiply By—
Acres	Square feet	4.356×10^4	Centipoise	Pounds-second/foot	6.72×10^{-1}
"	Square meters	4.047×10^3	"	Kilograms-hour/meters	3.6
"	Square miles	1.562×10^{-3}	Chain	Links	100
"	Square rods	1.6×10^2	"	Feet	66
Amperes	Faradays/second	1.04×10^{-5}	Circle	Degrees	360
Amperes/square foot	Amperes/square centimeter	0.00108	"	Radius	6.28319
Ampere-hours	Coulombs	3600	Circular mils	Square centimeters	5.07×10^{-6}
Amperes/square centimeter	Amperes/square foot	929	"	Square inches	7.854×10^{-7}
Angstrom units	Inches	3.94×10^{-9}	"	Square millimeters	5.067×10^{-4}
"	Meters	1×10^{-10}			
"	Microns	1×10^{-4}	"	Square mils	7.854×10^{-1}
Atmospheres	Centimeters of Hg at 0°C	76	Cord	Cubic feet	128
"	Inches Hg at 0°C	29.92	Cubic centimeters	Cubic feet	3.53×10^{-5}
"	Feet of water at 4°C	33.9	"	Cubic inches	0.061
			"	Gallons (U.S.)	2.64×10^{-4}
"	Kilograms/square centimeter	1.033	"	Liters	9.999×10^{-1}
			"	Fluid ounces	0.0338
"	Pounds/square inch	15.696	"	Pints	0.0021
"	Pounds/square foot	2116	"	Quarts (liquid)	0.0011
"	Bars, hectopieze	1.013	Cubic feet	Cubic centimeters	28,317
Barrels (U.S. liquid)	Gallons	31.5	"	Cubic inches	1728
Bars	Centimeters of Hg at 0°C	75.01	"	Gallons (U.S.)	7.48
			"	Liters	28.32
"	Pounds/square inch	14.5	"	Quarts	29.92
			"	Cubic yards	3.704×10^{-2}
Btu	Foot-pounds	778.26	Cubic feet of water (60°F)	Pounds	62.37
"	Horsepower hours	3.93×10^{-4}	Cubic feet/minute	Gallons/second	0.1247
"	Kilowatt hours	2.931×10^{-4}	"	Liters/second	4.719×10^{-1}
"	Kilogram calories	2.52×10^{-1}	"	Cubic meters/minute	2.832×10^{-2}
"	Kilogram meters	1.076×10^2	Cubic inches	Cubic centimeters	16.39
"	Joules	1055	"	Cubic feet	5.787×10^{-4}
Btu/second	Watts	1055	"	Gallons (U.S.)	0.0043
Centimeters	Inches	0.394	"	Liters	0.0164
"	Mils	393.7	"	Fluid ounces	0.554
"	Feet	0.0328	"	Quarts (liquid)	0.0173
Centimeters of Hg	Inches of water at 4°C	5.354	Cubic meters	Cubic inches	61,023
			"	Cubic yards	1.308
"	Feet of water at 4°C	4.46×10^{-1}	"	Cubic feet	35.31
"	Pounds/square inch	1.934×10^{-1}	"	Gallons (U.S.)	274.17
"	Pounds/square foot	27.85	Cubic yards	Cubic feet	27
			"	Cubic meters	7.646×10^{-1}
"	Kilograms/square meter	135.95	"	Gallons	2.022×10^2
"	Atmospheres	0.013158	"	Cubic inches	46,656
			Curies	Disintegrations/second	3.7×10^{10}
Centimeters/second	Feet/minute	1.9685	Degrees (arc)	Radians	1.745×10^{-2}
"	Feet/second	3.281×10^{-2}	Degrees/second	Revolutions/minute	0.1667
"	Miles/hour	2.237×10^{-2}	"	Radians/second	0.017453

To Convert From—	To—	Multiply By—
Dynes	Grams	1.020×10^{-3}
"	Pounds	2.248×10^{-6}
"	Poundals	7.233×10^{-5}
Electron volts	Ergs	1.602×10^{-12}
Ergs	Btu	9.478×10^{-11}
"	Dyne centimeters	1
"	Electron volts	6.242×10^{11}
"	Foot-pounds	7.376×10^{-8}
"	Gram centimeters	1.020×10^{-3}
"	Joules	10^{-7}
"	Kilogram calories	2.388×10^{-11}
Faradays	Coulombs	9.65×10^{-4}
Faradays/second	Amperes	96,500
Fathoms	Feet	6
Feet	Centimeters	30.48
"	Inches	12
"	Meters	0.3048
"	Yards	3.333×10^{-1}
"	Miles	1.894×10^{-4}
"	Nautical miles	1.646×10^{-4}
Feet of water at 4°C	Atmospheres	2.95×10^{-2}
"	Pounds/square inch	4.335×10^{-1}
"	Pounds/square foot	62.43
"	Kilograms/square meter	3.048×10^{2}
"	Inches of Hg at 0°C	8.826×10^{-1}
"	Centimeters of Hg at 0°C	2.24
Feet/minute	Miles/hour	1.136×10^{-2}
"	Kilometers/hour	1.829×10^{-2}
"	Centimeters/second	5.08×10^{-1}
Feet/second	Miles/hour	0.6818
"	Kilometers/hour	1.097
"	Meters/minute	18.29
"	Centimeters/second	30.48
"	Knots	0.5925
Foot-pounds	Btu	1.285×10^{-3}
"	Joules (absolute)	1.35582
"	Meter-kilograms	1.383×10^{-1}
"	Kilowatt hours	3.766×10^{-7}
Foot-pounds/minute	Horsepower	3.030×10^{-5}
Foot-pounds/second	Horsepower	1.818×10^{-3}
"	Kilowatts	1.356×10^{-3}
Fluid ounces	Drams	8
"	Cubic centimeters	29.6
Furlongs	Feet	660
"	Yards	220
"	Rods	40
Gallons (U.S. liquid)	Quarts (liquid)	4
"	Cubic centimeters	3785.4
Gallons (U.S. liquid)	Cubic inches	231
"	Cubic feet	1.337×10^{-1}
"	Liters	3.785
"	Imperial gallons	8.327×10^{-1}
"	Fluid ounces	1.28×10^{2}
"	Pints	8
"	Pounds (av) of water at 17°C	8.31
Gallons (U.S. dry)	Cubic inches	268.8
"	Cubic feet	1.556×10^{-1}
"	U.S. gallons (liquid)	1.164
"	Liters	4.405
Gallons (imperial)	Cubic inches	277.4
"	U.S. gallons	1.201
"	Liters	4.546
Grains	Grams	0.0648
"	Ounces (avoir.)	0.0023
"	Ounces (troy)	0.0021
"	Pennyweights (troy)	0.0417
"	Pounds (avoir.)	1/7000
"	Pounds (troy)	1/5760
Grams	Grains	15.43
"	Milligrams	1000
"	Ounces (avoir.)	0.0353
"	Ounces (troy)	0.0321
"	Pennyweights	0.643
"	Pounds (avoir.)	0.022
"	Pounds (troy)	0.0027
"	Kilograms	10^{-3}
"	Dynes	980.67
Grams/cubic centimeter	Pounds/cubic inch	0.03613
"	Pounds/cubic foot	62.43
"	Kilograms/cubic meter	1000
Grams/centimeter	Kilograms/meter	0.1
"	Pounds/foot	6.721×10^{-2}
"	Pounds/inch	5.601×10^{-3}
Grams/liter	Ounces/gallon (troy)	0.122
"	Ounces/gallon (avoir.)	0.134
"	Parts/million	1000
"	Pennyweights/gallon	2.44
Grams of U^{235} fissioned	Kilowatt hours heat gen.	23,000
Gram calories	Btu	3.969×10^{-3}
Hands	Inches	4
Hectares	Square meters	10^{4}
"	Acres	2.471
Hectopieze	Inches of Hg	29.53
Horsepower	Foot-pounds/minute	33,000
"	Foot-pounds/second	550
"	Meter-kilograms/second	76.04

To Convert From—	To—	Multiply By—
Horsepower	Metric horsepower	1.014
"	Kilowatts	7.457×10^{-1}
"	Btu/hour	2545.08
"	Btu/second	7.068×10^{-1}
Horsepower (metric)	Meters/kilogram second	75
"	Horsepower	9.863×10^{-1}
"	Kilowatts	7.355×10^{-1}
"	Btu/second	6.971×10^{-1}
Horsepower hours	Btu	2.545×10^{3}
"	Foot-pounds	1.98×10^{6}
"	Meter-kilograms	2.737×10^{5}
Inches	Centimeters	2.54
"	Feet	83.33×10^{-3}
"	Mils	1000
Inches of Hg at 0°C	Atmospheres	3.342×10^{-2}
"	Inches of water at 4°C	13.6
"	Feet of water	1.133
"	Pounds/square inch	4.912×10^{-1}
"	Pounds/square foot	70.73
"	Kilograms/square meter	3.453×10^{2}
Inches of water at 4 °C	Atmospheres	2.458×10^{-3}
"	Inches of Hg at 0°C	7.355×10^{-2}
"	Centimeters of Hg at 0°C	1.868×10^{-1}
"	Pounds/square inch	3.613×10^{-2}
"	Pounds/square foot	5.202
"	Kilograms/square meter	25.4
Joules	Btu	9.48×10^{-4}
"	Foot-pounds	7.376×10^{-1}
"	Kilogram calories	2.389×10^{-4}
"	Kilogram meters	1.020×10^{-1}
"	Watt hours	2.778×10^{-4}
"	Horsepower hours	3.725×10^{-7}
"	Ergs	10^{7}
Kilograms	Grains	15,432.4
"	Grams	1000
"	Ounces (avoir.)	35.27
"	Ounces (troy)	32.15
"	Pennyweights	643.01
"	Pounds (avoir.)	2.205
"	Pounds (troy)	2.679
Kilogram calories	Btu	3.968
"	Foot-pounds	3087
"	Meter-kilograms	4.269×10^{2}
Kilograms/cubic meter	Pounds/cubic foot	62.43×10^{-3}
"	Grams/cubic centimeter	10^{-3}
Kilograms/square centimeter	Pounds/square inch	14.22
"	Pounds/square foot	2.048×10^{3}
"	Inches of Hg at 0°C	28.96
"	Feet of water at 4°C	3.28×10^{-7}
Kilometers	Feet	3.281×10^{3}
"	Miles	6.214×10^{-1}
"	Nautical miles	5.4×10^{-1}
"	Centimeters	10^{5}
Kilometers/hour	Feet/second	9.113×10^{-1}
"	Knots	5.396×10^{-1}
"	Miles/hour	6.214×10^{-1}
"	Meters/second	2.778×10^{-1}
Kilowatt hours heat gen.	Grams of U^{235} fissioned	4.35×10^{-5}
Kilowatts	Btu/second	9.48×10^{-1}
"	Foot-pounds/second	7.376×10^{2}
"	Horsepower	1.341
"	Kilogram calories/second	2.389×10^{-1}
Knots	Nautical miles/hour	1
"	Feet/hour	6076.1033
"	Feet/second	1.688
"	Miles/hour	1.151
"	Kilometers/hour	1.853
"	Meters/second	5.148×10^{-1}
Leagues (U.S.)	Nautical miles	3
Link	Inches	7.92
Liters	Cubic centimeters	1000.027
"	Cubic feet	0.035
"	Cubic inches	61.025
"	Gallons (U.S.)	0.264
"	Gallons (imperial)	0.22
"	Ounces (fluid)	33.81
"	Pints	2.11
"	Quarts (liquid)	1.057
Meters	Centimeters	100
"	Miles	6.214×10^{-4}
"	Feet	3.281
"	Inches	39.37
"	Kilometers	0.001
"	Yards	1.094
Meter-kilograms	Foot-pounds	7.233
"	Joules	9.807
Meters/second	Feet/second	3.281
"	Miles/hour	2.237
"	Kilometers/hour	3.6
Microamperes	Unit charges/second	6.24×10^{12}
Microohms	Megaohms	1×10^{-12}
"	Ohms	1×10^{-6}
Microns	Inches	3.937×10^{-5}
"	Millimeters	0.001

To Convert From—	To—	Multiply By—	To Convert From—	To—	Multiply By—
Miles	Feet	5280	Ounces/gallon (fluid)	Cubic centimeters/liter	7.7
"	Kilometers	1.609			
"	Nautical miles	8.69×10^{-1}	Pennyweights	Grains	24
"	Furlongs	8	"	Grams	1.56
Miles/hour	Feet/second	1.467	"	Milligrams	1555
"	Meters/second	4.47×10^{-1}	"	Ounces (avoir.)	0.0549
"	Kilometers/hour	1.609	"	Ounces (troy)	0.05
"	Knots	8.69×10^{-1}	"	Pounds (avoir.)	0.0034
Miles/hour squared	Feet/second squared	2.151	"	Pounds (troy)	0.0042
Miles/hour/ second	Feet/second squared	1.4667	Pennyweights/ gallon	Grams/liter	0.41
Millibars	Inches of Hg at 0°C	2.953×10^{-2}	Pints	Cubic centimeters	473.2
Milligrams	Grains	0.0154	"	Cubic feet	0.017
"	Grams	0.001	"	Cubic inches	28.88
"	Kilograms	1×10^{-6}	"	Gallons	0.125
"	Ounces (avoir.)	3.5×10^{-5}	"	Liters	0.473
"	Ounces (troy)	3.215×10^{-5}	"	Ounces (fluid)	16
"	Pennyweights	6.43×10^{-4}	"	Quarts	0.5
"	Pounds (avoir.)	2.21×10^{-6}	Poles	Feet	16.5
"	Pounds (troy)	2.68×10^{-6}	"	Yards	5.5
Milligrams/liter	Parts/million	1	Pounds (avoir.)	Grains	7000
Milliliters	Cubic centimeters	1.000027	"	Grams	453.6
"	Cubic inches	0.061	"	Kilograms	0.454
"	Liters	0.001	"	Ounces (avoir.)	16
"	Ounces (fluid)	0.034	"	Ounces (troy)	14.58
Millimeters	Centimeters	0.1	"	Pennyweights	291.67
"	Inches	0.039	"	Pounds (troy)	1.215
"	Meters	0.001	"	Poundals	32.174
"	Microns	1000	"	Slugs	3.108×10^{-2}
Mils	Centimeters	0.0025	Pounds (troy)	Grains	5760
"	Inches	0.001	"	Grams	373.24
"	Microns	25.4	"	Kilograms	0.373
Nautical miles	Feet	6076.1	"	Ounces (avoir.)	13.17
"	Miles	1.151	"	Ounces (troy)	12
"	Meters	1852	"	Pennyweights	240
Ounces (avoir.)	Grains	437.5	"	Pounds (avoir.)	0.823
"	Grams	28.35	Pounds of water at 17°C	Cubic feet	0.016
"	Ounces (troy)	0.911	"	Cubic inches	27.68
"	Pennyweights	18.23	"	Gallons	0.1198
"	Pounds (avoir.)	1/16	Pounds/cubic foot	Kilograms/cubic meter	16.02
"	Pounds (troy)	0.076	Pounds/cubic inch	Pounds/cubic foot	1728
Ounces/gallon (avoir.)	Grams/liter	7.5	"	Grams/cubic centimeter	27.68
Ounces (troy)	Grains	480	Pounds/square inch	Inches of Hg at 0°C	2.036
"	Grams	31.1	"	Feet of water at 4°C	2.307
"	Milligrams	31103.5	"	Atmospheres	6.805×10^{-2}
"	Ounces (avoir.)	1.097	"	Kilograms/ square meter	7.031×10^{2}
"	Pennyweights	20			
"	Pounds (avoir)	0.069			
"	Pounds (troy)	1/12			
Ounces/gallon (troy)	Grams/liter	8.2	Quarts (liquid)	Cubic centimeters	946.4
Ounces (fluid)	Cubic centimeters	29.57	"	Cubic inches	57.75
"	Cubic inches	1.8	"	Cubic feet	0.033
"	Gallons	1/128	"	Gallons	0.25
"	Liters	0.0296	"	Liters	0.946
"	Milliliters	29.57	"	Ounces (fluid)	32
"	Pints	1/16	"	Pints	2
"	Quarts	0.031			

To Convert From—	To—	Multiply By—	To Convert From—	To—	Multiply By—
Radians	Degrees (arc)	57.3	Square inches	Square feet	1/144
Radians/second	Degrees/second	57.3	"	Square mils	1×10^{-6}
"	Revolutions/second	15.92×10^{-2}	"	Square yards	1/1296
"	Revolutions/minute	9.549	Square kilometers	Square miles	3.861×10^{-1}
Revolutions	Radians	6.283	Square meters	Square feet	10.76
Revolutions/minute	Radians/second	1.047×10^{-1}	"	Square yards	1.196
Rods	Foot	16.5	Square miles	Square kilometers	2.59
"	Yard	5.5	"	Acres	640
Slugs	Pounds	32.17405	Square rods	Square yards	30.25
Spans	Inches	9	Square yards	Square feet	9
Square centimeters	Circular millimeters	127.32	"	Square inches	1296
"	Circular mils	197,350	"	Square meters	8.361×10^{-1}
"	Square feet	0.001	Stones (British)	Pounds (avoir.)	14
"	Square inches	0.155	Tablespoons	Fluid ounces	0.5
"	Square millimeters	100	U^{235} fissions/second	Watts	3.21×10^{-11}
Square feet	Square centimeters	929.03	Unit charges/second	Microamperes	1.6×10^{-13}
"	Square inches	144	Watts	Btu/second	9.481×10^{-1}
"	Square yards	1.111×10^{-1}	"	U^{235} fissions/second	3.12×10^{10}
"	Acres	2.296×10^{-5}	Yards	Meters	9.144×10^{-1}
Square inches	Circular mils	1.2732	"	Feet	3
"	Square centimeters	6.45	"	Inches	36
			Years (sidereal)	Days (mean solar)	365.256

APPENDIX-B / Commonly Used Gas Turbine Engine Symbols and Abbreviations

Symbol	Definition	Units	Symbol	Definition	Units
A	Cross-sectional flow area	sq in., sq ft	eshp	Equivalent shaft horse-power (turboprop)	hp
a	Angle of attack	Degrees		$shp + \dfrac{F_n \times V_a}{550}$ (airplane moving)	
a	Linear or angular acceleration	ft/sec², rad/sec²		$shp + \dfrac{F_n}{2.5}$ (airplane static)	
C	Coefficient	None			
C-D	Converging-diverging	None	F	Thrust	lbt
CDP	Compressor discharge pressure	psi	F_g	Gross thrust	lbt
			F_j	Jet thrust	lbt
CDT	Compressor discharge temperature	°F, °C, °R	F_n	Net thrust $= F_g - F_r + F_j$	lb
CIT	Compressor inlet temperature	°F, °C, °R	F_r	Ram drag of engine airflow	lb
c	Speed of sound	ft/sec	f/a	Fuel/air ratio	lb/lb
c_p	Specific heat at a constant pressure	Btu/lb/°F	f	Frequency	1/sec
			g	Acceleration due to gravity	ft/sec²
c_v	Specific heat at a constant volume	Btu/lb/°F	g	Mass conversion factor	32.174
D	Diameter	ft, in.	H	Enthalpy, heating value	Btu/lb
EGT	Exhaust-gas temperature	°F, °C, °R	IGV	Inlet guide vane	None
			hp	Horsepower (shp, fhp)	hp
esfc	Equivalent specific fuel consumption $\dfrac{W_f}{eshp}$	lb/eshp/hr	J	Joule's constant $= 778.26$	ft-lb/Btu
			M	Mass W/g	lb
			M	Mach number $= V/c$	None
EPR	Engine pressure ratio	None	N	Engine rotational speed	rpm

Symbol	Definition	Units
OAT	Outside air temperature	°F, °C, °R
OGV	Outlet guide vane	None
P	Power	Btu/sec, hp
P	Absolute pressure (psia)	lb/sq in., in. Hg
PTA	Post-turbine augmentation	None
PTI	Pre-turbine injection	None
R	Gas constant for air = 53.35	ft-lb/(lb)(°F)
R	Reynolds number	None
SFC	Specific fuel consumption $\dfrac{W_f}{shp}$ (shaft engine)	lb/shp/hr
s	Specific entropy	Btu/lb-°R
shp	Shaft horsepower	hp
thp	Thrust horsepower	hp
TSFC	specific fuel consumption $\dfrac{W_f}{F_n}$ (jet engine)	lb/lbt/hr

Symbol	Definition	Units
T	Absolute temperature	°R
t	Time	sec
u	Specific internal energy	Btu/lb
u	Rotor linear velocity	ft/sec
V	Velocity	ft/sec
V	Volume	cu ft
v	Specific volume	cu ft/lb
W	Weight (force)	lb
w	Rate of flow	lb/sec, lb/hr
Δ	Difference	None
δ	Relative absolute pressure = P/P_o	None
η	Efficiency	percent
γ or κ	Ratio of specific heats c_p/c_v	None
μ	Microns	None
π	3.1416	None
θ	Relative absolute temperature T/T_o	None

APPENDIX-C / Glossary

ambient Refers to condition of atmosphere existing around the engine, such as ambient pressure or temperature.

ampere A unit of measurement of current flow. It is directly proportional to the voltage, but inversely proportional to the resistance to flow (ohm).

centrifugal force The outward force an object exerts on a restraining agent when the motion of the object is rotational.

centripetal force The inward force a restraining agent exerts on an object moving in a circle. It is the opposite of, and equal to, centrifugal force.

chocked nozzle A nozzle whole flow rate has reached the speed of sound.

density Mass per unit volume.

energy The capacity for doing work.

horsepower A man-made unit of power equal to 33,000 ft-lb of work per minute.

hot start Overtemperature during starting.

hung start Failure to reach normal-idling rpm during starting.

inertia The opposition of a body to have its state of rest or motion changed.

jam acceleration Rapid movement of the power lever, calling for maximum rate of rotor-speed increase.

jet silencer A device used to reduce and change the lower-frequency sound waves emitting from the engine's exhaust nozzle, and thus reducing the nose factor.

joule An electrical unit of energy or work.

mass The amount of matter contained within a substance.

molecule The smallest particle of a substance which can exist and still retain all of the characteristics of that substance.

momentum The tendency of a body to continue after being placed in motion.

ohm A unit which measures resistance to electrical current flow, and is equal to volts divided by amperes.

overspeed Temperature in excess of maximum allowable design temperature at the turbine exit.

power The rate of doing work; work per unit of time.

ram The amount of pressure buildup above ambient pressure at the engine's compressor inlet, due to forward motion of the engine through the air (air's initial momentum).

ram ratio The ratio of ram pressure to ambient pressure.

ram recovery The ability of an engine's air inlet duct to take advantage of ram pressure.

sonic speed Speed of sound under ambient or local conditions.

specific heat The ratio of the thermal capacity of a substance to the thermal capacity of water.

stable operation Condition where no appreciable fluctuation, intentional or unintentional, is occurring to any of the engine's variables such as rpm, temperature, or pressure.

subsonic speed Speed less than that of sound.

supersonic speed Speed in excess of that of sound.

thermocouple A pair of joints of two dissimilar metals across which a dc voltage is produced when one joint is warmer than the other.

thrust A reaction force in pounds.

thrust, gross The thrust developed by the engine, not taking into consideration any presence of initial-air-mass momentum.

thrust, net The effective thrust developed by the engine during flight, taking into consideration the initial momentum of the air mass prior to entering the influence of the engine.

thrust reverser A device used to partially reverse the flow of engine's nozzle discharge gases, and thus create a thrust force in the rearward direction.

thrust specific fuel consumption The fuel that the engine must burn per hour to generate any 1 lb of thrust.

thrust, static Same as gross thrust, without any initial air-mass momentum present due to engine's static condition.

torque A force, multiplied by its lever arm, acting at right angles to an axis.

transient conditions Conditions which may occur briefly while accelerating or decelerating, or while passing through a specific range of engine operation.

vector A line which, by scaled length, indicates magnitude, and whose arrow head represents direction of action.

volt A unit of measurement of electrical force. It is a function of the flow current (ampere) and the amount of resistance to flow (ohm) present.

watt A unit which measures power, and is equal to voltage multiplied by amperes.

work A force acting through a distance.

General properties of air

P = Absolute pressure — lb/ft^2
P_0 = Standard absolute pressure — lb/ft^2
T = Absolute temperature
T_0 = Standard absolute temperature
V = Velocity—ft/sec
g = Acceleration of gravity—ft/sec^2
n = Exponent of compression
q = Impact pressure—lb/ft^2
ρ = Density—lb sec^{2}ft^4
μ = Absolute viscosity lb sec/ft^2
ν = Kinematic viscosity—ft^2/sec
σ = Density ratio—ρ/ρ_0

$$P = \rho gRT$$

$$\frac{P}{P_0} = \frac{\rho}{\rho_0}\frac{T}{T_0} = \left(\frac{\rho}{\rho_0}\right)^n = \left(\frac{V_0}{V}\right)^n$$

Specific Weight of Air in lb/ft^3

$$g\rho = .07651\frac{P}{P_0}\frac{T_0}{T} = 1.325\frac{P \text{ in in. Hg}}{T \text{ in } F_{abs}}$$

Density of Air in lb sec^2/ft^4 or slugs/ft^3

$$\rho = .002378\frac{P}{P_0}\frac{T_0}{T} = .041187\frac{P \text{ in in. Hg.}}{T \text{ in } F_{abs}}$$

Air Density Ratio

$$\rho/\rho_0 = \frac{P}{P_0}\frac{T_0}{T} = 17.32\frac{P \text{ in in. Hg}}{T \text{ in } F_{abs}}$$

Impact Pressure

for incompressible
flow $q = \frac{1}{2}\rho V^2$

for compressible
flow $qc = \left(\dfrac{P_r}{P_{am}} - 1\right)^{P_{am}}$

$T_r = \left(\dfrac{T_r}{T_{am}} - 1\right)^{T_{am}}$

Approximate value (at sea level)

$$q = 25\left(\frac{V}{100}\right)^2 \text{ lb/sq ft}$$

$$= 5\left(\frac{V}{100}\right)^2 \text{ in. water}$$

Where V is in mph

Conversions (standard day condition)
1 cubic ft of air = .07651 lbs
1 pound of air = 13.07 cu. ft.

$$\frac{T}{T_0} = \left(\frac{P}{P_0}\right)^{\frac{n-1}{n}} = \left(\frac{V_0}{V}\right)^{n-1}$$

$$\frac{\rho}{\rho_0} = \left(\frac{T}{T_0}\right)^{\frac{1}{n-1}}$$

For adiabatic change
n = 1.39

Specific Heat of Air in Btu/lb/°F

at constant pressure, $C_p = .240$
at constant volume, $C_v = .1715$
for atmospheric temperature range
$\gamma = C_p/C_v = 1.40$

Gas Constant for Air

$R = 53.345$ ft-lb/lb F$_{abs}$

$= \dfrac{1545.4 \text{ ft-lb/lb - mole } F_{abs}}{\text{mol wt}}$

Molecular weight of air = 28.97

Speed of Sound in Air

$c_{fps} = \sqrt{\gamma gRT}$ $c_{fps} = 49.04\sqrt{T}$
$c_{knots} = 29.05\sqrt{T}$ $c_{mph} = 33.5\sqrt{T}$

where T is air temperature in °R
$c_{SL} = 661.74$ knots $= 761.52$ mph
$= 1117$ fps

Absolute Viscosity for Air

$\mu = \rho\nu$
$10^{10}\mu = 3538 + 9.870$ t in degrees C
$= 3408 + 5.483$ t in degrees F

Temperature rise resulting from adiabatic compression at impact

$$T = 1.792\left(\frac{V}{100}\right)^2 \text{ in degrees F}$$

Where V = True air speed in mph

Table for finding θ and $\sqrt{\theta}$ (temperature correction factor)

Relative Temperature

$°F = °R - 460$ $°C = °K - 273$ $°C = \dfrac{5}{9}(°F - 32)$ $°F = \dfrac{9}{5}°C + 32$ $\theta = \dfrac{T}{T_0} = \dfrac{T}{519}$ For interpolation, $1°C = 1.8°F$

T (°F)	θ	$\sqrt{\theta}$	T (°F)	θ	$\sqrt{\theta}$	T (°F)	θ	$\sqrt{\theta}$	T (°F)	θ	$\sqrt{\theta}$
69	1.020	1.010	49	0.982	0.990	29	0.943	0.971	9	0.905	0.951
68	1.018	1.009	48	0.980	0.989	28	0.941	0.970	8	0.903	0.950
67	1.016	1.008	47	0.978	0.988	27	0.939	0.969	7	0.901	0.949
66	1.014	1.007	46	0.976	0.987	26	0.937	0.968	6	0.899	0.948
65	1.012	1.006	45	0.974	0.986	25	0.935	0.967	5	0.897	0.947
64	1.010	1.005	44	0.972	0.985	24	0.934	0.966	4	0.895	0.946
63	1.008	1.004	43	0.970	0.984	23	0.932	0.965	3	0.893	0.945
62	1.006	1.003	42	0.968	0.984	22	0.930	0.964	2	0.891	0.944
61	1.004	1.002	41	0.966	0.983	21	0.928	0.963	1	0.889	0.943
60	1.002	1.001	40	0.964	0.982	20	0.926	0.962	0	0.887	0.942
59	1.000	1.000	39	0.962	0.981	19	0.924	0.961	−1	0.884	0.940
58	0.999	0.999	38	0.960	0.980	18	0.922	0.960	−2	0.883	0.939
57	0.997	0.998	37	0.959	0.979	17	0.920	0.959	−3	0.881	0.938
56	0.995	0.997	36	0.957	0.978	16	0.918	0.958	−4	0.879	0.937
55	0.993	0.996	35	0.955	0.977	15	0.916	0.957	−5	0.877	0.936
54	0.991	0.995	34	0.953	0.976	14	0.914	0.956	−6	0.875	0.935
53	0.989	0.994	33	0.951	0.975	13	0.912	0.955	−7	0.873	0.934
52	0.988	0.993	32	0.949	0.974	12	0.910	0.954	−8	0.871	0.933
51	0.986	0.992	31	0.947	0.973	11	0.908	0.953	−9	0.869	0.932
50	0.984	0.991	30	0.945	0.972	10	0.907	0.952			

Table for finding δ (pressure correction factor)

Relative Pressure

$$\text{InHg} = 0.07355 \times \text{in } H_2O \quad \text{InHg} = 0.1414 \times lbft^2 \quad \text{InHg} = 2.036 \times lb/in^2 \qquad \delta = \frac{P}{P_0} = \frac{P}{29.92}$$

P InHg ABS	δ	P InHg ABS	δ	P InHg ABS	δ	P InHg ABS	δ	P InHg ABS	δ
34.9	1.166	32.9	1.100	30.9	1.033	28.9	0.9659	26.9	0.8990
34.8	1.163	32.8	1.096	30.8	1.029	28.8	0.9626	26.8	0.8957
34.7	1.160	32.7	1.093	30.7	1.026	28.7	0.9592	26.7	0.8924
34.6	1.156	32.6	1.090	30.6	1.023	28.6	0.9559	26.6	0.8890
34.5	1.153	32.5	1.086	30.5	1.019	28.5	0.9525	26.5	0.8857
34.4	1.150	32.4	1.083	30.4	1.016	28.4	0.9492	26.4	0.8823
34.3	1.146	32.3	1.080	30.3	1.013	28.3	0.9458	26.3	0.8790
34.2	1.143	32.2	1.076	30.2	1.009	28.2	0.9425	26.2	0.8757
34.1	1.140	32.1	1.073	30.1	1.006	28.1	0.9392	26.1	0.8723
34.0	1.136	32.0	1.070	30.0	1.003	28.0	0.9358	26.0	0.8690
33.9	1.133	31.9	1.066	29.9	0.9993	27.9	0.9325	25.9	0.8656
33.8	1.130	31.8	1.063	29.8	0.9960	27.8	0.9291	25.8	0.8623
33.7	1.126	31.7	1.059	29.7	0.9926	27.7	0.9258	25.7	0.8586
33.6	1.123	31.6	1.056	29.6	0.9893	27.6	0.9224	25.6	0.8556
33.5	1.120	31.5	1.053	29.5	0.9859	27.5	0.9191	25.5	0.8523
33.4	1.116	31.4	1.049	29.4	0.9826	27.4	0.9158	25.4	0.8489
33.3	1.113	31.3	1.046	29.3	0.9793	27.3	0.9124	25.3	0.8456
33.2	1.110	31.2	1.043	29.2	0.9759	27.2	0.9091	25.2	0.8422
33.1	1.106	31.1	1.039	29.1	0.9726	27.1	0.9057	25.1	0.8389
33.0	1.103	31.0	1.036	29.0	0.9692	27.0	0.9024	25.0	0.8356

Fahrenheit–Celsius conversion table

Look up reading in middle column; if in degrees Celsius, read Fahrenheit equivalent in right-hand column; if in degrees Fahrenheit, read Celsius equivalent in left-hand column.

−60 to 43			44 to 93			94 to 510			520 to 1010			1020 to 1510			1520 to 2010			2020 to 2510			2520 to 3000		
C		F	C		F	C		F	C		F	C		F	C		F	C		F	C		F
−51	−60	−76	6.7	44	111.2	34.4	94	201.2	271	520	968	549	1020	1868	827	1520	2768	1104	2020	3668	1382	2520	4568
−46	−50	−58	7.2	45	113.0	35.0	95	203.0	277	530	986	554	1030	1886	832	1530	2786	1110	2030	3686	1388	2530	4586
−40	−40	−40	7.8	46	114.3	35.6	96	204.8	282	540	1004	560	1040	1904	838	1540	2804	1116	2040	3704	1393	2540	4604
−34	−30	−22	8.3	47	116.6	36.1	97	206.6	288	550	1022	566	1050	1922	843	1550	2822	1121	2050	3722	1399	2550	4622
−29	−20	−4	8.9	48	118.4	36.7	98	208.4	293	560	1040	571	1060	1940	849	1560	2840	1127	2060	3740	1404	2560	4640
−23	−10	14	9.4	49	120.2	37.2	99	210.2	299	570	1058	577	1070	1958	854	1570	2858	1132	2070	3758	1410	2570	4658
−17.8	0	32	10.0	50	122.0	37.8	100	212.0	304	580	1076	582	1080	1976	860	1580	2876	1138	2080	3776	1416	2580	4676
−17.2	1	33.8	10.6	51	123.8	38	100	212	310	590	1094	588	1090	1994	866	1590	2894	1143	2090	3794	1421	2590	4694
−16.7	2	35.6	11.1	52	125.6	43	110	230	316	600	1112	593	1100	2012	871	1600	2912	1149	2100	3812	1427	2600	4712
−16.1	3	37.4	11.7	53	127.4	49	120	248	321	610	1130	599	1110	2030	877	1610	2930	1154	2110	3830	1432	2610	4730
−15.6	4	39.2	12.2	54	129.2	54	130	266	327	620	1148	604	1120	2048	882	1620	2948	1160	2120	3848	1438	2620	4748
−15.0	5	41.0	12.8	55	131.0	60	140	284	332	630	1166	610	1130	2066	888	1630	2966	1166	2130	3866	1443	2630	4766
−14.4	6	42.8	13.3	56	132.8	66	150	302	338	640	1184	616	1140	2084	893	1640	2984	1171	2140	3884	1449	2640	4784
−13.9	7	44.6	13.9	57	134.6	71	160	320	343	650	1202	621	1150	2102	899	1650	3002	1177	2150	3902	1454	2650	4802
−13.3	8	46.4	14.4	58	136.4	77	170	338	349	660	1220	627	1160	2120	904	1660	3020	1182	2160	3920	1460	2660	4820
−12.8	9	48.2	15.0	59	138.2	82	180	356	354	670	1238	632	1170	2138	910	1670	3038	1188	2170	3938	1466	2670	4838
−12.2	10	50.0	15.6	60	140.0	88	190	374	360	680	1256	638	1180	2156	916	1680	3056	1193	2180	3956	1471	2680	4856
−11.7	11	51.8	16.1	61	141.8	93	200	392	366	690	1274	643	1190	2174	921	1690	3074	1199	2190	3974	1477	2690	4874
−11.1	12	53.6	16.7	62	143.6	99	210	410	371	700	1292	649	1200	2192	927	1700	3092	1204	2200	3992	1482	2700	4892
−10.6	13	55.4	17.2	63	145.4	100	212	413.6	377	710	1310	654	1210	2210	932	1710	3110	1210	2210	4010	1488	2710	4910
−10.0	14	57.2	17.8	64	147.2	104	220	428	382	720	1328	660	1220	2228	938	1720	3128	1216	2220	4028	1493	2720	4928
−9.4	15	59.0	18.3	65	149.0	110	230	446	388	730	1346	666	1230	2246	943	1730	3146	1221	2230	4046	1499	2730	4946
−8.9	16	60.8	18.9	66	150.8	116	240	464	393	740	1364	671	1240	2264	949	1740	3164	1227	2240	4064	1504	2740	4964
−8.3	17	62.6	19.4	67	152.6	121	250	482	399	750	1382	677	1250	2282	954	1750	3182	1232	2250	4082	1510	2750	4982
−7.8	18	64.4	20.0	68	154.4	127	260	500	404	760	1400	682	1260	2300	960	1760	3200	1238	2260	4100	1516	2760	5000
−7.2	19	66.2	20.6	69	156.2	132	270	518	410	770	1418	688	1270	2318	966	1770	3218	1243	2270	4118	1521	2770	5018
−6.7	20	68.0	21.1	70	158.0	138	280	536	416	780	1436	693	1280	2336	971	1780	3236	1249	2280	4136	1527	2780	5036
−6.1	21	69.8	21.7	71	159.8	143	290	554	421	790	1454	699	1290	2354	977	1790	3254	1254	2290	4154	1532	2790	5054
−5.6	22	71.6	22.2	72	161.6	149	300	572	427	800	1472	704	1300	2372	982	1800	3272	1260	2300	4172	1538	2800	5072
−5.0	23	73.4	22.8	73	163.4	154	310	590	432	810	1490	710	1310	2390	988	1810	3290	1266	2310	4190	1543	2810	5090
−4.4	24	75.2	23.3	74	165.2	160	320	608	438	820	1508	716	1320	2408	993	1820	3308	1271	2320	4208	1549	2820	5108
−3.9	25	77.0	23.9	75	167.0	166	330	626	443	830	1526	721	1330	2426	999	1830	3326	1277	2330	4226	1554	2830	5126
−3.3	26	78.8	24.4	76	168.8	171	340	644	449	840	1544	727	1340	2444	1004	1840	3344	1282	2340	4244	1560	2840	5144
−2.8	27	80.6	25.0	77	170.6	177	350	662	454	850	1562	732	1350	2462	1010	1850	3362	1288	2350	4262	1566	2850	5162
−2.3	28	82.4	25.6	78	172.4	182	360	680	460	860	1580	738	1360	2480	1016	1860	3380	1293	2360	4280	1571	2860	5180
−1.7	29	84.2	26.1	79	174.3	188	370	698	466	870	1598	743	1370	2498	1021	1870	3398	1299	2370	4298	1577	2870	5198
−1.1	30	86.0	26.7	80	176.0	193	380	716	471	880	1616	749	1380	2516	1027	1880	3416	1304	2380	4316	1582	2880	5216
−0.6	31	87.8	27.2	81	177.8	199	390	734	477	890	1634	754	1390	2534	1032	1890	3434	1310	2390	4334	1588	2890	5234
0.0	32	89.6	27.8	82	179.6	204	400	752	482	900	1652	760	1400	2552	1038	1900	3452	1316	2400	4352	1593	2900	5252
0.6	33	91.4	28.3	83	181.4	210	410	770	488	910	1670	766	1410	2570	1043	1910	3470	1321	2410	4370	1599	2910	5270
1.1	34	93.2	28.9	84	183.2	216	420	788	493	920	1688	771	1420	2588	1049	1920	3488	1327	2420	4388	1604	2920	5288
1.7	35	95.0	29.4	85	185.0	221	430	806	499	930	1706	777	1430	2606	1054	1930	3506	1332	2430	4406	1610	2930	5306
2.2	36	96.8	30.0	86	186.8	227	440	824	504	940	1724	782	1440	2624	1060	1940	3524	1338	2440	4424	1616	2940	5324
2.8	37	98.6	30.6	87	188.6	232	450	842	510	950	1742	788	1450	2642	1066	1950	3542	1343	2450	4442	1621	2950	5342
3.3	38	100.4	31.1	88	190.4	238	460	860	516	960	1760	793	1460	2660	1071	1960	3560	1349	2460	4460	1627	2960	5360
3.9	39	102.2	31.7	89	192.2	243	470	878	521	970	1778	799	1470	2678	1077	1970	3578	1354	2470	4478	1632	2970	5378
4.4	40	104.0	32.2	90	194.0	249	480	896	527	980	1796	804	1480	2696	1082	1980	3596	1360	2480	4496	1638	2980	5396
5.0	41	105.8	32.8	91	195.8	254	490	914	532	990	1814	810	1490	2714	1088	1990	3614	1366	2490	4514	1643	2990	5414
5.6	42	107.6	33.3	92	197.6	260	500	932	538	1000	1832	816	1500	2732	1093	2000	3632	1371	2500	4532	1649	3000	5432
6.1	43	109.4	33.9	93	199.4	266	510	950	543	1010	1850	821	1510	2750	1099	2010	3650	1377	2510	4550			

INDEX